Applications of Glass Capillary Gas Chromatography

CHROMATOGRAPHIC SCIENCE

A Series of Monographs

Editor: JACK CAZES

Waters Associates, Inc.
Milford, Massachusetts

Volume 1: Dynamics of Chromatography
J. Calvin Giddings

Volume 2: Gas Chromatographic Analysis of Drugs and Pesticides
Benjamin J. Gudzinowicz

Volume 3: Principles of Adsorption Chromatography: The Separation of Nonionic Organic Compounds (out of print)
Lloyd R. Snyder

Volume 4: Multicomponent Chromatography: Theory of Interference (out of print)
Friedrich Helfferich and Gerhard Klein

Volume 5: Quantitative Analysis by Gas Chromatography
Joseph Novák

Volume 6: High-Speed Liquid Chromatography
Peter M. Rajcsanyi and Elisabeth Rajcsanyi

Volume 7: Fundamentals of Integrated CG-MS (in three parts)
Benjamin J. Gudzinowicz, Michael J. Gudzinowicz, and Horace F. Martin

Volume 8: Liquid Chromtography of Polymers and Related Materials
Jack Cazes

Volume 9: GLC and HPLC Determination of Therapeutic Agents (in three parts)
Part 1 edited by Kiyoshi Tsuji and Walter Morozowich
Parts 2 and 3 edited by Kiyoshi Tsuji

Volume 10: Biological/Biomedical Applications of Liquid Chromatography
Edited by Gerald L. Hawk

Volume 11: Chromatography in Petroleum Analysis
Edited by Klaus H. Altgelt and T. H. Gouw

Volume 12: Biological/Biomedical Applications of Liquid Chromatography II
Edited by Gerald L. Hawk

Volume 13: Liquid Chromatography of Polymers and Related Materials II
Edited by Jack Cazes and Xavier Delamare

Volume 14: Introduction to Analytical Gas Chromatography: History, Principles, and Practice
John A. Perry

Volume 15: Applications of Glass Capillary Gas Chromatography
Edited by Walter G. Jennings

Volume 16: Steroid Analysis by HPLC: Recent Applications (in press)
Edited by Marie P. Kautsky

Other Volumes in Preparation

Applications of Glass Capillary Gas Chromatography

Edited by

WALTER G. JENNINGS

Department of Food Science and Technology
University of California
Davis, California

MARCEL DEKKER, INC. New York and Basel

Library of Congress Cataloging in Publication Data

Main entry under title:

Applications of glass capillary gas chromatography.

(Chromatographic science; v. 15)
Includes index.
1. Gas chromatography. I. Jennings, Walter G., [date]. II. Series.

QD79.C45A66 543'.0896 81-5436
ISBN 0-8247-1223-4 AACR2

MARCEL DEKKER, INC.
270 Madison Avenue, New York, New York 10016

Current printing (last digit):
10 9 8 7 6 5 4 3 2 1

PRINTED IN THE UNITED STATES OF AMERICA

PREFACE

Because it offers several advantages when compared with other methods of analysis, gas chromatography is today's most widely used analytical technique. The benefits--good resolution of complex mixtures, comparatively short analysis times, and relatively high sensitivities--have proven so attractive that considerable attention is currently directed to volatilizing, by way of derivatization, a wide variety of substances so that they too can be analyzed by this technique.

All these benefits are realized to a higher degree, and several others are engendered, by a transition from packed columns to open tubular glass capillary gas chromatography. Properly employed, these more inert columns can yield results on active compounds that react with (or entirely fail to negotiate) packed columns; in addition, increases in the degree of component separation and the sensitivity of detection, and a decrease in analysis time, can be realized. Packed columns retain three advantages: they possess larger sample capacities (but exhibit lower sensitivities), and they are more forgiving of defects in instrumental design and operator technique. Several equipment manufacturers now supply capillary-compatible instruments, and the continued growth of this new technology, largely at the expense of packed column applications, seems inevitable.

This volume has been compiled to allow the interested analyst an opportunity to explore the application of glass capillary gas chromatography to several diverse fields. The editor would like to thank each of the carefully selected contributors for his efforts, all of which combine to clearly demonstrate the dawn of what has been termed the "age of analytical renaissance."

Walter G. Jennings

CONTRIBUTORS

JEROME E. BAKKE *Metabolism and Radiation Research Laboratory, Agricultural Research, Science and Education Administration, U. S. Department of Agriculture, Fargo, North Dakota*

WOLFGANG BERTSCH *Department of Chemistry, University of Alabama, University (Tuscaloosa), Alabama*

LARS BLOMBERG *Department of Analytical Chemistry, Arrhenius Laboratory, University of Stockholm, Stockholm, Sweden*

HANS DITSCHUNEIT *Department of Internal Medicine, Division of Metabolism, Nutrition, and Gastroenterology, University of Ulm, Ulm, Federal Republic of Germany*

LESLIE S. ETTRE *Chromatography Division, The Perkin-Elmer Corporation, Norwalk, Connecticut*

HARTMUT FRANK *Institute for Toxicology, University of Tübingen, Tübingen, Federal Republic of Germany*

BRUCE W. HERMANN* *Department of Environmental Toxicology, University of California, Davis, California*

HALVOR JAEGER *Department of Internal Medicine, Division of Metabolism, Nutrition, and Gastroenterology, University of Ulm, Ulm, Federal Republic of Germany*

WALTER G. JENNINGS *Department of Food Science and Technology, University of California, Davis, California*

*Current Affiliation: *Department of Analytical Chemistry, Biological Sciences Research Center, Shell Development Company, Modesto, California*

HANS U. KLOER* *Department of Internal Medicine, Division of Metabolism, Nutrition, and Gastroenterology, University of Ulm, Ulm, Federal Republic of Germany*

DENIS C. K. LIN *Analytical Chemistry Section, Department of Chemistry, Battelle-Columbus Laboratories, Columbus, Ohio*

WILHELM NEBELUNG *Department of Internal Medicine, Division of Metabolism, Nutrition, and Gastroenterology, University of Ulm, Ulm, Federal Republic of Germany*

CARL D. PFAFFENBERGER *Chemical Epidemiology Division, Department of Epidemiology and Public Health, University of Miami School of Medicine, Miami, Florida*

A. RAPP *Department of Biochemistry and Physiology, Federal Research Institute of Viticulture, Siebeldingen, Federal Republic of Germany*

P. SANDRA *Laboratory for Organic Chemistry, State University of Ghent, Ghent, Belgium*

JAMES N. SEIBER *Department of Environmental Toxicology, University of California, Davis, California*

TAKAYUKI SHIBAMOTO *Department of Environmental Toxicology, University of California, Davis, California*

ERIC VANLUCHENE *Gynecologic and Obstetric Clinic, Academic Hospital, State University of Ghent, Ghent, Belgium*

M. VERZELE *Laboratory of Organic Chemistry, State University of Ghent, Ghent, Belgium*

GUNNAR WIDMARK *Department of Analytical Chemistry, Arrhenius Laboratory, University of Stockholm, Stockholm, Sweden*

---

*Current Affiliation: *Laboratory of Lipid and Lipoprotein Studies, Oklahoma Medical Research Foundation, Oklahoma City, Oklahoma*

CONTENTS

	Preface	iii
	Contributors	v
1	The Evolution of Open Tubular Columns *Leslie S. Ettre*	1
2	System Requirements *Walter G. Jennings*	49
3	Applications of Glass Capillary Gas Chromatography in Xenobiotic Research *Jerome E. Bakke, Lars Blomberg, and Gunnar Widmark*	55
4	Analysis of Air and Air Pollutants *Wolfgang Bertsch*	71
5	Analysis of Water and Water Pollutants *Denis C. K. Lin*	123
6	Glass Capillary Gas Chromatography of Pesticides *Bruce W. Hermann and James N. Seiber*	175
7	Glass Capillary Gas Chromatography in Clinical Medicine *Carl D. Pfaffenberger*	203
8	Analysis of Steroid Hormones *Eric Vanluchene and P. Sandra*	289
9	Glass Capillary Gas-Liquid Chromatography of Amino Acids *Halvor Jaeger, Hartmut Frank, Hans U. Kloer, and Hans Ditschuneit*	331
10	Glass Capillary Gas-Liquid Chromatography of Bile Acids *Halvor Jaeger, Hartmut Frank, Wilhelm Nebelung, Hans U. Kloer, and Hans Ditschuneit*	365
11	Glass Capillary Gas-Liquid Chromatography of Fatty Acids *Halvor Jaeger, Hartmut Frank, Hans U. Kloer, and Hans Ditschuneit*	395
12	Analysis of Essential Oils *Takayuki Shibamoto*	455
13	Analysis of Food and Flavors *Walter G. Jennings*	511

14 Analysis of Beer and Hops 535
M. Verzele and P. Sandra

15 Analysis of Grapes, Wines, and Brandies 579
A. Rapp

Index 623

THE EVOLUTION OF OPEN TUBULAR COLUMNS

LESLIE S. ETTRE

The Perkin-Elmer Corporation
Norwalk, Connecticut

1.	Introduction	2
	1.1 Nomenclature	2
2.	Beginnings	3
	2.1 Background	3
	2.2 Invention of Open Tubular Columns	4
	2.3 Work of Other Pioneers	9
3.	Evolution of the System	11
	3.1 Detection	12
	3.2 Sample Introduction	16
4.	Evolution of Open Tubular Columns	20
	4.1 The Coating Method	20
	4.2 Porous-layer Open Tubular Columns	22
	4.3 Material of the Tubing	27
	4.4 Liquid-phase Film Thickness	31
	4.5 Utilization of the Theoretical Best Efficiency	32
5.	Conclusions	37
	List of Symbols	38
	References	39

1. INTRODUCTION

Although the open tubular column was invented over 25 years ago by M. J. E. Golay, only now are we beginning to observe its wide acceptance. The delay was due mainly to technical problems in both instrumentation development (primarily suitable inlet systems and column connections), and in column preparation (mainly surface preparation and coating methodology). A survey of these developments provides an interesting introduction to the chapters that follow.

1.1 Nomenclature

In a lecture at the 1960 Edinburgh Symposium [1], Golay acknowledged that he "must take the blame" for "having used the word capillary instead of open-tubular columns" in his 1957 Lansing paper [2], thus creating the impression that it is the smallness of the column which is beneficial; he was careful to use the term *open tubular column* at the 1958 Amsterdam Symposium. As he emphasized in retrospect [1], "it is not the smallness, it is the 'openness' of the open-tubular column which permits us to realize a two orders of magnitude improvement over the packed column."

Continuing this theme, Golay stated that such columns with larger diameter could also be expected to have excellent performance for the proper sample such as, e.g., "the 24-inch diameter gas pipe with an oil-coated inner wall, stretching from Texas to Maine" for high-boiling hydrocarbons.

This clear distinction in the name of these columns is not only a question of semantics, it also helps avoid misunderstanding. *Packed* capillary columns are used in both gas and liquid chromatography, and the term *capillary* does not clearly specify whether reference is made to such columns or to open tubular columns.

These points have been well recognized by a number of leading scientists involved in the development of open tubular columns and also by national and international bodies dealing with the standardization of gas chromatography nomenclature such as the British Standards Association (BS), the American Society for Testing and Materials (ASTM), and the International Union of Pure and Applied Chemistry (IUPAC), *all of which clearly specify the term "open tube" or "open tubular" columns and not "capillary" columns.* Popular usage, however, has defied this distinction. The word *American* carries a worldwide connotation implying citizenship in the United States, to the justifiable distress of our Canadian and Mexican neighbors; it is regretable that except for a few scientists who like consistency, *capillary column* is widely substituted for the more precise term *open tubular column*. The offense is sometimes compounded by the use of "$(GC)^2$" for "Glass Capillary Gas Chromatography." Although this sloppiness should definitely be condemned, practitioners are regrettably still using this slang. It is this author's feeling that if such usage continues to expand, we soon will have to spell the name of the state representing the cradle of country music as Ten^{2}es^{2}e^2 and the next international symposium will be organized by a com^{2}it^{2}e^2.

2. BEGINNINGS

2.1 Background

We must go back in our considerations to 1955-1957, the period when gas chromatography started its exponential development. This is the period when van Deemter, Zuiderweg, and Klinkenberg published their basic paper [7] on the parameters influencing the performance of a packed column. Without wanting to diminish the merit of these authors we should realize that very few chromatographers have read the original publication, since the journal in which it was published, *Chemical Engineering Science,* is not often read by analysts. Thus, this theoretical treatment became widely known only following the 1957 publication of the proceedings of the 1956 London Symposium, in which Keulemans and Kwantes [8] presented a preview of this work, and after the publication, again in 1957, of Keulemanns' book [9]. This is demonstrated by the fact that, as mentioned by Dijkstra [10], most authors, while referring to the original publication [7], use that form of the van Deemter equation which was included in the discussion of Keulemans and Kwantes [8,9] although this differs slightly from the original.*

Some feel that the concept of open tubular columns can be traced back to the lecture of Martin, given at the 1956 London Symposium [11] in which, when speaking about the desirability of reducing the amount of sample needed for an analysis, he predicted that the diameter of the column would be decreased to about 0.2 mm. However, a careful reading of the corresponding part of Martin's lecture discloses no hint that he is speaking of "open tubular columns of capillary dimensions"; besides, as mentioned earlier, it

---

*E.g., the second term is given in the original publication as $\gamma D_G/\bar{u}$ and not as $2\gamma D_G/\bar{u}$ as listed by Keulemans and Kwantes.

is *not the small diameter but the openness* that constitutes the characteristic element of these columns. This was recognized by Desty [12] who pointed out that "at that stage his [Martin's] ideas had not progressed as far as the open coated capillary with what Golay was to point out a year later as its very important advantage—that of low resistance to gas flow and consequent low pressure drop."

In order to place Golay's work in the proper context, we should realize that at that time, column performance was far from exceptional. Indeed, Keulemans and Kwantes, in their 1956 paper [8] publicizing the theoretical treatment of van Deemter et al., set the modest goal of "a 20-metre column, having an average [height equivalent to one theoretical plate (HETP)] of 2 mm, and thus giving 10,000 theoretical plates."

2.2 Invention of Open Tubular Columns

Golay's activities in this area started in 1955, well before the disclosure of van Deemter et al. He first centered his investigations around the operation of the (packed) separation column from a theoretical point of view. His first results—a detailed theoretical discussion—were summarized in the paper presented at the first gas chromatography symposium organized by the American Chemical Society and held in spring 1956, in Dallas [13]. In subsequent months, Golay continued his investigations on the influence of gas chromatographic parameters on column efficiency and the speed of analysis. This work was summarized in a number of internal progress reports. The first [14], dated August 22, 1956, dealt with theoretical aspects, while the second [15], dated two weeks later, reported on experimental data obtained on various packed columns. These investigations showed the enormous discrepancy between theory and experimental data. In the second report Golay [15] outlined further experiments needed to understand this discrepancy and to improve the pracitcal results. One of the suggestions was "that experiments be undertaken with a single glass capillary, 0.05 cm or 0.1 cm in diameter, wetted with silicone oil." These experiments were carried out in September-October 1956 and first reported on November 15, 1956 [16]; they confirmed the theoretical predictions. Figure 1 shows a copy of Golay's report, while Fig. 2 shows the first chromatograms obtained on a short (12 ft = 366 cm), 0.055-in (1.375-mm) ID column analyzing Phillips 37 hydrocarbon mixture [18].

A major problem Golay and his colleagues had to face was the lack of detectors which would be sensitive enough without creating band spreading and of suitably fast recorders. The September 5, 1956 report [15] discussed

November 15, 1956

To: J. G. Atwood

From: Marcel J. E. Golay

Subject: Progress report of gas chromatographic experimental work for September and October 1956

An enormous discrepancy with the theory evolved up to that time was the circumstance that the HETP was ten times greater (0.1 cm) than the packing granule dimensions while the characteristic dimension controlling the pneumatic resistance to flow (0.001 cm) was only one-tenth as large. It had been anticipated that these two quantities, two orders of magnitude apart, would be of the same order of magnitude.

Experiments with open cylindrical columns with an oil coating on the wall were initiated in order to unravel these discrepancies and to obtain a better insight into the separation mechanisms.

An initial experiment with a glass capillary, 1 meter x 0.08 cm, gave inconclusive results because of the near equality of the volumes in the injection block, in the column itself and in the detector (0.8 cm^3). An experiment with an uncoated 10 meter x 0.3 cm length of Tygon tubing gave a $\Delta t/t$ ratio corresponding very closely to the calculated number of theoretical plates of this column for the air peak (about 14,000).

Experiments were then initiated with two steel tubings, 12 feet x 0.055", wall coated with ethylene glycol by filling and wiping with a cotton wad drawn through the tubing. For these experiments, a new detector was constructed in order to reduce its volume.

Figure 1. Copy of Golay's first report on his investigations concerning open tubular columns. (Source: Ref. 16.)

the question of detectors and outlined the concept of a small-volume thermal conductivity detector that was built soon after. The chromatograms shown in Fig. 2 were obtained with this detector. Fast recording represented another problem. At that time, very fast recorders were difficult to find, and the chromatograms shown in Fig. 2 were obtained with the help of a Sanborn high-speed recording galvanometer. Besides this, Polaroid photographs of oscilloscope displays were also utilized. Figure 3 shows such a chromatogram, in which the retention time of the last peak is only 15 seconds. Similar chromatograms, also obtained on oscilloscopes, were included in the patent application filed June 24, 1957 [19].

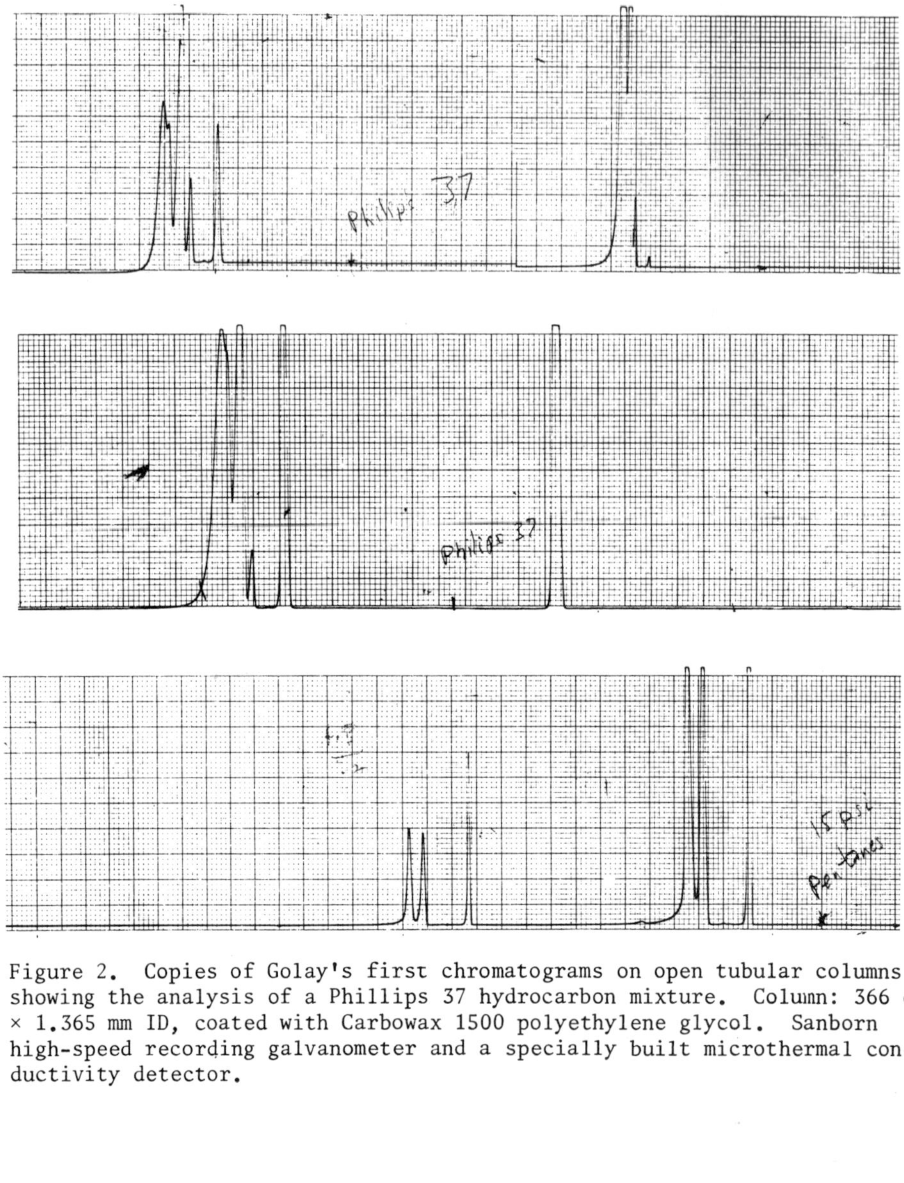

Figure 2. Copies of Golay's first chromatograms on open tubular columns showing the analysis of a Phillips 37 hydrocarbon mixture. Column: 366 cm × 1.365 mm ID, coated with Carbowax 1500 polyethylene glycol. Sanborn high-speed recording galvanometer and a specially built microthermal conductivity detector.

We should also refer to Golay's theoretical treatments that resulted in the development of the *performance index* [20]. Although this term had not been generally accepted as a basic efficiency term, it had an important role in the development of open tubular columns because it provided a way for Golay and his associates to judge the relative performance of various columns. For example, Golay, when noting the enormous discrepancy (representing a factor of 10,000) between actual and theoretical values for

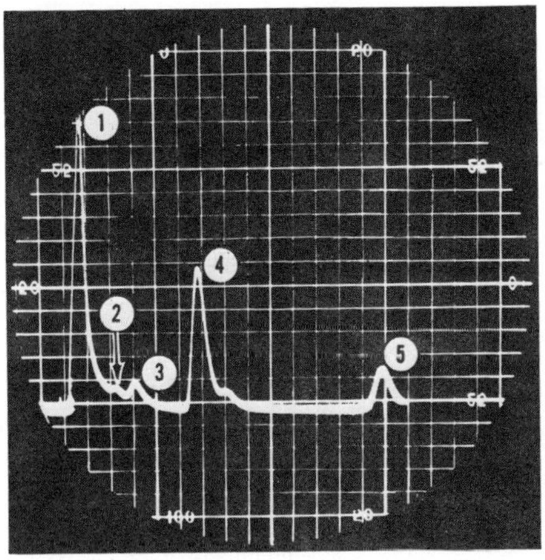

Figure 3. One of the early chromatograms of Golay obtained on an open tubular column. Column: 488 cm × 0.25 mm ID, coated with Carbowax 1500 polyethylene glycol. Oscilloscope readout and a specially built microthermal conductivity detector. Peaks: 1—air, 2—acetone, 3—carbon disulfide, 4—chloroform, 5—methylene chloride. Retention time of last peak: 15 sec.

packed columns, stated that "this very large ratio between the actual and ideal performance indexes constitutes a measure of how much room for improvement there is in basic column design."

Golay's first—and mainly theoretical—discussion on open tubular columns was presented at the first international gas chromatography symposium organized by the Instrument Society of America on the Michigan State University campus in East Lansing, August 28-30, 1957 [2]. This theoretical discussion was still incomplete mainly because some terms were not fully considered according to gas kinetics. Most interesting, however, was Golay's presentation of three chromatograms of a three-component mixture, one on a 4-m packed column, the other on a 91-m × 0.25-mm ID open tubular column, and the third on a 30-m × 0.25-mm ID open tubular column: the respective theoretical plate numbers were 1700, 12,000, and 2,500, while the analysis times (the retention time of the third peak) were about 25, 25, and a little over 3 minutes, respectively. Golay emphasized in his presentation that of the development neither the theory nor the technique was yet finished and the paper should really be considered as the first report on these columns.

Martin, during the discussion, congratulated Golay for "beating him thoroughly" in this concept and drew his attention to the work of Taylor who dealt with the gas kinetics problems. In the next months, while his colleagues—mostly R. D. Condon—continued working on the technical questions related to column preparation, Golay* completed the development of the theory considering the previous treatments of Taylor [22-24] on the dispersion of a solute (k = 0) in a fluid moving through an open tube, and particularly of Westhaver who studied the conditions influencing the HETP in an open tube distillation column for both cases of $k = \infty$ [25] and k = 0 [26]. The conclusions of this work were reported at the Amsterdam Symposium, May 19-23, 1958 [13], in which the famous Golay equations describing the influence of various parameters on column performance were derived:†

$$h = \frac{B}{\bar{u}} + C_G\bar{u} + C_L\bar{u} \tag{1}$$

$$B = 2D_G \tag{2}$$

$$C = \frac{1 + 6k + 11k^2}{(1 + k)^2} \frac{r^2}{24D_G} \tag{3}$$

$$C_L = \frac{k^3}{6(1 + k)^2} \frac{r^2}{K^2D_L} = \frac{2k}{3(1 + k)^2} \frac{d_f^2}{D_L} \tag{4a}$$

If the actual inside tube area is larger than that of a smooth wall tube, then the C_L term can be written in the following form:

$$C_L = \frac{k^3}{6(1 + k)^2} \frac{r^2}{F^2K^2D_L} = \frac{2k}{3(1 + k)^2} \frac{d_f^2}{F^2D_L} \tag{4b}$$

where F represents the ratio of the actual inside surface area coated with the liquid phase and the inside surface area of a smooth wall tube.

It is noteworthy to mention that the form of the C_L term as given in Eq. (4b) has been largely forgotten, although it has very important implications on the influence of the inside surface on column efficiency.

In addition to these detailed theoretical treatments (which also dealt with the possibility of rectangular columns) Golay, in his Amsterdam paper, also showed two chromatograms obtained on a 45-m × 0.25-mm ID open tubular column coated with didecyl phthalate liquid phase. The first showed the separation of p/m-xylene in 58 min while the other—reproduced in Fig. 4—illustrated the separation of C_8 hydrocarbons in 9 min.

* For his own recollections on this work, see Ref. 21.

† The equations are written here using present-day symbols corresponding to the cited nomenclatures. For explanation of the symbols, see the List of Symbols at the end of the chapter.

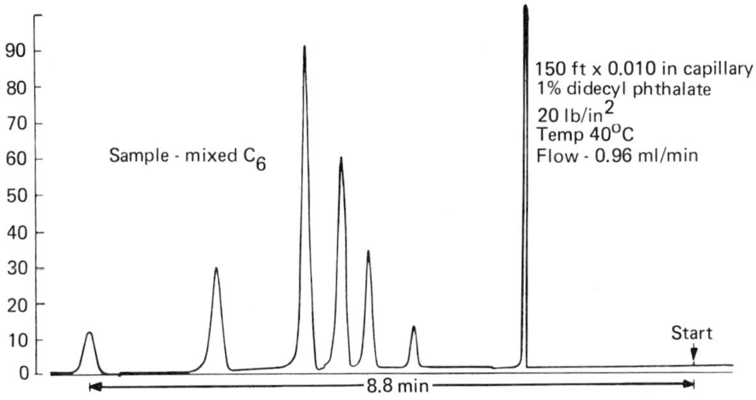

Figure 4. Analysis of C_6 hydrocarbon isomers. Column: 45.7 m × 0.25 mm ID open tubular, coated with didecyl phthalate liquid phase. Film thickness: 0.63 μm. Column temperature: 40°C. Carrier gas (He) inlet pressure: 20 psig. Flowrate: 0.96 μl/min. Specially built thermal conductivity detector. (Source: Ref. 3.)

2.3 Work of Other Pioneers

At the 1958 Amsterdam Symposium, Dijkstra and DeGoey of Unilever also presented results obtained on similar columns [27]. The results showed by them were inferior to those shown by Golay, which was probably due to a number of factors: too thick liquid-phase coatings, dead volumes in the injector and detector, and gas velocities that were much too high. However, as mentioned below, their coating technique soon became the standard procedure for the preparation of open tubular columns.*

Immediately after Golay's presentation in Amsterdam, a number of researchers picked up his ideas and started to apply them in practice. One of the first was D. H. Desty at British Petroleum, who in the fall of 1958, at the First Gas Chromatography Symposium organized in Leipzig, October 9-11, presented a paper on the application of open tubular columns in the petroleum industry [28] which was then followed soon by a more detailed paper on this subject [29], a short paper [30], and, finally, by two basic papers [31,32] on the influence of various parameters on column performance.

*For the sake of historical accuracy it should be mentioned that the columns prepared by Condon for the two chromatograms were coated using practically the same dynamic method. This was, however, not mentioned earlier in Golay's paper or in the following discussion [27a].

Among other early European pioneers in open tubular column gas chromatography who should be mentioned are, R. P. W. Scott [33-36], Grant [37,38], and Adlard [39,40] in England; Halász [41-44], Kaiser [45,46], Jentzsch [47-49], and Schomburg [50] in Germany; and the group of Liberti, Cartoni, and Bruner, in Italy [51-54].

In the United States, Zlatkis at the University of Houston, Lipsky at Yale University, and Lovelock, who had just invented the argon ionization detector, were probably the first scientists outside the laboratories of Perkin-Elmer to immediately recognize the importance of open tubular columns. Figure 5 shows the first chromatogram obtained by the Houston group on November 16, 1958. In the lower left-hand corner of the chart paper appear the signatures of the people who witnessed this historical run. Zlatkis continued to have a leading role in the applications of open tubular columns [56-58]. Lipsky and co-workers [59-63] utilized these columns for one of the most difficult analyses to which gas chromatography was applied: fatty acid analysis. Other American pioneers who should be mentioned here are Hollis [64,65]; scientists at the Shell laboratories in Houston [66-70] who pioneered in hydrocarbon analysis; and those at the Western Regional Research Center of the U.S. Department of Agriculture in Albany, California [71-77] who pioneered in the application of open tubular columns to the analysis of food flavors and gas chromatography-mass spectrometry (GC-MS) systems; and, last but not least, those associated with the Perkin-Elmer Corporation [78-93].

This author has tried to list and refer to those who immediately recognized Golay's work and, through their publications in the first five years, contributed to the development of open tubular columns. In order to show the high level reached by 1963, four chromatograms are randomly selected from that year which are shown in Figs. 6-9.

Naturally, this listing cannot be complete; citations were restricted to those who had a number of publications and do not include those chromatographers who had a single paper, however important. Consideration has been restricted to gas-liquid *partition* chromatography, ignoring work on adsorption-type open tubular columns, although significant work was also accomplished in this field during this period. Furthermore, let us not forget that a number of analysts, although active in the use of open tubular columns, did not publish their results due to the nature of their work; however, some participated in the discussions of the frequently held symposia, and we should realize that these discussions were equally important

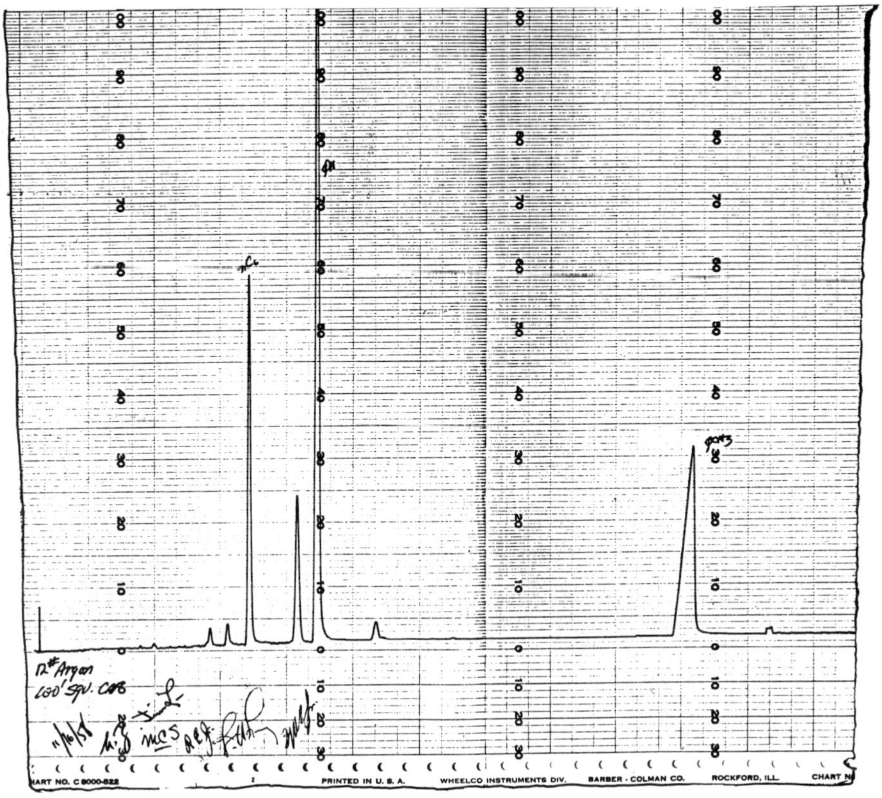

Figure 5. The first chromatogram obtained at the University of Houston, on November 16, 1958. A 30.5-m open tubular column coated with squalane was used, with an argon ionization detector. The three last peaks are n-hexane, benzene, and toluene. Identified signatures are those of A. Zlatkis, J. E. Lovelock, M. C. Simmons, R. E. Johnson and H. Lilly. (Source: Ref. 55.)

to the papers presented there. The collective contribution of all these pioneers gave us the base on which future development could be built and the results of which we are enjoying today.

3. EVOLUTION OF THE SYSTEM

Early work with open tubular columns was plagued by a number of problems, making the results of the pioneers even more remarkable when applied to our present-day standards. These problems continued for a number of years, and no discussion of the evolution of open tubular columns could be complete without mentioning them and their solutions.

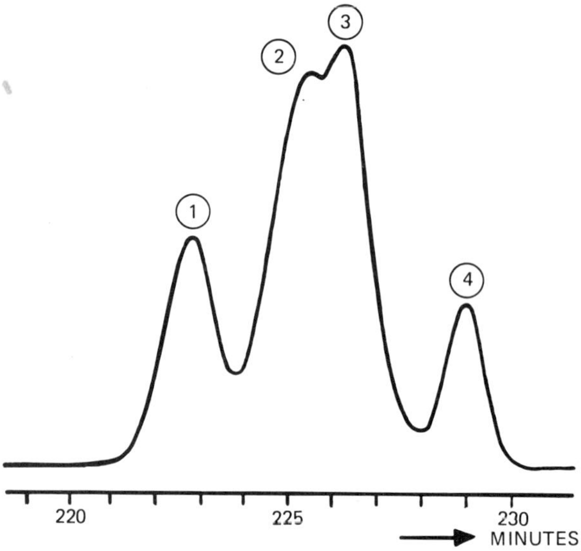

Figure 6. Separation of benzene and deuterobenzene. Column: 250 m × 0.28mm ID open tubular coated with squalane liquid phase. Column temperature: 10°C. Carrier gas (N_2) flowrate at outlet: 0.25 µl/min. Flame ionization detector. Peaks: 1—C_6H_6, 2—$C_6H_3D_3$, 3—$C_6H_2D_4$, 4—C_6D_6. (Source: Ref. 53.)

3.1 Detection

The first problem was related to detection. When Golay started his work, only thermal conductivity detectors were available and their relatively high volumes as compared to the low flowrates used with open tubular columns gave such a degree of band broadening that their practical use was precluded. This band broadening was even more evident because of the large sample sizes needed; many of the early publications clearly show column overloading. In the work of Dijkstra, already referred to, this is quite obvious. Golay designed a special small-volume thermal conductivity detector which could be used with these columns, but some band broadening still occurred in this detector, and one may question whether the general application of open tubular columns would have been possible without significant advances in detectors.

Fortunately, the development of open tubular columns coincided with the development of highly selective, small-volume ionization detectors. One of these, Lovelock's argon ionization detector [49] (in its small-volume version), was used in the quoted investigations of Zlatkis and Lipsky and represented a significant improvement; however, there were still problems

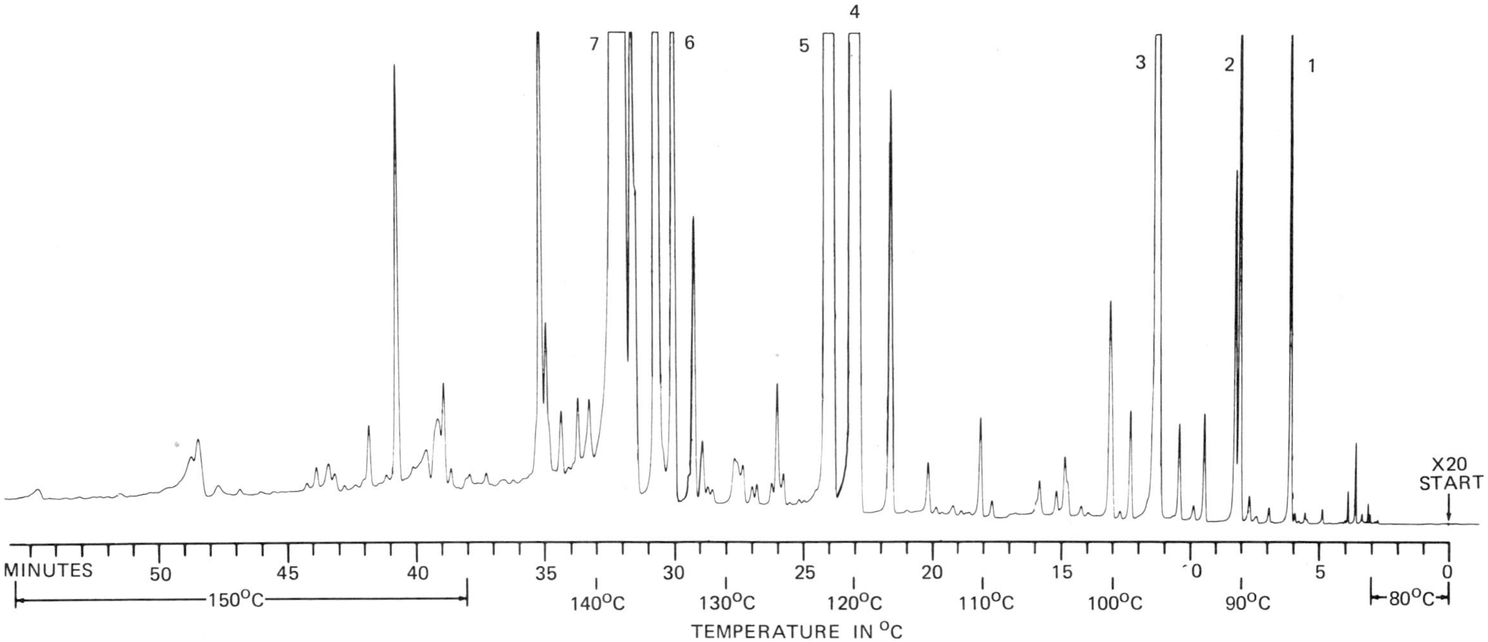

Figure 7. Analysis of a natural peppermint oil sample from Yakima Valley, Wash. Column: 45.7 m × 0.25 mm ID open tubular coated with Ucon Oil 50 HB 200 polypropylene glycol liquid phase. Column temperature: as given. Program rate: 2°C min^{-1}. Carrier gas (He) inlet pressure: 20 psig. Identified peaks: 1—α-pinene, 2—β-pinene, 3—eucalyptol, 4—menthone, 5—menthofuran, 6—menthyl acetate, 7—menthol. (Source: Ref. 91.)

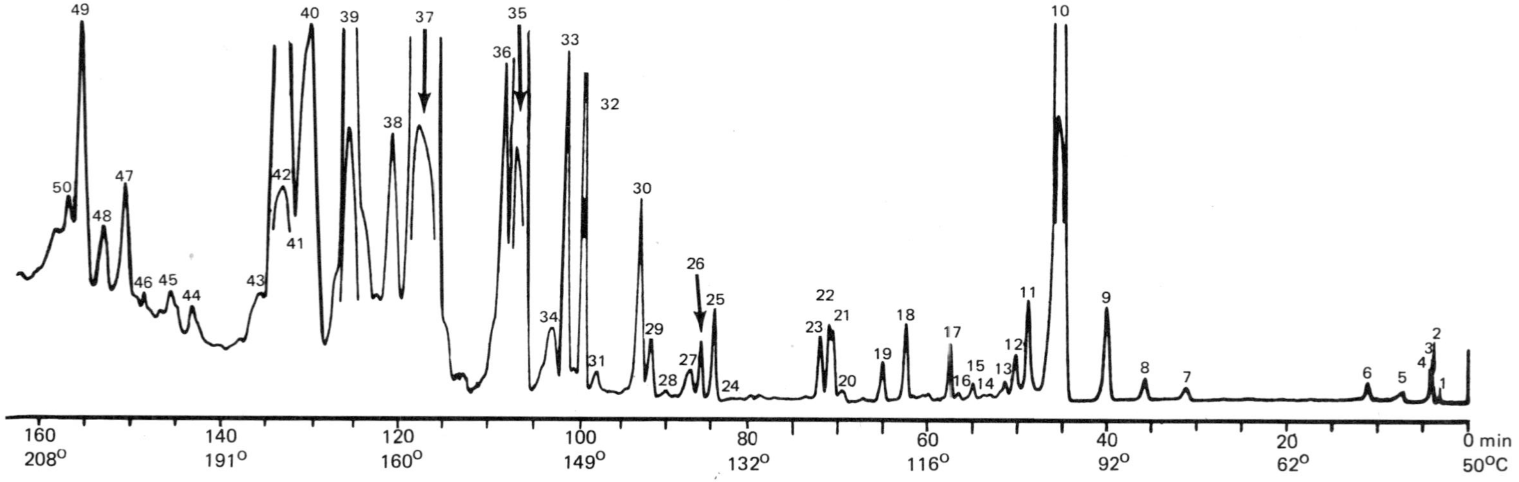

Figure 8. Analysis of a "Gebrig" hop oil. Column: 30.5 m × 0.25 mm ID open tubular coated with Tween 20 polyoxyethylene sorbitan monolaurate liquid phase. Column temperature: programmed, as given, nonlinear rates. Carrier gas (He) inlet pressure: 12 psig. Flame ionization detector. In the original chromatogram most of the peaks are identified; here, only a few are listed: 5—acetone, 7—α-pinene, 10—myrcene, 17—methyl heptanoate, 26—linalöol, 32—undecanone-2, 35—β-caryophyllene, 37—humulene. (Source: Ref. 73.)

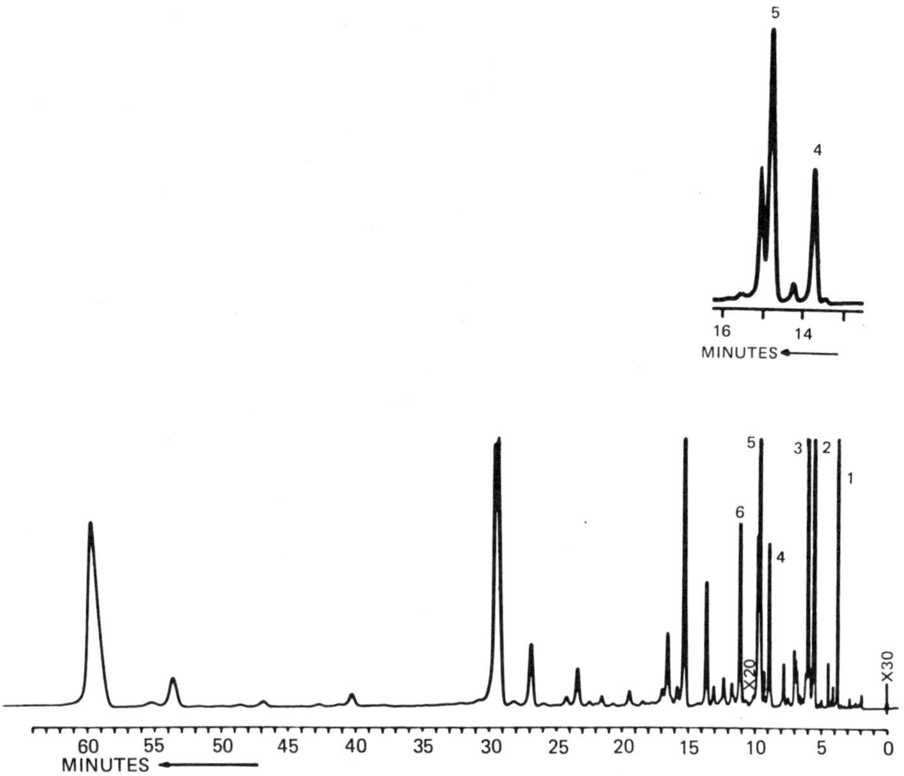

Figure 9. Analysis of the fatty acid methyl ester mixture obtained from menhaden oil. Column: 45.7 m × 0.25 mm ID open tubular coated with butanediol succinate liquid phase. Column temperature: 185°C. Carrier gas: nitrogen for the full chromatogram and helium for the cutout chromatogram. Carrier gas inlet pressure: 40 and 24 psig, respectively. Flame ionization detector. Peaks: methyl esters of 1—myristic, 2—palmitic, 3—palmitoleic, 4—stearic, 5—oleic, and 6—linoleic acids. (Source: Ref. 93.)

related to the need for a scavenger gas with this detector [59]. A major advance was achieved with the flame ionization detector of Harley, Nel, and Pretorius [95] and McWilliam and Dewar [96-97] which was introduced at this time. This detector is very sensitive and has no dead volume, and it is not surprising that it was applied almost immediately to open tubular columns. Figure 10 shows Desty's first instrument that was built a few weeks after the Amsterdam symposium. As can be seen, the end of the column was directly connected to the detector jet. Of course, Golay also immediately realized the potential of the flame ionization detector; this author well remembers that, upon joining Perkin-Elmer on October 6, 1958, R. D. Condon was al-

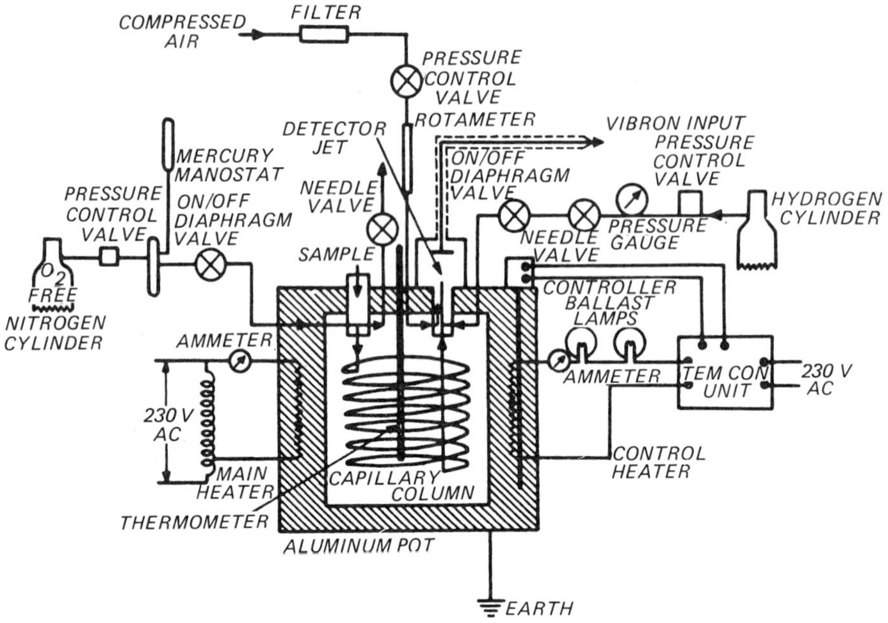

Figure 10. The schematic of the first gas chromatograph built by Desty, employing a flame ionization detector, open tubular column and split-type sample introduction. (Source: Ref. 28.)

ready obtaining one excellent chromatogram after the other on a working prototype of an instrument with open tubular columns and a flame ionization detector. This instrument was introduced at the Tenth Pittsburgh Conference on Analytical Chemistry and Applied Spectroscopy, March 2-6, 1959 [78]. Today, practically all gas chromatographs employing open tubular columns utilize flame ionization detection as the standard mode of operation.

3.2 Sample Introduction

The second problem involved sample introduction. Due to the very thin liquid-phase film, resulting in a decrease in the available liquid phase by at least two orders of magnitude as compared with the conventional packed columns used at that time, the sample capacity of open tubular columns was naturally limited. We have already mentioned the problem of column overloading, and this was not only due to the fact that one simply could not introduce a small enough sample. Desty was probably the first to routinely use split sampling; the schematic of his instrument is shown in Fig. 10. The following quotation from his paper, presented in October 1958 [28], clearly describes the problem and its solution:

> The importance of introducing a sharp plug of sample of small enough size has been stressed by many authors in connection with conventional packed columns. With capillary columns, where plate capacities are of the order of a thousand times lower than packed columns and efficiencies ten or more times higher, the introduction of extremely small samples very sharply becomes of paramount importance. We have achieved this by a dynamic division, by far the largest proportion of the sample escaping through the external adjustable needle valve.

Early split systems represented a simple tee design (see Fig. 11A), but it was soon found that such systems do not always split the sample linearly.* Thus, split design was subsequently refined, changing it to the so-called two-concentric-tube design (see Fig. 11B). It is safe to say that by 1961-1962, reliable systems were available. Halász and Schneider [43,44], Clarke [98], as well as Ettre, Gill, and Averill [99,100] dealt in detail with the various designs and the investigation of the different systems, while Condon and Ettre [101] have summarized the theory and practice of split sampling.

In the most popular design (see Fig. 11B) the split ratio is governed by the relative flow restrictions of the column and the vent line. Usually a needle valve or an interchangeable vent restrictor (capillary tube or needle) is used in the vent line to permit the variation of the split ratio. Naturally, it is important to have these at a sufficiently high temperature to prevent sample condensation in the restriction which would change the vent flow. As a precaution a small tube filled with an adsorbent and placed between the split point and vent restrictor (needle valve) was proposed as early as 1962 [102]; in this case the restrictor sees only pure carrier gas. Such systems have not been used widely in practice but there is some renewed interest in them.

In the classical design, with a restrictor (needle valve) in the vent line, the total flow (inlet pressure) is controlled. Thus, when changing the split ratio by changing the vent restrictor, the column flow will

---

*"Linear" splitting has three basic criteria: (a) every sample component is split in the same ratio without any loss; (b) when analyzing mixtures with different concentrations, the individual peak areas are proportional to concentration; (c) when changing the analytical conditions (temperature, column flow rate, split ratio, etc.), the relative sizes of the peaks in the chromatogram remain constant.

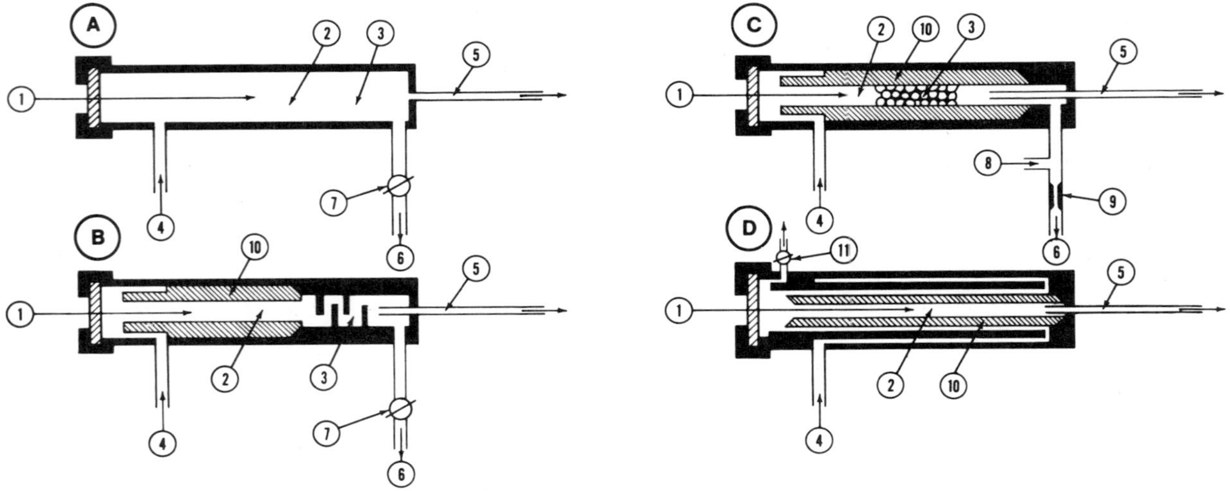

Figure 11. Four modes of sample introduction into open tubular columns. A--Bypass tee; B--concentric tube design with mixing chamber; C—inlet system developed by German and Horning; D—splitless injector with septum purge. 1—Sample injection through the septum, 2—evaporation zone, 3—mixing chamber, 4—carrier gas inlet, 5—open tubular column, 6—split vent, 7—vent restrictor (capillary tube, capillary needle, or needle valve), 8—auxiliary carrier gas inlet, 9—restrictor in the vent line, 10—(glass) insert, 11—septum purge line with needle valve.

automatically also change and readjustment is necessary. In order to avoid this problem, German and Horning in 1972 [103] developed a new system (see Fig. 11C) in which the split ratio is controlled by two gas flows. Here the inlet pressure, and hence flow velocity of the column, is controlled by the auxiliary gas flow while the vent flow is controlled by the main carrier gas flow through the whole system. If the latter is changed while keeping the auxiliary gas flow (inlet pressure) constant, the column flow will remain constant but the vent flow, and hence split ratio, is changed.

In split-type sample-introduction systems, the proper mixing of carrier gas plus sample vapor prior to splitting is important. The classical designs contained properly designed mixing chambers representing either a tortous path or baffles. In the design of German and Horning, this was replaced by a tube containing glass beads or glass wool (to provide a tortuous path). More recently, systems with a simple straight tube have also been recommended. However, these recommendations are usually given without the proper data on split linearity, dynamic range, and reproducibility and thus one should not blindly follow them: Schomburg et al. [104] clearly demonstrated the poor repeatability of peak area ratios with such a design.

The shortcomings of split sampling can be summarized in two points: part of the sample is lost and, if sample components in small concentrations are to be analyzed, the absolute amounts of these components introduced into the column might be insufficient for detection. In order to overcome this problem, the Grob family in Switzerland has introduced the so-called splitless sample-introduction technique [105-107] (see Fig. 11D). In this mode special requirements exist concerning the selection of the solvent, and thus this method cannot be considered as something for general application. After evaporation in the inlet system, the solvent vapor reaching the column is utilized to "sharpen up" the vapor band of the sample proper, thus creating sharp, narrow peaks, while that part of the solvent vapor which temporarily accumulates at the front of the inlet system (next to the septum) and which would create an extended tailing of the solvent peak is vented directly from the injector ("septum purge"). The criteria of such sampling systems have been discussed in detail in the literature [104,108, 109].

In splitless sample introduction, practically the whole volume of the injected sample is introduced into the column (except some part of the solvent). However, even here, the sample is first evaporated, and this is undesirable in some cases. It would be better if the liquid (or solid) sample could be deposited directly onto the top of the column. Such an on-column sample-introduction system was developed in 1972 at Eindhoven Technical University [110]; in this, the sample solution was coated on a wire, the solvent (having much lower boiling point than the sample proper) evaporated, and then the wire with the solid sample forwarded to the top of the column where it was then evaporated. This system, however, had only limited applications. Recently Schomburg [104] and Galli et al. [111] described new systems which are less restricted in their use. However, these inlet systems are too new to judge on their general applicability and acceptance.

4. EVOLUTION OF OPEN TUBULAR COLUMNS

The previous discussion dealt with the development of two important components of the system in which open tubular columns are applied. Naturally, without the optimum system, the superior performance possibilities of the open tubular columns could not be fully utilized. However, even the best system would mean nothing without the availability of high-efficiency reliable columns. Thus, in this last section of the chapter, I will discuss five subjects: the coating method, porous-layer open tubular columns, the material of column tubing, the variation of the liquid-phase film thickness, and, finally, the utilization of the theoretical efficiency.

4.1 The Coating Method

In the first weeks of column development, Golay coated the inside of the relatively short tubes by drawing through them a cotton wad wetted by the liquid-phase solution. Obviously, this was a very primitive method and it was difficult to control the thickness of the liquid-phase film on the inside tube wall. As the next, a coating method practically identical to the dynamic method of Dijkstra and deGoey [27] was applied, and in fact the columns on which the chromatograms shown by Golay at the 1958 Amsterdam Symposium [3] were obtained were made by Condon using this technique [27a]. However, this was soon replaced by a new method [3] which we today call the *static coating method*. The column is first completely filled with a solution of the liquid phase. One end of the tubing is then closed, following which the tubing is slowly drawn through an oven, open end first, in order to evaporate the solvent from the column.

The significant advantage of this method is that the phase ratio of the column (β), and hence the average liquid-phase film thickness, can be calculated directly from the volume concentration of the liquid-phase solution. For example, if a 0.5% by volume solution is used for coating a 0.25-mm ID (0.125 mm radius) column, then

$$\beta = \frac{V_G}{V_L} = \frac{99.5}{0.5} = 199 \tag{5}$$

and

$$d_f = \frac{r}{2\beta} = \frac{125}{398} = 0.31\ \mu m \tag{6}$$

The disadvantage of this method is that, in the original form, it can only be applied to metal columns since one cannot draw a glass tube through an oven and then recoil it. In order to overcome this problem, Ilkova and Mistryukov [112] conceived the idea of filling a precoiled glass column with the dilute coating solution, closing its lower end, and then treating the filled column as if it were a gigantic threaded screw, "screwing" it into the oven, forcing the solvent to evaporate through the open end.

This method was later modified by Jennings [113,114] by using an arc of stainless steel tubing, heated electrically to a high temperature, as the entrance to the oven. When the solution-filled column enters this tube, the solvent is abruptly converted to a superheated vapor containing an aerosol of the liquid phase developing high pressures in the column. It is claimed that these conditions are particularly favorable for uniform and stable coating.

A simpler procedure for static coating was developed by Verzele and his group [113,114]. The coiled column is completely filled with the solution of the liquid phase in a suitable low-boiling solvent and one end sealed as usual. However, now the filled column is placed under vacuum through the open end and the solvent evaporated very slowly, leaving a thin liquid-phase film on the inside wall. Obviously, this is a lengthy procedure and is difficult to apply to columns with small diameter and greater length.

The second coating method--the so-called *dynamic method*--had been first described by Dijkstra and DeGoey [27]. Here a more concentrated (about 10%) solution of the liquid phase is forced through the column under closely controlled flow conditions. The volume of the coating solution may be such that it completely fills the column*; however, more frequently, it only represents an aliquot of the total tube volume. In either case, the solution is forced through the column at low velocity (about 1 to 2 cm sec^{-1}) with nitrogen pressure.

A special modification of the dynamic technique was developed by Schomburg and Husmann [117]; this is the so-called *mercury plug technique*. After the plug of the liquid-phase solution is introduced into the column, it is followed by a plug consisting of liquid mercury, and then both are forced through the column by gas pressure. The mercury plug serves as a kind of road roller, pressing down the liquid phase to a thin film and pushing excess liquid phase out of the column. This technique, the conditions of which have been investigated in detail by Alexander and Lipsky [118], can be particularly recommended for viscous coating solutions where it is not easy to prepare a thin film.

Among these techniques, the dynamic method is used most frequently, mainly because it is simpler. Some researchers claim that the static method gives somewhat better columns, but opinions are divided.

4.2 Porous-layer Open Tubular Columns

As stated earlier, one version of Golay's equations, derived at the 1958 Amsterdam Symposium (see Eq. 4b) described the resistance to mass transfer in the liquid phase for the case when the actual inside tube area is larger than that of a smooth wall tube. As stated by Golay, further improvement in column efficiency could be obtained by replacing the smooth inner wall by a roughened surface or by a porous layer, thus further reducing liquid-phase thickness. This was then further elaborated by Golay two years later, in the lecture given at the 1960 Edinburgh Symposium [1]:

> If a coated tubular column owes its somewhat too high performance index to the thickness of its oil coating and if a packed column owes its over a hundred-fold too high performance index to the smallness of its interstitial passages, why not make a semi-packed column with a large open passage in the centre, say nine-tenths as large as the column inside diameter and with a thin-layer of packing material in the remaining space on the periphery? The answer is why not indeed? I do verily believe that such columns consti-

* This was the technique used by Condon for the 1958 Amsterdam chromatograms. [27a].

tute a nearly ideal column for a wider range of analysis than present-day smooth tubular columns.

This proposal was the basis for the porous-layer open tubular (PLOT) columns which were first realized in the work of Horvath [119-121] in the laboratory of Professor Halasz at the University of Frankfurt, West Germany; since these columns are prepared by coating the inside tube wall with a porous support layer, they are also called the support-coated open tubular (SCOT) columns. These columns were extensively studied from both the scientific and practical points of view and made commercially available by Perkin-Elmer's Applications Group in 1965-1968 [122-130].

SCOT columns have a number of advantages as compared with conventional smooth wall columns. Their sample capacity is higher, permitting smaller split ratios or even splitless sample introduction [129-130]. As a conclusion these columns are eminently suited for trace analysis, in the sub-ppm region. Figure 12 shows the analysis of "pure" n-heptane in which the concentration of the n-hexane impurity was 0.1 ppm; the peak in the chromatogram corresponds to 0.34×10^{-10}g of n-hexane [127a,129].

A further advantage of these columns is the possibility of higher flowrates. While the optimum flowrate for 0.25-mm ID columns is about 0.5 ml min^{-1} and one can rarely tolerate more than 1.5 ml min^{-1}, and flowrates up to 10 ml min^{-1} may be used. In practice, the possibility of higher flowrates means that such columns do not require the ultimate in gas chromatographic systems.

Finally, we should emphasize that the efficiency of these columns is excellent, both from the practical and the theoretical point of view. Figure 13, as an example, shows the separation of disastereoisomers on a Carbowax 1540 column; these analyses were carried out by us in the first part of 1966 on the request of B. C. Karger of Northeastern University, Boston [127b].

Due to these advantages and their convenience, SCOT columns have been and are extensively used in a large number of laboratories for day-to-day, routine applications. In fact, such columns have also been developed and are presently produced in the People's Republic of China [131].

Two shortcomings of the SCOT columns are that one cannot select completely inert support particles (just like in packed columns) and that, due to the high viscosity of the coating suspension, the length of columns which can be prepared is somewhat limited. These are, however, not major points in most of the applications in which these columns are used.

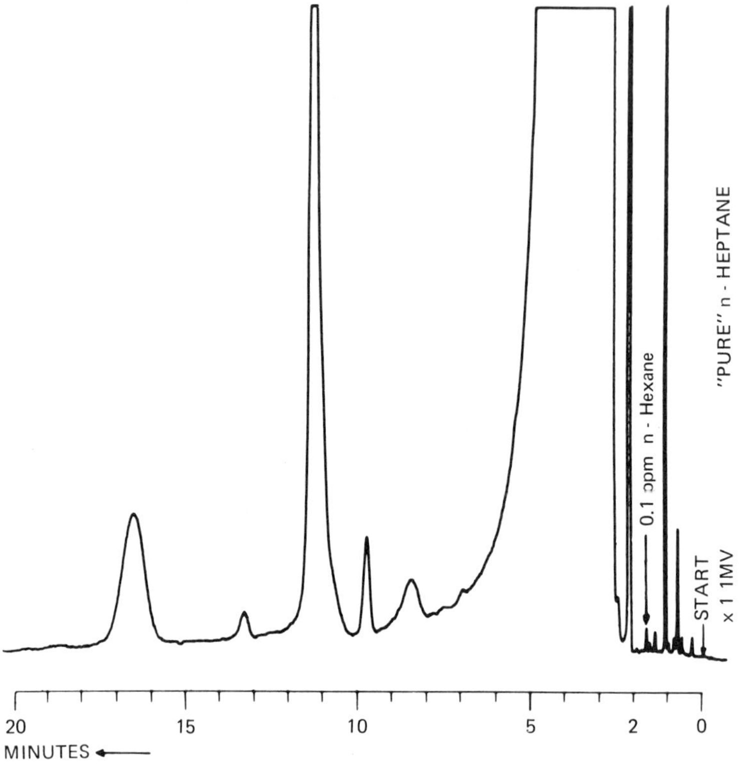

Figure 12. Analysis of traces of n-hexane in "pure" n-heptane. Column: 15.2 m × 0.50 mm ID support-coated open tubular coated with squalane liquid phase. Phase ratio: 67. Column temperature: 60°C. Carrier gas (He) flowrate: 9 ml min^{-1}. Sample volume: 0.5 μl, no split. Flame ionization detector. (Source: Ref. 127a.)

Horvath and our group at Perkin-Elmer prepared the SCOT columns with the help of the static coating method of Golay, by filling the (metal) tube with the suspension of the very fine support particles in which the liquid phase is also dissolved and then drawing the tube--with the open end up--through an oven. This method is still used by Perkin-Elmer in the manufacture of these columns. This technique results in the formation of a ringlike deposit of support particles coated with liquid phase [129,130]. Some authors [132,133,133a] used the dynamic method for the preparation of metal SCOT columns, and this technique is also partly utilized by others in the manufacturing of SCOT columns made of glass tubes. There is, however, no evidence that the dynamic technique also provides a truly ringlike coating.

We should mention here that the method of Horning et al. [134,135] for

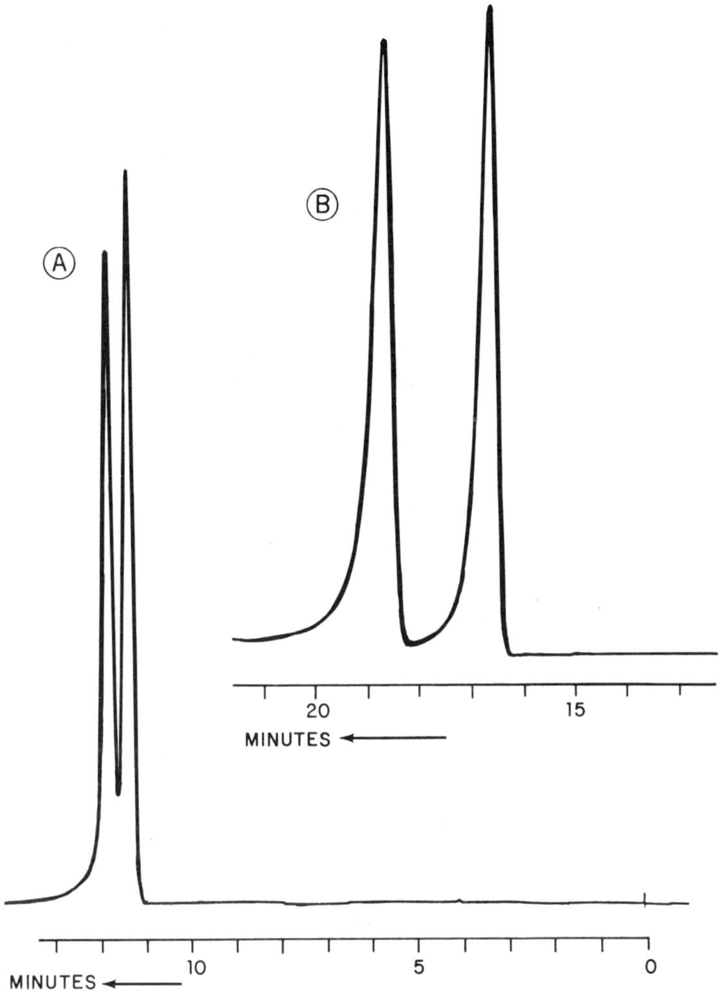

Figure 13. Analysis of diastereoisomers of (A) 2-butyl lactate and (B) 3,3-dimethyl-2-butyl lactate. Column: 15.2 m × 0.50 mm ID support-coated open tubular coated with Carbowax 1540 polyethylene glycol liquid phase. Phase ratio: 50. Column temperature: 130°C. Carrier gas (He) average linear gas velocity: 19.6 cm sec^{-1}. Flame ionization detector. (Source: Ref. 127b.)

the preparation of a stable liquid-phase film on glass open tubes by adding Silanox 101 (a silanized silica) to the liquid-phase solution may be considered similar to the above methods, and the columns prepared in this way represent a kind of SCOT column, having, however, relatively less support particles in them. A number of other authors [135a-d] also utilized a Silanox addition to prepare support-coated open tubular columns.

Other methods described for the preparation of PLOT columns include drawing the capillary from glass tubing loosely packed with support particles and providing a way to maintain an "open, unrestricted path" in the middle. Although these methods are attributed to Grant [136] and Cronin [137], they were actually first described by Jentzsch [138] and Ness [139]; for details see Ref. 129. Another very interesting method is represented by the growth of "whiskers" on the inside surface of a glass tube as developed by the group of Pretorius [140,140a].

It should finally be noted that the original Golay equations are not fully applicable to PLOT columns. Recognizing this problem, Golay reinvestigated the various terms which may have to be modified. His first brief treatment was published in 1963 [141], but in this he did not yet consider the resistance to mass transfer in the liquid phase (the CL term). Expressing the thickness of the porous layer (d_p) as a fraction of the radius of the unobstructed gas path ($r_g = r - d_p$),

$$d_p = ar_g \tag{7}$$

and neglecting the C_L term, the relationship between the HETP and the various parameters can be written as

$$h = \frac{2D_G}{u} + \left[\frac{1 + 6k + 11k^2}{(1 + k)^2} + 8a + \frac{16ka}{(1 + k)^2}\right]\frac{r_g^2\bar{u}}{24D_G} \tag{8}$$

In subsequent years, he elaborated this equation in more detail considering both the C_G and C_L terms. As derived in a paper published in 1968 [142], for porous-layer open tubular columns, these two terms can be written as*

$$C_G = \left[\frac{1 + 6k + 11k^2}{(1 + k)^2} + \frac{8 + 32k}{(1 + k)^2}a^2 + \frac{8k^2}{(1 + k)^2}\frac{a_1^2}{a_2}\right]\frac{r^2}{24D_G} \tag{9}$$

$$C_L = \frac{k^3}{6\,(1 + k)^2}\,\frac{1 + 2a_2}{F^2}\,\frac{r^2}{K^2D_L} = \frac{k}{6\,(1 + k)^2}\,\frac{1 + 2a_2}{F^2\beta^2}\,\frac{r^2}{D_L} \tag{10}$$

In these equations, r is the inside tube radius, β is the phase ratio of the column, and F is the same as already indicated in connection with Eq. (4b). The terms a_1 and a_2 are explained as follows:

---

* It should be noted that in Ref. 142 Eq. (9) was incorrectly printed: in the last fraction (after the bracket) the second power is missing.

a_1: We assume that because of the porous particles, the molecules in the gas phase of the porous layer traveling to and from the liquid phase cannot travel a straight line; the length of their tortous path is a_1r.

a_2: We assume that the thickness of the average free gas volume is a_2r. Comparing these two terms with the term used in Eq. (8), we can write that

$$a_1 > a \text{ and } a_2 < a \tag{11a-b}$$

It is also true that

$$a_1 << 1 \text{ and } a_2 << 1 \tag{11c-d}$$

4.3 Material of the Tubing

Golay's first open tubular column was made of a Tygon tubing (see the report reproduced in Fig. 1). Plastic tubing was also used by others in a few cases (see, e.g., Refs. 33 and 34), but there are obvious disadvantages with such materials: temperature limitations, poor coatability, and short life due to plasticizer migration. Still, we should mention these columns here because a very interesting report from this period demonstrated the ultimate possibilities of open tubular columns: the 1-mile-long column made of nylon tubing and coated with n-hexadecane described in 1959 by Zlatkis and Kaufman [57]. Table 1 summarizes the available information; as seen, the internal diameter was 1.7 mm and an HETP of 1.61 mm was achieved, representing less than twice the tube radius. This is an excellent result for such a column, particularly since the carrier gas velocity was obviously much higher than optimum. It is interesting to note that the gas holdup time was about 36 min and a peak with a capacity ratio of 2 emerged in about 110 min after sample introduction.

Most of Golay's early developmental work was carried out on glass columns, but at that time serious problems arose from poor coatability and the short lifetime of these columns. For this reason, metal open tubular columns were soon introduced and used almost universally for at least a decade: stainless steel was the material of choice. These tubes had a number of advantages: easy handling, mechanical strength, fairly good coatability, and stability. However, two problems soon became evident. The

Table 1. Information Concerning the 1-mile-long Open Tubular Column Described by Zlatkis and Kaufman [57]

Length	1 mile (1609 m)
Internal diameter	1.676 mm
Liquid phase	n-Hexadecane
Temperature	25°C
Carrier gas	Nitrogen
Flowrate at outlet	250 μl min^{-1}
Average linear carrier gas velocity	73.3 cm sec^{-1}
Inlet pressure	34.3 psig [2.3 atm (gauge)]
Number of theoretical plates	1,000,000
HETP	1.61 mm
Gas holdup time	36.6 min
Retention time for k = 2	109.7 min

Source: Data from Ref. 57. Values not listed in Ref. 57 were calculated by utilizing basic gas chromatographic equations.

tubes were not clean enough*, and therefore fairly elaborate cleaning procedures have been described in the literature to prepare the inside column wall for coating. The second problem was related to the activity of the metal surface. For many sample types, the use of metal tubing resulted in peak tailing, and complete sample retardation or even decomposition was sometimes experienced. A significant breakthrough occurred when Averill, in a paper presented at the 1961 Lansing Symposium [83], proposed to use long-chain surface-active additives to the liquid phase. A typical additive is Atpet 80 (sorbitan monooleate); the five hydroxyl groups at one end are assumed to attach to the column wall and from the velvetlike structure visualized in Fig. 14. The liquid-phase coating will wet this "velvet," and the sample molecules are prevented from contacting the active sites of the tube wall.

During this period glass columns were tried a number of times, and probably the most spectacular chromatogram ever achieved was obtained on

* In fact, the foreign substances present in the original tube may represent a fairly high volume "coating." For example, Porcaro, in 1963 [143], found hydrocarbon-type substances equivalent to 0.32-μm coating in the untreated metal capillary tubing.

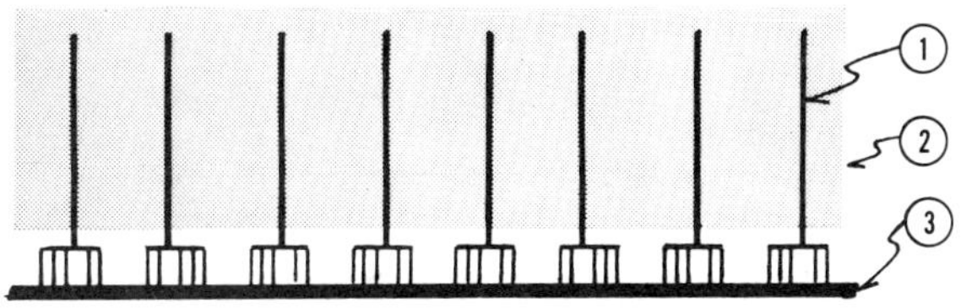

Figure 14. Idealized schematic diagram of the coated column inside wall using Averill's surface-active additives to the liquid phase. 1--Additive molecule; 2--liquid phase coating; 3--column wall. (Source: Ref. 17b.)

a 273-m × 0.152-mm ID glass open tubular column coated with squalane liquid phase. This is the famous chromatogram shown by Desty at the 1961 Lansing Symposium, illustrating the separation of the components of a Ponca crude oil [32]. Figure 15 shows a portion of the chromatogram representing the first 3 hr, 27 min. The total chromatogram occupies a whole roll of chart paper. However, Desty was one of few able to make and use glass open tubular columns at that time. The general opinion was that these columns would not last, if indeed one could make them at all. The following quotation from the discussion on open tubular columns during the 1961 Lansing Symposium is typical of the opinion at that time [144]:

> With a copper column coated with squalane, we worked for about 7 months, eight to ten, sometimes 24 hr daily. With glass capillary columns, however, coated also with squalane, we were unable to work longer than 2 or 3 days. On glass columns coated with squalane, you can see with your eyes after two days that your film is not in one piece.

Finally, starting in the latter half of the 1960s a number of researchers slowly and painstakingly developed methods to modify the inside wall of the glass tubing in order to assure a better coatability. Numerous methods have been tried, e.g., "etching" with HCL and HF; silylation; deposition of $BaCO_2$; NaCl, or silanized silica on the glass surface; undercoating with polyethylene glycol; etc. This work is still continuing, and thus only the names of the most important contributors with references to their more significant publications [134,135,145-180] are listed.

The newest development in high-resolution open tubular columns is the use of fused silica—i.e., quartz—as the material of the tubing. The preliminary report [181] looked extremely interesting; more recent developments are making it clear that this new material will be very important in the preparation of columns with stable and uniform coating.

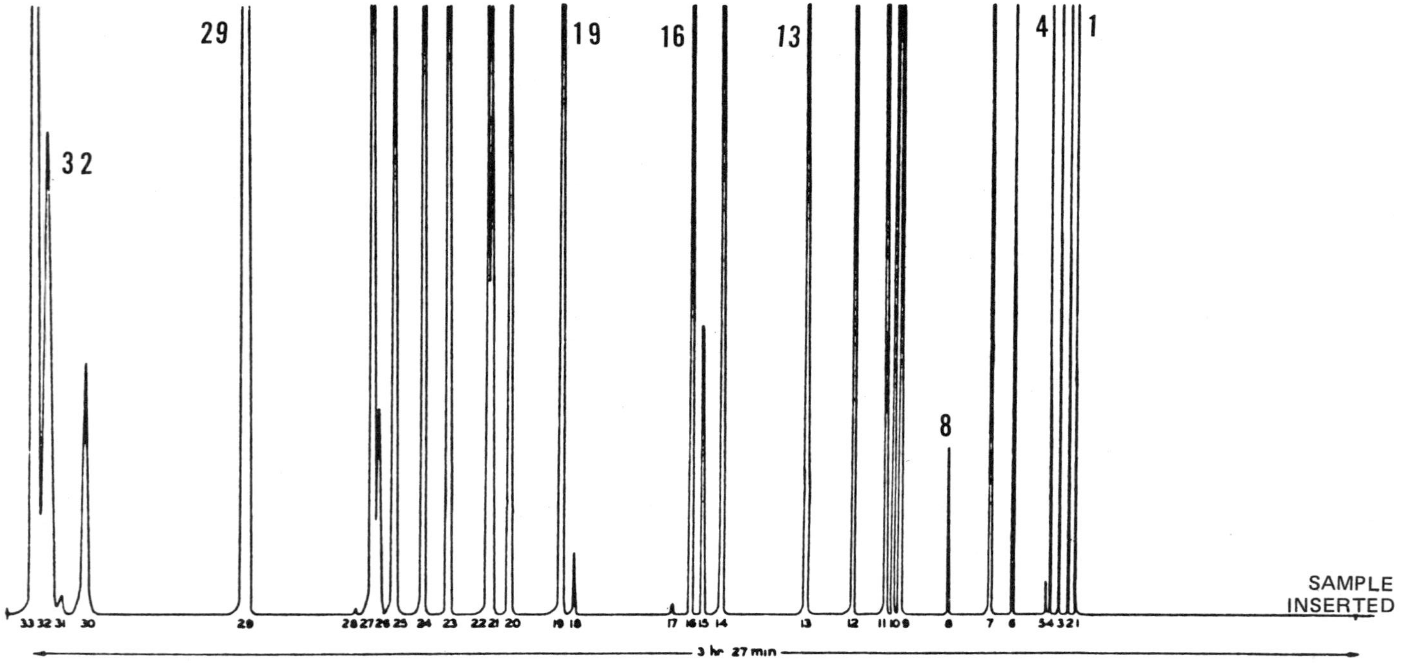

Figure 15. Desty's chromatogram of the light (-42 to 150°C cut) fraction in Ponca crude. Column: 273 m × 0.152 mm ID open tubular (glass) coated with squalane liquid phase. Column temperature: 25°C. Carrier gas (N_2) flowrate: 0.5 ml min^{-1}. Flame ionization detector. Efficiency calculated for the n-heptane peak: 300,000 effective plates. In the original publication each peak is identified; here only a few are listed: 1—methane (marker), 4—n-butane, 8—2,2-dimethylbutane, 13—n-hexane, 16—benzene, 19—cyclohexane, 29—n-heptane, 32—1,1,3-trimethylcyclopentane. (Source: Ref. 32.)

4.4 Liquid-phase Film Thickness

The need of achieving a very thin liquid-phase coating is evident from the basic GC theory. As shown by Eq. (4b), the term describing the resistance to mass transfer in the liquid phase (C_L) is proportional to the *square* of the thickness of the liquid-phase film. This had already been emphasized by Golay in his presentation at the 1958 Amsterdam Symposium [3], and Dijkstra and deGoey at the same meeting [27] predicted that the optimum film thickness for a 0.5-mm ID open tubular column for high-boiling compounds would be about 0.1 μm. On the other hand, the actual values in their columns were 1 to 2 μm, and the metal wall-coated open tubular columns used in practice for some time had an average thickness of the liquid-phase film of about 0.5 to 0.6 μm. The reason for this was twofold. It was not easy to achieve very thin films together with uniform coating, and the surface activity of the tubing became more evident when the film thickness was reduced below about 0.5 μm.

Apart from the need expressed in Eq. (4b), there is one additional reason why the adjustment of the liquid-phase film thickness is preferred. This is related to the adjustment of the capacity ratio (k). If we express the phase ratio (β) as

$$\beta = \frac{r}{2d_f} \tag{12}$$

and substitute it into the expression of the partition coefficient

$$K = \beta k \tag{13}$$

then the following relationship is obtained:

$$K = \frac{rk}{2d_f} \tag{14}$$

According to basic GC theory (see e.g., Refs. 17, 182, 183) compounds with smaller capacity ratio values are more difficult to separate than those with large-capacity ratio values. Thus, for low-boiling compounds, it is desirable to increase the film thickness, thereby decreasing the phase ratio of the column, and so increasing capacity ratios. On the other hand, for high-boiling compounds, the opposite is true. Here, we would like to reducc the capacity ratio (i.e., increase the phase ratio) in order to reduce the time the compounds spend in the column:

$$t_R = t_M (1 + k) = \frac{L}{\bar{u}} (1 + k) \tag{15}$$

This adjustment of the phase ratio is possible with SCOT columns, and the quoted literature [122-130] gives many examples. However, for the reasons mentioned, such an adjustment is not possible with metal columns; there the only practical possibility to adjust the phase ratio is by changing the column radius (see Eq. 12).

The new glass and quartz open tubular columns with significantly reduced surface activity finally permit us to adjust the thickness of the liquid-phase film according to the sample analyzed. This question has been discussed in detail in the recent literature [184-186]. Thus, while the so-called general-purpose columns have an average liquid-phase film thickness of about 0.2 to 0.3 μm, columns with both thinner and thicker films are used according to the particular need. The two extremes are illustrated here by two chromatograms. Figure 16 represents the analysis of amino acid derivatives on a column with a calculated film thickness well below 0.1 μm [187], while Fig. 17 illustrates the use of a column with an about 0.9 μm-thick film for the analysis of low-boiling hydrocarbons [188]. Some characteristic data for this column as compared to a similar column with 0.2-μm film thickness and operated at the same conditions are given below:

		0.2 μm	9.0 μm
Capacity ratio:	n-hexane	0.23	1.04
	n-octane	1.13	5.16
HETP of n-hexane		0.25 mm	0.33 mm
Resolution of n-butane/isobutane		1.66	4.50

4.5 Utilization of the Theoretical Best Efficiency

As expressed by the Golay equation (see Eq. 1), the relationship between the HETP and the average linear gas velocity is described by a hyperbolic curve. This hyperbola has a minimum at

$$\mathrm{HETP}_{min} = 2\sqrt{B(C_G + C_L)} \tag{16}$$

For high values of β, C_G dominates the resistance to mass transfer and C_L can be considered very small compared to C_G. If C_L is negligibly small, Eq. (16) can be written by substituting the proper expressions for B and C_G from Eqs. (2) and (3):

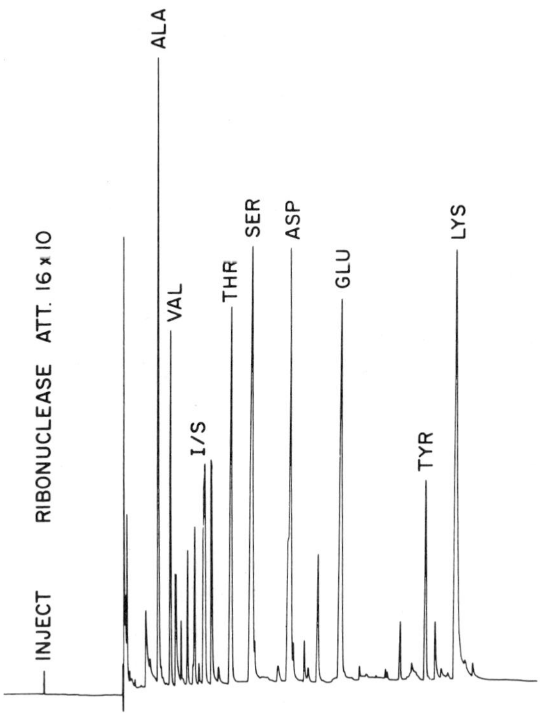

Figure 16. Analysis of the amino acids in ribunuclease hydrolyzate, in the form of the n-propyl, N-acetyl derivatives. Column: 50 m × 0.27 mm ID open tubular (glass) coated with a 1:1 mixture of Carbowax 20M and Silar 5CP. Film thickness: <0.1 μm. Column temperature: programmed at 8°C min^{-1} from 110 to 190°C then ballasted to 250°C. Carrier gas (He) flowrate: 0.75 μl min^{-1}. Split sample injection: 1:20 split. Thermionic detector sensitive for nitrogen compounds. Peaks: ALA--alanine, VAL--valine, I/S—norleucine (internal standard), THR—threonine, SER—serine, ASP—aspartic acid, GLU—glutamic acid, TYR—tyrosine, LYS—lysine. The retention time of the last peak is less than 30 min. (Source: Ref. 187.)

$$HETP_{min} = 2\ BC_G = r\sqrt{\frac{1 + 6k + 11k^2}{3(1 + k)^2}} \tag{17}$$

In other words, $HETP_{min}$ is proportional to the inside tube radius and the two limiting values are

$$\text{For } k = 0: \quad HETP_{min} = 0.58r \tag{18a}$$

$$\text{For } k = \infty: \quad HETP_{min} = 1.9r \tag{18b}$$

The considerations on $HETP_{min}$ were first mentioned by Golay in his 1958 Amsterdam paper [3] and one year later by Desty et al. [29], who also

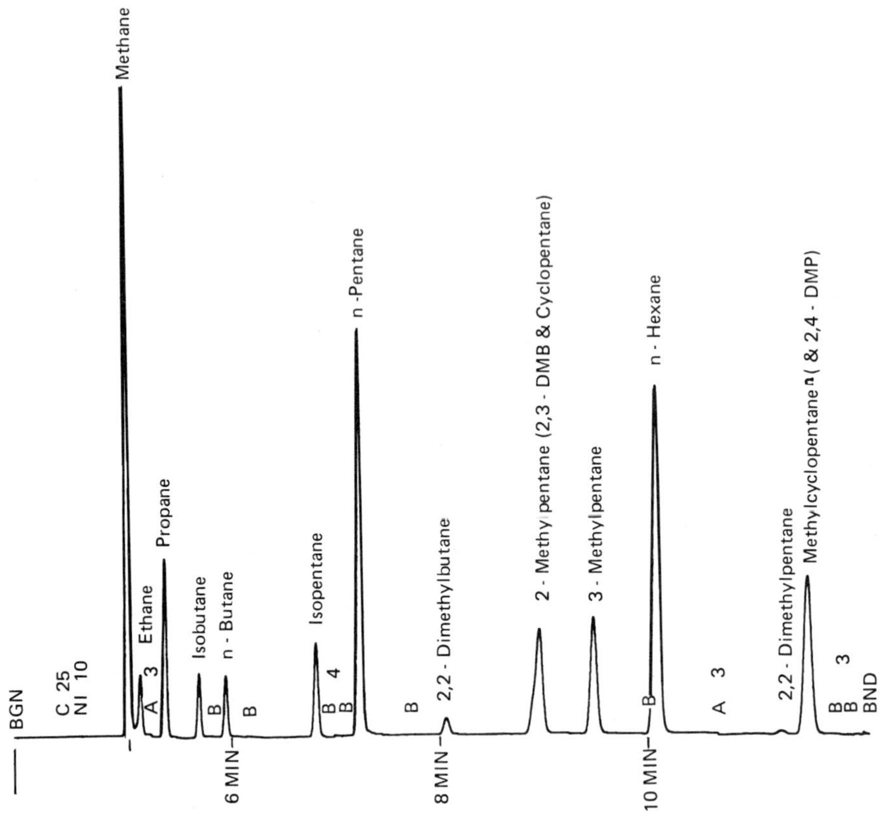

Figure 17. Analysis of C_1 through C_7 hydrocarbons. Column: 55 m × 0.27 mm ID open tubular (glass) coated with OV-101 methyl silicone liquid phase. Film thickness: 0.9 μm. Phase ratio: 75. Column temperature: 69°C. Carrier gas (He) average linear gas velocity: 19.0 cm sec^{-1}. (Source: Ref. 188.)

mentioned that if one compares actual HETP values with the corresponding values of $HETP_{min}$ calculated from Eq. (17), the actual values for earlier peaks will be closer (about 50 to 60%) to the theoretical values, while for later peaks the actual HETP values might be 3 times those of $HETP_{min}$ theoretical. This meaning of the theoretical $HETP_{min}$ was further discussed by M. A. Khan during the discussion of Desty's paper presented at the 1960 Edinburgh Symposium [31].

We have found this concept very important, and in 1963 [90] we investigated the relationship between actual HETP and theoretical $HETP_{min}$ values for a number of wall-coated open tubular columns having different diameters. Our numerical results supported Desty's rule and showed that, in general, the larger the column diameter, the poorer the agreement between the theoretical and actual values. A few years later, when investigating SCOT columns, we again used this concept to see how close these columns approached the ideal theoretical values.

In a book published in 1965 [17c] this author elaborated on this concept, exploring the reasons why only a fraction of the theoretical maximum plate number is obtained:

1. In Eq. (17), the C_L term is neglected although it is not equal to zero.
2. Columns have been assumed to be ideal, i.e., with uniform film thickness and constant diameter along the entire length.
3. It has been assumed that ideal gas chromatographic systems are used in which the width of the peak is a function only of partitioning in the column.
4. It has been assumed that one is working at optimum gas velocities, although this is almost never the case during practical analysis.

Nevertheless, this comparison, which in 1966 [126] I termed as the *utilization of the theoretical best efficiency* (UTE) and expressed as a percentage value,

$$\text{UTE\%} = \frac{\text{HETP}_{\min(\text{theor})}}{\text{HETP}_{\text{act}}} 100 \tag{19}$$

is a convenient value and is not in contradiction to chromatographic theory since it can be written as

$$\text{UTE\%} = \frac{\sigma^2_{\min(\text{theor})}}{\sigma_{\text{act}}} \tag{20}$$

and, after all, band spreading in the column is proportional to the variance.

Three conclusions can be drawn from the previous discussion:

1. The value of UTE% will depend on the column *and* the chromatographic system and not only on the column itself, since HETP_{act} will also reflect, e.g., band broadening in the injector.
2. Even if we disregard the chromatographic system, HETP_{act} embraces the *whole column* and is not restricted to the liquid phase. For example, it is very sensitive to variation in the column radius along the column (which influences *both k and r*) or to any adsorption on active sites of the glass surface which does not necessarily have to result in observable peak tailing but can simply increase the bandwidth.

3. Obviously, if $HETP_{act}$ is not measured at exactly the optimum average linear gas velocity, the UTE% value will be lower than could be obtained on that column.

These points mean that the value of UTE%, while indicative of the *whole column and the system*, cannot be related only to the liquid-phase coating. In spite of this, however, to the surprise of this author, in the years following the 1966 Rome Symposium, others renamed this term the "coating efficiency" and used it to show how good the *liquid-phase coating* is; even worse, companies producing open tubular columns for commercial sale used this term to specify their column efficiencies.

These points were emphasized by the author at the 1973 Bratislava Symposium during the discussion of Alexander's paper [158]. However, use of the term *coating efficiency*, like the terms *capillary column* and *(GC)*^{2}, has continued.

It is also important to realize that Eq. (19)—which expresses the deviation of experimental performance from theoretical ideality—is *only* valid for wall-coated open tubular columns *having a smooth inside wall.* For support-coated open tubular columns, neglecting again the C_L term, one may derive an $HETP_{min(theor)}$ expression from Eq. (8) [126]:

$$HETP_{min(theor)} = r_g \sqrt{\frac{k + 6k + 11k^2}{3(1 + k)^2} + \frac{8a}{3} + \frac{16ka}{3(1 + k)^2}} \qquad (21)$$

where a and r_g are as specified in Eq. (7).

For example, at k = 3 and a = 0.1, the value of the square root term in Eq. (21) is 1.6808 versus a value of 1.5679 for the square root term in Eq. (17). This means that, for a column with an equivalent radius, this term increased by 7.2%, that is, the $HETP_{min(theor)}$ is 7.2% larger than it would be in an ideal wall-coated open tubular column assuming that the resistance to mass transfer in the liquid phase is negligible in both.

At the Third Hindelang Symposium on High Resolution Glass Columns (April 29 to May 3, 1979) the incorrect meaning of the coating efficiency term was discussed in detail. It was also pointed out that, for present-day glass open tubular columns with roughened surface which, however, still cannot be termed as having a "porous layer", most likely neither form of the $HETP_{min(theor)}$ equations is fully valid and some correction will have to be added to properly describe the diffusion phenomena at this roughened surface. This possibility underlines even more the fallacy of the "coating efficiency" term to characterize the quality of these columns and the skill of the people preparing them.*

5. CONCLUSIONS

In the 20 years which have passed since their invention, high-resolution open tubular columns have undergone remarkable development, and the analysis of complex mixtures would be unthinkable today without them.

A renaissance of high-resolution gas chromatography was made possible by the most recent developments, particularly in the preparation of the proper glass tubing for coating by the liquid phase. However, we should not forget that these results are part of a continuing evolution and that none of them would have been possible without the activities of M. J. E. Golay and those other pioneers who, in the first decade, laid down the foundations of open tubular column gas chromatography.

ACKNOWLEDGMENTS

The author would like to express his gratitude to Academic Press, New York, Butterworths Publishers, London, and the Deutsche Akademie der Wissenschaften, as well as to the editors and publishers of the *Journal of Chromatography, Journal of Chromatographic Science,* and *Nature* for permission to use copyrighted figures; to Messrs. N. Johansen, F. J. Kabot, and E. W. March for unpublished chromatograms; to the management of the Perkin-Elmer Corporation for permission to quote internal reports and use chromatograms from company publications; and last but not least, to Professor J. P. Friend and Mr. D. R. Rushnek for placing at his disposal the historical chromatograms shown in Figs. 2 and 5.

* A most recent paper published after submitting the text of this chapter [189] dealt with this question in detail, emphasizing the importance of also considering the pressure drop.

LIST OF SYMBOLS

a	Ratio of porous layer thickness to the radius of the unobstructed gas path
a_1, a_2	Terms in the Golay equation describing the influence of porous layer in PLOT columns
B	Term in the Golay equation describing longitudianl diffusion
C_G	Term in the Golay equation describing the resistance to mass transfer in the gas phase
C_L	Term in the Golay equation describing the resistance to mass transfer in the liquid phase
d_f	Average thickness of the liquid-phase film
d_p	Average thickness of the porous layer in PLOT columns
D_G	Gaseous diffusion coefficient
D_L	Liquid diffusion coefficient
F	Ratio of the actual inside tube area and the geometric inside area of a smooth wall tube
h	Height equivalent to one theoretical plate (HETP)
$HETP_{min}$	Minimum of the HETP vs $\overline{u}$ plot
$HETP_{min(theor)}$	Calculated theoretically achievable best HETP value
$HETP_{act}$	Actual HETP value
k	Capacity ratio
K	Partition coefficient
L	Column length
r	Inside tube radius
r_g	Radius of the unobstructed gas path in the column
t_M	Gas holdup time
t_R	Retention time
$\overline{u}$	Average linear carrier gas velocity
UTE	Utilization of the theoretical best efficiency
β	Phase ratio
σ	Standard deviation of a Gaussian peak
σ^2	Variance of a Gaussian peak

REFERENCES

1. Golay, M. J. E., in *Gas Chromatography 1960* (Edinburgh Symposium), Butterworths, London, 1960, pp. 139-143.

2. Golay, M. J. E., in *Gas Chromatography* (1957 Lansing Symposium) Coates, V. J., H. J. Noebels, and I. S. Fagerson, eds., Academic Press, New York, 1958, pp. 1-13.

3. Golay, M. J. E., in *Gas Chromatography 1958* (Amsterdam Symposium) Desty, D. H., ed., Butterworths, London, 1958, pp. 36-55.

4. *British Standard 3282*, rev. ed., Glossary of Terms to Gas Chromatography, British Standards Institution, London, 1969; first published 1963.

5. *ASTM E 355-77*, Gas Chromatography Terms and Relationships, American Society for Testing and Materials, Philadelphia, 1977; first published 1968.

6. Recommendations on Nomenclature, *Pure Appl. Chem.*, 37; 447 (1974).

7. van Deemter, J. J., Zuiderweg, F. J., and Klinkenberg, A., *Chem. Eng. Sci.* 5;271 (1956).

8. Keulemans, and Kwantes, A., in *Vapor Phase Chromatography* (1956 London Symposium) (D. H. Desty, ed.), Butterworths, London, 1957, pp. 15-34.

9. Keulemans, A. I. M., *Gas Chromatography*, Reinhold, New York, 1957.

10. Dijkstra, G., in *Seventy-five Years of Chromatography: A Historical Dialogue* L. S. Ettre and A. Zlatkis, eds., Elsevier, Amsterdam, 1979, pp. 43-51.

11. Martin, A. J. P., in *Vapour Phase Chromatography* (1956 London Symposium) (D. H. Desty, ed.), Butterworths, London, 1957, pp. 1-4.

12. Desty, D. H., in *Advances in Chromatography*, Vol. 1., (J. C. Giddings and R. A. Keller, eds.), Dekker, New York, 1965, pp. 199-228.

13. Golay, M. J. E., 129th National American Chemical Society Meeting, Dallas, April 8-13, 1956; *Anal. Chem.*, 29:928 (1957).

14. Golay, M. J. E., Notes on Gas Chromatographic Parameters, *Engineering Report No. 522*, August 22, 1956, Perkin-Elmer Corp., Norwalk, Conn.

15. Golay, M. J. E., Discussion of Results with Three Experimental Gas-Liquid Chromatographic Columns, Engineering Report No. 523, September 5, 1956, Perkin-Elmer Corp., Norwalk, Conn.

16. Golay, M. J. E., Progress Report on Gas Chromatographic Experimental Work for September and October 1956, November 15, 1956, Perkin-Elmer Corp., Norwalk, Conn; partly quoted in Ref. 17a.

17. Ettre, L. S., *Open Tubular Columns in Gas Chromatography*, Plenum Press, New York, 1965; a p.3, b p.86, c p.15.

18. Courtesy of Prof. J. P. Friend, Drexel University.

19. Golay, M. J. E., (to Perkin-Elmer Corp.), U. S. Patent 2,920,478, filed June 24, 1957, issued January 12, 1960.

20. Golay, M. J. E., *Nature*, 180:435 (1957).

21. Golay, M. J. E., in *Seventy-five Years of Chromatography: A Historical Dialogue* (L. S. Ettre and A. Zlatkis, eds.), Elsevier, Amsterdam, 1978, pp. 109-114.

22. Taylor, G., *Proc. Roy. Soc.*, 219A:186 (1953).

23. Taylor, G., *Proc. Roy. Soc.*, 223A:446 (1954).

24. Taylor, G., *Proc. Roy. Soc.*, 225A:473 (1954).

25. Westhaver, J. W., *Ind. Eng. Chem.*, 34:126 (1942).

26. Westhaver, J. W., *J. Res. Nat. Bur. Stand.*, 38:169 (1947).

27. Dijkstra, G., and DeGoey, J., in *Gas Chromatography 1958* (Amsterdam Symposium) (D. H. Desty, ed.), Butterworths, London, 1958, pp. 56-68.

27a. Condon, R. D., personal communication to the author.

28. Desty, D. H., *Abh. Deut. Akad. Wiss. Berlin, Kl. Chem. Geol. Biol.*, No. 9, pp. 176-184 (1959).

29. Desty, D. H., Goldup, A., and Whyman, B. H. F., *J. Inst. Petrol.*, 45:287 (1959).

30. Desty, D. H., Goldup, A., and Swanton, W. T., *Nature*, 183:107 (1959).

31. Desty, D. H., and Goldup, A., in *Gas Chromatography 1960* (Edinburgh Symposium) (R. P. W. Scott, ed.), Butterworths, London, 1960 pp. 162-183.

32. Desty, D. H., Goldup, A., and Swanton, W. T., in *Gas Chromatography* (1961 Lansing Symposium) (N. Brenner, J. E. Callen, and M. D. Weiss, eds.), Academic Press, New York, 1962, pp. 105-138.

33. Scott, R. P. W., *Nature*, 183:1753 (1959).

34. Scott, R. P. W., and Hazeldean, G. S. F., in *Gas Chromatography 1960* (Edinburgh Symposium) (R. P. W. Scott, ed.), Butterworths, London, 1960, pp. 144-161.

35. Scott, R. P. W., *J. Inst. Petrol.*, 47:284 (1961).

36. Hazeldean, G. S. F., and Scott, R. P. W., *J. Inst. Petrol.*, 48:380 (1962).

37. Grant, D. W., Potentialities of Capillary Columns in the Analysis of Coal Tar, *Coal Tar Res. Assoc. Report* No. 627, 1960.

38. Grant, D. W., and Vaugham, G. A., in *Gas Chromatography 1962* (Hamburg Symposium) (M. van Swaay, ed.), Butterworths, London, 1962, pp. 305-314.

39. Adlard, E. R., and Whitham, B. T., in *Gas Chromatography* (1961 Lansing Symposium) (N. Brenner, J. E. Callen, and M. D. Weiss, eds.), Academic Press, New York, 1962, pp. 371-390.

40. Adlard, E. R., Khan, M. A., and Whitham, B. T., in *Gas Chromatography 1962* (Hamburg Symposium) (M. van Swaay, ed.), Butterworths, London, 1962, pp. 84-101.

41. Halász, I., and Schreyer, G., *Chem.-Ing.-Tech.*, 32:675 (1960).

42. Halász, I., and Schreyer, G., *Z. Anal. Chem.*, 181:367 (1961).

43. Halász, I., and Schneider, W., *Anal. Chem.*, 33:978 (1961).

44. Halász, I., and Schneider, W., in *Gas Chromatography* (1961 Lansing Symposium) (N. Brenner, J. E. Callen, and M. D. Weiss, eds.), Academic Press, New York, 1962, pp. 287-306.

45. Kaiser, R. E., and Struppe, H. G., in *Gas-Chromatographie 1959* (Böhlen Symposium) (R. E. Kaiser and H. G. Struppe, eds.), Akademie Verlag, Berlin, 1959, pp. 177-194.

46. Kaiser, R. E., Struppe, H. G., Holzhäuser, H., and Kuhl, H., *Freiberger Forschungsh.*, A192:205 (1961).

47. Jentzsch, D., in *DECHEMA Monographien*, Vol. 43 (1961 ACHEMA Symposium, Frankfurt, Main) (H. Bretschneider and K. Fischbeck, eds.) Verlag Chemie, Weinheim, 1962, pp. 231-240.

48. Jentzsch, D., and Hövermann, W., in *Gas Chromatography 1962* (Hamburg Symposium) (M. van Swaay, ed.), Butterworths, London, 1962, pp. 204-215.

49. Jentzsch, D., and Hövermann, W., *J. Chromatogr.* 11:440 (1963).

50. Henneberg, D., and Schomburg, G., in *Gas Chromatography 1962* (Hamburg Symposium) (M. van Swaay, ed.), Butterworths, London, 1962, pp. 191-203.

51. Bruner, F., Cartoni, G., and Liberti, A., *Chem. Ind. (Milan)*, 44:999 (1962).

52. Cartoni, G., Liberti, A., Pallotta, U., and Palombardi, R., *Riv. Ital. Sostanze Grasse*, 40:653 (1963).

53. Bruner, F., and Cartoni, G. P., *J. Chromatogr.*, 10:396 (1963).

54. Liberti, A., Cartoni, G. P., and Bruner, F., *J. Chromatogr.*, 12:8 (1963).

55. Courtesy of D. R. Rushnek, Interface, Inc., Fort Collins, Colo.

56. Zlatkis, A., and Lovelock, J. E., *Anal. Chem.*, 31:620 (1959).

57. Zlatkis, A., and Kaufman, H. R., *Nature*, 184:2010 (1959).

58. Zlatkis, A., in *Lectures on Gas Chromatography 1962* (H. A. Szymanski, ed.), Plenum Press, New York, 1963, pp. 87-104.

59. Lipsky, S. R., Landowne, R. A., and Lovelock, J. E., *Anal. Chem.* 31:852 (1959).

60. Lipsky, S. R., Lovelock, J. E., and Landowne, R. A., *J. Amer. Chem. Soc.*, 81:1010 (1959).

61. Lipsky, S. R., Lovelock, J. E., and Landowne, R. A., *Anal. Chem.*, 33:818 (1961).

62. Landowne, R. A., and Lipsky, S. R., *Biochim. Biophys. Acta*, 46:1 (1961).

63. Landowne, R. A., and Lipsky, S. R., *Biochim. Biophys. Acta*, 47:589 (1961).

64. Hollis, O. L., *Anal. Chem.*, 33:352 (1961).

65. Hollis, O. L., and Hayes, W. V., *Anal. Chem.*, 34:1223 (1962).

66. Simmons, M. C., Richardson, D. B., and Dvoretzky, I., in *Gas Chromatography 1960* (Edinburgh Symposium) (R. P. W. Scott, ed.), Butterworths, London, 1960, pp. 211-223.

67. Richardson, D. B., Simmons, M. C., and Dvoretzky, *J. Amer. Chem. Soc.*, 83:1934 (1961).

68. Durrett, L. R., Taylor, L. M., Wantland, C. F., and Dvoretzky, I., *Preprints, Div. Petrol. Chem. Amer. Chem. Soc.*, 7(3):5 (1962); *Anal. Chem.*, 35:637 (1963).

69. Schwartz, R. D., and Brasseaux, D. J., *Preprints, Div. Petrol. Chem. Amer. Chem. Soc.*, 8(4C):77 (1963); *Anal. Chem.*, 35:1374 (1963).

70. Schwartz, R. D., Brasseaux, D. J., and Shoemake, G. R., *J. Gas Chromatogr.*, 1(1):32 1(2):17 (1963).

71. Teranishi, R. C., Nimmo, C. C., and Corse, J., *Anal. Chem.*, 32:1384 (1960).

72. Buttery, R. G., Hendel, C. E., and Boggs, M. M., *J. Agr. Food Chem.*, 9:245 (1961).

73. Buttery, R. G., McFadden, W. H., Teranishi, R., Kealy, M. P., and Mon, T. R., *Nature*, 200:435 (1963).

74. Buttery, R. G., and Teranishi, R., *J. Agr. Food Chem.* 11:504 (1963).

75. McFadden, W. H., and Teranishi, R., *Nature*, 200:329 (1963).

76. McFadden, W. H., Teranishi, R., Black, D. R., and Day, J. C., *J. Food Sci.*, 28:316 (1963).

77. Teranishi, R., Corse, J. W., McFadden, W. H., Black, D. R., and Morgan Jr., A. I., *J. Food Sci.*, 28:478 (1963).

78. Condon, R. D., *Anal. Chem.*, 31:1717 (1959).

79. Averill, W., *Instrument News*, 12(1):6 (1960).

80. Averill, W., in *Progress in Industrial Gas Chromatography*, (H. A. Szymanski, ed.), Plenum Press, New York, 1961, pp. 31-36, 225-230.

81. Ettre, L. S., *Instrument News*, 13(1a):1 (1961).

82. Brenner, N., and Ettre, L. S., *Acta. Chim. Acad. Sci. Hung.*, 27:205 (1961).

83. Averill, W., in *Gas Chromatography* (1961 Lansing Symposium) (N. Brenner, J. E. Callen, and M. D. Weiss, eds.), Academic Press, New York, 1962, pp. 1-6.

84. Ettre, L. S., Cieplinski, E. W., and Brenner, N., *ISA Reprints* No. 78-La/61 (1961); *ISA Trans.*, 2:134 (1963).

85. Norem, S. D., *Anal. Chem.*, 34:40 (1962).

86. Averill, W., and Ettre, L. S., *Nature*, 196:1198 (1962).

87. Ettre, L. S., and Kabot, F. J., *Anal. Chem.*, 34:1431 (1962).

88. Ettre, L. S., Averill, W., and Kabot, F. J., Gas Chromatographic Analysis of Fatty Acids, GC Applications No. GC-AP-001, Perkin-Elmer Corp., Norwalk, Conn., 1962

89. Averill, W., *J. Gas Chromatogr.*,1(1):22 (1963).

90. Ettre, L. S., Cieplinski, E. W., and Averill, W., *J. Gas Chromatogr.*, 1(2):7 (1963).

91. Cieplinski, E. W., and Averill, W., Gas Chromatographic Analysis of Essential Oils Using Golay Columns and a Flame Ionization Detector,

GC Applications No. GC-AP-002, Perkin-Elmer Corp., Norwalk, Conn., 1963.

92. Ettre, L. S., Cieplinski, E. W., and Kabot, F. J., *J. Gas Chromatogr.*, 1(11):38 (1963).

93. Kabot, F. J., unpublished results, Perkin-Elmer Corp., Norwalk, Conn., 1963; see Ref 17, Figs. 14 and 33.

94. Lovelock, L. E., *Nature*, 182:1663 (1958).

95. Harley, J., Nel, W., and Pretorius, V., *Nature*, 181:117 (1958).

96. McWilliam, I. G., and Dewar, R. A., *Nature*, 181:760 (1958).

97. McWilliam, I. G., and Dewar, R. A., in *Gas Chromatography 1958* (Amsterdam Symposium) (D. H. Desty, ed.), Butterworths, London, 1958, pp. 142-152.

98. Clarke, D. R., *Nature*, 198:681 (1963).

99. Ettre, L. S., and Averill, W., *Anal. Chem.*, 33:680 (1961).

100. Gill, H. A., and Averill, W., 13th Pittsburgh Conf. Anal. Chem. and Appl. Spectroscopy, Pittsburgh, March 1962; reprint; Perkin-Elmer Corp., Norwalk, Conn., 1962

101. Condon, R. D., and Ettre, L. S., in *Instrumentation in Gas Chromatography* (J. Krugers, ed.), Centrex, Eindhoven, 1968, pp. 87-105.

102. Jentzsch, D., and Hövermann, W., (to Perkin-Elmer), U.S. Patent 3,257,775, filed August 5, 1963, issued June 28, 1966; German patent issued October 20, 1962.

103. German, A. L., and Horning, E. C., *Anal. Lett.*, 5:619 (1972).

104. Schomburg, G., Behlau, H., Dielmann, R., Weeke, G., and Husmann, H., *J. Chromatogr.*, 142:87 (1977).

105. Grob, K., and Grob, G., *Chromatographia*, 5:3 (1972).

106. Grob, K., and Grob, Jr., K., *J. Chromatogr.*, 94:53 (1974).

107. Grob, K., and Grob, Jr., K., *J. High Resolution Chromatogr.*, 1:57 (1978).

108. Hartigan, M. J., Giordano, B., and Ettre, L. S., *Chromatogr. Newslett.*, 5:24 (1977).

109. Yang, F. J., Brown III, A. C., and Cram, S. P., *J. Chromatogr.*, 158:91 (1978).

110. Vanden Berg, P. M. J., and Cox, T., *Chromatographia*, 5:301 (1972).

111. Galli, M., Trestianu, S., and Grob, Jr., K., *J. High Resolution Chromatogr.*, 2:366 (1979).

112. Ilkova, E. L., and Mistryukov, E. A., *J. Chromatogr. Sci.*, 9:569 (1971).

113. Jennings, W. G., and Wohleb, R., *Chem. Mikrobiol. Technol. Lebesm.*, 3:33 (1974).

114. Jennings, W. G., Yabamoto, K., and Wohleb, R. H., *J. Chromatogr. Sci.* 12:344 (1974).

115. Bouche, J., and Verzele, M., *J. Gas Chromatogr.*, 6:569 (1971).

116. Verzele, M., Verstappe, M., Sandra, P., van Lüchene, E., and Vuye, A., *J. Chromatogr. Sci.*, 10:668 (1972).

117. Schomburg, G., and Husmann, H., *Chromatographia*, 8:517 (1975).

118. Alexander, G. and Lipsky, S. R., *Chromatographia*, 10:487 (1977).

119. Horváth, C., inaugural dissertation, J. W. Goethe University, Frankfurt, Main, 1963

120. Halasz, I., and Horvath, C., *Anal. Chem.*, 35:499 (1963).

121. Halasz, I., and Horvath, C., *Nature*, 197:71 (1963).

122. Ettre, L. S., Purcell, J. E., and Norem, S. D., *J. Gas Chromatogr.*, 3:181 (1965).

123. Purcell, J. E., and Ettre, L. S., *J. Gas Chromatogr.*, 4:23 (1966).

124. Ettre, L. S., and Purcell, J. E., in *Chromatographie et Méthods de Séparation Immediate*, Vol. 1, (G. Parrisakis, ed.), Union of Greek Chemists, Athens, 1966, pp. 67-84.

125. Ettre, L. S., Purcell, J. E., and Billeb, K., *J. Chromatogr.*, 24:335 (1966).

126. Ettre, L. S., in *Gas Chromatography 1966* (Rome Symposium) (A. B. Littlewood, ed.), Institute of Petroleum, London, 1967, pp. 115-118.

127. Ettre, L. S., Purcell, J. E., and Billeb, K., *Separation Sci.*, 1:777 (1966).

127a. March, E. W., unpublished results, Perkin-Elmer Corp., Norwalk, Conn., 1968: see Ref. 129, Fig. 11.

127b. Ettre, L. S., and Billeb, K., unpublished results, Perkin-Elmer Corp., Norwalk, Conn., 1966.

128. Averill, W., *Chromatogr. Newslett.*, 1:27 (1972).

129. Ettre, L. S., and Purcell, J. E., in *Advances in Chromatography, Vol. 10* (J. C. Giddings and R. A. Keller, eds.), Dekker, New York, 1974, pp. 1-94.

130. Horvath, C., Ettre, L. S., and Purcell, J. E., *Amer. Lab.*, 6(8):75 (August 1974).

131. Lue, Z. F., Tung, Y. Y., Chao, R. M., Ou, Q. Y., Yang, P. Y., Dong, K. N., and Yu, W. L., *J. High Resolution Chromatogr.*, 2:429 (1979).

132. Kaiser, R. E., *Chromatographia*, 1:34 (1968).

133. Nikelly, J., *Anal. Chem.*, 44:623, 625 (1972).

133a. Nikelly, J., *Anal. Chem.*, 45:2280 (1973).

133b. Blumer, M., *Anal. Chem.*, 45:980 (1973).

133c. Nikelly, J. G., and Blumer, M., *Amer. Lab.*, 6(1):12 (January 1974).

134. German, A. L., Pfaffenberger, C. D., Thenot, J. P., Horning, M. G., and Horning, E. C., *Anal. Chem.*, 45:930 (1973).

135. van Hout, P., Szafranek, J., Pfaffenberfer, C. D., and Horning, E. C., *J. Chromatogr.*, 99:103 (1974).

135a. Blakesley, C. N., and Torline, P. A., *J. Chromatogr.*, 105:385 (1975).

135b. McKeag, R. G., and Hougen, F. W., *J. Chromatogr.*, 136:308 (1977).

135c. Deelder, R. S., Ramaekers, J. J. M., van den Berg, J. H. M., and Wetzels, M. L., *J. Chromatogr.*, 119:99 (1976).

135d. Thiecke, R. J., van den Berg, J. J. M., Deelder, R. S., and Ramaekers, J. J. M., *J. Chromatogr.*, 160:264 (1978).

136. Grant, D. W., *J. Gas Chromatogr.*, 6:18 (1969).

137. Cronin, D. A., *J. Chromatogr.*, 48:406 (1970).

138. Jentzsch, D., (to Perkin-Elmer Corp.), U.S. Patent 3,182,394, filed December 1, 1962, issued May 11, 1965.

139. Ness, H. H., (to Perkin-Elmer Corp.), U.S. Patent 3,238,759, applied November 12, 1963, issued March 8, 1966.

140. Schieke, J. D., Comins, N. R., and Pretorius, V., *Chromatographia*, 8:354 (1975).

140a. Schieke, J. D., Comins, N. R., and Pretorius, V., *J. Chromatogr.*, 112:97 (1975).

140b. Schieke, J. D., and Pretorius, V., *J. Chromatogr.*, 132:217 (1977).

140c. Schieke, J. D., and Pretorius, V., *J. Chromatogr.*, 132:223 (1977).

140d. Schieke, J. D., and Pretorius, V., *J. Chromatogr.*, 132:231 (1977).

141. Golay, M. J. E., *Nature*, 199:370 (1963).

142. Golay, M. J. E., *Anal. Chem.*, 40:382 (1968).

143. Porcaro, P. J., *J. Gas Chromatogr.*, 1(6):17 (1963).

144. Discussion on Open Tubular Columns, in *Gas Chromatography* (1961 Lansing Symposium) (N. Brenner, J. E. Callem, and M. D. Weiss, eds.), Academic Press, New York, 1962, p. 560.

145. Grob, K., in *Gas Chromatography 1966* (Rome Symposium) (A. B. Littlewood, ed.), Institute of Petroleum, London, 1967, pp. 113-114.

146. Grob, K., and Grob, G., *Chromatographia*, 5:3 (1972).

147. Grob, K., and Jaeggi, J., *Chromatographia*, 5:382 (1972).

148. Grob, K., *Chromatographia*, 7:94 (1974).

149. Grob, K., *Chromatographia*, 8:423 (1975).

150. Grob, K., and Grob, G., *J. Chromatogr.*, 125:471 (1976).

151. Grob, K., Grob, G., and Grob, Jr., K., *Chromatographia*, 10:181 (1977).

152. Grob, K., Grob, G., and Grob, Jr., K., *J. High Resolution Chromatogr.*, 2:31 (1979).

153. Tesarik, K., and Novotny, M., in *Gas-Chromatographie 1968* (Berlin Symposium) (H. G. Struppe, ed.), Akademie Verlag, Berlin, 1968, pp. 575-584.

154. Novotny, M., and Tesarik, K., *Chromatographia*, 1:332 (1968).

155. Novotny, M., Blomberg, L., and Bartle, K. D., *J. Chromatogr. Sci.*, 8:390 (1970).

156. Novotny, M., Bartle, K. D., *Chromatographia*, 7:122 (1974).

157. Alexander, G., and Rutten, G. A. F. M., *Chromatographia*, 6:231 (1973).

158. Alexander, G., Garzo, G., and Palyi, G., *J. Chromatogr.*, 91:25 (1974).

159. Alexander, G., and Rutten, G. A. F. M., *J. Chromatogr.*, 99:81 (1974).

160. Rutten, G. A. F. M., and Luyten, J. A., *J. Chromatogr.*, 74:177 (1972).

161. Franken, J. J., Rutten, G. A. F. M., and Rijks, J. A., *J. Chromatogr.*, 126:117 (1976).

162. Franken, J. J., de Nijs, R. C. M., and Schulting, F. L., *J. Chromatogr.*, 144:253 (1977).

163. de Nijs, R. C. M., Franken, J. J., Dooper, R. P. M., Rijks, J. A., de Ruwe, H. J. J. M., and Schulting, F. L., *J. Chromatogr.*, 167:337 (1978).

164. Rutten, G. A. F. M., and Rijks, J. A., *J. High Resolution Chromatogr.*, 1:279 (1978).

165. de Nijs, R. C. M., Rutten, G. A. F. M., Franken, J. J., Dooper, R. P. M., and Rijks, J. A., *J. High Resolution Chromatogr.*, 2:447 (1979).

166. Schomburg, G., Husmann, H., and Weeke, F., *J. Chromatogr.*, 99:63 (1974).

167. Schomburg, G., Husmann, H., and Weeke, F., *Chromatographia*, 10:580 (1977).

168. Roerade, J., *Chromatographia*, 8:511 (1975).

169. Watanabe, C., and Tomita, H., *J. Chromatogr. Sci.*, 13:123 (1975).

170. Badings, H. T., van der Pol, J. J. G., and Wassink, J. G., *Chromatographia*, 8:440 (1975).

171. Badings, H. T., van der Pol, J. J. G., and Schmidt, D. G., *Chromatographia*, 10:404 (1977).

172. Badings, H. T., van der Pol, J. J. G., and Wassink, J. G., *J. High Resolution Chromatogr.*, 2:297 (1979).

173. Welsch, T., Engewald, W., and Klancke, C., *Chromatographia*, 10:22 (1977).

174. Sandra, P., and Verzele, M., *Chromatographia*, 10:419 (1977).

175. Sandra, P., Verstappe, M., and Verzele, M., *Chromatographia*, 11:223 (1978).

176. Sandra, P., Verstappe, M., and Verzele, M., *J. High Resolution Chromatogr.*, 1:28 (1978).

177. Verzele, M., and Sandra, P., *J. High Resolution Chromatogr.*, 2:303 (1979).

178. Merle d'Aubigne, J., Landault, C., and Guiochon, G., *Chromatographia*, 4:309 (1971).

179. Hartigan, M. J., and Ettre, L. S., *J. Chromatogr.*, 119:187 (1976).

180. Hartigan, M. J., Billeb, K., and Ettre, L. S., *Chromatographia*, 10:571 (1977).

181. Dandeneau, R. D., and Zerenner, E. H., *J. High Resolution Chromatogr.*, 2:351 (1979).

182. Jennings, W., *Gas Chromatography with Glass Capillary Columns*, Academic Press, New York, 1978.

183. Ettre, L. S., *Introduction to Open Tubular Columns*, Perkin-Elmer Corp., Norwalk, Conn., 1978.

184. Averill, W., and March, E. W., *Chromatogr. Newslett.*, 4:20 (1976).

185. Ettre, L. S., All-Union Symposium on Chromatography, Samarkand, U.S. S.R., September 20-22, 1976.

186. Grob, Jr., K., and Grob, K., *Chromatographia*, 10:250 (1977).

187. Adams, R. F., Vandemark, F. L., and Schmidt, G. J., *J. Chromatogr. Sci.*, 15:63 (1977).

188. Johansen, N., to be published

189. Cramers, C. A., Wijnheijmer, F. A., and Rijks, J. A., *Chromatographia*, 12:643 (1979).

2

SYSTEM REQUIREMENTS

WALTER G. JENNINGS

University of California
Davis, California

1. Introduction 49

2. Inlets 50

3. Oven Considerations 51

4. Detector Considerations 51

5. Carrier Gas 52

6. Signal Handling 53

7. Column Installation 53

References 54

1. INTRODUCTION

Of all available analytical techniques, gas chromatography (GC) is the most widely used; indeed, its popularity continues to increase, and it has become the analytical method of preference for a wide number of applications. In addition, nonvolatile materials from many analytical areas are converted to volatile derivatives and analyzed by GC. This is largely because GC offers several advantages over other methods of analyzing volatile compounds: it normally achieves better separations, analysis times are short, and the sensitivity is usually high. Glass capillary columns promise to eventually capture the major portion of GC applications because, as compared to packed, support-coated, or porous-layer columns, they are capable of much more efficient separations, even shorter analysis times, and even higher sensitivities [1]. In addition, they can be much more inert, and some compounds that suffer attrition or total disappearance in packed or metal (including nickel) capillary columns can be analyzed with glass capillaries. Recent developments have produced columns of fused quartz and fused silica that not only can be even more inert, but also offer certain other advantages (vide infra). These considerations of glass capillary gas chromatography have been explored [1].

The glass capillary system, however, is less forgiving and more demanding of equipment design and operator technique. Instrumental defects or sloppy injection techniques that cause no noticeable problems with packed column instruments can lead to disastrous results on glass capillary chromatograms. In this chapter we will explore some of the system and operator requirements for good results in glass capillary GC.

2. INLETS

Four general methods are most widely used to introduce the sample plug onto the glass capillary column: cold trapping, split [2,3], "splitless" [4-12], and on-column injection [13,14]. Other injection techniques that offer advantages for special application--e.g., capsule insertion, falling needle, thermal focusing--have also been suggested. These various inlet systems have been recently reviewed [1]. The choice of injection technique is usually governed by the type of equipment available, the amount of sample obtainable, the concentration of components to be detected, component lability, and operator skill. Schomburg [15] compared the results of various split designs with splitless and direct injections.

The whole idea of the injection process is to introduce detectable quantities of the substances to be detected on the beginning of the column in the shortest possible length of sample plug. With cold trapping, splitless, and on-column injection techniques, a finite period is allowed for the deposition of sample on the column before the chromatographic process begins. With split injections, on the other hand, the high flowrate through the inlet chamber, most of which is split to atmosphere, must rapidly flush that chamber clean so that the short plug deposited on column is followed by pure carrier gas. When the initial column temperature is sufficiently low and/or the sample injected contains considerable solvent, cold trapping and/or splitless injection may also play roles in split injections. Split injection is the simplest technique, and even the novice can obtain good results provided that the concentrations of the components to be detected fall roughly within the range 0.2 to 10% of the sample. Syringe technique can have a profound effect on the subsequent chromatography. With an inlet splitter operating at split ratios to 1:50 to 1:100, injection sizes of 0.2 to 1 μl are usually used; injections as large as 5 to 8 μl may be required for the detection of trace components, but this may result in overloads for some peaks. After drawing the sample into the syringe, the syringe needle should be wiped down and the plunger drawn back an additional 1 to 2 cm. This withdraws sample from the needle and into the body of the syringe before the needle is inserted into the heated inlet. Otherwise that sample flashing from the needle as it is inserted into the heated inlet begins the chromatographic process before the main sample is injected. This can result in "doublets" (more noticeable on early peaks) for each peak in the chromatogram [1].

3. OVEN CONSIDERATIONS

Ideally, the oven should possess a low thermal mass, a high-velocity fan, and precise temperature control; there should be no possibility of radiant heat reaching any portion of the column. A number of manufacturers encase the glass capillary column in a protective cage, but with experience one soon learns that these columns are much less fragile than the uninitiated would believe. The author prefers a free-hanging installation on a rack located centrally in the oven; five to ten coils on each end of the column are not placed on the hanger, but are left free to provide flexibility to the inlet and detector connections.

4. DETECTOR CONSIDERATIONS

Most commercial chromatographs are fitted with detectors designed for use with packed columns, and the detector may not operate properly at the low carrier gas flows emerging from a capillary column. An auxillary or makeup gas is added to compensate for the difference between the optimum carrier gas flow for that detector (consult the manufacturer's specifications) and that delivered by the capillary column (normally 1 to 3 cm^3 min^{-1}). The makeup gas used can be different from the carrier, and should be that providing the greatest detector sensitivity, i.e., nitrogen for flame ionization detectors (FIDs) and argon/methane for electron capture detectors. The outlet end of the glass capillary column should project beyond the point of makeup (and hydrogen, for FID) gas addition, so that the end of the column is in a high-velocity gas zone [1,2]. Most modern chromatographs contain a built-in provision for makeup gas, and various manufacturers offer makeup gas kits for the conversion of older designs [1].

5. CARRIER GAS

Particle filters should always be used on carrier gas lines; entrained dust or particulate matter will quickly destroy the glass capillary, evidenced by peak broadening and a rapid loss of column efficiency. Some liquid phases (e.g., Carbowax 20M, SP 1000, FFAP) are very sensitive to trace amounts of water and oxygen, and the carrier gas supplying these should be passed through gas driers and oxygen scrubbers to prevent their rapid deterioration. These devices should preferably be located upstream from the dust or particle filters. The methyl silicone columns supplied by some manufacturers seem to be quite resistant to water and oxygen [3], and the use of gas driers and oxygen scrubbers is usually a needless expense for these columns.

Inadequate interpretations of the classical van Deemter plot have been responsible for many errors in the choice of carrier gas and in setting the carrier gas velocity. Nitrogen is frequently selected because it gives a slightly lower value for h at u_{opt}; such a selection overlooks several salient points. First, the increase in h as one proceeds from nitrogen to helium to hydrogen is disproportionately small when compared to the increase in u_{opt} values; i.e., even for operation at u_{opt}, analysis times can be made much shorter with the less dense carrier gases [1,2]. Secondly, the curve for hydrogen is much flatter than that for helium, which in turn is flatter than that for nitrogen. This means that the less dense carrier can accommodate a wider range of partition ratios (i.e., a greater array of solute retention times) at or near optimum conditions than can the denser gases. Thirdly, the rate at which h increases with increasing u is highest for nitrogen and lowest for hydrogen; where time of analysis is a consideration (or temperature of elution in programmed runs), hydrogen is a vastly superior carrier gas. Because of the explosic hazard, many investigators prefer (or in the more safety-conscious United States, are required) to use helium.

6. SIGNAL HANDLING

Even if all other factors that can affect the chromatographic results--injection technique, inlet temperature, dead space or excessive volumes in the flow path, column efficiency, makeup gas considerations, detector design--have been resolved, results (particularly where older instruments have been converted to capillary capability) may be less than satisfactory. Ultimately, the signal produced by the detector must be amplified, and it is then usually fed to a strip chart recorder. With older instruments, the electrometer may prove limiting, in that it simply cannot respond fast enough to follow the very narrow peaks generated by a glass capillary column operating under ideal conditions. Similarly, the recorder (and/or integrator) should be capable of full-scale response in less than ½ sec; electrical or mechanical sluggishness can result in the lowering and broadening of peaks.

7. COLUMN INSTALLATION

Where an inlet splitter is used, it is important that both the inlet and outlet ends of the capillary column lie in high-velocity gas zones. This can be achieved by straightening the column ends or by using special fittings that achieve the same goal with unstraightened columns [2]. A major advantage of the thin wall flexible fused quartz and fused silica columns is that they are inherently straight; one need merely free from the column ties whatever length is required for the inlet and outlet ends [1].

Where splitless, cold trapping, or on-column injection techniques are used, the gas velocity at the inlet end of the column is less important, but that at the outlet end remains critical.

Column connections commonly use graphite or Vespel ferrules, or in some cases the column may be fused to a length of platinum-iridium tubing [2]. Ferrule connections should be designed so that the graphite or Vespel does not come in contact with the flow path, i.e., the column should extend through the ferrule and terminate well beyond it.

REFERENCES

1. Jennings, W., *Gas Chromatography with Glass Capillary Columns*, 2nd. ed., Academic Press, New York, 1980.

2. Jennings, W., and Shibamoto, T., *Qualitative Analysis of Flavor and Fragrance Volatiles with Glass Capillary Gas Chromatography*, Academic Press, New York, 1980.

3. Jennings, W., *J. Chromatogr. Sci.*, 13:185 (1975).

4. Grob, K., and Grob, G., *J. Chromatogr. Sci.*, 7:584,587 (1969).

5. Grob, K., and Grob, G., *Chromatographia*, 5:3 (1972).

6. Grob, K., *J. Chromatogr.*, 90:303 (1974).

7. Grob, K., and Grob, Jr., K., *J. Chromatogr.*, 94:53 (1974).

8. Grob, K., and Grob, Jr., K., *HRC&CC*, 1:57 (1978).

9. Jennings, W. G., Freeman, R. R., and Rooney, T. A., *HRC&CC*, 1:275 (1978).

10. Schulte, E., and Acker, L., *Z. Anal. Chem.*, 268:260 (1974).

11. Schulte, E., *Chromatographia*, 7:138 (1974).

12. Miller, R. J., and Jennings, W. G., *HRC&CC*, 2:72 (1979).

13. Grob, K., and Grob, Jr., K., *J. Chromatogr.*, 151:311 (1978).

14. Grob, K., *HRC&CC*, 1:263 (1978).

15. Schomburg, G., Behlau, H., Dielmann, R., Weeke, F., and Husmann, H., *J. Chromatogr.*, 142:87 (1977).

3

APPLICATIONS OF GLASS CAPILLARY GAS CHROMATOGRAPHY IN XENOBIOTIC RESEARCH

JEROME E. BAKKE

United States Department of Agriculture
Fargo, North Dakota

LARS BLOMBERG
GUNNAR WIDMARK

University of Stockholm
Stockholm, Sweden

1. Introduction 55
2. Sample Preparation and Injection 56
3. Columns 60
4. Detectors 62
5. Applications 63
6. Two-Dimensional Gas Chromatography 66
References 68

1. INTRODUCTION

One of the most significant advances in gas chromatography has been the high resolving power that can be obtained with the wall-coated open tubular glass capillary column (WCOT). The practical use of this major improvement in column technology has required simultaneous developments in the design of detectors and simultaneous changes in the concepts of sample preparation and sample introduction. The problems associated with sample preparation and sample introduction, we believe, are the main reasons for the limited use that has been made of WCOTs in biological research.

Even though the principles of glass capillary gas chromatography are the same as for packed column gas chromatography, the technology in actual practice is very different. Therefore, a review of the progress that has been made in the use of WCOT, stressing the applications and needs of the xenobiotic* scientist, is appropriate because many of the technological problems associated with their use have been overcome with the commercial availability of prepared columns and the instruments designed to accomodate them.

It is our intent not only to familiarize xenobiotic scientists with the uses that have been made of the WCOT in xenobiotic studies, but also to present some of the xenobiotic researcher's problems to the chromatography specialists. The latter, we believe, is necessary because the development of columns and instrumentation capable of performing these separations is far ahead of the xenobiotic scientist's ability to prepare samples that can directly utilize this extremely efficient and sensitive separation methodology. Those attempting to isolate, identify, and quantify trace amounts of metabolites from biological souces have enough difficulty to purify their samples to a point where high-capacity packed columns can be utilized without considering the sample preparation required for the low-capacity WCOTs. Thus, we would hope to stimulate studies on new sample-injection systems and more contact between biological scientists and gas chromatography specialists.

2. SAMPLE PREPARATION AND INJECTION

Most of the natural forces that act upon xenobiotics result in metabolites with a wide range of polarities and, in cases where an increase in polarity results from conjugation reactions, an increase in molecular weight. The polarities of the metabolites from any one xenobiotic can range from very nonpolar (for example, the methylsulfides and methylsulfones from polychlorinated biphenyls [1,2]) to the very polar glycosides, glucuronides, amino acid conjugates, and sulfate esters. In general, metabolic processes convert xenobiotics to compounds that exhibit the broad range of chemical properties similar to the natural products from which they are to be ex-

---

* For this chapter, a *xenobiotic* may be defined as any compound, whether man-made or natural, that has been added by man to some compartment of the biosphere.

tracted. Therefore, the wide variations in concentration, volatility, polarity, and stability of both the metabolites and the accompanying natural products made sample preparation for ultimate use of any of the modern high-performance chromatographic methods very difficult. All of the above-mentioned sample characteristics must be overcome for the analysis of these extracts by gas chromatography. Furthermore, to utilize the high resolution of WCOTs, which are easily overloaded, the ranges in concentration become of paramount importance.

The difficulties that arise from sample polarity and volatility are usually overcome by appropriate derivatization of the sample components. The derivatization techniques used in packed column methodology are applicable for WCOTs. These methods have been adequately reviewed [3,4,5]. For the use of WCOTs, the derivatization reaction mixtures may require additional processing to eliminate excess reagents, overwhelming concentrations of derivatized natural products, and nonvolatile natural products and/or reagents. These nonvolatile compounds can often be tolerated by the high-capacity packed column but cannot be tolerated by the WCOT systems. The reactive derivatization reagents can also destroy the integrity of the phase film in the WCOT.

To utilize the high resolving power of WCOTs, it is imperative that the sample enter the column in a narrow band. The appropriate bandwidth cannot be achieved using the conventional packed column injection technique because the evaporation of the sample and the solvent requires a relatively large injector volume to contain the vapor formed on rapid injection. The subsequent transfer of this gas volume into the column, at the slow carrier gas flowrates used, may take several minutes. This slow transfer would result in a broad sample band. To avoid this problem, the carrier gas flow through the injector can be divided (split injection). One exit of this split is through the column; the other exit is out of the injector. By the proper temperature control of the splitter and adjustment of the splitting ratio, the desired sample bandwidth on the column is obtained. The use of the split injection requires that the compounds of interest be present in high enough concentrations so that, for example, 1/50 the sample injected is enough for their detection. For trace analysis and for many biological samples, this is not the case. Another disadvantage of the split injection technique is that most splitters do not split the sample components uniformly, i.e., low- and high-boiling components are split at different ratios (sample discrimination). This discrimination complicates the quantitation of sample components that have widely differing volatilities.

A splitless sample-injection technique was introduced by Grob and Grob [6]. In this procedure, the sample is vaporized in the injector and the sample and solvent are carried into a column that is maintained at a temperature low enough to collect the components in a narrow band in the first part of the column while the solvent is passed through the column. After a predetermined sample transfer time, the materials remaining in the injector are vented out of the instrument. The injection parameters (solvent, injector, and column temperature and transfer interval) must be determined for each sample. The sample that was trapped at the beginning of the column is then chromatographed by increasing the column temperature.

In addition to the obvious volatility differences required between the sample components and the solvent, several other sample conditions must be present to use the splitless injection technique. The sample cannot contain excessive amounts of nonvolatile or thermally unstable materials that will decompose in the injector. If these decomposition products do not affect the first elution, they can lead to artifacts that will appear as ghost peaks on subsequent runs. It is therefore necessary to routinely clean the glass liner of the injector to remove these materials. It is also necessary that the large amounts of solvents and/or derivatization reagents that pass through the column do not destroy the integrity of the stationary-phase film on the column wall. The exposed glass surface that can result can give rise to adsorption of sample components and broadening and tailing of peaks, and, with labile compounds, can result in the catalytic formation of artifacts.

The physical design of the components of the instrument is also important for the successful use of splitless injection. It is essential that no portion of the injection system, including the exit port through which the excess sample is vented, be in thermal contact with the column oven. When the column oven is cold for the sample transfer, sample will condense on these cooled surfaces and will be subsequently volatilized when the oven temperature is increased. Materials volatilized from these surfaces will appear as ghost peaks and/or column bleed. This is true for any condensable compounds bleeding into the injector from sources such as valves, gaskets, septa, and the carrier gas. These latter sources of contaminants are a problem even with thermally isolated injectors.

The retention of the sample components at the beginning of the column, when splitless injection is used, not only is a function of the column temperature but also is aided by the solvent plug which precedes the sample. For a discussion of this effect, see Ref. 7. The utility of the effect of the solvent plug in splitless injection has been demonstrated by Buser [8] in the analysis of polychlorinated dibenzo-p-dioxins.

Another useful injection system that is commercially available is the all-glass falling needle injector that was first described by Ros [9] and given its present configuration by van den Berg and Cox [10]. With this injector, the sample is applied to a quartz probe that is initially contained outside the heated injection port. The solvent is evaporated and removed from the injector using a back flush of carrier gas. To inject the sample, the probe is moved into the heated zone at the head of the column where the sample is volatilized and carried into the column. The advantages of this system are that the solvent does not enter the column, nonvolatile constituents in the sample will remain on the probe, and isothermal chromatographic conditions can be used. A limitation of the system is that the volatility of the components of interest must be such that they will not be evaporated with the solvent.

It is possible to form certain derivatives in situ on the probe of the all-glass solid sample injector described above. Lauwereys and Vercruysse [11] described the in situ formation of the trimethylsilyl derivatives of methylprylon and pyrithyldion. The test compounds were applied to the probe in chloroform. After the solvent was evaporated, the derivatization was accomplished by adding 0.5 mm^3 of N,0-bis (trimethylsilyl) acetamide to the probe.

Another injection device has been reported that may be applicable to the determination of volatile residues in biological samples. Karlsen [12] designed a precolumn reactor that enabled him to prepare volatiles from plant foliage for direct injection into the gas chromatograph. The device required no modification of the injection system, and he demonstrated its applicability with a capillary column system equipped for split injections.

Another, more sophisticated method of introduction of pesticide samples into WCOTs has been demonstrated by Kirsten et al. [13]. This process, which involves the diversion of selected fractions eluted from a packed column into a WCOT, will be discussed later.

Schomburg et al. [14] and Grob and Grob [15] have described a direct injection technique that can overcome many of the problems that could be encountered. The direct injection technique is in essence a scaled-down version of the packed column injection technique. When direct injection can be used, i.e., when the sample composition is satisfactory and can be contained in a small volume, it will be the method of choice.

3. COLUMNS

In this section, some general properties of WCOTs will be discussed to aid those unfamiliar with WCOT chromatography in their transition from packed columns to this high-resolution technique. A detailed discussion of the methods for the preparation and coating of glass capillary columns will not be presented for the following reasons: High-quality WCOTs with several different stationary phases are now commercially available. The methodology for preparation of these columns has been described in text books [16,17]. The methodology for the preparation of more sophisticated column (such as bonded phases) is, at present, a continually evolving art that varies from laboratory to laboratory. Furthermore, the literature in this field is written by experts for experts and therefore not very descriptive. These latter factors make it difficult for the uninitiated to prepare these columns that will reproduce the reported separations.

An ideal column for any particular separation would contain a stationary phase with sufficient selectivity to effect the separation of the components in the sample. This phase should be deposited as a continous film having a uniform thickness along the entire length of the column, and this film structure should be stable at the temperature required for the spearation. The uniformity of the film thickness and the film thickness along with the physical dimensions of the column together determine the efficiency (degree of band broadening) of the column. The thicker the film and the more non-uniform the film deposition, the broader the peaks will be because of the varying rates of diffusion of the sample molecules into and out from the film. Discontinuity in the film results in the exposure of glass surfaces. These exposed surfaces contribute to adsorption of some sample components and artifact formation with labile compounds. Most capillary columns elicit some adsorption of typically sensitive compounds, such as endrin and p,p'-DDT [18]. In most cases, these catalytic and adsorption effects can be tolerated when sample sizes in the nanogram range are analyzed; however, in the picogram range, a condition often encountered in biological analysis, more extensive column deactivation is necessary. Franken and Rutten [18] have demonstrated this in the analysis of chlorinated pesticides. They effected the necessary deactivation by coating the capillaries with relatively thick layers of stationary phase. However, this thicker film led to increased retention times of the components, and therefore higher elution temperatures were needed to effect the efficient elution of the samples. Thicker-phase films can therefore be used to deactivate the glass surface at the expense of band broadening.

The search for ways to deactivate the glass surface without increasing the thickness of the phase film is a continuing problem. Some useful methods include covering the active sites with Carbowax* 20M [19], nitrogen-containing compounds [20], surface-active agents [21], and the application of barium carbonate to the surface [22,23].

The problems of obtaining and maintaining a continuous film become even greater as the polarity of the coating material is increased because the ability of the polymer to wet the glass is decreased as the polarity of the polymer is increased. The wettability of the glass surface for these polar-phase materials can be improved by corrosion (etching) of the glass with reagents such as hydrochloric and hydroflouric acids or by the deposition of a thin layer of fine particles, e.g., Silanox [24] or Carbo-O-Slil [25]. Even though it is more difficult to get an even-phase film deposition on the roughened surface than it is on smooth glass, the film formation with polar phases is improved. The amount of etching of the glass surface should not exceed that which is required to maintain the polar film in position because the etching increases the surface activity which will enhance sample losses by adsorption and decomposition.

For the separation of the components in samples of biological origin, the capacity of the stationary phase for sample components must also be considered. With samples in which the compounds of interest are present in concentrations much lower than the interfering natural components, which is usually the case, the capacity of the column will become of primary importance. This is especially true when the most abundant components are eluted in juxtaposition to the components of interest. The capacity of the column

---

*Mention of this or any other trade name, proprietary product, or specific equipment does not constitute a guarantee or warranty by the U.S. Department of Agriculture and does not imply its approval to the exclusion of other products that may be suitable.

is determined by the film thickness and/or column inside diameter, i.e., the greater the amount of stationary phase per unit length of the column, the greater will be the capacity. However, as was pointed out above, this increased capacity will be gained at the expense of column efficiency due to peak broadening. The greater film thickness may, however, yield a column with decreased adsorptive and catalytic activity. At present, the only other apparent solution to the above capacity problem is further purification of the sample before attempting to use WCOTs. This purification of the sample can be accomplished by trapping the compounds of interest from a packed column for subsequent reinjection into the WCOT. This process has been accomplished with only one injection of the sample with the two-dimensional gas chromatograph built and tested by Kirsten et al. [13]. This process will be discussed later.

4. DETECTORS

Most detectors used in gas chromatography have been developed for connection to packed columns and therefore their geometry was designed for relatively high gas flowrates. Because the carrier gas flowrates are much lower through WCOTs, and because components eluted from properly prepared WCOTs appear as very narrow bands that are eluted in very rapid succession, detectors must have very low dead volumes to avoid remixing of the peaks prior to detection and must have short response times. In general, it is possible to add some purge gas to the carrier gas as it exits from the capillary column to overcome the remixing of the peaks. In this way, most short-response-time detectors can be adapted to WCOTs; however, for some detectors, such as electron capture [26], the addition of a purge gas results in some loss of sensitivity. Present-day microcoulometric, thermal conductivity, and radioactivity detectors do not have response times compatible with WCOTs.

A microelectron capture detector has been designed for use with WCOTs [27]. For the nondestructive detection and possible subsequent trapping of components eluted from WCOTs, the development of the photoionization detector should be followed [28,29]. Jennings et al. [30] have tested this detector for the nondestructive detection of flavor constituents and found it to be only slightly less sensitive than a flame ionization detector. Trapping fractions from this detector, followed by liquid scintillation counting, may be a method for detection of radioactivity during metabolic studies.

The use of a mass spectrometer as a detector with WCOTs can now be considered a routine procedure becasue of the commercial availability of instruments and columns. Complete electron impact and chemical ionization mass spectra can be obtained for the identification of compounds as well as specific ion monitoring for detection purposes.

The laboratory interfacing of WCOTs with existing mass spectrometers has also been accomplished. The general types of coupling systems have been discussed by Suñol and Gelpi [31]. The major additional requirement for the successful use of capillary gas chromatography/mass spectrometry (GC/MS) is that both the mass spectrometer and the data system are able to scan the mass range at a rapid rate. This is necessary because of the narrow bandwidths of the peaks that are eluted from the WCOTs. Grob [32] has studied the activities of various noble metals used for the direct coupling of columns to the ion source and has found that none are ideal. The best surface he found was deactivated glass.

5. APPLICATIONS

The series of papers by the Swiss and Swedish groups (e.g., Refs. 8,33-39) on the assay and identification of polychlorinated dibenzofurans (PCDFS) and polychlorinated dibenzodioxins (PCDDS) in commercial polychlorinated biphenyl (PCB) mixtures, herbicide formulations, and environmental samples are a good example of the use of the resolving power of WCOTs combined with mass fragmentography for the detection and quantitation of these environmental pollutants. The high resolution was required not only to separate the myriad of positional isomers that were present due to the various degrees of chlorination, but also was required to separate these components from interfering substances. Three different stationary phases were tested for theses separations: OV-101, OV-17, and Silar 10c. Of these, OV-17 gave the most satisfactory separations.

The compositions of PCB mixtures have also been extensively studied using WCOTs [40-51] using electron capture, flame ionization, and mass spectrometric detection. Ballschmiter et al. [50] have studied the formation of chlorinated benzoic acids as bacterial metabolites of PCBs. The acids were separated as their methylesters on columns coated with SE-30 or Carbowax 20M and identified by retention indices and/or mass spectrometry.

Test mixtures of chlorinated hydrocarbon pesticides have been used extensively for the testing of WCOTs. However, WCOTs have rarely been used for the routine analysis of these pesticides. One reason is that some of these pesticides are subject to adsorption or decomposition if the columns are not sufficiently deactivated. A typical example of this problem is the formation of p,p'-DDD from p,p'-DDT. Pentachloronitrobenzene, pentachloroaniline, and penta-and hexachlorobezene have been quantitated in extracts from lettuce using a WCOT coated with OV-17 [27], and chlorinated hydrocarbons have been quantitated in extracts from blood plasma using a SE-30 column [52].

Atallah et al. [53] used a WCOT coated with SE-30 to support evidence that a packed column analytical method prescribed for the quantitation of some residues from technical chlordane in human fat samples does not give adequate separation for the monitoring of oxychlordane, heptachlor epoxide, and trans-nonachlor in these samples. The packed columns gave only 4 components, while the WCOT separated the same sample into at least 12 components.

Holmstead [54] used an SE-30=coated column for the identification of degradation products from Mirex. Three reductive dechlorination products produced by the reaction of Mirex with reduced hematin were identified by cochromatography.

Jansson and Bergman [55] used WCOTs for the isolation and identification of rat metabolites from hexachlorobenzene. They isolated and identified a series of methylsulfide-containing metabolites from the urine and feces using columns coated with SE-30, OV-225, or OV-101. Capillary GC/MS was performed with the SE-30 column.

Technical toxaphene is another complex mixture that has been studied using WCOTs [56,58]. Recently, the distribution of toxaphene components in various tissues has been studied using a WCOT [59].

Test mixtures of organophosphorus insecticides have been separated on WCOTs coated with SE-30 [60], SE-54 [61], and diethylene glycoladipate (DEGA) [62], and a test mixture of s-triazine herbicides has been separated on a column coated with Carbowax 20M [63]. A WCOT was used for the assay of plant residues of an experimental insect growth regulator [64]. The column was coated with SF-96 silicone oil and DDT was used as an internal reference standard.

Badings et al. [65] have used wide-bore (0.7- to 0.9-mm) WCOTs for the separation of test mixtures of pesticides. These columns, SE-30 or diethylene glycol succinate (DEGS)/phosphoric acid (3:1, film thickness 1 μm), had capacities for up to 0.9 μg of any one component. They also described a column effluent splitting and trapping system that enabled them to trap separate components for further analysis. The trapping efficiencies for both butanal (0.8 μg) and nonanal (0.8 μg) were 70 and 75%, respectively, using a 10:1 effluent splitter. The trapping method probably represents the only method at present for the detection of radioactivity in fractions eluted from WCOTs.

One important group of compounds, the conjugates (glycosides, glucuronides, and amino acid conjugates) of metabolites of xenobiotics have not, to our knowledge, been studied using WCOTs. However, the groundwork is being laid for these separations. Examples for reference are the separations of derivatives of biologically active amines [66] and the analysis of serotonin and its metabolites by GC/MS [67], the separation of the trifluoroacetyl derivatives of monosaccharides [68], the quantitation of permethylated urinary tetrasaccharides by GC/MS [69], and the separation of derivatized amino acids and diastereoisomeric amino acid derivatives [70].

Ward et al. [71] have separated the organic acids (as trimethylsilyl derivatives) present in saliva by capillary GC/MS. In this study, they showed the presence of the food preservative BHT (2,6-di-t-butyleresol) in human saliva samples.

Several other uses that have been made of capillary GC/MS that have potential use as an aid in the identification of metabolites have been reported. Blum et al. [72] performed on-line hydrogen-deuterium exchange in a capillary GC/MS system using CH_3O^2H as the reagent gas for chemical ionization. This method was used to determine the number of replacable hydrogens in compounds by the comparison of the mass differences between the quasi-molecular ions in two successive runs. In one run, the reagent gas was undeuterated methanol.

On-line hydrogenation of compounds has been performed using WCOT GC/MS [73]. Glass capillary columns were used for both the separation and for the reactor that contained the catalyst. This reactor was placed after the WCOT and prior to the connection to the mass spectrometer. This form of reaction GC/MS could be of great value in structural determinations.

Rapp et al. [74] have demonstrated that exact mass determinations can be obtained for ions in the mass spectra from compounds separated in a WCOT GC/MS. At a resolution of $M/\Delta M$ of 3000, they obtained exact mass values within 3 millimass units from chromatographic peaks that contained 1 to 2 ng of the compound.

The addition of an internal standard labeled with a heavy isotope (by either derivatization or synthesis of the compound) followed by GC/MS is a very sensitive method for quantitative analysis (for examples, see Refs. 75 and 76). Where applicable--i.e., when an appropriately labeled internal standard can be prepared--this method compensates not only for the losses that occur during sample processing but also for sample-injection problems.

6. TWO-DIMENSIONAL GAS CHROMATOGRAPHY

The limited use of the resolving power of WCOTs in xenobiotic assay and in metabolite isolation and identification studies is very apparent from the examples given above. From discussions with colleagues, it is evident that one of the major reasons for this limited use is the amount of added sample preparation that is necessary to get the concentrations of the interfering components present in the biological samples into the range that can be accommodated by the limited capacity of the WCOT. The process that is termed two-dimensional gas chromatography (2D-GC) presents a good method to overcome this sample preparation problem.

In 2D-GC two gas chromatographic columns are connected in series. In its more sophisticated forms, it is possible to divert only selected fractions eluted from the first column into the second column. The other components in the sample are either vented out of the system prior to the second column or back-flushed out of the first column. Several designs for 2D-GC instruments that utilize a packed precolumn and a WCOT as the main column have been reported [13,77-79]; one of these instruments was designed, built, and tested for pesticide residue detection [13].

To utilize a WCOT in series with a packed precolumn, the introduction of the samples from the effluent stream from the precolumn must be accomplished in a manner such that the sample bandwidth is not excessive. In one instrument, the effluent stream that contains the components to be diverted to the capillary column is split so that only a portion of the flow is passed into the capillary [77]. Another method is to use a cooled trap between the two columns to concentrate the sample before it is subsequently volatilized into the WCOT [13]. The other designs employ a combination of of these [78,79]. The method that traps the entire effluent from the precolumn is the most attractive for trace analysis. The efficiencies of these intermediate traps for both high- and low-boiling compounds have been tested and found to be very good, and Kirsten et al. have discussed the problems that can be encountered with aerosol formation.

An extremely simplified block diagram of a 2D-GC is given in Figure 1. For details of the various systems, one should consult the original literature.

The 2D-GC instruments that have been discussed are equipped with two detectors (D_1 and D_2). D_1 is used to identify and subsequently locate the fractions that are to be diverted to the capillary column. In metabolism studies where the state of the art usually involves the use of ^{14}C- or ^{3}H-labeled compounds, the metabolites that contain radioactivity could be detected in the first run with a radioactivity/flame ionization detector combination at D_1. In the analytical run, the fractions containing radioactivity could then be diverted to the trap for subsequent separation on the WCOT. The ultimate design, for metabolite identification, would include a mass spectrometer for D_2.

Two-dimensional GC offers the following advantages for xenobiotic studies. The higher sample capacity of the packed precolumn makes it possible to inject sample compositions with which the xenobiotic scientist is familiar without overloading the WCOT. The WCOT is protected from sample components that can shroten column life, such as water, solvents, and excess derivatization reagents. Minor components can be collected in the trap from several runs on the precolumn and then be subsequently analyzed. Shomburg et al. [78] have demonstrated this. When 2D-GC/MS is used, the mass spectrometer is protected from sample components that can decrease the source life.

All of the components for such a 2D-GC system have been designed, built, and tested; it only remains for the profession to create the demand for this equipment to have it become available commercially. Indeed, a 2D-GC built on the balanced-flow principles of Deans [77] is already available.

Figure 1. Block diagram of the two-dimensional gas chromatographic process utilizing intermediate trapping and two detectors.

REFERENCES

1. Jensen, S., and Jansson, B., *Ambio.*, 5:257 (1976).
2. Takaya, M., Sumino, K., and Mizutani, T., *Chem. Pharm. Bull.*, 24:1958 (1976).
3. Drozd, J., *J. Chromatogr.*, 113:303 (1975).
4. Pierce, A. E., Silylation of Organic Compounds, Pierce Chemical Co., Ill., 1968
5. *Handbook of Derivatives for Chromatography* (K. Blow and G. S. King,eds.), Heyden, London, 1977.
6. Grob, K., and Grob, G., *J. Chromatogr. Sci.*, 7:584 (1969).
7. Grob, K., and Grob, Jr., K., *J. Chromatogr.*, 94:53 (1974).
8. Buser, H.-R., *Anal. Chem.*, 48:1553 (1976).
9. Ros, A., *Gas Chromatogr.*, 3:252 (1965).
10. van den Berg, P. M. J., and Cox, T. P. H., *Chromatographia*, 5:301 (1972).
11. Lauwereys, M., and Vercruysse, A., *Chromatographia*, 9:520 (1976).
12. Karlsen, J., *J. Chromatogr. Sci.*, 10:642 (1972).
13. Kirsten, W. J., Mattsson, P. E., and Alfons, H., *Anal. Chem.*, 47:1974 (1975).
14. Schomburg, G., Dielmann, R., Husmann, H., and Weeke, F., *Chromatographia*, 10:383 (1977).
15. Grob, K., and Grob, Jr., K., *J. Chromatogr.*, 151:311 (1978).
16. Ettre, L. S., *Open Tubular Columns in Gas Chromatography*, Plenum Press New York, 1965.
17. Jennings, W., *Gas Chromatography with Glass Capillary Columns*, Academic Press, New York, 1978.
18. Franken, J. J., and Rutten, G. A. F. M., *Chromatography 1972* (S. G. Perry, ed.), Applied Science, Barking,
19. Blomberg, L., *J. Chromatogr.*, 115:365 (1975).
20. Sandra, P., and Verzelem M., *Chromatographia*, 10:419 (1977).
21. Rutten, G. A. F. M., and Juyden, J. A., *J. Chromatogr.*, 74:177 (1972).
22. Grob, K., and Grob, G., *J. Chromatogr.*, 125:471 (1976).
23. Grob, K., Grob, G., and Grob, Jr., K., *Chromatographia*, 10:181 (1977).
24. Van Hout, P., Szafranek, J., Pfaffenberger, C. D., and Horning, E. C., *J. Chromatogr.*, 99:103 (1974).
25. Cramers, C. A., Vermeer, E. A., and Franken, J. J., *Chromatographia*, 10:412 (1977).
26. Rajthar, L., and Tesarik, K., *J. Chromatogr.*, 131:404 (1977).
27. Brechbühler, B., Gay, L., and Jaeger, H., *Chromatographia*, 10:478 (1977).
28. Ostojić, M., and Sternberg, Z., *Chromatographia*, 7:3 (1974).
29. Driscoll, J. N., and Clarici, J. B., *Chromatographia*, 9:567 (1976).

30. Jennings, W. G., Wyllie, S. G., and Alves, S., *Chromatographia*, 10:426 (1977).

31. Suñol, C., and Gelpi, E. N., *J. Chromatogr.*, 142:559 (1977).

32. Grob, K., *Chromatographia*, 9:509 (1976).

33. Buser, H.-R., *J. Chromatogr.*, 114:95 (1975).

34. Buser, H.-R., *Anal. Chem.*, 49:018 (1977).

35. Rappe, C., Gara, A., Buser, H.-R., and Bosshardt, H.-P., *Chemosphere*, 6:231 (1977).

36. Buser, H.-R., Bosshardt, H.-P., and Rappe, C., *Chemosphere*, 7:109 (1978).

37. Buser, H.-R., Bosshardt, H.-P., and Rappe, C., *Chemosphere*, 7:165 (1978).

38. Buser, H.-R., and Rappe, C., *Chemosphere*, 7:199 (1978).

39. Buser, H.-R., Bosshardt, H.-P., Rappe, C., and Lindahl, R., *Chemosphere*, 7:419 (1978).

40. Schulte, E., and Acker, L., *J. Anal. Chem.*, 268:260 (1974).

41. Ballschmiter, K., Niederschulte, U., Thamm, H., and Neu, J. H., *Chemosphere*, 5:367 (1976).

42. Seidl, G., and Ballschmiter, K., *Chemosphere*, 5:373 (1976).

43. Nygren, S., and Mattsson, P. E., *J. Chromatogr.*, 123:101 (1976).

44. Mattsson, P. E., and Nygren, S., *J. Chromatogr.*, 124:265 (1976).

45. Onuska, F. J., and Comba, M. E., *J. Chromatogr.*, 126:133 (1976).

46. Krupcík, J., Kristin, M., Valachovicová, M., and Janiga, S., *J. Chromatogr.*, 126:147 (1976).

47. Krupcík, J., Leclericq, P. A., Símová, A., Suchánek, P., Collák, M., and Hrivnák, J., *J. Chromatogr.*, 119:271 (1976).

48. Krupcík, J., Kríz, J., Prusová, D., Suchánek, P., and Cervenka, Z., *J. Chromatogr.*, 142:797 (1977).

49. Nygren, S., *J. Chromatogr.*, 142:109 (1977).

50. Ballschmiter, K., Unglert, Ch., and Nerr, H. J., *Chemosphere*, 6:51 (1977).

51. Zell, M., Neu, H. J., and Ballschmiter, K., *Chemosphere*, 6:69 (1977).

52. Franken, J. J., and Vader, H. L., *Chromatographia*, 6:22 (1973).

53. Atallah, Y. H., Whitacre, D. M., and Polen, P. B., *Chemosphere*, 6:17 (1977).

54. Holmstead, R. L., *J. Agr. Food Chem.*, 24:620 (1976).

55. Jansson, B., and Bergman, A., *Chemosphere*, 7:257 (1978).

56. Holmstead, R. L., Khalifa, S., and Casida, J. E., *J. Agr. Food Chem.*, 22:939 (1974).

57. Seiber, J. M., Landrum, P. F., Madden, S. C., Nugent, K. D., and Winterlin, W. L., *J. Chromatogr.*, 114:361 (1975).

58. Saleh, M. A., and Casida, J. E., *J. Agr. Food Chem.*, 25:63 (1977).

59. IUPAC Meeting

60. Krijgsman, W., and van de Kamp, C. G., *J. Chromatogr.*, 117:201 (1976).

61. Stan, J. H., *Chromatographia*, 10:233 (1977).

62. Hild, J., Schulte, E., and Thier, H. P., *Chromatographia*, 11:397 (1978).

63. Matisová, E., and Krupcik, J., *J. Chromatogr.*, 142:597 (1977).

64. Hangartner, W. W., Suchý, M., Wipf, H.-K., and Zurfluch, R. C., *J. Agr. Food Chem.*, 24:169 (1976).

65. Badings, H. T., van der Pol, J. J. G., and Wassink, J. G., *Chromatographia*, 8:440 (1975).

66. Franken, J. J., and Trijbels, M. M. F., *J. Chromatogr.*, 91:425 (1974).

67. Suñol, C., and Gelpi, E., *J. Chromatogr.*, 142:559 (1977).

68. Eklund, G., Josefsson, B., and Roos, C., *J. Chromatogr.*, 142:575 (1977).

69. Lennartson, G., Lundblad, A., Sjöblad, S., Svensson, S., and Öckerman, P. A., *Biomed. Mass Spectrom.*, 3:51 (1976).

70. König, W. A., and Rahn, W., *J. Chromatogr.*, 133:141 (1977).

71. Ward, M. E., Atom, S. Q., Politzer, I. R., and Laseter, J. L., *Biomed. Mass Spectrom.*, 3:77 (1976).

72. Blum, W., Schlumpf, E., Liehr, J. G., and Richter, W. J., *Tetrahedron Lett.*,

73. Henneberg, D., Henrichs, U., and Schomburg, G., *J. Chromatogr.*, 112:343 (1975).

74. Rapp, U., Schröder, U., Meier, S., and Elmenhorst, H., *Chromatographia*, 8:474 (1975).

75. Summons, W. E., Pereira, W. E., Reynolds, W. E., Rindfleisch, T. C., and Duffield, A. M., *Anal. Chem.*, 46:582 (1974).

76. Chapman, J. R., and Bailey, E., *J. Chromatogr.*, 89:215 (1974).

77. Deans, D. R., *Chromatographia*, 1:18 (1968).

78. Shomburg, G., Husman, H., and Weeke, F., *J. Chromatogr.*,

79. Ducass, A., Gonnord, M. F., Arpino, P., and Guiochon, G., *J. Chromatogr.*, 148:321 (1978).

4

ANALYSIS OF AIR AND AIR POLLUTANTS

WOLFGANG BERTSCH

University of Alabama
Tuscaloosa, Alabama

1. Introduction 71

2. Sampling 74

   2.1 General Considerations 74
   2.2 Sampling Procedures for Highly Volatile Substances 76
   2.3 Methods for Precolumn Concentration of the Intermediate Volatility Range 77
   2.4 Sample Preparation of Air Particulate Matter 84
   2.5 Calibration and Verification 84

3. Instrumentation 85

   3.1 Sample Introduction 85
   3.2 Glass Capillary Columns 87
   3.3 Detectors 88

4. Applications 89

   4.1 General Characterization of Volatiles 89
   4.2 Determination of Specific Volatile Compounds 99
   4.3 Analysis of Polynuclear Aromatic Hydrocarbons and Similar Substances Associated with Air Particulate Matter 104

5. Summary 114

References 115

1. INTRODUCTION

Air analysis is a multifaceted discipline which incorporates a wide range of topics. While much data has been available for many years on matters such as the composition of inorganic components in air particulate matter or on atmospheric photochemistry, relatively little has been known about the nature of organic substances in air, especially of organics in the intermediate- to high-molecular-weight range which are not associated with air particulates. Classical air analysis for organic materials, conducted prior to 1965, has mostly been focused on the determination of polynuclear hydrocarbons in air particulate matter. The review literature on the analysis of general organics in air by gas chromatography is surprisingly sparse [1-4], although there are numerous texts which cover air analysis by other methods. Capillary column chromatography has been applied to air analysis in only a few isolated instances. Only starting in the 1970s have data surfaced which deal with the "intermediate range of volatility" of such organics in more detail and provide a more comprehensive picture. This is undoubtedly linked to advances in the development of both sampling methods and analytical methodology, making possible the determination and quantitation of individual components in exceedingly complex samples at the parts per billion (ppb) level and below. The sheer number of organics which potentially may be present in air is enormous. It has been estimated that some 10^5 compounds [5] may be present at the level of about 10 ppt (parts per trillion). In practice, however, the picture is less complicated and quite manageable, at least to the ppb level.

The total number of substances which can be observed at the state of the art (ppb level or slightly lower) does not amount to more than a few hundred. At the ppt level, a much higher number should be expected. The degree of complexity which can be dealt with is, in principle, limited by the ability of the sampling and analytical system to process the minute traces of sometimes reactive components without loss. Thus, resolving power of the analytical system, sample size, and detector sensitivity set practical limits for the analyst.

It appears that many substances which are active at the low concentration levels where biocatalysis is effected often escape analysis due to the constraints cited above. Several cases have been described in the literature where volatile organics in air exert toxic effects at ppb levels or below. A large number of substances which act as mutagens or carcinogens in humans is known. Recently, the U.S. Occupational Safety and Health Administration classified important industrial chemicals into three categories: confirmed carcinogens, suspect carcinogens, and unconfirmed substances

[6]. Some 269 chemicals fall into category 1, and 218 substances fall into category 2. Approximately half of these substances are organics and thus constitute a formidable challenge to the analyst. Clearly, additional research is required to both establish optimal analytical methodology and determine safe levels of exposure. It is undisputed that mortality rates are directly linked to various measures of air pollution. Linear regression analysis clearly proves this [2]. Future advances in both sampling and analytical methodology will progressively make more and more of these substances accessible and eventually bring them under scrutiny.

There are two principal approaches to the analysis of organics in air [7]. The particular route to be taken primarily depends on the type of information expected. If it is necessary to fully characterize all organics in air, different methodology needs to be applied in comparison to a case where only a specific substance or group of substances need to be analyzed. In the second case, full advantage can be taken of specific detectors and selective sampling methods.

Rapid progress has been made in the development of separation methods capable of high resolution. In particular, the combination of glass capillary column chromatography with sensitive detectors and ancillary methods of identification have significantly contributed to the understanding of the many different processes which take place in the environment. The mass spectrometer occupies a central position as an almost ideal partner for capillary columns. Gas chromatography (GC) has played a particularly significant role in air analysis, since most (but not all) organics in air are volatile and thermally stable. Glass capillary column GC is thus ideally suited for this type of analysis, lending itself to trace determination of complex samples better than any other separation procedure.

In spite of the emergence of capillary column GC as the almost ideal tool for trace analysis of complex samples some 20 years ago, the technique has hardly been used in air analysis until quite recently. Unquestionably, this delay has to be attributed to the lack of proper sampling methods. Reliable procedures for collecting and isolating the volatile portion of organics in air have only been developed within the last few years. Sampling methodology is one of the more difficult aspects, and its correct combination with gas chromatographic instrumentation still remains a major hindrance to successful analytical work. Another point deserving equal consideration is the ability to assemble a chromatographic system, where secondary reactions or irreversible adsorption are minimized. The situation may not be quite as severe as in water analysis, where, by one estimate, only 5 percent of the organics amenable to GC are presently analyzed by this technique [8]. It is very likely, however, that a large proportion of organics in air which in principle can be analyzed by GC escape analysis, due to limitations in both sampling and analytical technology.

This chapter is concerned with the analysis of organics in air only. Organics found in air range from low-molecular-weight materials as simple as methane to very complex structures of considerable molecular weight, e.g., substituted polynuclear aromatic hydrocarbons. Organics may be present in the form of true vapors or may be associated with aerosols or solid particles. Depending on such variables, the analytical approach must be adjusted to the particular problem. No single sampling method or chromatographic instrument is capable of resolving the entire spectrum of organic components in air. Analytical methodology is thus conventionally divided into procedures which are geared for the determination of gases, of organics in the intermediate range of volatility, and of materials which are commonly associated with air particulate matter. These areas overlap in practice but will be treated separately in the following chapters, due to the different sampling and instrumental methods applied.

2. SAMPLING

2.1 General Considerations

Water is a major component of interference in the analysis of trace organics in air. The water content of air can vary considerably, depending on sampling conditions. One liter of air at 25°C and 50% relative humidity contains approximately 11 mg of water. In comparison, the substances of interest are usually only present at the ppb level, corresponding to nanogram amounts and less. Water thus constitutes a significant interference, and sampling methods must either incorporate a mechanism to deal with large quantites of water or eliminate it.

One of the inherent difficulties in obtaining meaningful samples is related to the possible generation of artifacts. If trace organics are removed from their natural matrix, moist air, and are concentrated either in a solvent or on a surface, secondary reactions may take place. Compounds which are reasonably stable in air may thus be lost or altered. A study which was undertaken to assess the potential for the formation of n-nitrosodimethylamine (DMN) from dimethylamine nitrous oxides under atmospheric conditions may serve as an example [9]. When brought together in cryogenic traps, the concentration of n-nitrosodimethylamine increased by 1 to 2 orders of magnitude in comparison to traps which operated at ambient temperature. Another question is whether samples should be preconcentrated in the field. Without enrichment, the probability of losing sample is quite high, especially for reactive or polar trace components. With effective and carefully controlled preconcentration methods, such losses are less likely [5].

A wide variety of sampling systems have been described in the literature. The major ones will briefly be discussed in this chapter. Since gases and other low-molecular-weight substances are usually not analyzed by glass capillary column GC, specific sampling methods for this molecular weight region will not be discussed.

The major sample collection and sample preconcentration methods may be categorized according to the physical processes involved. The primary methods for gases and volatiles in the intermediate boiling range (substances which are liquids at standard conditions) are:

1. Grab samples (plastic bags or metal or glass containers preevacuated or used with pumps)
2. Cryogenic methods
    a. Containers where organics from air, including water, are condensed at the temperature of liquid oxygen.
    b. Packed columns at subambient temperature.
    The air may be passed over drying agents prior to collection. The method is most frequently applied in conjunction with bags or containers in the "on-column enrichment" mode.
3. Adsorbents at ambient temperature
    a. Moderately active materials, mostly organic polymers, used in conjunction with heat desorption.
    b. Highly active materials, usually of carbonaceous nature, employed with solvent extraction.
4. Miscellaneous methods such as impingers, specific reactants, etc.

For organics associated with air particulate matter, extraction of the collection media (mostly glass fiber filters) with a series of solvents usually represents the first step. The extracted material is frequently subjected to further fractionation, such as by liquid/liquid partitioning, gel permeation chromatography, liquid chromatography, or a combination of such methods.

Due to the tremendous importance of sampling in air analysis, the principal techniques are briefly scrutinized in this chapter.

2.2 Sampling Procedures for Highly Volatile Substances

Grab samples are rarely taken in the analysis of organics in air. There are two primary reasons: (1) The concentration of the substances of interest is very low and most sample containers have relatively large surface areas; thus adsorption and catalysis may take place or contaminants may be introduced. (2) Prior to analysis, enrichment of the trace organics is almost always necessary or desirable due to limitations in detector sensitivity. Exceptions are possible for substances of high chemical stability if truly sensitive and selective detectors are available for their analysis. Halogenated compounds, especially fluorocarbons, fall into this category. Containers made of glass [10] or specially treated and electropolished steel [11] have been used for such purposes. Plastic bags are particularly popular [12-15] but may cause secondary reactions or losses [16,17]. Generally, selective enrichment is advisable to improve detectability and thus analytical precision. A noticeable exception to this general rule is the analysis of hydrocarbons at the ppb level determined by glass capillary column GC with flame ionization detection [18]. The sample volume introduced into the preceding preconcentration column was less than 10 ml. A chromatographic system of extreme stability and sensitivity, however, was required.

Grab samples are often enriched by cryogenic condensation from relatively modest volumes (10 to 100 ml) at the head of the analytical column. Either valves [19,20] or syringes [21] commonly serve as means of sample introduction. Graphitized carbon and similar supports have been found to be suitable supports for cryogenic condensation of trace organics from air [19,22]. Refined versions of this principle, using a temperature gradient, have also been described [5,23-25]. Figure 1 shows an elegant technical solution [26]. Cryogenic methods are preferred but not limited to the volatile portion of organics in air, i.e., for substances which are in the gaseous state at ambient temperature. Samplers suitable for field work have been described [26,27]. Liquid oxygen is the preferred coolant for such cryocondensation devices, since its condensation point lies above

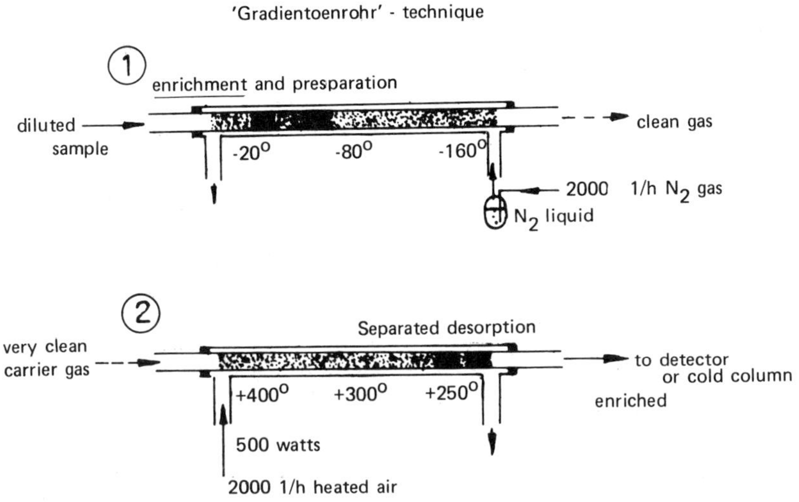

Figure 1. Schematic diagram of temperature gradient tube, with enrichment and elution steps. (Source: Ref. 26.)

liquid nitrogen. These samplers are not in widespread use, since water will also condense to form ice, making further sample processing difficult. Air samples which have been collected by cryogenic methods are mostly processed with packed columns or micropacked columns. Low-molecular-weight organics can thus be effectively separated with little overlap.

2.3 Methods for Precolumn Concentration of the Intermediate Volatility Range

Glass capillary columns are currently employed in air analysis primarily for substances in the intermediate range of volatility and for the determination of polynuclear aromatic hydrocarbons or related substances in air particulate matter. Glass capillary column technology has greatly improved within the last few years, and column wall surface modification procedures are now available which are relatively resistant to water. Nevertheless, the volume of the sample which can be introduced into a capillary column is in the low-microliter range. This sets a natural limit to the overall sensitivity of methods which do not exclude water, e.g., for cryogenic procedures. There are at least three possibilities to overcome this problem: (1) Water can be excluded by drying agents, preceding the sampler; (2) hydrophobic adsorbents can be chosen to selectively enrich the organics; or (3) water which has been collected with the organic material can physically be removed prior to analysis.

Numerous drying agents have been evaluated for air sampling [21,24,26, 28-31]. The primary prerequisite of a suitable medium is that only water is removed, while organic compounds are unaffected, regardless of molecular structure or volatility. Most drying agents which have been proposed either chemically react with water or, in the case of inorganic salts, incorporate it into their crystalline structure. With the exception of stable components such as fluorocarbons or alkanes [32], most substances passing through beds of drying agents are likely to be affected either by chemical reaction, selective adsorption, or discrimination according to boiling points. Such methods are problematic in general trace analysis.

Methods have been reported where water is removed and vented by preparative scale chromatography before the organic concentrate is allowed to enter the capillary column [33]. This approach can be classified as precolumn fractionation and leads to high enrichment, but the procedure is rather complex. More recently adsorbents have been used at ambient temperature which are capable of retaining organics from air but which have a low affinity for water.

Two slightly different approaches can be distinguished. In the first case, rather active materials, usually of carbonaeous nature [34-38], are employed in conjunction with solvent extraction. The other method is based on adsorbents of only moderate activity, mostly in combination with heat desorption. Organic resins [39-69] or graphitized carbon [70] have become extremely popular. The use of one organic polymer of particularly good thermal stability, Tenax GC, has been reported by at least 20 different laboratories. Support-bonded phases have also been used [71]. Methods have been suggested to modify adsorbents and tailor their characteristics to special needs [72,73], in particular the improvement of temperature stability [74]. For general applicability, an adsorbent should preferably exhibit nonspecific interaction, have large sample capacity, and be free from the generation of artifacts.

Some controversy has arisen around the use of highly active carbons which are often suspected of causing irreversible adsorption and of organic polymers which are prone to artifact formation by outgassing. It has been postulated that many of the problems caused by irreversible adsorption on highly active carbon can largely be eliminated if care is taken to control the quality of the charcoal [75] and if only a very small quantity of the adsorbent is applied. The relative merits can be illustrated with the following example: Some of the early work in monitoring air quality in enclosed atmospheres like space capsules or submarines involved samplers with large volumes of charcoal [76-80]. In spite of stringent recovery methods such as solvent extraction, thermal elution under vacuum, and steam distillation, recoveries were incomplete [81], especially of high-molecular-weight compounds. Grob and Grob [82], on the other side, have used as little as 25 mg of wood charcoal in a microtechnique with excellent results. The key lies in the optimization of the ratio of sample to adsorbent and in the quality of the charcoal. Another important factor is the extraction technique which permits the effective removal of the absorbate with a minimum volume of solvent. Figures 2 and 3 show the sampler and the extraction apparatus.

As with any dynamic enrichment process, breakthrough does occur, even for such highly active adsorbents. In the following, it will be illustrated that the degree of breakthrough can be calculated from experiments with model substances with reasonable accuracy. Figure 4 illustrates both the degree of breakthrough and the effectiveness of the extraction procedure. Components 4 (benzene) and 10 (toluene) are completely removed by extraction. Due to their relatively high volatility, however, they are continuously revaporized/readsorbed in the extraction process, leading to a stable equilibrium. The Grob microextraction method is superior to conventional carbon extraction procedures in several respects: surface effects of the adsorbent are minimized, extraction is more complete, and the disadvantages associated with removal of excess solvent--a partial loss of the volatile portion of the sample and concentration of solvent impurities--are largely eliminated.

Organic polymers are more limited in respect to sample capacity. Since adsorbed samples are usually regenerated by heat desorption, adsorbents must be chemically and thermally stable. Typical adsorbent behavior may best be illustrated with an example: Four glass tubes, each containing 50 mg of Tenax, were connected in series and several liters of air were pulled through this sampling train. Upon thermal desorption into a short capillary column, the four chromatograms shown in Fig. 5 were obtained. It can be seen that considerable breakthrough occurs under these conditions. It is also apparent that several competing processes take place on the adsorbent, including displacement reactions. The substance in peak a is reduced in intensity, substance b stays approximately constant, and substance c shows an opposite tendency to substance a. The first compound to be quantitatively

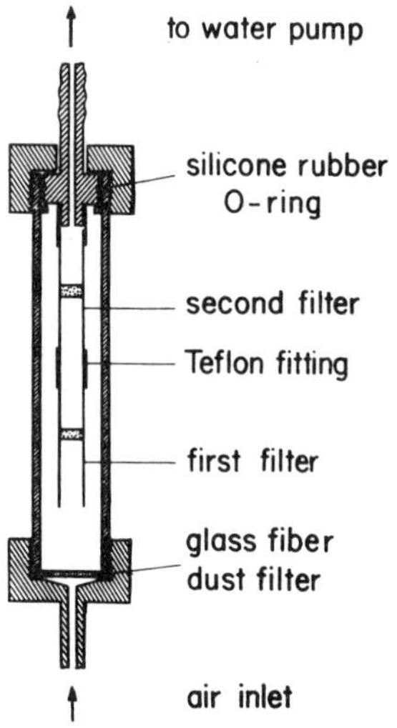

Figure 2. Filter holder containing two charcoal filters in series. For sampling, the holder is placed directly into the atmosphere. (Source: Ref. 82.)

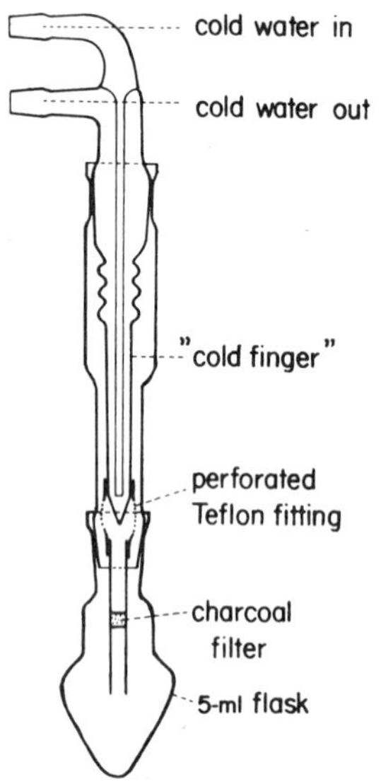

Figure 3. Glass apparatus allowing continuous extraction of charcoal filters using a very low volume of solvent. (Source: Ref. 82.)

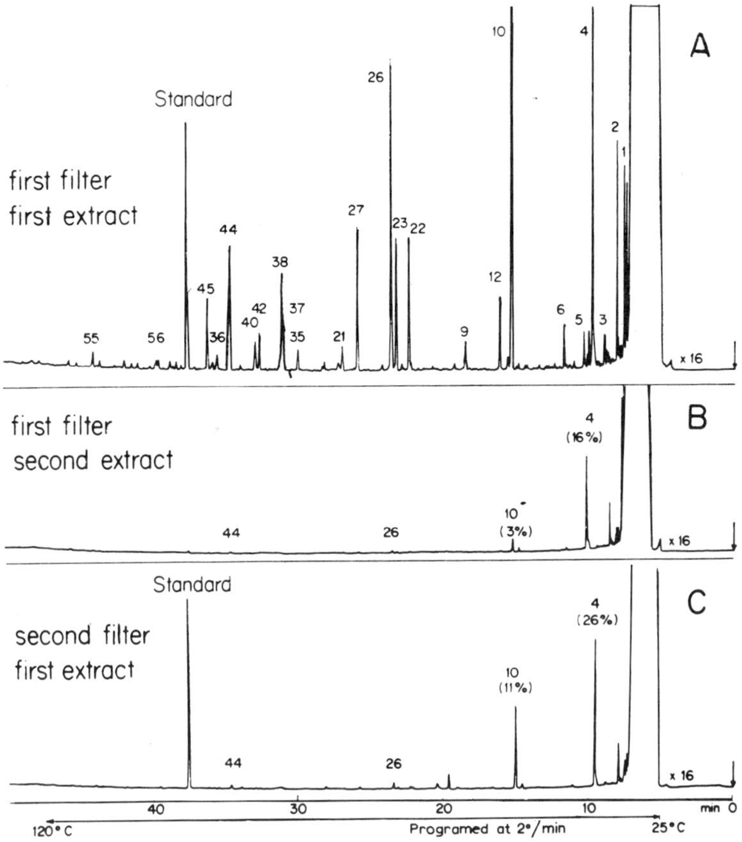

Figure 4. Quantitative analysis of lower-molecular-weight organic compounds in the atmosphere. (Source: Ref. 82.)

retained under these conditions is toluene, peak d. The substance in peak e which is present at approximately equal quantities in all four sections represents an artifact. At first look, the low capacity of most adsorbents which also fulfill the requirements of thermal stability seem to be an insurmountable problem. If breakthrough capacities are, however, determined for model substances, the problem becomes more manageable, and the range of application for a particular adsorbent can be defined. The most convenient way of determining capacities is by extrapolating retention data obtained at elevated temperatures to the conditions which prevail at sampling. A wealth of retention data have been reported in the literature, mostly for organic polymers. Table 1 lists a few retention data for some selected adsorbents. Surprisingly wide discrepancies can be observed when data reported in the literature are compared for the same substances and adsorbents.

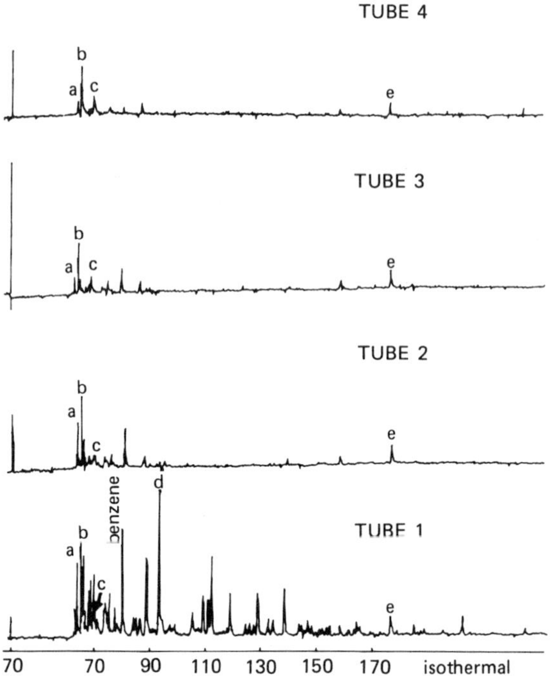

Figure 5. Breakthrough of organics in air on four adsorbent tubes at ambient temperature. Sample volume 6 liters, 50 mg of Tenax in each tube. (Source: Ref. 83.)

Retention data for benzene on Tenax have been reported by five different laboratories. They range from about 20 to approximately 75 liters g^{-1} of adsorbent at 25°C (data taken from tables or extrapolated from figures by this writer). In practical sampling situations variables such as physical dimension of the adsorbent bed, sampling rate, humidity, adsorbent particle size, density of packing, etc. play some role, and it is necessary to include a safety factor or accurately determine losses due to breakthrough by arranging several adsorbent beds in series. Fortunately, humidity has only a minor influence on retention volume [55].

An important advantage of organic polymers is related to the ease at which substances can be regenerated thermally. Recovery of substances is generally complete to the nanogram level [83-87]. Thermal elution thus offers the potential of complete transfer of the entire adsorbate into a glass capillary column without splitting. High sensitivity is one of the prime advantages of this method [3,65].

Table 1. Relative Trapping Efficiencies for Coated Versus Noncoated Tenax GC and Carbopak BHT

	Tenax 60-80		Tenax 5% OV-101		Tenax 25% OV-101		Carbopack BHT		Carbopack 5% OV-101		Carbopack 25% OV-101	
	Outer	Inner	Outer	Inner	Outer	Inner	Outer	Inner	Outer	Inner	Outer	Inner
C_6H_5	100	100	100	45	100	80	100	100	100	170	∿100	∿100
C_6H_5-CH_3	100	<2	100	4	100	7	100	<5	100	<5	100	0
C_6H_5-C_2H_5	100	0	100	0	100	<3	100	∿0	100	30	100	∿0

Source: Ref. 65.

2.4 Sample Preparation of Air Particulate Matter

Polynuclear aromatic hydrocarbons (PAHs) and similar heterocyclic compounds are usually associated with air particulate matter. Their vapor pressures at ambient temperature are negligible. Benzo(a)pyrene (BaP) is a recognized carcinogen and often used as an important indicator. Polynuclear aromatic hydrocarbons are usually sampled by pulling large volumes of air through glass fiber filters.

A large variety of industrial samplers are available [88]. It is, however, not certain that samples obtained by such high-volume samplers represent a true picture of the compounds present. Small particles can easily be lost when large volumes of air pass through the filters. Several sample losses may also occur during storage; in addition, samples have to be kept in the dark and refrigerated [89].

Soxhlet extraction of the filters with solvents such as benzene, methanol, cyclohexane, and chlorinated hydrocarbons is standard practice. Although such solvents are rather effective in removing aromatic materials, large amounts of polar compounds, including oxygenated hydrocarbons, are also extracted [90,91]. Removal of solvent at moderate temperature, e.g., by freeze-drying, largely avoids losses, degradation, and polymerization [92]. Heterocyclics can be extracted with strong acids such as $HClO_4$ [93] and H_2SO_4 [94].

Some rather elaborate schemes for precolumn fractionations of air particulate extracts have been adapted from tobacco smoke [95-100] and automobile exhaust work [101,102] where much research is focused on the same substances. Most of these procedures are based on a series of extractions, followed by conventional preparative-scale liquid chromatography. Several solvent combinations are available which permit rather selective enrichment of the relatively polar PAH fraction in solvent such as DMSO, DMF, or nitromethane. Although the separation power of glass capillary columns is very high, elaborate precolumn separation schemes are very helpful in reducing interferences at a later stage.

2.5 Calibration and Verification

In practical analysis it is necessary to experimentally verify system integrity. This is a difficult task at very low concentration levels, and accurate generation of standards in a major task. Sampling bags, permeation tubes, and exponential dilution devices have been the main tools to produce such trace standards. While reproducibility is generally excellent with permeation tubes [103,104], frequent reproducibility checks are advisable, since permeation rates may vary over extended periods. Errors with dynamic dilution methods are primarily caused by inadequate mixing, and optimization of flask geometry is important [105]. A wide range of concentrations can, however, be covered conveniently [107]. Bags or containers have the disadvantage of exposing the sample to a relatively large surface area, but are popular because of convenience and low cost.

The prediction of substance behavior by extrapolation from data observed for a single compound or a small number of homologs is a frequent source of systematic errors. Such errors, caused by preferential depletion, partial condensation, selective chemical reactions, or general adsorption of one or more components in the multicomponent standard are difficult to detect. Kaiser [23] has discussed methods for recognizing systematic errors.

Individual substances can accurately be determined by isotope dilution methods [108]. Since isotopic substances are chemically equivalent to their nonisotopic counterparts, both suffer the same interferences. The mass spectrometer is the most convenient detection device to distinguish between both, even if the degree of resolution is minimal. Isotope dilution analysis is expensive and therefore rarely applied to air analysis. It is, however, particularly useful for the verification of sampling procedures if only a few critical substances are of interest.

3. INSTRUMENTATION

3.1 Sample Introduction

The gas chromatographic system of choice to be used for a particular problem in air analysis must have the proper combination of the three principle components; sample inlet, chromatographic column, and detector/readout system. For work with capillary columns it is especially important that inlet and detector geometry are suitable for such columns. The response time of the system must be sufficiently fast.

The importance of properly introducing samples into glass capillary columns cannot be overemphasized. Since the enrichment factor which can be obtained during sampling is usually limited, it is important to be able to introduce a large volume of the highly dilute solution into the glass capillary without splitting. Grob [109-111] and others [112-114] have discussed the principles of introducing very large samples into glass capillary columns with no loss of efficiency. The underlying mechanisms were discussed in a series of recent papers.

In the splitless mode, samples undergo conventional flash vaporization. It has been pointed out that large systematic errors can result unless the system is well controlled. Transfer losses, especially of high-molecular-weight compounds, must be experimentally determined. It has been shown that substances in the boiling point range of polynuclear hydrocarbons are not quantitatively transferred from the syringe needle [115]. Elegant technical solutions for "on-column" injection have recently been proposed. Samples are directly introduced into the glass capillary columns without the need for flash vaporization [116-118]. On-column injection largely avoids sample discrimination and removes some of the hazards of flash vaporization to the more labile sample components. In any of these methods in which a large volume of solvent is introduced into a glass capillary, it is advisable to remove the phase from the first few turns. Another improvement is the use of precolumns, which insure consistent performance with samples containing a significant portion of nonvolatiles. Practical recommendations have been made [119].

Direct introduction of substances, trapped on adsorbents, is also possible by thermal elution. Since no solvent is involved, the sample is not diluted and higher sensitivities can be obtained. The first part of the capillary column has to be cryogenically cooled. This type of arrangement can be rather simple, consisting of a U-shaped section of the column or a spiral precolumn submerged in liquid nitrogen [65,120,121,149]. A few minor problems may have to be dealt with on occasion: (1) if excessive amounts of water are driven off the adsorbent tube, the trap may become plugged and the sample transfer from the adsorbent stops; (2) breakthrough may occur due to microfog formation; (3) losses may occur when the adsorbent tube is introduced into a hot desorption chamber. Most of these effects can be eliminated by appropriate experimental techniques such as selection of suitable column dimensions and geometry or by appropriate hardware design and sample handling. It is definitely feasible to transfer substances in the nanogram range from adsorbents directly into capillary columns by this method, despite some claims to the contrary in the literature.

It is the area of sample introduction in which some of the most significant advances toward practical utilization of capillary columns have been made over the last few years. Environmental analysis in general and air analysis in particular stand to benefit the most from these developments, since enrichment is always a critical factor.

3.2 Glass Capillary Columns

The success of methods based on the introduction of large sample volumes depends on the availability of processes to produce stable films of stationary phases in glass capillary columns. Deactivation techniques which are capable of maintaining the inertness of glass surfaces despite the injection of large volumes of solvents are of great importance. Numerous procedures have been discussed for the production and testing of columns, and improved methods of preparation are now appearing almost monthly in the literature.

The selection of an optimal glass capillary column strongly depends on the particular application envisioned. Glass capillary column technology as such cannot be discussed in this context, but a few important considerations should be stressed.

It is important to establish the adsorptive behavior of a glass capillary column on a continuous basis, since perfectly deactivated columns may become active after prolonged use. This can only be done by investigation of the quantitative recovery of suitable test components [8]. The acid/base characteristics of glass surfaces are also important [122]. Efficiency considerations, which are always somewhat overemphasized with glass capillary columns, should be considered secondary.

For the analysis of air samples in the intermediate range of volatility, phases of moderate polarity such as the Ucons or Emulphor have traditionally been preferred. Such phases generally offer better selectivity than the nonpolar ones and are sufficiently temperature-stable. For polynuclear aromatic hydrocarbons, nonpolar (OV-1, OV-101 etc.) or slightly polar silicone phases (SE-52, SE-54, etc.) are preferred. Liquid crystal substrates [123] inherently have excellent selectivity for isomeric substances; e.g., selectivities of between 1.2 and 1.4 have been reported for the separation of anthracene and phenanthrene [124]. Their temperature stability, however, is not sufficient for practical work. Primary prerequisites for glass capillary column in conjuction with high temperature GC or GC/mass spectrometry (GC/MS) work are low bleed and stability of deactivation at high temperature. Up to a few years ago practical high-temperature GC on glass capillary columns has been limited to approximately 260°C even with phases which are specially designed for this type of work, e.g., the carboranes. It appears now that the inherent temperature stability of the phase is not the limiting factor but rather the catalytic action of the column wall. Significent progress has been made to reduce such effects, e.g., by deposition of barium carbonate [125], exhaustive silanization [126-128], and direct chemical bonding of crosslinked polysiloxane polymers onto the glass surface [129]. Temperatures of 300°C and above can now be reached without excessive bleed or generation of active sites. One important variable for the optimization of separations which has been neglected by many chromatographers is the selection of the appropriate film thickness [130]. Relatively thin films are not usually required for high-temperature work.

3.3 Detectors

The flame ionization detector is the most widely used detection device in air analysis. In cases where specific substances are analyzed for, selective detectors are preferred. Selective detectors arranged in parallel with a universal detector [131] are especially powerful. In many cases, relative response factors can be used as additional means of identification [6,132]. This is especially useful for the determination of polynuclear aromatic hydrocarbons which are responsive to electron capture detection [133]. Several systems which allow effluent splitting from glass capillary columns into two or more detectors have been described [134-136]. The mass spectrometer, however, remains the detector of choice.

Quadrapole GC/MS has made a large impact, although magnetic sector instruments are generally more sensitive. GC/MS in general has developed to a rather high degree of sophistication. Computerized capillary column GC/MS is now routinely performed in many applications laboratories. Almost all manufacturers of mass spectrometers offer instruments which have been designed for capillary column work. In most GC/MS combinations, the entire effluent is directed into the ion source. A variety of different technical solutions have been suggested for direct coupling [137-140].

A computer adds still another dimension to GC/MS. Besides its recognized capability in the area of data reduction and file search, components which are only partially resolved can be deconvoluted. The computer permits the use off mass fragmentography as a method to gain additional selectivity. Virtually at the touch of a button, selected ions can be extracted from a complex chromatogram and the presence of a class of compounds buried under large amounts of other substances can be deducted from characteristic mass fragments. Further specificity can be procided with high-resolution mass determinations. Different species having the same nominal mass can thus be readily distinguished [7]. The sensitivity of modern double-focusing instruments permints elemental composition analysis with as little as a few nanograms of a substance [141]. One of the newest tools in GC/MS is the use of chemical ionization, which permits easier determination of molecular weights and may greatly simplify the appearance of a chromatogram [142]. GC/MS units, capable of both CI and EI, are common but have rarely been applied to air analysis in this capacity. Since a large volume of data accumulates, data systems are often an intergral part of such instruments. Figure 6 shows a diagram of a comprehensive setup. It appears that the mass spectrometer will remain the most universal detector for instrumentation dealing with air analysis by glass capillary columns for many years to come.

4. APPLICATIONS

4.1 General Characterization of Volatiles

It was stated in the introductory chapter that glass capillary columns were rarely used for the study of organics in air prior to 1970. Only since the mid-1970s have a sizable number of publications surfaced with glass capillary columns as the primary tool of analysis. Some of these papers deal with the general characterization of air in selected environments; others report on the analysis of specific substances or single components. Several appli-

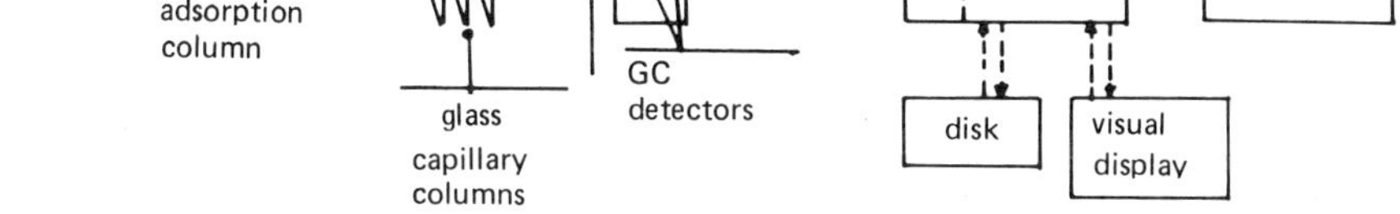

Figure 6. Flow scheme of a comprehensive analytical unit. (Source: Ref. 82.)

cations have been reported in which metal capillary columns were used [39, 40,42,144,180]. These will not be covered in this context.

One of the first successful attempts to thoroughly characterize the volatile components in urban air was reported in 1971 [82]. Figure 7 shows a chromatogram and Table 2 reports the identity of the major components from this study. The sample extract was subjected to treatment with sulfuric acid as additional means of identification. This treatment eliminates or reduces all compounds except alkanes and arylalkanes. Less than 20 out of the 460 compounds observed were influenced by this treatment. Similar results have been obtained in a study of South African cities [37,38]. An on-line GC/infrared and subtractive precolumns provided additional confirmation and identification in this case. The general conclusion has to be that the largest proportion of organics in urban air consists of relatively stable hydrocarbons which are derived from fuel. These observations have been confirmed by others [22,24,34,36,41]. The predominance of these compounds is hardly surprising, since most emissions in urban areas are likely to derive from automobiles or from heating units. Typical concentrations of organics in air are in the ppb range, with the exception of methane, which is present at a much higher concentration. Toluene is almost always the single largest component. Louw et al. [38] have tabulated literature data for six large cities in five different countries. Table 3 represents an abbreviated version from their work. It should be pointed out that concentrations may easily vary by a factor of 10 at the same location, depending on the prevailing meterological conditions [144].

As expected, there are significant variations in both the absolute amount of organics in air as well as their relative composition [145,146]. Figure 8 shows the distribution for nonmethane hydrocarbons in a city during a 24-hour cycle. Apparently, the concentration in an urban environment is highest during the early morning rush hour, whereas it peaks later in a suburban environment. It is likely that this shift is caused by a combination of two factors: (1) the time necessary to transport the pollution from the urban to the suburban areas; and (2) a difference in urban vs. suburban traffic patterns. Figure 9 shows three chromatograms of a sample taken in the suburban area during the trapping period indicated in Fig. 8. The advantages of selective detectors are indicated. The electron-capture-detector-responsive substances are low-molecular-weight chlorinated hydrocarbons, and peak 41, giving a signal in the flame photometric detector

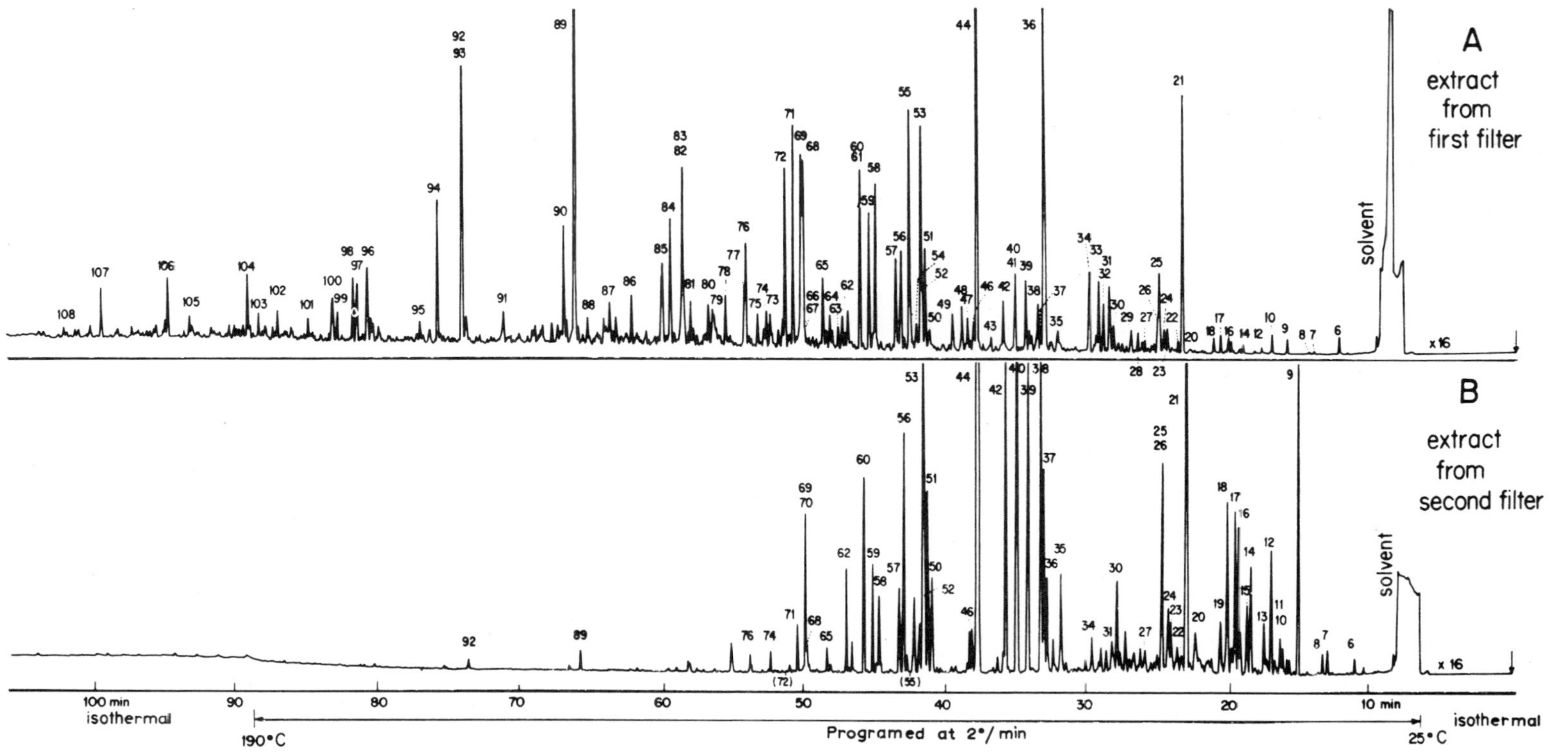

Figure 7. Qualitative analysis of organic compounds in the atmosphere. Large sample volume (25 m^3) to allow detection of heavier substances, with corresponding loss of lighter ones. Numbers refer to substances listed in Table 2. (Source: Ref. 82.)

Table 2. Substances Identified in the Air of Zurich, Switzerland[a]

Peak	Retention time (min)	Identified as (MS)	Amount (ppb CH_4)
1	0.72	Butene-1 + cis-butene-2 + isobutane	25
2	1.13	n-Butane + trans-butene-2	16
3	1.35	1,3-Butadiene	7
4	2.02	Methyl mercaptan	4
5	2.37	Acetone + CH_2CI_2[b]	16
6	3.42	C_5 unsaturated hydrocarbon	2
7	3.88	i-C_5 hydrocarbon	1
8	4.48	n-C_5 hydrocarbon	40
9	6.45	Cyclohexanc	17
10	10.0	—	50
11	11.20	—.[b]	5
12	11.80	Carbon tetrachloride[b]	5
13	14.66	Dietyl ketone	11
14	15.64	Benzene	74
15	16.88	C_6 unsaturated hydrocarbon	33
16	18.00	Trichloroethylene[b]	<1
17	19.54	n-C_6 hydrocarbon	8
18	25.03	Butanol	6
19	27.63	Methylcyclohexane	19
20	27.99	C_7 unsaturated hydrocarbon	90
21	28.55	C_7 unsaturated hydrocarbon	21
22	29.06	Toluene	20
23	29.61	C_7 hydrocarbon	100
24	29.80	1,3-Dichloropropane[b]	6
25	33.59	Dimethylcyclohexane	30
26	35.19	C_8 unsaturated	26
27	35.91	Ethyl butyl ketone	15
28	36.25	Styrene	13
29	36.96	Ethylbenzene	21
30	37.50	m-Xylene	11

Table 2 (cont'd.)

Peak	Retention time (min)	Identified as (MS)	Amount (ppb CH_4)
31	37.98	p-Xylene + isopropylbenzene	41
32	38.44	o-Xylene	3
33	39.81	—[b]	4
34	41.04	n-C_9 hydrocarbon	20
35	42.07	Ethyltoluene	19
36	42.83	1,3,5-Trimethylbenzene	18
37	44.38	1,2,4-Trimethylbenzene	54
38	45.46	1,2,3-Trimethylbenzene	22
39	46.05	Ethyldimethylbenzene	18
40	47.67	Methylisopropylbenzene	18
41	48.67	Benzothiophene[b]	3
42	50.77	1,2,3,5-Tetramethylbenzene	13
43	51.57	1,2,3,4- + 1,2,4,5-Tetramethylbenzene	4
44	52.59	—	18
45	57.74	—	18
46	58.99	—	4
47	65.05	—	19
		Total:	988

[a] Numbers correspond to those in Fig. 9.
[b] Detected with specific detectors and identified with standards. Source: Ref. 82.

(FPD), is benzothiophene. Selective detectors thus not only point out specific compounds but are also almost always more sensitive than a flame ionization detector (FID). It is noteworthy that the chromatograms of Fig. 8 were taken on an efficient micropacked column rather than a capillary column.

Figure 10 shows a direct comparison between fuels (gasoline plus diesel) and air sampled at a parking lot and in a wooded area. At first sight

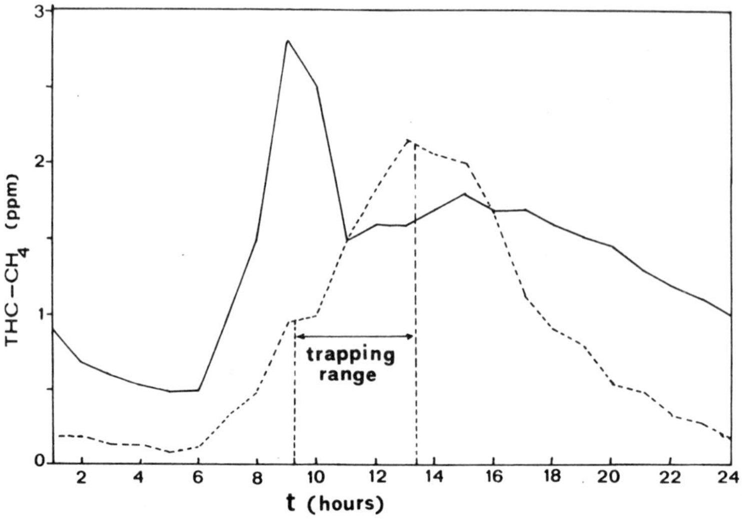

Figure 8. Total hydrocarbon content during 24 hr. monitored in typical urban (---) and suburban (- - -) areas. (Source: Ref. 147.)

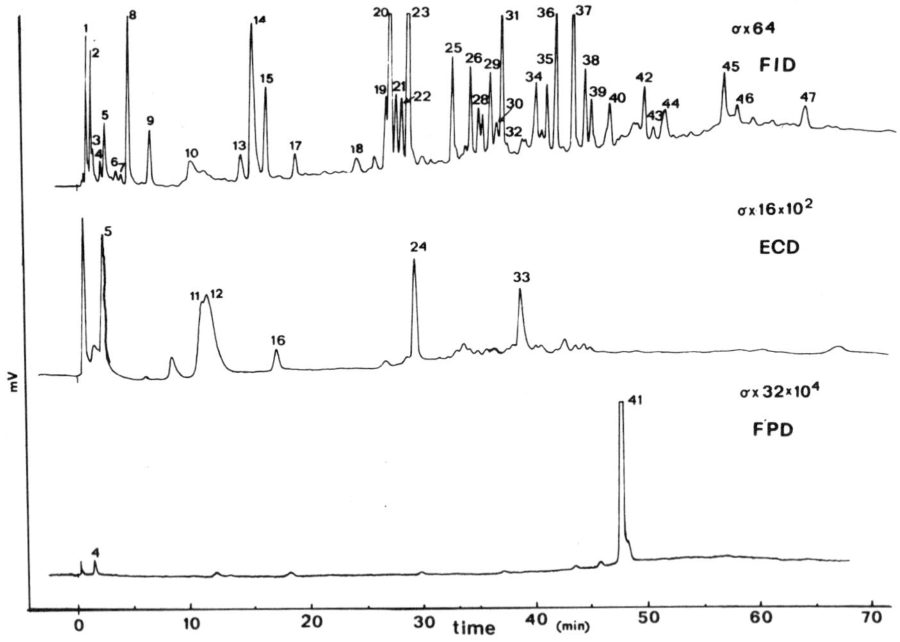

Figure 9. Triple detector chromatogram showing the analysis of a sample of air taken in suburban open atmosphere. Sampling material: Carbopack B. (Source: Ref. 147.)

Table 3. Concentrations of Selected Volatile Organic Compounds in the Air of Some Major Cities (ppb)[a]

Compound	Johannesburg	Paris	Houston	Zurich	Los Angeles	The Hague
Paraffinic hydrocarbons						
n-Octane	1.3			3.4		
n-Nonane	0.6	0.1-0.6	1.6-4.4	1.7		
n-Decane	0.7	0.7-1.9	1.0-2.7	1.6		
Aromatic hydrocarbons						
Benzene	3.4[b]		1.3-15	54	15	8.9
Toluene	10.3		0.3-9.7	39	37	17.9
Ethylebenzene	1.9		3.1-4.5	8.7	6	3.1
m- + p-Xylene	4.5	0.5-1.5	7.5-11.2	30	22	12.4
o-Xylene	1.6	0.2-0.6	3.0-4.8	9.3	8	3.1
n-Propylbenzene	0.4			1.5	2	
1-Methyl-3-ethylbenzene + 1-methyl-4-ethylbenzene	1.6	0.2-0.8		9.0	8	
1,3,5-Trimethylbenzene	0.6	0.3-1.0		2.1	3	1.6
1,2,4-Trimethylbenzene	1.6	0.9-3.1		9.0		7.3
Chlorinated hydrocarbons						
Tetrachloroethylene	2.1			10		

[a] Johannesburg: average of 9 samples taken during summer and winter 1975-1976; Paris: Raymond and Guiochon (1974); Houston: Bertsch et al. (1974); Zurich: Grob and Grob (1971); Los Angeles: Lonneman et al. (1968); The Hague: Burghardt and Jeltes (1975)

[b] Values for benzene are approximate because they have been corrected (Grob and Grob, 1971) for losses due to inefficient collection.

Source: Ref. 38.

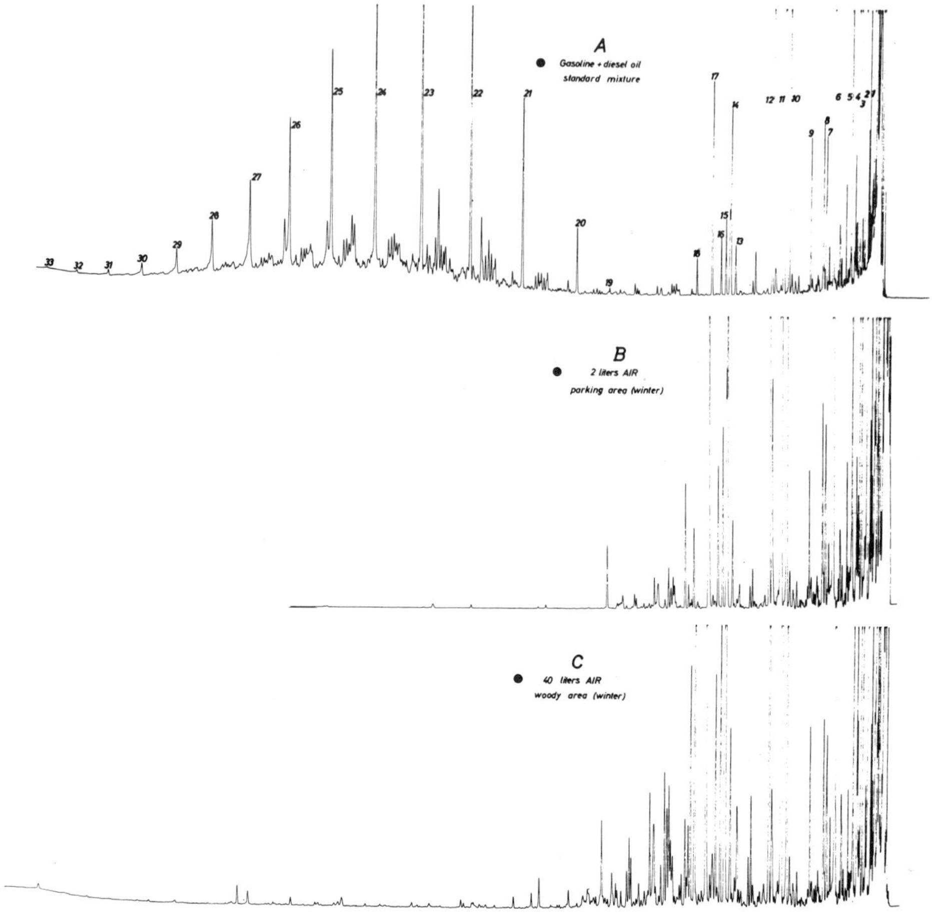

Figure 10. Profiles of organic volatiles in air: A—standard mixture of gasoline plus diesel oil; B—2 liters of air from a car parking area (winter); C—40 liters of air from a woody area (winter). Glass capillary column; 60 m × 0.3 mm ID, OV-101. Helium flowrate: ∿4 ml min^{-1}. Temperature program: 40°C for 5 min, 40 to 250°C at 2° min^{-1}. Detector: FID, 270°C. (Source: Ref. 143.)

it appears that chromatogram C, which represents the nonurban profile, has many additional components in the high-molecular-weight range in comparison to chromatogram B. This, however, is a result of the much larger sample size (40 liters vs. 2 liters) which magnifies the heavy portion. A closer look shows that the early parts of chromatograms B and C are almost identical. This condition is indicative of a stable equilibrium which is reached after the adsorbent capacity for these substances has been exceeded. It has been shown, however, by the same authors that additional components are generated from vegetation. A sample taken during the growing season clearly shows significant variations [134]. Regardless of such natural sources of emission, it is clear that the fingerprint pattern of urban activities, mainly the profile of gasoline, is superimposed on such samples even when these are taken at relatively large distances from such point sources. The concentration of the fuel constituents decreases, mostly due to dilution. Emissions observed in nonurban areas are thus dependent on the meterological conditions. Holzer et al. [65] have shown that fuel constituents are ambiguous components in rural air, although at greatly reduced concentration levels. Significant amounts of terpenes and some oxygenated substances are added from natural sources even in winter when the contribution from vegetation is greatly reduced.

In routine analysis, the process of sample preparation itself takes up a considerable amount of time, since most preenrichment processes are rather work intensive. A partially automated design of a comprehensive system, including sample processing, has been described [25]. The main benefits of automation are improvements in the precision of data and long-term performance. The system was designed to remove most of the manual manipulations which are frequent sources of error. Samples were dynamically enriched in a temperature gradient tube. Since sample volumes were in the order of 30 liters, a drying tube had to precede the gradient tube. Magnesium perchlorate was applied. It is reported that no losses or interferences were observed with the exception of short-chain alcohols. Retention indices were reproduced to better than 1 index unit, and a standard deviation of better than 5% was reported for substances in the ppb range. In another application, some of the effects of tobacco smoke on air quality were quantitatively investigated. This is an important problem in light of recent legislation which requires that separate facilities be provided for nonsmokers in public places. Samples of only 3.5 liters were taken in a room before and after the smoking of cigarettes. As expected, the air polluted by cigarette smoke contains significant portions of substances which are derived from smoking. Figure 11 shows the chromatograms. Peak 130 represents nicotine. Some 133 compounds were identified down to ppb levels. An OV-101 glass capillary column coupled to a magnetic-sector GC/MS was used.

Some of the most extensive studies of volatile organics in air have been undertaken by Pellizzari and co-workers [148-150]. Several hundred samples from many major industrial locations in the United States were collected on Tenax and analyzed by computerized GC/MS. Besides the usual hydrocarbon-type substances, a variety of oxygen-, nitrogen-, sulfur-, and

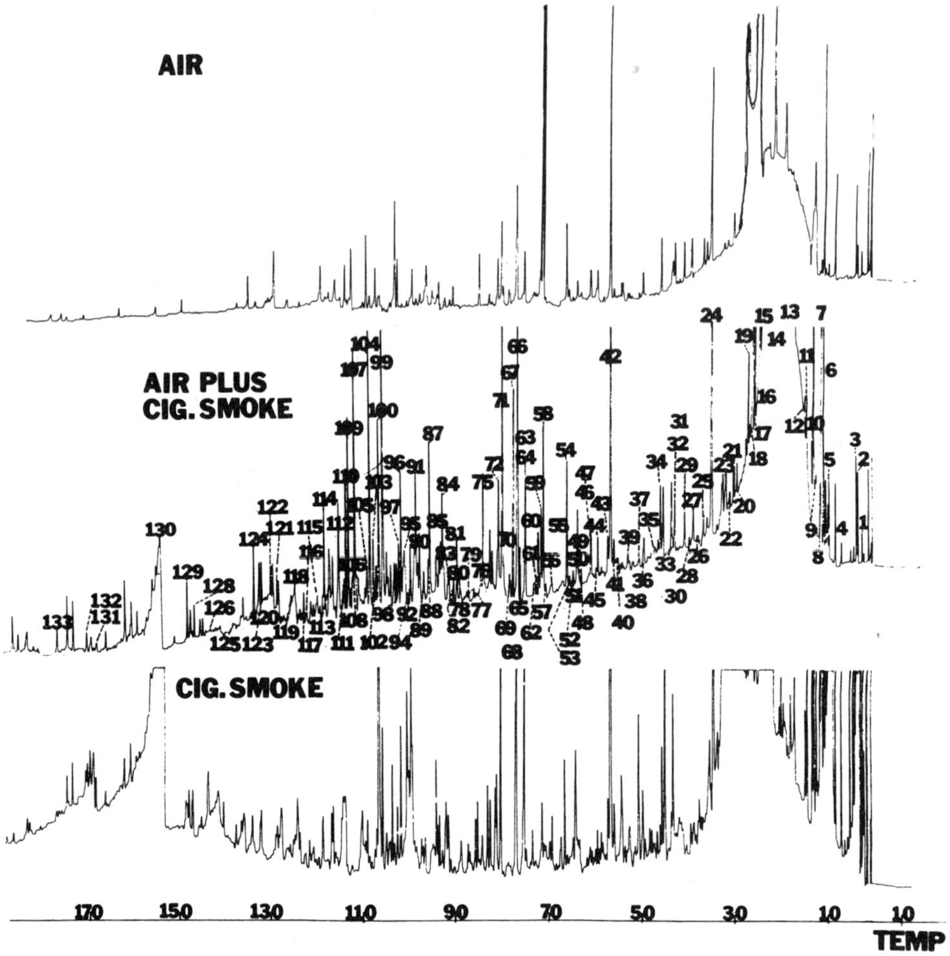

Figure 11. Total ion current chromatograms of urban air, air contaminated by cigarette smoke, and cigarette smoke. Sample size of air sampling: 3.5; Sample size of cigarette puff; 3 ml; Glass capillary column: 38 m × 0.35 mm ID, coated with OV-101; Carrier gas (helium) flowrate: 3 ml min^{-1}. Temperature program: 20°C for 8 min, 20 to 200°C at 2° min^{-1}. (Source: Ref. 65.)

silicon-containing substances were detected. This was possible through the use of mass fragmentography. Characteristic features were extracted, in spite of the relatively modest resolving power of the support-coated open tubular column. Figure 12 shows the plots of three characteristic mass fragments, corresponding to chlorinated hydrocarbons [148]. Table 4 is a summary of the quantitative distribution of halogenated organics in a heavily industrialized area around Houston, Texas. Many of the substances identified in this nationwide program were characteristic of the type of industrial activity which was surveyed. Such information forms the basis

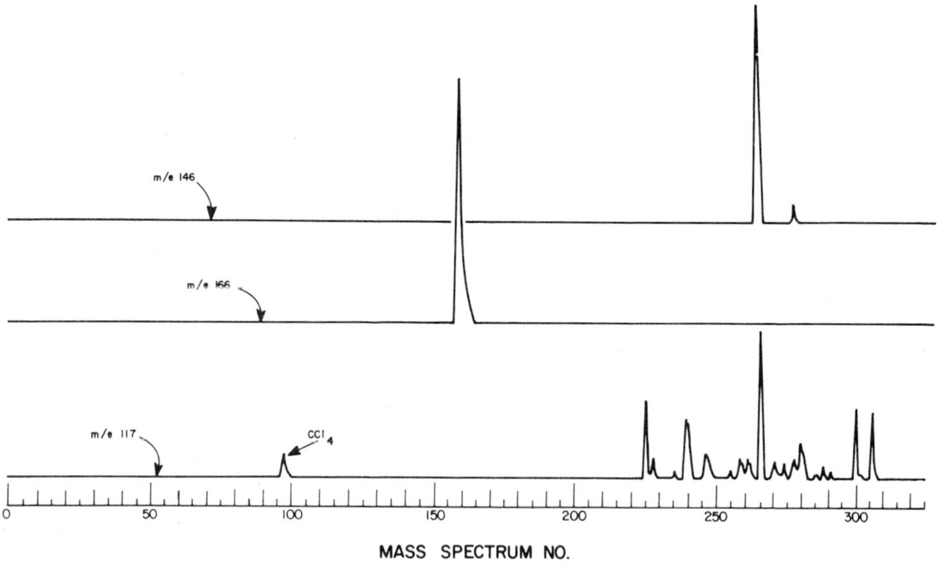

Figure 12. Mass fragmentograms of characteristic ions representing carbon tetrachloride (m/e 117), tetrachloroethylene (m/e 166), and m-dichlorobenzene (m/e 146). From an air sample collected in Glendora, Calif. (Source: Ref. 148.)

of correlation studies for correlating mortality data with the chemical composition of air at various industrial and nonindustrial sampling sites. Detailed "cancer maps" are available for most of the United States, and it is important to deconvolute the many variables which cause local accumulation of cancer deaths.

Certain types of organics in air may well be linked to such statistics [2]. It has been shown that certain kinds of cancer are particularly predominant in the city of Baltimore and an adjacent county [151]. N-nitrosodimethylamine has been detected in ambient air by glass capillary column GC/MS at levels as high as 10 ppb [152]. It is quite unlikely that this single substance alone is responsible for this high incidence of cancer, but it may significantly contribute. It is thus necessary to subject chemical data to multivariant analysis to unravel underlying patterns. Detailed data, such as obtained by high-resolution methods, are necessary for such analysis.

4.2 Determination of Specific Volatile Compounds

The examples demonstrated in the previous chapter pointed out that glass capillary columns are indeed capable of handling the complexity of organics in air, at least down to the ppb level. In many cases, where only one or a

Table 4. Ambient Air Levels of Halogenated Organics in the Houston, Pasadena, Deer Park, Freeport, and LaPorte, Texas areas (ng m^{-3})[a]

Chemical class	HL1	HL2	HL3	PL1	PL2	DSL1
Halogenated hydrocarbons						
2-Chloro-1,3-Butadiene (chloroprene)	4,000	266	—	—	—	—
Dichloropropene isomer (tent.)	—	—	—	—	—	—
Chloroform	11,539	—	11,538	T	T	53,846
Vinylidene chloride	—	—	—	—	—	—
1,1-Dichloroethane	—	—	—	—	—	555
1,2-Dichloroethane	T	—	—	—	158	66,300
Dibromoethane	—	—	—	158	—	—
1,1,1-Trichloroethane	522	—	900	—	T	—
Carbon tetrachloride	238	114	T	T	146	846
Dichloropropane isomer	—	—	—	—	—	T
Trichloroethylene	75	—	39	76	5,071	321
Dichlorobutane isomer	—	700	—	—	—	—
Dichlorobutene isomer	262	—	—	—	—	—
1,1,2-Trichloroethane	—	—	—	—	—	6,700
Tetrachloroathylene	29	T	21	20	18	—
Dichlorobutane isomer	52	—	—	—	—	—
Dichloropropene isomer(s)	—	—	—	—	—	180 90 90
1,1,1,2-Tetrachloroethane	—	—	—	—	—	—
Bis-(2-chloroisopropyl)-ether	—	—	—	—	—	—
Hexachloro-1,3-butadiene	—	—	—	—	—	—
1,2-Dibromopropane	—	—	—	—	—	—
1,1,2,2-Tatrachloroethane	—	—	—	—	—	—
1,2,3-Trichloropropane	—	—	—	—	—	—
Pentachloroethane	—	—	—	—	—	—
Perchloroethane	—	—	—	—	—	—
Tetrachlorobutadiene isomer(s)	—	—	—	—	—	—
Pentachlorobutadiene isomer(s)	—	—	—	—	—	—
Vinyl chloride	—	—	—	—	—	—
Trichloropropene isomer(s)	—	—	—	—	—	—

[a] In this table, T = trace. Column headings refer to the various sampling sites. Source: Ref. 41.

Table 4 (cont'd.)

DSL2	DDL1	DTL1	DTL2	DTL3	DTL4	FL2	FL3	LL1	LL2	LL3
—	—	—	—	—	—	—	—	—	—	—
—	—	—	—	—	—	—	—	—	—	—
6,420	7,692	1,923	8,846	T	15,384	280	T	8,461	8,850	7,692
—	—	—	—	—	—	—	531	—	—	—
—	—	—	—	—	—	—	—	—	—	—
—	6,722	T	4,055	T	T	3,300	4,500	778	—	—
—	—	—	—	—	—	—	T	—	—	—
144	—	—	1,000	400	400	16,600	15,200	3,889	27,700	T
T	T	T	T	276	69	11,538	T	T	1,230	T
—	—	—	2,586	—	—	69	1,478	—	—	T
—	2,535	—	—	—	—	107	200	—	43	T
—	—	—	—	—	—	—	—	—	—	—
—	—	—	T	—	—	—	—	—	—	—
—	—	—	3,334	—	—	—	3,821	—	—	—
T	2,019	75	68	—	72	94	1,585	17	83	T
—	—	—	—	—	—	—	—	—	—	—
—	—	T,T	241 1,293 72 1,293 345	—	T	—	—	—	—	—
—	—	—	—	—	—	—	21	—	—	—
—	—	—	—	—	—	27	333	—	—	T
—	334	—	2,066	—	25	13.3	8.3	—	T	T
—	—	—	—	—	—	—	T	—	—	—
—	—	—	19	—	—	—	33	—	—	—
—	—	—	—	—	—	—	298	—	—	—
—	—	—	—	—	—	—	3,984	—	—	—
—	—	—	—	—	—	—	2,903	—	—	—
—	—	—	T 100	—	—	—	—	—	—	—
—	67	—	T	—	—	—	—	—	—	—
—	T	—	—	—	—	—	—	—	—	—
—	—	—	T,T	—	—	—	—	—	—	—

few substances are of interest, it is not necessary to resolve all components in a sample. An ideal situation exists if the substance of interest can selectively be derivatized and detected by a selective detector. GC in packed columns with highly specific detectors is the predominant approach to the analysis of such special compounds, and glass capillary columns are less frequently used. Analysis of specific substances is of considerable interest because of potential health hazards or because they may serve as indicators of illegal activities. Portable mass spectrometers capable of selective ion monitoring or plasma chromatographs have been used to detect explosives and drugs in air [153,154].

In cases where only one or a few chemicals are of interest, nonchromatographic determinations based on specific colorimetric reactions may also be quite successful. Such methods are, however, prone to interferences, and large errors are sometimes introduced into such measurements. For critical determinations, methods based on gas chromatography are still preferred because of high selectivity and specificity. In truly complex air samples ambiguities may, however, still arise if the substances are present at only very low concentration levels. The analysis of bis-chloromethyl ether (BCME) may serve as an example.

Chloromethyl ether (CME) is an important chloromethylating reagent which is widely used in the manufacture of ion exchange resins, membranes, and similar products. BCME, a stable and extremely toxic and carcinogenic substance [155,156], may be present as an impurity in CME at high levels. It is thus important to find a very sensitive method for its detection. Many analytical methods based on different principles have been offered. Some approaches are based on the extraordinary selectivity of chemical derivatization reactions or on detector specificity and largely neglect the separation aspects. Others require a high degree of separation and only moderate detector specificity. Both methods have their pitfalls, pointing out the need to carefully verify an analytical system for the correct determination of critical compounds.

BCME can inherently be readily derivatized by replacement of its chlorine group with a substituent, which is sensitive to electron capture detection. One particular approach based on this principle of selective derivatization does not require expensive instrumentation [157] but depends on many variables which are difficult to control [158,159]. It appears that the extra steps required for derivatization need to be checked carefully with respect to reproducibility of the reaction, stability of products, and interferences.

Although high-resolution GC with glass capillary columns is a very powerful method in separating single substances from a large number of interferences, overlap may occur occasionally, even under such conditions. An alternative method, two-dimensional GC with heartcutting and back-flushing, may be prefereable in some cases, especially if air is to be analyzed automatically on a continuous basis [51]. A system has been described in which a section containing partially resolved BCME is transferred from one column into a second one of different selectivity. Although packed columns were used, complete separation was easily achieved [87]. A short precolumn, in which derivatization takes place for subsequent selective detection, may also be considered a two-dimensional system. An automated system based on this principle has been described [160]. It appears that two-dimensional GC is an ideal method for isolation and characterization of specific compounds. The basis of this technique which is eminently suitable for glass capillary columns has recently been reviewed [161-163].

In another approach a method has been described [164] which largely depends upon the superb selectivity of a high-resolution mass spectrometer to precisely monitor a specific fragment, $C_2H_4OCl^+$. The chromatographic column was merely used as a convenient way of sample introduction. For misidentification, a substance producing the same mass fragment would thus have to be eluted at the same retention time. The specificity of the method was still further increased with a system in which two mass fragments comprising a chlorine isotope pair were simultaneously monitored [165]. Three conditions thus had to be filled for a substance to be identified as BCME: correct mass, correct chlorine isotope ratio, and correct retention time. Misidentification, however, may still occur under unfavorable circumstances. In a further step, these requirements were further stiffened by substitution of a packed column with a very efficient glass capillary column [166]. BCME and 1-chloro-2-propanol have almost identical retention times on a SE-30 column and cannot be distinguished chromatographically, even on a glass capillary column. These two substances also have similar fragmentation patterns. Both produce identical chlorinated mass fragments at the masses which have been suggested for selective ion monitoring [165].

From the above example it is clear that neither extraordinary chromatographic resolving power nor high detector specificity are necessarily a guarantee of success. This applies especially for truly complex samples at very low concentration levels. At the present state of technology, GC/MS with glass capillary column unquestionably is the most powerful tool for the analytical chemist charged with the determination of toxic trace components. Tetrachlorodibenzodioxins (TCDDs) are extremely toxic, teratogenic, and mutagenic substances [167] at levels far below the concentrations at which pesticides are active. An incident at an industrial site in Italy in which large amounts of these substances were accidentally released into air forced a close look at the analytical options.

The analysis scheme calls for an extremely selective and sensitive method. The problem is compounded by potential interferences from polychlorinated biphenyls and by the fact that many isomers of TCDD exist but that only a few, most notably the 2,3,7,8-tetrochlorodibenzo-p-dioxin isomer, are of interest. Buser [168] was able to determine this specific substance in soil down to ppt levels using mass fragmentography with glass capillary columns. Figure 13 shows partial mass fragmentograms of 2,3,7,8-TCDD before a background of polychlorinated biphenyls (PCBs), demonstrating both excellent resolution and specificity. As expected, the soil around the site of the accident was heavily contaminated, but traces of 2 to 5 ppt were also found at a sampling location which is more than 100 miles removed from the site of this industrial accident. It is very likely that the substance had been transported through air, although some claims have recently been made by Dow chemists that TCDDs are formed in normal combustion processes [169, 170].

4.3 Analysis of Polynuclear Aromatic Hydrocarbons and Similar Substances Associated with Air Particulate Matter

The formation of polynuclear aromatoc hydrocarbons (PAHs) and similar large-ring compounds are a consequence of a series of high-temperature reactions. The resulting mixtures are very complex. It is well known that many of these structures are highly carcinogenic, depending on factors such as molecular shape, size, steric factors, substitution patterns, etc. Alkyl substitution and displacement of ring carbon compounds by heteroatoms may either increase or decrease the biological activity of PAHs [171,172]. A wealth of data on toxicological effects are available, and it is clearly important that the composition and level of PAH in air particulate matter be known to assess potential health risks.

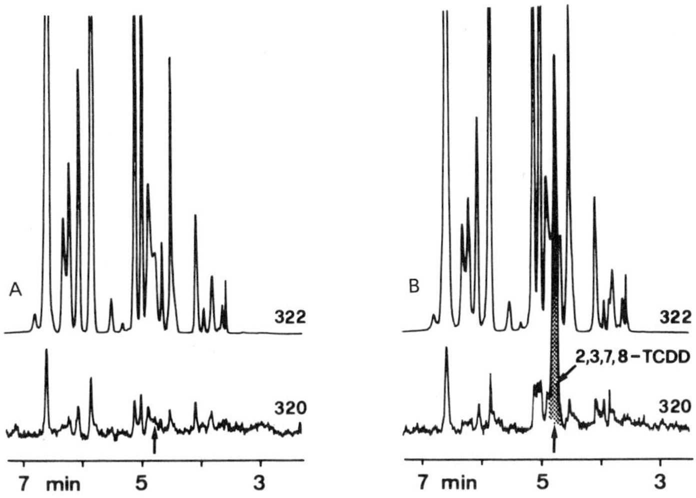

Figure 13. Partial mass fragmentograms at m/e 320 and 322 of (A) Aroclor 1254 (8 μg) with 2,3,7,8-TCDD (200 pg) added. (Source: Ref. 168.)

Polynuclear aromatic hydrocarbons have been under intense investigation for several decades. The classical analytical methods are centered around liquid chromatography and various kinds of molecular spectroscopy. A huge volume of literature has appeared, mainly describing methods of characterization for PAH by thin-layer chromatography and fluorescence spectroscopy. Gas chromatography has been applied to the analysis of PAH almost from its onset. Most of this early work has been based on packed columns. Gas-solid chromatography on inorganic salts [173-179] was more widespread than gas-liquid chromatography [180-182], reflecting the lack of suitable high-temperature phases these days. Some of the first attempts with glass capillary columns have been reported by several Italian workers [183,184] who used gum phases.

The advantages of combining a flame ionization detector with an electron capture detector for the analysis of PAH were recognized early [185]. Condensed ring systems such as polynuclear aromatic hydrocarbons can sustain a negative charge and thus produce a relatively weak but characteristic response in an ECD. Figure 14 and Table 5 represent a recent application of this principle in which predetermined response factors were applied in the identification and confirmation of PAH from air particulate matter [133].

The introduction of high-temperature phases, notably Dexsil, and general availability of computerized GC/MS gave a strong impetus to the analy-

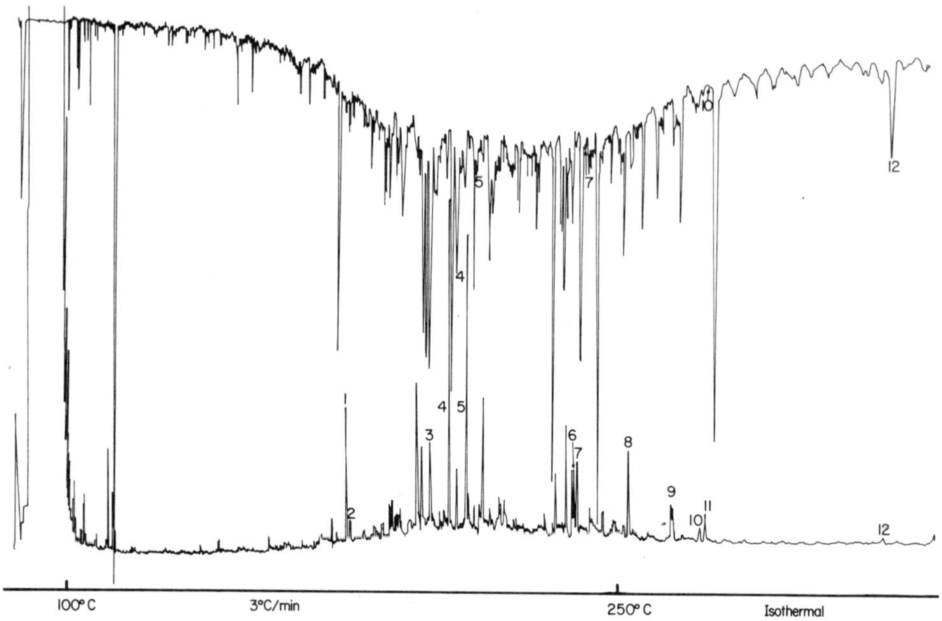

Figure 14. Glass capillary gas chromatogram of PAHs in urban air particulates. Lower trace--FID, upper trace--ECD. For peak identities, see Table 5. (Source: Ref. 133.)

sis of PAH by gas-phase methods. Lao et al. [186,187] applied both packed and capillary columns. Many chromatographically unresolved compounds were identified with the aid of the background-subtraction and ion-extraction features of the computer-controlled GC/MS. Figure 15 shows the chromatogram of a PAH extract from a large Canadian city and Table 6 lists the identified substances. Since the mass spectra of such large condensed systems are all quite similar and relatively insensitive to substitution patterns, extensive use was made of retention data. Figure 16 shows the mass spectra of BaP and BeP. It is clear that BaP, a carcinogen, cannot be distinguished from its benign counterpart on this basis.

In recent work, emphasis has been placed on extensive precolumn cleanup methods. These involve a multitude of different processes, all designed to minimize later interferences and isolate chemically well-defined groups [188]. A rather detailed liquid/liquid extraction scheme including chromatography on Sephadex [189] was described in 1972. This method was subsequently improved by addition of a further step, consisting of conventional liquid chromatography on silica gel [190] or high-pressure liquid chroma-

Table 5. Ratios of ECD and FID Responses for PAHs in Air Particulates in Oslo, Norway

Peak Number[a]	Compound	ECD/FID	Determined from standard[b]	Comment
1	Phenanthrene	c	0.090	
2	Anthracene	c	0.348	
3	3,6-Dimethylphenanthrene	c	c	Internal standard
4	Fluoranthrene	1.51	1.664	
5	Pyrene	0.17	0.154	
6	Benzo(a)anthracene	d	0.073	e
7	Chrysen/Triphenylene	0.34	0.371	
8	β,β-Binaphthalene	c	c	Internal standard
9	Benzo(b&k)fluoranthene	12.3	d	
10	Benzo(e)pyrene	0.26	0.272	
11	Benzo(a)pyrene	47.1	d	e
12	o-Phenylenepyrene	50.0	48.5	

[a] See Fig. 14.
[b] See table in original text [133].
[c] No or very small EC response.
[d] Not determined.
[e] Possibly interference from another compound with strong EC response.
Source: Ref. 133.

tography. The entire process thus consisted of a Soxhlet extraction step, followed by solvent partitioning, lipophilic gel chromatography, high-pressure liquid-solid chromatography, and finally gas-liquid chromatography on glass capillary columns [191]. Some 122 compounds were tentatively identified, mostly by mass spectrometry. The difficulty in assignment of structures on the basis of mass spectral fragmentation patterns became apparent. Some seven different isomers of ethylchrysene or ethylbenzo (a) anthracene all produced very similar mass spectra. Unequivocal identification was thus not possible in the absence of authentic standards.

A great deal of data have accumulated on the distribution of PAH in urban areas, but not much information is available on human exposure to these chemicals at specific industrial production sites. It is well known that workers involved in the handling of petroleum and coal products are in contact with such materials. Significantly higher rates of cancer have been

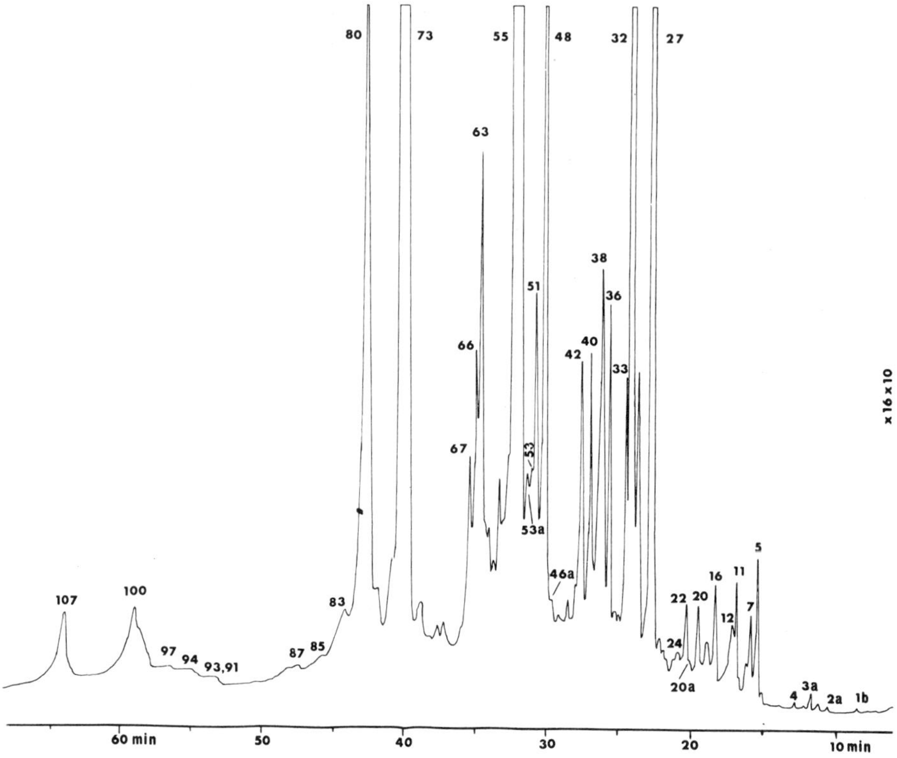

Figure 15. Gas chromatogram of air sample on Dexsil-300 SCOT column. Numbers correspond to Table 6. (Source: Ref. 186.)

reported for personnel exposed to pitch volatiles in coke oven batteries [192]. Björseth and co-workers have undertaken extensive studies of several working environments in Norway [193,194]. They concluded that coke oven workers are exposed to between 5 and 1000 mg m^{-3} of polynuclear hydrocarbons [195]. To include both respirable and nonrespirable material, air was drawn through a conventional filter, followed by an impinger. Figures 17 and 18 show the chromatograms of an air filter extract and the content of the impinger, respectively. Table 7 represents data from one sampling station. It can be seen that the more volatile components are primarily found in the wash bottle, whereas substances of higher molecular weight such as BaP are quantitatively retained on the glass fiber filter preceding the impinger. The PAH profiles generated from the coke plant and from an aluminum production facility were similar, but another aluminum smelter using a different process produced a larger proportion of the more volatile material [196]. This demonstrates that type and distribution of such substances is linked to the particular industrial processes involved.

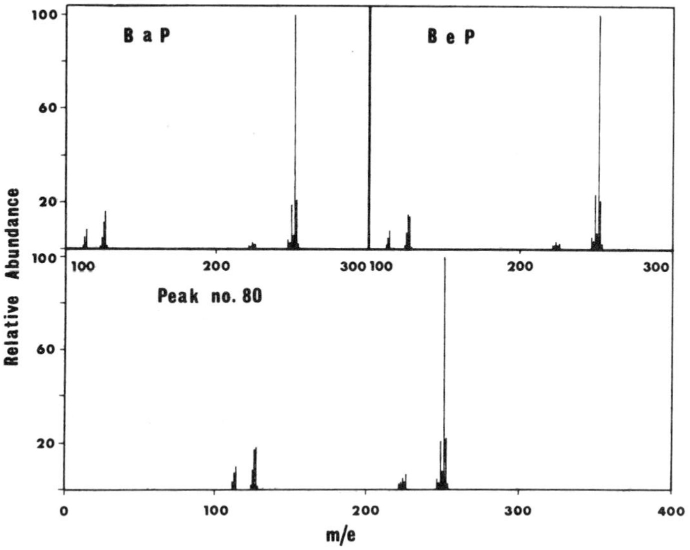

Figure 16. Mass spectra of GC effluent peak no. 80 of air sample, represented in Fig. 15, and the corresponding reference compounds. (Source: Ref. 186.)

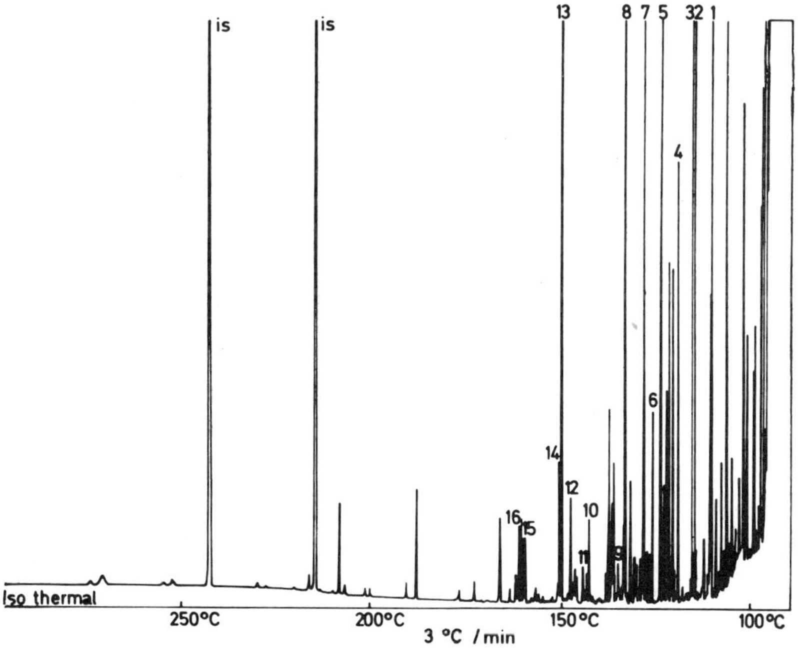

Figure 17. Gas chromatogram of PAHs in absorption bottles. The peak identities are (1) naphthalene, (2) 2-methylnahpthalene, (3) 1-methylnaphthalene, (4) biphenyl, (5) acenaphthylene, (6) acenaphtene, (7) dibenzofuran, (8) Fluorene, (9) 9-methylfluorene, (10) 2-methylfluorene, (11) 1-methylfluorene, (12) dibenzothiophene, (13) phenanthrene, (14) anthracene, (15) carbazole, and (16) 2-methylanthracene. Numbers refer to substances listed in Table 7. (Source: Ref. 195.)

Table 6. PAH Concentration of Air Sample on Dexsil-300 Capillary Column (Numbers correspond to Fig. 15.)

Time, min	Area	RRT	RF	Concn.[a] μg	Peak No.	Compound name
5.29	0.0051	0.233	0.751	0.0013	1	Biphenyl[b]
7.96	0.0067	0.350	0.776	0.0018	1a	Octahydro-phenanthrene[c] and octahydro-anthracene[b,c]
8.51	0.0177	0.375	0.758	0.0046	1b	Dihydro-fluorene[c]
8.93	0.0011	0.393	0.758	0.0003	1c	Dihydro-fluorene[c]
9.35	0.0029	0.412	0.796	0.0008	2	Methyl-biphenyl
10.52	0.0320	0.463	0.796	0.0088	2a	Methyl-biphenyl
10.96	0.0038	0.483	0.810	0.0011	3	Benzindene[b]
11.74	0.0937	0.517	0.810	0.0262	3a	Benzindene[b]
12.36	0.0148	0.544	0.864	0.0044	3b	Fluorene[b]
12.71	0.0359	0.560	0.827	0.0102	4	Dihydro-phenanthrene[b,c]
13.18	0.0082	0.580	0.803	0.0023	4a	Dihydro-anthracene[b,c]
13.87	0.0085	0.611	0.916	0.0027	4b	2-Methyl-fluorene[b]
14.60	0.0017	0.643	0.916	0.0005	4c	1-Methyl-fluorene[b]
14.94	0.0220	0.658	0.916	0.0070	4d	9-Methyl-fluorene[b]
15.34	0.6918	0.676	0.920	0.2199	5	Phenanthrene[b]
15.86	0.3479	0.698	0.880	0.1058	7	Anthracene[b]
16.15	0.1903	0.711	0.926	0.0609	11	Benzoquinoline[b]
16.63	0.0496	0.732	0.926	0.0159	11a	Benzoquinoline[b]
16.85	0.6011	0.742	0.918	0.1907	12	Acridine[b]
17.65	0.0929	0.777	0.926	0.0297	13a, 14	Fluorene carbonitrile[b,c]
18.22	0.5736	0.802	1.042	0.2065	16	Methyl-phenanthrene[b]
18.83	0.3037	0.829	0.998	0.1048	19	Methyl-anthracene[b]
19.46	0.3479	0.857	1.185	0.1424	20	Ethyl phenanthrene[c] and dimethyl-phenanthrene[c]
20.13	0.0461	0.887	1.185	0.0199	20a	Ethyl-phenanthrene[c] and ethyl-anthracene[c]
20.34	0.3252	0.896	1.132	0.1272	22	Ethyl-anthracene[c] and dimethyl-anthracene[c]
20.82	0.1264	0.917	0.882	0.0385	24	Octahydro-fluoranthene[c]
21.13	0.0919	0.931	0.941	0.0299	24a	Octahydro-pyrene[c]
21.80	0.1263	0.960	0.902	0.0394	25	Dihydro-fluoranthene
22.11	0.1904	0.974	0.962	0.0633	25a	Dihydro-pyrene[b,c]
22.69	14.3616	1.000	1.000	4.9620	27	Fluoranthene[b]
23.54	1.5204	1.037	1.025	0.5384	29	Dihydro-benzo [*a*] fluorene[c] and dihydro-benzo [*b*] fluorene[c]
24.09	13.0534	1.061	1.067	4.8121	32	Pyrene[b]
24.39	1.1789	1.074	0.996	0.4057	35	Dihydro-benzo [*c*] fluorene[c]
25.15	0.0965	1.108	0.996	0.0332	35a	Dihydro-benzo [*c*] fluorene[c]
25.52	1.5112	1.124	1.137	0.5937	36	Benzo [*a*] fluorene[b]
26.06	3.5523	1.148	1.137	1.3954	38	Benzo [*b*] fluorene[b] and benzo [*c*] fluorene[b]
26.84	1.4097	1.182	1.070	0.5211	40	Methyl-fluoranthene[b]
27.42	1.4279	1.208	1.142	0.5634	42	Methyl-pyrene[b]
27.91	0.0780	1.230	1.142	0.0308	43	Methyl-pyrene[b]
28.38	0.1408	1.250	1.291	0.0628	44	Trimethyl-fluoranthene[c] and trimethyl-pyrene[c]
28.79	0.0579	1.268	1.291	0.0258	44a	Trimethyl-fluoranthene[c] and trimethyl-pyrene[c]
29.06	0.0367	1.280	1.090	0.0138	46	Dihydro-benzo [*c*] phenanthrene[c]
29.55	0.1349	1.302	1.090	0.0508	46a	Dihydro-benzo [*c*] phenanthrene[c]
30.03	7.3187	1.323	1.238	3.1300	48	Benzo [*c*] phenanthrene[b] and hexahydro-chrysene[c]
30.60	1.9386	1.348	1.250	0.8372	51	Benzo [*ghi*] fluoranthene[b]
30.92	0.5301	1.362	1.108	0.2029	53	Dihydro-benzo [*a*] anthracene[c] dihydro-chrysene,[c] and dihydro-triphenylene[c]

Table 6 (cont'd.)

Time, min	Area	RRT	RF	Concn.[a] µg	Peak No.	Compound name
31.19	0.9648	1.374	1.108	0.3694	53a	Dihydro-benzo [*a*] anthracene[c] dihydro-chrysene,[c] and dihydro-triphenylene[c]
32.30	30.6304	1.423	1.241	13.1333	55	Benzo [*a*] anthracene,[b] chrysene,[b] and triphenylene[b]
32.93	0.2359	1.451	1.212	0.0988	58	Tetrahydro-methyl-benzo [*a*] anthracene,[c] chrysene,[c] and triphenylene[c]
33.16	0.5427	1.460	1.153	0.2162	59	Dihydro-methyl-benzo [*ghi*] fluoranthene[c]
33.55	0.0552	1.478	1.328	0.0253	61	Methyl-benzo [*a*] anthracene[b]
33.92	0.7511	1.494	1.328	0.3446	62	Methyl-triphenylene
34.44	3.1790	1.517	1.328	1.4590	64	Methyl-chrysene[b]
34.79	1.8216	1.533	1.267	0.7974	66	β,β'-Binaphthyl[b]
35.24	1.1726	1.553	1.269	0.5141	67	Dihydro-methyl-benzo [*k&b*] fluoranthenes[c] and dihydro-methyl-benzo [*a&e*] pyrenes[c]
37.09	0.3416	1.634	1.356	0.1600	68	Methyl-β,β'-binaphthyl[c]
37.50	0.1994	1.652	1.421	0.979	68a	Dimethyl-benzo [*a*] anthracene[b,c] and triphenylene[c]
38.02	0.0861	1.675	1.420	0.0423	69	Dimethyl-chrysene[c]
38.67	0.9308	1.704	1.293	0.4158	70	Benzo [*j*] fluoranthene[b]
40.42	24.9984	1.781	1.333	11.5132	72	Benzo [*k*] fluoranthene[b] and benzo [*b*] fluoranthene[b]
41.82	0.9861	1.843	1.426	0.4858	77	Methyl-benzo [*k*] fluoranthene and methyl-benzo [*b*] fluoranthene
42.84	11.1724	1.888	1.327	5.1223	80	Benzo [*a*] pyrene,[b] benzo [*e*] pyrene,[b]
44.08	1.4384	1.942	1.332	0.6619	83	Perylene[b]
45.38	0.2320	2.000	1.337	0.1072	85	3-Methyl-cholanthrene[b]
47.11	0.2356	2.076	1.419	0.1155	87	Methyl-benzo [*a*] pyrene and methyl-benzo [*e*] pyrene
49.77	0.0211	2.193	1.252	0.0092	90	*o*-Phenylene-fluoranthene
52.98	0.2679	2.334	1.522	0.1409	91,93	Dimethyl-benzo [*k*] fluoranthene,[c] dimethyl-benzo [*b*] fluoranthene,[c] dimethyl-benzo [*a*] pyrene,[c] and dimethyl-benzo [*e*] pyrene[c]
54.52	0.9655	2.402	1.342	0.4477	94	1,2,3,4-Dibenzanthracene[b]
56.16	0.8556	2.475	1.352	0.3397	97	2,3,6,7-Dibenzanthracene[b]
59.01	4.1676	2.600	1.352	1.9482	100	Benzo [*b*] chrysene[b] and *o*-phenylenepyrene[b]
64.07	2.4232	2.823	1.353	1.1328	107	Benzo [*ghi*] perylene[b] and anthanthrene[b]
72.45	0.0415	3.193	1.444	0.0207	117	Methyl-benzo [*b*] chrysene[c] and methyl-benzo [*c*] tetraphene[c]
76.72	0.0223	3.381	1.447	0.0111	119	Methyl-*o*-phenylene-pyrene[c] and methyl-anthanthrene[c]

[a] Calibrated with fluoranthene the specific responses 0.3455 µg/unit area.
[b] Determined by comparison with a primary standard compound of known purity with respect to relative retention time and mass spectrum. [c] Compounds found in the air sample which have not been previously reported. Source: Ref. 186

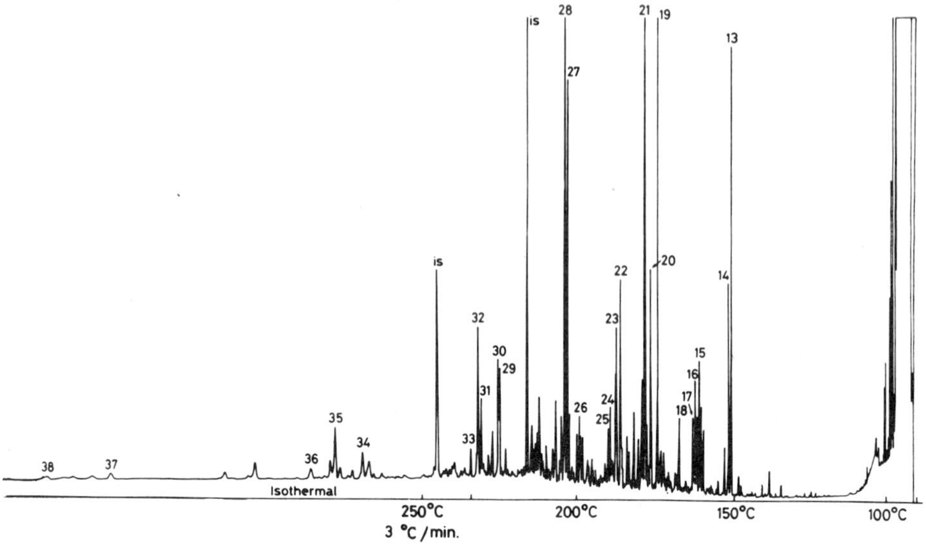

Figure 18. Gas chromatogram of particulate PAHs. For peak identities of nos. 13 through 16 see Fig. 17. The rest are (17) 1-methylphenanthrene, (18) 9-methylanthracene, (19) fluoranthene, (20) dihydrobenzo(a,b)fluorene, (21) pyrene, (22) benzo(a)fluorene, (23) benzo(b)fluorene, (24) 4-methyl pyrene, (25) 1-methylpyrene, (26) benzo(c)phenanthrene, (27) benz(a)anthracene, (28) chrysene/Triphenylene, (29) benzo(b)fluoranthrene, (30) benzo(k)-fluoranthene, (31) benzo(e)pyrene, (32) benzo(a)pyrene, (33) perylene, (34) o-phenylenepyrene, (35) benzo(ghi)perylene, (36) anthanthene, (37) 1,2,3,4-dibenzopyrene, and (38) coronene. Numbers refer to substances listed in Table 7. (Source: Ref. 195.)

Sulfur, nitrogen, and oxygen heterocyclics have also been reported [195] in air and also in the manufacture of carbon black, a major ingredient in the production of tires [197]. Studies recently conducted in Norway have shown that polynuclear aromatic hydrocarbons may be transported over long ranges. The information is based on analysis of snow and water samples which act as convenient collection media for organics from the atmosphere, in particular for particulate matter. Figure 19 shows a comprehensive analytical scheme for the study of such precipitated material [198]. It can be seen that gas chromatography occupies a central position in the overall scheme, but other methods are necessary to obtain a more complete picture.

These studies revealed that several classes of compounds can be found in such precipitates, including alkanes, PAH, phthalic esters, and fatty acid ethylesters in the neutral fraction and fatty acids and dicarboxylic acids in the acidic fraction [198,199]. Phthalates are widely associated

Table 7. PAHs in the Working Atmosphere of a Coke Plant, Mobile Sampling, Fall 1976

PAH	Sampling site			
	Battery top[a]		Battery top[b]	
	Particulates	Gaseous	Particulates	Gaseous
Naphthalene	--	294.46	--	459.91
2-Methylnaphthalene	--	41.03	--	126.34
1-Methylnaphthalene	--	20.78	--	68.35
Biphenyl	--	7.62	--	17.93
Acenaphthylene	--	47.55	--	113.14
Acenaphthene	--	10.86	--	25.19
Dibenzofuran	--	17.33	--	35.24
Fluorene	--	24.92	--	37.14
Dibenzothiophene	0.77	5.35	4.07	8.43
Phenanthrene	11.24	71.39	59.40	71.82
Anthracene	1.76	16.85	25.83	18.66
Carbazole	1.89	--	4.22	0.46
2-Methylanthracene	0.76	--	9.12	--
1-Methylphenanthrene	1.15	1.00	7.54	1.02
Fluoranthene	32.38	7.60	71.29	1.62
Dihydrobenzo(a&b) fluorene	3.25	--	16.30	--
Pyrene	24.87	3.98	51.56	0.77
Benzo(a)fluorene	6.40	--	21.10	--
Benzo(b)fluorene	0.92	--	28.27	--
4-Methylpyrene	0.68	--	--	--
1-Methylpyrene	1.27	--	5.81	--
Benzo(c)phenanthrene	5.47	--	5.21	0.38
Benzo(a)anthracene	11.32	--	31.09	--
Chrysene/triphenylene	14.49	--	35.55	--
Benzo(b,j,&k) fluoranthene	4.70	--	7.20	--
Benzo(e)pyrene	3.88	--	6.41	--
Benzo(a) pyrene	8.02	--	14.51	--
Perylene	1.80	--	3.15	--
o-Phenylenepyrene	3.83	--	3.14	--
Benzo(ghi)perylene	4.20	--	7.53	--
Anthanthrene	1.70	--	1.48	--
Total PAH:	146.75	569.72	419.78	986.40

[a] Particle concentration 5.76 mg m^{-3}. [b] Particle concentration 7.21 mg m^{-3}.
Source: Ref. 195.

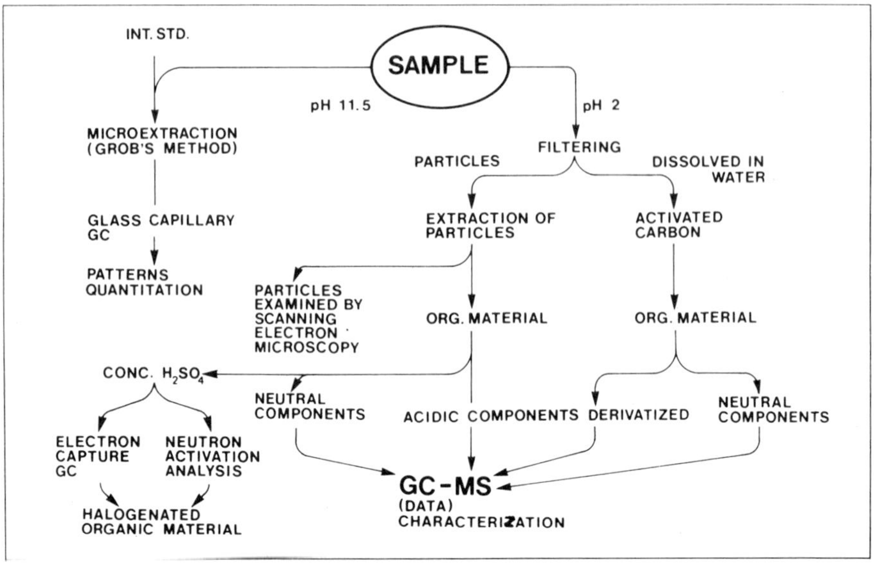

Figure 19. Outline of a comprehensive analytical procedure for organics from the atmosphere, trapped in snow. (Source: Ref. 199.)

with industrial activities; some of the acids or esters may, however, stem from natural sources. It is interesting to note that the PAH levels in these samples, collected in southern Norway, varied by as much as a factor of 20, depending on meterological conditions. High levels of PAH were observed when aerosols were transported from England and France, whereas concentrations were much lower when the air around the sampling site was stagnant or came from northern Norway [200,201].

It is clear that such data not only aid in proper assessment of occupational health risks but also precipitate much-needed further study in other fields of environmental analysis.

5. SUMMARY

Sampling and analytical methodology were discussed for the analysis of organic substances in air. It was shown that glass capillary column GC, especially in conjunction with mass spectoometry, presently represents the most powerful method toward qualitative and quantitative determination of trace organics in air. This is particularly true of highly complex samples such as encountered in the high-molecular-weight range of organics in air.

The importance of suitable sampling techniques was discussed in some detail and the significance of proper sampling-introduction methods into high-performance glass capillaries was underlined. One of the most important factors which often is not sufficiently recognized, the chemical inertness of the glass capillary column, was stressed. Several examples of applications were presented. Some of these involved the comprehensive characterization of organics in air; others were targeted toward specific substances or groups of substances.

Glass capillary column GC is the preferred method of separation for all organics in air, with the exception of the most volatile section. Advantages of such technology are not limited to improved resolution but also to increased speed and better sensitivity.

Glass capillary columns are the heart of separation systems, but they must always be seen in conjunction with the proper sample-introduction methods, detection systems, and data-handling capabilities. The enormous amount of information generated by such systems is greatly facilitated by on-line data processing. High-precision GC with glass capillary columns may become an important extension which in current practice has not been utilized to its full potential.

REFERENCES

1. Altshuller, A. P., in *Advances in Chromatography*, Vol. 5 (J. C. Giddings, and R. A. Keller, eds.), Marcel Dekker, New York, 1968.
2. Thring, M. W., *Int. J. Environ. Stud.*, 5:251 (1974).
3. Bunn, W. W., Deane, E. R., Klein, D. W., and Kloepfer, R. D., *Water, Air, and Soil Pollution*, 4:376 (1975).
4. Villalobos, R., in *Analytical Methods Applied to Air Pollution Measurements*, (R. K. Stevens and W. F. Herget, eds.), Ann Arbor Science, Ann Arbor, Mich., 1977.
5. Kaiser, R. E., *J. Chromatogr. Sci.*, 12:36 (1974).
6. *Chem. Eng. News*, p. 20 (July 1978).
7. Keith, L. H., *J. Chromatogr. Sci.*, 17:48 (1979).
8. Grob, K., Jr., Grob, G., and Grob, K., *J. Chromatogr.*, 156:1 (1978).
9. Fisher, R. L., Reiser, R. W., and Lasoski, B. A., *Anal. Chem.*, 49:1821 (1977).
10. Schneider, W., and Frohne, J. C., *Staub-Reinhalt Luft*, 35:275 (1975).
11. Cronn, D. R., and Harsch, D. E., *Anal. Lett.*, 9:1015 (1976).
12. Drasche, H., Funk, L., and Herbolsheimer, R., *Staub-Reinhalt Luft*, 32:20 (1972).

13. Polasek, J. C., and Bullin, J. A., *Environ. Sci. Technol.*, 12:708 (1978).

14. Kopczynski, S. L., Kuntz, R. L., and Bufalini, J. J., *Environ. Sci. Technol.*, 9:648 (1975).

15. Crabtree, J. H., and Mayrsohn, H., *Environ. Sci. Technol.*, 11:97 (1977).

16. Calvert, J. G., *Environ. Sci. Technol.*, 11:99 (1977).

17. Ludlum, K. H., and Bailey, B. S., *Environ. Sci. Technol.*, 13:235 (1979).

18. Schneider, W., Frohne, J. C., and Bruderreck, H., *J. Chromatogr.*, 155: 311 (1978).

19. Rasmussen, R. A., *Amer. Lab.*, 10:43 (1971).

20. Hoshika, Y., *J. Chromatogr.*, 137:455 (1977).

21. Westberg, H. H., Rasmussen, R. A., and Holdren, M., *Anal. Chem.*, 46: 1852 (1974).

22. Bruner, F., Ciccioli, P., and DiNardo, F., *J. Chromatogr.*, 99:661 (1974).

23. Kaiser, R. E., *J. Chromatogr. Sci.*, 9:227 (1971).

24. Bergert, K. H., Betz, V., and Pruggmayer, D., *Chromatographia*, 7:115 (1974).

25. Bergert, K. H., and Betz, V., *Chromatographia*, 7:681 (1974).

26. Kaiser, R. E., *Anal. Chem.*, 45:965 (1973).

27. Rasmussen, R. A., *Amer. Lab*, 7:19 (1972).

28. Rohrschneider, L., Jaeschke, A., and Kubik, W., *Chem.-Ing.-Tech.*, 43: 1010 (1971).

29. Nicotra, C., Cornu, A., Massot, R., and Perilhon, P., in *Gas Chromatography, 1972* (S. G. Perry and E. R. Adlard, eds.), Applied Science, Barking, England 1973, p. 9.

30. Simmons, J. H., in *Gas Chromatography, 1972* (S. G. Perry and E. R. Adlard, eds.), Applied Science, Barking, England 1973, p. 17.

31. McEwan, D. J., *Anal. Chem.*, 38:1047 (1966).

32. Russell, J. W., and Shadoff, L. A., *J. Chromatogr.*, 134:375 (1977).

33. Tyson, B. J., and Carle, G. C., *Anal. Chem.*, 46:611 (1974).

34. Raymond, A., and Guiochon, G., *Analysis*, 1:357 (1973).

35. Raymond, A., and Guichon, G., *Environ. Sci. Technol.*, 8:143 (1974).

36. Jeltes, R., Burghardt, E., Thijsse, T. R., and den Tonkelaar, W. A. M., *Chromatographia*, 10:430 (1977).

37. Louw, C. W., and Richards, J. F., *S. Afr. J. Sci.*, 73:240 (1977).

38. Louw, C. W., Richards, J. F., and Faure, P. K., *Atm. Environ.*, 11:703 (1977).

39. Ioffe, B. V., Isidorov, V. A., and Zenkevich, I. G., *J. Chromatogr.*, 142:787 (1977).

40. Bertsch, W., Chang, R. C., and Zlatkis, A., *J. Chromatogr. Sci.*, 12:175 (1974).

41. Bursey, J. T., Smith, D., Bunch, J. E., Williams, R. N., Berkley, R. E., and Pellizzari, E. D., *Amer. Lab.*, p. 35. (Dec. 1977).

42. Liebich, H. M., Bertsch, W., Zlatkis, A., and Schneider, H., *Aviat. Space and Environ. Med.*, p. 1002 (Aug. 1973).

43. Butler, L. D., and Burke, M. F., *J. Chromatogr. Sci.*, 14:117 (1976).

44. Gearhart, H. L., and Burke, M. F., *J. Chromatogr. Sci.*15:2 (1977).

45. Bertsch, W., Zlatkis, A., Liebich, H. M., and Schneider, H. T., *J. Chromatogr.*, 99:673 (1974).

46. Berkley, R. E., and Pellizzari, E. D., *Anal. Lett.*, 4:327 (1978).

47. Hoshika, Y., and Muto, G., *J. Chromatogr.*, 157:277 (1978).

48. Novotny, M., Lee, M. L., and Bartle, K. D., *Chromatographia*, 7:333 (1974).

49. Mieure, J. P., and Dietrich, M. W., *J. Chromatogr. Sci.*, 11:559 (1973).

50. Russell, J. W., *Environ. Sci. Technol.*, 9:1175 (1975).

51. Pellizzari, E. D., Carpenter, B. H., Bunch, J. E., and Sawicki, E., *Environ. Sci. Technol.*, 9:556 (1975).

52. Parsons, J. S., and Mitzner, S., *Environ. Sci. Technol.*, 9:1053 (1975).

53. Neher, M. B., and Jones, P. W., *Anal. Chem.*, 49:512 (1977).

54. Murray, K. E., *J. Chromatogr.*, 135:49 (1977).

55. Janák, J., Růzicková, J., and Novák, J., *J. Chromatogr.*, 99:689 (1974).

56. Tanaka, T., *J. Chromatogr.*, 153:7 (1978).

57. Sydor, R., and Pietrzyk, D. J., *Anal. Chem.*, 50:1842 (1978).

58. Novotny, M., and Lee, M. L., *Experientia*, 29:1038 (1973).

59. Vanhaelen, M., Vanhaelen-Fastre, R., and Geeraerts, J., *J. Chromatogr.*, 144:108 (1977).

60. Mann, J. B., Enos, H. F., Gonzalez, J., and Thompson, J. F., *Environ. Sci. Technol.*, 8:584 (1974).

61. Williams, F. W., and Umstead, M. E., *Anal. Chem.*, 40:2232 (1969).

62. Black, M. S., Rehg, W. R., Sievers, R. E., and Brooks, J. J., *J. Chromatogr.*, 142:809 (1977).

63. Versino, B., de Groot, M., and Geiss, F., *Chromatographia*, 7:302 (1974).

64. Holzer, G., Shanfield, H., Zlatkis, A., Bertsch, W., Juarez, P., Mayfield, H., and Liebich, H. M., *J. Chromatogr.*, 142:755 (1977).

65. Holzer, G., Oro, J., and Bertsch, W., *J. Chromatogr.*, 126:771 (1976).

66. Pellizzari, E. D., Bunch, J. E., Berkeley, R. E., and McRae, J., *Anal. Lett.*, 9:45 (1976).

67. Bowen, B. E., *Anal. Chem.*, 48:1584 (1976).

68. Dravnieks, A., Krotoszynski, B. K., Whitfield, J., O'Donnell, A., and Burgwald, T., *Environ. Sci. Technol.*, 5:1220 (1971).

69. Ellgehausen, D., *Z. Anal. Chem.*, 272:284 (1974).

70. Bruner, F., Ciccioli, P., and Di Nardo, F., *J. Chromatogr.*, 99:661 (1974).

71. Aue, W. A., and Teli, P., *J. Chromatogr.*, 62:15 (1971).

72. Komers, R., Kopecká, H., and Kraus, M., *J. Chromatogr.*, 148:43 (1978).

73. Smith, J. R. L., Tameesh, A. H. H., and Waddington, D. J., *J. Chromatogr.*, 148:353 (1978).

74. Smith, J. R. L., Tameesh, A. H. H., and Waddington, D. J., *J. Chromatogr.*, 148:365 (1978).

75. Yasuda, S. K., and Loughran, E. D., *J. Chromatogr.*, 137:283 (1977).

76. Saalfield, F. A., Williams, F. W., and Saunders, R., *Amer. Lab.*, 3:8 (1971).

77. Saunders, R. A., Atmospheric Contamination in SEALAB 1, Pro. Con. Atmospheric Contamination in Confined Spaces, AMRL-TR-65-230 (1965).

78. Deuel, C. L., and Moberg, M. L., NASA, Langley Research Center, NASA CR-111851 (1970).

79. Saalfield, F. E., Williams, F. W., and Saunders, R. A., *Amer. Lab.*, 3:8 (1971).

80. Saunders, R. A., Umstead, M. E., and Gammon, R. H., Pro. 3rd An. Con. Atmospheric Contamination in Confined Spaces, AMRL-TR-67-200 (1967).

81. Chiantella, A. J., Smith, W. D., Umstead, M. E., and Johnson, J. E., *Amer. Ind. Hyg. Ass. J.*, 3:186 (1966).

82. Grob, K., and Grob, G., *J. Chromatogr.*, 62:1 (1971).

83. Bertsch, W., and Zlatkis, A., unpublished results.

84. Juarez, P., M.S. thesis, University of Alabama, Spring 1979.

85. Bowen, B. E., *Anal. Chem.*, 48:1584 (1976).

86. Pellizzari, E. D., The Measurement of Carcinogenic Vapors in Ambient Temperatures, U.S. Environmental Protection Agency, Contract #68-02-1228, 1977.

87. Frankel, L. S., and Black, R. F., *Anal. Chem.*, 48:732 (1976).

88. Air Sampling Instruments for Evaluation of Atmospheric Contaminants, American Conference of Government, Industrial, Hygienists, Cincinnati, 1972.

89. Lindsted, G., *Atm. Environ.*, 2:1-11 (1968).

90. Cautreels, W., and Van Cauwenberghe, K., *J. Chromatogr.*, 131:253 (1977).

91. Hill, H. H., Jr., Chan, K. W., and Karasek, F. W., *J. Chromatogr.*, 131:245 (1977).

92. Schuette, F. J., *Atm. Environ.*, 1:515-19 (1967).

93. Sawicki, E., Meeker, J. E., and Morgan, M. J., *Int. J. Air Wat. Poll.*, 9:291-298 (1965).

94. Sawicki, E., McPherson, S. P., Stanley, T. W., Meeker, J., and Elbert, W. C., *Int. J. Air. Wat. Poll.*, 9:515-524 (1965).

95. Grimmer, G., *Beitr. Tabakforsch.*, 7:291 (1962).

96. Elmenhorst, H., Stadler, L., and Goertz, E., *J. Chromatogr.*, 40:264(1969)

97. Klimisch, J., *Anal. Chem.*, 264:275 (1973).

98. Swain, A. P., Cooper, J. E., and Stedman, R. L., *Cancer Res.*, 29:579 (1969).

99. Stedman, R. L., Miller, R. L., Lakrtiz, L., and Chamberlain, W. J., *Chem. Ind.*, 3:394 (1968).

100. Schmeltz, I., Dooley, C. J., Stedman, R. L., and Chamberlain, W. J., *Phytochem.*, 6:33 (1967).

101. Haenni, E. O., Howard, J. W., and Joe, F. L., Jr., *J. AOAC*, 45:67 (1962).

102. Grimmer, V. G., Hildebrandt, A., and Böhnke, H., *Erdöl Kohle*, 25:442 (1972).

103. O'Keefe, A. E., and Ortmann, G. C., *Anal. Chem.*, 38:760 (1966).

104. Singh, G. B., Salas, L., Lillian, D., Arnts, R. R., and Appleby, A., *Environ. Sci. Technol.*, 11:511 (1977).

105. Saltzmann, B. E., Burg, W. R., and Ramaswamy, G., *Environ. Sci. Technol.*, 5:1121 (1971).

106. Burg, W. R., Birch, S. R., Cuddeback, J. E., and Saltzman, B. E., *Environ. Sci. Technol.*, 10:1233 (1976).

107. Bruner, G., Canulli, C., and Possanzini, M., *Anal. Chem.*, 45:1790 (1973).

108. Tsang, W., and Wasik, S. P., *J. Chromatogr. Sci.*, 9:567 (1971).

109. Grob, K., and Grob, G., *J. Chromatogr. Sci.*, 7:584 (1969).

110. Grob, K., and Grob, K., Jr., *J. Chromatogr.*, 94:53 (1974).

111. Grob, K., Jr., *HRC & CC*, 1:57 (1978).

112. Jennings, W. G., Freemann, R. R., and Rooney, T. A., *HRC & CC*, 1:275 (1978).

113. Schomburg, G., Belan, H., Dielmann, R., Weeke, F., and Husmann, H., *J. Chromatogr.*, 142:87 (1977).

114. Miller, R., and Jennings, W., *HRC & CC*, 2:72 (1979).

115. Grob, K., Jr., and Neukom, H. P., *HRC & CC*, 2:15 (1979).

116. Grob, K., and Grob, K., Jr., *J. Chromatogr.*, 151:311 (1978).

117. Schomburg, G., Husmann, H., and Weeke, F., *Chromatographia*, 10:580 (1977).

118. Grob, K., *HRC & CC*, p. 263 (Nov. 1978).

119. Kaiser, R. E., Chromcard Information, Precolumns in Gas Chromatography, *HRC & CC*, 2:95 (1979).

120. Teranishi, R., Mon, T. R., Robinson, A. B., Cary, P., and Pauling, L., *Anal. Chem.*, 44:18 (1972).

121. Zlatkis, A., Bertsch, W., Lichtenstein, H. A., Tishbee, A., Shunbo, F., Liebich, H. M., Coscia, A. M., and Fleischer, N., *Anal. Chem.*, 45:763 (1973).

122. Grob, K., and Grob, G., *Chromatographia*, 4:422 (1971).

123. Janini, G. M., Johnson, K., and Fielinski, W., *Anal. Chem.*, 47:670 (1975).

124. Janini, G. M., Muschik, G. M., and Zielinski, W. L., *Anal. Chem.*, 48:809 (1976).

125. Grob, K., Grob, G., and Grob, K., Jr., *Chromatographia*, 10:181 (1977).

126. Welsch, T., Engewald, W., and Klaucke, C., *Chromatographia*, 10:22 (1977).

127. Grob, K., Jr., Grob, G., and Grob, K., *HRC & CC*, p. 149 (Sept. 1978).

128. Grob, K., Grob, G., and Grob, K., Jr., *HRC & CC*, p. 31 (Jan. 1979).

129. Blomberg, L., and Wännman, T., *J. Chromatogr.*, 168:81 (1979).

130. Grob, K., and Grob, G., *HRC & CC*, 2:109 (1979).

131. Sevcik, J., Kaiser, R. E., and Rieder, R., *J. Chromatogr.*, 126:263 (1976).

132. Bächmann, K., Ewig, W., Rudolph, J., and Tsotsos, D., *Chromatographia*, 10:684 (1977).

133. Björseth, A., and Eklund, G., *HRC & CC*, 2:22 (1979).

134. Etzweiler, F., and Neuner-Jehle, N., *Chromatographia*, 6:503 (1973).

135. Goretti, G., and Pozzanzini, M., *J. Chromatogr.*, 77:317 (1973).

136. Anderson, E. L., and Bertsch, W., *HRC & CC*, 1:13 (1978).

137. Henneberg, D., Henrichs, U., and Schomburg, G., *J. Chromatogr.*, 122:343 (1975).

138. Köhler, M., and Höhn, M., *Chromatographia*, 9:611 (1976).

139. Stan, J., and Abraham, B., *Anal. Chem.*, 50:2161 (1978).

140. Dürbeck, W., Büker, I., and Leymann, W., *Chromatographia*, 11:295 (1978).

141. Rapp, U., Schröder, U., Meier, S., and Elmenhorst, H., *Chromatographia*, 8:474 (1975).

142. Blum, W., and Richter, W. J., *J. Chromatogr.*, 132:249 (1977).

143. Versino, B., Knöppel, H., de Groot, M., Peil, A., Poelman, J., Schauenburg, H., Vissers, H., and Geiss, F., *J. Chromatogr.*, 122:373 (1976).

144. Bertsch, W., unpublished results.

145. Altshuller, A. P., Lonneman, W. A., Sutterfield, F. D., and Kopczynski, S. L., *Environ. Sci. Technol.*, 5:1009 (1971).

146. Lonneman, W. A., Kopczynski, S. L., Darley, P. E., and Sutterfield, F. D., *Environ. Sci. Technol.*, 8:229 (1974).

147. Ciccioli, P., Bertoni, G., Bancaleoni, E., Fratarcangeli, R., and Bruner, F., *J. Chromatogr.*, 126:757 (1976).

148. Pellizzari, E. D., Bunch, J. E., Berkley, R. E., and McRae, J., *Anal. Chem.*, 48:803 (1976).

149. Office of Research and Development, U.S. Environmental Protection Agency, Development of Method for Carcinogenic Vapor Analysis in Ambient Atmospheres, EPA-650/2-74-121, July 1974.

150. U.S. Environmental Protection Agency, Development of Analytical Techniques for Measuring Ambient Atmospheric Carcinogenic Vapors, EPA-600/2-75-076, Nov. 1975.

151. U.S. Dept. HEW, U.S. Cancer Mortality by County, 1950-1969, HEW/NIH/NCI DHEW Publication No. (NIH) 74-614.

152. Pellizzari, E. D., Bunch, J. E., Bursey, J. T., and Berkley, R. E., *Anal. Lett.*, 9:579 (1976).

153. Karasek, F., *Res. Develop.*, 5:32 (1974).

154. Evans, J. E., and Arnold, J. T., *Environ. Sci. Technol.*, 9:1134 (1975).

155. VanDuren, B. L., Goldschmidt, B. M., Katz, C., Langseth, L., Mercado, G., and Sivac, A., *Arch. Environ. Health*, 16:472 (1968).

156. Laskin, G., Drew, R. T., Capiello, V. P., and Nelsol, N., *Arch. Environ. Health*, 23:135 (1971).

157. Solomon, R. A., and Kallos, G. J., *Anal. Chem.*, 47:955 (1975).

158. Yao, C. C., and Zollinger, H., *Anal. Chem.*, 51:299 (1979).

159. Kallos, G. J., Solomon, R. A., and Tou, J. C., *Anal. Chem.*, 51:301 (1979).

160. Kallos, G. J., Albe, W. R., and Solomon, R. A., *Anal. Chem.*, 49:1817 (1977).

161. Bertsch, W., *HRC & CC*, 1:85 (1978).

162. Bertsch, W., *HRC & CC*, 1:187 (1978).

163. Bertsch, W., *HRC & CC*, 1:289 (1978).

164. Evans, K. P., Mathias, A., Mellor, N., Silvester, R., and Williams, A. E., *Anal. Chem.*, 47:821 (1975).

165. Shadoff, L. A., Kallos, G. J., and Woods, J. S., *Anal. Chem.*, 45:2341 (1973).

166. Schulting, F. L., and Wils, E. R. J., *Anal. Chem.*, 49:2365 (1977).

167. Schwetz, B. A., Norris, J. M., Sparschu, G. L., Rowe, V. K., Gering, P. J., Emerson, J. L., and Gerbig, C. G., *Environ. Health Perspect.*, 5:87 (1973).

168. Buser, H. R., *Anal. Chem.*, 44:918 (1977).

169. *Chem. Eng. News*, p. 23, Feb. 12, 1979

170. *Chem. Eng. News*, p. 6, Feb. 26, 1979

171. Hoffmann, D., Bondinall, W. E., and Wynder, E. L., *Science*, 183:215 (1974).

172. Tilak, B. D., *J. Occup. Med.*, 9:76 (1960).

173. Hanneman, W. W., Spencer, C. F., and Johnson, J. F., *Anal. Chem.*, 32:1386 (1960).

174. Chortyk, O. T., Schlotzhauer, W. S., and Stedman, R. L., *J. Gas Chromatogr.*, 11:394 (1965).

175. Sauerland, H. D., and Zander, M., *Erdoel und Kohle Erdgas Petrochemie*, 19:502 (1966).

176. Solomon, P. W., *Anal. Chem.*, 36:476 (1964).

177. Favre, J. A., and Kallenbach, R. L., *Anal. Chem.*, 36:63 (1964).

178. Gump, B. H., *J. Chromatogr. Sci.*, 7:755 (1969).

179. Vernon, F., *J. Chromatogr.*, 60:406 (1971).

180. Robb, E. W., Guvernator, G. C., III, Edmonds, M. D., and Bavley, A., *Beitr. zur Tabakforsch.*, 3:278 (1965).

181. DeMaio, L., and Corn, M., *Air Poll. Contr. Ass. J.*, 16:67 (1966).

182. DeMaio, L., and Corn, M., *Anal. Chem.*, 38:131 (1966).

183. Liberti, A., Cartoni, G., and Cantuti, V., *J. Chromatogr.*, 15:141 (1964).

184. Albertini, G., Cantuti, V., and Cartoni, G. P., Int. Symp. on GC, Paper 17, Rome, 1966.

185. Cantuti, V., Cartoni, G. P., Liberty, A., and Torri, A. G., *J. Chromatogr.*, 17:60 (1965).

186. Lao, R. C., Thomas, R. S., Oja, H., and Dubois, L., *Anal. Chem.*, 45:908 (1973).

187. Lao, R. C., Thomas, R. S., and Monkman, J. L., *J. Chromatogr.*, 112:681 (1975).

188. Novotny, M., Lee, M. L., and Bartle, K. D., *J. Chromatogr. Sci.*, 12:606 (1974).

189. Grimmer, G., and Böhnke, H., *Z. Anal. Chem.*, 261:310 (1972).

190. Grimmer, G., and Böhnke, H., *Chromatographia*, 9:30 (1976).

191. Lee, M. L., Novotny, M., and Bartle, K. D., *Anal. Chem.*, 48:1566 (1976).

192. Lloyd, J. W., *J. Occup. Med.*, 13:53 (1971).

193. Björseth, A., *Anal. Chim. Acta*, 94:21 (1977).

194. Björseth, A., and Eklund, G., *Anal. Chim. Acta*, 105:119 (1978).

195. Björseth, A., Björseth, O., and Fjeldstad, F. E., *Scand. J. Environ. Health*, 4:224-236 (1978).

196. Björseth, A., and Lunde, G., *Amer. Ind. Hyg. Ass. J.*, 38:224 (1977).

197. Lee, M. L., and Hites, R. A., *Anal. Chem.*, 48:1890 (1976).

198. Gether, J., Gjös, N., and Lunde, G., Isolation and Characterization in Precipitation of Organic Micropollutants Associated with Long-Range Transport in Air, SNSF Project FR 8/77, Oslo.

199. Lunde, G., Gether, J., Gjös, N., and Lande, M. B., Organic Micropollutants in Precipitation in Norway, SNSF Project FR 9/76, Oslo.

200. Björseth, A., Lunde, G., and Lindskog, A., *Atmospheric Environment*, 13:45 (1979).

201. Lunde, G., and Björseth, A., *Nature*, 268:518 (1977).

5

ANALYSIS OF WATER AND WATER POLLUTANTS

DENIS C. K. LIN

Battelle-Columbus Laboratories
Columbus, Ohio

1. Introduction 123

2. Design of the Analytical Scheme 125

 2.1 Sampling Site and Sampling Frequency 126
 2.2 Sample Collection and Preservation 127
 2.3 Sample Preparation 129
 2.4 Instrumentation 134
 2.5 Columns 137
 2.6 Data Interpretation 138
 2.7 Data Management 139
 2.8 Quality Assurance 139

3. Application 140

 3.1 Groundwater 141
 3.2 Industrial Effluents 146
 3.3 Domestic Effluents 153
 3.4 River, Lake, Ocean, and Related Compartments 157
 3.5 Drinking Water 162

4. Summary and Future Needs 166

References 169

1. INTRODUCTION

With the number of existing organic compounds estimated at more than 3 million [1], a key question asked by scientists in water research as well as many governmental and industrial decision makers is: How many of these compounds affect human health or the environment if they are present in water? There are three basic approaches to answering this very important question; seeking to find either what compounds are in water, whether specific compounds are in water, or what compounds in the water are known to be of health and/or environmental significance.

Each approach has its own merits. The first should give a complete picture; the second gives specific answers; and the third should provide the relevant answers. Inevitably, scientists involved in water analysis will find themselves using all three approaches. Regardless of the approach, however, they generally find many constraints in the path of producing analytical data. Some constraints are scientific limitations, others are extraneous in nature, but all must be considered in designing an analytical approach. The discussion in this chapter attempts to demonstrate how glass capillary gas chromatography with appropriate detectors can provide the most reliable data within time and cost constraints and to show where it is applicable.

Involatile organic material can at times amount to 95% of the total organics present in water, and most of this material cannot be analyzed by gas chromatography. However, it has often been postulated that from 5 to 25% of the volatile organic compounds that can be analyzed by gas chromatography are responsible for the odor, taste, and toxicity of water [1]. Perhaps because of this and the limited methods available for analyzing the involatile organic compounds, most of the analysis on water reported thus far has involved the use of gas chromatography. For example, in water analyses reported in some 222 literature publications on water pollution in 1977 [2], over 90% were performed using gas chromatography. The surprising thing is that the number of these analyses using glass capillary gas chromatography is relatively small. In 1975, Grob and Grob [3] injected the same river water extract on both the packed and capillary column (Fig. 1, and Table 1) and demonstrated clearly the distinct advantage of capillary gas chromatography. It is clear that the packed column results can be deemed unacceptable in light of the capillary column results. Why, then, is there so much reluctance to switch from packed columns to capillary columns?

There have never been any good scientific reasons for this, but until recently there were several good scientific excuses: the lack of high-quality capillary columns, proper equipment, a basic modern text on the subject, and access to expertise. This book is intended to help meet those

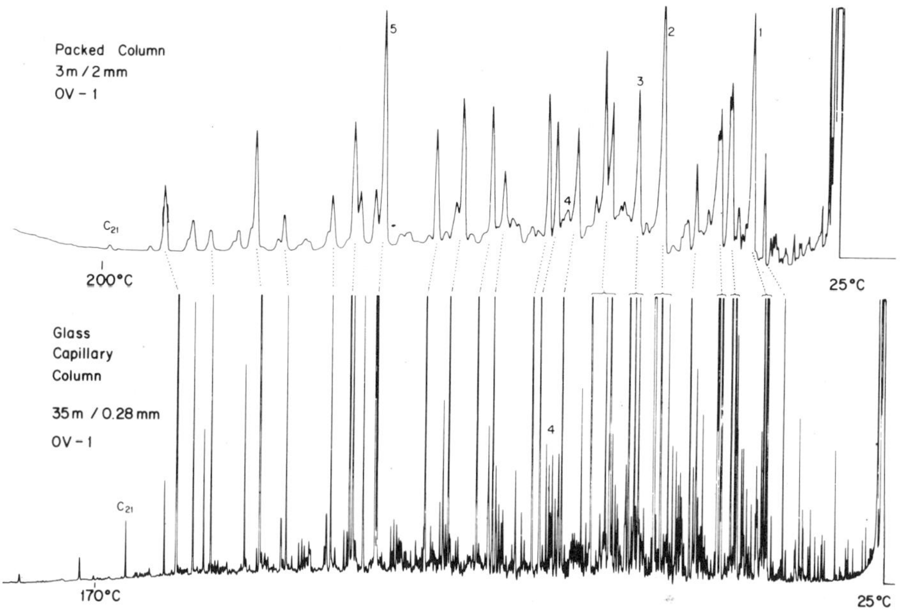

Figure 1. Comparison of typical separation from packed and capillary columns. Sample (extract from river water) and liquid phase identical. For characteristics of columns and chromatographic conditions, see Table 1. Dotted lines indicate corresponding sample components. (Source: Ref. 3.)

needs by providing the reader with a good understanding of the basic techniques of capillary gas chromatography. This chapter furnishes more specific information on the analyses of water and water pollutants.

2. DESIGN OF THE ANALYTICAL SCHEME

There are several factors to consider in designing an analytical scheme for the analysis of water. Each factor can be considered separately, but it is important to determine which is "rate-determinant" and to design the experiments around it.

Frequently the "rate-determinant" factor is the sensitivity and specificity of the analytical method. For a general-purpose qualitative and quantitative analysis, the choice instrumentation is a combination of glass capillary gas chromatography and mass spectrometry (GC/MS). Using such instrumentation on a routine basis, it is relatively safe to say that a good, low-resolution electron-impact mass spectrum may be obtained if 1 to 5 ng of a compound is put on a capillary column. Based on this kind of

Table 1. Column Characteristics and Chromatographic Conditions

	Packed	Capillary: Narrow-bore	Capillary: Wide-bore
Length	3 m	35 m	60 m
Inner Diameter	2 mm	0.28 mm	0.60 mm
Coating	3% OV-1 on Gaschrom Q, 80-100 mesh	OV-1 film 1.0×10^{-5} cm	OV-1 film 1.8×10^{-5} cm
Liquid-phase load	120 mg	1.5 mg	14 mg
Carrier gas flow	He 10-22 ml min^{-1} (progr)	H_2 2.2 ml min^{-1}	H_2 14 ml min^{-1}
Temperature range	50-200°C	25-170°C	25-200°C
Temperature Program rate	2°C min^{-1}	3.5°C min^{-1}	2.5°C min^{-1}
Analysis time	90 min	45 min	65 min
Sample Size	0.15 μl direct	0.6 μl splitting 1:25	0.6 μl splitting 1:8
Attenuation	×32	×4	×16
Number of Peaks	118	490	320

information, one can judge how the other factors fit into the design of the analytical scheme. The other factors to be considered, in the general order in which they occur in the analytical process, are:

- Sampling site and sampling frequency
- Sample collection and preservation
- Sample preparation
- Instrumentation
- Columns
- Data interpretation
- Data management
- Quality assurance

Each of these factors is discussed in the following sections.

2.1 Sampling Site and Sampling Frequency

The selection of the proper sampling site and most appropriate sampling frequency is perhaps one of the trickiest areas of water analysis. Sampling frequency is very important statistically in assuring that the analytical data will be meaningful. Consideration of sampling requires expert knowledge and a more extensive background than can be provided in the context of this chapter. The reader is referred to numerous articles on the subject cited in Ref. 4.

2.2 Sample Collection and Preservation

Water samples are generally collected in appropriate containers and sent back to the laboratory for analysis. This sounds simple enough but requires precaution to ensure that the water actually analyzed truly represents the water at the sampling site at the time of collection. Amber glass bottles, generally accepted as the most inert vessels for water storage, must be thoroughly cleaned. This is best accomplished by oven-heating the bottles at 400°C or higher for at least 3 hr. The bottles should be cooled in a contamination-free atmosphere and sealed with Teflon-lined screw caps. At the sampling site, these bottles are filled to overflow with the water of interest and resealed with the Teflon-lined screw caps, being sure that no air space remains. At this same time, other bottles are filled with an acceptable "organic-free" water to serve as sample blanks. The sample volume depends on the quality of the water, the sample preparation technique, and the desired level of sensitivity. Using liquid-liquid extraction for the organics at levels of 1 $\mu g\ liter^{-1}$ (1 ppb) or above, a 1.5- to 5-liter sample is usually collected. Either grab samples, composite samples, or both may be taken. A grab sample is taken at a particular location and time. A composite sample is made up of water at a particular site but taken over a period of time.

At times, it is desirable to add preservatives to the water sample to stablize certain classes of compounds of interest. For example, residual chlorine should be eliminated if a sample is to be analyzed for volatile halogenated compounds. Potassium ferrocyanide is recommended over sodium thiosulfate and sodium sulfite for this purpose, because the sodium compounds may react with alkyl halides [5]. However, potassium ferrocyanide should be used only if the pH of the water is high. In any case, chemical preservatives should be made organic-free before use. This can be achieved by muffling the chemicals or extracting the chemical preservative solutions several times with high-purity solvent. In any event, regardless of whether or not chemical preservatives are added, the samples and blanks should be refrigerated at 4°C for return to the laboratory and analyzed as soon as possible after return.

Another method of sampling uses accumulative columns. This increasingly popular method actually combines the sampling and extraction into a single step. It is based on the principle that the organic compounds in water can be adsorbed on solids such as the macroreticular resins Amberlite XAD-2, XAD-4, XAD-7, and XAD-8 [6-13]; the porous polymer Tenax [13,14]; and carbon [15-17]. The most commonly used adsorbent is XAD-2, although recent literature has described an equal-weight mixture of XAD-4 and XAD-8 [9] and XAD-8 by itself [10-11] as superior. All these resins must be cleaned thoroughly by successive Soxhlet extractions with various solvents to eliminate organic impurities before use.

The apparatus described by Stephan et al. [13] appears to be ideally suited for on-site sampling and extraction of organics in water. To minimize sampling contamination, the resin-adsorbent column is placed at the inlet of the apparatus, which consists of peristaltic pumps and electronic control devices, and the column is submerged in the water to be sampled. After sampling, the column is returned to the laboratory for further processing. Accumulative columns are also frequently used under laboratory situations to extract organics from water sent from sampling sites. Under these circumstances, the pH of the water can be tightly controlled, which is very difficult to do in on-site sampling.

As mentioned in the introduction to this chapter, one approach to determining the effects of organic compounds in water is to find out what compounds in the water have been shown to be of health or environmental significance. This approach is perhaps the most cost-effective way to identify the problem-causing compounds that should be monitored and regulated. To show the significance of a water sample, the organics from a large volume (up to hundreds or even thousands of liters of drinking water) have to be concentrated for bioassay [18,19]. This has been done using the reverse osmosis process [20-22]. This process is often carried out at the sampling site. Pressure applied to a water sample in contact with a semipermeable membrane allows the water to pass through the membrane leaving an aqueous reject behind, which contains the dissolved compounds. The reject is refrigerated at 4°C and sent to the laboratory for further sample work-up [20]. Note that the very volatile components are lost in this process and an additional water sample should be collected specifically for the analysis of those components.

An alternative path to direct sampling of the water is to sample appropriate living organisms in the particular aquatic environment. Shellfish, fin fish, and plankton are often employed [23-28]. They are natural bioaccumulators and in that respect the concentrations of significant pollutants are higher than in the surrounding waters. The biota samples should be frozen as soon as they are taken, and stored at subfreezing temperature, preferably about -80°C. If this sampling approach is used, it is advisable to work with an appropriate biologist.

2.3 Sample Preparation

An analytical scheme should be designed to obtain maximum resolution, i.e., separation of each component from the others so that it can be detected, identified, and quantified. The resolution obtainable from the instrumentation and the capillary column is predetermined to a certain degree. It is in sample preparation that the knowledge of separation science is best applied to optimize the analysis. Sample preparation for water analysis generally requires two basic steps, extraction and concentration, although a third step—sample cleanup—may be necessary. Before considering extraction and concentration, a brief discussion of internal standards is appropriate.

In water analysis, internal standards are used to monitor recovery of compounds, to calculate relative retention times, and for quantification [29,30]. Furthermore, internal standards may exhibit a "carrier effect" and enhance the recovery of some compounds. Obviously, the compounds used for these purposes cannot be present in the sample. And they must be stable in the sample matrix and remain unaltered through the analytical scheme. Isotopically labeled compounds of different classes [29-31] and a series of 1-chloroalkanes [32] added before the extraction step have been used as internal standards. In most cases where the isotopically labeled compounds are used it is necessary to use the mass spectrometer as the detector. The amount of internal standards added should be compatible with the amount of organics present in the sample.

After internal standards are added, organics are extracted from the water. The extraction method chosen is highly dependent on the quality of water and on whether only a certain class of compound, such as polycyclic aromatic hydrocarbons, pesticides, phenols, polychlorinated dibenzo-p-dioxins, or purgeable volatiles, is of interest.

The common liquid-liquid extraction method is usually used to examine pollutants as low as 1 μg liter^{-1} (1 ppb). It usually requires a fairly large volume of organic solvent, e.g., 100 ml of methylene chloride to extract 1 to 5 liters of water after it has been adjusted to a suitable pH [33-36]. The organic solvent is separated from the aqueous layer and concentrated with a Kuderna-Danish evaporator. Because of the concentration step, it is imperative that clean glassware and ultrahigh-purity solvents be used. To avoid the large-volume concentration step, a small volume of solvent in hundreds of μl of n-pentane has been used successfully to extract organics from water [37]. Actually, it was shown that the microsolvent extraction technique with capillary gas chromatography can extend the detection limit of organics in water at the ng liter^{-1} (ppt) level.

The discussion on sample collection and preservation touched on the use of solid adsorbent to extract organics from water. Macroreticular XAD resins are more commonly used as an adsorbent in accumulative columns than carbon. This is because the compounds that adsorb on the resins can be desorbed by solvent elution, whereas Soxhlet extraction often must be employed for carbon. Furthermore, impurities in carbon are difficult to eliminate. On the other hand, it is generally accepted that carbon has better adsorption properties. In the closed-loop stripping method for purgeable organics (to be discussed later), a small amount of carbon is used as an elution solvent. The use of carbon disulfide to desorb a regular-size accumulative column has not been reported, but liquid carbon dioxide has been used successfully to desorb compounds on fly ash [38]. If liquid carbon dioxide can be used successfully to elute compounds from carbon accumulative columns, it will be a breakthrough in that the extracted sample can be concentrated without effort by just allowing the carbon dioxide to evaporate.

Diethyl ether is the solvent most often used to elute compounds from XAD resins [6,7,12,39]. This is achieved by adding ether to the wet column. After the ether penetrates the resins for 10 min, it is allowed to elute from the column. Recently, the use of two-bed volumes of acetone followed by eight-bed volumes of methylene chloride has been recommended [9]. This avoids the impurities, peroxides, and preservatives in the ether and also makes it unnecessary to dry the organic solvent after it is separated from the water. The water is removed in the Kuderna-Danish concentration step apparently as an azeotrope. It has been demonstrated that using a 50:50

dry weight mixture of XAD-4 and XAD-8 as adsorbent for an accumulative column, it is possible to extract compounds for analysis from a 1000-liter water sample at the ng liter^{-1} (ppt) level [9].

In using the XAD resins as adsorbent, minicolumns (1.2 to 1.8 mm × 25 mm) have been employed to avoid evaporation of large volumes of elution solvent [8]. These columns can be effectively used to analyze organics in a 100-ml water sample at the μg liter^{-1} (ppb) level. The advantages of this analysis lie in the simplicity and quickness.

The reverse osmosis reject, mentioned in the collection discussion, can be extracted and concentrated by liquid-liquid extraction or accumulative columns after the sample arrives in the laboratory [20].

With biota samples, preparation is perhaps the most challenging aspect of the analysis. Sludge samples, and to a lesser degree reverse osmosis and highly polluted samples, are similar. They are extremely complex, as illustrated by Fig. 2 [30], and contain large amounts of material that cannot be separated by gas chromatography. In these cases, a sample cleanup step is necessary. The biota and sludge samples are extracted with or without saponification after homogenization. Concentrates of these or other extracts are usually cleaned up by fractionation with silica gel, alumina, gel permeation chromatography, or a combination of these techniques [27,28, 30,40-42]. An example of the fractionation scheme for reverse osmosis water concentrates is shown in Fig. 3 [30]. If removal of elemental sulfur is necessary, a granulated copper column should be used before the other types of column chromatography [42]. If the sample is made up of a large amount of high-molecular-weight compounds, gel permeation chromatography is highly recommended [28]. High-performance liquid chromatography with various stationary phases has also been employed effectively for sample cleanup [28, 43-45].

At times some of the fractions obtained from sample preparation are best handled by capillary gas chromatography after derivatization. There are many reviews and books on the subject [46,47], and thus there is no need to elaborate here except perhaps on the methylation of acids and phenols. Acids and phenols are frequently found in water and their analyses without derivatization require specialized columns [48]; even then the tailing of some of the compounds are not acceptable. There are many ways to derivatize acids and phenols, but the most effective appears to be methylation with diazomethane. Diazomethane is often considered to be hazardous and diffi-

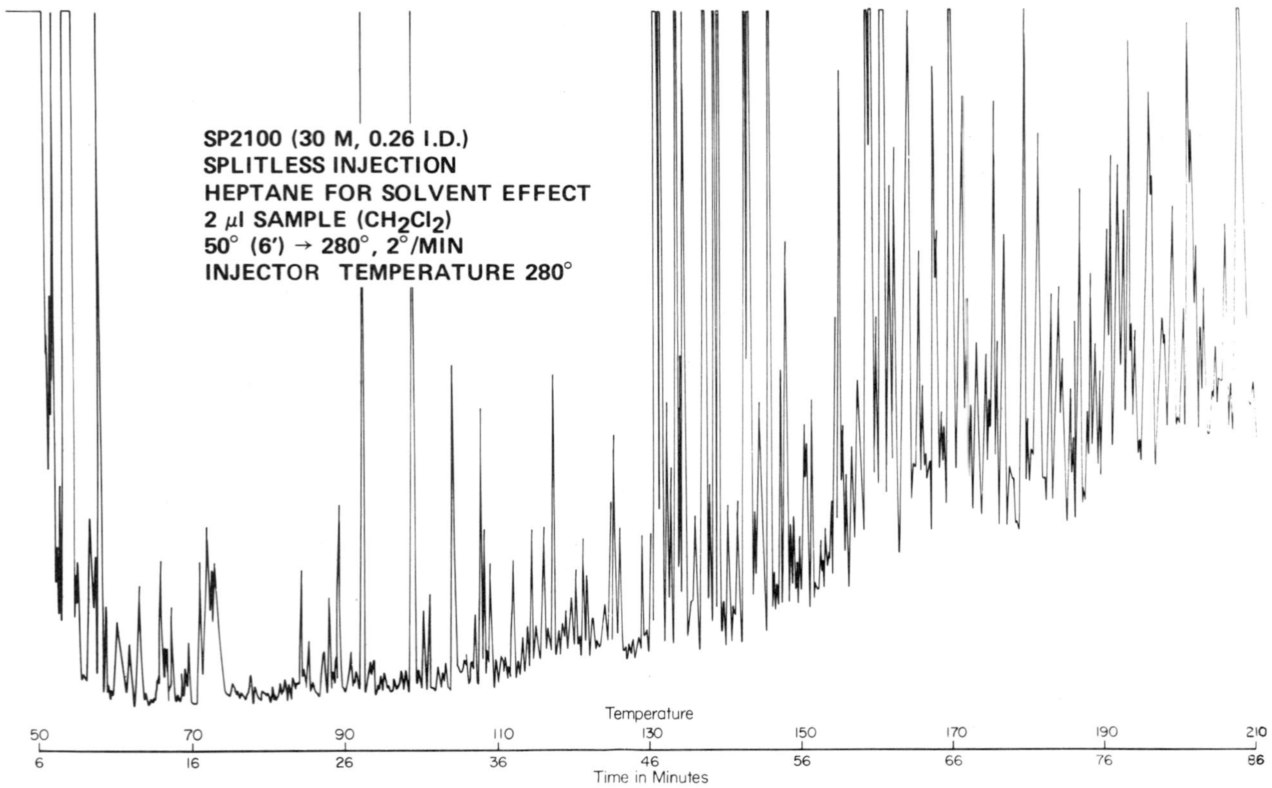

Figure 2. Partial gas chromatogram (FID) of a typical water concentrate sample; 30$\overline{M}$ SP2100 glass capillary column (ID 0.26 mm), 2 μℓ splitless injection with heptane for "solvent effect," injector temperature 280°C, temperature program from 50 (6 min) to 280°C, 2°C min^{-1}. (Sources: Refs. 30,80.)

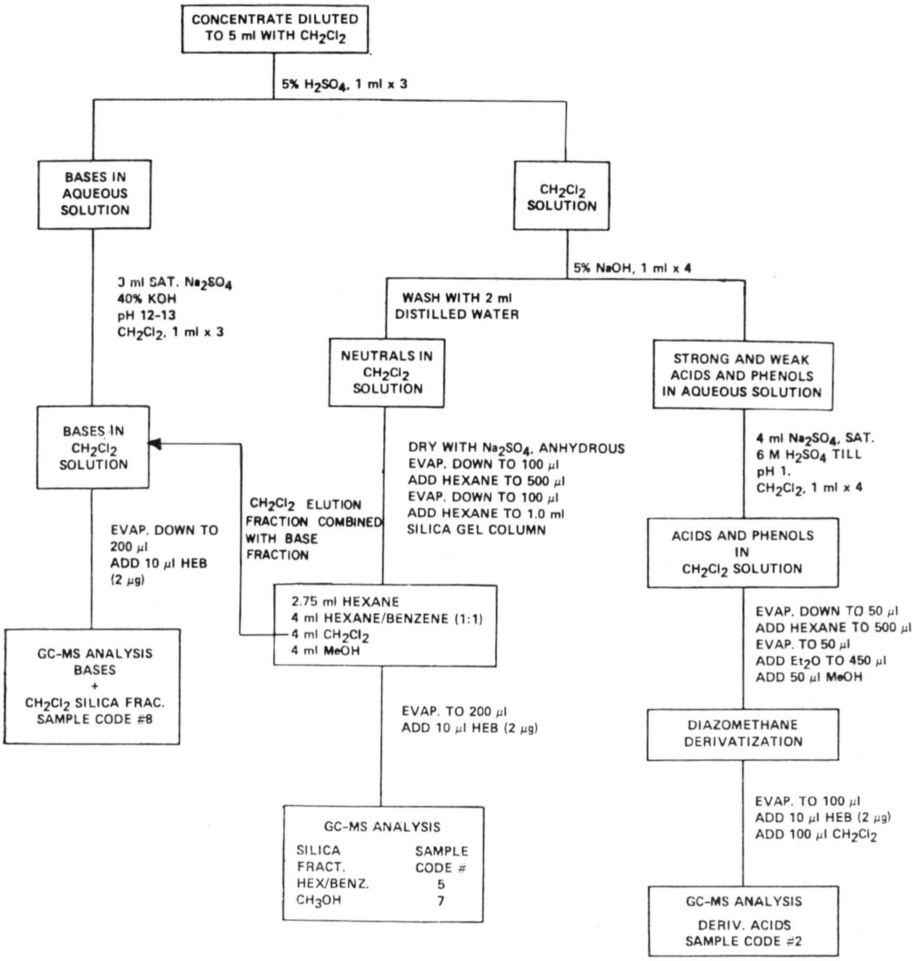

Figure 3. Fractionation scheme for reverse osmosis. (Sources: Refs. 30, 80.)

cult to use, but there is a simple miniature generator available to safely generate just enough diazomethane for the reaction [49]. The generator can be made in a glassblowing shop. In the reaction, it is important to remove methylene chloride as solvent (Fig. 3). Diazomethane reacts with methylene chloride to generate a homologous series of chlorinated hydrocarbons, which complicate the analysis of the acids and phenols.

Before moving on to the next factor in the analytical scheme, it is important to consider the two most commonly used extraction and concentration methods for the organic analysis of purgeable volatile compounds in water. They are the Bellar-Lichtenberg "purge-and-trap" method [32,37,56-59].

When first developed, the purge-and-trap method was used exclusively with a packed column. Since then it has been modified in various laboratories for use with capillary columns. One of the versions used in my laboratory is shown schematically in Fig. 4 [60]. Basically it involves purging the volatile organics out of the water onto a Tenax trap. The organics are later desorbed by heat onto the head of the capillary column for analysis.

The closed-loop stripping apparatus shown in Fig. 5 uses a specially designed filter containing 1.5 mg of specially pure activated carbon adsorbent. The headspace which is used to continuously purge the water passes through the filter in its path. The most commonly employed stripping time is 2 hr. The carbon filter is extracted with 10 to 30 μℓ of ultrapure carbon disulfide and eluted into a special collector. Usually, the extract does not need further concentration before analysis.

No direct comparison has been made between the two methods. It appears that the purge-and-trap method was designed for μg liter^{-1} (ppb) levels [50] and the closed-loop method for sub-ng liter^{-1} (sub-ppt) levels [56]. In both cases, the apparatus can be easily assembled in the laboratory with all the parts commercially available.

Finally in regard to sample preparation, note that the sample blanks collected at the sampling sites should be processed in parallel with the samples. Contributions from glassware, reagents, etc. can then be differentiated from the real data obtained on the sample.

2.4 Instrumentation

Earlier chapters in this book describe how to obtain optimum separation from glass capillary gas chromatography with respect to time. The proper pressure regulator, injector, column (see following section), and detector must be chosen. In any water analysis seeking to identify and quantify the unknown components with the highest degree of sensitivity and specificity, the mass spectrometer is definitely the best detector. If the object is to analyze for a specific class of compounds, other, less expensive specific detectors may suffice. However, in cases where results will shape important decisions, the mass spectrometer is highly recommended. If cost is the limiting fact, perhaps environmental monitoring or field laboratories should only screen samples with the highest resolution one can obtain using dual flame ionization and electron capture detectors. The results should be checked constantly against a certain standard. If any variations occur,

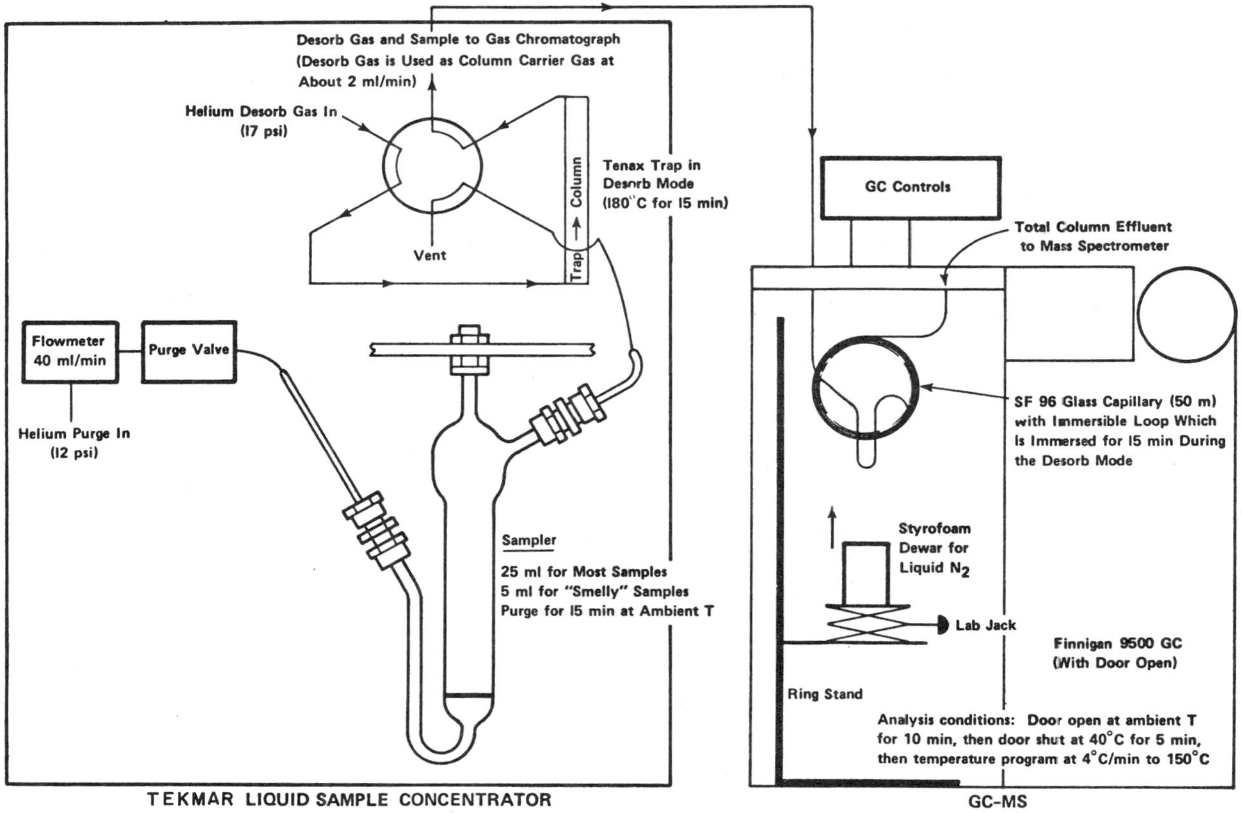

Figure 4. Instrumentation used for the purge-and-trap analysis of volatile organic compounds in water and diluted samples by capillary GC/MS.

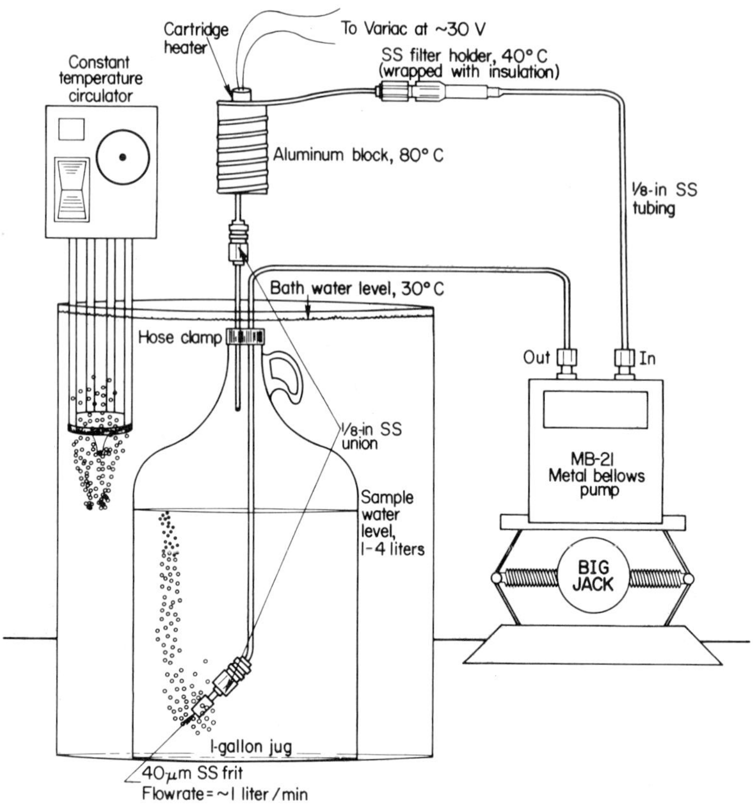

Figure 5. Diagram of Grob closed-loop stripping apparatus.

the samples should be sent to one of the central laboratories with expertise in mass spectrometry and other techniques for more detailed analysis. The idea of comparing the GC profiles by computer has already been developed [61,62].

Requirements of a mass spectrometer for glass capillary water analysis are that:

The GC/MS interface is proper for capillary column use.
The scan time is fast enough (2 sec or less per cycle).
The computer is capable of handling the amount of data generated by the capillary column.
The operator is trained to use it properly.

Most of the modern GC/MS systems satisfy the second and third requirements. In a central laboratory situation, the fourth requirement can be satisfied;

but when it comes to the GC/MS interface, some systems are not ideally suitable for capillary column work.

There are many articles published on capillary GC/MS interfaces [63-72], but a slightly modified version of the one described by Blum et al. [68] appears to be the best (Fig. 6). The coaxial interface allows the MS operator freedom to use electron impact (EI) as well as positive or negative chemical ionization (CI) conditions. The original version [68,73] uses a platinum capillary to connect the column onto the ion source of the MS. Later versions use a 0.1-mm internal diameter (ID) glass capillary. There is still controversy over the use of a platinum capillary, which is avoided by simply using a glass capillary. Interfaces, as shown in Fig. 6, have been successfully adapted in other laboratories (e.g., Battelle Columbus Laboratories and Woods Hole Oceanographic Institution) for the use of glass capillary GC/MS.

2.5 Columns

The question of which column to use in an analysis should and always does come up. Previous chapters have already discussed the type (WCOT, SCOT, PLOT, WWOT, micropacked, etc.), the diameter, the length, the film thickness, the efficiency, the deactivation method, the adsorptive sites, the acidity, the basicity, the optimum flow velocity, and the evaluation of glass capillary columns. Obviously, the liquid phase to be used depends on the type of analysis to be performed, which in water analysis will be for either specific compounds or for all the components that can be separated by gas chromatography. The choice of liquid phase is important in spite of suggestions to the contrary. This has recently been demonstrated theoretically [74].

Generally speaking, analysis of polycyclic aromatic hydrocarbons is performed best with SE-54 [74]. For polychlorinated dibenzo-p-dioxins and dibenzofurans, it has been demonstrated that OV-17 gives the best results [76]. For acetylated chlorophenols, SE-30 has been recommended [77], while many underivatized phenols, tricresols, dicresols, and xylenols have been successfully analyzed on a 50-wall=coated open tubular (WCOT) column with tri-(2,4-xylenyl)phosphate as the liquid phase [48]. Supelco, Inc. recently reported that they have a new phase (SP-1240DA) that can separate the free phenol and various substituted phenols. For purgeable volatile organics, a low-temperature methysilicon, such as SF-96, appears to be the best choice [37,60]. Two new types of phases have recently been introduced:

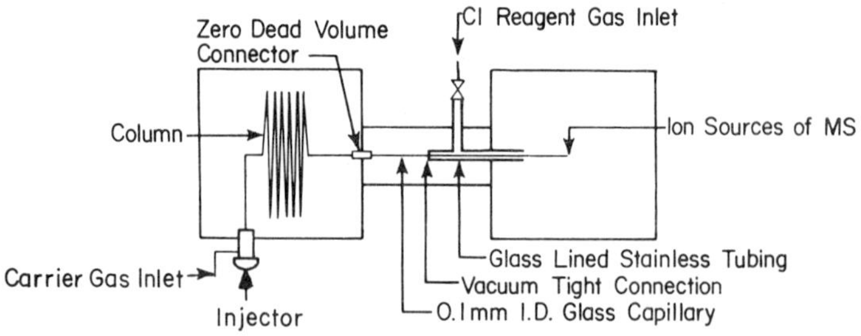

Figure 6. Schematic diagram of a working capillary GC/MS interface. (Source: Ref. 68.)

Pluronics [78] and Superox-4 [79]. Both are polar phases that behave similarly to polyethylene glycol, Carbowax 20M, except they have much better stability at high temperatures. They seem to have excellent potential for solving a variety of separation problems.

In analyzing water for all the components that can be separated on a gas chromatograph, use of two columns with different types of liquid phase on the same sample has been suggested [30,80]—a nonpolar phase (SP2100), which allows the elution of as many components as possible, and a polar phase (SP1000), which gives good separation on most of the components. If only one column is permitted in the analysis, a slightly polar column is the natural choice; perhaps an SE-54 column should be used.

Regardless of which columns are chosen, each should be tested and should meet predetermined standards before use. Furthermore, the performance of a column should be monitored frequently, perhaps daily, with a test mixture to ensure that valuable samples will not be wasted. In my opinion, each laboratory should have a simple gas chromatograph into which a test mixture with a predetermined split ratio can be injected and from which the components of the test mixture can elute directly into the bottom of the flame tip of the detector (flame ionization detector) via the column. This instrument should be used to test columns. The performance of the tested columns on other more sophisticated instruments can then be compared to results obtained on the simple "baseline" instrument.

2.6 Data Interpretation

Once data are obtained with a specified column in a specified instrument, the analyst has to analyze the data. If capillary GC/MS is used, there are two pieces of information on which an identification can be based: the retention time or relative retention time and the mass spectrum itself.

Since this is not a mass spectrometry text I shall not elaborate on the interpretation of mass spectra. If other detectors are used, other information may be useful for the compound identification. Quantification can be achieved in the usual manner with the aid of the various internal standards added. (Hexaethylbenzene is frequently added to a sample prior to injection as an additional internal standard.) Naturally, the use of authentic reference compounds is highly recommended. The question of the use of retention time for information has already been discussed in previous chapters.

2.7 Data Management

Frequently, important decisions with social, economic, political, environmental, and health impacts are based on the results of water analyses. Data from capillary gas chromatography are often massive and their significance may be masked to the untrained observer, who will ask, "What are a few more peaks here and there?" In that respect, packed column data may be more appealing, but the picture has been oversimplified to a certain degree (Fig. 1). Scientists have an obligation to see that the most accurate information is obtained and disseminated. To that end, the more accurate data from capillary columns should be organized and interpreted in a manner that will not mask their significance. For example, an attempt has been made to computerize capillary column chemical data so that correlation can be made with other factors, such as toxicity [81].

2.8 Quality Assurance

Quality assurance is an integral part of any water analysis experimental design. Without provision for quality assurance, any data generated are simply unacceptable. Many aspects of quality assurance have already been alluded to. More specifically, the experimental design must consider quality assurance in the following:

Sampling sites and frequence selection. The final design must be statistically sound.

Sample collection and preservation. Glasswares must be meticulously clean and preservatives free of organics. Sample blanks should be created at the same time that samples are collected.

Sample Preparation. Internal standards should be used to monitor the fractionation recoveries. The sample blanks should be processed in parallel with the samples.

Instrumentation. The instruments used should be tuned reproducibly to reach certain specified standards before data are acquired.

Column selection. The columns used must meet certain requirements and their performance must be faithfully monitored with test mixtures.

Data interpretation. If at all possible, authentic reference compounds should be used for compound identification and quantification. Components that occur in similar quantities in both the blanks and the samples will be considered as arising from contaminations during the course of analysis. Finally, an accurate log must be kept on every step of the analysis.

3. APPLICATION

Like other branches of analytical chemistry, water analysis is only a tool to help answer questions. It has become important because of the recent recognition that organic water pollutants play a significant role in the earth's ecosystem. If we are to preserve a viable ecosystem for mankind, we have to understand how, where, when, why and to what degree water pollutants affect the system. An understanding of the dynamics and hazards of the water pollutants will enable us to institute measures to avoid any immediate or long-term disastrous effects.

The ecosystem is complex and has many compartments; water itself is one of them. Water can be further subdivided according to its origin--e.g., the ground, industrial effluents, domestic or sewage effluents, rivers, lakes, oceans, and drinking water sources. A study of organic pollutants in these types of water is certainly appropriate but it is equally important to study the organic pollutants in other compartments (the air, sediments, etc.) for they are all interrelated. The following discussion attempts to illustrate the use of glass capillary gas chromatography in analyzing different types of water and also touches on the analysis of water pollutants in the other compartments of our ecosystem that have not been covered in other chapters.

Generally speaking, the degree of pollution in different types of water varies according to the sampling site. In wilderness areas, pollution from synthetic organics has to be transported from other sources. In agricultural areas, farm chemicals inevitably get into the water supply. For heavily populated and industrial areas, pollution can come from many sources. To analyze any type of water, it is useful to know as much as possible about the water sample itself. It is not cost-effective to work with a large sample volume when the organics in a small sample would suffice for the analytical technique.

3.1 Groundwater

Groundwater, as a rule, contains low levels of organic compounds; consequently, it is often used as a drinking water supply. But with the use of landfill sites for waste disposal (including chemicals), underground activities (coal gasification etc.), and the occurrence of accidental spills, this source of drinking water is threatened by pollution of synthetic organic chemicals.

Grob and Grob [57] in their investigation of the closed-loop stripping method reported results of groundwater analysis from Hardhof, Switzerland, and from an accidental diesel oil spill site. The Hardhof groundwater (Fig. 7B) is used as part of the drinking water supply in Zurich. Its components were tentatively identified by their mass spectra from data obtained on a capillary GC/MS system. In the identification process, two columns, Ucon HB550 and Emulphor, were used to facilitate separation of components at the early and late part of the chromatogram respectively. However, the chromatograms shown (Fig. 7) were obtained with a Ucon HB5100 column, which allowed the separation of most of the components from the room temperature, closed-loop stripping method. In a more recent publication [37], the use of SF-96 columns was reported. In Fig. 7, components represented by peaks 2, 9, 127, 128, and 129 are heptane, benzene, tri-n-butyl phosphate, octadecane, and acenaphthene, respectively. Most of the components present were identified as products in gasoline. In comparison with the springwater from Dolder (Fig. 7A), the groundwater contained 3 times more gasoline constituents. In addition, the groundwater contained components 18, 25, and 48--trichloroethylene, tetrachloroethylene, and chlorobenzene, respectively. The springwater came from an unpopulated, forested hill close to a city. It was known that gasoline vapor surrounded the hill. It is not surprising then that the components in the springwater are either of gasoline origin or have terpenoid structures. The groundwater passed through a densely populated and partly industrialized area before it was collected. Knowing these facts, the springwater and groundwater analytical results seem very reasonable. It is interesting to note that the Rämibühl tap water (Fig. 7C), which came from the Dolder springwater and Hardhof groundwater, appeared to contain more organic pollutants.

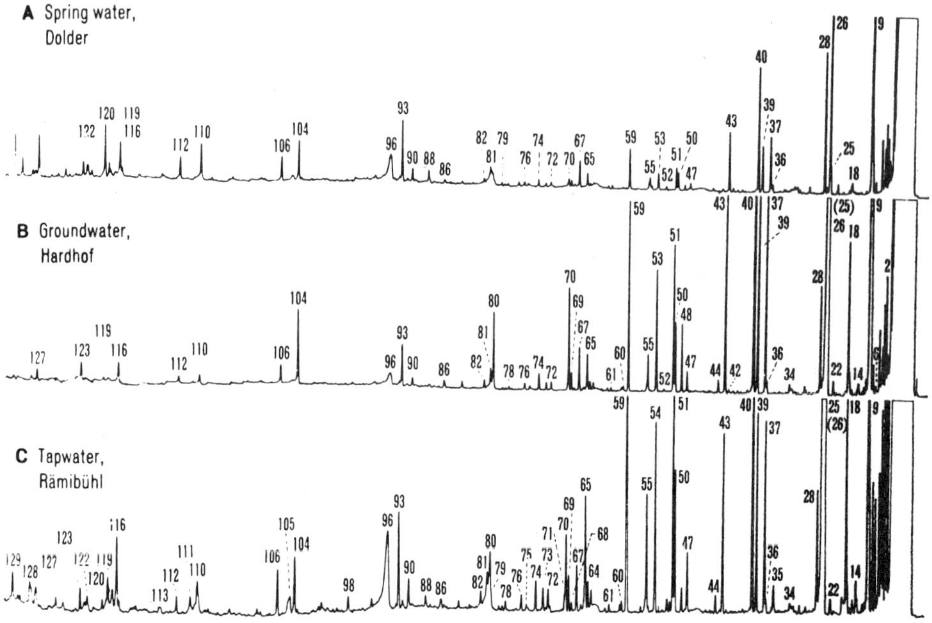

Figure 7. Chromatographic conditions: glass capillary column, 120 m × 0.32 mm, coated with Ucon HB 5100; hydrogen carrier gas at a flowrate of 4.5 ml min^{-1}. A 4.0-µl amount of carbon disulfide solution was injected into the column at room temperature and the split valve closed for 30 sec. The temperature was programmed at 1.5° min^{-1} from 25 to 175°C. The Model GI flame ionization detector, X8 was from Carlo Erba, Milan, Italy. (Source: Ref. 57.)

In the study of groundwater from an accidental diesel oil spill site by the closed-loop stripping method, samples were collected after conventional infrared analysis detected no contamination. The samples had to be diluted 10 times to obtain the chromatograms shown (Fig. 8B). Figure 8A is a chromatogram obtained on pure water spiked with 0.5 ppb gasoline, while Fig. 8C is from pure water spiked with 1 ppb diesel oil. It was surprising that Fig. 8B resembled Fig. 8A rather than Fig. 8C. The authors [57] theorized that the source of pollution was an old gasoline contamination.

Grob et al. [37] also reported results obtained on contaminated groundwater using the microsolvent extraction technique (Fig. 9). Components 1, 2, 47, 49, and 52 were identified as toluene, tetrachloroethylene, fluoranthene, dioctyl phthalate, and C_{27}-alkane, respectively.

These authors also compared the closed-loop stripping method with the n-pentane microsolvent extraction method using a test mixture containing n-alkanes, aromatics, chlorinated hydrocarbons, ketones, alcohols, pyridine,

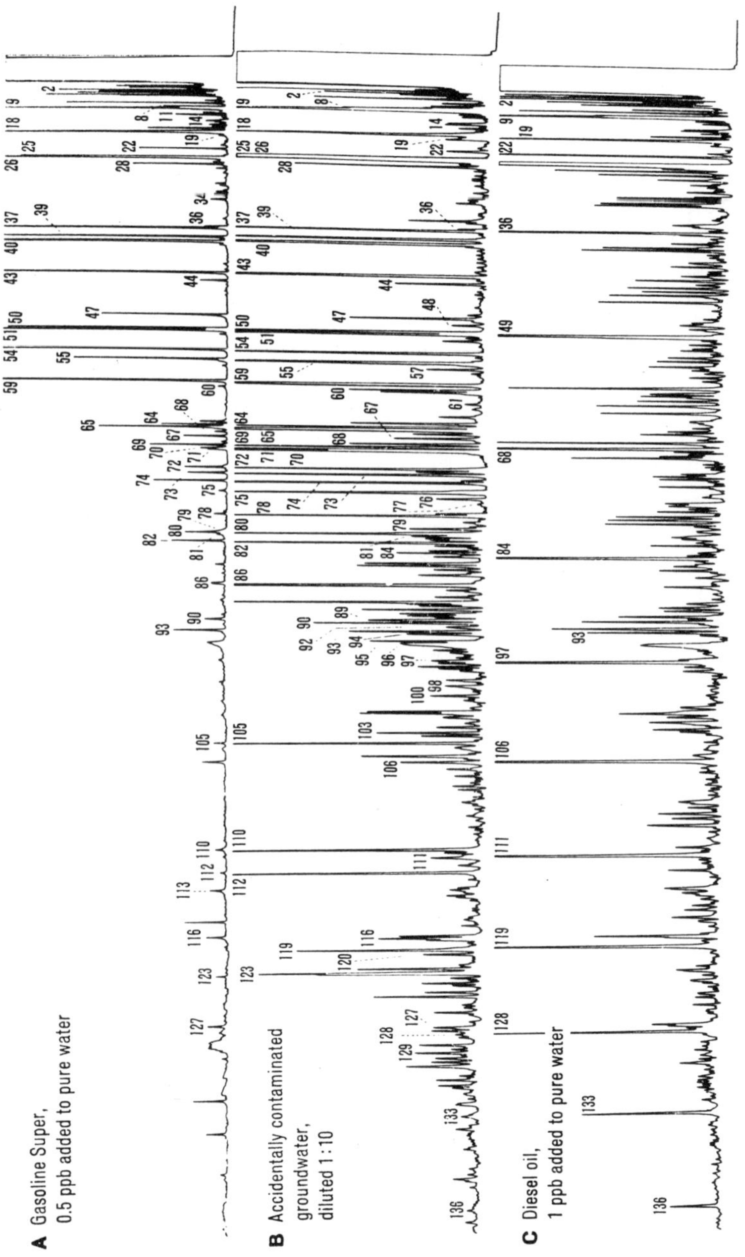

Figure 8. Comparison of an accidentally contaminated water sample (B) with pure water spiked with known amounts of gasoline (A) and diesel oil (C). For direct quantitative comparison the peak areas in B have to be increased 10 times to compensate for the 1:10 dilution. Note the exact coincidence of peaks in A and B, but almost no coincidence between B and C. A and C are also used as calibration runs for the estimation of gasoline and diesel oil mixtures, respectively. (Source: Ref. 57.)

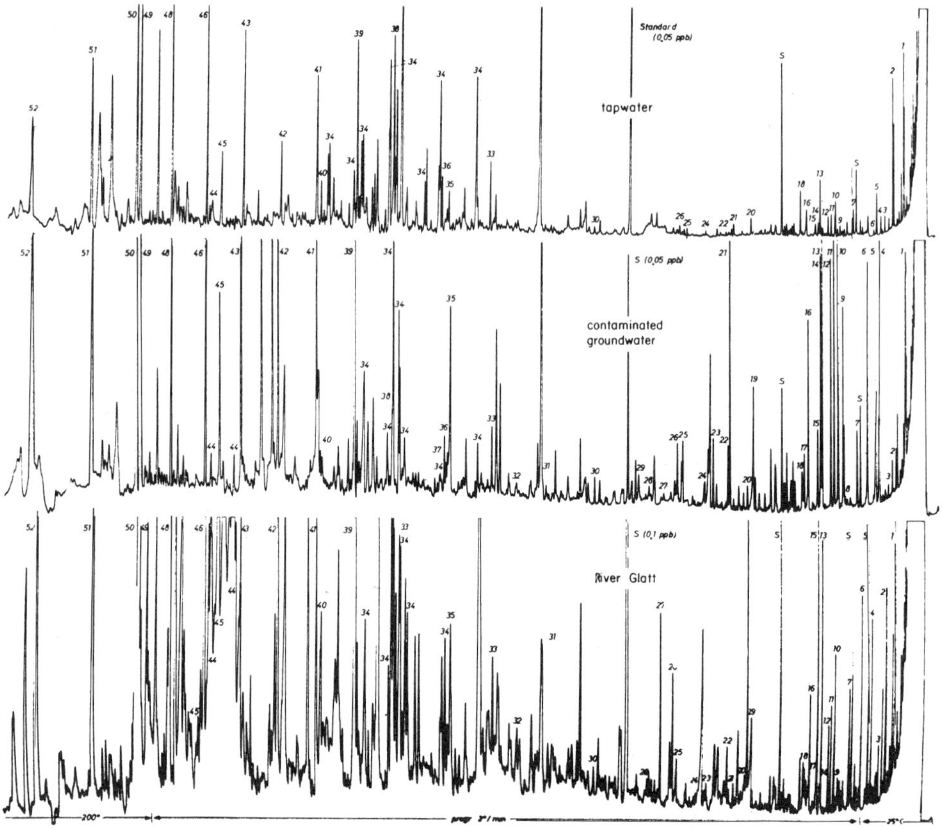

Figure 9. Three water analyses as examples of practical routine work, n-pentane extract ∿200 μl, concentrated to 3 μl for one injection. Column: glass, 40 m × 0.32 mm; SF-96; carrier gas (hydrogen) flowrate, 3.2 ml min^{-1}, Carlo Erba (Milan, Italy) Model 2100; FID attenuation ×4. Column immersed in ice water during injection and elution of solvent, then programmed at 2°C min^{-1} from 5 to 200°C. (Source: Ref. 37. Reproduced courtesy of Dr. E. Romann, Official Chemist of Zurich.)

phenols, a 4-methoxytoluene (Fig. 10, Table 2). Their general conclusion was that closed-loop stripping was superior in the more volatile range while microsolvent extraction showed increased efficiency with decreasing volatility. Furthermore, for pyridine and its derivatives, or phenols, the pH of the water had to be adjusted appropriately to facilitate recovery by either method.

It is interesting to compare the above results on groundwater to other investigations of unpolluted and polluted groundwater using packed columns. Figure 11A is a chromatogram from packed column analysis of groundwater prior to coal gasification, and Fig. 11B is the chromatogram after gasification [82]. Obviously, the authors did not intend to study the pollutants

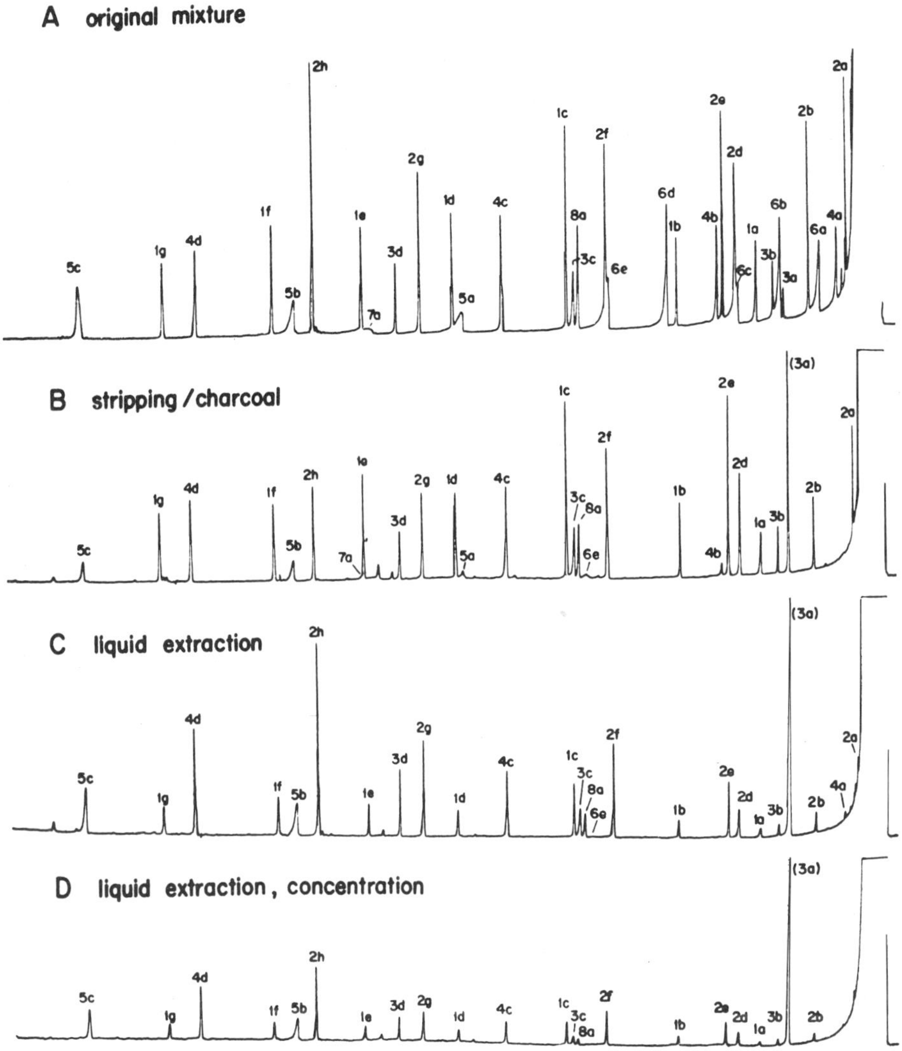

Figure 10. A test mixture using a 50 m × 0.29 mm, SF-96 slightly basic column, carrier gas (hydrogen), flowrate 3.0 ml min^{-1}, column programmed at 2°C min^{-1} from 5 to 140°C, 3 μl splitless injection at 5°C. (A) original mixture, representing 100% recovery. (B) Same amount of mixture added to tap water and extracted by stripping. (C) Identical sample as used for B, extracted with n-pentane. Splitless injection of 3 μl of extracts made up to 250 μl (no concentration step). (D) n-pentane extract as obtained in C, concentrated to 3 μl and diluted to original volume of 250 μl to show losses from concentration. (Source: Ref. 37.)

Table 2. Constituents of Recovery Test Mixture as Shown in Fig. 10

No.	Type of substance	Member	No.	Type of substance	Member
1	n-Alkanes		4	Ketones	
1a		n-C_8	4a		3-Pentanone
1b		n-C_9	4b		4-Heptanone
1c		n-C_{10}	4c		5-Nonanone
1d		n-C_{11}	4d		2-Dodecanone
1e		n-C_{12}			
1f		n-C_{13}	5	Alcohols	
1g		n-C_{14}	5a		1-Octanol
			5b		1-Decanol
2	Aromatics		5c		1-Dodecanol
2a		Benzene			
2b		Toluene	6	Pyridines	
2d		Ethylbenzene	6a		Pyridine
2e		m-Xylene	6b		2-Picoline
2f		1,2,3-Trimethylbenzene	6c		4-Picoline
2g		Naphthalene	6d		2,3-Lutidine
2h		1-Methylnaphthalene	6e		3,4-Lutidine
3	Chlorinated hydrocarbons		7	Phenols	
			7a		Phenol
3a		Tetrachloroethylene	7b		p-Cresol
3b		Chlorobenzene			
3c		1,2-Dichlorobenzene	8	Miscellaneous	
3d		1,2,3-Trichlorobenzene	8a		4-Methoxytoluene

in detail. Nevertheless, their analytical task would have been much better served with capillary columns. In another study of groundwater at and in the vicinity of a landfill site, carbon accumulative columns were used [17]. The carbon was Soxhlet-extracted with either chloroform (CCE) or ethanol (CAE, A for alcohol). Figure 12 shows the chromatograms obtained on packed columns. Again, the analytical task would have been more fruitful if capillary columns were used, perhaps in conjunction with other extraction techniques.

3.2 Industrial Effluents

Many studies are being conducted on effluents from the 22 categories of industry in the United States. Unfortunately, (apparently due to time and

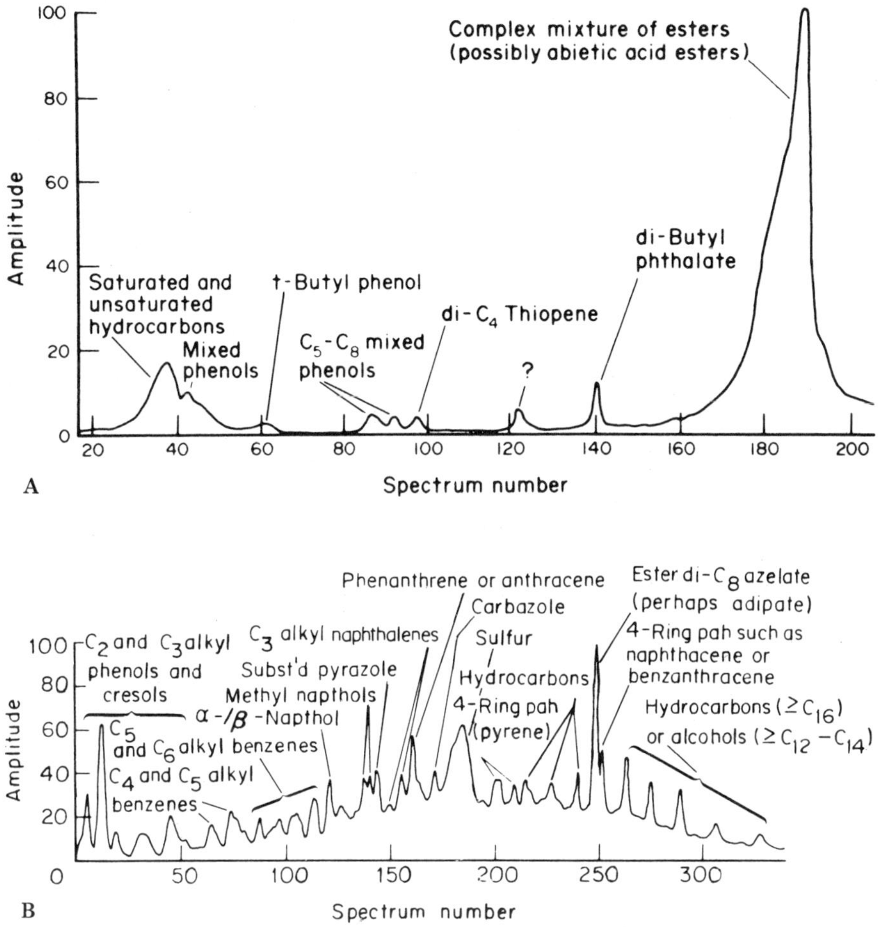

Figure 11. (A) GC/MS analysis; baseline contamination. (B) GC/MS analysis; contamination after gasification. (Source: Ref. 82.)

cost constraints), glass capillary gas chromatography was not considered the best available technology for these studies and, in the recommended method, is offered only as an alternative to packed column analysis [83]. However, in a recent study on energy-related wastes and effluents, support-coated open tubular (SCOT) columns were used extensively [84]. In that particular study, purgeable volatile organics were analyzed by a modified purge-and-trap method [50, 85-87]. A 100-m glass SCOT OV-101 column programmed from 20 to 240°C at 4°C min^{-1}, with a helium flowrate of 3 ml min^{-1}, was employed for the GC/MS analysis. The other volatile organics were extracted with large volumes of Freon-TF at various pH levels to separate the components into acidic, basic, and neutral fractions. For apparent health

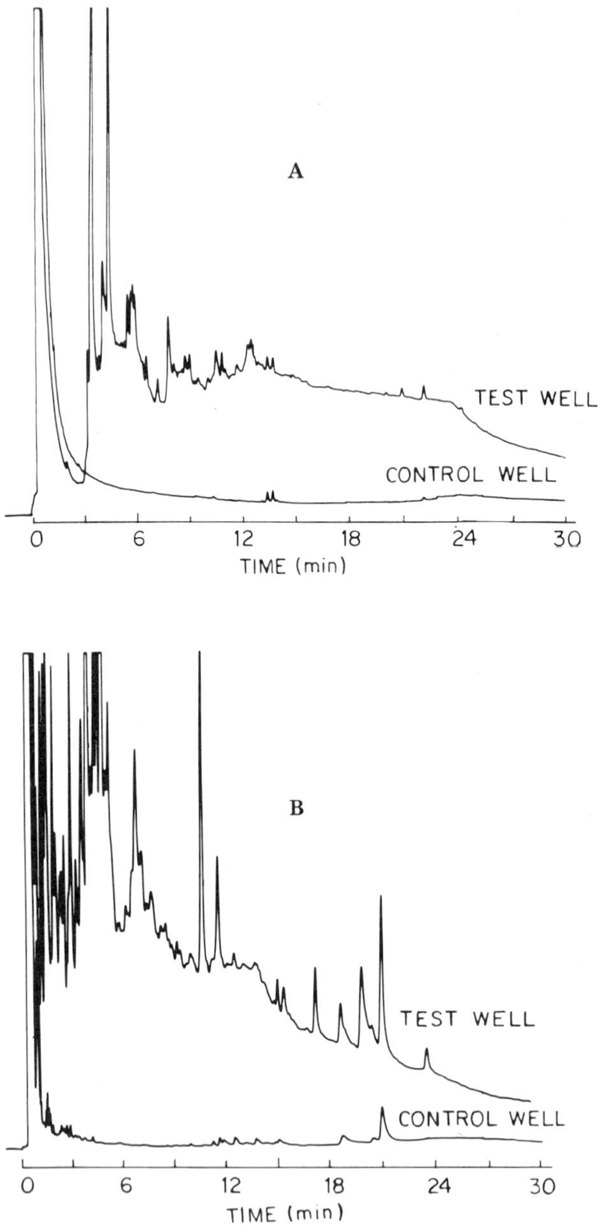

Figure 12. (A) FID chromatogram of test and control well CAE. (B) FID chromatogram of test and control well CCE. (Source: Ref. 17.)

reasons, Freon-TF, which had to be purified by florasil column, was chosen as the solvent. The basic fraction was derivatized with perfluoropropionic anhydride, while the acidic fraction was derivatized with diazomethane or dimethyl sulfate. A 50-m SCOT Carbowax 20 M column programmed from 80 to 240°C at 4°C min^{-1} with a helium flowrate of 3 ml min^{-1} was employed in the GC/MS analysis of the acidic, basic, and neutral fractions. SCOT columns were chosen over WCOT columns in these analyses, probably because the complete scan cycle of the mass spectrometer used was approximately 6 sec. A SCOT column is much more compatible with a slow-scanning instrument. With this experimental design, the number of components analyzed in each sample (over 50 were reported) usually exceeded 200. Most analyses were at the ppb levels. Many classes of compounds were found, including alkanes, alkenes, aromatics, alkyl aromatics, polycyclic aromatics, and other sulfur-, nitrogen-, and oxygen-containing compounds. Those interested in oil-shale processing and coal gasification should refer directly to the report [84].

A recent study of the effluents from the bleach plant of a sulfite pulp mill provides another example of using high-resolution capillary columns to solve a complex analytical problem. A 50-m OV-1 WCOT column programmed from 25 to 260°C at 3.5°C min^{-1} with a helium flowrate of 2 ml min^{-1} was employed in the GC/MS analysis to examine the nonpolar chlorinated and brominated compounds [88,89]. The mass spectrometer used could record a spectrum every 2 sec. The nonpolar organic compounds were extracted from a 4.5-liter sample with 2-ml volumes of cyclohexane after the pH of the effluent was adjusted to 11. The solvent was eventually concentrated to 5 ml. Nineteen halogenated compounds were identified (Fig. 13, Table 3), some of which were synthesized. It is interesting to note that the extract and the synthesized compounds exhibited mutagenic activity in the *Salmonella* test system [90].

An earlier study on paper mill effluent was performed with a 50-ft Carbowax 20M/terephthalic acid SCOT column programmed from 100 to 200°C at 4°C min^{-1} [1,91]. Chloroform was used in the large-volume solvent extraction method on 1- to 5-liter samples. Such extractions were found to be as effective as the carbon adsorption method on a 1800-gallon effluent sample. The acid fraction was of special interest and was methylated with dimethyl sulfate before analysis. With GC/MS, about 15 fatty acids, 12 resin acids, 10 phenols, and several other compounds were identified.

Another study of industrial effluents was conducted at a factory that operated in a weekly batch mode to produce pharmaceuticals, herbicides,

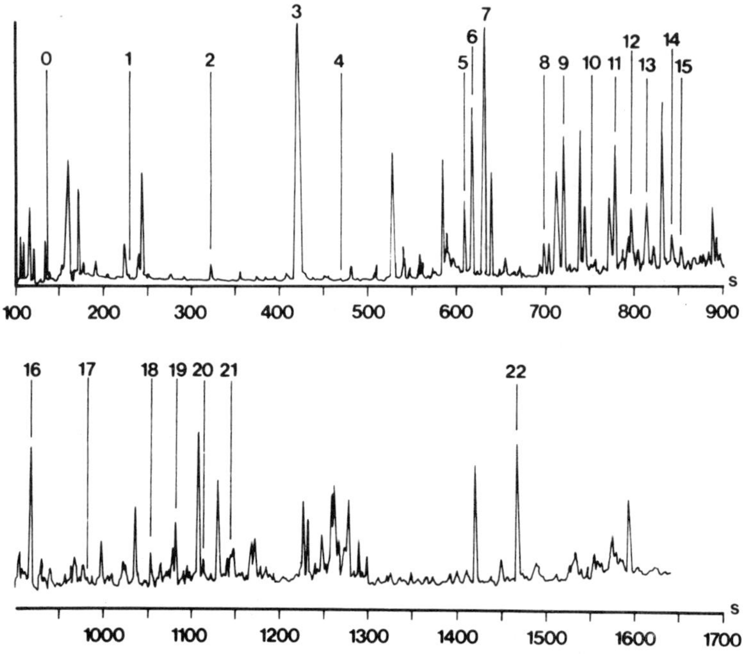

Figure 13. Computer-reconstructed chromatogram, showing sum of the ion intensities versus the scan number. The halogenated compounds are listed in Table 3. In addition, peak 3 is cymene, 5 is cymene-α-ol, 16 is calamenene, and 22 is methyl dehydroabietate. (Source: Ref. 88.)

antioxidants, thermal stabilizers, ultraviolet light absorbers, optical brighteners, and surfactants [36]. A variation of the purge-and-trap method was used to analyze the purgeable volatile organic compounds. To study the other volatile organics, large-volume solvent extractions were performed on 4-liter samples with 300 ml of methylene chloride at an initial acidic pH. Then the aqueous layer was extracted again with 200 ml of methylene chloride after the pH had been adjusted to extract the basic compounds. One of the columns used in the GC/MS analysis was a 20-m SE-52 WCOT column programmed from 60 to 240°C at 4°C min^{-1}. With the aid of other techniques, 123 compounds were identified and quantified in the wastewater and the other compartments of that particular ecosystem. These compounds included 15 nitrogen-containing heterocyclics, 5 nitrogen-containing compounds, 40 oxygen-containing compounds, 24 halogenated compounds, 31 aromatic hydrocarbons, and 8 alkanes. Several compounds were of known biological activity (herbicides, bacteriostats, and disinfectants), whereas others, such as dichlorodibenzodioxin, were potentially hazardous. The concentration of compounds in the wastewater ranged between 1 ppb and 15 ppm.

Table 3. Halogenated Compounds Found in Sulfite Bleachery Effluents

Compound	Peak number[a]
Dibromochloromethane	0
Bromoform	1
Chlorotoluene	2
Hexachloroethane	4
Chlorocymene (2 isomers)	6, 7
Dichlorocymene (3 isomers)	10, 11, 13
Trichlorocymene	17
Bromocymene (2 isomers)	8, 9
Bromochlorocymene	15
Chlorocymene-α-ol	12
Bromocymene-α-ol	14
Chlorocalamenene (2 isomers)	18, 19
Bromocalamenene (2 isomers)	20, 21

[a] Refer to Fig. 13

All of the examples of industrial effluent analysis cited above used liquid-liquid extraction and/or the purge-and-trap method. Waggott recently reported the analysis of industrial effluents using the closed-loop stripping method [58]. He pointed out that the capacity of the carbon filter was an important consideration in ensuring good recovery throughout the range of compounds that can be analyzed with this technique. He further demonstrated that the absolute capacity of a 1.5-mg filter for a homologous series of 1-chloro-n-alkanes was 6 μg for complete recovery, and within the series maximum recoveries were obtained for compounds between C_8 and C_{12}. Thus the volume of the industrial effluent sample had to be balanced with the capacity of the carbon filter. An example was given for an effluent containing flavor and fragrance compounds (Fig. 14).

This discussion would not be complete without citing the use of a macroreticular resin column. In this analysis, polycyclic aromatic hydrocarbons were the compounds of interest in industrial effluents [92]. Tenax,

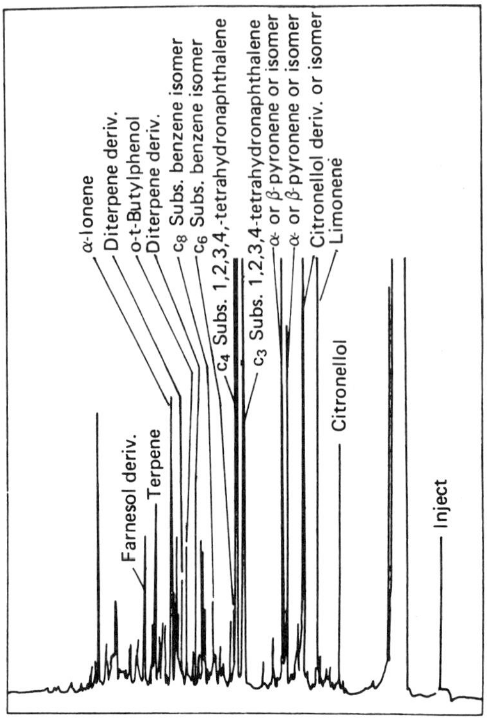

Figure 14. Gas-liquid chromatographic capillary column profile of a headspace analysis extract obtained from an industrial column using closed-loop stripping method. Sample size, 2.5 μl; solvent, dichloromethane; column, 50 m glass capillary containing OV-1; detector, flame ionization; conditions, 30 to 200°C at 4°C min^{-1}. (Source: Ref. 58.)

XAD resins, and polyurethane foam [93] were examined as the adsorbent material for the accumulative column, which was designed specially for polycyclic aromatic hydrocarbons. The XAD-2 resin was chosen over the others. Like other researchers [12,13], the authors recommended that the column should be slurry-packed and stored after it had been filled with methanol. The effluent sample was pumped through the column under field conditions and the accumulated organics were eluted by Soxhlet extraction with methylene chloride. The methylene chloride solution was concentrated using Kuderna-Danish evaporation apparatus before analysis. A 20-m SE-54 WCOT column programmed from 150 to 275°C at 3°C min^{-1} was used in the analysis (Fig. 15). The identification was done with capillary chemical ionization GC/MS.

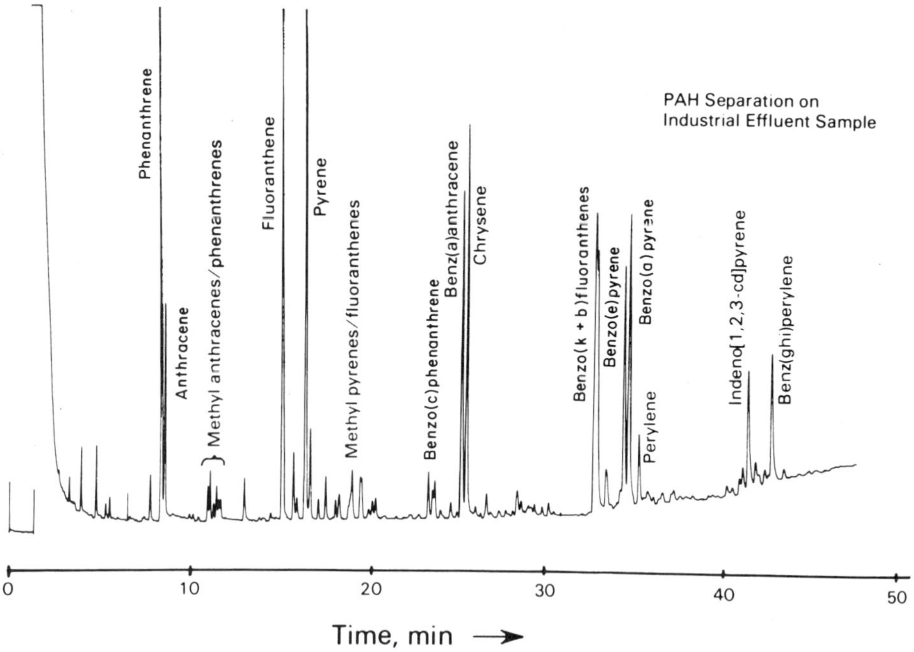

Figure 15. Polycyclic aromatic hydrocarbons found in an industrial effluent sample.

3.3 Domestic Effluents

Nowadays, most of the domestic effluent is treated at a plant before it is discharged. One of the most closely studied treatment plants is Water Factory 21 of the Orange County Water District in California [59]. Factory 21 is a 15,000,000 gal day^{-1} tertiary (advanced) waste treatment plant. Its effluent is injected into the ground to act as a seawater barrier system. The organics in the plant influent and effluents of individual treatment processes have been analyzed using a variety of techniques including purge-and-trap, closed-loop stripping, liquid-liquid extraction, Hall electroyltic conductivity detection, flame ionization detection, electron capture detection, and GC/MS.

Both grab and composite samples were used for the purge-and-trap method to study halomethanes with sodium thiosulfate added to the collected samples. For closed-loop stripping analysis, a 25-m Ucon HB WCOT column was used with simultaneous flame ionization and electron capture detection. The identification was achieved by GC/MS using a 50-m Ucon HB 5100 WCOT column programmed from 30 to 180°C at 3°C min^{-1}. The liquid-liquid extraction was performed on 1-liter samples with two volumes of 15 ml of hexane. The solvent was dried with sodium sulfate and concentrated to 2 ml. A cleanup step was performed on the concentrate using a florisil column. The sample was analyzed by an electron capture detector with a 20-m SE-54 WCOT column programmed from 90 to 230°C at 4°C min^{-1}. Figures 16 and 17 demonstrate how the effectiveness of the treatment processes can be closely monitored.

Earlier work had examined the effectiveness of primary and secondary sewage treatments [94,95]. The effluents were filtered through a glass-fritted funnel before extraction with methylene chloride. The methylene chloride extract was then separated into pentane, methylene chloride, and methanol fractions by silica gel column chromatography. After concentration, the fractions were analyzed on a 50-m OV-101 WCOT column programmed from 30 to 240°C at 2.8°C min^{-1} with a hydrogen flowrate of 4 ml min^{-1}. Figure 18 shows the flame ionization chromatogram of the pentane (A) and methylene chloride (B) fractions of the effluent after secondary treatment. The compounds were identified by GC/MS.

Advanced waste treatment effluents, as well as drinking water, have been processed with reverse osmosis to extract and concentrate the organics from the water for bioassay and analysis [18-20,29,30,81]. The concentrates obtained were complex (Fig. 2) and had to be fractionated (Fig. 3) before GC/MS analysis. Purgeable volatile organics were analyzed separately on the original water samples [96]. Deuterated compounds were added before fractionation and the various fractions obtained from the scheme (Fig. 3) were analyzed by flame ionization detection and GC/MS. Two columns were used on the analysis of each fraction: a 40-m SP1000 WCOT column programmed from 50 to 225°C at 2°C min^{-1} with a helium flow velocity of 23 to 25 cm sec^{-1} at 125°C and a 20-m SP2100 WCOT column programmed from room temperature (3 min) quickly to 50°C (3 min) then to 280°C at 2°C min^{-1} with a helium flow velocity of 23 to 25 cm sec^{-1} at 125°C. For Lake Tahoe (pentane) concentrate 264 compounds were tentatively identified by their mass spectra and 152 of these were confirmed by comparison with the retention times, relative retention times, and mass spectra of authentic reference material. In the analyses, an extensive quality assurance program was

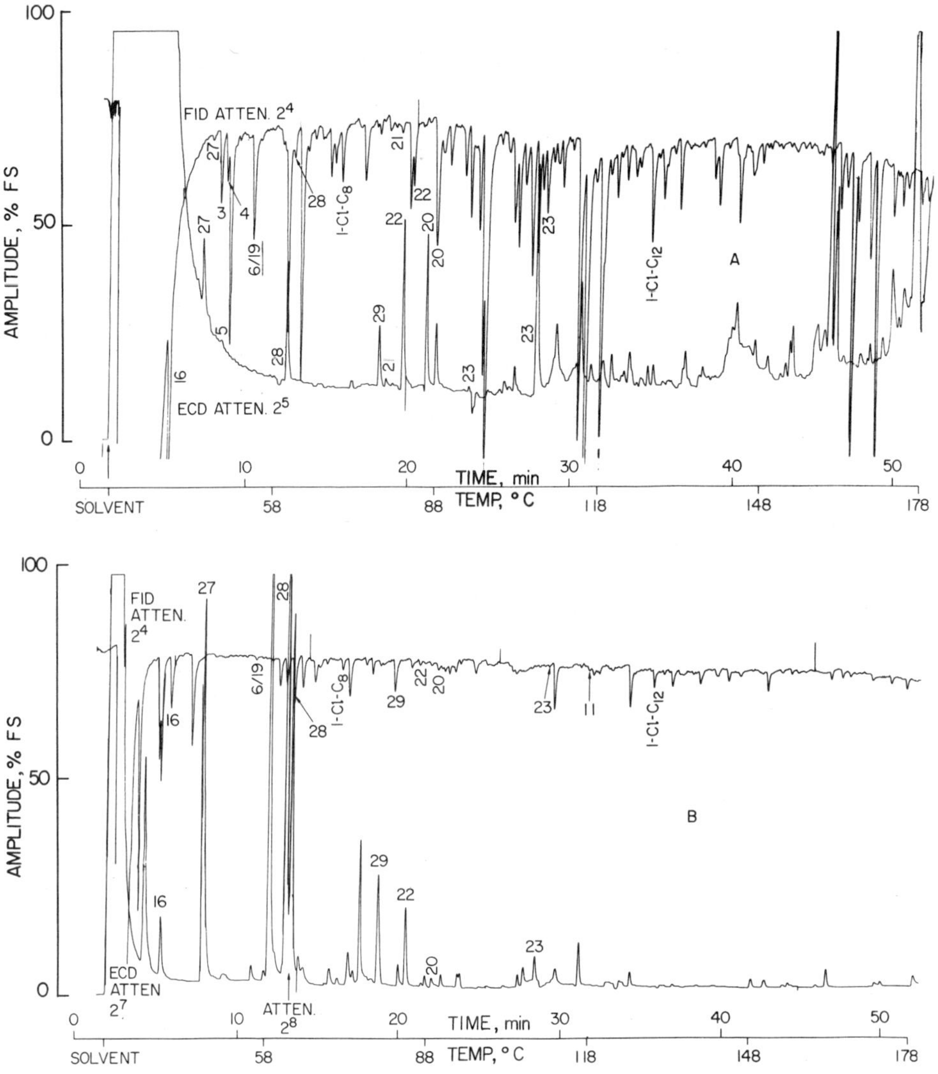

Figure 16. (A) Chromatograms obtained on closed-loop stripping sample at the initial stage of the treatment processes. Upper trace: flame ionization detection. Lower trace: electron capture detection. Identification was obtained by GC/MS: 3—ethylbenzene; 4—p-xylene; 5—m-xylene; 6—o-xylene; 16—tetrachloroethylene; 19—chlorobenzene; 20—1,2-dichlorobenzene; 21—1,3-dichlorobenzene; 22—1,4-dichlorobenzene; 23—trichlorobenzene; 27—dichlorobromomethane; 28—chlorodibromomethane; and 29—bromoform. Internal standard 1-C_1-C_8 and 1-C_1-C_{12} alkanes. (B) Chromatograms obtained on closed-loop stripping sample at the end of the treatment processes. Identification: see Fig. 16(A), and peak 11 is naphthalene. (Source: Ref. 59.)

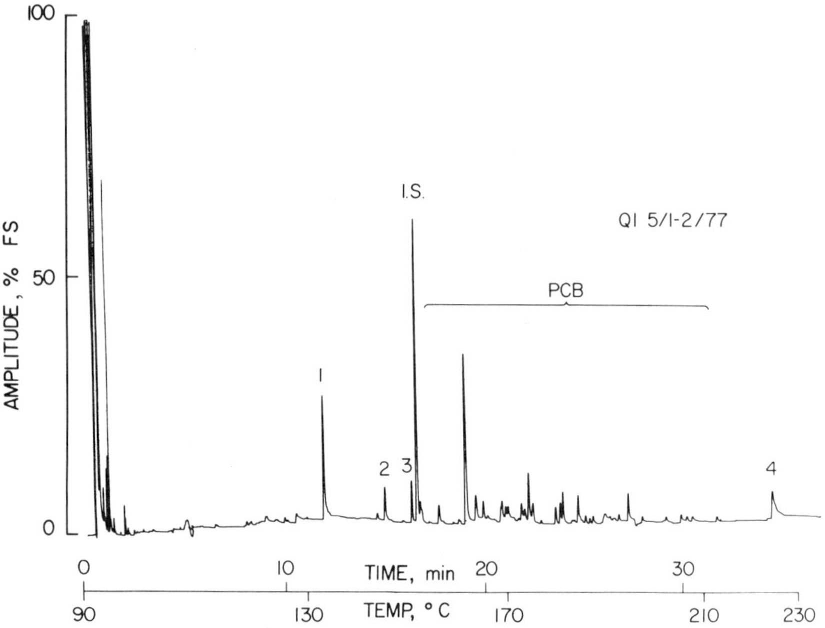

Figure 17. ECD chromatogram (setting $337/2^4$) of the pesticide fraction of the liquid-liquid extract of the influent internal standard 2.34 μg $liter^{-1}$. 1—diethylphthalate; 4—dioctylphthalate; 3—retention time as lindane; 2—unknown. (Source: Ref. 59.)

instituted. Those interested in the advanced waste treatment effluent and drinking water concentrates should refer directly to the report [81].

Using the modified purge-and-trap apparatus (Fig. 4) effluents of a secondary waste treatment plant have been studied. The experimental conditions are shown in Fig. 4 and an example of the results in Fig. 19. In this study the main objective was to search for the so-called priority pollutants.* They were analyzed as three fractions by GC/MS: the purgeable volatile organics, the basic components combined with neutrals, and the phenols and acids. The column used for the nonpurgeable organics was a 30-m SE-30 WCOT. The compounds were searched for with their characteristic ions within a certain retention window in the GC/MS data. Those identified were quantified with response factors and compared to an internal standard. The compounds identified in Fig. 19 are some of the purgeable priority pollutants.

* A list of chemicals in which the U.S. Environmental Protection Agency has special interest. Of the 129 compounds now on the list, 114 are organic.

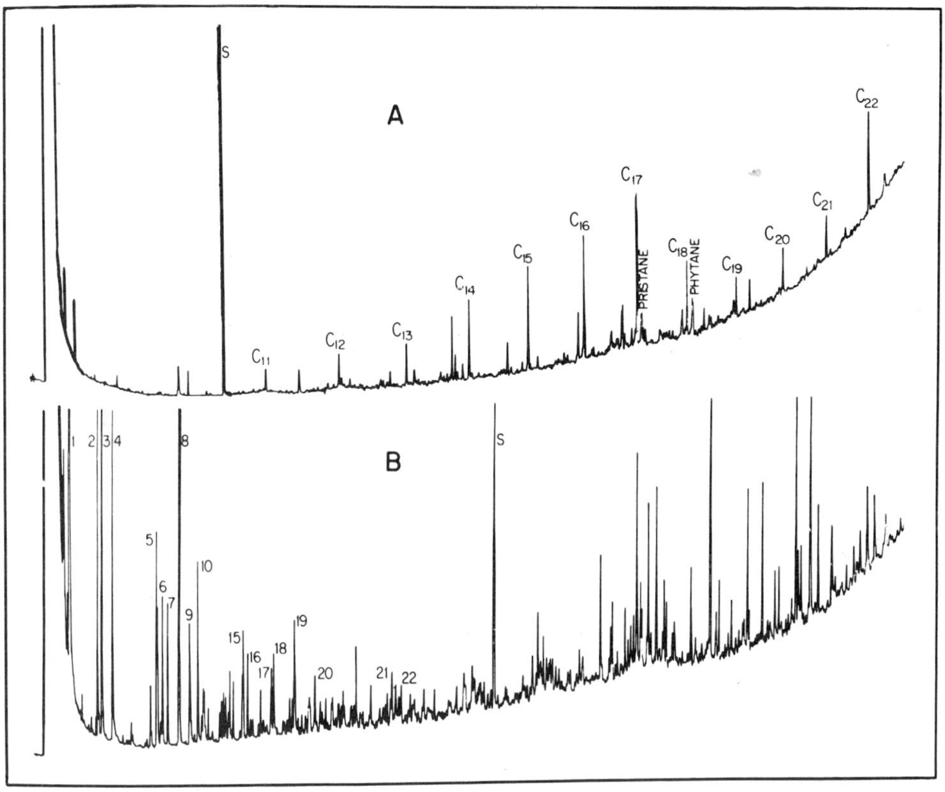

Figure 18. Capillary gas chromatograms of hydrocarbon fractions of an activated sludge effluent: (A) saturated hydrocarbons; (B) aromatic hydrocarbons; (S) internal standards; C_{11}-C_{22}, n-alkanes; 1—toluene; 2—ethylbenzene; 3—m- and p-xylene; 4—o-xylene; 5—1-ethyl-4-methylbenzene and 1-ethyl-3-methylbenzene; 6—1,3,5-trimethylbenzene; 7—1-ethyl-2-methylbenzene; 8—1,2,4-trimethylbenzene; 9—1,4-dichlorobenzene; 10—1,2,3-trimethylbenzene; 11—1-isopropyl-4-methylbenzene; 12—indan; 13—1,4-diethylbenzene; 14—1,3-dimethyl-5-ethylbenzene; 15—1,3-dimethyl-2-ethylbenzene and 1,3-dimethyl-4-ethylbenzene; 16—1,2-dimethyl-4-ethylbenzene; 17—1,2,4,5-tetramethylbenzene; 18—1,2,3,5-tetramethylbenzene; 19—tetralin; 20—naphthalene; 21—2-methylnaphthalene; 22—1-methylnaphthalene. (Sources: Refs. 94, 95.)

3.4 River, Lake, Ocean, and Related Compartments

The amount of pollutants in river, lake, and ocean water depends to a large extent on the groundwater and on industrial and domestic effluents feeding into the system. Depending on the level of contamination, the techniques described for these other types of water can be applied. For example, liquid-liquid extraction and the purge-and-trap method were used prior to capillary GC/MS analysis to study samples from the Delaware River [97] and an unnamed river [36]. Using capillary columns, River Glatt (Switzerland)

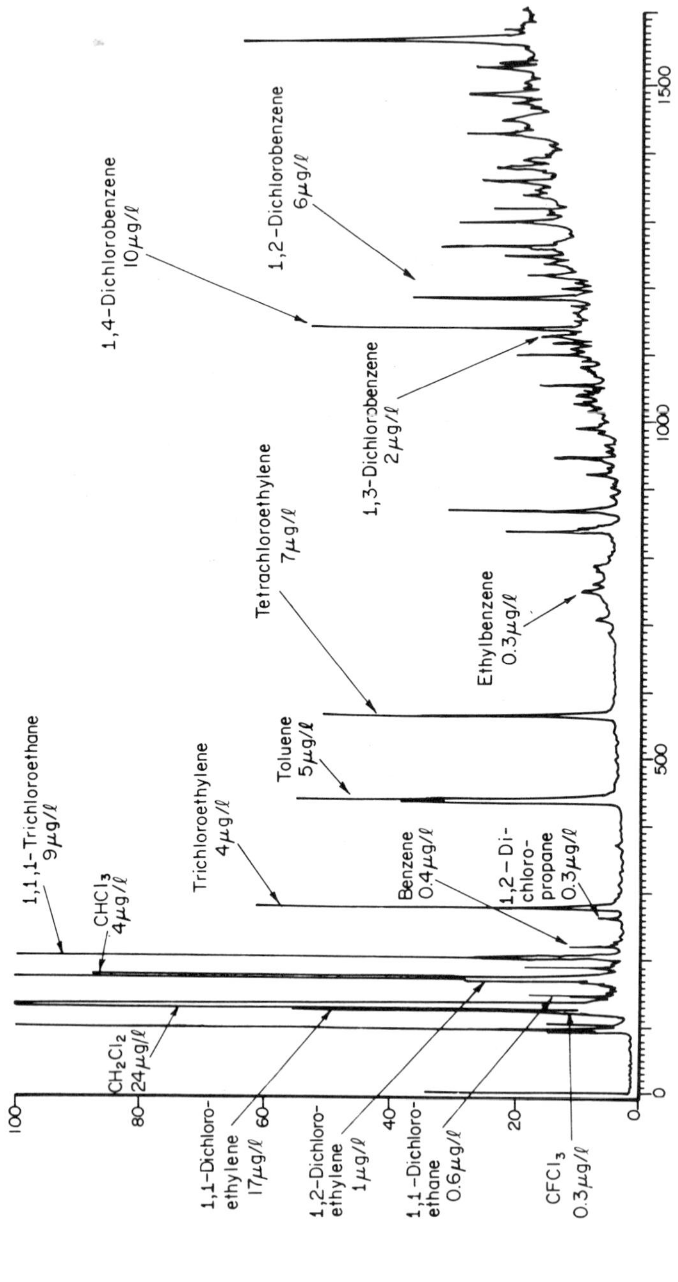

Figure 19. Reconstructed total ion chromatogram of purgeable organic compounds from a sewage sample with a few representative compounds identified and quantitated.

samples were studied by pentane microsolvent extraction [37] (Fig. 9) as well as closed-loop stripping [94,95,98]. Organics in the Columbia River (200 to 500-liter samples) were collected with an XAD-2 accumulative column and eluted with ether [99]. The sample was fractionated with gel permeation chromatography to eliminate compounds above a molecular weight of 800. Then it was further separated by silica gel chromatography into a hexane fraction and an ether (16%) hexane fraction before analysis on a 30-m OV-101 WCOT column. Flame ionization, electron capture, and mass spectrometry were used for detection. As for lake water, surface water and water from a 30-m depth of Lake Zurich were analyzed by capillary columns using the closed-loop stripping method of organic extraction [56,57]. The surface water of Lake Maggiore, Italy, was sampled with a Tenax accumulative column and analyzed with capillary GC/MS [14].

The volatile halogenated hydrocarbons in seawater on the west coast of Sweden were analyzed on a 33-m SE-52 WCOT column using a small-volume solvent extraction technique. Using electron capture detection, sensitivity of 1 fg (10^{-15} g) for carbon tetrachloride was reported [100]. Seawater from Sequim, Washington was examined in the same way as the Columbia River water [99,101]. The objective of this ongoing investigation is to study the effect of low-level chlorination of seawater, as for cooling towers. Surprisingly, many of the compounds found appeared to contain bromine instead of chlorine. These brominated compounds are usually at the μg liter^{-1} (ppb) level, with the exception of haloforms, and many of them remain unidentified [101].

Since rivers, lakes, and oceans usually support living organisms, this is an appropriate place to discuss the analysis of water pollutants in biota samples or other compartments. In an effort to prepare for any drilling activities in the outer continental shelf, the U.S. Bureau of Land Management had a number of water, surface, film, particulate, sediment, macrofauna, and zooplankton samples analyzed for hydrocarbon. There were 942 samples in all from the South Atlantic outer continental shelf [28,102]. The analytical approach involved extraction, saponification if necessary, silica gel column chromatography, and glass capillary gas chromatography. Two fractions from each sample, a saturated hydrocarbon fraction and an unsaturated and aromatic hydrocarbon fraction, were analyzed separately. About 8% of the samples were also analyzed by GC/MS.

For the analysis, the water was extracted with methylene chloride. The particulates were filtered and the filters were extracted for 4 hr in a Soxhlet extractor using methylene chloride as solvent. The frozen sediment samples were thawed, freeze-dried before Soxhlet extraction for 96 hr with benzene/methanol (3:2 ratio) as the solvent. The solvent concentrate was saponified before silica gel chromatography. The frozen macrofauna and zooplankton samples were thawed, homogenized, and saponified before solvent extraction with hexane. The sample extracts were fractionated with silica gel chromatography. Two fractions, petroleum ether and benzene/petroleum ether (1:1), were collected from all the samples. The benzene/petroleum ether fraction from zooplankton was further separated into two fractions, $>C_{12}$ and $>C_{12}$, with gel permeation chromatography using a high-performance liquid chromatograph. The $>C_{12}$ fractions were analyzed along with the petroleum ether fractions (Fig. 20) using a 30-m SE-30 WCOT column programmed from 80 to 260°C at 4°C min^{-1} with a flowrate of 2 ml min^{-1} at 80°C. Extensive quality assurance and method validation studies were made to insure the reliability of the data.

The results of the above study showed that although numerous hydrocarbon components were detected, especially biogenic components in sediment, macrofauna, and zooplankton samples, there was little indication of contamination by petroleum aromatic hydrocarbons. A few samples, mainly water particulate and surface film samples, contained petroleum contamination that was likely acquired during sample collection. Most of the hydrocarbons characterized by GC/MS analyses were biogenic components. They were prevalent in sediment as well as in macrofauna and zooplankton and seriously interfered with the detection of petroleum aromatic hydrocarbons at low levels.

Sediments have been studied rather extensively for water pollutants. Studies have been made in the Delaware River and an unnamed river for general industrial pollutants [97,36]; in the River Monchaltorfer, Aa, Switzerland [42], Lake Greifensee, Switzerland [42], and the Norwegian Sea [75] for polycyclic aromatic hydrocarbons; in the region of Buccaneer oilfield in the Mexican Gulf for alkanes [103]; in the Strait of Juan de Fuca, Washington for hydrocarbons [104]; and in Nantucket Shoals after the grounding of the oil tanker *Argo Merchant* for hydrocarbons [105]. These are just a few examples of such studies using capillary column chromatography. Polycyclic aromatic hydrocarbons in sediments and soils appear to be of intense

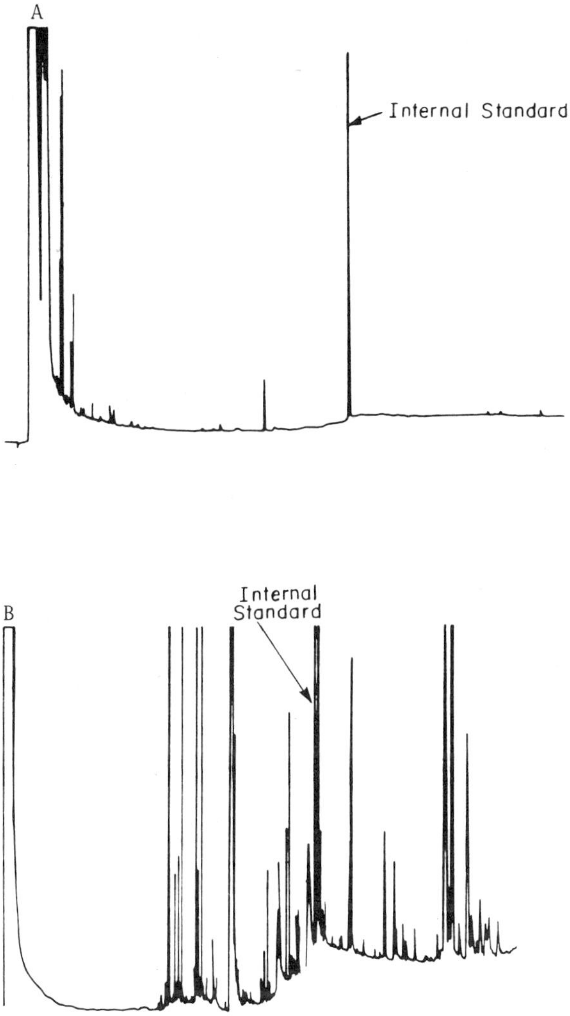

Figure 20. (A) Gas chromatogram of "aromatic" fraction from a zooplankton sample after GPC cleanup. (B) Gas chromatogram of "aromatic" fraction from a zooplankton sample before GPC cleanup. (Source: Ref. 28.)

interest and their global distribution was discussed in a recent review article [106]. Also of interest is the analysis of polychlorinated dibenzo-p-dioxins. Using capillary column chromatography with GC/MS, concentrations as low as 3 ppt and 5 ppt of 2,3,7,8-tetrachlorodibenzo-p-dioxin have been quantified in the soil and grass in Switzerland [76,107].

Analysis of biota samples requires the most elaborate preparation. Frequently only capillary columns can provide acceptable sensitivity and specificity even after extensive cleanup. Sample preparation requires multiple steps and is similar to the preparation of the zooplankton sample described earlier. In analyzing the oil spill by the tanker *Argo Merchant*, 37 samples were homogenized, saponified, extracted, chromatographed on silica gel, and then the saturated and aromatic fractions were analyzed by capillary gas chromatography and GC/MS, respectively [108]. One sample (a codfish) exhibited the two gas chromatographic profiles similar to those of the *Argo Merchant* cargo.

In other studies, mussels form Norwegian fjords have been examined for polycyclic hydrocarbons [75]. Accumulation of hydrocarbons in tissues of mallard ducks have also been examined [109,110]. And accumulation of chlorinated hydrocarbons in human tissues have been studied with the aid of capillary gas chromatography [111-113].

3.5 Drinking Water

The importance of drinking is obvious. There are no data on the direct or indirect human consumption of potable water, but implications are that the potable water should not contain any material that will cause acute or chronic adverse effects. Undesirable material can come mainly from two sources: raw water and water treatment processes. However, regardless of the sources of contamination, the final water supply usually has to meet certain standards. For example, in the United States the interim primary drinking water regulation of February 1978 [114] requires that total trihalomethanes be below 100 ppb and that if the raw water supply contains some of the chemicals (59 compounds and compound types) known as chemical contamination indicators, the drinking water treatment system has to install a granular activated-carbon treatment device. Regulations such as this and the general concern over the health effects of potable water generate immense interest in every aspect of drinking water. Most of the investigations involve some form of chemical analysis of water. However, as indicated previously, the use of capillary columns is not as widespread as their separation power deserves.

A recent study investigated the purgeable volatile organics through each step of the treatment process of the Tuscaloosa, Alabama waterworks [54]. The plant uses a double chlorination process and draws its water supply from an artificial lake. In the same study, single chlorination of water from a deep well was also investigated. A variation of the purge-and-trap method was used to extract the organics. The organics were analyzed on either a glass capillary OV-101 WCOT column or a nickel capillary column coated with Witconol LA25 [115]. Flame ionization detection (FID), electron capture detection (ECD), and GC/MS were used in the analysis. The series of chromatograms (Fig. 21) followed the fate of the purgeable volatile organics through the treatment processes. Figure 22 shows a simultaneous FID and ECD chromatogram of the finished drinking water.

As an alternative to the chlorination process for disinfection, ozonolysis has been used. The ozonolysis process has not been approved for use in the United States because it lacks a residual disinfection capability. However, this process has been studied at the Upper Thompson Sanitation District wastewater treatment plant at Estes Park, Colorado [53]. The purgeable volatile organics again were extracted by a variation of the purge-and-trap method. The columns used were either a 60-m OV-101 WCOT column with FID or a 50-m OV-101 SCOT column in the GC/MS. Figure 23 shows the chromatograms obtained on the GC/MS before and after ozonolysis. As expected, the process did not generate additional halogenated compounds in numbers or quantities. The principal volatile products were n-hexanal, n-heptanal, n-octanal, and n-nonanal. This is not surprising since substances such as oleic acid, unsaturated fats, olefin, and other unsaturated compounds can act as aldehyde precursors.

To illustrate analytical methods, various tap water samples have been studied. Figures 7 and 9 show chromatograms of tap water extracted with closed-loop stripping [57] and n-pentane microsolvent techniques [37], respectively. A quick, simple, and sensitive method has also been reported for low concentrations of volatile halogenated hydrocarbons in tap water. It uses small-volume solvent extraction, a capillary column, and electron capture detection [100]. Analysis of tap water using Tenax as the accumulative column has also been reported [14]. In that study, the authors theorized that mild chlorination, with a pH of 8, might explain why more chlorinated compounds were not found in the tap water than the raw water supply. They used the same method to study samples of bottled mineral water and in those cases at least a phthalate was identified as a major component.

Attempts to use capillary columns to study the nonpurgeable volatile organics in potable water are relatively few. A mixture of XAD-4 and XAD-8

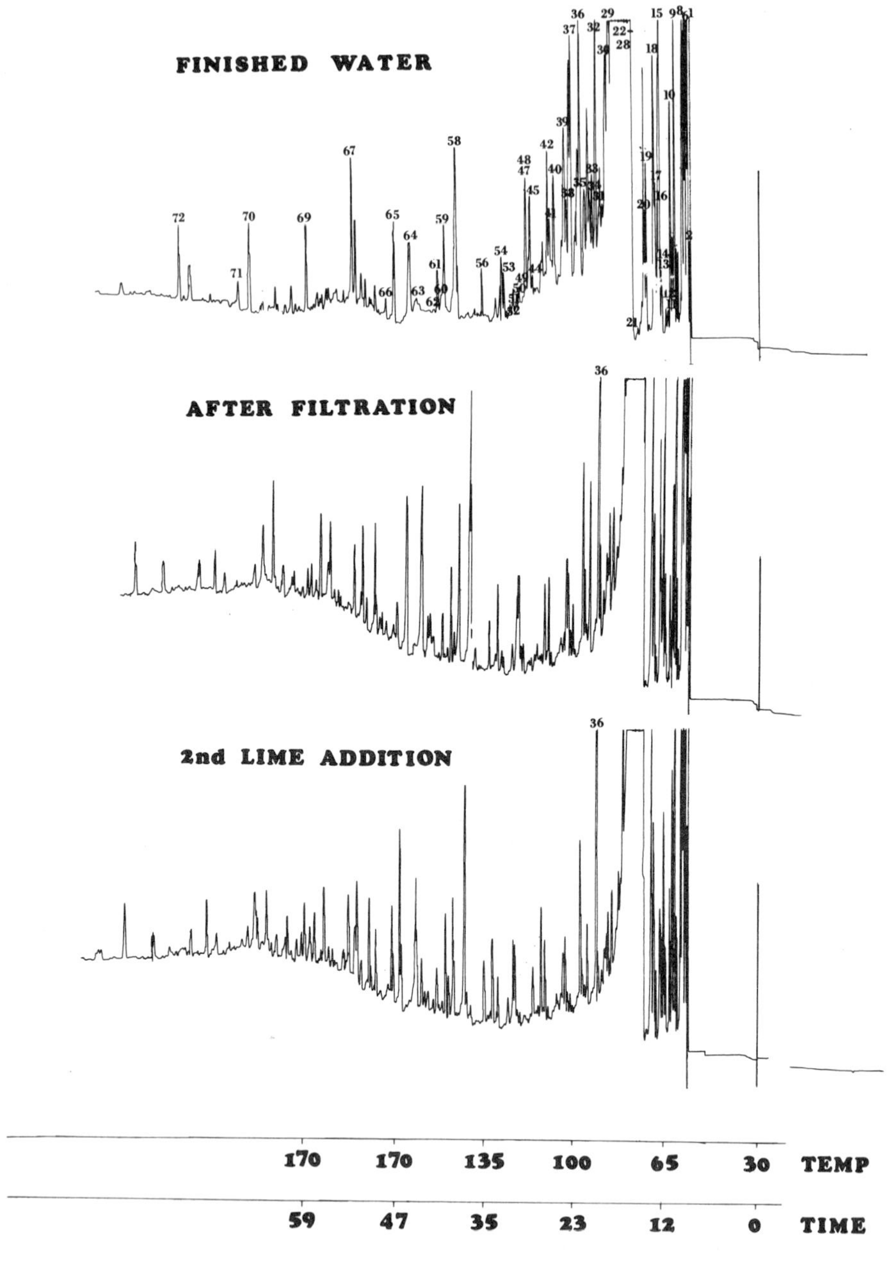

Figure 21. Total ion current chromatograms of volatiles in surface water after each treatment step. Sampling: sample size 250 ml, purge at 50 ml min^{-1} for 1 hr, room temperature. Analysis: nickel capillary column, 80 m × 0.5 mm coated with Witconol LA 23. (Source: Ref. 54.)

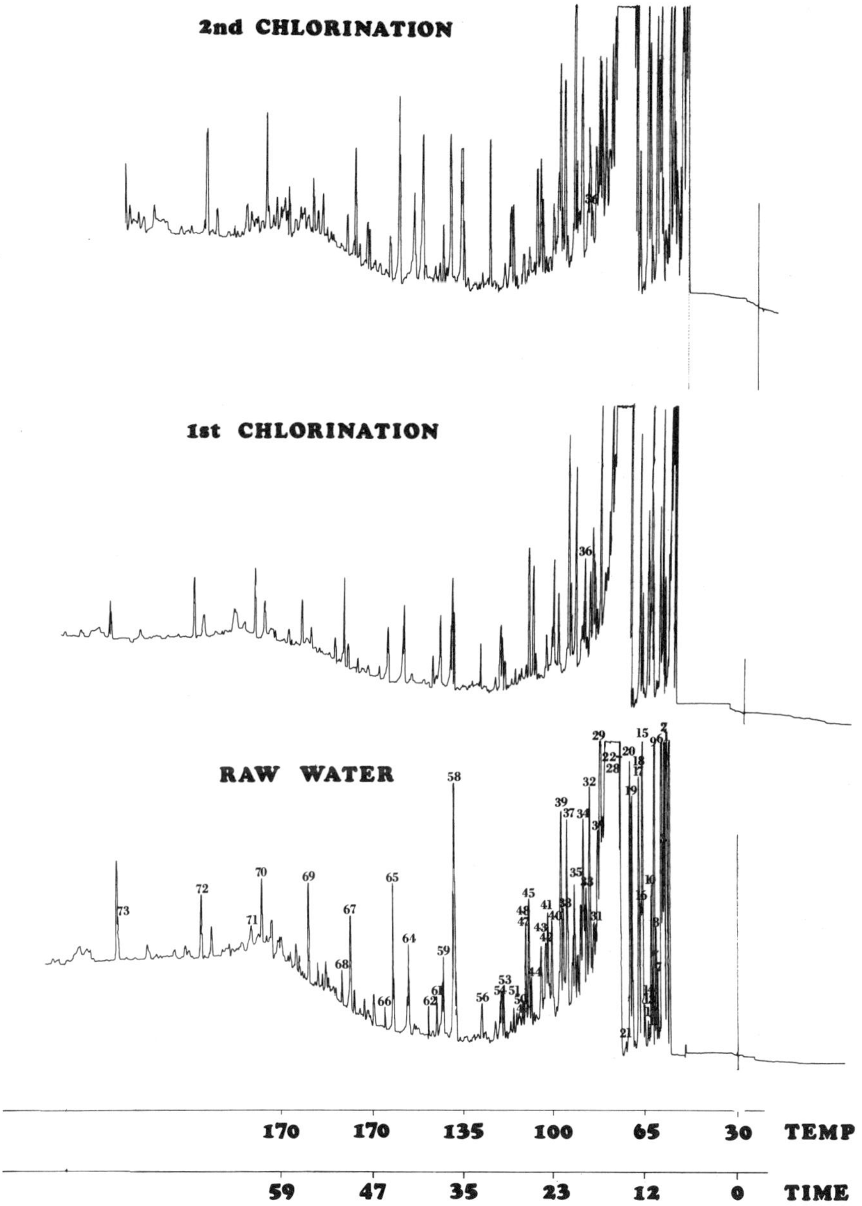

Figure 21 (cont'd.)

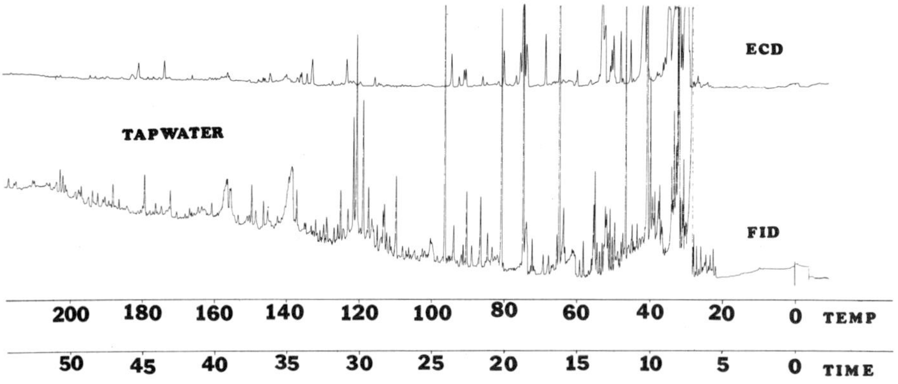

Figure 22. Volatiles in drinking water by FID and ECD in parallel. Sampling: sample size 250 ml, purge at 50 ml min^{-1} for 1 hr, room temperature. Analysis: glass capillary column, 45 m × 0.4 mm ID coated with OV-101. (Source: Ref. 54.)

resin as adsorbent in a column was used to extract the organics from 1000 liters of tap water from Athens, Georgia [9]. The sample was analyzed on a SP2100 capillary column and several polycyclic aromatic hydrocarbons (fluorene, phenanthrene, and fluoranthene) were identified. By far the major concerted effort in the capillary GC analysis of the nonpurgeable volatile organics in drinking water involves water samples from five metropolitan areas that have different types of raw water supplies [30,81]. The method being used, which involved reverse osmosis, extraction, bioassay, fractionation, and capillary GC/MS, is similar to that described for analysis of domestic effluents at Lake Tahoe. The ultimate goal of this ongoing project is to pinpoint the compounds that are responsible for the positive results in the drinking water concentrate bioassay.

4. SUMMARY AND FUTURE NEEDS

This chapter has reviewed various commonly practiced options available in building an experimental design around the superior separation power of capillary columns. Examples illustrate how some scientists utilize capillary columns to solve their analytical problems in a variety of water matrices and other compartments. More stress could have been placed on the constraints of cost and time, but the arguments are relatively simple: packed columns cannot give the kind of resolution capillary columns can offer [3] (Fig. 1), and if one operates capillary columns at the same resolution as packed columns, the analysis time will be much shorter. For example, using a 2-m

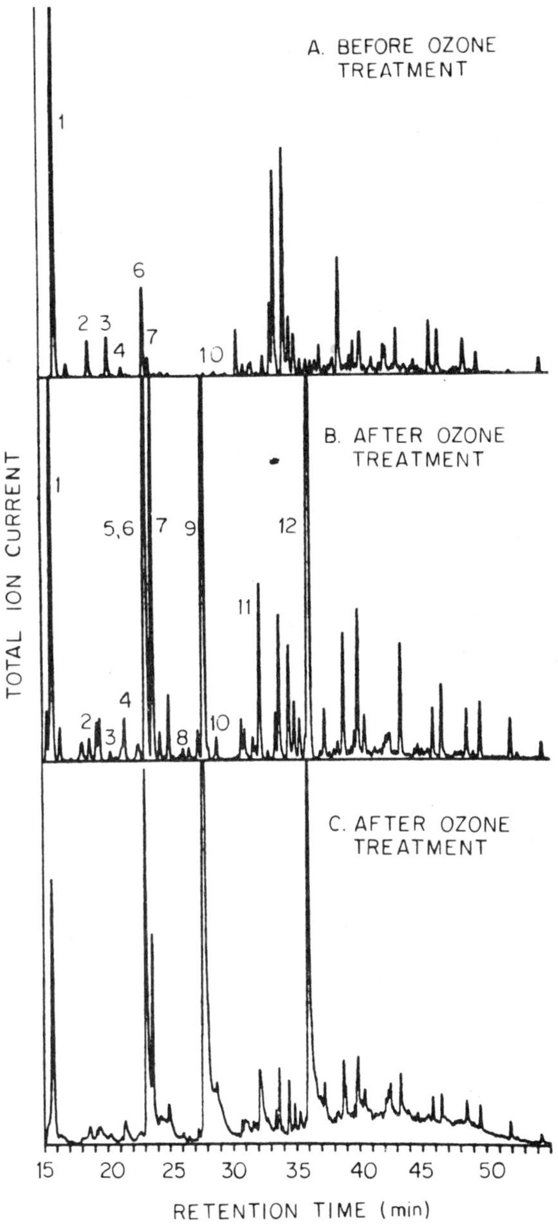

Figure 23. Chromatograms of wastewater obtained from the Upper Thompson Sanitation District wastewater treatment plant at Estes Park, Colorado. The sample was desorbed onto the capillary column while the GC oven was maintained at -20°C for 8 min. The oven was then temperature-programmed at 170°C at 4°C min^{-1}. Helium carrier flowrate was 4 ml min^{-1}. Mass spectra from m/e 35 to 400 were scanned and stored every 2.2 sec. in the electron impact mode. (A) Mass-resolved chromatogram prior to ozonolysis. (B) Mass-resolved chromatogram subsequent to ozonolysis. (C) Same data as B prior to data reduction by Biller and Biemann technique. 1—chloroform; 2—bromodichloromethane; 3—dimethyldisulfide; 4—toluene; 5—n-hexanal; 6—tetrachloroethylene; 7—n-octane; 8—xylene isomers; 9—heptanal; 10—dichlorobenzene isomer; 11—n-octanal; 12—n-nonanal. (Source: Ref. 52.)

capillary column to analyze hydrocarbons from n-$C_{10}H_{22}$ to n-$C_{34}H_{70}$ requires 7 min, whereas the same analysis on packed column requires 40 min [107]. The cost saving in terms of time alone is tremendous. Furthermore, where important decisions hinge on results, can one afford not to have the most reliable data?

So far as the technical content is concerned, this contribution is certainly not meant to be an all-inclusive review. It does attempt to familiarize the reader with all factors in an analysis and considerations in building an analytical scheme around them, to briefly review the applications of water analysis, and to direct the reader to sources of more detailed information. As a final consideration, then, what are the future needs in this area?

Ideally, we can dream of having an apparatus that will continuously sample water and instantaneously identify and quantify components that can cause acute or chronic adverse effects. In the meantime, we need to optimize all the parameters to give the best possible results. Several areas present themselves in terms of expected improvement. Direct on-column injectors have recently been introduced for narrow-bore WCOT columns, but they have not been applied in water analysis [116,117]. These injectors will extend the range of compounds that capillary column chromatography can analyze. The injectors do not need the evaporation process to deposite the sample components into the column. In terms of column technology, the newly introduced phases will improve column performance further, but it is unlikely any major breakthrough will occur in the near future. Insofar as extraction of organics is concerned, success in desorption of organics from carbon accumulative columns with carbon dioxide will play an important role in identifying additional components. It may also make extraction of nonpurgeable organics simpler and more effective. Compound identification will be easier if water scientists can agree on producing a set or two (for a polar and a nonpolar phase) of relative retention times of different classes of compounds found in water. Is this a close-to-impossible task? And what about an efficient data management method? Perhaps we can answer these questions more easily and meet some of the future needs outlined above if we ponder a final question--can the ecosystem of this planet wait?

ACKNOWLEDGEMENT

I would like to thank Drs. Rodger L. Foltz and Gregory A. Jungclaus for reading this manuscript.

REFERENCES

1. Garrison, A. W., *Ann. N.Y. Acad. Sci.*, 298:2 (1977).
2. Chian, E. S. K., and DeWalle, F. B., *J. Water Poll. Control Fed.*, 50: 1026 (1978).
3. Grob, K., and Grob, G., in *Identification and Analysis of Organic Pollutants in Water*, (L. H. Keith, Ed.), Ann Arbor Science, Ann Arbor, Mich., 1976, p. 75.
4. Eberhardt, L. L., Gilbert, R. O., Hollister, H. L., and Thomas, J. M., *Environ. Sci. Technol.*, 10:917 (1976) and references cited therein.
5. Kopfler, F. C., Melton, R. G., Lingg, R. D., and Coleman, W. E., in *Identification and Analysis of Organic Pollutants in Water* (L. H. Keith ed.), Ann Arbor Science, Ann Arbor, Mich. 1976, p. 87.
6. Junk, G. A., Richard, J. J., Griesen, M. D., Witiak, D., Witiak, J. L., Argnello, M. D., Vick, R., Snee, H. J., Fritz, J. S., and Galder, G. U., *J. Chromatogr.*, 99:745 (1974).
7. Junk, G. A., Richard, J. J., Fritz, J. S., and Svec, H. J., in *Identification and Analysis of Organic Pollutants in Water*, (L. H. Keith ed.) Ann Arbor Science, Ann Arbor, Mich., 1976, p. 136.
8. Tateda, A., and Fritz, J. S., *J. Chromatogr.*, 152:329 (1978).
9. Rossum, P. V., and Webb, R. G., *J. Chromatogr.*, 150:381 (1978).
10. Thurman, E. M., Aiken, G. R., and Malcolm, R. L., *Proc. 4th Joint Conf. on the Sensing of Environmental Pollutants, New Orleans, 1977*
11. Malcolm, R. L., Thurman, E. M., and Aiken, R. G., *Proc. 11th An. Conf. on Trace Substances in Environmental Health, 1977*
12. Stephan, S. F., and Smith, J. F., *Water Res.*, 11:339 (1977).
13. Stephan, S. F., Smith, J. F., Flego, M., and Reukers, J., *Water Res.*, 12:447 (1978).
14. Versino, B., Knoppel, H., DeGroot, M., Peil, A., Poelman, J., Schauenburg, H., Visser, H., and Geiss, F., *J. Chromatogr.*, 122:373 (1976).
15. Huffman, Jr., E. N. D., Farrier, D. S., and Poulson, R. E., "Proceedings of Symposium on Measurement of Organic Pollutants in Water and Wastewater, 1978", ASTM Technical Publication (in press)
16. Environmental Protection Agency, *J. Amer. Water Works Assoc.*, 67:281 (1975).
17. Dunlap, W. J., Shew, D. C., Scalf, M. R., Cosby, R. L., and Robertson, J. M., in *Identification and Analysis of Organic Pollutants in Water*, (L. H. Keith, ed.), Ann Arbor Science, Ann Arbor, Mich., 1976, p. 453.
18. Loper, J. C., Schoeny, R. S., and Tardiff, R. G., *Mutat. Res.*, 52:223 (1978).
19. Loper, J. C., Lang, D. R., Smith, C. C., Schoeny, R. S., Kopfler, F. C., and Tardiff, R. G., *Proc. Int. Symp. on Aquatic Pollutants*, Pergamon Press, Oxford and New York
20. Kopfler, F. C., Coleman, W. E., Melton, R. G., Tardiff, R. G., Lynch, S. C., and Smith, J. K., *Ann. N.Y. Acad. Sci.*, 298:20 (1977).
21. Deinzer, M., Melton, R., and Mitchell, D., *Water Res.*, 9:799 (1975).

22. Kopfler, F. C., Melton, R. G., Mullaney, J. L., and Tardiff, R. G., *Fate of Pollutants in Air and Water Environments*, Wiley, New York

23. Harless, R. L., Harris, D. E., Sovocool, G. W., Zehr, R. D., Wilson, N. K., and Oswald, E. O., *Biomed. Mass Spectrometry*, 5:232 (1978).

24. Bedford, J. W., Roelfs, E. W., and Zabick, M. J., *Limnol. Oceanogr.*, 13:112 (1968).

25. Goldberg, E. D., *Marine Poll. Bull.*, 6:111 (1975).

26. Dunn, B. P., *Environ. Sci. Technol.*, 10:1018 (1976).

27. Farrington, J. W., *Preprints on Symp. on Analytical Chemistry of Petroleum Hydrocarbons in Marine/Aquatic Environment*, 23:818 (1978).

28. Warner, J. S., Riggin, R. M., and Graffeo, A. P., *Preprints on Symp. on Analytical Chemistry of Petroleum Hydrocarbons in Marine/Aquatic Environment*, 23:835 (1978).

29. Lin, D. C. K., Foltz, R. L., Lucas, S. V., Petersen, B. A., and Slivon, L. E., *Proc. 3rd Int. Conf. on Stable Isotopes*, (P. Klein, and R. Klein, eds.), Academic Press, New York

30. Lin, D. C. K., Foltz, R. L., Lucas, S. V., Petersen, B. A., Slivon, L. E., and Melton, R. G., *Proc. Symp. on Measurement of Organic Pollutants in Water and Wastewater, Denver, 1978*, ASTM Technical Publication

31. Middleditch, B. S., Basile, B., and Chang, E. S., *J. Chromatogr.*, 142: 777 (1977).

32. Grob, K., and Zurcher, F., *J. Chromatogr.*, 117:285 (1976).

33. Ahnoff, M., and Josefsson, B., *Anal. Chem.*, 46:658 (1974).

34. Goldberg, M. C., Delong, L., and Sinclair, M., *Anal. Chem.*, 89:45 (1973).

35. Hites, R. A., *J. Chromatogr. Sci.*, 11:570 (1973).

36. Jungclaus, G. A., Lopez-Avila, V., and Hites, R. A., *Environ. Sci. Technol.*, 12:88 (1978).

37. Grob, K., Grob Jr., K., and Grob G., *J. Chromatogr.*, 106:299 (1975).

38. Sucre, L., Jennings, W., Fisher, G. L., Raabe, O. G., and Olechno, J., *Proc. Symp. on Trace Organic Analysis, March 1978*, National Bureau of Standards, Washington, D.C.

39. Osterroht, C., *J. Chromatogr.*, 101:289 (1974).

40. Warner, J. S., *Anal. Chem.*, 48:578 (1976).

41. Lawler, G. C., Loong, W. A., and Laseter, J. L., *Environ. Sci. Technol.*, 11:47 (1977).

42. Giger, W., and Schaffner, C., *Anal. Chem.*, 50:243 (1978).

43. Thruston, Jr., A. D., *J. Chromatogr. Sci.*, 16:254 (1978).

44. DeRidder, J. J., Hoppens, P. C. J. M., and VanHal, H. J. M., *J. Chromatogr.*, 143:281 (1977).

45. Graffeo, A. P., Lin, D. C. K., and Foltz, R. L., *J. Chromatogr.*, 126: 717 (1976).

46. Ahuja, S., *J. Pharm. Sci.*, 65:163 (1976).

47. Blau, K., and King, G. S., *Handbook of Derivatives for Chromatography*, Heyden, London, 1977.

48. Buryan, P., Macak, J., and Nabivach, V. M., *J. Chromatogr.*, 148:203 (1978).

49. Miniature diazomethane generator, Panton Woods Glass Shop, 7500 Brill Road, Cincinnati, Ohio.

50. Bellar, T., and Lichtenberg, J. J., *J. Amer. Water Works Assoc.*, 66: 729 (1974).

51. Lingg, R. D., Melton, R. G., Kopfler, F. C., Coleman, W. E., and Mitchell, D. E., *J. Amer. Water Works Assoc.*, 69:605 (1977).

52. Sievers, R. E., Barkley, R. M., Eiceman, G. A., Shapiro, R. H., Walton, H. F., Kolonko, K. J., and Field, L. R., *J. Chromatogr.*, 142:745 (1977).

53. Bertsch, W., Anderson, E., and Holzer, G., *J. Chromatogr.*, 112:701 (1975).

54. Thomason, M., Shoults, M., Bertsch, W., and Holger, G., *J. Chromatogr.*, 158:437 (1978).

55. Dowty, B. J., Antoine, S. R., and Laneter, J. L., *Proc. Symp. on Measurement of Organic Pollutant in Water and Wastewater, December 1978*, ASTM Technical Publication

56. Grob, K., *J. Chromatogr.*, 84:255 (1973).

57. Grob, K., and Grob, G., *J. Chromatogr.*, 90:303 (1974).

58. Waggott, A., *Proc. Analyt. Div. Chem. Soc.*, p. 232 (August 1978).

59. Reinhard, M., Dolce, C. J., McCarty, P. L., and Argo, D. G., *J. Amer. Soc. Civ. Eng.*

60. Jungclaus, G. A., personal communication, Battelle Columbus Laboratories, King Avenue, Columbus, Ohio.

61. Glaser, E. R., Silver, B., and Suffet, I. H., *J. Chromatogr. Sci.*, 15: 22 (1977).

62. Suffet, I. H., and Glaser, E. R., *J. Chromatogr. Sci.*, 16:12 (1978).

63. Ryhage, R., Wikstrom, S., and Waller, G. R., *Anal. Chem.*, 37:435 (1965).

64. Grob, K., and Jaeggi, H., *Anal. Chem.*, 45:1778 (1973).

65. Leferink, J. G., and Leclercq, P. A., *J. Chromatogr.*, 91:385 (1974).

66. Bergert, K. H., Betz, V., and Pruggmayer, D., *Chromatographia*, 7:115 (1974).

67. Thome, F. A., and Young, G. W., *Anal. Chem.*, 48:1423 (1976).

68. Blum, W., Schlumpf, E., Liehr, J. G., and Richter, W. J., *Tetrahedron Lett.*, p. 565 (1976).

69. Blum, W., and Richter, W. J., *J. Chromatogr.*, 132:249 (1977).

70. Matthiensen, U., Kump, B., and Staib, W., *Chromatographia*, 10:303 (1977).

71. Sunol, C., and Gelpi, E., *J. Chromatogr.*, 142:559 (1977).

72. Durbeck, H. W., Baker, I., and Leymann, W., *Chromatographia*, 11:295 (1978).

73. Neuner-Jehle, N., Etzweiler, F., and Zarske, G., *Chromatographia*, 6:211 (1973).

74. Guiochon, G., *Anal. Chem.*, 50:1812 (1978).

75. Bjorseth, A., in *Carcinogenesis*, Vol. 3: *Polynuclear Aromatic Hydrocarbons* (P. W. Jones and R. I. Freudenthal, eds.), Raven Press, New York, 1978, p. 75.

76. Buser, H. R., *Anal. Chem.*, 48:1553 (1976).

77. Krijgsman, W., and Van De Kamp, C. G., *J. Chromatogr.*, 131:412 (1977).

78. Grob, Jr., K., and Grob, K., *J. Chromatogr.*, 140:257 (1977).

79. Verzele, M., and Sandra, P., *J. Chromatogr.*, 158:111 (1978).

80. Lin, D. C. K., Petersen, B. A., Melton, R. G., and Coleman, E., paper 16, Am. Chem. Soc. National Meeting, Symposium on High Resolution Gas Chromatography, August 28 to September 2, 1977, Chicago, IL.

81. Lin, D. C. K., Contes, D. A., Foltz, R. L., Goff, V. R., Hayes, T. L., Kelly, P. P., Petersen, B. A., Slivon, L. E., Wallen, R., and Watson, S. C., GC/MS Analysis of Organics in Drinking Water Concentrates and Advanced Waste Treatment Concentrates, Lake Tahoe II Pentane Concentrate Analysis Report, Project No. 68-03-2548, Health Effect Research Laboratory, U.S. Environmental Protection Agency, 1978.

82. Humenick, J. J., and Mattox, C. F., *Water Res.*, 12:463 (1978).

83. U.S. Environmental Protection Agency, *Sampling and Analysis Procedures for Screening of Industrial Effluents for Priority Pollutants*, Cincinnati, April 1977.

84. Pellizzari, E. D., Identification of Components of Energy-Related Wastes and Effluents, Project No. 68-03-2368, Environmental Research Laboratory, U.S. Environmental Protection Agency, 1978.

85. Zlatkis, A., Lichenstein, H. A., and Tishbee, A., *Chromatographia*, 6:67 (1973).

86. Dowty, B., Carlisle, D., Laseter, J. L., and Storer, J., *Science*, 187:75 (1975).

87. Pellizzari, E. D., Bunch, J. E., Carpenter, B. H., and Sawicki, E., *Environ. Sci. Technol.*, 9:552 (1975).

88. Eklund, G., Josefsson, B., and Bjorseth, A., *J. Chromatogr.*, 150:161 (1978).

89. Bjorseth, A., Lunde, G., and Gjos, N., *Acta Chem. Scand., Ser. B*, 31:797 (1977).

90. Bjorseth, A., Carlberg, G. E., and Moller, M., *Sci. Total Environ.*

91. Keith, L. H., in *Identification and Analysis of Organic Pollutants in Water* (L. H. Keith, ed.), Ann Arbor Science, Ann Arbor, Mich., 1976, p. 671.

92. Strup, P. E., Wilkinson, J. E., and Jones, P. W., in *Carcinogenisis*, Vol. 3: *Polynuclear Aromatic Hydrocarbons* (P. W. Jones and R. I. Freudenthal, eds.), Raven Press, New York, 1978, p. 131.

93. Saxena, J., Kozuchonski, J., and Baser, D. K., *Environ. Sci. Technol.*, 11:682 (1977).

94. Giger, W., Reinhard, M., Schaffner, C., and Zurcher, F., in *Identification and Analysis of Organic Pollutants in Water* (L. H. Keith, ed.), Ann Arbor Science, Ann Arbor, Mich., 1976, p. 135.

95. Giger, W., and Roberts, P. V., *Water Pollution Microbiology*, Vol. 2 (R. Mitchell, ed.), Wiley, New York, 1978, p. 135

96. Melton, R. G., private communication, Health Effect Research Laboratory, U.S. Environmental Protection Agency, Cincinnati.

97. Sheldon, L. S., and Hites, R. A., *Environ. Sci. Technol.* 12:1188 (1978).

98. Zucker, F., and Giger, W., *Von Wasser*, 47:37 (1976).

99. Bean, R. M., Ryan, P. W., and Riley, R. G., in *High Resolution Gas Chromatography* (S. P. Cram, ed.), Academic Press, New York

100. Eklund, G., Josefsson, B., and Roos, C., *J. High Res. Chromatogr.*, 1:34 (1978).

101. Bean, R. M., Riley, R. G., and Ryan, P. W., *Water Chlorination: Environmental Impact and Health Effects*, Vol. 2, Ann Arbor Science, Ann Arbor, Mich., 1978.

102. Warner, J. S., Hydrocarbon Analysis of Marine Samples from the South Atlantic Outer Continental Shelf, report submitted to Bureau of Land Management via Texas Instruments, Inc., June 1978.

103. Middledith, B. S., and Basile, B., *J. Chromatogr.*, 158:449 (1978).

104. MacLeod, Jr., W. D., Brown, D. W., Jenkins, R. G., and Ramos, L. S., in *Fate and Effects of Petroleum Hydrocarbons in Marine Organisms and Ecosystems* (A. D. Wolfe, ed.), Pergamon Press, New York, 1977.

105. MacLeod, W. D., Uyeda, M. Y., Thomas, L. C., and Brown, D. W., *Proc. Symp.: In the Wake of the Argo Merchant, 1978*, University of Rhode Island Press, Kingston

106. Laflamme, R. E., and Hites, R. A., *Geochim. Cosmochim. Acta*, 42:289 (1978).

107. Buser, H. R., *Proc. 2nd Int. Symp. on Glass Capillary Chromatography, 1977* (R. E. Kaiser, ed.), Institute of Chromatography, Bad Burkeim, Germany, 1977; *Anal. Chem.*, 49:918 (1977).

108. MacLeod, Jr., W. D., Thomas, L. C., Uyeda, M. Y., and Jenkins, R. G., *Proc. Conf. on Assessment of Ecological Impacts of Oil Spills, Colorado, 1978*, American Institute of Biological Science, Arlington, Va.

109. Lawler, G. C., Holms, J. P., Fiorito, B. J., Laseter, J. L., and Szaro, R. C., *Proc. Conf. on Assessment of Ecological Impacts of Oil Spills, Colorado, 1978*, American Institute of Biological Science, Arlington, Va.

110. Lawler, G. C., Loong, W. A., and Laseter, J. L., *Environ. Sci. Technol.* 12:47,51 (1978).

111. Luyten, B. J., and Franken, J. J., *J. Assoc. Off. Anal. Chem.*, 59: 1279 (1976).

112. Bjorseth, A., Lunde, G., and Dybing, E., *Bull. Environ. Contamin. Toxicol.*, 18:581 (1977).

113. Bjorseth, A., and Lunde, G., *Sci. Total Environ.*, 8:241 (1977).

114. Federal Interim Primary Drinking Water Regulations, *Federal Register*, 43(28):5756 (February 9, 1978).

115. Bertsch, W., Shumbo, F., Chang, R. C., and Zlatkis, A., *Chromatographia*, 7:128 (1974).

116. Grob, K., and Grob, Jr., K., *J. Chromatogr.*, 151:311 (1978).

117. Schomburg, G., Behlan, H., Dielmann, R., Weeke, F., and Husmann, H., *J. Chromatogr.*, 142:87 (1977).

6

GLASS CAPILLARY GAS CHROMATOGRAPHY OF PESTICIDES

BRUCE W. HERMANN AND JAMES N. SEIBER

University of California
Davis, California

1. Introduction 175

2. Analytical Considerations 179
   - 2.1 Sample Preparation 179
   - 2.2 Sample Injection 180
   - 2.3 Columns 181
   - 2.4 Detectors 181

3. Applications 182
   - 3.1 Organochlorine Insecticides 182
   - 3.2 Organophosphorus Insecticides 189
   - 3.3 Herbicides 191
   - 3.4 Fungicides and Bacteriocides 193
   - 3.5 Pheromones and Insect Growth Regulators 195

4. Conclusions 197

References 199

1. INTRODUCTION

Pesticides have played a major role in increasing worldwide food production and in protecting man's health and natural resources [1]. New chemicals are continually under development which provide better control for a wide variety of pests while minimizing any adverse effects on the environment. The first major class of synthetic organic pesticides, of which DDT represents the best-known member, were the organochlorine insecticides [2]. Though still widely used, they have now been largely replaced by the generally less persistant and more selective organophosphate and carbamate insecticides. The other major pesticides include: herbicides, used to elimate plants that compete with the farmer's crops for nutrients, water, and sun; fungicides and bacteriocides, which control the agents responsible for many plant diseases; and the pheromones and insect growth regulators, new and highly specific groups of insecticides now under active development.

These pest control tools are not without shortcomings. Too much dependence on pesticides results in selection pressures which may give rise to resistant strains of insects or alter natural balances [3]. Also, it is now known that chemicals generally do not remain where they are applied, but instead may partition between the major environmental compartments in accordance with their physicochemical properties and the ways in which they are deployed [4]. These properties can also interract with the various environmental transport mechanisms, moving the more stable pesticides hundreds of miles from the point of their original release. Such environmental distribution may lead to exposure of living organisms--including man--that are far removed from the intended target.

Pesticides may also be chemically modified—i.e., by oxidation, reduction, hydrolysis, and photodecomposition—when present in the environment, frequently forming products which may bear little resemblance to the ones originally applied [5,6]. Reactions may lead to detoxification of the poison, such as in the hydrolysis of parathion, or to a compound more toxic than the parent, as in the oxidation of parathion to paraoxon. It is important that both the parent and products be identified and quantitated, so that environmental exposure to these potentially toxic chemicals can be assessed.

Another problem in deciphering the fate and relative residual hazard associated with any individual pesticide formulation lies in knowing the chemical identity of its components. Many pesticide formulations, such as toxaphene, are complex mixtures whose components may vary in both their environmental behavior and biological activity [7] (Fig. 1). Furthermore, trace contaminants such as the "dioxin" (2,3,6,7-tetrachlorodibenzo-p-dioxin) in 2,4,5-T may be far more of a hazard than the major formulation component [8].

Thus the environmental analytical chemist is faced with the enormous task not only of analyzing for a chemical where it is applied, but also in

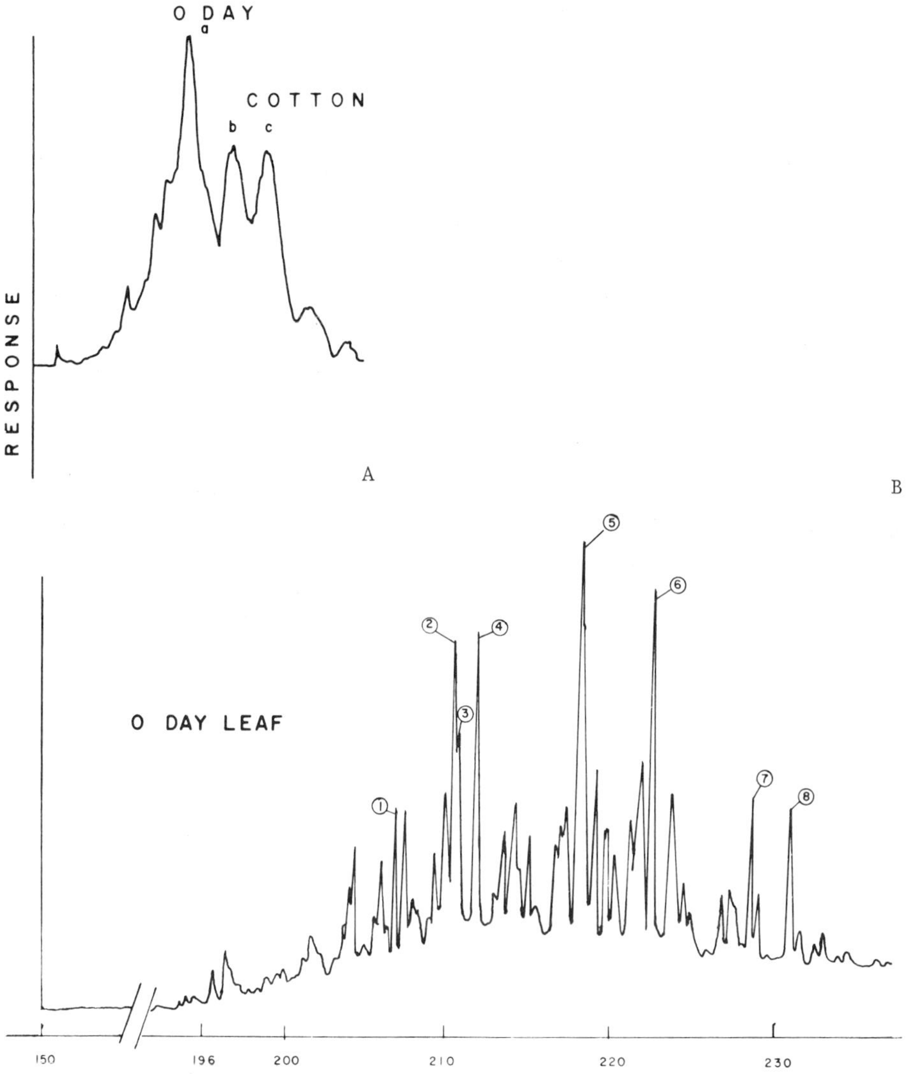

Figure 1. Chromatogram of toxaphene: (A) by packed column gas chromatography; (B) by capillary gas chromatography. (Source: Ref. 58.)

determining how and in what form the chemical will be transported and eventually deposited in the environment. The tool most frequently used for these determinations has been packed column gas chromatography. This technique has received wide acceptance due to its applicability to the analysis of a wide variety of pesticides and their environmental products. With the proper selection of liquid phase and detector, separations and quantitative determinations can often be made with very little sample preparation. Even when the compounds of interest are nonvolatile or thermally labile, derivatives can often be made which overcome these undesirable chromatographic properties. Derivatization can also be used to make the compound responsive to selective detectors, such as in the formation of halogenated derivatives for enhanced electron capture detection [9]. High-performance liquid chromatography (HPLC) has recently made some inroads into pesticide residue analysis, but its development has been slowed by the lack of detection capability matching the sensitivity and selectivity available for gas chromatography.

Packed column gas chromatography does have limitations. As effective as high-performance columns are, they may provide insufficient resolution for the complex mixtures frequently encountered in environmental samples and result in mistaken assignments or overestimation of pesticide residues [10]. This problem can be overcome by the use of high-resolution glass capillary gas chromatography, a technique not widely used in pesticide residue analysis until just a few years ago.

The slow adoption of glass capillary gas chromatography to pesticide residue analyses may be traced to several factors. Initially there was an inability to make or purchase the stable, inert columns required to avoid loss of sensitive pesticides under the required gas chromatographic conditions. Instead, columns were produced which had unpredictable and/or unreproducible characteristics, as demonstrated in the many publications dealing with the coating and deactivation of capillary columns. Detectors had to be modified to minimize dead volumes, and to compensate for the low flowrates typical of narrow-bore capillary columns. Artifacts caused by impure carrier gas, septum bleed, or back-diffused samples were a frequent problem. Some felt that splitting of a large portion of the sample upon injection lowered the sensitivity below that attainable with packed columns. But perhaps of greatest importance, residue chemists frequently attempted to directly transfer the techniques learned from packed columns, not realizing that capillary chromatography has a technology all its own [11]. A combination of these problems often led the residue analyst to abandon the promise of high-resolution separations for the security of reproducible results using the more familiar packed column technology.

Most of these problems have been minimized or eliminated. It is the author's belief that, once the pesticide analyst becomes familiar with the techniques necessary for the successful use of glass capillary chromatography, he will take advantage of their increased powers of resolution in everyday applications to residue analysis.

2. ANALYTICAL CONSIDERATIONS

In many ways, samples for capillary chromatography may be handled with the same basic procedures used for packed column gas chromatography [12]. There are important differences, however, and overlooking these differences may cause serious difficulties.

2.1 Sample Preparation

The major limitation in sample preparation has been the small capacity of capillary systems. Even with the sensitive detectors now available, samples must be concentrated many orders of magnitude before a meaningful analysis can be made. Air samples are concentrated by the use of liquid impingers or by sorption to charcoal or polymeric resins. Pesticides in water are similarly concentrated by adsorption, or by partitioning with organic solvents [13]. A variety of methods exist for the extraction of pesticides from solid matrices and their subsequent concentration. These procedures are detailed in standard treatises dealing with pesticide residue methodology, or discussed in detail in the applications which follow in this chapter.

Once collected, the sample must be cleaned for analysis. This is usually of less importance for the volatile coextractives in the sample due to the higher resolving power of capillary columns than for nonvolatile or thermally unstable components. The latter tend to concentrate in the injector or at the head of the column, with the possibility of absorbing or catalyzing decomposition of the compounds of interest. In addition, these contaminants can lead to the formation of ghost peaks as they eventually make their way through the column or begin to degrade the column liquid phase. With packed columns, minimal cleanup was required, but with the smaller sample sizes involved in capillary chromatography, any adsorption or degradation of the compound may be intolerable. Many times it is adsorption problems in the injector that have given glass capillary gas chromatography the reputation, among pesticide analysts, of being tricky and unreliable [14]. Many a "bad" column may be "repaired" by cleaning or replacing the injection port, or by discarding the first few coils of a damaged column [15] (Fig. 2).

Besides the conventional methods of cleanup using solvent partition and/or column adsorption chromatography, refined cleanup may involve fractionation by preparative HPLC or GC. Two-dimensional gas chromatography, with a packed or wide-bore capillary precolumn coupled to the capillary

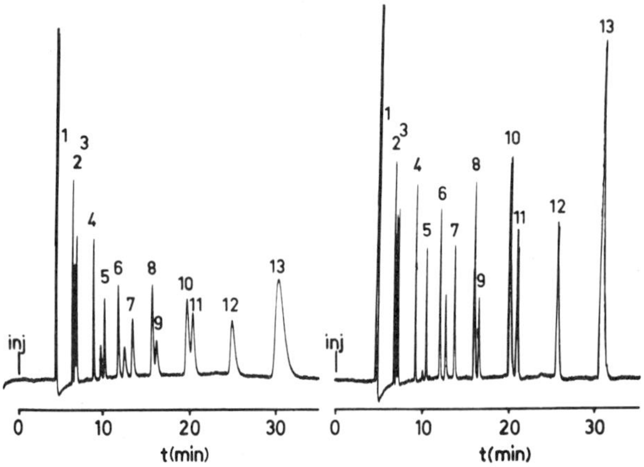

Figure 2. Illustration of the effect of changing injector glass tube. Left: standard solution run after about 100 injections of sewage sludge extracts. Right: the same solution run after changing the injector glass tube. Peaks: 1—α-BHC; 2—β-BHC; 3—lindane; 4—heptachlor; 5—aldrin; 6—heptachlor epoxide; 7—p,p-DDMU; 8—dieldrin; 9—p,p'-DDE; 10—p,p'-DDD; 11—o,p-DDT; 12—p,p'-DDT; 13—hexabromobenzene. (Source: Ref. 15.)

system, provides both cleanup and improved resolution with a single chromatographic pass.

2.2 Sample Injection

Even with clean samples, the analyst is faced with the problem of introducing the sample into a small-bore column with a minimum of band spreading, especially considering the large injector volume that usually accompanies the need for enough heated surface area to completely volatilize the sample. One strategy is to split off most of the sample [16,17]. This method not only restricts the sample size, but it continually flushes the injector, minimizing the problem of extended injections entering the column. Splitless injection is becoming a popular technique in trace analysis where large sample losses cannot be tolerated. A capsule insertion technique, with an injector dead volume of only 0.15 μl, that places the sample directly at the head of the column has been described [18]. Other injector techniques involve the use of the falling needle, Curie point injector [19], a precolumn with a cold trap for concentrating the analyte [13,20,21], and direct injection. With pesticides in particular, it is of paramount importance that the injector remain free of any residual organic compounds that can cause breakdown or absorption of the sample, resulting in tailing or loss of sensitivity.

2.3 Columns

Pesticide analysis generally requires high efficiency and inert columns with low surface activity that are stable toward high temperatures. A great deal of attention is still directed toward the creation of such columns. Approaches have included [22]:

1. Roughening (etching) the glass surface with the use of HCl or HF gas [23].
2. Coating the surface with a layer of salt crystals (either NaCl [23] or $BaCO_3$ [24]).
3. Growing whiskers on the walls (WWOT [25,26]).
4. Coating the column with a solid support coated with the liquid phase (SCOT).
5. Precoating the column with a monolayer of a polar phase such as Carbowax 20M [27].
6. Adding surface active or wetting agents to the liquid phase (Ionox etc. [28]).
7. Using a thicker coating of liquid phase [29].

Some of these modifications increase the surface area of the column and therefore increase its sample capacity. Wide-bore WCOT columns have been created for the same purpose.

The loss of resolution or shortened lifetime of glass capillaries is generally caused by a breakdown of the liquid phase. The breakdown is usually more a result of mistreatment than inherent instability. Causes include exceeding the upper temperature limit of the liquid phase, large injections of crude extracts containing polar, nonvolatile components [11], the presence of oxygen or water in the carrier gas, or overloading the column with solvent. Properly made columns should compare well in lifetime with packed columns when precautions are taken to avoid these abuses.

2.4 Detectors

Capillary systems require detectors that have a low dead volume, high sensitivity, and fast response time. Among the commercial detectors, these criteria are met by the flame ionization (FID), alkali flame ionization (AFID), flame photometric (FPD), and thermionic nitrogen-phosphorus (NPD) detectors [8,30]. The latter three have the selectivity and sensitivity needed for residue determination of several classes of pesticides. There are few constraints on their geometry, allowing them to be interfaced directly or easily scaled down to capillary GC. These detectors can be flushed with components of the combustion mixture or the makeup gas, minimizing dead volume problems.

Other detectors have been more difficult to adapt. The electron capture detector is hindered by a minimum-volume requirement due to the energy of the β particles emitted by the foil [31]. However, by placing the column outlet at the base of the detector cell and efficiently flushing the cell with a purge gas, the response time can approach that of the other ionization detectors [32].

The interfacing of the capillary column and the mass spectrometer has proven to be a landmark achievement in residue analysis, making possible cost-effective survey analyses for all pesticides present in the sample [33-36]. This has been achieved through the parallel development of computerized data-reduction systems and the use of mass spectral libraries [37]. In addition, such techniques as single-ion monitoring (SIM) and chemical ionization (CI) make the mass spectrometer one of the most specific and sensitive detectors available. Though mass spectral data often cannot distinguish between isomeric structures, the capillary column frequently has the ability to resolve them. Therefore, when retention time and mass spectra are used together, many chemicals may be positively identified.

3. APPLICATIONS

3.1 Organochlorine Insecticides

Organochlorine insecticides (OCIs) demonstrate both the utility and the problems in the use of glass capillary GC. Mixtures of OCIs that can be difficult or impossible to separate with packed columns can be resolved, though pg amounts of compounds sensitive to adsorption or decomposition cannot be analyzed except on the most inert systems. For these reasons, the OCIs have been used by many investigators to demonstrate the adequacy of their coating and deactivation techniques, the utility of different surface treatments, detector and injector efficiencies, and the utility of such novel techniques as flow programming [38,39] and precolumn interfacing.

Franken and Rutten tried a variety of liquid phases and surface treatments for such analyses [29]. They statically coated several columns with SE-30 according to the methods of Bouch and Verzele (thickness 0.12 to 1.13 μm [40]). With an electron capture detector and sample concentrations in the ng and pg ranges, they determined that active sites were decomposing the aromatic OCIs and that, despite attempts at deactivation, this effect was minimized only with the use of thick (>0.5μm) liquid loadings.

Subsequently, these same authors [23] treated a variety of columns with different surface treatments, with the best results given by NaCl SCOT columns. In this method the drawn column was treated with a NaCl solution in a relatively simple procedure that may prove to be a good alternative to HCl etching—a method of column preparation which accomplishes a similar advantage in increased surface area and liquid-phase retention (Fig. 3).

Low-resolution columns were prepared by Einig and MacDonald by chemically bonding octadecyltrichlorosilane to a borosilicate glass column after the column had been cleaned, etched, and dried [41]. This column (20 m × 0.5 mm ID) gave resolutions similar to that of a mixed phase (10% DC-200/15% QF-1) packed column and was stable up to 300°C. Unidentified peaks were noted on the chromatograms, demonstrating the possibility of on-column degradation.

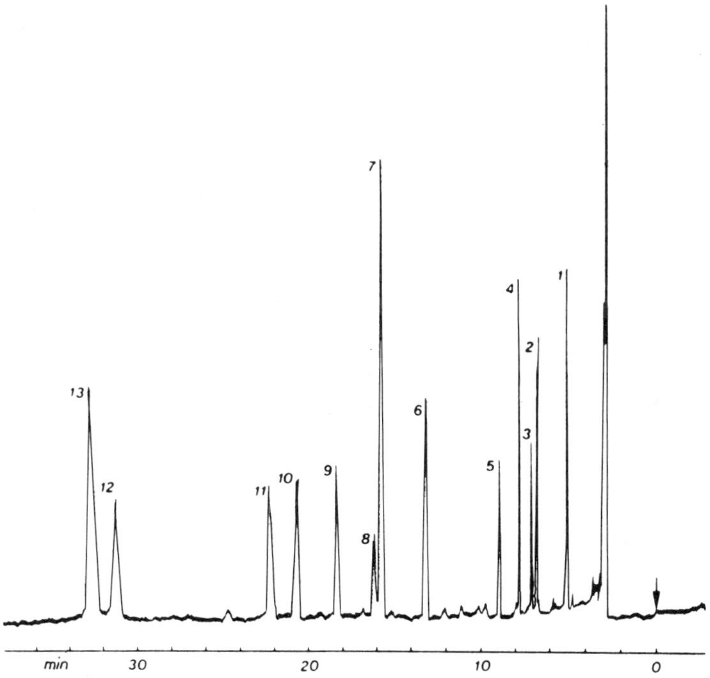

Figure 3. Separation of a mixture of chlorinated pesticides on an NaCl SCOT column. Peaks: 1—hexachlorobenzene; 2—2,2'-dibromobiphenyl; 3—α-BHC; 4—aldrin; 10—o,p'-DDT; 11—o,p'-DDD; 12—p,p'-DDT; 13—p,p'-DDD. (Source: Ref. 23.)

Whisker-walled open tubular columns (WWOT) were prepared to enable the handling of larger sample sizes (1-5 μg [25]). Wide-bore columns were prepared by Badings et al. in an effort to ease handling of the columns as well as to increase sample capacity [42]. These lower-resolution columns are easier to use in instruments designed for packed column, since they allow higher flowrates and do not suffer from the problems associated with detector and injector dead volumes. In addition, fractions can be collected for possible reinjection into high-resolution capillary columns. Polar support-coated open tubular columns (SCOT) were prepared by Thiecke et al. [43]. These columns proved to be stable under prolonged use at 220°C, though a small amount of catalytic decomposition of DDT to DDD was observed.

Highly efficient WCOT columns were prepared by Onuska and Comba by growing filamentary crystals in the column [44]. The column was suitable for such sensitive compounds as endrin and photodieldrin (Fig. 4). Endrin is a good compound for measuring a column's activity; de Nijs et al. showed that endrin would decompose on imperfectly deactivated columns coated with a methyl silicone liquid phase (Fig. 5 [45]). To prevent this decomposition, they recommended a final coating with the bleed from a Carbowax-packed precolumn. Schomburg demonstrated that columns prepared from borosilicate glass gave superior results to columns of alkali glass [46].

Flow programming has been suggested as an alternative to temperature programming [38] since changes in flowrates are masked in the detector by relatively large makeup gas flows. These authors used ECD, a temperature-sensitive detector, and varied the flowrate by changing the outlet pressure of the injector splitter. Although there is some loss in column efficiency, the detector response was not affected. This technique can also be effectively used with temperature programming (Fig. 6 [39]).

Kirsten et al. developed a two-dimensional GC system with a packed precolumn for analyzing OCIs in fish [20]. The crude extracts were injected on the precolumn and the analyte trapped in a cooled capillary trap, which was then warmed to elute the sample onto the capillary column. The important feature of this system is that it is all glass with no valves or other metal parts in contact with the sample (Fig. 7).

Spencer used a mixture of OCIs in hexane to demonstrate the utility of the splitless mode of a new injector system [47]. The initial temperature (100°C) condensed the pesticides at the head of the column. Once the solvent (hexane) had eluted, the column oven was temperature programmed to 160°C.

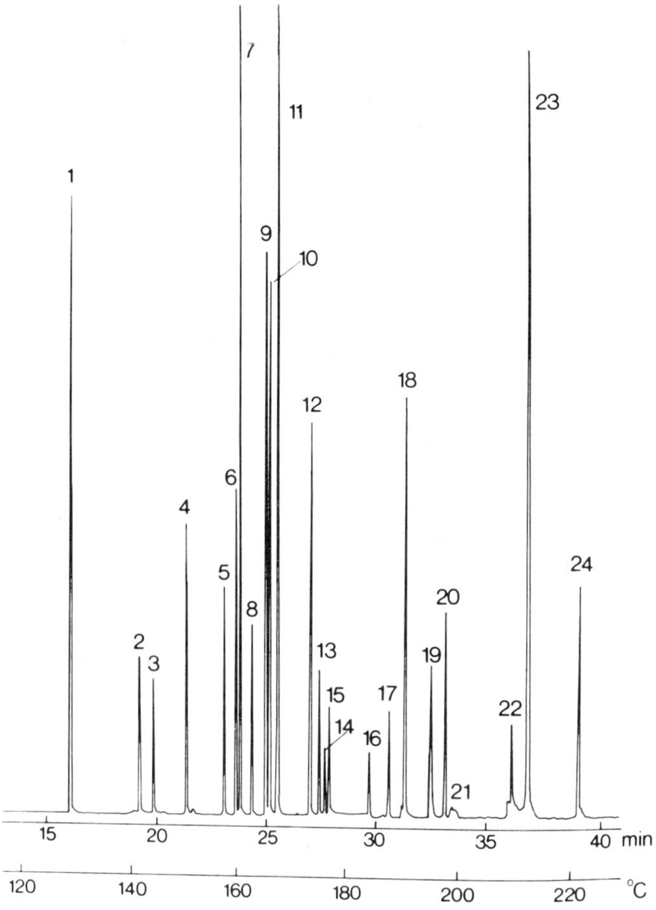

Figure 4. Chromatogram of a OCI mixture on the WCOT column prepared by Onuska and Comba [44]. Peaks: 1—chlordene; 2—heptachlor; 3—aldrin; 4—telodrin; 5—oxychlordane; 6—photoheptachlor; 7—photochlordene; 8—heptachlor epoxide; 9—trans-chlordane; 10—trans-nonachlor; 11—cis-chlordane; 12—dieldrin; 13—endrin; 14—dihydromirex; 15—p,p'-DDE; 16—monohydromirex; 17—kepone; 18—photoaldrin; 19—p,p'DDT; 20—mirex; 21—endrin aldehyde; 22—endrin ketone; 23—methoxychlor; 24—photodieldrin. (Source: Ref. 44.)

Mixtures of OCIs were used to compare the resolution of various WCOT and SCOT capillary columns and packed columns (Suzuki et al. [48]). Of the capillary columns, both of which were superior to the packed column, the SCOT column had the better resolution. However, the SCOT column promoted the decomposition of photodieldrin in the sample. Nevertheless, it was used successfully by the authors to analyze for other OCIs in river water and sediments and in soil from a spinach field.

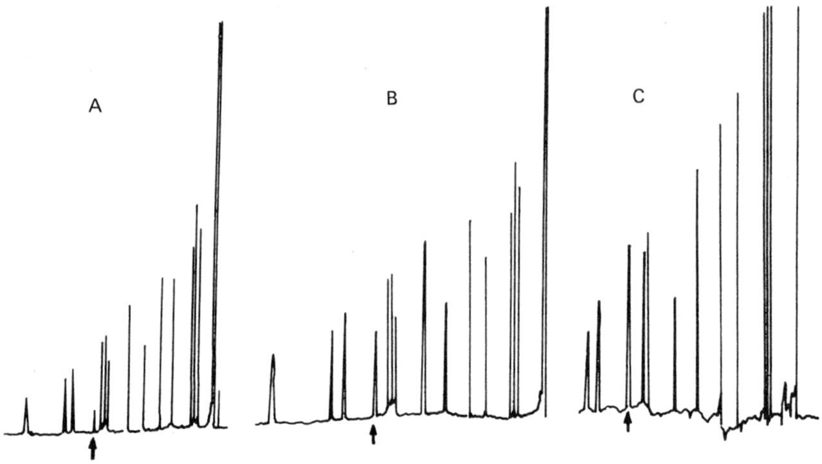

Figure 5. Separation of OCI mixture showing endrin adsorption [4] on deactivated borosilicate and alkali glass surfaces: (A) 20-m OV-1 (borosilicate), 180°C isothermal, 1 bar H_2 (carrier), 1.5 μl injection (0.1 μg μl^{-1} each compound), showing partial adsorption of endrin; (B) alkali glass column, same GC parameters, 2 μl injection (0.1 μg μl^{-1} each compound), showing no endrin adsorption; (C) same as B, 2.5 μl injection (10 pg μl^{-1} each compound) with ECD, showing no adsorption of endrin. (Source: Ref. 45.)

Brodtman tested a 50-m SE-30 WCOT column for analysis of pesticides in water [49]. He measured initial separation numbers for several OCIs and the loss of resolution after 3 months' use (12%). Despite the longer elution time compared to a 6-ft packed column, analysis time was actually reduced due to the avoidance of a cleanup step. Some degradation was observed, which increased with use. This was reduced by cleaning the injector.

Mirex, formerly used extensively in the southeastern United States to control fire ants, is an unusually stable compound. Investigations into its spread through the environment has included reports of low concentrations in fish from Lake Ontario. Glass capillary systems have been used by two authors to confirm these findings. An SE-52 capillary column was used by Hallet et al. to resolve mirex, 0-monohydromirex, and DDE in Coho salmon using FID detection [50]. This was done without the need for the separation and derivatization techniques usually employed, despite the presence of polychlorinated byphenyl (PCB) isomers and other organochlorine compounds. Laseter et al. showed by glass capillary GC/MS that the levels of mirex in Lake Ontario fish were 3 to 6 times lower than reported previously in the literature [51]. They demonstrated that use of the 272 m/e ion for low-

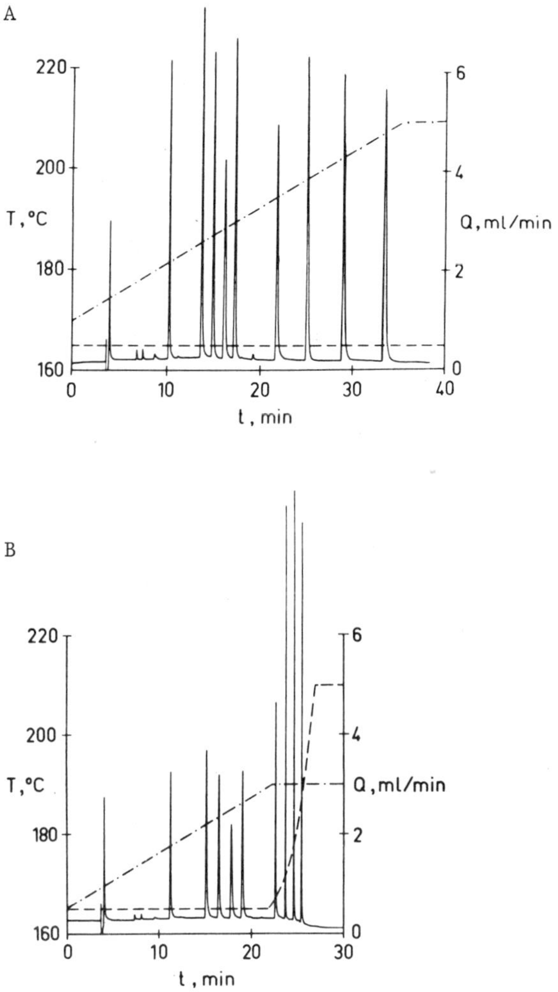

Figure 6. Chromatograms showing the utility of flow programming: (A) temperature programming only; (B) temperature programming with flow programming. (Source: Ref. 39.)

resolution SIM was subject to interferences through the use of alternate detectors (showing coelution of the PCBs with the mirex) and high-resolution MS (showing that only PCBs were adding to the interferences).

Holmstead used commercial glass capillary columns (15.5 m × 0.25 mm ID, WCOT, 220°C, ^{63}Ni ECD) to confirm 3 hematin reduction products from mirex [52]. This work demonstrated possible mirex degradation pathways under anaerobic conditions. Parlar et al. combined a glass capillary with ECD to show that technical chlordane was a mixture of at least 50 components [53]. The authors identified at least 6 different C_{10} skeletons.

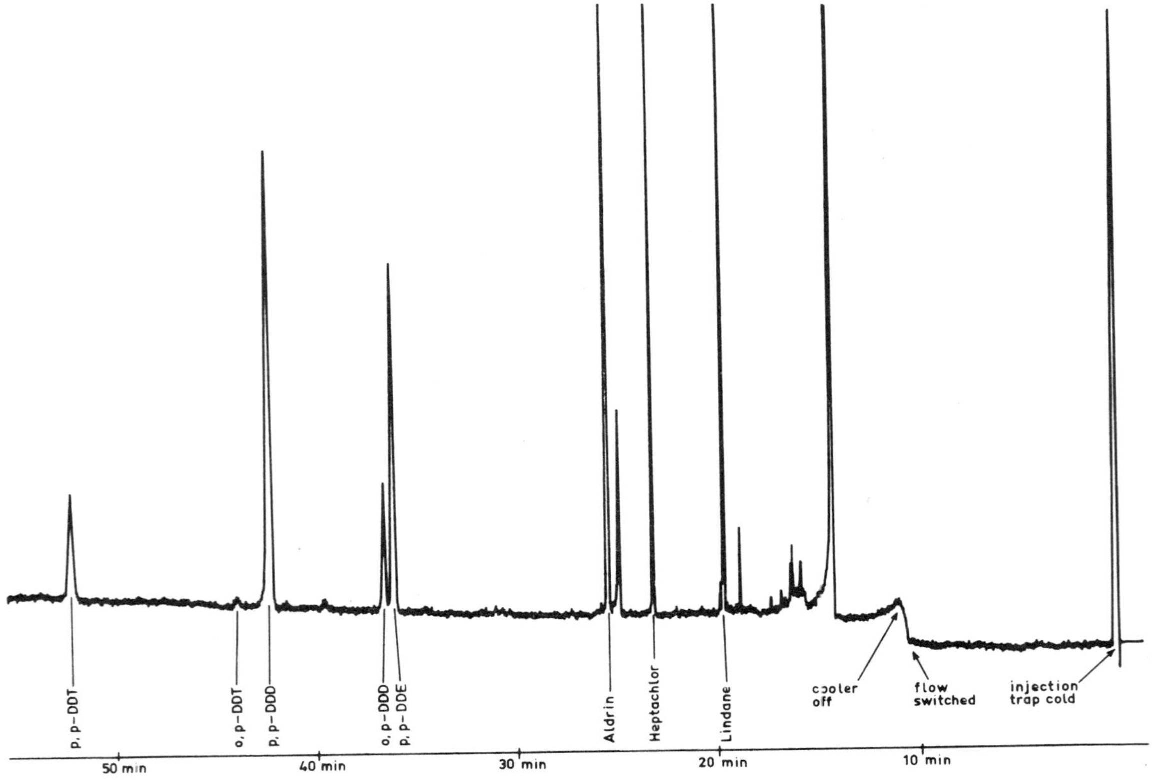

Figure 7. Analysis of fish extract fortified with an OCI mixture using a packed precolumn with a cold capillary trap interfaced with a capillary analytical column. (Source: Ref. 20.)

Toxaphene is a widely used contact insecticide whose determination is complicated by its multicomponent nature (Fig. 1 [54]). Seiber et al. used capillary GC/ECD to locate components in the toxaphene profile which had been isolated with liquid and preparative gas-liquid chromatography [55]. Saleh used similar conditions to compare the composition of different toxaphene formulations, interfacing an on-line computer to normalize peak areas and provide precise T_r values [56]. The same investigators monitored the reductive dechlorination of one toxaphene component in various chemical, photochemical, and metabolic systems [57].

Seiber et al. followed the environmental dissipation of toxaphene from a treated cotton field [58]. Peak-by-peak comparison of capillary chromatograms showed that no significant decomposition occurred in foliage, topsoil, or airborne residues, but that anaerobic decomposition in tilled soil or ditch sediment favored the formation of products of shorter retention times. The same authors employed HPLC fractionation preceding capillary GC analysis to follow the residual fate of toxicants A and B, two major toxaphene components, in topsoil samples [59].

An extracted ion current profile (EICP) was used to identify dieldrin in New Orleans drinking water [37]. In this procedure, a single peak or group of peaks was extracted from the total ion current profile (TICP), plotting its change in relative abundance as a function of time. This greatly enhances the sensitivity and selectivity of the detector toward certain compounds or classes of compounds. It is more flexible than single-ion monitoring, since different m/es can be extracted from the TICP after a run has been completed.

3.2 Organophosphorus Insecticides

Krijgsman and van de Camp used capillary systems to separate organophosphorous (OP) esters [60]. Using dynamically coated WCOT SE-30 columns and FPD detection, they reported the relative retention times of 59 OP pesticides in the isothermal and temperature programming modes.

Stan applied glass capillary GC/MS to the detection and identification of mixtures of up to 23 organophosphorus pesticides, with nearly complete separation [61,62]. He initially used CI-MS for identification of the pesticides and later added mass fragmentography (MF), which allowed detection limits in the low ppb range. This method proved more sensitive than electron impact (EI) mass fragmentography. The m/e fragments used in MS are listed in Table 1.

Table 1. Ions Used in Mass Fragmentographic Analysis of Organophosphorous Insecticides and Their Relative Intensities in the Mass Spectrum

Compound	MW	Ion(s)
Dichlorvos	220	221(100), 223(66)
Mevinphos	224	225(100)
Trichlorfon	256	111(100), 257(42), 259(40)
Demeton-S-methyl	230	231(100)
Dibrom	378	221(100)
Phorate	260	261(100)
Demeton	258	259(100)
Diazinon	304	305(100)
Dimethoate	229	230(100)
Phosphamidon	299	300(100)
Fenchlorphos	320	321(100), 323(98), 111(99)
Methyl parathion	263	264(100)
Malathion	330	331(100)
Fenthion	278	279(100)
Parathion	291	292(100)
Bromophos	364	367(100), 365(60), 111(73)
Demeton-S-methylsulfone	262	263(100)
Chlorofenvinphos	358	359(100), 361(88)
Ethyl bromophos	392	395(100), 393(60)
Methidathion	302	145(100), 303(50)
Carbophenothion	342	343(100)
Azinphosmethyl	317	160(100), 318(42)
Azinphosethyl	345	160(100), 346(32), 132(36)

Alkali flame thermionic detection has been used for OPs by two groups of investigators. DePotter et al., using a simplified extraction procedure and wide-bore columns (for fast analysis times), followed the serum levels of patients who had ingested commercial solutions of organophosphorus insecticides [63]. Useful quantitative results were attained to levels of 10 ng ml^{-1}. In the extraction procedure, serum was extracted with benzene, centrifuged, dried, the residue dissolved in ethyl acetate and injected on-column. Hild et al. [64] separated mixtures of OPs and their metabolites from spinach and resolved a mixture of parathion ring-subsitiuted isomers using a 25 m × 0.3 mm diethylene glycoladipate (DEGA) dynamically coated SCOT column (Fig. 8).

3.3 Herbicides

S-triazine herbicides were first subjected to glass capillary analysis by Matisova and Krupcik [65]. These workers used different coating methods (static vs. dynamic) and liquid phases for this analysis. Liquid phases, including OV-101, SE-30, and Carbowax 20M, were coated on soda lime capillary columns. The most suitable columns were coated with Carbowax 20M, with both the static and dynamic coating procedures providing adequate results (Fig. 9).

Hattula and Raeisanenen were able to analyze as little as 100 pg of the methyl ester of 2-methyl-4-chlorophenoxyacetic acid (MCPA) from water samples using FID detection and glass capillary columns [66]. Although the authors

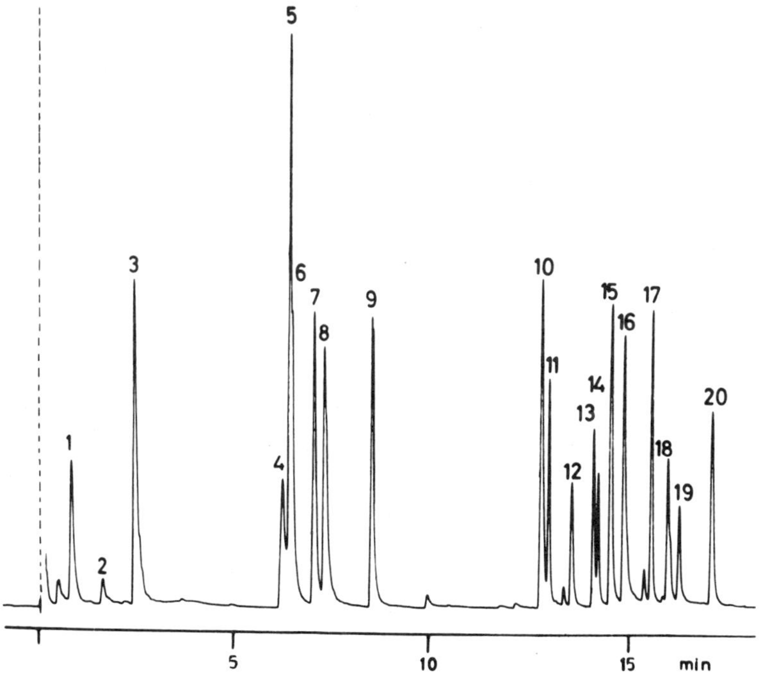

Figure 8. Extract from spinach fortified with organophosphrous pesticides and corresponding oxons. Peaks: 1—dimefox; 2—naled; 3—dichlorvos; 4—mevinphos; 5—sulfotepp; 6—thionazin; 7—fonofos; 8—dioxathion; 9—trichloronat; 10—bromophos; 11—malaoxon; 12—formothion; 13—methyl parathion; 14—malathion; 15—fenitrothion; 16—paraoxon; 17—parathion; 18—chlorofenvinphos; 19—tetrachlorvinphos; 20—ethion. (Source: Ref. 64.)

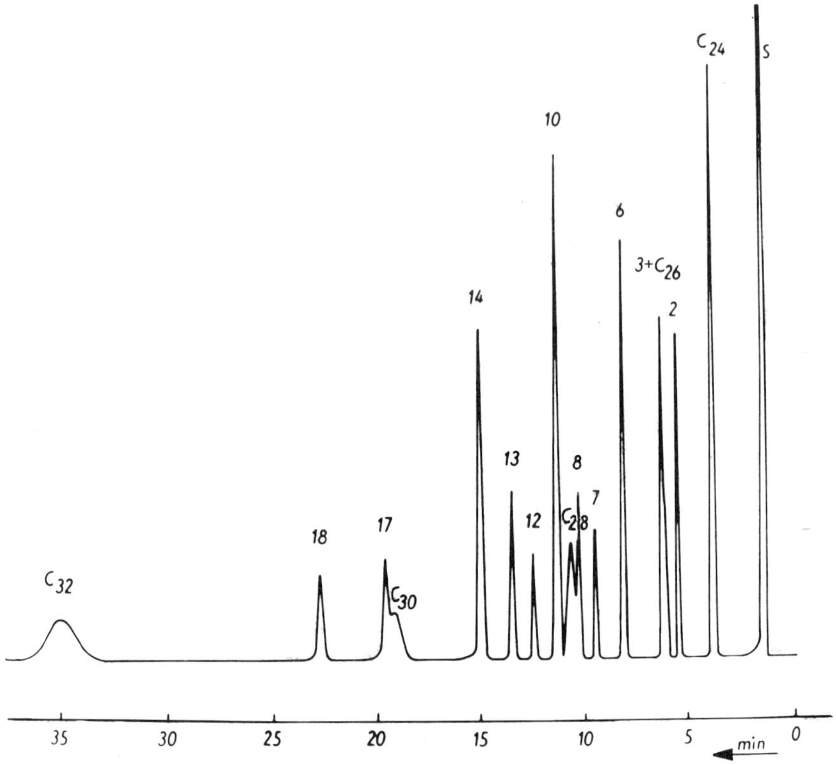

Figure 9. Separation of S-triazines and n-alkanes on statically coated Carbowax 20M. Peaks: 2—ipazine; 3—2-chloro-4,6-bis(tert-butylamino)-S-triazine; 6—terbuton; 7—propazine; 8—terbutylazine; 10—Aratan; 12—atrazine; 13—prometryn; 14—terbutryn; 17—desmetryn; 18—simetryn. (Source: Ref. 65.)

used a free fatty acid phase (FFAP) column, they recommended instead a nonpolar column such as SE-30 or OV-101. Methylation of the isoctyl ester with boron trifluoride reagents was also recommended. Using an extraction-derivatization scheme analagous to that for MCPA, 2,4-D was determined in water samples by the use of mass spectral matching following capillary GC resolution [67].

Dichloroaniline, a compound used in herbicide production, has been implicated as a cause of chloracne. Six dichloraniline isomers and related mono-, tri-, tetra-, and pentachloroaniline derivatives were resolved by capillary chromatography and identified by coinjection techniques.

3.4 Fungicides and Bacteriocides

Triphenyltin hydroxide (Duter) is used to control rice diseases in the United States. Wright et al. developed a method for determining Duter as well as some of its possible metabolic breakdown products as the methyl derivatives using a tin-selective FPD [68]. Methyl triphenyl-, dimethyl diphenyl-, and tetraphenyltin were resolved with no significant interference at levels to 10 ppb in water. Ethylenethiourea (ETU) is a trace contaminant in and a decomposition product of the ethylenebis (dithiocarbamate) (EBDC) fungicides [69]. This compound is acutely toxic and has carcinogenic and teratogenic properties. It can be determined by conventional GC with or without derivatization at nanogram levels. In either case, the analytical procedure is long and requires a column chromatography cleanup. With capillary gas chromatography, sensitivity in the picogram range was possible for the underivatized ETU and the column cleanup was elminiated. The method can also be used to quantify the parent EBDC following conversion to ETU. For example, maneb was decomposed in 75% yield to ETU in 2 hr at 90°C when dissolved in H_2O at pH 9.

A number of columns (OV-101, OV-17, Carbowax 20M, and FFAP prepared according to Grob and Grob [24]) have been used in the analytical determination of ETU at temperatures as high as 260°C. H_2 and ^{63}Ni ECD were utilized in these analyses. Sample preparation was performed on a number of matrices: formulations were simply extracted with ethyl acetate; rat urine was subjected to vacuum evaporation and dissolved in ethyl acetate; rat liver tissue was homogenized with MeOH, filtered, dried, and dissolved in ethyl acetate. For foodstuff residues, the dried MeOH extract was boiled in EtOH for 30 min to release bound residues, dried, and dissolved in ethyl acetate.

Chlorophenols are introduced into the environment as industrial effluents, herbicide breakdown products, and through direct use of fungicides. Many chlorophenols can be analyzed in a single run with the use of glass capillary columns. Krijgsman and van de Camp analyzed a mixture of chlorophenols using an SE-30 column (WCOT, 25 × 0.35 mm) with electron capture detection following acetylation [70]. The compounds were extracted from H_2O with toluene, shaken with K_2CO_3 solution, acetic anhydride was added, and the solution extracted with hexane. Recoveries were 80 to 100% for fortifications down to pg levels, and the minimum detection limit was 1 pg for pentachlorophenol (Fig. 10).

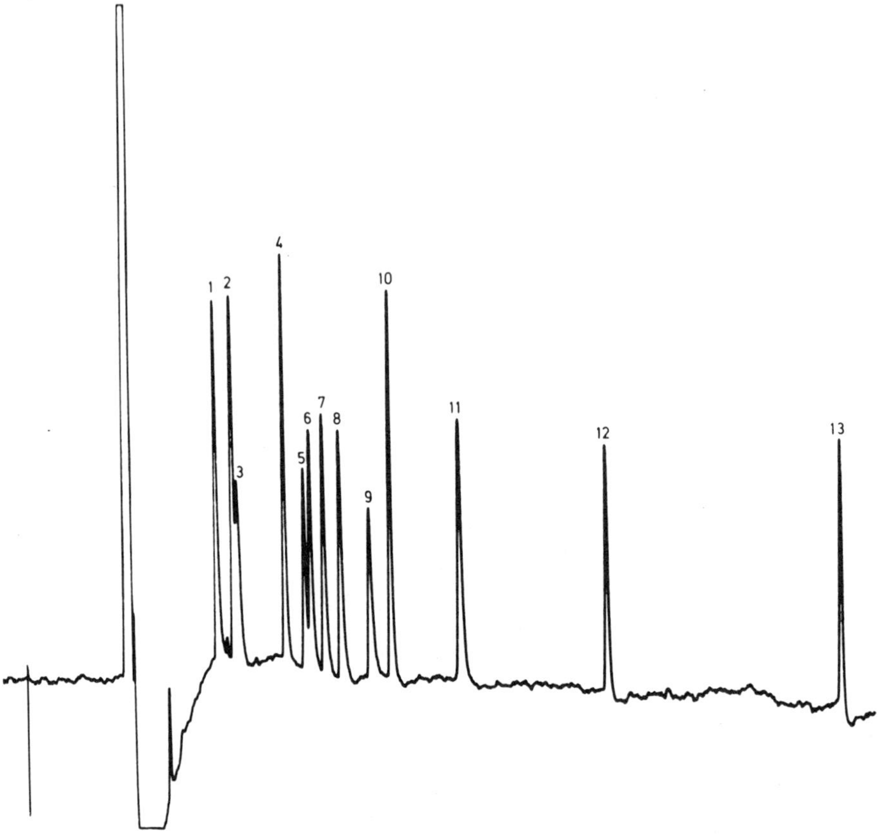

Figure 10. Separation of chlorophenol acetates with linear programming from 95 to 180°C at 3° min^{-1}. Peaks: 1—2-chlorophenol acetate; 2—3-chlorophenol acetate; 3—4-chlorophenol acetate; 4—2,6-dichlorophenol acetate; 5—2,5-dichlorophenol acetate; 6—2,4-dichlorophenol; 7—3,4-dichlorophenol acetate; 8—2,3-dichlorophenol acetate; 9—3,5-dichlorophenol acetate; 10—2,4,6-trichlorophenol acetate; 11—2,4,5-trichlorophenol acetate; 12—2,3,4,6-tetrachlorophenol acetate; 13—pentachlorophenol acetate. (Source: Ref. 70.)

Onuska and Comba (see OCI applications) monitored levels of pentachlorophenol (PCP) and its primary degradation product, 2,3,5,6-tetrachlorophenol (TCP), in paper mill effluent in a watershed. The farther the samples were taken from the plant, the smaller should have been the ratio of PCP to TCP. With packed columns, 2,3,5,6-TCP could not be separated from a formulation impurity, 2,3,4,6-TCP, thus preventing application of this method. The use of glass capillary GC resolved these two compounds, allowing investigators to determine the paper mill's influence on watershed residues, as well as pinpointing other sources of environmental contamination by PCP.

An extract of Quintozene, pentachloronitrobenzene fungicide, its formulation contaminant, pentachlorobenzene, and two plant and soil metabolites, pentachloroaniline and pentachlorothioanisole, were analyzed in lettuce extracts by Brechbuehler et al. [54] using capillary GC.

Polychlorinated dibenzo-p-dioxins (PCDD) are found as contaminants in clorinated phenols and some herbicide formulations, being produced at high temperatures under alkaline manufacturing conditions. Some isomers are extremely teratogenic, mutagenic, and/or acnegenic. While glass capillary GC would seem to be the obvious choice for analysis, surprisingly little work has been done in this area, perhaps due to the propensity of these chemicals to degrade at the high temperatures required for analysis.

Most of the work in glass capillary determinations of PCDDs has been done by Buser and co-workers with mass fragmentographic detection [71-76]. These studies have included the comparison of various liquid phases [72], preparation of PCDD standards by pyrolysis of potassium chlorophenates [71], the effects of ultraviolet and gamma irradiation on the octachloro-DD compounds [70], their identification in fly ash [74] and other industrial and environmental samples [69], and the use of MS to identify PCDD substitution patterns [73] (Fig. 11).

3.5 Pheromones and Insect Growth Regulators

Pheromones and insect growth regulators (IGRs) are under active development as pest control agents, as they may be effective in very small quantities and apparently have no adverse effects on humans or wildlife. Pheromones

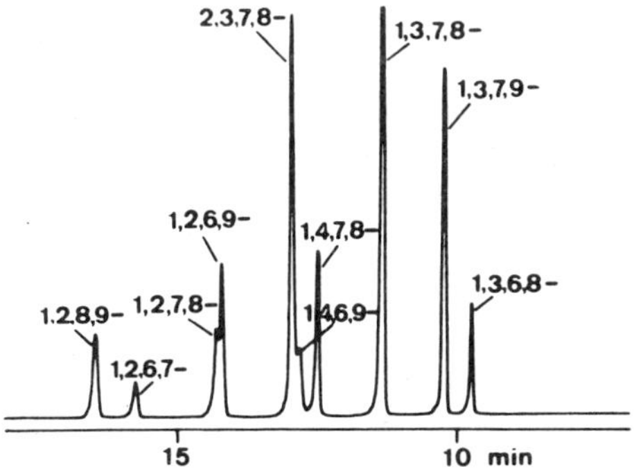

Figure 11. Mass fragmentogram (m/e 320) showing elution of 10 tetrachlorodibenzodioxin isomers with 2:2 chlorine substitution patterns prepared by microscale pyrolysis of different trichlorophenates; 50-m OV-17 glass capillary column, 200°C, isothermal. (Source: Ref. 76.)

elicit communication responses ranging from sexual excitation and oviposition to alarm and dispersion. The sex pheromones seem to offer the greatest promise in pest control, either as survey tools or in the suppression (by confusion) of emerging insect pest populations. Juvenile hormones (JHs), one class of IGRs, inhibit or prevent metamorphosis, causing the insects to die before reaching the adult stage. Both pheromones and IGRs are stereospecific natural products, making their separation and identification difficult—a problem suited for capillary chromatography.

The isolation and structure elucidation of sex pheromones is a difficult endeavor due to the minute quantities emitted from, or contained within, the female insect. A pheromone may also be a blend of closely related chemicals rather than a single compound.

Mass fragmentography (MF) detection used in conjunction with capillary GC retention times was the basis for the analysis of the codling moth and European grapevine moth pheromones [77]. Methods for sample cleanup and analysis were developed with the use of model synthetic pheromones. The samples were prepared by grinding 500 abdominal tips with diethyl ether and extracting them in an ultrasonic bath. After filtering through glass wool, the filtrates were concentrated and fractionated by column chromatography. Splitless injections up to 10 μl were made on a Ucon 50 HB 5100 glass capillary column, directly interfaced with the ion source. Complete mass spectra required approximately 200 ng (50 moth equivalents of the codling moth or 150 moth equivalents of the European grapevine moth). Once the spectra were characterized, detection was made by SIM which lowered the minimum sensitivity to below 200 pg. Glass capillary GC has also been used as confirmation in identifying the primary attractant pheromone of the Douglas fir tussock moth [78].

The degradation pathway of the IGR 6,7-epoxy-1-(p-ethylphenoxy)-3-ethyl-7-methylnonane (RO 10-3108) in water was followed by Dorn et al. [79]. The possible breakdown products were synthesized in the laboratory and used to confirm their identity in environmental samples by capillary cochromatography and GC/MS. Automated glass capillary radio gas chromatography (RGC) in conjunction with HPLC was used by Hamnett and Pratt to identify biosynthesized JHs and proximate precursors [80]. Conventional methods of analysis were not possible, as insects contain a nanogram or less of the compounds. By using higher-specific-activity labels, the authors were able to improve on the sensitivity. The RGC was done on a Pye 104 GC fitted with a Ucon

HB 5100 column (20 m WCOT). The injector was an all-glass solid injector operated by depositing the sample on a movable glass needle. The carrier gas back-flushed solvent to the atmosphere, and the sample was injected through an injector block at 200°C. FID detection was used, with precautions taken to avoid hot metal surfaces. The RGC system was an adaption of that used by Wels [81], removing the ^3H_2O trapping tube and using a single absorption fluid (5% 2-phenylethylamine in 2-ethoxyethanol) for trapping both ^3H and ^{14}C from the FID effluent. All-glass lined tubing was used. Average recoveries were 40% for ^3H and 80% for ^{14}C from double-labeled (^{14}C, ^3H) methyl palmitate standard. Two-ml fractions of effluent were collected automatically every 12 sec and analyzed in a scintillation spectrometer.

The JHs were biosynthesized in an organic culture medium and labeled with the high-specific-activity precursor, ^3H-methyl methionine. The glands were extracted, cleaned using TLC, and then analyzed by RGC. Cochromatography with standards was used for confirmation. This technique (cochromatography) used with radio HPLC serves as a further means of confirmation, or as an alternative method of identification when working with thermally labile JHs (Fig. 12).

Capillary GC/MS identification was also used by McGovern and Redfern to investigate the biological effects associated with changes in the terpenoid side chains of JH analogs [82].

4. CONCLUSIONS

Pest control chemicals span a wide range of structures, from simple halogenated alkanes such as methyl bromide to complex synthetic and natural products. As a consequence, their polarities, volatilities, and stabilities vary considerably, so that no single analytical technique or set of conditions is universally appropriate. It is noteworthy, as the examples in this chapter illustrate, that glass capillary GC has found many applications in the pesticide analytical literature. Nevertheless, many areas have remained virturally unexplored and further examples can be anticipated in the future. The availability of directions in the preparation of reliable columns—and in fact their commercial availability at reasonable cost—coupled with the developments in injector and detector adaptations tailored to the need of residue analysts should hasten progress in this exciting field. The problems of detector modifications and interfacing have been largely

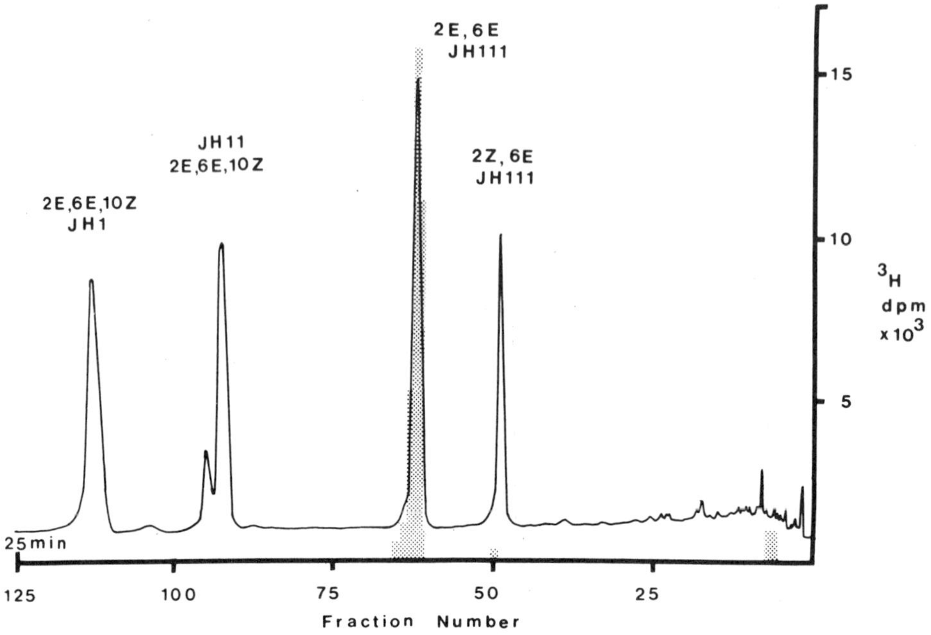

Figure 12. Radio GC of products synthesized by *Perinplaneta americana* incubated with (^{3}H-methyl)methionine ([80] 80 Ci/mmol). The solid line is the FID response and the shaded area the radioactivity of the FID combustion products. The sample is fortified with internal reference marker compounds. (Source: Ref. 80.)

solved, except in the case of electrochemical devices such as the Hall electrolytic and the microcoulometric detectors. These detectors would add a dimension of selectivity to capillary GC, but their slow response and recovery times must first be overcome.

The most immediate need of pesticide capillary GC is to adopt techniques, such as those demonstrated in many of the papers cited, to routine analysis. One way to hasten this adoption would be for authors to use a set of standard terms and a complete set of parameters when reporting their applications. Many references omit analytical parameters used in isolation, such as the percentage deactivation of a Florisil cleanup column or information on injector split ratios and detector interfaces associated with capillary chromatography. This information is of vital interest to any analyst attempting to quantitatively duplicate a procedure, and its omission contributes to the uncertainties regarding the potential of capillary GC among practitioners of residue analysis.

REFERENCES

1. Ware, G., *The Pesticide Book*, W. H. Freeman, San Francisco, 1978.
2. Brooks, G. T., in *Pesticide Chemistry in the 20th Century* (J. R. Plimmer, ed.), American Chemical Society, Washington, D.C., 1977.
3. Mulla, M. S., in *Pesticide Management and Insecticide Resistance* (D. L. Watson and A. W. A. Brown, eds.), Academic Press, New York, 1977.
4. Freed, V. H., and Haque, R., *Environmental Dynamics of Pesticides*, Plenum Press, New York, 1975.
5. Ford, J. H., McDaniel, C. A., White, F. C., Vest, R. E., and Roberts, R. E., *J. Chromatogr. Sci.*, 13:291 (1975).
6. Crosby, D. G., *An. Rev. Plant Physiol.*, 24:467 (1973).
7. Holmstead, R. L., Khalifa, S., and Casida, J. E., *J. Agr. Food Chem.*, 22:939 (1974).
8. Novotny, M., *Anal. Chem.*, 50:16A (1978).
9. Blau, K., and King, G. S., *Handbook of Derivatives for Chromatography*, Heyden, London, 1977.
10. Atallah, Y. H., Whitacre, D. M., and Polen, P. B., *Chemosphere*, 1:17 (1977).
11. Grob, K., *Chromatographia*, 8:423 (1975).
12. U.S. Department of Health, Education, and Welfare, *Pesticide Analytical Manual*, FDA/ACA-77/114A, 1977.
13. Bertsch, W., Anderson, E., and Holzer, G., *J. Chromatogr.*, 112:701 (1975).
14. Grob, K., and Grob, G., in *Identification and Analysis of Organic Pollutants in Water* (L. Keith, ed.), Ann Arbor Science, Ann Arbor, Mich., 1976.
15. Mattson, P. E., and Nygren, S., *J. Chromatogr.*, 124:265 (1976).
16. Grob, K., and Grob, G., *J. Chromatogr. Sci.*, 7:587 (1969).
17. Grob, K., and Grob, G., *J. Chromatogr. Sci.*, 7:584 (1969).
18. Buser, H. V., and Widmer, H. M., *J. High Res. Chromatogr. Chromatogr. Comm.*, 1:177 (1979).
19. de Leeuw, J. W., Maters, W. L., van der Meent, D., and Boon, J. J., *Anal. Chem.*, 49:1881 (1977).
20. Kirsten, W. J., Mattson, P. E., and Alfons, H., *Anal. Chem.*, 47:1974 (1975).
21. Bertsch, W., and Anderson, E., Proc. 1st Int. Symp. on Glass Capillary Chromatography, Hindelang Conference, Institute of Chromatography, p.9 (1977).
22. Onuska, F. I., and Comba, M. E., Proc. 1st Int. Symp. on Glass Capillary Chromatography, Hendelang Conference, Institute of Chromatography, p. 283 (1977).
23. Franken, J. J., Rutten, G. A. F. M., and Rijks, J. A., *J. Chromatogr.*, 126:117 (1976).

24. Grob, K., and Grob, G., *J. Chromatogr.*, 125:471 (1976).

25. Schieke, J. D., and Pretorius, V., *J. Chromatogr.*, 132:231 (1977).

26. Schieke, J. D., Comins, N. R., and Pretorius, V., *J. Chromatogr.*, 112: 97 (1975).

27. Schomberg, G., Husmann, H., and Weeke, F., *J. Chromatogr.*, 99:63 (1974).

28. Jennings, W., *Gas Chromatography with Glass Capillary Columns*, Academic Press, New York, 1978.

29. Franken, J. J., and Rutten, G. A. F. M., *Gas Chromatogr.*, 1972 (S. G. Perry ed.), Applied Science Publishers LTD., Essey, England.

30. Hartigan, M. J., Purcell, J. E., Navotny, M., McConnel, M. I., and Lee, M. L., *J. Chromatogr.*, 99:339 (1974).

31. Franken, J. J., and Vader, H. L., *Chromatographia*, 6:22 (1973).

32. Rejthar, L., and Tesarik, K., *J. Chromatogr.*, 131:404 (1977).

33. Christiansen, B., *Chromatographia*, 8:463 (1975).

34. Gudzinowicz, B. J., Gudzinowicz, M. J., and Martin, H. F., *Fundamentals of Integrated GC-MS*, Marcel Dekker, New York, 1977.

35. Sheldon, L. S., and Hites, R. A., *Environ. Sci. Technol.*, 12:1188 (1978).

36. Jungclaus, G. A., Lopez-Avila, V., and Hites, R. A., *Environ. Sci. Technol.*, 12:88 (1978).

37. Keith, L. H., *J. Chromatogr. Sci.*, 17:48 (1979).

38. Nygren, S., and Mattson, P. E., *J. Chromatogr.*, 123:101 (1976).

39. Nygren, S., *J. Chromatogr.*, 142:109 (1977).

40. Bouche, J., and Verzele, M., *J. Gas Chromatogr.*, 6:501 (1968).

41. Einig, R. G., and MacDonald, J. L., *Anal. Chem.*, 48:2281 (1976).

42. Badings, H. T., van der Pol, J. J. G., and Wassink, J. G., *Chromatographia*, 8:440 (1975).

43. Thiecke, R. J., Van den Berg, J. H. M., Deelder, R. S., and Ramaekers, J. J. M., *J. Chromatogr.*, 160:264 (1978).

44. Onuska, F. K., and Comba, M. E., *J. High Res. Chromatogr. Chromatogr. Comm.*, 1:209 (1978).

45. de Nijs, R. C. M., Franken, J. J., Dooper, R. P. M., Rijks, J. A., de Ruwe, H. J. J., and Schultig, F. L., *J. Chromatogr.*, 167:231 (1978).

46. Schomburg, G., Proc. 2nd Int. Symp. on Glass Capillary Chromatography, Hindelang Conference, Institute of Chromatography, p. 393 (1979).

47. Spencer, S. F., *Amer. Lab.*, p. 69 (October 1977).

48. Suzuki, M., Yamato, Y., and Watanabe, T., *Environ. Sci. Technol.*, 11:1109 (1977).

49. Brodtman, N. V., and Koffskey, W. E., *J. Chromatogr. Sci.*, 17:97 (1979).

50. Hallet, D. J., Norstrom, R. J., Onuska, F. I., and Comba, M. E., Proc. 1st Int. Symp. on Glass Capillary Chromatography, Hindelang Conference, Institute of Chromatography, p. 115 (1977).

51. Laseter, J. L., de Leon, I. R., and Remele, P. C., *Anal. Chem.*, 50:1169 (1978).

52. Holmstead, R. L., *J. Agr. Food Chem.*, 24:620 (1976).

53. Parlar, H., Hustert, K., Gaeb, S., and Korte, F., *J. Agr. Food Chem.*, 27:278 (1979).

54. Brechbuehler, B., Gay, L., and Jaeger, H., *Chromatographia*, 10:478 (1977).

55. Seiber, J. N., Landrum, P. F., Madden, S. C., Nugent, K. D., and Winterlin, W. L., *J. Chromatogr.*, 114:361 (1975).

56. Saleh, M. A., and Casida, J. E., *J. Agr. Food Chem.*, 25:63 (1977).

57. Saleh, M. A., and Casida, J. E., *J. Agr. Food Chem.*, 26:583 (1978).

58. Seiber, J. N., Madden, S. C., McChesney, M. M., and Winterlin, W. L., *J. Agr. Food Chem.*, 27:284 (1979).

59. Madden, S. C., Winterlin, W. L., Seiber, J. N., and Mast, T. J., Pittsburgh Conference, Paper #574, March 2, 1978.

60. Krijgsman, W., and van de Kamp, C. G., *J. Chromatogr.*, 117:201 (1976).

61. Stan, H. J., *Z. Lebensm.-Unters. Forsch.*, 164:153 (1977).

62. Stan, H. J., *Chromatographia*, 10:233 (1977).

63. DePotter, M., Mueller, R., and Willems, J., *Chromatographia*, 11:220 (1978).

64. Hild, J., Schulte, E., and Thier, H.-P., *Chromatographia*, 11:397 (1978).

65. Matisova, E., and Krupcik, J., *J. Chromatogr.*, 142:597 (1977).

66. Hattula, M. L., and Raeisanenen, S., *Bull. Environ. Contam. Toxicol.*, 16:355 (1976).

67. Versino, B., Knoeppel, H., de Groot, M., Peil, A., Poelman, J., Schauenburg, H., Vissers, H., and Geiss, F., *J. Chromatogr.*, 122:373 (1976).

68. Wright, B. W., Lee, M. L., and Booth, G. M., *J. High Res. Chromatogr. Chromatogr. Comm.*, 1:189 (1978).

69. Hirvi, T., Pyysalo, H., and Savolainen, K., *J. Agr. Food Chem.*, 27:194 (1979).

70. Krijgsman, W., and Van de Kamp, C. G., *J. Chromatogr.*, 131:412 (1977).

71. Buser, H. -R., *Anal. Chem.*, 49:918 (1977).

72. Buser, H. -R., *J. Chromatogr.*, 129:303 (1976).

73. Buser, H. -R., *J. Chromatogr.*, 114:95 (1975).

74. Buser, H. -R., *Anal. Chem.*, 48:1553 (1976).

75. Buser, H. -R., and Rappe, C., *Chemosphere*, 2:199 (1978).

76. Buser, H. -R., Bosshardt, H. -P., and Rappe, C., *Chemosphere*, 2:165 (1978).

77. Buser, H. -R., and Arn, H., *J. Chromatogr.*, 106:83 (1975).

78. Smith, L. M., Smith, R. G., Loehr, T. M., Daves, Jr., G. D., Daterman, G. E., and Wohleb, R. H., Paper 19, Symposium on High Resolution Chromatography, American Chemical Society Annual Meeting, August 28-September 2, 1977, Chicago, Illinois.

79. Dorn, S., Oesterhelt, G., Suchy, M., Trautman, K. H., and Wipf, H. K., *J. Agr. Food Chem.*, 24:637 (1976).

80. Hamnet, A. F., and Pratt, G. E., *J. Chromatogr.*, 158:387 (1978).

81. Wels, C. M., *J. Chromatogr.*, 142:459 (1977).

82. McGovern, T. P., and Redfern, R. E., *J. Agr. Food Chem.*, 27:82 (1979).

7

GLASS CAPILLARY GAS CHROMATOGRAPHY IN CLINICAL MEDICINE

CARL D. PFAFFENBERGER

University of Miami School of Medicine
Miami, Florida

1. Introduction 204
   1.1 Survey of Uses, 1974 to the Present 205
   1.2 The Concept of Metabolic Profiling 206

2. Urinary Steroid Metabolic Profiles 207
   2.1 Functions of Major Steroidal Hormones 212
   2.2 Obtaining and Using the Profiles 217
   2.3 Derivatives for Gas Chromatography of Steroids 221
   2.4 Representative Urinary Steroid Metabolic Profiles 223
   2.5 Hepatic Enzyme Imprinting Studies 233
   2.6 Breast Lesion Pathology Studies 244

3. Aldose and Alditol Metabolic Profiles 250
   3.1 Associated Clinical Pathology 250
   3.2 Derivatives for Gas Chromatography of Aldoses and Alditols 253
   3.3 Obtaining and Using the Profiles 259
   3.4 Representative Metabolic Profiles of Urinary Aldoses and Alditols 261
   3.5 Separation of Sugar Acids and Lactones 266
   3.6 Studies of Human Cataractous Lens Tissue 267

4. Additional Biomedical Applications 273
   4.1 Rat Fecal Steroid Studies 273
   4.2 Drugs and Drug Metabolites 273
   4.3 Neonatal Urinary Steroid Metabolic Profiles 276
   4.4 Plasma α-Tocopherol and Phosphatidylcholines 276

5. Concluding Remarks 279

References 279

1. INTRODUCTION

The use of glass capillary gas chromatography while investigating problems of a clinical nature is just beginning to accelerate. For that matter, use of any type of gas chromatography in clinical medicine is still relatively rare, and such studies are usually described by the more general term *biomedical applications*. Ultimately however, clinical use of the remarkable ability of glass capillary columns to resolve complex mixtures into their component parts will probably rank among the more significant utilizations of the entire field of glass capillary gas chromatography.

Presently several developments are leading clinical medicine closer to glass capillary gas chromatography laboratories: (1) greatly improved and more easily employed glass capillary gas chromatographs are now available; (2) increasingly more precoated glass capillary columns are being made commercially available; (3) more glass capillary gas chromatographers have entered the scientific community; (4) more investigators in clinical medicine are aware of just how powerful an analytical tool glass capillary gas chromatography is; and (5) an ever-increasing number of clinical laboratories are purchasing glass capillary gas chromatography/mass spectrometry analytical systems.

The overall reason that clinically based analyses have lagged behind other areas of application such as water and air pollution studies and petroleum product analyses etc. is probably related to the overriding fact that almost all clinical chemistry laboratories are headed by pathologists, individuals who generally receive little (if any) exposure to gas chromatography during their formal didactic training. Moreover, many gas chromatographers obtain degrees through analytical chemistry divisions of chemistry departments and by studying completely different curricula than do pathologists.

Presentations of bioanalytical studies at some of the more medically oriented symposia and inclusion of a biomedical applications section in the program of such events as the annual meeting Advances in Chromatography have served to bring these two groups of scientists with diversely differing backgrounds into closer proximity and somewhat more mutual understanding.

An indication of where the area of biomedical applications of gas chromatography now stands as well as the directions in which it is heading may be obtained from a summary of the program content of the past several sessions of Advances in Chromatography [1-6], an international symposium organized each year by Professor Albert Zlatkis of the University of Houston, Houston, Texas (Table 1).

Table 1. Number and Type of Gas Chromatographic Studies Presented by the Biomedical Applications Section at Advances in Chromatography during the Years 1973 to 1978

Number of biomedical investigations which ...	Meeting year					
	1973	1974	1975	1976	1977	1978
Were presented	13	11	16	17	14	8
Used packed columns[a]	11(8)	11(8)	9(7)	10(6)	7(4)	3(3)
Used metal capillaries[a]	1	0	3(2)	3(2)	1(1)	1(0)
Used glass capillaries[a]	1	0	4(3)	4(4)	6(3)	4(2)

[a] The number of studies which included structural confirmations by combined gas chromatography/mass spectrometry are given in parentheses. Source: Refs. 1,2,3,4,5, and 6.

It is thus readily apparent that the use of glass capillary gas chromatography in clinical medicine has been slower to materialize than in other areas of application, but Cram and Juvet did point out in a 1976 review [7] that there was a substantial increase in the number and the diversity of applications of glass capillary gas chromatography during 1974 and 1975. And over the past several years the number of analytical separation problems solved by capillary columns has continued to proliferate [8].

1.1 Survey of Uses, 1974 to the Present

Although many of the capillary separations performed between 1974 and the present have not been biomedical in nature, the methodologies and technological advances which the work represents created a firm analytical foundation upon which the more clinically based investigations were built. And during this period satisfactory separations of many biomedically related mixtures were achieved such as those of lactone trimethylsilyl ethers [9], aldoses and alditols [10-12], disaccharides [13], fatty acids [14-23], keto acids [24], fats and vegetable oils [25], nitrosamines [26], steroids and steroidal metabolites [27-41], tobacco smoke [42,43], physiological fluids [44], amines [45], amino acids and amino acid enantiomers [46-48], diastereomeric amino acids [49], polychlorinated dibenzo-p-dioxins and dibenzofurans [50], chlorinated pesticides [51,52], chlorophenols [53], organophosphorous pesticides [54], barbituates and other pharmaceuticals [55-59], marijuana constituents [60,61], and volatile constituents of human serum, urine, and cerebrospinal fluid [62]. This diverse list of capillary column gas chromatographic analyses represents the separation of samples ranging in complexity from two amino acids [47] to the thousands of components present in tobacco smoke [43].

Many of the above applications of biomedical gas chromatography are covered elsewhere in this book. This chapter is directed mainly toward some aspects of urinary steroid, sugar, and sugar alcohol analyses as well as to the overall concept and utility of metabolic profiling introduced by Horning and Horning [63] and extensively reviewed by Jellum et al. [64] in 1976.

1.2 The Concept of Metabolic Profiling

Kutsky [65] pointed out in 1973 that with the new discoveries in molecular biology, biochemistry, physiology, and endocrinology, the emphasis in research in life sciences is turning from descriptive biochemistry and electron microscopy to the functional, dynamic aspects of vital metabolic processes. Moreover, particularly great interest has been generated in the actual controlling agents which accurately regulate all of the cellular enzyme systems and organelles to produce a living cell and, from that, a living multicellular organism with many highly differentiated tissues. Steroidal hormones play a key role in this area in that they are controlling agents; and because many of them are chemically and structurally similar and exist in a constant state of physiological flux, they are prime candidates for study by metabolic profiling approaches. Glass capillary gas chromatographic columns, which now routinely possess theoretical plate efficiencies in the range of 100,000 to 150,000, have sufficient resolving power to afford baseline separation of the component parts of the complex mixture of steroidal metabolites normally present in human urine [10,11]. Again the complexity of the sample, viz., many components which are quite similar structurally, demand the use of high-resolution columns such as glass capillaries. Dietary and physiological fluctuations of these components can be readily determined by following an individual over any convenient investigative period of time. Many other biomedical applications of the metabolic profiling technique have appeared in the literature, including analyses of human blood serum and cerebrospinal, amniotic, and seminal fluids [66]. As mentioned earlier, an extensive review of this area has been published [64]. Jellum and colleagues [67] have published organic acid metabolic profiles of percutaneous biopsy samples of human hepatic, pancreatic, kidney, and muscle tissue (Figs. 1-4) as well as those of serum and sinovial fluid from a healthy control subject and a patient with rheumatoid arthritis (Figs. 5 and 6) [68].

Conceptually, metabolic profiling is simple; technologically, it can tax the imagination. One isolates a group of chemically and structurally similar substances, usually from a physiological fluid or tissue, converts the biological sample into an analytical sample, and performs an analysis by using a high-resolution technique. The biological sample may be a single "point-in-time" collection (a grab sample) or a more averaged one such as an aliquot of a 24-hr urine collection etc. The obvious utility of the profiling approach is its multicomponent nature and, in the case of grab sampling, its inherent characteristic of indicating the physiological level of each component at one point in time. (Many more metabolic profiles are included in the sections which follow.)

2. URINARY STEROID METABOLIC PROFILES

According to Kutsky [65], a hormone is a biologically active organic compound, a controlling agent essential for normal health and growth (its absence causing a deficiency disease or disorder), synthesized within the organism (human body) in ductless glands which release the agent in very small concentrations into the circulatory system to act on the target organs or tissues. Among the hormones are the lipid-soluble steroidal series (testosterone, cortisol, estradiol, progesterone, and aldosterone) which seems (1) to act by virtue of lipid solubility to stabilize and augment cell membrane permeability; (2) to regulate enzyme activity and membrane polarization; (3) to regulate redox potential; and (4) to affect ribonucleic acid transcription in the nucleus (aldosterone, estradiol, and testosterone).

If one considers the physiological forms, active analogs, and related compounds of each of the five major steroidal hormones listed above and studies the physiological functions and associated deficiency symptoms, diseases, and disorders as well as the effects of overdosage or excessive production (hyperplasia) of these compounds, one readily appreciates why metabolic profiling has been and will continue to be extensively applied to studies of them and their corresponding urinary metabolites.

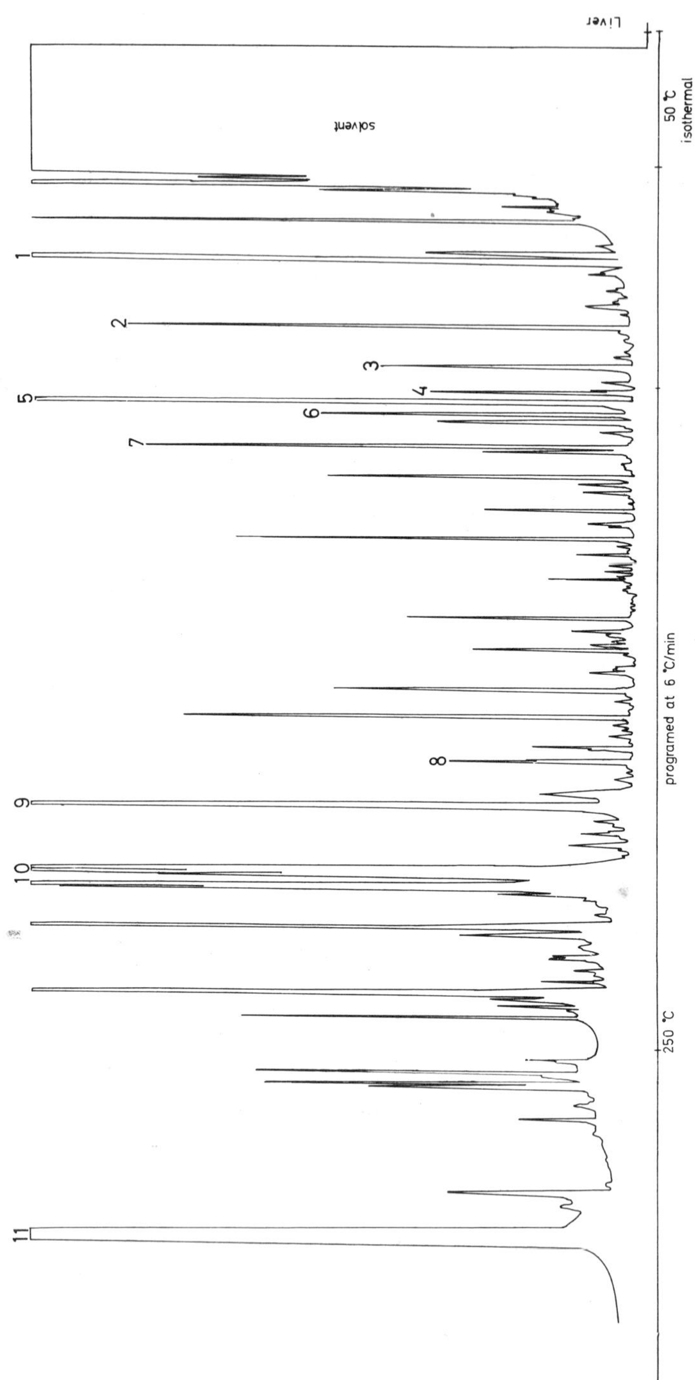

Figure 1. Organic acid profile of human liver. The biopsy sample was taken from a patient undergoing surgery in connection with a kidney transplant. Sample work-up and separation on a 25-m glass open tubular capillary column in a combined GC/MS instrument are described in Ref. 67. Identified components include: 1—lactic acid, 2—3-hydroxybutyric acid, 3—benzoic acid, 4—glyoxylic acid, 5—phosphoric acid, 6—glycerol, 7—fumaric acid, 8—phthalate plasticizer, 9—palmitic acid, 10—C_{18} fatty acids (steric, linoleic), and 11—cholesterol. (Source: Ref. 67. Reproduced courtesy of Elsevier Scientific Publishing Co., Amsterdam, and the major author.)

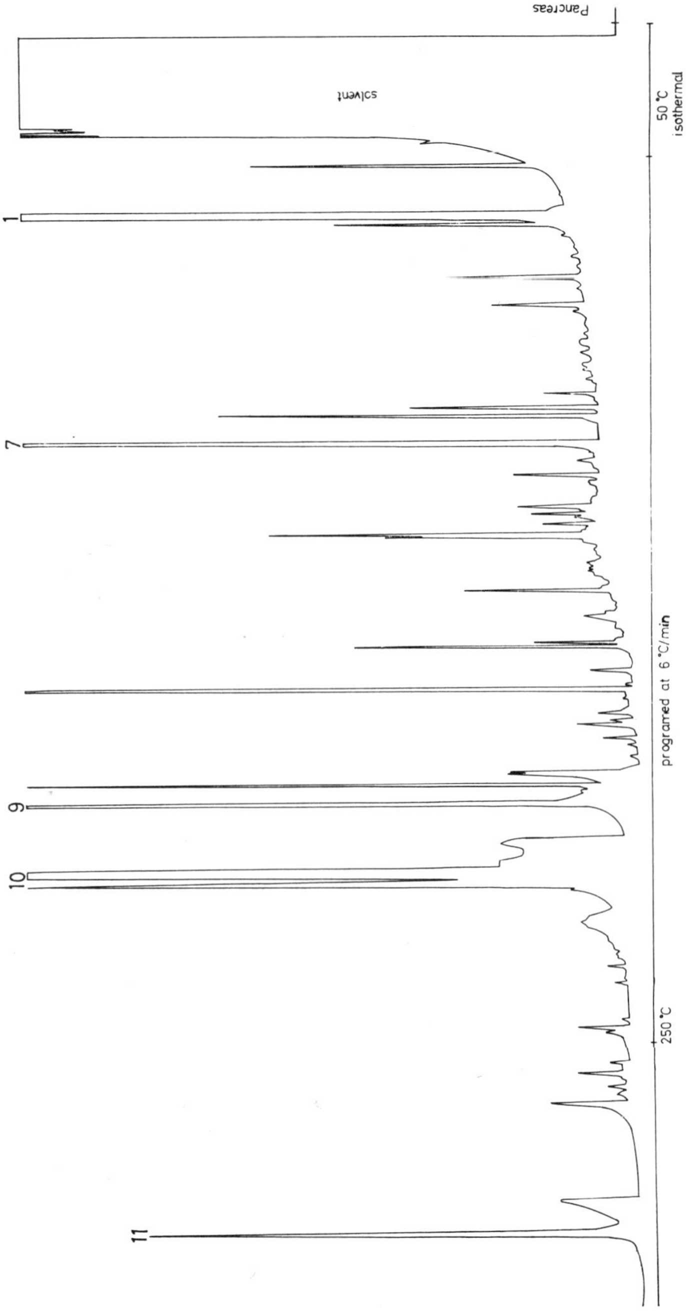

Figure 2. Organic acid profile of human pancreas. Patient, experimental conditions, and numbering for the identified components are as in Fig. 1. (Source: Ref. 67. Reproduced courtesy of Elsevier Scientific Publishing Co., Amsterdam, and the major author.)

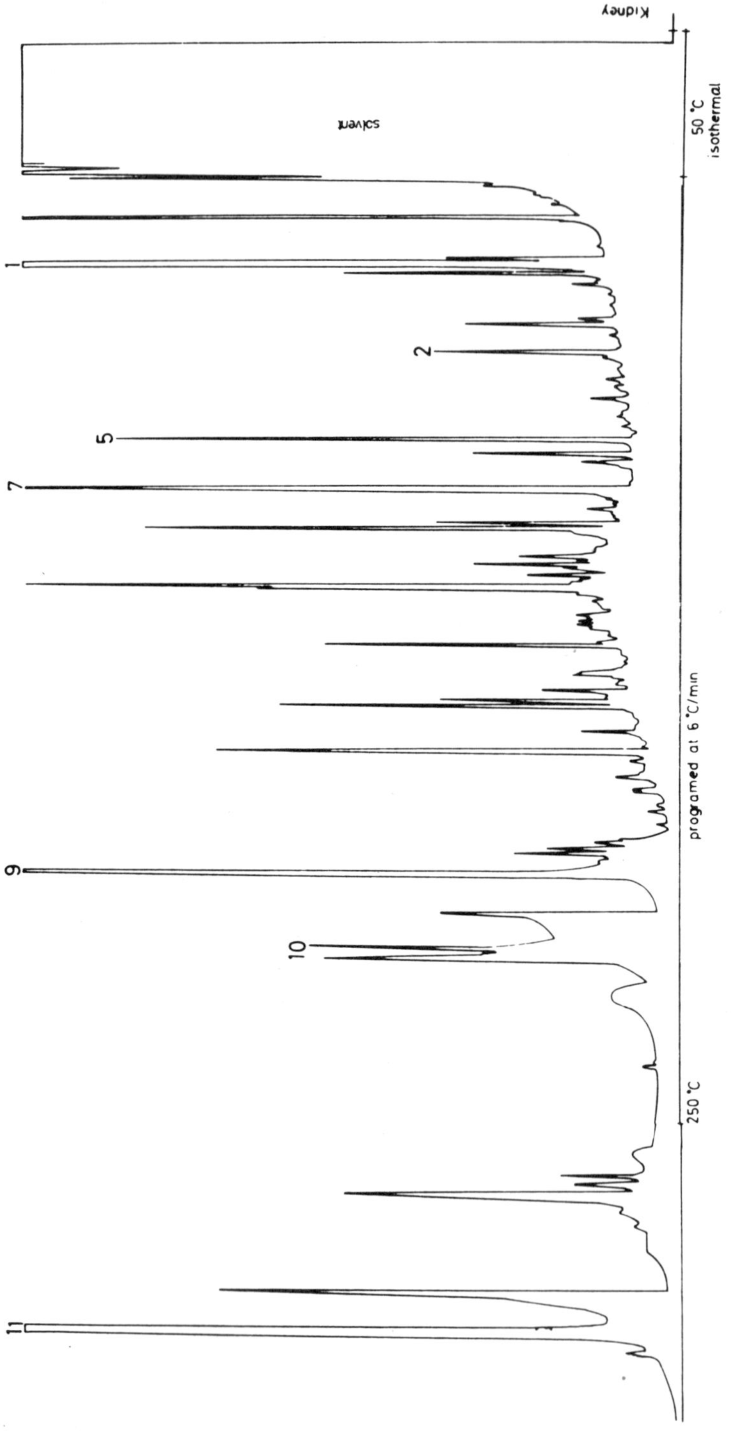

Figure 3. Organic acid profile of human kidney. The biopsy sample was taken from a patient undergoing abdominal surgery. Experimental conditions and peak identifications are as in Fig. 1. (Source: Ref. 67. Reproduced courtesy of Elsevier Scientific Publishing Co., Amsterdam, and the major author.)

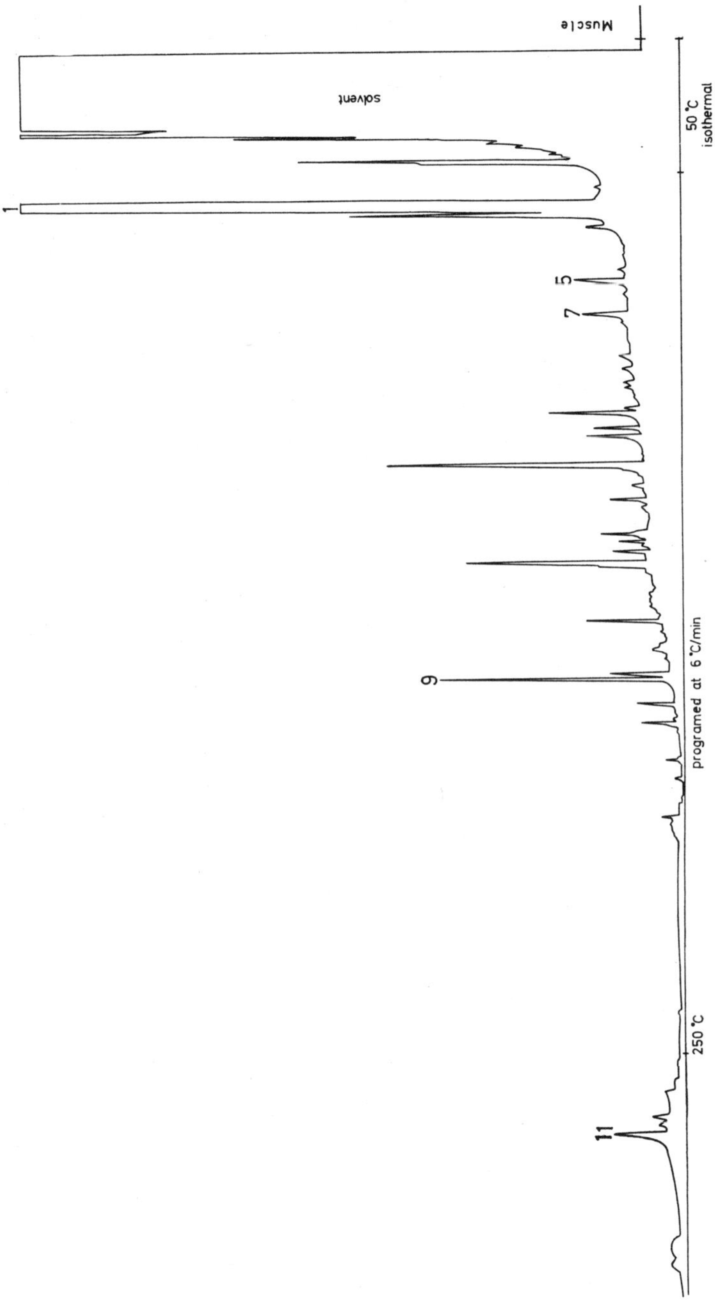

Figure 4. Organic acid profile of human muscle. Experimental conditions and peak identifications are as in Fig. 1. (Source: Ref. 67. Reproduced courtesy of Elsevier Scientific Publishing Co., Amsterdam, and the major author.)

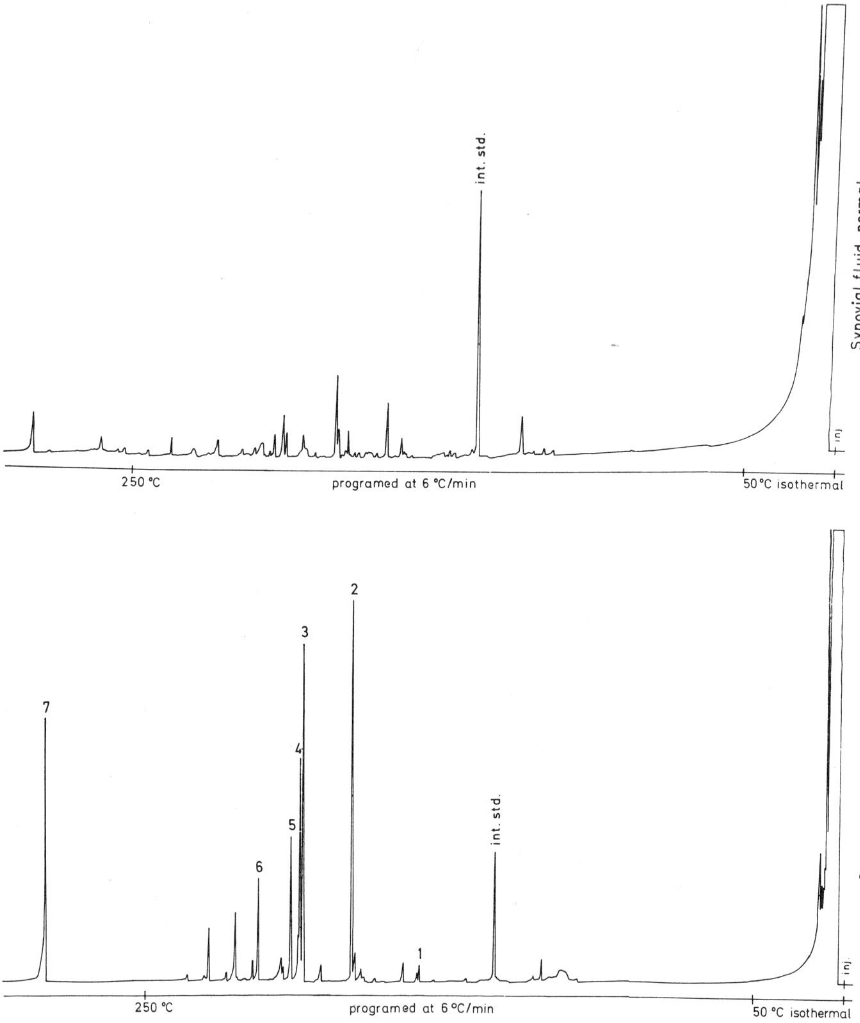

Figure 5. Organic acid profile of sinovial fluid (top) and blood plasma (bottom) from a healthy control subject. The experimental conditions are described in Ref. 68. Identified components include: 1—myristic acid, 2—palmitic acid, 3—linoleic acid, 4—oleic acid, 5—stearic acid, 6—arachidonic acids, and 7—cholesterol. Internal standard:nonanedioic acid. (Source: Ref. 68. Reproduced courtesy of Elsevier Scientific Publishing Co., Amsterdam, and the major author.)

2.1 Functions of Major Steroidal Hormones

Testosterone [69-76]

The average plasma level of testosterone (synonyms: 17B-hydroxy-4-androsten-3-one, Δ^4-androsten-17β-ol-3-one) is 0.60 μg 100 ml^{-1} of plasma in healthy males and 0.05 μg 100 ml^{-1} in healthy females. Active analogs and

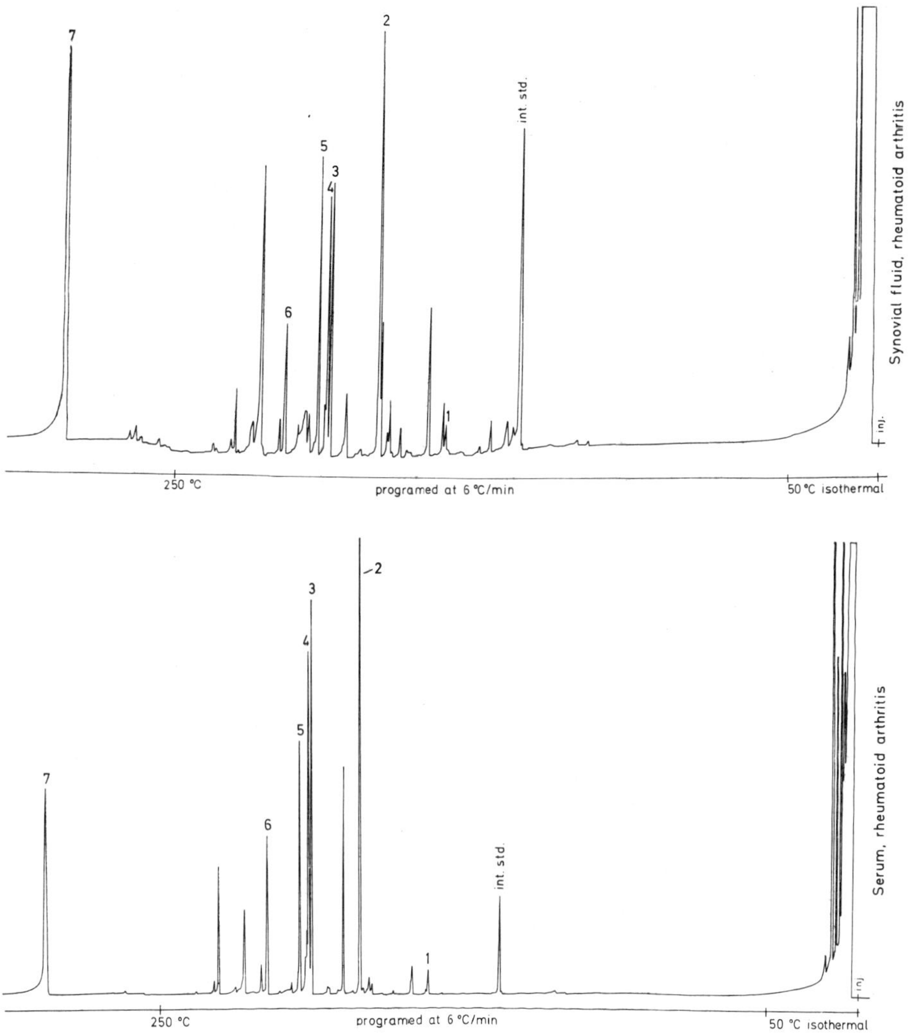

Figure 6. Organic acid profile of sinovial fluid (top) and blood serum (bottom) from a patient with rheumatoid arthritis. Experimental conditions and peak identifications are as in Fig. 5. (Source: Ref. 68. Reproduced courtesy of Elsevier Scientific Publishing Co., Amsterdam, and the major author.)

related compounds (androgens) include natural substances like 11-hydroxyandrosterone, adrenosterone, 7-hydroxyprogesterone, 11-keto-androsterone, and 11β-hydroxyandrostenedione and synthetics like ethynyl testosterone, methyltestosterone, 6α-chlorotestosterone, 19-nortestosterone, and 17α-ethyl-19-nortestosterone. The physiological forms include androstenedione, testosterone, 11β-hydroxyandrostenedione and adrenosterone. Inactive analogs and related compounds (for androgenic activity) include cortisone, cortisol, 17-hydroxyprogesterone, lumiandrosterone, 17α-hydroxy-11-desoxycorticosterone, 17β-methyl epitestosterone, and etiocholanolone.

Normal plasma levels (including fluctuations) of testosterone are sufficient to: control secondary male sex characteristics, maintain functional competence of male reproductive ducts and glands, increase protein anabolism, maintain spermatogenesis, inhibit gonadotropin, increase male sexual behavior, and increase the closure of epiphyseal plates. A deficiency of testosterone can lead to one or more of these symptoms: involution of accessory organs such as the prostate or seminal vesicles, decreased male behavior patterns and libido, decreased secondary sex traits, poor muscle development and function, delayed closure of epiphyses, and decreased urinary excretion of 17-keto-steroids.

Sustained testosterone deficiency can result in male hypogonadism, eunuchoidism, feminizing testes, and hyperplasia of the adrenals and the testes. Excessive amounts of testosterone can cause increases in libido, virilization, acne, increased lipid catabolism, increased androgen and estrogen excretion (17-keto-steroids), precocious sex development, hypertropy of accessory sex organs, increased skeletal growth until the epiphyses close, increased muscle mass, hirsutism, and decreased scalp hair growth.

Cortisol [69-76]

The average blood level of cortisol (synonyms: hydrocortisone, compound F, 17-hydroxycorticosterone, substance M) is 10 μg 100 ml^{-1} whole blood. Active analogs and related compounds (glucocorticoids) include natural substances like deoxycorticosterone, cortexolone, cortisone E, corticosterone B, and dehydrocorticosterone A and synthetics like dexamethasone, 9α-F-cortisol, prednisone, and prednisolone. The physiological forms are cortisol and cortisone. Inactive analogs and related compounds include estrone, progesterone, 17-α-hydroxyprogesterone, cortexolone, and adrenosterone.

Normal plasma levels (including fluctuations) are sufficient to increase all of the following: protein catabolism, gluconeogensis, carbohydrate anabolism (liver), blood sugar, glucose absorption, brain excitation, spread of infections, urinary glucose and nitrogen, stress tolerance, lactation, and water diuresis. Cortisol decreases fat anabolism, growth rate, inflammation, eosinophils, lymphocytes, antigen sensitivity, respiratory quotient, ketosis, wound healing, skin pigmentation, and red blood cell hemolysis; it regulates general adaptation syndrome, water balance, blood pressure, and hormone release. Cortisol deficiency symptoms include decreases in growth, secondary sex characteristics, blood pressure, body temperature, kidney function (leading to death), liver glycogen, gluconeogenesis, intestinal absorption, blood sugar, and stress response (ultimately death). Deficiency also leads to increased glucose oxidation, adrenocorticotropic hormone levels, respiratory quotient, fat anabolism, hemoconcentration, muscular weakness, skin pigmentation, and insulin sensitivity.

Sustained cortisol deficiency can result in Addison's disease, adrenal insufficiency, adrenogenital syndrome, rheumatic arthritis, and inflammation. Excessive amounts of cortisol can cause Cushing's syndrome, buffalo obesity, bruisability, moonface, osteoporosis, adrenal regression, anesthesia, atherosclerosis, hypercholesterolemia, lipemia, diabetes, alkalosis, decreased growth, and inhibition of inflammatory responses and wound healing.

Estradiol [69-76]

The average blood level of estradiol (synonyms: estradiol-17β, female hormone, β-estradiol, dihydrotheelin, dihydrofollicular hormone, dihydrofolliculin) is 0.008 μg 100 ml^{-1} in healthy males and 0.2 μg 100 ml^{-1} or less in healthy females. Pregnant females have average blood levels of 2.1 μg 100 ml^{-1} estradiol, 6.5 μg 100 ml^{-1} estrone, and 10.9 μg 100 ml^{-1} estriol, the actual levels varying with the stage of pregnancy. Active analogs and related compounds (estrogens) include natural substances like estriol, estrone, and equilin and synthetics like diethylstilbesterol, hexestrol, dienestrol, benzestrol, ethinylestradiol, and chlorotrianisene. The three physiological forms are estrone, estriol, and estradiol. Inactive analogs and related compounds (for estrogen function) include pregnanediol, lumiestrone, progesterone, 17α-hydroxyprogesterone, and 17α-hydroxypregnenolone.

Normal blood levels (including fluctuations) of estradiol are sufficient to: regulate the menstrual cycle and female sexual behavior, maintain secondary sex characteristics, affect antibody properties, induce estrus, bring about uterine hypertrophy and vaginal cornification, and potentiate and stimulate calcitonin secretion. A deficiency of estradiol can lead to delayed maturation, regression of female accessory and reproductive organs, decreased female behavioral patterns, senescence, and menopause.

Sustained estradiol deficiency can result in menopause, gonadal dysgenesis, and delayed maturation. Excessive amounts of estradiol can cause tumors and inhibition of the gonads (decrease in follicle-stimulating and luteinizing hormones) which can lead to permanent sterility.

Progesterone [69-76]

The average blood level of progesterone (synonyms: progestin, luteosterone, corpus luteum hormone) is 0.03 μg 100 ml^{-1} plasma in healthy males and 0.1 to 0.3 μg 100 ml^{-1} in healthy females. Pregnant females have levels between 10 and 28 μg 100 ml^{-1} blood plasma. Active analogs and related compounds (progestins, progestogens) include natural substances like 20α-hydroxypregnenone, 20β-hydroxypregnenone, 11-dehydroprogesterone, cortexone, and 17α-hydroxyprogesterone and synthetics like A-norprogesterone, 19-norprogesterone, 21-norprogesterone, ethisterone, 17α-methyltestosterone, 6α-methyl-17α-acetoxyprogesterone, and Δ^1-dehydro-6α-methyl-17α-acetoxyprogesterone. Inactive analogs and related compounds include 5α-pregnane-3β-ol-20-one, 21-ethylprogesterone, 11β-hydroxyprogesterone, pregnanediol, and pregnanetriol.

Low plasma levels of progesterone are sufficient to prepare the uterus for blastocyst implantation and promote ovulation and mammary gland development as well as regulate female sex accessory organs, function with weak corticosteroid properties, and serve as a precursor to sex hormones. Higher plasma levels are required to maintain pregnancy, repress ovulation and sexual activity, inhibit vaginal cornification and parturition, and decrease myometrial excitation. A deficiency of progesterone can lead to termination of pregnancy, a decreased production of steroids, decreased ovulation, loss of normal menstrual cycle changes, and decreased development for embryonic implantation and gestation.

Sustained progesterone deficiency can result in acne and dysfunctional uterine bleeding and pseudopregnancy in laboratory animals. Excessive amounts can cause progestational changes, pregnancy prolongation, inhibition of uterine growth, and increased sodium and potassium ion excretion.

Aldosterone [69-76]

The average blood level of aldosterone (synonyms: electrocortin, mineralocorticoid, aldocortin, 18-oxocorticosterone) is 0.03 μg 100 ml^{-1} in healthy individuals. Active analogs and related compounds (mineralocorticoids) include natural substances like 11-deoxycorticosterone and corticosterone and synthetics like 2α-methylcortisol, 9α-fluorocortisol, and 2α-methyl-9α-fluorcortisol. The physiological forms include the 11,18-semiacetal form as well as the true aldehyde. Inactive analogs and related compounds (for aldosterone function) include cortexolone, dexamethasone, and prednisolone.

Normal blood levels of aldosterone are sufficient to maintain the normal electrolyte balance of blood, prolong survival of adrenalectomized animals, accelerate gluconeogenesis, and regulate kidney function. A deficiency of aldosterone can lead to one or more of these symptoms: decreased blood pressure, blood sugar and pH, decreased weight, liver glycogen, and urinary potassium ion, decreased body temperature and reproductive function; increased urinary sodium, chloride, and bicarbonate ions; and kidney failure, muscular weakness, gastrointestinal disturbances, hemoconcentration, stress intolerance, and acidosis.

Sustained aldosterone deficiency can result in Addison's disease, adrenocortical insufficiency, and primary hyperaldosteronism. Excessive amounts of aldosterone can cause Cushing's syndrome, hypertension, congestive heart failure, increased sodium ion and water in blood and muscle tissue, hemodilution, hypokalemia, edema, alkalosis, and a type of diabetes insipidus.

Brooks and colleagues [77-83] have published extensively on the use of cyclic alkaneboronates for the gas-phase characterization of various corticosteroids and the several antomeric forms of aldosterone.

2.2 Obtaining and Using the Profiles

With the preceding review of steroidal hormones and steroidal metabolites we can now begin a step-by-step description of how urinary steroid metabolic profiles such as the one given as Fig. 7 are obtained.

Isolation of "Steroid Profile" Sample by Enzymic Hydrolysis

Urine Sample. A 24-hr urine sample is collected in a sterile container. The volume is recorded, and if the collection is less than 2 liters it is diluted to 2 liters using deionized water which has been glass-distilled from basic potassium permanganate. If not immediately used, the urine is stored at -14°C. Analysis should be made within 72 hr after collection. A creatinine value is obtained from an aliquot by any accepted clinical chemistry procedure.

Enzymic Hydrolysis. Sodium acetate trihydrate (1.0 g) is added to 50 ml of diluted urine in a sterile 125-ml screw-capped Erlenmeyer flask. The pH is adjusted to 4.5 using acetic acid, and 0.5 ml of enzymic solution (Glusulase, Endo Labs., Garden City, N.Y.; 1 ml contains 100,000 units of β-glucuronidase and 50,000 units of sulfatase) is added. After incubation at 37°C with gentle motion for 24 hr, the pH is adjusted to 5.5 using a few drops of 40% aqueous potassium hydroxide and a second 0.5-ml portion of Glusulase is added. The incubation is continued for a second 24-hr period

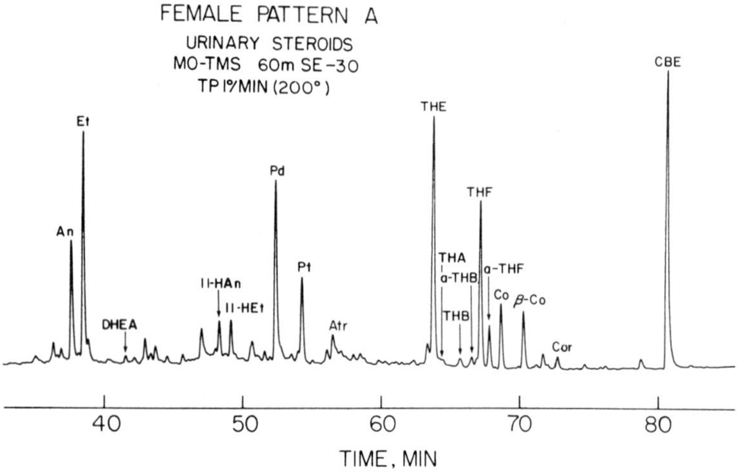

Figure 7. Urinary steroid metabolic profile of a premenopausal female showing pattern A ratios of etiocholanolone to androsterone (Et/An) and tetrahydrocortisol to allo-tetrahydrocortisol (THF/a-THF) of about 2:1 and 3:1, respectively. The separation of the methoxime-trimethylsilyl (MO-TMS) ether and trimethylsilyl (TMS) ether derivatives was performed on a 60-m glass open tubular capillary column temperature programmed 1°C min^{-1} from 200°C. All of the usual components are present; see the text for an explanation of the abbreviations. (Source: Ref. 36. Reproduced courtesy of Academic Press, Inc.)

at 37°C. If the hydrolysate cannot be extracted immediately, the flask is stored at 7°C.

Extraction of Steroids. Into a 1-liter rotary extractor are placed 75 ml of dichloromethane and 100 µg of an internal reference compound, obtained from a stock solution (1 mg ml^{-1}) of cholesteryl butyl ether (Sigma, St. Louis, Mo.) in pyridine. The hydrolysate is then added. The extraction step requires 10 min of the swirling action of the rotary extractor. Another extraction with 75 ml of dicloromethane and a final extraction with 75 ml of ethyl acetate (all solvents are Nanograde from Mallinckrodt, St. Louis, Mo.) leads to a combined organic phase of 225 ml which is collected in a 500-ml round-bottomed flask and evaporated (Rotovap; temperature maintained below 40°C). The residue is transferred to a 125-ml separatory funnel using a total of 15 ml of ethyl acetate. The organic layer is washed three times with 10-ml portions of aqueous 5% sodium bicarbonate/10% sodium chloride, and then three times with saturated aqueous sodium chloride. The water used to make the wash solutions is deionized and glass-distilled from basic potassium permanganate. The ethyl acetate solution is drained into a 50-ml

Erlenmeyer flask, dried over 0.5 g of anhydrous magnesium sulfate, filtered into a 15-ml conical test tube, and evaporated using a nitrogen stream. The residue is transferred using ethyl acetate to a 1-ml Reacti-vial (Pierce, Rockford, Ill.) and stored at -14°C if not immediately processed.

Derivative Formation

Preparation of Methoxime-Trimethylsilyl Ether (MO-TMS) Derivatives. After the ethyl acetate solution of steroids is evaporated with the aid of a nitrogen stream, 100 μl of a stock solution of methoxyamine hydrochloride (Eastman-Kodak, Rochester, N.Y.) in pyridine, 100 mg ml^{-1}, is added, and the vial is fitted with a Teflon-lined screw cap. The solution is heated at 70°C for 15 mins, then evaporated using a nitrogen stream. After the addition of 100 μl of N-trimethylsilylimidazole (TSIM, Pierce), the solution is heated at 100°C for 2 hr. The final solution is used directly for glass capillary gas chromatographic analysis, using 2- to 5-μl samples in most instances.

Gas Chromatography

Column Preparation. Sixty-meter borosilicate glass capillaries (1.0 mm OD, 0.3 mm ID) are drawn from 1.25-m tubes (7.8 mm OD, 3.8 mm ID) using a Shimadzu model GDM-1 glass-drawing-and-coiling apparatus. Pyrex tubes to be drawn are successively rinsed with acetone, dichloromethane, 1% aqueous potassium hydroxide, and methanol, then dried under vacuum. The diameter of the capillary coil is 10.5 cm.

Glass capillaries silanized with dimethyldichlorosilane and coated with SE-30 containing Silanox 101 (Cabot, Boston, Mass.) are prepared according to the method of German and Horning [84]. These are conditioned under carrier gas (nitrogen) flow by temperature programming at 1°C min^{-1} from 25°C to 280°C, holding at 280°C for 2 hr, lowering to 200°C, programming at 1°C min^{-1} to 300°C, and holding at 300°C for 1 hr. The theoretical plate efficiencies of the resulting columns (for n-tetracosane at 250°C) range from 125,000 to 150,000.

Commercial Availability. Similar high-performance columns are currently available from several suppliers for around $200 to $300 per graded column.

Instrumentation. Separation of the MO-TMS derivatives of urinary steroid metabolites is carried out on any gas chromatograph equipped with temperature programming, good gas flow control, an all-glass inlet splitter system, and the ability to accept glass capillary columns. Most of the separations reported herein were performed on a Tracor model 550 gas chromatograph which had been modified to include a glass inlet splitter system [85] and to accept glass capillary columns. A Fisher Recordall series 5000 recorder was employed. Hydrogen was supplied by a Hewlett-Packard model 18591A hydrogen generator. All gas flows except air were controlled by Brooks model 5840 flow controllers. The flame ionization detector was extensively modified [86] for use with glass capillaries. Many acceptable glass capillary gas chromatographs are now being marketed.

Retention Behavior. It is convenient to determine methylene unit (MU) [87,88] values through use of n-alkanes coinjected with the sample using an initial temperature of 200°C and programming at 1°C min^{-1}. Other GC conditions include: split ratio, 5:1; temperature of precolumn inlet splitter, 270°C; detector bath temperature, 300°C; column inlet pressure, 10 lb in^{-2}, resulting in a carrier gas (helium) linear velocity of 18-20 cm sec^{-1} (200°C); hydrogen flowrate, 37.5 ml min^{-1}; air flowrate, 200 ml min^{-1}; nitrogen makeup gas to the detector, 50 ml min^{-1}. The order of elution of steroid derivatives with an SE-30 phase is the same for packed and open tubular capillary columns, but the observed MU values are usually slightly different and should be determined for each column.

Quantitative Analyses. Calculations can be made using a peak integrator; those herein were made by hand. Area measurements are made by multiplying the height of the peak by the width at half-height. The chart speed and pen line width are variables that affect the results. The height of the peak is measured to the outside of the pen line; the width of the peak at half-height is measured to the inside of the pen line. A magnifying scale is used for the width measurement. Usually a chart speed of 1.25 cm min^{-1} is used for quantitative calculation strip charts and 9.5 cm min^{-1} for illustrations. The precision of this method is about 2% standard deviation. The calculations are based on individual peak area responses, corrected by experimentally determined mass response factors, and compared to the area response exhibited by a known amount of internal reference compound.

Mass Spectrometry

The identity of each steroid indicated in the illustrations was confirmed by comparisons of retention behavior and mass spectra fragmentation patterns of authentic samples. The mass spectra were obtained with an LKB 9000/PDP 12 analytical system in the usual way, using an SE-30 column; it is not advisable to depend upon retention data alone for the identification of individual steroids.

2.3 Derivatives for Gas Chromatography of Steroids

MO-TMS derivatives are the most satisfactory derivatives now known for gas-phase analytical studies of steroids. For special purposes, n-butyloxime-trimethylsilyl ethers [89], benzyloxime-trimethylsilyl ethers [90], and trimethylsilyloxime-trimethylsilyl ethers [91] are also useful; and persilylated trimethylsilyl [92] (hydroxyl and enol) ethers have been used in some applications. Very few problems are associated with the preparation of derivatives of steroids with two or three functional groups; the major difficulties lie in working with steroid mixtures of biologic origin containing compounds with as many as six functional groups.

Three principal problems are present. The purpose of converting ketone groups to methoximes [93] is to prevent enol ether formation during the silylation procedure. Ketone groups in the usual unhindered postions (3-, 16-, 17-, 20-) react rapidly under a variety of conditions, but a hindered ketone (11-oxo) group will react very slowly. The reaction condition described here was established through an investigation of conditions designed to result in rapid derivative formation for 3-, 16-, 17-, and 20-oxo groups, with negligible reaction of the 11-oxo group in tetrahydrocortisone and related compounds [93-96]. A second problem involves the fact that conversion of the 17α-hydroxyl group of tetrahydrocortisol (and of related compounds with the same type of functional groups) to an ether is required to provide thermal stability. Otherwise the side chain is lost during the vaporization process. Conditions for the preparation of derivatives of steroids with a tertiary 17-hydroxyl group were developed by Gardiner and Horning [97] and for some time this procedure was used routinely. An additional problem however, is posed by the fact that steroids with a free hydroxyl group usually show evidence of adsorption or dehydration when separated under conditions involving relatively high temperatures (280 to 320°C). For this reason, studies were carried out of relative rates of reaction of steroidal hydroxyl groups with different silylating reagents and different reaction conditions. This work

[96,98,99] resulted in the establishment of conditions for the conversion of all hydroxyl groups to trimethylsilyl ether groups. The rates of reaction of unhindered, moderately hindered, and highly hindered hydroxyl groups are quite different; the condition described here was established by Thenot and Horning [96] and was chosen in order to convert all hydroxyl groups in human urinary steroids, alone and in combination, to trimethylsilyl ethers while restricting enol ether formation for the 11-oxo group of tetrahydrocortisone to 2% or less. (This method has been modified for use with dexamethasone [100].)

The resulting MO-TMS derivatives are thermally stable, easily volatilized, and show no indication of dehydration or compound loss when good gas chromatographic analytical practices are followed. The methods can be used for many types of steroid mixtures of biologic or synthetic origin, providing suitable precautions are taken if steroidal hormones with a 3-oxo-4-ene moiety are involved. All 17-oxo steroids that have been studied yield a single MO derivative, but 3-oxo-4-ene steroids form syn- and antiisomers which may or may not separate when packed columns are employed [94,95]. The isomeric derivatives are stable and usually form in characteristic ratios for each steroid. It is usually preferable to employ methods yielding a single derivative whenever possible; in this instance it has been found that isomer formation is acceptable for compounds of interest including costisone [101], cortisol [101], and 6β-hydroxycortisol [102].

Caution should be exercised in the separation of all MO-TMS derivatives with respect to column conditions. In a study of the problem of acid-catalyzed decomposition or alteration of MO derivatives, Thenot and Horning [101,103] found that under acidic conditions 17-MO steroids are converted to a nitrile by D-ring opening and 3-MO derivatives are lost by decomposition except for 4,4-dimethyl steroids, which are transformed to other products by A-ring opening.

When the MO-TMS procedure described here is employed, there is no evidence of residual acidity (due to hydrogen chloride) in the reagent-sample mixture, and the use of a short (1-cm) initial segment of 10% SE-30 column packing is recommended [101] to prevent the development of acidic active sites at the head of the column. When the inlet system of German and Horning [85] is used, this segment is placed in the forecolumn. This derivatization procedure is suitable for many types of urinary steroid samples. When excessive amounts of steroids have been excreted, however, the urine sample should be diluted as required to approximate the range of steroid concentrations usually encountered.

2.4 Representative Urinary Steroid Metabolic Profiles

Metabolic profiles of human urinary steroids obtained by glass capillary gas chromatography reflect the production and metabolism of steroidal hormones. Their present use is largely for research purposes, but in time they will be used in diagnosis, long-term therapy, and possibly in establishing risk in disorders involving steroid biosynthesis and steroid metabolism.

Profiles as Possible Aids in Clinical Diagnosis

Figure 7 is a metabolic profile of a normal postpubertal premenopausal female. All of the usual components are present: androsterone (An), etiocholanolone (Et), dehydroepiandrosterone (DHEA), 11β-hydroxyandrosterone (11-HAn), 11β-hydroxyetiocholanolone (11-HEt), pregnanediol (Pd), pregnanetriol (Pt), 5-androsten-3β,16α,17β-triol (Atr), tetrahydrocortisone (THE), tetrahydro-11-dehydrocorticosterone, 5β-pregnan-3α21-diol-11,20-dione (THA), tetrahydrocorticosterone, 5β-pregnan-3α,11β,21-triol-20-one (THB), allo-tetrahydrocorticosterone, 5α-pregnan-3α,11β,21-triol-20-one (a-THB), tetrahydrocortisol (THF), allo-tetrahydrocortisol (a-THF), cortolone, 5β-pregnan-3α,-17α,20α,21-tetrol-11-one (Co), β-cortolone, 5β-pregnan-3α,17α,20β,21-tetrol-11-one (β-Co), and cortol, 5β-pregnan-3α,11β,17α,20α,21-pentol (Cor). The last component appearing in the chromatogram is the internal reference compound, cholesteryl butyl ether (CBE).

Androsterone, etiocholanolone, allo-tetrahydrocostisol, and tetrahydrocortisol are all products of liver enzymic reduction of the 3-oxo-4-ene moiety of the steroidal A/B ring system. 5α-oxidoreductases are involved in the conversion of testosterone to androsterone and cortisol to allo-tetrahydrocortisol; 5β-oxidoredutases are involved in the conversion of testosterone to etiocholanolone and cortisol to tetrahydrocortisol. Fig. 8 indicates the steroidal A/B ring enzymic oxidoreductive conversions which lead to the observed metabolites. Ratios of these metabolites, viz., Et/An (5β-H/5α-H) and THF/a-THF (5β-H/5α-H), may be taken as a reflection of the relative activities of the corresponding oxidoreductases that act on testosterone and cortisol.

Many females as the one shown in Fig. 7 excrete etiocholanolone and androsterone in a ratio (5β-H/5α-H) of about 2:1; their usual ratio of tetrahydrocortisol to allo-tetrahydrocortisol (5β-H/5α-H) is about 3:1. This point will bc expanded later, but for now suffice it to say that women who have these Et/An and THF/a-THF ratios are called "pattern A" females. (No healthy male has ever been found to exhibit pattern A ratios.) Women

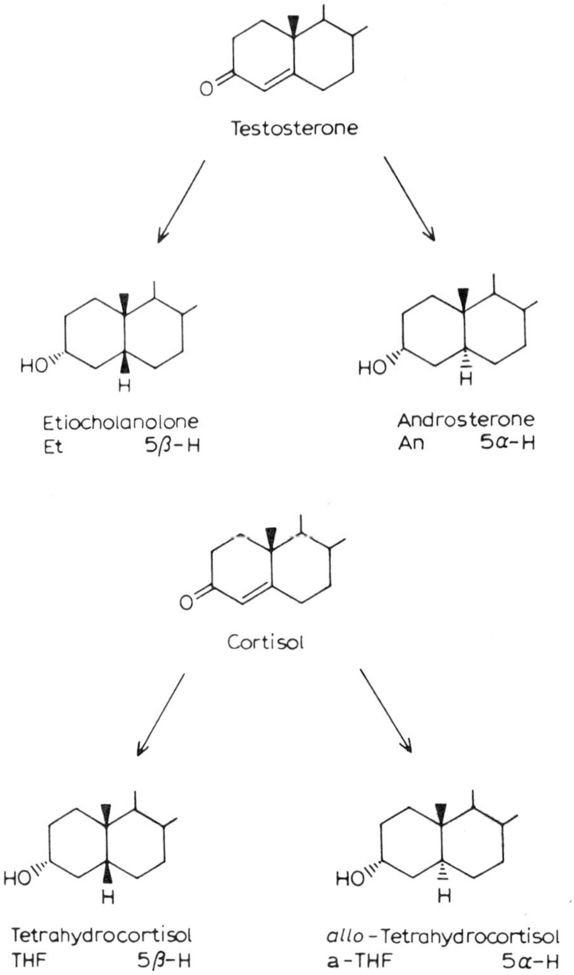

Figure 8. Structural representations of the steroidal A/B ring segments of testosterone, cortisol, and the respective products which result from enzymic oxidoreduction of the 3-oxo-4-ene moiety. 5β-H oxidoreductases are involved in the formation of etiocholanolone and tetrahydrocortisol; 5α-H oxidoreductases, in the formation of androsterone and allo-tetrahydrocortisol. The metabolite ratios, Et/An and THF/a-THF, reflect the relative activities of the corresponding hepatic enzymes. (Source: Ref. 40. Reproduced courtesy of Elsevier Scientific Publishing Co., Amsterdam.)

with marked deviations from the usually observed ratios may have a higher state of risk of cystic disease, either polycystic or sclerocystic disease of the ovary or fibrocystic disease of the breast. In Fig. 7 the ratios and the amounts of both androgen and andrenocorticosteroid metabolites are typical.

Figure 9 shows a metabolic profile of a normal postmenopausal pattern A female. Although the Et/An ratio is about 3:1 and the THF/a-THF ratio is about 5:1, this is not regarded as a significant deviation from the usual female ratio. The lower amount of androgen metabolite excretion observed is typical for the postmenopausal female; a steady decline in androgen metabolites can be observed with advancing years. The amounts of adrenocorticosteroid metabolites seen here are comparable to those of the premenopausal female.

Figure 10 is a composite of four metabolic profiles taken over the menstrual cycle of a female with bilateral polycystic ovary syndrome. One can follow the cyclical rise and fall of pregnanediol (Pd) and pregnanetriol (Pt), the major urinary metabolites of progesterone which are increasingly secreted as a woman passes through the follicular, ovulatory, and luteal phases of the cycle. This female has quite different ratios of Et/An and THF/a-THF and falls within the group designated "pattern B" females, those who have metabolite ratios similar to males (discussed below). Over half of the subjects involved in an unpublished study of polycystic and sclerocystic ovary syndrome had pattern B ratios, whereas only 1 in 3 in the corresponding control group had pattern B ratios.

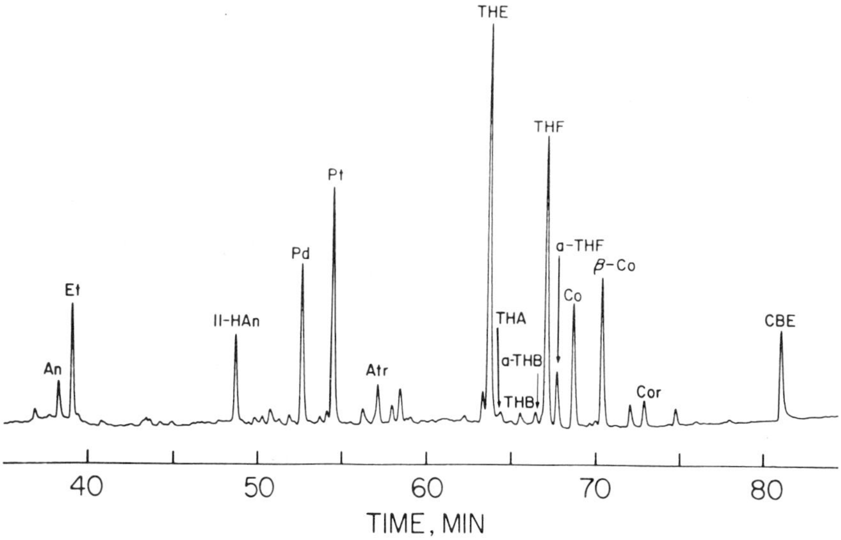

Figure 9. Urinary steroid metabolic profile of a healthy pattern A postmenopausal female. Analysis conditions and abbreviations are as in Fig. 7.

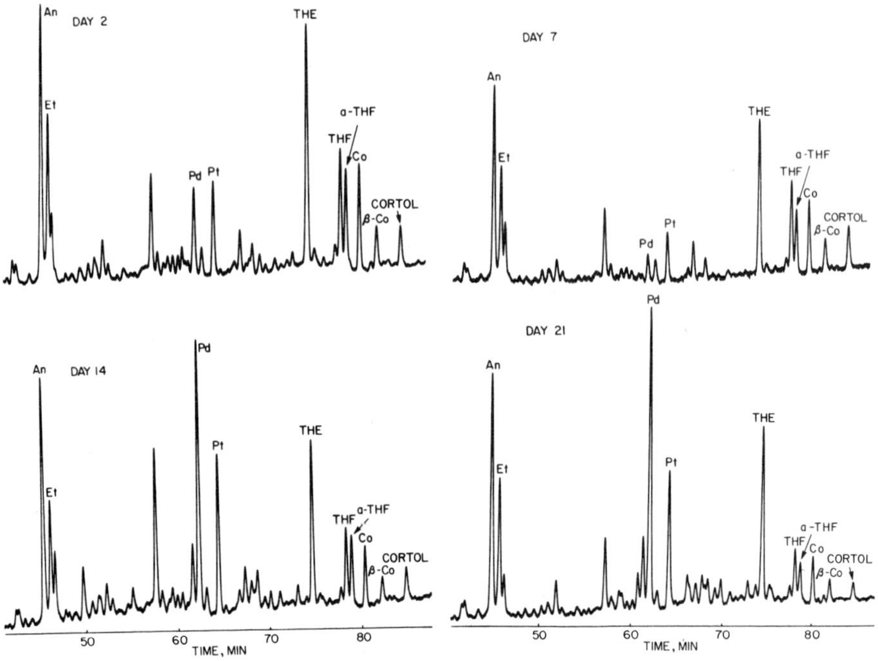

Figure 10. Composite of four urinary steroid metabolic profiles taken over the menstrual cycle of a female with bilateral polycystic ovary syndrome. Analytical conditions are as in Fig. 7. The cyclical rise and fall of pregnanediol (Pd) and pregnanetriol (Pt) can be observed as the subject passes through the follicular, ovulatory, and luteal phases of the cycle. The Et/An and THF/a-THF ratios fall within pattern B values which resemble those of males.

Figure 11 shows a metabolic profile of a female entering the second trimester of pregnancy. Significant increases in the excretion of 11β-hydroxyandrosterone (11-HAn), 11β-hydroxyetiocholanolone (11-HEt), 16α-hydroxydehydroepiandrosterone (16-HDHEA), pregnanediol (Pd), androstenetriol (Atr), tetrahydrocortexolone, 5β-pregnan-3α,17α,21-triol-20-one (THS), estriol (E_3), and allo-tetrahydrocortexolone, 5α-pregnan-3α,17α,21-triol-20-one (a-THS) are present. Small amounts of 11-ketoandrosterone (11-KAn), 11-ketoetiocholanolone (11-KEt), estrone (E_1), and estradiol (E_2) are also excreted.

As early as 1946 Venning [104] found only a slight rise in 17-ketosteroid excretion during pregnancy. Her observations were later confirmed by both Schuller [105] and Birke et al. [106] who detected no appreciable rise in androsterone or etiocholanolone excretion until late in the third trimester. A steady increase in the excretion of 11-oxygenated 17-ketosteroids, primarily 11-hydroxyandrosterone and 11-hydroxyetiocholanolone, was observed.

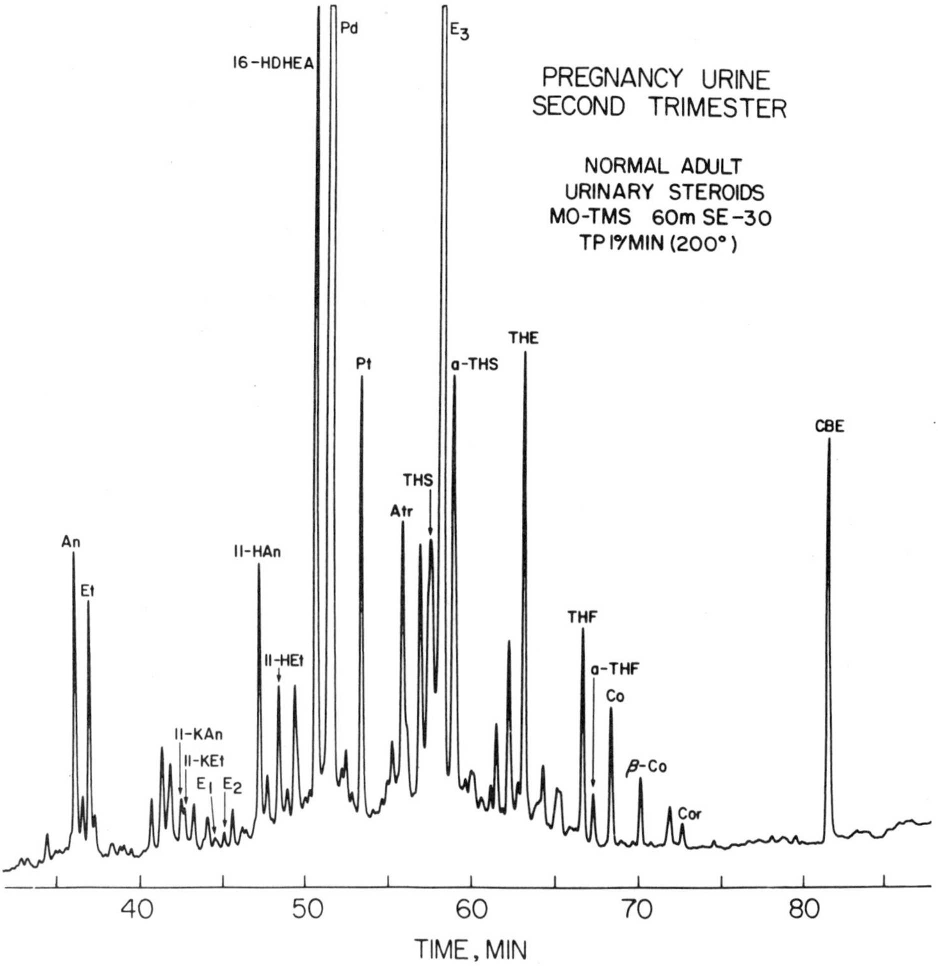

Figure 11. Urinary steroid metabolic profile of a female entering the second trimester of pregnancy. Analysis conditions are as in Fig. 7; see the text for an explanation of the abbreviations. (Source: Ref. 32. Reproduced courtesy of Elsevier Scientific Publishing Co., Amsterdam.)

The greatly elevated excretion of estriol during pregnancy is due to fetoplacental processes [107]. Fetal 16α-hydroxylation, which is required for estriol formation, is also responsible for increased production of 16 -hydroxydehydroepiandrosterone. Because 16α-hydroxylation is primarily a fetal adrenal (rather than a placental) hydroxylation [108], estriol has been regarded as a specific indicator of fetal well-being. The excretion of less than 1 mg daily in late pregnancy usually indicates fetal death. Klopper [109] regards less than 3 mg daily indicative of probable fetal death; Green et al. [110] considered an excretion of 4 to 12 mg daily a warning that the fetus is at risk.

Increased pregnanediol excretion (always found in pregnancy) results from maternal hepatic reduction of progesterone produced by placental tissue. Dässler [111] found significant rises in cortexolone (4-pregnen-17α, 21-diol-3,20-dione) excretion during pregnancy, an observation consistent with the elevated excretion of tetrahydrocortexolone and allo-tetrahydrocortexolone indicated in the profile. These metabolites (THS and a-THS) are usually present in only trace amounts under ordinary circumstances.

Figure 12 is a metabolic profile of a normal adult male. As in Fig. 7, all of the usual steroid metabolites are present. Normal males excrete etiocholanolone and androsterone in a ratio (5β-H/5α-H) of about 1:2; the usual ratio of tetrahydrocortisol to allo-tetrahydrocortisol (5β-H/5α-H) is about 1:1. These ratios reflect the hepatic 3-oxo-4-ene oxidoreductase distribution found in males.

Figure 13 shows a metabolic profile of a premenopausal female with a testosterone-secreting ovarian tumor. (The urine was diluted 20 times with glass-distilled water before the enzymic hydrolysis step.) Large amounts of etiocholanolone and androsterone were excreted (ratio about 2:3). That an ovarian tumor can produce very large quantities of testosterone (leading to greatly elevated excretion of androsterone and etiocholanolone) was demonstrated in 1964 by Rosner et al. [112]. Their patient excreted 45 times the amount of urinary testosterone glucuronide normally found for females. Greatly elevated excretion of 11β-hydroxyandrosterone and 11β-hydroxyetiocholanolone, and of pregnanetriol, is evident from the profile. Adrenocorticosteroid metabolites are present in larger amounts than are observed for normal subjects.

Figure 14 is a metabolic profile of an adult male with congenital adrenal insufficiency. In the absence of adrenal cortex function, adrenocorticosteroid metabolites are not observed as excretion products. Androgen excretion levels are normal in this patient; the Et/An ratio is about 2:3. Pregnenolone, the precursor of progesterone, leads to pregnanetriol via 17α-hydroxylation and subsequent reduction of both the 20-oxo- and the 3-oxo-4-ene moieties of 17α-hydroxyprogesterone. The component which elutes about 53 mins after injection is a pregnanediolone, presumably 5β-pregnan-3α,17α-diol-20-one, which would result from reduction of the 3-oxo-4-ene system of 17α-hydroxyprogesterone [113].

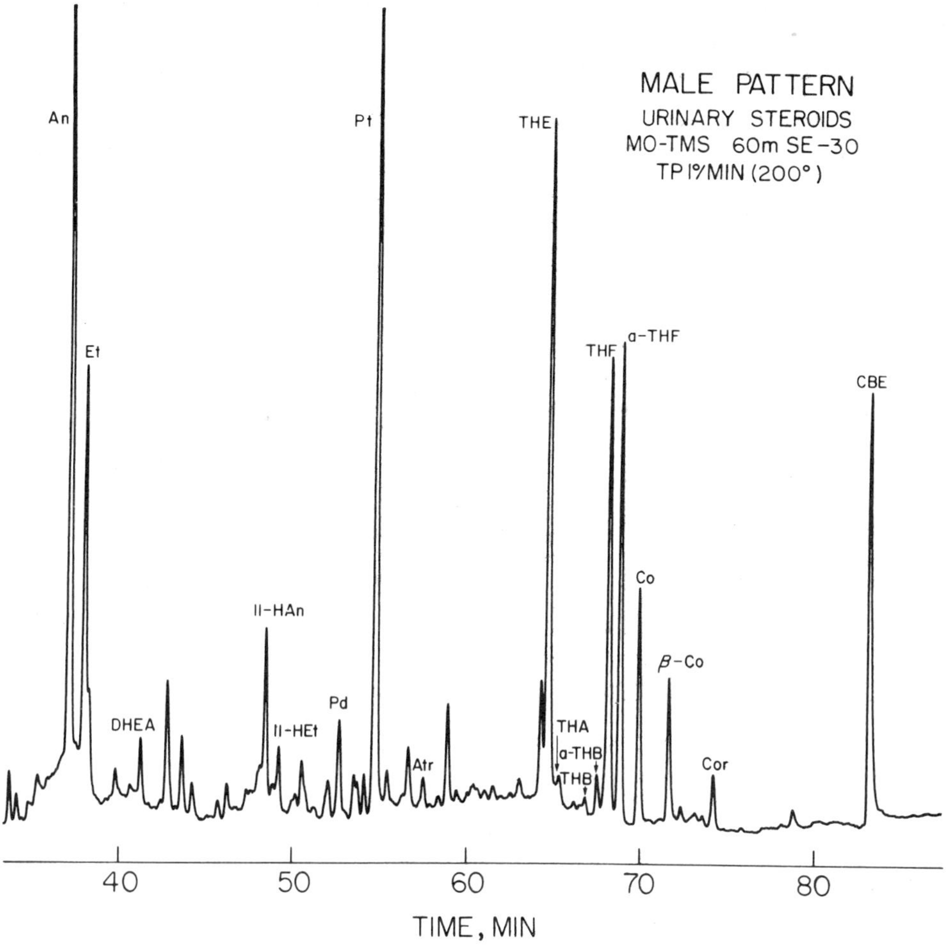

Figure 12. Urinary steroid metabolic profile of a healthy adult male. The analytical conditions are as in Fig. 7. All of the usual androgen and adrenocorticosteroid metabolites are present. The Et/An and THF/a-THF ratios are both about 1:2; see the text for an explanation of the abbreviations. (Source: Ref. 32. Reproduced courtesy of Elsevier Scientific Publishing Co., Amsterdam.)

Figure 15 shows the metabolic profile of a postmenopausal female with a dehydroepiandrosterone-secreting ovarian tumor.

Dehydroepiandrosterone was isolated from adrenal tumor tissue by Plantin et al. [114] in 1957, and is found in increased amounts in the urine of patients with adrenal carcinoma. Dorfman and Shipley [115] found greatly elevated excretions of 3β-hydroxy-17-ketosteroids in 20 of 24 cases of adrenal cortical tumors.

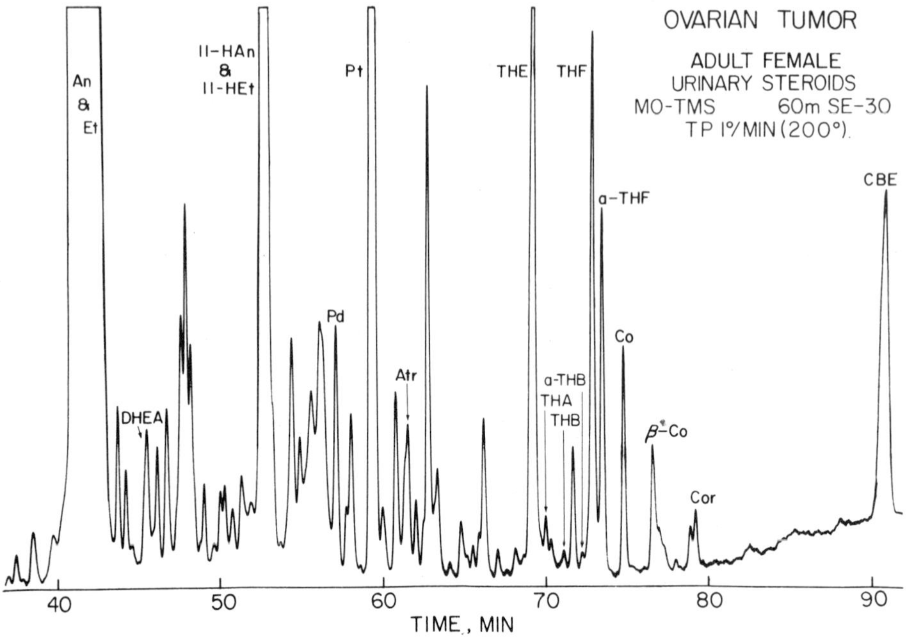

Figure 13. Urinary steroid metabolic profile of a premenopausal female with a testosterone-secreting ovarian tumor. Greatly elevated levels of androsterone, etiocholanolone, 11β-hydroxyandrosterone, 11β-hydroxyetiocholanolone, and pregnanetriol are evident. Analysis conditions and abbriviations are as in Fig. 7. (Source: Ref. 32. Reproduced courtesy of Elsevier Scientific Publishing Co., Amsterdam.)

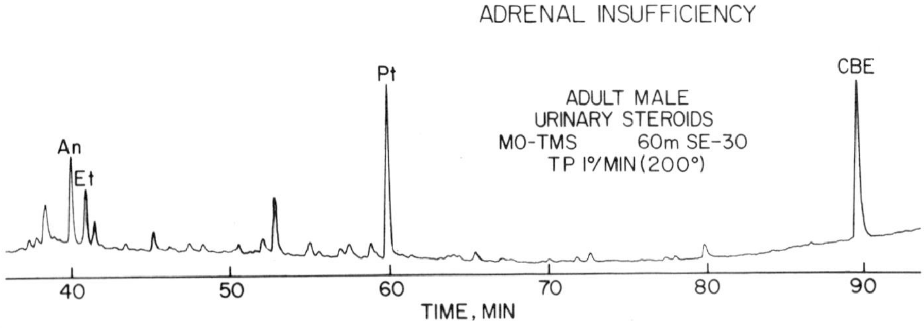

Figure 14. Urinary steroid metabolic profile of an adult male with congenital adrenal insufficiency. The usual adrenocorticosteroid metabolites are missing. Analysis conditions and abbreviations are as in Fig. 7. (Source: Ref. 32. Reproduced courtesy of Elsevier Scientific Publishing Co., Amsterdam.)

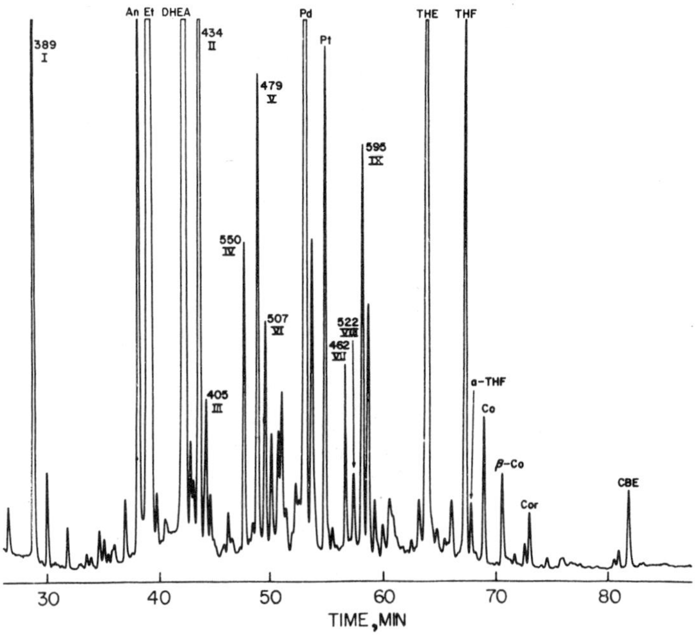

Figure 15. Urinary steroid metabolic profile of a postmenopausal female with a DHEA-secreting adrenal tumor. Components I to IX are discussed in the text. Analysis conditions and abbreviations are as in Fig. 7. (Source: Ref. 32. Reproduced courtesy of Elsevier Scientific Publishing Co., Amsterdam.)

A GC/MS analysis of this sample was carried out to identify the components numbered I to XI. Methylene unit and fragmentation data led to the following identifications, MO-TMS derivatives of: I (M = 389), 9,(5β)-androsten-3α-ol-17-one; II (M = 434), 5-androsten-3β,17β-diol; III (M = 405), 11-ketoetiocholanolone; IV (M = 550), 5-pregnen-3β,17α,20α-triol; V (M = 479), 11β-hydroxyandrosterone; VI (M = 507), 5β-pregnan-3α,17α-diol-20-one; VII (M = 462), 5-pregnen-3β,20α-diol; VIII (M = 522), 5-androsten-3β,16α,17β-triol; and IX (M = 595), 5β-pregnan-3α,17α,20α-triol-11-one.

The open tubular glass capillary columns used to obtain the above urinary steroid metabolic profiles are capable of handling rather large amounts of components. Thus, as illustrated in Fig. 11, estriol (E_3) is well separated from allo-tetrahydrocortexolone (a-THS). In Fig. 13, the descending side of the etiocholanolone peak remains sharp with little trailing; the tetrahydrocortisone (THE) peak does not obscure the tetrahydro-11-dehydrocorticosterone (THA) component.

Values used for Identification and Quantification Purposes

Table 2 contains mass response factors (RF) and methylene unit (MU) values for the metabolites which are often quantified using urinary steroid metabolic profiling. Methylene unit values are given as obtained from two different open tubular glass capillary columns coated with SE-30 containing dispersed Silanox 101. The values are highly reproducible on any given column but must be determined for each column.

Table 2. Mass Response Factors (RF)[a] and Methylene Unit (MU) Values for Quantifying Urinary Steroid Metabolites[b]

Metabolite	Abbreviation	RF	MU values[c] on column I	MU values[c] on column II
Androsterone	An	0.40	25.32	25.31
Etiocholanolone	Et	0.41	25.46	25.45
Dehydroepiandrosterone	DHEA	0.43	26.06	26.03
11β-Hydroxyandrosterone	11-HAn	0.63	27.16	27.12
11β-Hydroxyetiocholanolone	11-HEt	0.63	27.29	27.25
Pregnanediol	Pd	0.55	27.88	27.80
Pregnanetriol	Pt	1.16	28.16	28.12
5-Androsten-3β,16α,17β-triol	Atr	0.82	28.55	28.51
Tetrahydrocortisone	THE	0.61	29.75	29.72
Tetrahydro-11-dehydrocorticosterone	THA	d	29.85	29.82
Tetrahydrocorticosterone	THB	d	30.06	30.03
Allo-tetrahydrocorticosterone	a-THB	d	30.24	30.21
Tetrahydrocortisol	THF	0.74	30.34	30.30
Allo-tetrahydrocortisol	a-THF	0.72	30.45	30.41
Cortolone	Co	0.81	30.62	30.59
β-Cortolone	β-Co	0.93	30.93	30.90
Cortol	Cor	1.15	31.37	31.32

[a]Determined in triplicate and averaged. The reference compound was cholesteryl butyl ether.

[b]SE-30 glass open tubular capillary column, temperature programmed at 1°C min^{-1} from 200°C.

[c]Of MO-TMS or TMS derivatives.

[d]RF assumed = 1.00.

2.5 Hepatic Enzyme Imprinting Studies

Studies of sexual differences in various enzyme distributions have been underway since the initial observations of Forchielli et al. [116] in 1958. Most investigations have been made on the rat, and they have included liver cell culture studies [117-119] as well as those involving physiological modification (castration, oophorectomy, hypophysectomy, and adrenalectomy) [120-131]. Studies with Rhesus monkeys have also been made [132]. Most data corroborate with the concept that both "male" and "female" liver cells and similarly differentiated hypothalamic brain cells [132-142] can and commonly do exist.

Supporting Data from Urinary Steroid Metabolic Profiles

Over the past several years, Pfaffenberger et al. [36,39,40] have been seeking evidence for sex-related differences in human hepatic metabolism. Their evidence, which is derived from urinary steroid metabolic profiles, indicates that quite different enzyme distributions can occur with liver cells of different females. Their conclusions are based upon the urinary metabolites of two of the major steroidal hormones, testosterone and cortisol (see Fig. 8).

As mentioned earlier, androsterone, etiocholanolone, allo-tetrahydor-cortisol, and tetrahydrocortisol are all products of liver enzymic reduction of the 3-oxo-4-ene moiety of the steroidal A/B ring system. 5α-oxidoreductases are involved in the conversion of testosterone to androsterone and cortisol to allo-tetrahydrocortisol; 5β-oxidoreductases are involved in the conversion of testosterone to etiocholanolone and cortisol to tetrahydrocortisol. The relative activities of these enzymes leads to the ratios of metabolites observed for each individual. Some females possess decidedly male distributions of the 5α- and 5β-oxidoreductases.

Females who have low etiocholanolone/androsterone (5β-H/5α-H) ratios also have low tetrahydrocortisol/allo-tetrahydrocortisol (5β-H/5α-H) ratios. The classification of normal pre- and postmenopausal females into two subgroups, one with values similar to those of males, was established through two sets of data for each individual. Pfaffenberger and Horning believe these findings are indicative of events which occur early in human development although the consequences of the events are not fully manifested until postpubertal life. Moreover, they believe that these events are probably related to the phenomenon of either late fetal or neonatal biologic imprinting of the aforementioned hepatic enzyme systems.

Postpubertal Males. Profiles were obtained for 21 healthy postpubertal males who were all less than 36 years of age. Normal males were found to usually excrete etiocholanolone (5β-H) and androsterone (5α-H) in a ratio of of about 1:2; the usual ratio of tetrahydrocortisol (5β-H) to allo-tetrahydrocortisol (5α-H) was about 1:1. The range which was actually observed for males was from 0.31 to 0.89 for Et/An (Fig. 16) and from 0.91 to 1.99 for THF/a-THF (Fig. 17).

Premenopausal Females. Profiles were obtained for 31 premenopausal females who had no clinical symptom of disease. Each participant was carefully screened to exclude anyone with hypothyroidism, hyperthyroidism, idiopathic hirsutism, weight problems, polycystic or sclerocystic ovary syndrome, ovarian problems of any other significant nature, hepatic pathology, fibrocystic breast disease, breast cancer, or any other type of lesion. Both parous and nulliparous women, all less than 36 years of age, were included; none had taken oral contraceptives over at least the preceding year.

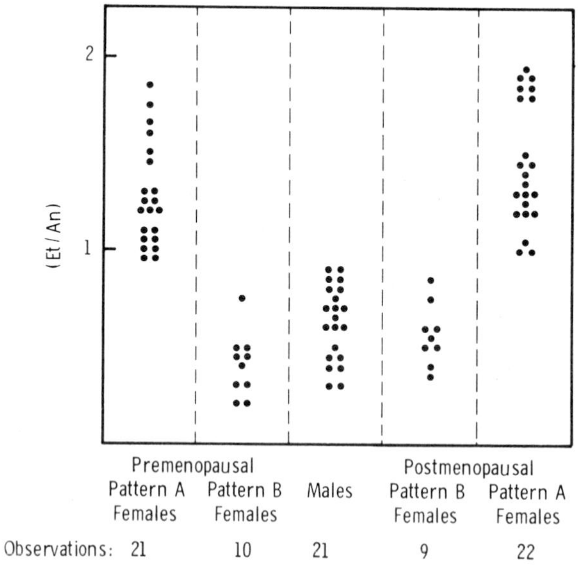

Figure 16. Values of the etiocholanolone/androsterone (Et/An) ratio found for pre- and postmenopausal female and male subjects. The females show a 2:1 distribution between pattern A values (greater than 0.90) and pattern B values (less than 0.90). Pattern B values are the same as male ratio values for these testosterone metabolites. (Source: Ref. 39. Reproduced courtesy of Academic Press, Inc.)

Many females were found to excrete etiocholanolone and androsterone in a ratio (5β-H/5α-H) of about 2:1; their usual ratio of tetrahydrocortisol to allo-tetrahydrocortisol (5β-H/5α-H) was about 3:1. Women who had these ratios were designated "pattern A" females. The observed ranges within the pattern A category were from 0.92 to 1.83 for Et/An (Fig. 16) and from 2.00 to 4.57 for THF/a-THF (Fig. 17). No healthy male was found to exhibit a pattern A ratio.

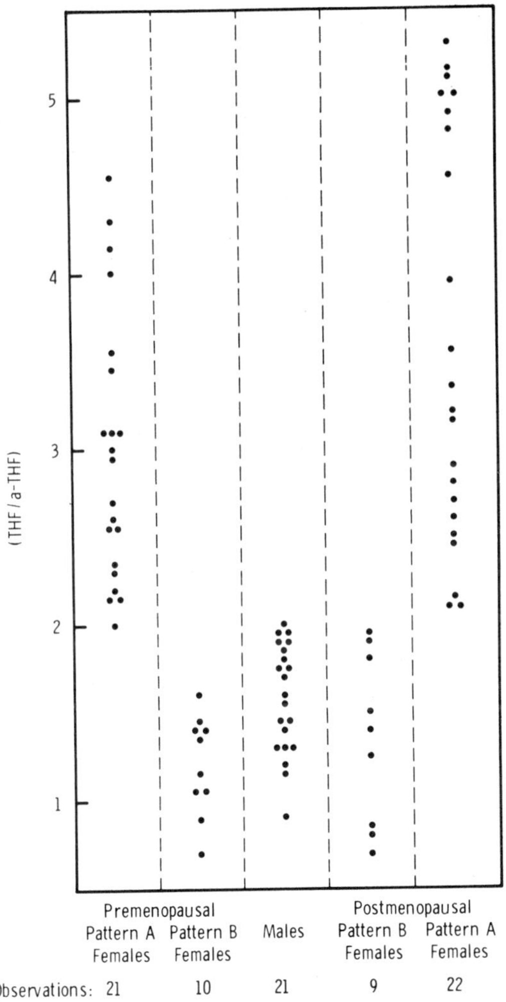

Figure 17. Values of the tetrahydrocortisol/allo-tetrahydrocortisol (THF/a-THF) ratio found for pre- and postmenopausal female and male subjects. The same individuals are distributed in the same ratios shown in Fig. 16. Twice as many females have pattern A values (greater than 2.00) as pattern B values (less than 2.00). Again, pattern B values are the same as male ratio values for these cortisol metabolites. (Source: Ref. 39. Reproduced courtesy of Academic Press, Inc.)

Some females, about 1 in 3, were found to have Et/An excretion ratios of about 1:2. These ratios reflect the hepatic 3-oxo-4-ene oxidoreductase distributions commonly found in males. These women were designated "pattern B" females. The observed ranges within the pattern B category were from 0.21 to 0.75 for Et/An (Fig. 16) and from 0.70 to 1.59 for THF/a-THF (Fig. 17).

Table 3 contains the excretion ranges and average excretion values of the urinary steroid metabolites from the postpubertal group of males and the premenopausal females.

Postmenopausal Females. Profiles were obtained for 31 postmenopausal females who likewise had no clinical symptom of disease. Each participant was screened in the same manner as the premenopausal group. Again both parous and nulliparous women were included. All were over 55 years of age and none were on conjugated estrogen therapy. The same distribution of about 2:1 pattern A to pattern B females was found (Figs. 16 and 17). The values of the ratios were quite similar: pattern A, from 1.01 to 1.95 for Et/An and 2.10 to 5.32 for THF/a-THF; pattern B, from 0.35 to 0.86 for Et/An and 0.69 to 1.98 for THF/a-THF.

Cellular Imprinting in Animals. It has been known for many years that there are differences in male and female liver metabolism, and that the course of development of animals can be altered by neonatal exposure to androgens or estrogens [116]. Liver steroid oxidoreductases and steroid hydroxylases have been studied in detail in the rat by Gustafsson and co-workers [120,121,124-131,143]; these have been classified according to whether their activity is hormone-dependent or hormone-independent, and whether hormone-dependent activity is determined during the neonatal period. Several types of enzyme activity were found to be typically male or typically female, and to be determined by processes requiring pituitary action during the neonatal period.

One particularly interesting example of an enzyme activity which appears irreversibly imprinted by androgens neonatally yet is reversibly influenced by sex hormones postpubertally is the 15β-hydroxylase active on 5α-androstane-3α,17β-diol 3,17-disulfate. Gustafsson and Ingleman-Sundberg [127] could not detect the 15β-hydroxylase system in liver microsomes from male rats; the system is thus at least 3000 times more active in female rats. The 15β-hydroxylase activity cannot be measured in rats younger than 20 days of age, and at this time no sexual difference in enzyme activity exists.

Table 3. Excretion Ranges and Average Excretion Values of Urinary Steroid Metabolites of Postpubertal Males and Premenopausal Females[a]

Metabolite	Pattern A females			Pattern B females			Males		
	Min	Max	Avg	Min	Max	Avg	Min	Max	Avg
An	0.73	6.86	2.36	0.80	7.29	3.21	1.71	10.46	5.49
Et	0.75	9.13	2.95	0.35	3.31	1.29	1.16	8.47	3.58
DHEA	0.01	2.45	0.47	0.01	1.14	0.24	0.01	3.64	0.66
11-HAn	0.04	0.99	0.49	0.12	0.99	0.63	0.23	2.83	1.11
11-HEt	0.01	0.24	0.08	0.01	0.24	0.08	0.01	1.32	0.31
Pd	0.30	4.59	1.74	0.28	3.67	1.50	0.07	1.54	0.55
Pt	0.21	3.60	1.04	0.35	1.07	0.65	0.08	2.34	1.07
Atr	0.01	0.24	0.13	0.02	0.17	0.07	0.01	0.59	0.16
THE	0.90	6.04	2.21	0.84	5.04	2.17	1.07	8.09	3.73
THA	0.01	0.09	0.04	0.02	0.08	0.03	0.01	0.12	0.05
THB	0.01	0.09	0.05	0.01	0.05	0.03	0.01	0.08	0.04
a-THB	0.01	0.09	0.04	0.01	0.09	0.05	0.01	0.14	0.05
THF	0.64	2.95	1.19	0.44	2.11	1.10	0.44	2.89	1.62
a-THF	0.14	1.07	0.42	0.30	2.04	0.97	0.30	1.74	1.10
Co	0.28	1.22	0.50	0.31	0.72	0.44	0.27	1.62	0.67
β-Co[b]	0.16	0.69	0.34	0.14	0.93	0.35	0.18	1.25	0.48
Cor	0.05	0.37	0.12	0.05	0.19	0.10	0.03	0.38	0.14

[a]Values in mg g^{-1} of creatinine.
[b]Values not adjusted for a trace of β-cortol which coelutes.

At 30 days of age, female rats tend to hydroxylate (15β) more effectively than male rats, and after this time the 15β-hydroxylase activity increases rapidly in female rats, almost reaching adult levels in 40 days. Over this same interval, the hydroxylase activity of male rats decreases; from 45 days onward it is no longer detectable. It has been speculated that the physiological role of this sex-specific enzyme system is female deactivation of potentially androgenic compounds.

Further experiments suggest that although the activity of this enzyme system is neonatally imprinted, it remains reversibly influenced by sex hormones. Postpubertal gonadectomy does not affect the 15β-hydroxylase activity in female rats and does not lead to the appearance of detectable enzyme levels in the liver microsomes of male rats. Neonatal castration of male rats results in completely feminized levels of 15β-hydroxylase in the adult animals, yet this feminization is totally inhibited by a single dose of 1.45 μmol of testosterone propionate administered on the day following neonatal testectomy. Treatment of postpubertally castrated male rats with estradiol benzoate leads to a transient, partial feminization of the liver with a 15β-hydroxylase activity of about 30% of that present in normal female rats. When postpubertally castrated female rats are treated with testosterone propionate, the activity of the 15β-hydroxylase system is suppressed [127].

Denef and colleagues [122,123,144] have made similar observations by studying the effect of hypophysectomy and pituitary implants at puberty on the sexual differentiation of testosterone metabolism in rat liver. Their results suggest that sexual differentiation of hepatic testosterone metabolism requires the presence of the pituitary gland in situ and that it depends on a sex difference in pituitary hormone secretion, the regulation of which is probably located in the hypothalamus.

In a series of studies of rat liver steroid reductase enzymes carried out with cell culture and mass fragmentography microanalytical techniques, Padieu and co-workers [117-119] found typical male and typical female patterns of reductase activity for corticosterone, the principal corticosteroid of rats. They established [117] that the female rat mainly has an enzyme which reduces the 4-5 double bond of corticosterone, 11β,21-dihydroxy-4-pregnene-3,20-dione to two 5α-H isomers, 11β,21-dihydroxy-5α-pregnane-3,20-dione and 3α,11β,21-trihydroxy-5α-pregnane-20-one. Moreover, the female rat apparently has no enzyme to convert tetrahydrocorticosterone into hexahydro compounds. The male rat liver mainly reduces corticosterone to tetrahydrocorticosterones with a 3β-hydroxy structure, then further metabolizes these compounds to form hexahydrocorticosterones. In male rats, the intermediary dihydrocorticosterone metabolites do not appreciably accumulate. Because these patterns do not change through successive cell cultures, this system appears to be firmly imprinted and hormone-independent once established.

Biologic imprinting also occurs during the development of the central nervous system [132-142] as supported recently by techniques more advanced than mere observation of behavioral patterns. Using autoradiography, McEwen and colleagues [133,137-140] have confirmed that during the neonatal period hormonal action in rats results in the development of a "male type" or "female type" of brain cell, particularly those of the preoptic area, the hypothalamus, and an adjacent area, the amygdala. All of these are areas of the "primitive" brain, the archipallum, which has long been known to play a role in sexual behavior.

If a hormone is to have a permanent effect on cells, it must be present during the period of early development when the brain is particularly sensitive to hormones. The critical period in the human occurs during fetal life. Fortunately for investigators, this same period for the laboratory rat occurs during the first week following birth.

Testes of the newborn male rat secrete testosterone; the testosterone initiates events which lead to sexual differentiation of the brain. Paradoxically, the "male" brain pattern appears to be imprinted in newborn rats by one of the female hormones, estradiol. Of key importance is the fact that the estradiol which is involved in the differentiation acts within the nuclei of brain cells [137,138,140]. Estradiol of newborn genetically female rats is prevented from entering the cell nucleus by α-fetoprotein, which strongly binds the estrogen and prevents its entry into the nucleus [145]. α-Fetoprotein does not bind testosterone; the testosterone of genetically male rats enters the nucleus where it is converted into estradiol by steroidal A-ring aromatizing enzymes [146,147]. It is intranuclear estradiol which permanently differentiates brain cells into "male" types. Even if castrated and given estradiol, adult male rats continue to show strong male sexual behavior and only slightly exhibit the female lordosis response (the female mating posture in which both the rump and the head are raised and held rigid, making the back concave). The permanent nature of cell differentiation suggests that testosterone (via nuclear estradiol) influences the development of neurons and the formation of synaptic contracts with other neurons which differ from those found in the female brain. The events are thought to augment both cell replication (DNA synthesis) and gene expression (RNA and protein synthesis). Because genetic material (DNA) is the same in all cells of an organism, the process of cell differentiation must involve the selective "turning on" of certain genes and the "turning off" of others. Cell differentiation is believed to be irreversible for the lifetime of the tissue involved. This hypothesis is strengthened by the observation that newborn male rats deprived of testosterone by castration show a female pattern of brain organization as adults. Moreover, if newborn female rats are given testosterone, they exhibit male sexual behavior as adults [135,136].

Corroborating Clinical Syndromes. Human experimentation utilizing the methods of animal work are obviously precluded, but a study of several clinical syndromes are relevant to the idea of biologic imprinting: (1) adrenogenital syndrome; (2) progestin-induced hermaphroditism; (3) androgen-insensitivity syndrome (testicular feminization); and (4) Turner's syndrome—all recently reviewed by Reinisch [141].

Adrenogenital syndrome is an endocrine anomaly; in the homozygous individual, an enzymatic defect in cortisol biosynthesis causes excessive production of adrenal androgens during fetal life. The excess androgen has a masculinizing effect of the genitalia, often leading to clitoral hypertrophy and partial fusion of the labia. Analyses of the personality and social behavior of 15 females with adrenogenital syndrome who had received steroid treatment from early childhood [148] described them as tomboys throughout childhood. The girls displayed high energy levels with a strong interest and participation in outdoor physical activity and minimal interest in doll play, dresses, and the usual girls' activities. All displayed less concern about marriage and child care and much more interest in careers than did the corresponding control subjects. Moreover, some of the patients, in adulthood, reported experiences more typical of normal males, namely erotic arousal with a strong genitopelvic component from the stimulation of visual and narrative perceptual material. Ehrdhardt et al. [148,149] and Money and Meredith [150] thought it possible that the cause of these behavioral patterns was traceable to the effects of excessive fetal androgens on the hypothalamus or related areas of the brain.

Gravid women are sometimes given synthetic progestins to prevent miscarriage. Some of these patients give birth to partially virilized females (progestin-induced hemaphrodites), but masculinization of the external genitalia is usually not to the same extent as with adrenogenital syndrome. This condition, exogenously produced, more closely resembles animal experiments in which androgens are prenatally administered to female fetuses. These patients differ in two ways from those with endogenously produced adrenogenital syndrome: they do not have elevated levels of 17-ketosteroids and they do not show progressive virilization. They exhibit normal feminization at puberty; menstruation and ovulation occur without treatment. Nine of ten girls studied [151] were labeled tomboys. The girls also displayed (mothers' judgments) a high frequency of self-assertive independence and self-reliance.

In androgen-insensitivity syndrome, there is a total discrepancy between genetic and gonadal sex (XY male) on the one hand and morphological sex appearance and sex of rearing (female) on the other. These genetic males with testes are born with normal-appearing external female genitalia. According to Money et al. [152], the body of such individuals cannot utilize androgen from fetal life onward. The testes produce normal amounts of both androgen and estrogen for a male but the body responds only to testicular estrogens. At puberty there is enough estrogen produced to develop the female secondary sex characteristics. These individuals rate high in verbal ability over performance tests which is typical of normal females. The 10 patients studied strongly desired marriage and had only minimal interest in outside jobs. Seven enjoyed taking care of the home and 8 played primarily with dolls and other girls' toys during childhood. None was described or described herself as a tomboy. Money et al. [152] reported that patients with androgen-insensitivity syndrome are exclusively feminine in their gender role and identity. They found no evidence of masculinized behavior that could be attributed to masculine genetics. Evidently the XY chromosomal constitution does not carry a genetic code directly responsible for masculine psychology and behavior. They concluded that the mediation between genetics and psychological behavior is hormonal. Presumably the failure of fetal androgens to influence tissue on a cellular level enhanced the capacity of the nervous system to comply with the demands of female sex assignment and rearing.

Patients with Turner's syndrome lack gonads and are thus not exposed to any gonadal hormones during the critical period. They differentiate morphologically as phenotypic females. Mascia et al. [153] compared 15 girls with Turner's syndrome to closely matched control subjects. All of the patients were more passive than their controls with respect to childhood fights. None tended to initiate fights and they showed less interest and liking for outdoor activities. They showed strong interest in doll play, were feminine in appearance, and scored very low on athletic interests, rough outdoor games, and energy expenditure. All had uniformly differentiated psychosexual identity as girls. Money and Ehrhardt [154] reported that Turner's syndrome represents the fact that feminine differentiation takes place in the absence of any added masculine factor.

The bulk of the data from careful observation of the clinical syndromes just described suggests that the presence or absence of sex hormones in utero is involved in the organization and differentiation of certain patterns of human cognition, intellectual functioning, and energy expenditure as well as in influencing the acquisition of other sexually dimorphic behaviors and interests found among humans. The effect of fetal androgen on the central nervous system has been demonstrated to continue to exert its influence into adulthood in both animals and humans, suggesting that the modification, organization, and differentiation which occur in its presence during the cirtical period remain fixed [142,149,152,155].

Imprinting Human Liver Cells. The precise mechanism(s) involved in the late expression of early events in human development remains unknown. Anatomical and biochemical differences resulting from steroidal hormone action on undifferentiated tissues have been well documented in instances where changes in organ structure occur. When the changes are intracellular, as in the case for liver steroid oxidoreductases, the "anatomical" change in the liver involves subcellular assemblies rather than gross organ structure.

There is no direct proof at this time that the observations of Pfaffenberger and Horning of biologic imprinting of liver enzyme activities are based on processes similar to those observed in animal studies, but this appears to be the most likely explanation. It must be emphasized that no female participant in their studies was either externally or behaviorally virilized. Figure 18 includes both sets of metabolite ratios for the 62 females studies. The region labeled "A" contains no male ratio values. Figure 19 indicates that apparently whatever the causative imprinting agent is, it simultaneously brings about increased 5α-oxidoreduction of both testosterone and cortisol.

The currently adopted hypothesis, by analogy with organizational mechanisms observed in animals and especially the recent work of McEwen and colleagues [140], is that pattern B steroid oxidoreductase activities in females may be the result of late fetal or neonatal exposure of developing hepatic tissue to intranuclear estradiol. The degree of exposure, both in intensity and duration, is unknown. For males, the secreted hormone is presumably testosterone, and the period of exposure probably corresponds to the period of Leydig cell activity [132]. For females, the time of exposure, the origin of the hormone(s) and the concentrations required to produce the effect are unknown. If, however, this interpretation is correct, several lines of investigation should be pursued. Obvious questions are whether

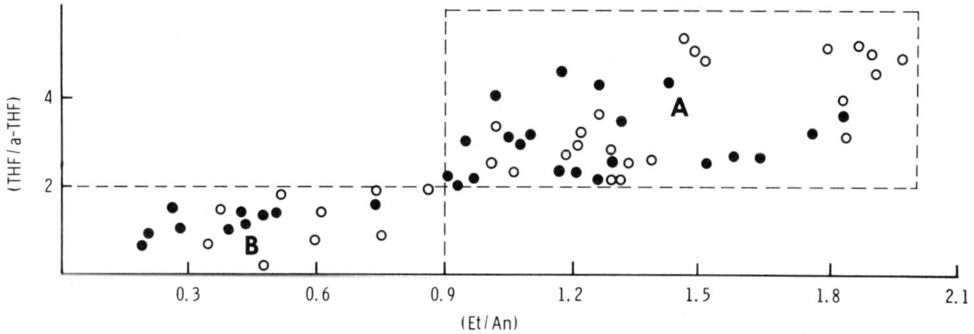

Figure 18. Ratio values (Et/An and THF/a-THF) for premenopausal females (closed circles) and postmenopausal females (open circles). Two female groupings, A and B, are evident; the block labeled "B" would contain all of the male values if they were plotted. Each individual is characterized by two sets of data determined from two separate ratios of metabolites, one from testosterone and one from cortisol. Low values are associated with block B, higher values with block A. (Source: Ref. 39. Reproduced courtesy of Academic Press, Inc.)

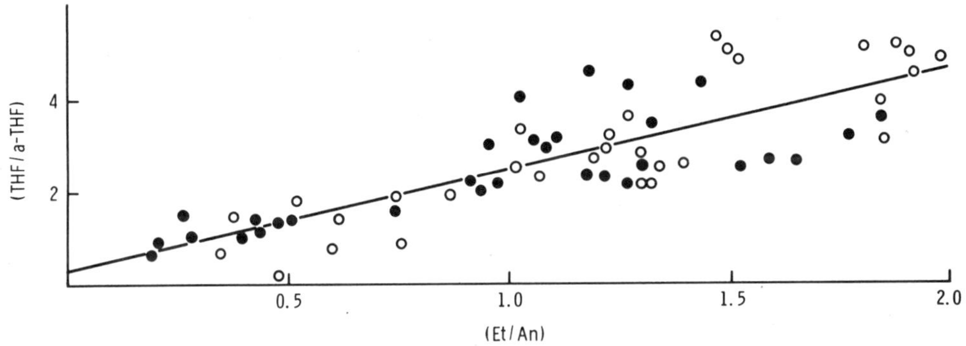

Figure 19. Regression analysis curve of the data presented in Fig. 18. The curve supports the contention that a single chemical agent simultaneously imprints the relative activities of certain 5β-H and 5α-H oxidoreductive enzymes.

male examples of pattern A will be found (corresponding to a lack of fetal hormonal stimulation); and if they are, will they be obtained from normal, healthy subjects or ones with recognized pathologies. It would also be interesting to learn if fetal or neonatal exposure of females to some hormonal conditions would lead to an exaggerated female pattern with greatly repressed 5α-H steroid oxidoreductase activity. Another question is whether or not additional enzymic activities are determined by organizational mechanisms. It seems unlikely that only 5α-H/5β-H steroid oxidoreductases would be involved in a basic mechanism of considerable bioligic importance.

2.6 Breast Lesion Pathology Studies

A major question which was raised by the above investigations was whether or not there is any difference in susceptibility to specific disorders associated with pattern A or B females. Unpublished observations of profiles of women with polycystic or sclerocystic ovary syndrome had indicated that more than one out of three of these subjects had pattern B ratios. A more well-defined study was made of groups of pre- and postmenopausal women with either fibrocystic breast disease or malignant breast lesion pathology. Earlier observations and the firm convictions of Allen et al. [156], Bulbrook et al. [157,158], and other endocrinologists that the concentrations of urinary androgen and adrenocorticosteroid metabolites should be useful in predicting the results of adrenalectomy or hypophysectomy in advanced cases of inoperable breast cancer added support to this endeavor.

The objectives of the study were to ascertain if each of the three pathological groups would exhibit the same two patterns, A and B, and in the same distribution, 2:1. If this were not the case, it would be of interest to know which patterns and distributions would present themselves and whether any unusual metabolite ratios would be particularly associated with one or more of the defined pathological states. The possibility of any of the observations being diagnostically useful was considered, specifically for early differentiation between malignant and nonmalignant pathology; and finally the possibility of any pattern bing associated with or being a presumptive causative agent of any type of breast lesion was examined.

Androsterone Sulfate Hydrolysis

A separate solvolysis procedure is used to determine values for the portion of androsterone which is excreted as the corresponding sulfate conjugate (which remains intact during Glusulase treatment). A 75-ml aliquot of the diluted urine is placed in a 125-ml screw-capped Erlenmeyer flask and the pH is adusted to 1 using 16N sulfuric acid. Then 15 g of sodium chloride are added and dissolved; the solution is poured into a 1-1 rotary extractor that contains 75 ml of ethyl acetate. The extraction step requires 10 min of the swirling action of the extractor. A further two extractions of the aqueous layer with 75-ml portions of ethyl acetate lead to a combined organic phase of 225 ml, which is reduced to 60 ml (Rotovap; temperature maintained below 40°C). This solution and 15 ml of tetrahydrofuran are placed in a 125-ml water bath at 37°C for 48 hr. The solvolysis mixture is poured into a rotary extractor and 25 μl of an internal reference compound, obtained from a stock solution (1 mg ml^{-1}) of 5α-androstan-17β-ol (methylene unit value, MU = 23.10 for the corresponding TMS ether) in pyridine, are added. A 15-ml portion of saturated sodium hydrogen carbonate solution is slowly added and the mixture is swirled for 10 min. The organic layer is washed once with 75 ml of glass-distilled water and the volume reduced to 3 to 4 ml. This residue is transferred into a 125-ml separatory funnel using 15 ml of ethyl acetate. From this stage on, the solvolysis sample is treated according to the latter part of the extraction procedure given earlier in Sec. 2.2 (under "Extraction of Steroids"). The urinary excretion values for androsterone are calculated using the sum of the androsterone liberated from the corresponding glucuronide during hydrolysis and that liberated from the sulfate during solvolysis.

Urinary Steroid Metabolic Profiles of Pathological Females

Premenopausal Benign Breast Lesions. Figure 20 shows a metabolic profile of a postpubertal, premenopausal female with bilateral benign fibrocystic

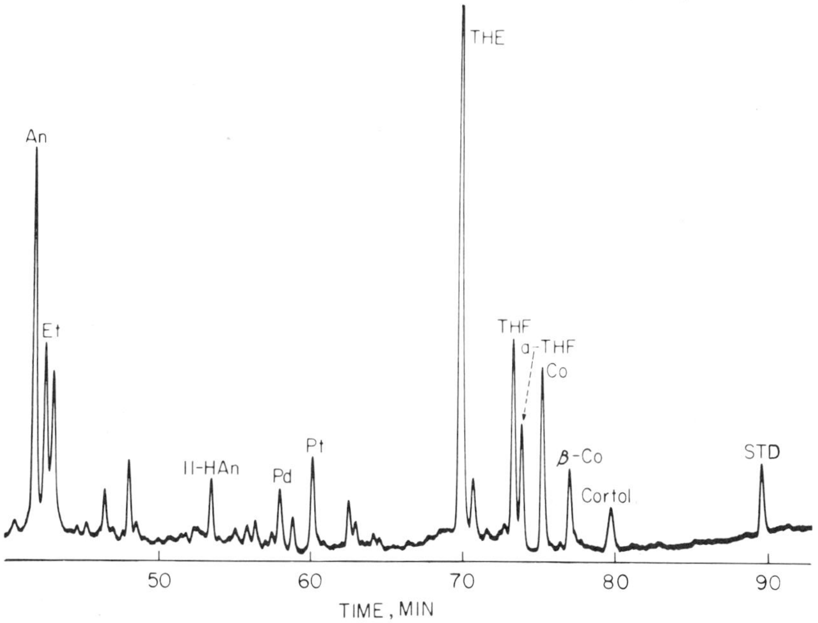

Figure 20. A representative pattern B urinary steroid metabolic profile of a postpubertal premenopausal female with bilateral benign fibrocystic disease of the breast. Analysis conditions and abbreviations are as in Fig. 7. STD is the internal reference compound, cholesteryl butyl ether. (Source: Ref. 40. Reproduced courtesy of Elsevier Scientific Publishing Co., Amsterdam.)

disease of the breast. The Et/An and THF/a-THF ratios are typical of the pattern B females mentioned earlier, i.e., they reflect decidedly male distributions of hepatic 5α-H and 5β-H oxidoreductases. Pattern B ratios appear to be associated with premenopausal females with benign breast lesions. Of the 37 women in this category who were studied, three-quarters had pattern B ratios although only one-third of those in the control group had pattern B ratios. It would be difficult to accept that the excretion of certain metabolites in certain ratios can be a direct cause of fibrocystic breast disease, but there does appear to be an association of the disease with pattern B ratios. (Abnormal levels of metabolite excretion were not observed for this group.) Figure 21 summarizes the findings and includes the data obtained from the control group. Even the pattern A ratios obtained from this pathological group appear to be unevenly distributed within area A, being shifted in the direction of pattern B values.

Postmenopausal Benign Breast Lesions. Figure 22 is a metabolic profile of a postmenopausal female with benign breast lesions; pattern A ratios are evident. The lower amount of androgen metabolite excretion observed here is typical for the postmenopausal female; a steady decline in androgen metabolite production can be observed with advancing years. The amounts of adrenocorticosteroid metabolites seen here are slightly elevated, probably as a stress reaction to being hospitalized.

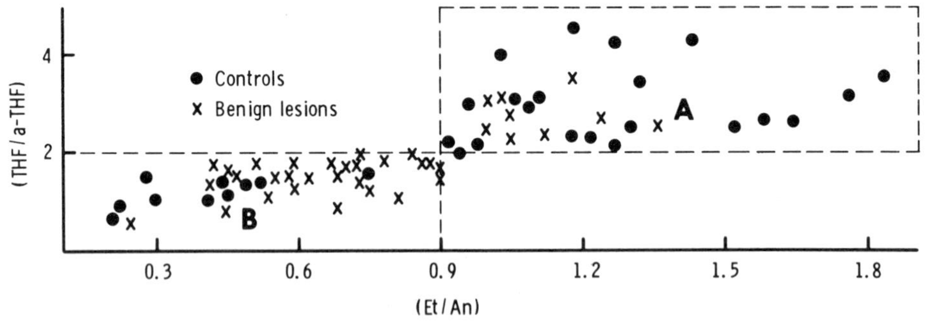

Figure 21. Et/An and THF/a-THF ratios for healthy premenopausal control subjects (; distributed 2:1 between pattern A and pattern B) and premenopausal subjects with benign breast lesions (X; distributed 1:3 between pattern A and pattern B). Even the pattern A ratios obtained from the pathological group appear unevenly distributed within area A, being shifted in the direction of area B. (Source: Ref. 40. Reproduced courtesy of Elsevier Scientific Publishing Co., Amsterdam.)

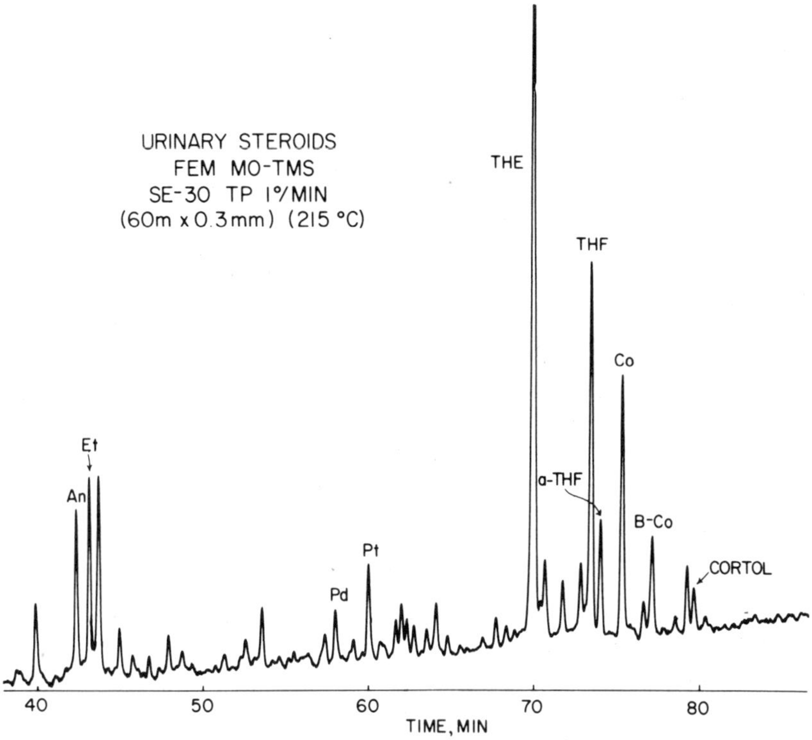

Figure 22. A pattern A urinary steroid metabolic profile of a postmenopausal female with benign breast lesions. The lower amount of androgen metabolite excretion observed here is typical of the postmenopausal female; a steady decline in androgen metabolites can be observed with advancing years. The amounts of adrenocorticosteroid metabolites seen here are comparable to those of the premenopausal female. (Source: Ref. 40. Reproduced courtesy of Elsevier Scientific Publishing Co., Amsterdam.)

During the investigation of this pathological group and the group with malignant breast lesions, a third pattern, which was designated pattern C, emerged. Pattern C had never been observed in any of the control groups or in the group with premenopausal benign lesions. Pattern C females exhibit elevated values for the Et/An ratio, viz., greater than 1.95. Figure 23 is an example of pattern C ratios with Et/An = 3.72 and THF/a-THF = 3.61.

Of the 30 postmenopausal females with benign breast lesions, 12 had pattern A, 10 had pattern B, and 8 had pattern C ratios. It was not completely surprising that the results of studying this group differed from those obtained from the premenopausal group. Pre- and postmenopausal fibrocystic breast disease are recognized by physicians to be clinically different

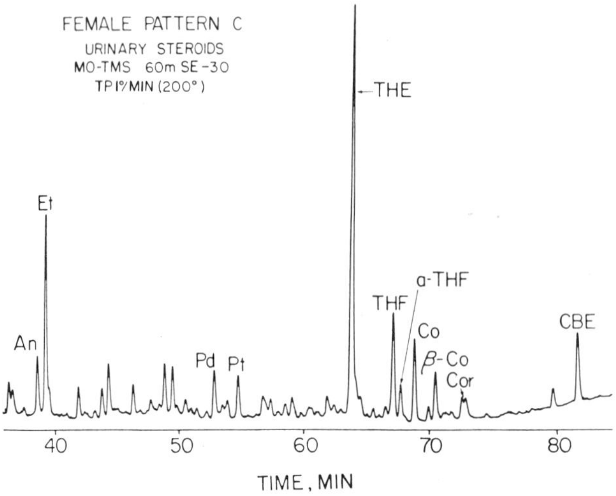

Figure 23. A pattern C urinary steroid metabolic profile of a postmenopausal female with benign breast lesions. Et/An = 3.72 and THF/a-THF = 3.61. Analysis conditions and abbreviations are as in Fig. 7. (Source: Ref. 40. Reproduced courtesy of Elsevier Scientific Publishing Co., Amsterdam.)

and to exhibit differing therapeutic responses during treatment. Figure 24 summarizes the testosterone reduction data for both types of benign breast lesions and compares the ratio distributions with those obtained from the two control groups.

Postmenopausal Malignant Breast Disease. Figure 25 summarizes the testosterone reduction data for both postmenopausal pathological groups and compares the ratio distribution with the one obtained for the postmenopausal control group. Of the 30 women with malignant breast lesions, 16 had pattern A, 7 had pattern B, and 7 had pattern C. In Fig. 26 all of the postmenopausal ratio data are compiled. Patterns A, B, and C fall within the respective areas A, B, and C.

Consideration of the pattern C tissue type is of interest in that no control subject or any premenopausal benign lesion subject ever exhibited pattern C ratios. Again, it appears that there is some associative relationship between pattern C ratios and breast lesion pathology, although a wider study would be required to justify acceptance of this suggestion.

The distribution of the postmenopausal females with breast lesions between pattern A and pattern B tissue types is similar to that obtained from the postmenopausal control group, viz., about 2:1. This was interpreted

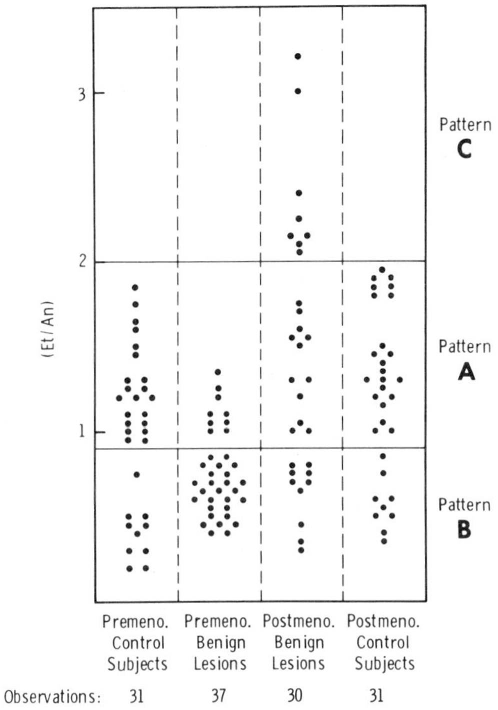

Figure 24. Testosterone reduction data (Et/An) for both categories of benign breast lesions and the ratio distributions obtained for the two female control groups. Only some of the postmenopausal subjects with benign lesions had pattern C ratios. (Source: Ref. 40. Reproduced courtesy of Elsevier Scientific Publishing Co., Amsterdam.)

as indicating that neither pattern A nor pattern B is selectively associated with either postmenopausal pathological state. This is in contrast to the results of the study of premenopausal women with benign breast lesions, in which an association of pattern B ratios with the disease is suggested. It was thus concluded that in the postmenopausal pathological subjects, pattern A and pattern B ratios are not diagnostically useful for differentiating between malignant and nonmalignant pathologies.

Table 4 recapitulates all of the data obtained for each group studied. It includes the averages of the urinary steroidal metabolite ratios, Et/An and THF/a-THF, according to pattern type, for males, both female control groups and the three pathological female populations. A total of 179 subjects were involved during the course of the investigation.

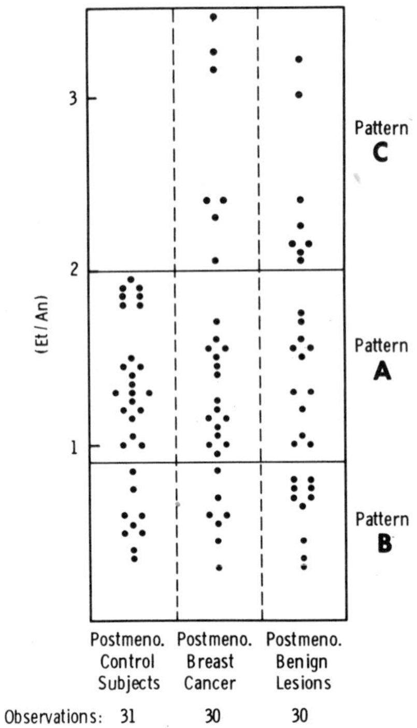

Figure 25. Testosterone reduction data for both categories of postmenopausal females with breast lesions and the ratio distribution obtained for the postmenopausal control group. Both pathological groups contained subjects who exhibited pattern C ratios. (Source: Ref. 40. Reproduced courtesy of Elsevier Scientific Publishing Co., Amsterdam.)

3. ALDOSE AND ALDITOL METABOLIC PROFILES

3.1 Associated Clinical Pathology

Although carbohydrates are required as energy-producing substrates for certain organs such as the central nervous system, and although they are essential components of the nucleic acids and some structural compounds such as hyaluronic acid, all of the necessary sugar units can be biosynthesized in the human body from precursors derived from the amino acids and the glycerol of various lipids. As a result, no carbohydrate is required in the diet; indeed, it has been shown experimentally that humans can survive in good health for months on a diet of fats and meats which provides only trace amounts of sugar. But many diets provide considerable amounts of carbohydrate, both as components of vegetables and grains and as purified

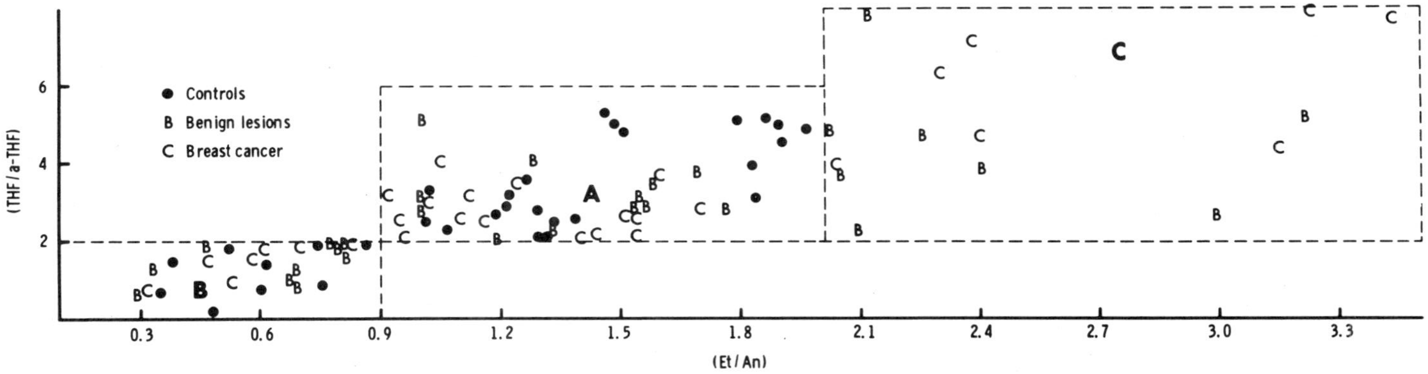

Figure 26. Testosterone and cortisol metabolite ratio data for both postmenopausal pathologic groups and contro subjects. Benign lesion patients are represented by B, cancer patients by C, and control subjects by closed circles. Patterns A, B, and C fall within the respective areas A, B, and C. The distribution of postmenopausal females with breast lesions between pattern A and pattern B is similar to that obtained for the postmenopausal control group. No member of the control group exhibited a pattern C ratio. (Source: Ref. 40. Reproduced courtesy of Elsevier Scientific Publishing Co., Amsterdam.)

Table 4. Average Urinary Steroidal Metabolite Ratios by Groups and Pattern

Group and pattern	Et/An	THF/a-THF
I. Premenopausal females		
A. Controls (31)		
1. Pattern A[a] (21/31)	1.39	2.98
2. Pattern B[b] (10/31)	0.60	1.29
B. Benign lesions (37)		
1. Pattern A (9/37)	1.28	2.81
2. Pattern B (28/37)	0.66	1.58
II. Males (= pattern B) (21)	0.67	1.57
III. Postmenopausal females		
A. Controls (31)		
1. Pattern A (22/31)	1.35	3.29
2. Pattern B (9/31)	0.61	1.32
B. Benign lesions (30)		
1. Pattern A (12/30)	1.48	3.12
2. Pattern B (10/30)	0.67	1.46
3. Pattern C[c] (8/30)	3.03	4.09
C. Breast cancer (30)		
1. Pattern A (16/30)	1.37	2.85
2. Pattern B (7/30)	0.64	1.48
3. Pattern C (7/30)	3.19	5.58

[a]Pattern A: $2.0 > \text{Et/An} \geq 0.9$; $6.0 > \text{THF/a-THF} \geq 2.0$.

[b]Pattern B: $\text{Et/An} < 0.9$; $\text{THF/a-THF} < 2.0$.

[c]Pattern C: $\text{Et/An} > 2.0$; $8.0 > \text{THF/a-THF} \geq 2.0$.

substances such as refined sugar. Often one-third to one-half of an individual's caloric intake is derived from carbohydrate sources; and because it is more efficient and less expensive to use vegetables and grains for food, the proportion of carbohydrate in the diet tends to be higher in lower socioeconomic groups. Carbohydrate metabolism is thus of central importance to a large portion of the world's population.

Disturbances in carbohydrate metabolism, either congenital or developed, lead to a host of diseases which should lend themselves to investigation by metabolic profiling techniques.

Von Mering and Minkowski [159] first demonstrated that if a dog's pancreas is removed, serious disturbances of glucose metabolism result with sudden elevation of blood glucose concentration and development of the clinical symptoms of diabetes mellitus. Insulin, the major hormone involved in carbohydrate metabolism, affects cell membrane activity, amino acid transport, intracellular glucose metabolism, lipid metabolism, protein synthesis, and growth. Carbohydrate metabolism is also affected by epinephrine, glucagon, adrenocorticosteroids, and pituitary growth hormone.

Diabetes mellitus is a generalized, chronic metabolic disorder which manifests itself, in its fully developed form, by hyperglycemia, glycosuria, increased protein breakdown, ketosis, acidosis, and hyperlipemia. Other diseases associated with improper carbohydrate metabolism include acromegaly, hypoglycemia, fructose intolerance, congenital galactosemia, pentosuria, congenital lactose intolerance, and congenital and acquired renal glycosuria.

An array of at least seven glycogen-storage diseases exist, including type I (von Gierke's disease or hepatorenal glycogne-storage disease), type II (Pompe's disease), type III (limit dextrinosis, debrancher glycogen-storage disease, or Cori's disease), type IV (amylopectinosis, brancher glycogen-storage disease, or Andersen's disease), type V (muscle phosphorylase deficiency or McArdle's disease), type VI (liver phosphorylase dificiency or Hers' disease), and type VII (muscle phosphoglucomutase deficiency, another type of McArdle's disease).

The formation of cataractous human lens tissue has also been linked to carbohydrate metabolism, particularly to the accumulation of myo-inositol within the aging lens [12].

Jellum et al. [64] have published a carbohydrate profile for normal seminal fluid. Their chromatogram includes derivatives of glycerol, erythritol, fucose, ribose, fructose (the major component), glucose, sorbitol, a relatively large amount of myo-inositol, and 11 components which were not identified.

Aldose and alditol metabolic profiles have been useful in studies of both normal and unusual carbohydrate metabolism. Many more uses should materialize, especially in studies of glycogen storage disease.

3.2 Derivatives for Gas Chromatography of Aldoses and Alditols

Two major complications have always presented themselves in analytical procedures devised to separate complex mixtures of aldoses and related polyols. One is the well-known problem of the formation of several compounds from any one aldose, viz., an anomeric pair with furano and pyrano derivatives of each form. The other problem is associated with the fact that all the members of a family of aldoses, e.g., the aldohexoses, are quite similar in structure and thus inherently difficult to separate. Invariably, any GC method employing packed columns leads to the result of at least two derivatives having identical methylene unit (MU) values [87,88,160].

With the advent of 60-m SE-30 open tubular glass capillary columns having thermal stability to 300°C and possessing between 100,000 and 125,000 theoretical plate efficiencies, it became feasible to attempt to separate complex mixtures of aldoses and related polyols. The procedure described here was developed by Szafranek, Pfaffenberger, and Horning [10] and is based on the first step of the classical Wohl degradation of aldoses [161-163]. Open-chain poly-O-acetylaldonic nitriles are formed, one from each aldose, in high yield and with retention of terminal dissymmetry [10]. The Wohl derivatives give good GC peaks; and polyols, when subjected to the Wohl conditions, lead to polyacetyl esters which also chromatograph well. Figure 27 is a separation of 13 common aldoses as poly-O-acetylaldonic nitriles. Methylene unit (MU) values of the derivatives are given in Table 5.

Complete separation of these aldoses is possible. The order of elution of the derivatives depends not only upon the molecular weight but also upon the configuration of the asymmertric carbon atoms [164]. The only pair of compounds not separable under the experimental conditions is D-xylose and 6-deoxy-L-galactose. The latter compound was not included in the study;

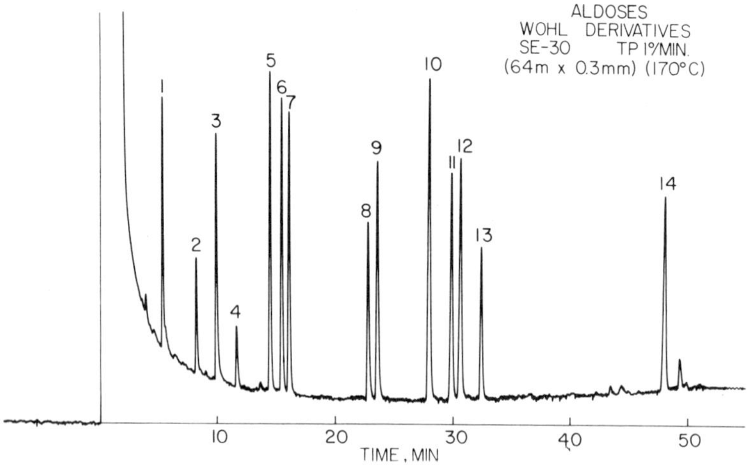

Figure 27. Separation of 13 aldoses as Wohl derivatives on a 64-m SE-30 glass open tubular capillary column temperature programmed 1°C min^{-1} from 170°C. The components include: 1—D-erythrose, 2—L-threose, 3—2-deoxy-D-ribose, 4—L-threitol, 5—D-ribose, 6—arabinose, 7—D-xylose, 8—2-deoxy-D-glucose, 9—2-deoxy-D-galactose, 10—3-O-methyl-D-glucose, 11—mannose, 12—D-glucose, 13—D-galactose and 14—D-glycero-D-gulo-heptose.

Table 5. Methylene Unit (MU) Values of Wohl Derivatives of Aldoses[a]

Aldose	MU value
D-erythrose	14.07
L-threose	15.10
2-Deoxy-D-ribose	15.57
D-ribose	16.63
D-arabinose	16.80
D-xylose	16.94
2-Deoxy-D-glucose	18.07
2-Deoxy-D-galactose	18.15
3-O-methyl-D-glucose	18.81
D-mannose	19.06[b]
D-glucose	19.17[b]
D-galactose	19.40[b]
D-glycero-D-gulo-heptose	21.37[b]

[a]Initial temperature, 105°C.
[b]Initial temperature, 170°C.

its MU value is 16.91 (Wohl derivative). Figure 28 is a separation of 22 polyols as polyacetyl esters; Table 6 contains the methylene unit (MU) values for these derivatives.

Because the molecular weight of a Wohl derivative of an aldose is always lower than that of the corresponding alditol acetate, it is possible to separate complex mixtures of polyols and aldoses. From Fig. 29 it may be observed that only two pairs of a 32-component mixture were incompletely separated, viz., the separation of D-galactose from 2-deoxy-D-glucitol and L-glycero-D-gulo-heptitol from meso-glycero-gulo-heptitol.

Methyl and trimethylsilyl ethers or acetyl and trifluoroacetyl esters have been used in GC analyses of aldoses, but then it is possible for the aldoses to form anomeric pairs of cyclic furano and pyrano isomers when derivatized. This fact leads to the well-known multiplicity of derivatives for aldoses, and a mixture containing multiple derivatives of several aldoses is difficult to separate, even with capillary columns [165]. Although the reduction of the carbonyl group of a sugar removes the possibility of forming isomers, retention of terminal dissymmetry is likewise lost. For example, only six polyols are obtained from the eight aldohexoses. Therefore, re-

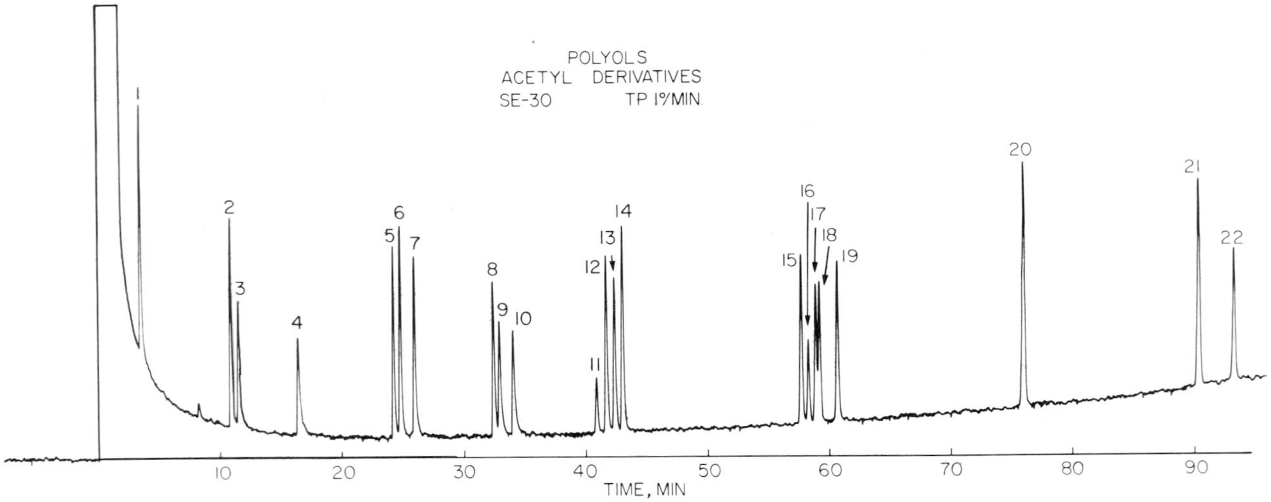

Figure 28. Separation of 22 polyols as polyacetyl esters; see Fig. 27 for column description and analytical conditions. The components include: 1—glycerol, 2—erythritol, 3—L-threitol, 4—2-deoxy-D-ribitol, 5—ribitol, 6—D-arabinitol, 7—xylitol, 8—2-deoxy-D-glucitol, 9—2-deoxy-D-galactitol, 10—3-deoxy-D-glucitol, 11—myo-inositol, 12—D-mannitol, 13—D-glucitol, 14—galactitol, 15—D-glycero-D-gluco-heptitol, 16—D-glycero-D-manno-heptitol, 17—L-glycero-D-gulo-heptitol, 18—meso-glycero-gulo-heptitol, 19—D-glycero-D-galacto-heptitol, 20—D-erythro-L-galacto-octitol, 21—D-arabino-D-manno-nonitol, and 22—D-arabino-L-galacto-nonitol.

Table 6. Methylene Unit (MU) Values of Polyacetyl Esters

Polyol	MU value
Glycerol	13.08[a]
Erythritol	15.88
L-threitol	15.98
2-Deoxy-D-ribitol	16.99
Ribitol	18.29
D-arabinitol	18.37
Xylitol	18.55
2-Deoxy-D-glucitol	19.40
2-Deoxy-D-galactitol	19.48
3-Deoxy-D-glucitol	19.65
Myo-inositol	20.47[b]
D-mannitol	20.59
D-glucitol	20.67
Galactitol	20.76
D-glycero-D-gluco-heptitol	22.63
D-glycero-D-manno-heptitol	22.69
L-glycero-D-gulo-heptitol	22.78
Meso-glycero-gulo-heptitol	22.82
D-glycero-D-galacto-heptitol	23.00
D-erythro-L-galacto-octitol	25.01
D-arabino-D-manno-nonitol	26.99
D-arabino-L-galacto-nonitol	27.40

[a]For glycerol through 3-deoxy-D-glucitol an initial temperature of 105°C was employed.

[b]For myo-inositol through D-arabino-L-galacto-nonitol, the initial temperature was 170°C.

duction of the carbonyl group should not be applied to a mixture containing both sugars and polyols. Conversion of aldoses into aldonolactone-tri-methylsilyl ethers [166] retains terminal dissymmetry, but from D-arabinose, D-xylose, D-mannose, and D-glycero-D-gulo-heptose more than one derivative is obtained [167]. The procedure described here is based upon formation of poly-O-acetylaldonic nitriles (Wohl derivatives of aldoses) [161,162]. The GC analysis of mixtures of polyols and aldoses thus becomes satisfactory because the nitriles have open-chain structures and molecular weights lower than the acetyl derivatives of the corresponding polyols.

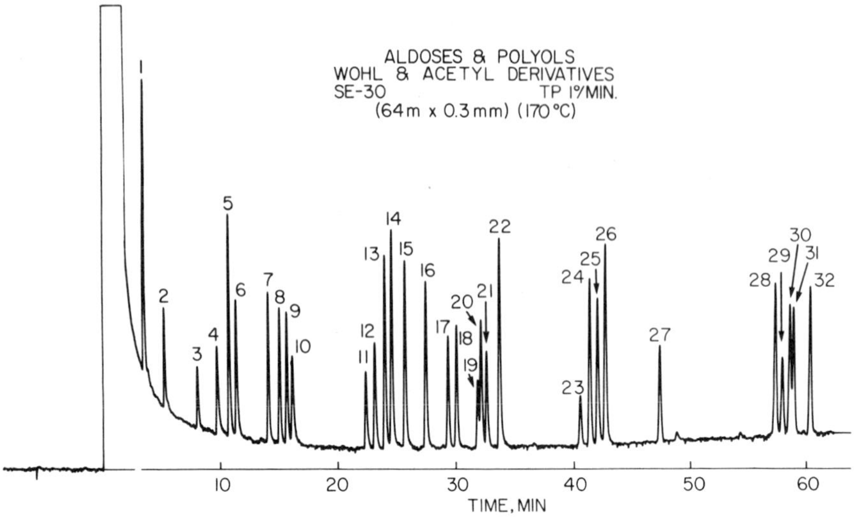

Figure 29. Separation of polyols and aldoses as the corresponding polyacetyl and Wohl derivatives; see Fig. 27 for column description and analytical conditions. The components include: 1—glycerol, 2—D-erythrose, 3—L-threose, 4—2-deoxy-D-ribose, 5—erythritol, 6—L-threitol, 7—D-ribose, 8—D-arabinose, 9—D-xylose, 10—2-deoxy-D-ribitol, 11—2-deoxy-D-glucose, 12—2-deoxy-D-galactose, 13—ribitol, 14—D-arabinitol, 15—xylitol, 16—3-O-methyl-D-glucose, 17—D-mannose, 18—D-glucose, 19—D-galactose, 20—2-deoxy-D-glucitol, 21—2-deoxy-D-galactitol, 22—3-deoxy-D-glucitol, 23—myo-inositol, 24—D-mannitol, 25—D-glucitol, 26—galactitol, 27—D-glycero-D-gulo-heptose, 28—D-glycero-D-gluco-heptitol, 29—D-glycero-D-manno-heptitol, 30—L-glycero-D-gulo-heptitol, 31—meso-glycero-gulo-heptitol, and 32—D-glycero-D-galacto-heptitol.

The reactions which occur in the Wohl synthesis are included in Ref. 10 for β-D-glucose. The high temperature of acetylation used in the procedure brings about the exclusive formation [162] of poly-O-acetylaldonic nitriles. A cyclic by-product which is formed at low reaction temperatures is not a final product under the reaction conditions employed. Ketoses, including D-fructose, form only nonvolatile products which are not detected under the analysis conditions. Eklund et al. [168] have reported an acceptable procedure based on the formation of trifluoroacetyl esters of sugars; this method should be used when ketoses are important components of the mixture being analyzed. The method is extremely sensitive, the minimum detectable amounts of monosaccharides being at the subpicogram level when an electron capture detector is employed.

3.3 Obtaining and Using the Profiles

The recommended procedure for obtaining Wohl derivatives of aldoses converts polyols to polyacetyl esters without complication. The method may therefore be applied in studies of biologic samples containing both types of compounds.

Preparation of Derivatives

Acetyl Derivatives of Reference Alditols [*10*]. One mg of a reference alditol in 200 μl of acetic anhydride containing 2 mg of dry sodium acetate is heated at 120°C for 2 hr. It is necessary to heat longer (5 hr) only in the case of D-arabino-L-galacto-nonitol. Sodium acetate is preferable to pyridine catalyst; the reaction is much faster.

Wohl Derivatives of Reference Aldoses [*10*]. A mixture of 1 mg of a reference aldose, 1 ml of hydroxylamine hydrochloride, and 2.5 mg of dry sodium acetate is dissolved in 200 μl of dry methanol and heated at 60°C for 1 hr. The methanol is removed with a stream of dry nitrogen, and the residue is treated with 200 μl of acetic anhydride at 120°C for 1 hr. The product is the corresponding poly-O-acetylaldonic nitrile.

Metabolic Profile Samples from Urine [*11*]. Urine samples from female and male human neonates, juveniles, and adults on unrestricted diets (excepting neonates in the nursery service) are collected in sterilized containers over a 24-hr period unless only one void is obtainable. No preservative is used, but the sample is kept at 4°C during the collection period and stored under deep-freeze conditions as soon as possible after the volume has been measured. A creatinine determination is made on one aliquot by any standard clinical laboratory procedure.

The pH of a 20-ml aliquot of urine is adjusted to 4.5 using 10% aqueous acetic acid. This sample is poured onto a packing of DEAE-Sephadex A-25 Medium (Pharmacia, Uppsala, Sweden), contained in a 10 cm × 1.3 cm ID glass column. Basic and neutral compounds are eluted by washing the packing with 100 ml of deionized water which has been glass-distilled from basic potassium permanganate. The eluate is collected in a 250-ml round-bottomed flask and lyophilized to constant weight. The dry residue is dissolved in 10 ml of methanol. A 1-ml aliquot of the methanolic solution is heated with 5 mg of hydroxylamine hydrochloride and 12.5 mg of sodium acetate at 60°C for 30 min in a PTFE-lined screw-cap vial. The product is dried at 40°C using a nitrogen stream, and derivatized in 1 hr using 300 μl of acetic anhydride at 100°C. When an internal reference compound is required, 200 μl of a solution of glyceryl-1-decyl ether (6.30 mg 10.0 ml^{-1} of methanol) is added prior to heating with hydroxylamine hydrochloride. Also, for samples of biologic origin, 100 μl of pyridine is used as catalyst (instead of sodium acetate). The dry residue from these reactions is treated with 300 μl of acetic anhydride and 50 μl of pyridine at 100°C for 1 hr.

Gas Chromatography. In the work reported here [11], separation of the urinary polyols and aldoses as polyacetyl esters and poly-O-acetylaldonic nitriles was carried out with a Barber Colman gas chromatograph (model 5000 with flame ionization detection) which had been modified to accept 60-m glass capillary columns (0.3 mm ID) coated with SE-30 containing Silanox 101 (Cabot Corp., Boston, Mass.) prepared according to German and Horning [84]. The theoretical plate efficiencies of the capillary columns were between 100,000 and 125,000 when measured isothermally at 250°C for n-tetracosane. A Keithley model 417 picoammeter and a Texas Instruments recorder were employed. The precolumn helium [85] and flame hydrogen flowrates were controlled by Brooks model 8744 flow controllers and measured with Matheson model 8110-0121 mass flow meters.

Methylene unit (MU) values [87,88] were measured through the use of n-alkanes coinjected with the samples. Temperature programming was employed at 1°C min^{-1} from 150°C. Other GC conditions included: sample volume 2 μl; split ratio, 5:1; precolumn-inlet splitter, 240°C; detector bath, 300°C; column inlet pressure, 18 lb in^{-2}, resulting in a carrier gas (helium) linear velocity of 16 to 18 cm sec^{-1} (200°C); hydrogen flow, 10 ml min^{-1}; air flow, 30 ml min^{-1}.

Table 7 lists methylene unit (MU) values and molar detector response factors for derivatives of the 17 polyols and aldoses identified in a study of normal human urine by Pfaffenberger et al. [11]. A typical GC separation of these compounds, using a 60-m glass capillary column with SE-30 liquid phase, temperature-programmed 1°C min^{-1} from 165°C, is shown in Fig. 30.

The methylene unit (MU) values listed in Table 7 are highly reproducible (±0.03 MU) for each column (column I), but a shift of up to about 0.1 MU can occur when a different column (e.g., column II) is used. Values for all components are shifted in the same direction when this effect occurs, and no change in the order of elution of urinary components has been observed.

Table 7. Methylene Unit (MU) Values for Acetyl and Wohl Derivatives of Polyols and Aldoses Identified in Human Urine[a]

Peak no.	Assignment	MU value Col. I	Col. II	Molar response factor (MRF)[b]
1	Erythritol	15.78	15.89	0.62 ± 0.08
2	Threitol	15.95	16.08	0.62 ± 0.08[c]
3	Ribose	--	16.66	0.45 ± 0.02
4	Arabinose	--	16.83	0.40 ± 0.03
5	Fucose	--	16.93	0.40 ± 0.03
6	Xylose	--	16.96	0.40 ± 0.02
7	Ribitol	18.24	18.35	0.65 ± 0.03
8	Fucitol	--	18.38	--
9	Arabinitol	18.33	18.42	0.61 ± 0.08
10	Xylitol	18.50	18.61	0.62 ± 0.08
11	Mannose	--	19.12	0.52 ± 0.03
12	Glucose	19.11	19.20	0.53 ± 0.03
13	Galactose	--	19.46	0.38 ± 0.02
IS	Glyceryl-1-decyl ether	--	19.88	1.00[d]
14	Myo-inositol	20.54	20.63	0.66 ± 0.04
15	Mannitol	20.63	20.70	0.84 ± 0.10
16	Glucitol	20.70	20.79	0.82 ± 0.09
17	Galactitol	20.79	20.87	0.76 ± 0.02

[a]SE-30 glass open tubular capillary column, temperature-programmed at 1°C min^{-1} from 150°C.

[b]Determined in triplicate using column II.

[c]Assumed to be the same as for erythritol.

[d]Internal reference compound (IS).

A typical GC analysis requires about 1 hr. Molar response factors (MRF) are calculated in relation to the detector response observed for the diacetyl derivative of glyceryl-1-decyl ether.

3.4 Representative Metabolic Profiles of Urinary Aldoses and Alditols

Many human pathologic circumstances result in alteration of urinary profiles of polyols and aldoses. Common examples include diabetes mellitus [159],

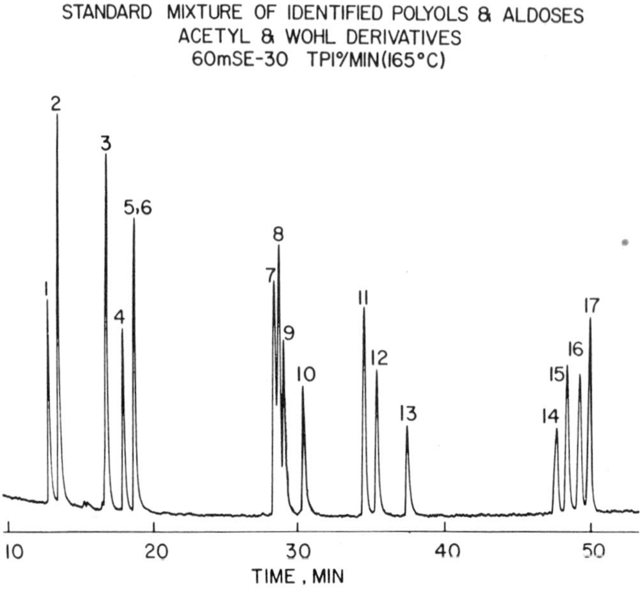

Figure 30. Separation of the reference mixture of 17 polyols and aldoses which were quantified as acetyl and Wohl derivatives during a study of human urine from healthy neonates, juveniles, and adults. A 60-m SE-30 glass open tubular glass capillary column was temperature programmed 1°C min^{-1} from 165°C. The components include: 1—erythritol, 2—threitol, 3—ribose, 4—arabinose, 5—fucose, 6—xylose, 7—ribitol, 8—fucitol, 9—arabinitol, 10—xylitol, 11—mannose, 12—glucose, 13—galactose, 14—myo-inositol, 15—mannitol, 16—glucitol, and 17—galactitol. (Source: Ref. 11. Reproduced courtesy of Academic Press, Inc.)

other glucosurias [169], galactosemia [170], pentosuria [171], and other disturbances or inborn errors in the usual carbohydrate metabolic pathways. Pfaffenberger et al. [11] chose three age groups for their initial studies. Neonates, aged less than 6 months, were generally in the first few days of life. Juveniles, 4 to 10 years of age, were on unrestricted diets. Adults, also on unrestricted diets, were 20 years or more in age.

According to Fig. 30, baseline separation of each of the components quantified was obtained except between derivatives of the fucose-xylose pair and the ribitol-fucitol-arabinitol trio. This complication proved to be unimportant because fucose and fucitol are rarely present above trace levels, and derivatives of ribitol and arabinitol are well separated. The internal reference compound (IS), the diacetyl ester of glyceryl-1-decyl ether, elutes in the otherwise clear region between derivatives of galactose and myo-inositol.

Excretion Profiles of Neonates

Figure 31 is a typical chromatogram observed for a neonate. The major components are 1—erythritol, 2—threitol, 9—arabinitol, and 14—myo-inositol. Erythritol and threitol are invariably present in neonatal urine, usually in a ratio of about 6:1. Average excretion values for these compounds are, respectively, 120 and 21 μg mg^{-1} creatinine. Pentoses are usually not detectable, or are present in trace amounts (<4 μg mg^{-1} creatinine). Fucitol is usually present in small amounts. Mannose, glucose, and galactose are often present in small amounts, with glucose predominating. Average excretion values for these three hexoses are respectively 4, 24, and 8 μg mg^{-1} creatine. By far the largest component of urinary samples from neonates is myo-inositol. The average excretion observed for myo-inositol is 2.2 mg mg^{-1} creatinine. The range of observed values is wide: 0.3 to 4.9 mg mg^{-1} creatinine for the 10 neonates reported [11]. The presence of comparatively

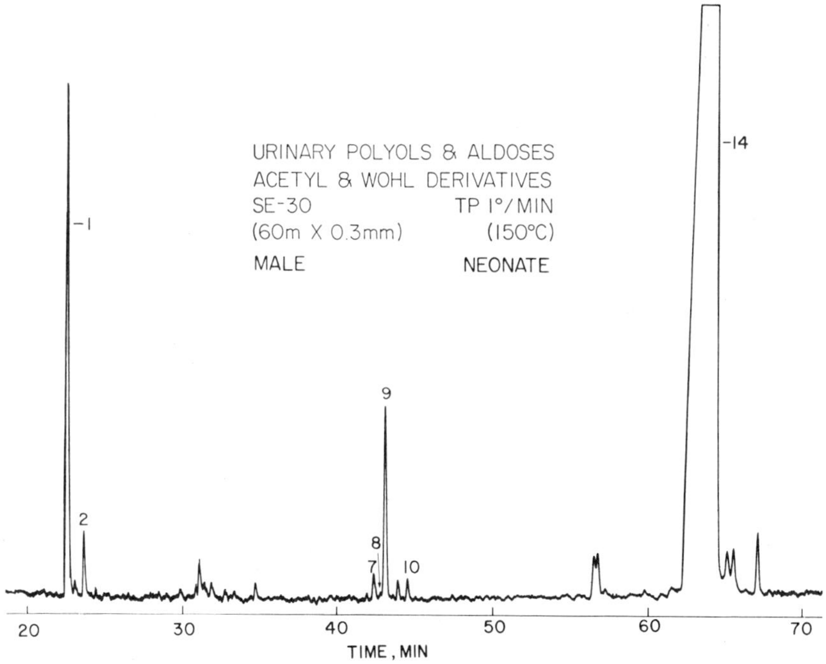

Figure 31. Profile of human (male neonate) urinary polyols and aldoses separated as in Fig. 30. The identified components include: 1--erythritol, 2—threitol, 7—ribitol, 8—fucitol, 9—arabinitol, 10—xylitol, and 14—myo-inositol, usually in relatively large amounts in neonatal urine. (Source: Ref. 11. Reproduced courtesy of Academic Press, Inc.)

large amounts of myo-inositol in the urine may be related to the continuing synthesis of central nervous system tissue during the first few days of life. The samples which had the lowest myo-inositol levels were from the oldest infants (1 month of age).

The presence of large amounts of myo-inositol will mask or delay the elution of derivatives of mannitol, glucitol, and galactitol. When the amount of myo-inositol is low, other hexitols are readily detected.

No signigicant difference was noted [11] on the basis of the sex of neonates, but females appeared to excrete about 2.5 times as much glucose as males. Female juveniles and female adults also usually excrete more glucose than do males; the ratio is about 10:1 for adults. (This comparison is based upon μg mg^{-1} creatinine.)

Excretion Profiles of Juveniles

Figure 32 is a typical chromatogram for a juvenile. The major components are 1—erythritol, 2—threitol, 7—ribitol, 9—arabinitol, 10—xylitol, 14—myo-inositol, 15—mannitol, 116—glucitol, and 17—galactitol. Erythritol and threitol are always present in a ratio of about 4:1. The average excretion values for these compounds are 110 and 25 μg mg^{-1} creatinine, respectively. Pentoses are usually present only in trace amounts. The unidentified compound which elutes around 30 min (MU = 17.04) is not a simple sugar or alditol. Ribitol, arabinitol, and xylitol are always present; the average excretion values are respectively 16, 75, and 17 μg mg^{-1} creatinine. Arabinitol predominates; a trace of fucitol is usually present. Mannose, glucose, and galactose are often present in small amounts. The four hexitols—myo-inositol, mannitol, glucitol, and galactitol—are present in varying ratios. The average excretion values are respectively 81, 93, 25, and 20 μg mg^{-1} creatinine. No significant difference based on sex was observed [11] except for the glucose ratios already discussed.

Excretion Profiles of Adults

Figure 33 is a typical chromatogram observed for an adult. The major components are those found in samples from juveniles, but with larger amounts of mannose. Adults excrete erythritol and threitol in a ratio of about 6:1. Average excretion values were found to be 63 and 11 μg mg^{-1} creatinine, respectively [11]. Pentoses are usually found in only trace amounts. Ribitol, arabinitol, and xylitol are always present. The average excretion values are respectively 10, 53, and 22 μg mg^{-1} creatinine. A trace of

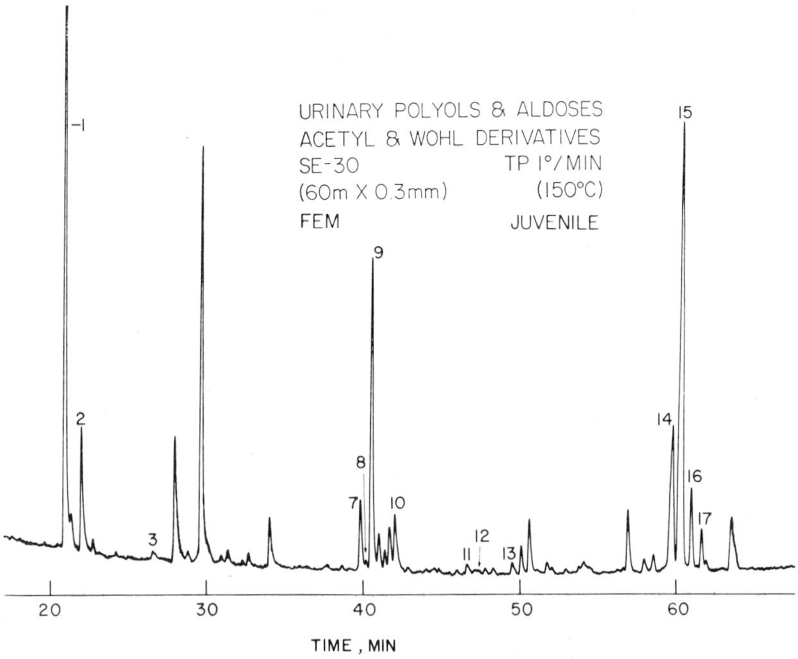

Figure 32. Profile of human (female juvenile) urinary polyols and aldoses separated as in Fig. 30. The identified components include: 1—erythritol, 2—threitol, 3—ribose, 7—ribitol, 8—fucitol, 9—arabinitol, 10—xylitol, 11—mannose, 12—glucose, 13—galactose, 14—myo-inositol, 15—mannitol, 16—glucitol, and 17—galactitol. (Source: Ref. 11. Reproduced courtesy of Academic Press, Inc.)

fucitol is usually found. Mannose, glucose and galactose are usually present in small amounts.

Myo-inositol, mannitol, glucitol, and galactitol were found [11], with respective average excretion values of 33, 26, 23, and 6 μg mg^{-1} creatinine. No sex-related differences were observed, except for the aforementioned glucose values.

Other Polyols and Aldoses

The region of the GC separation corresponding to MU values from 21 to 28 is usually not of interest and is often not included in the profile analysis to conserve time. No major urinary polyol has been observed in this region to date; however, the region is readily accessible. MU values for five heptitols, one octitol, and two nonitols have been reported [10].

Table 8 contains values for the maximum, minimum, and average amounts of urinary polyols and aldoses excreted according to Pfaffenberger et al. [11].

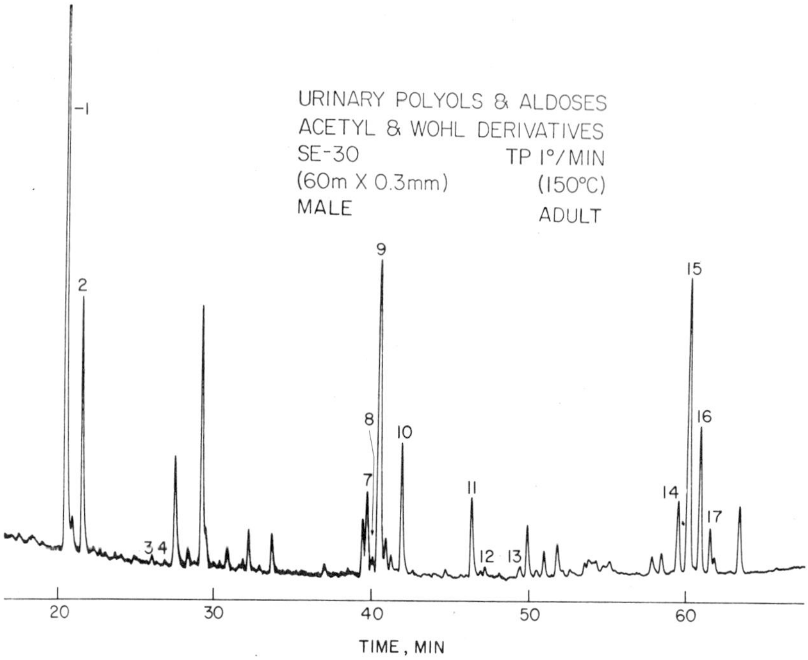

Figure 33. Profile of human (male adult) urinary polyols and aldoses separated as in Fig. 30. The identified components include: 1—erythritol, 2—threitol, 3—ribose, 4—arabinose, 7—ribitol, 8—fucitol, 9—arabinitol, 10—xylitol, 11—mannose, 12—glucose, 13—galactose, 14—myo-inositol, 15—mannitol, 16—glucitol, and 17—galactitol. (Source: Ref. 11. Reproduced courtesy of Academic Press, Inc.)

3.5 Separation of Sugar Acids and Lactones

In 1974, Szafranek et al. [172] reported that attempts to analyze aldonic or uronic acids as the corresponding alditol acetates were successful, but terminal dissymmetry was forfeited as evidence by four hexuronic acids yielding only two hexitols. Moreover, aldonic and uronic acids led to identical reaction products [173]. Mixtures containing aldonic, hexuronic, and hexaric acids cannot be analyzed as the corresponding alditols because the reduction step destroys family identity [173-175]. The substances do, however, lend themselves to analysis as trimethylsilyl derivatives, the formation of which does not involve prior chemical modification. At present these compounds are not known to be of large biological significance; but this fact may change with thorough investigations of the various glycogen storage diseases. Figure 34 demonstrates the great utility of glass capillary gas chromatographic separations of rather complex mixtures.

Table 8. Excretion Values for Urinary Polyols and Aldoses (μg mg^{-1} creatinine)

	Neonate			Juvenile			Adult		
	Min	Max	Avg	Min	Max	Avg	Min	Max	Avg
Erythritol	50	200	120	58	200	110	33	110	63
Threitol	5	48	21	11	72	25	6	16	11
Ribose	X[a]	T[b]	X	X	5	T	X	T	T
Arabinose	X	T	X	X	T	T	X	T	T
Fucose	X	T	X	X	T	T	X	T	T
Xylose	X	T	X	X	T	T	X	T	T
Ribitol	5	24	14	8	29	16	5	40	10
Fucitol	X	T	T	X	T	T	T	T	T
Arabinitol	51	170	98	47	110	75	31	140	53
Xylitol	9	28	15	9	30	17	6	100	22
Mannose	X	20	4	X	T	T	X	T	T
Glucose	T	80	24	X	9	T	T	61	8
Galactose	X	30	8	X	17	6	X	50	7
Myo-inositol	300	4900	c	11	250	81	9	72	33
Mannitol	X	60	d	T	360	93	X	58	26
Glucitol	X	45	d	10	44	25	T	120	23
Galactitol	7	51	d	7	64	20	T	19	6

[a]X = not present in the chromatogram.

[b]T = trace component (<4 μg mg^{-1} creatinine).

[c]Not averaged; decreases during neonatal period.

[d]Not averaged for neonatal samples.

3.6 Studies of Human Cataractous Lens Tissue

Cataract formation results in a loss of transparency of the lens of the eye. No cure or means of preventing senile cataract, the most common form of the disorder, is known. Most investigations of this circumstance have indicated that cataract formation, both in man [176-179] and in experimental animals [180-188], involves the accumulation of one of several polyols within the tissue of the lens. The osmotic balance of the lens must be carefully maintained for proper vision. Although amino acids are mainly responsible for maintaining osmotic balance [185], potassium and chloride ions make osmotic contributions [178]. Free polyols and aldoses also affect the osmotic bal-

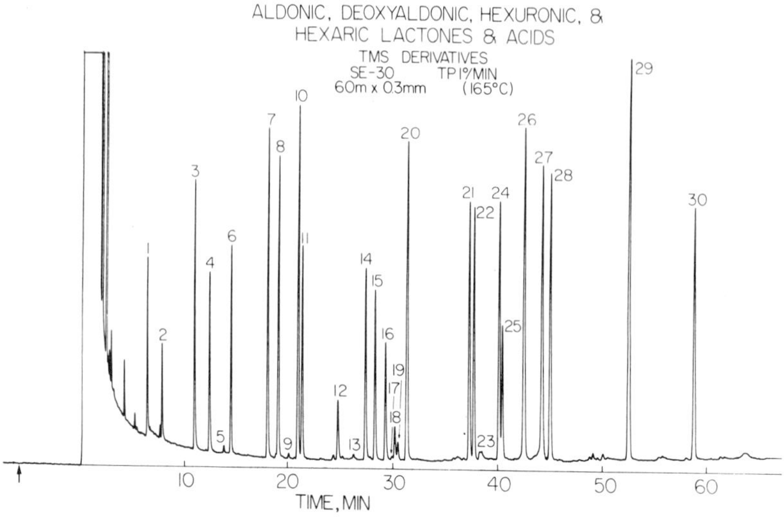

Figure 34. Separation of TMS derivatives of aldonic, deoxyaldonic, hexuronic, and hexaric lactones and acids as in Fig. 30. The 30 components include: 1,6—L-threonic; 2,5—D-erythronic; 3—2-deoxy-D-ribonic; 4—2,6 -dideoxy-D-allonic; 7,9—D-arabonic; 8—L-xylonic; 10—D-ribonic; 11,13—6-deoxy-L-galactonic; 12,19—D-lyxonic; 14—3-deoxy-D-gluconic; 15—2-deoxy-D-galactonic; 16,17—2-deoxy-D-gluconic; 18,20—D-glucuronic; 21,23—D-gluconic; 22—D-galactonic; 24—D-gulonic; 25—D-talonic; 26—D-glucaric; 27—D-galacturonic; 28—L-mannonic; 29—D-galactaric; and 30—D-glycero-D-gulo-heptonic. (Source: Ref. 9. Reproduced courtesy Elsevier Scientific Publishing Co., Amsterdam.)

ance. Accumulation of polyols within the lens is believed to modify the usual osmotic balance [185] and to initiate loss of lens transparency.

When a sugar alcohol such as myo-inositol accumulates in the lens, the resulting hypertonicity is corrected by an influx of water [178]. The osmotic changes which occur initially from polyol accumulation do not seriously alter the state of viability of the lens, and in the early stages of sugar cataract formation the process is reversible. If normal osmotic balance is not restored, lens opacity develops. At this stage there is a gradual breakdown of the permeability barriers and the electrolytes and polyols, including myo-inositol [189,190], become freely permeable and only the larger proteins are retained. After this stage becomes fully established, quantities of free sugar alcohol found in the cataract are insignificant and free myo-inositol levels are significantly altered from normal. This agrees with the wide of values obtained experimentally for myo-inositol in the study by Pfaffenberger et al. [12].

A diabetic is about 5 times more likely to have an initial cataract operation than is a nondiabetic of the same age and sex [191]; the polyol in these instances is believed to be glucitol.

Preparation of Analytical Sample from Lens Tissue

Adherent tissue (zonular ligament, ciliary muscle) is removed from surgically extracted human lenses. The intact lens is quickly weighed and frozen under an atmosphere of nitrogen. Lyophilization if effected over a 48-hr period at 0.5 torr using a Virtis model 10-300CF lyophilizer. The dry lens is weighed (51 to 88 mg), pulverized to a homogenous powder using a glass rod, and stored under nitrogen, under deep-freeze conditions, until analyzed.

Analytical samples are prepared by adding 10 mg of powdered lens, 10 mg of hydroxylamine hydrochloride, 0.0315 mg of glyceryl-1-decyl ether (internal reference compound, from a solution of 6.30 mg 10.0 ml^{-1} methanol) dissolved in 50 μl of methanol, 23 mg of sodium acetate, and 1 ml of methanol. The mixture is heated at 60°C for 2 hr, evaporated to dryness using a nitrogen stream, and treated with 1 ml of acetic anhydride. After this mixture has been heated for 2 hr at 100°C, it is allowed to stand overnight in a desiccator for complete conversion of the carbohydrates, which react more slowly in tissue preparations than they do as pure reference compounds [192]. Precipitated protein is removed by centrifugation. GC samples are removed from the clear supernatant solution and coinjected with n-alkenes when methylene unit (MU) [87,88] values are being determined.

Preparation of Analytical Sample from Urine

Urine samples from patients are collected over a 24-hr period just prior to cataract extraction. An analytical sample is prepared from a 20-ml aliquot of urine following exactly the procedure described in Sec. 3.3, above.

Representative Lens Profile

By far the largest free polyol component of human lens tissue is myo-inositol [12]. This is evident in Fig. 35. Amounts of this component range (in μg mg^{-1} lyophilized lens) from 2 to 19 for noncataractous lenses, 3 to 23 for uncomplicated senile cataractous lenses, and 7 to 29 for senile cataractous lenses from patients with diabetes [12].

All 20 lenses in the study by Pfaffenberger et al. [12] contained a trace of glucose, and all but two of the senile cataractous lenses from patients with diabetes contained at least a trace of glucitol.

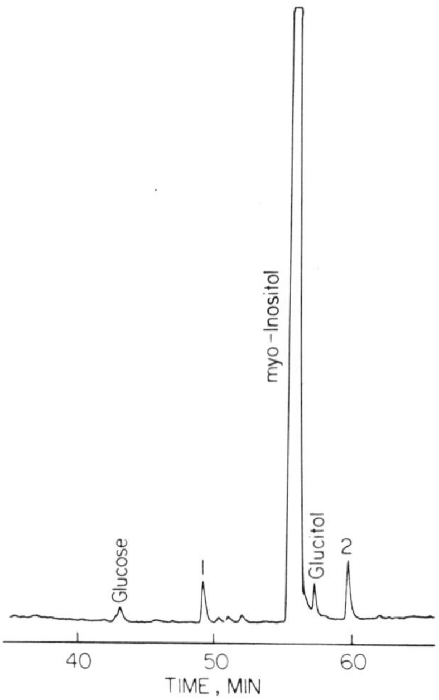

Figure 35. Profile of the free polyols and aldoses found in human cataractous lens tissue. Separation of the components as acetyl and Wohl derivatives was carried out on a 60-m SE-30 open tubular glass capillary column temperature programmed 1°C min^{-1} from 150°C. For further information see text. (Source: Ref. 12. Reproduced courtesy Elsevier Scientific Publishing Co., Amsterdam.)

All except two lenses (noncataractous) contained a trace of an unidentified compound (labeled 1 in Fig. 35) with MU = 19.79, and all except four lenses contained an unidentified compound (labeled 2) with MU = 20.95. The first unknown was not one of the simple aldohexoses, which elute within an MU range of 19.02 to 19.56 under the analytical conditions employed [11]. The second unknown was not mannitol [10], galactitol [11], scyllo-inositol [193], or 1,5-anhydroglucitol [194].

Urinary Profile of a Controlled Diabetic

Blood sugar levels fluctuate in diabetes, and the control achieved by therapy may vary over both short and long periods. Caird et al. [195] reported that control of diabetes is often below average in effectiveness in diabetics who require cataract extraction. He and his co-workers [196] later demonstrated a direct relationship between degree of control of diabetes and frequency of lens opacity in men of middle age.

That Pfaffenberger et al. [12] found only low concentrations of glucitol in the cataractous lenses they examined was explained by the facts that diabetes is usually well controlled before surgery and the cataractous lens (advanced) is abnormally permeable and loses glucitol to the aqueous humor by diffusion [178].

The original reference [12] contains excretion values (in mg g^{-1} creatinine) for 11 polyols and aldoses found in the urine of five of the subjects whose lenses were analyzed. The values are typical of the adult human over 50 years of age. No obvious differences were reported between noncataractous subjects and patients having senile cataracts [12].

Figure 36 is a urinary metabolic profile for a controlled diabetic patient with senile cataract. Excretion values for glucose, myo-inositol, and glucitol in this sample were 56, 6, and 4 mg g^{-1} creatinine, respectively.

Figure 37 is a composite which includes both urinary metabolic profiles of a controlled diabetic mother and her newborn son on the day of delivery. Virtually every aldose and polyol found in the mother's urine is also present in the son's. A relatively large amount of an unknown compound appears in the infant's profile between components 9 and 10 which correspond to arabinitol and xylitol.

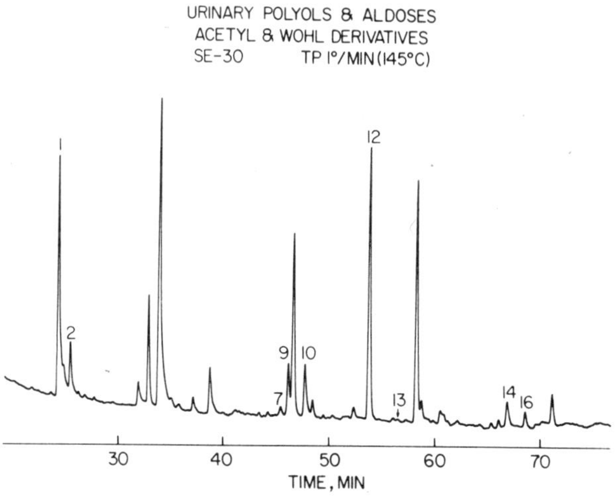

Figure 36. Metabolic profile of the urinary polyols and aldoses excreted by a controlled diabetic patient from whom the senile cataractous lens analyzed in Fig. 35 was obtained. Experimental conditions are as in Fig. 35. The identified components include: 1—erythritol, 2—threitol, 7—ribitol, 9—arabinitol, 10—xylitol, 12—glucose, 13—galactose, 14—myo-inositol, and 16—glucitol. (Source: Ref. 12. Reproduced courtesy of Elsevier Scientific Publishing Co., Amsterdam.)

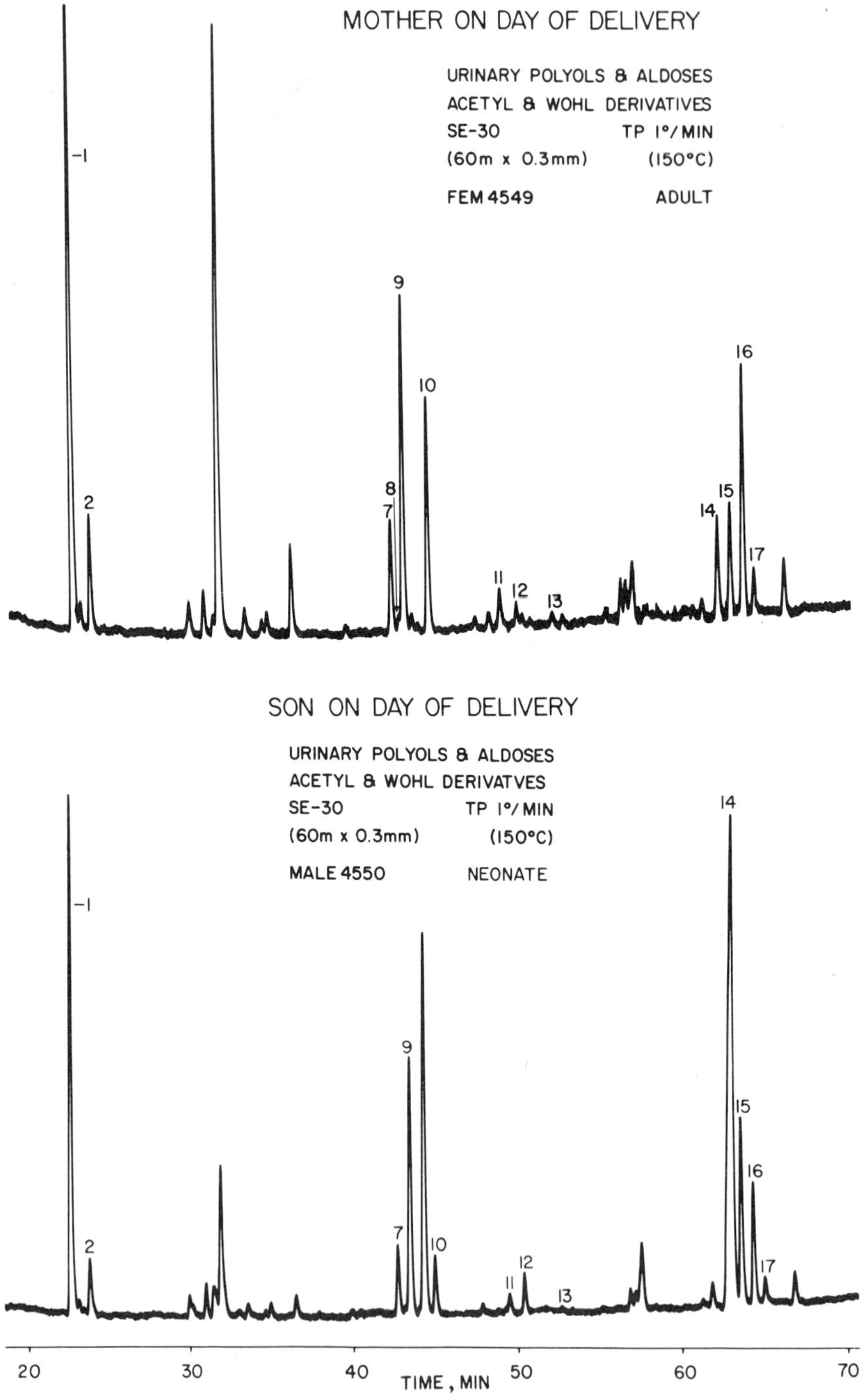

Figure 37. Metabolic profiles of urinary polyols and aldoses of a newborn male infant and his mother on the day of delivery. Experimental conditions are as in Fig. 35. Each identified component was in both samples: 1—erythritol, 2—threitol, 7—ribitol, 9—arabinitol, 10—xylitol, 11—mannose, 12—glucose, 13—galactose, 14—myo-inositol, 15—mannitol, 16—glucitol, and 17—galactitol.

4. ADDITIONAL BIOMEDICAL APPLICATIONS

Several additional biomedical applications of glass capillary gas chromatography are briefly considered here. This is not to imply that these areas are of lesser importance, but most of these studies are either still in progress or were located by the author too late to be incorporated into earlier sections.

4.1 Rat Fecal Steroid Studies

Fecal collections from experimental rats maintained on well-defined diets can be used to study the ability of various compounds to lower absorption of dietary cholesterol from the gastrointestinal (GI) tract. Microorganisms in the GI tract metabolized dietary cholesterol; thus litter mates must always be used for these investigations. Unchanged cholesterol, cholesterol metabolites, and other dietary sterols can be extracted from the fecal material, converted to the corresponding trimethylsilyl derivatives, and separated on SE-30 open tubular glass capillary columns. Such separations tax the upper temperature limits of the columns, but good profiles are obtainable. Figure 38 is one such profile in which component 4 is the internal reference compound, cholesteryl butyl ether. Tentative identifications of the other components include: Sq--squalene, 1--coprostanol, 2--cholesterol plus cholestanol, 3--campesterol, 5--β-sitosterol and 6--reduced β-sitosterol.

Complete separation of the four isomeric cholestanols as trimethylsilyl (TMS) ethers has been reported [86] using a 20-m glass capillary column coated with a polyphenyl ether sulfone phase, PZ-176 (now marketed as Poly-S 176). Figure 39 is a separation of the TMS derivatives of: 1--5β-cholestan-3β-ol, 2--5β-cholestan-3α-ol, 3--5α-cholestan-3α-ol, 4--5α-cholestan-3β-ol, 5--stigmasterol, and 6--cholesteryl butyl ether, the internal reference compound. The column was temperature programmed 1°C min^{-1} from 200°C.

4.2 Drugs and Drug Metabolites

Most drugs form a number of metabolites; characteristic reactions include aliphatic chain hydroxylation, aromatic epoxide formation followed by dihydrodiol formation, or conversion to phenols, epoxidation and diol formation from unsaturated carbon-carbon bonds, demethylation, oxidation of alcohols to ketones as well as the reverse reaction, and a variety of conjugation reactions [30]. The relative amount of each product in a multiproduct reaction scheme is determined by relative rates of enzymic action; these rates

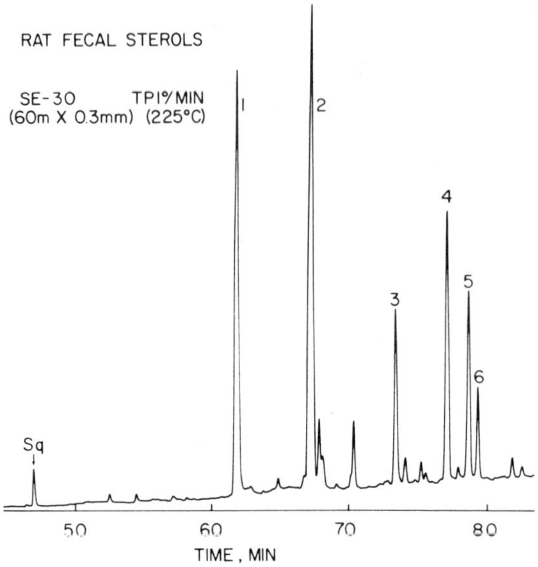

Figure 38. Rat fecal sterol profile of the corresponding TMS derivatives separated on a 60-m SE-30 glass capillary column temperature programmed 1°C min^{-1} from 225°C. Component 4—cholesteryl butyl ether, the internal reference compound. Tentative, identifications of the other components include: Sq—squalene, 1—coprostanol, 2—cholesterol plus cholestanol, 3—campesterol, 5—β-sitosterol, and 6—reduced β-sitosterol.

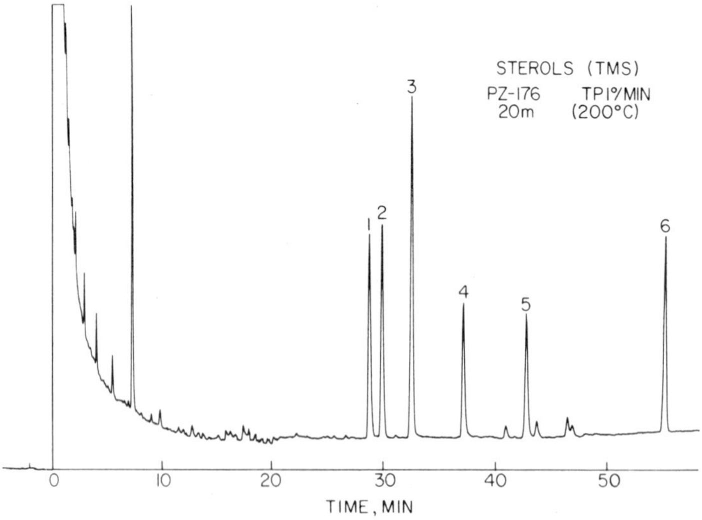

Figure 39. Separation of the trimethylsilyl (TMS) ether derivatives of four isomeric cholestanols and stigmasterol plus an internal reference compound on a 20-m PZ-176 (now Poly-S 176) open tubular glass capillary column temperature programmed 1°C min^{-1} from 200°C. 1—5β-cholestan-3β-ol, 2—5β-cholestan-3α-ol, 3—5α-cholestan-3α-ol, 4—5α-cholestan-3β-ol, 5—stigmasterol, and 6—cholesteryl butyl ether. (Source: Ref. 86. Reproduced courtesy of Elsevier Scientific Publishing Co., Amsterdam.)

in turn vary with each species and with individual considerations which include age, sex, genetic background, nutritional state, and extent of drug-drug interaction. Glass capillary gas chromatographic methods, because of their capability of providing multicomponent analyses, are highly valuable in studies of drug metabolism as well as toxicological studies.

Figure 40 is a urinary profile of methyl and methyl-trimethylsilyl derivatives of drugs and drug metabolites from a human receiving Dilantin, phenobarbital, and methsuximide. The urine was worked up and derivatized according to Horning et al. [197], and the components which have been identified include the parent drugs and a variety of metabolites including dihydrodiols formed through the epoxide-diol pathway. (In this figure, the order of elution is from right to left.)

German et al. [198] reported an instance of accidental poisoning of a child with a pine oil preparation. A separation of urinary acids as methyl ester-trimethylsilyl derivatives, prepared according to Horning and Horning

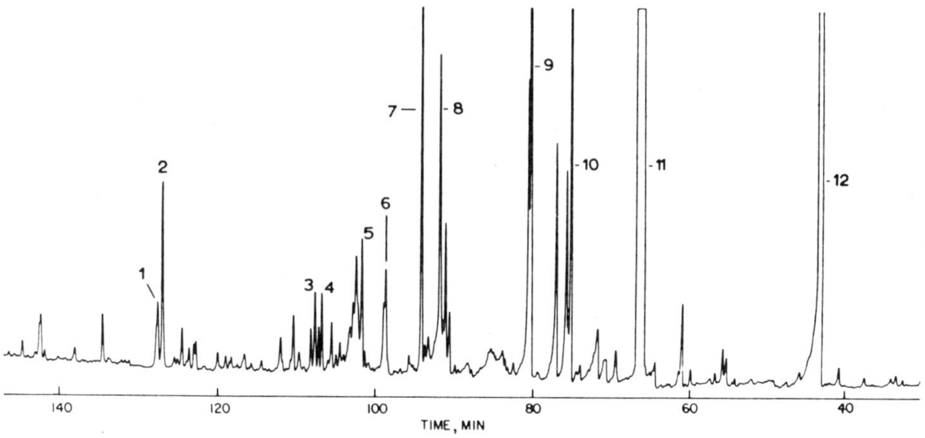

Figure 40. Profile (obtained using a 60-m SE-30 glass capillary column) of methyl ether (ME) and methyl-trimethylsilyl ehter (ME-TMS) derivatives of drugs and urinary drug metabolites from a human receiving dilantin, phenobarbital, and methsuximide. The identified components include the parent drugs and a variety of metabolites including dihydrodiols formed through the epoxide-diol pathway. (The order of elution in this chart is from right to left.) 1—dihydrodiol derivatives of dilantin; 2—hydroxydilantin; 3,4—dihydrodiol derivatives of hydroxymethsuximide; 5—dilantin; 6—ρ-hydroxyphenobarbital; 7—diol derivatives of methsuximide; 8—dihydrodiol derivative of methsuximide; 9—ρ-hydroxymethsuximide; 10—n-hexadecane; 11—phenobarbital, and 12—methsuximide. (Source: Ref. 30. Reproduced courtesy of Elsevier Scientific Publishing Co., Amsterdam.)

[199], employed a 60-m SE-30 glass capillary column with temperature programming at 2°C min^{-1} from 100°C. Urinary acid profiles contain many components, but abnormal peaks can usually be detected in the chromatogram without difficulty.

4.3 Neonatal Urinary Steroid Metabolic Profiles

Metabolic profiles may also be used in studies of developmental changes occurring during the first few days or weeks of life. Horning et al. [30] prefer the use of benzyloximetrimethylsilyl (BO-TMS) derivatives [90,99] for this type of investigation because they afford better separation of some of the predominant components of the profile. Most of the steroidal compounds in neonatal urine are 3β-ol-5-ene steroids which originate in cells of the fetal adrenal zone; these gradually disappear and are replaced by cells characteristic of adult adrenals.

4.4 Plasma α-Tocopherol and Phosphatidylcholines

In 1975, Lin and Horning [19] developed a procedure for the concurrent analysis of mixtures of long-chain acids (as methyl esters), cholesterol (as the trimethylsilyl ether), and the tocopherols (as trimethylsilyl ethers). The method was used in a study of free fatty acids and α-tocopherol in the blood plasma of normal subjects and stroke patients. The chromatographic separations were based upon the use of thermostable glass open tubular capillary columns prepared with Silanox 101 according to Lin et al. [200] using a polyphenyl ether sulfone phase, PZ-176 (now Poly S-176) [201]. The plasma lipids were extracted essentially according to Dole and Meinertz [202], with the substitution of isooctane for heptane.

Derivatization was carried out in two steps. The acidic components were converted to methyl esters by short treatment with diazomethane; this was accomplished by dissolving the lipid sample in 0.5 ml of methanol and adding 2 ml of a solution of diazomethane in ether (prepared from Diazald, Aldrich, Milwaukee, Wisc.). After 10 to 15 min, the solvents and excess reagent were removed with the aid of a nitrogen stream. To the residue there were added 50 μl of pyridine, 50 μl of bis(trimethylsilyl) acetamide and 25 μl of trimethylchlorosilane. The mixture was heated at 90°C for 1 hr; and the resulting solution was used directly for GC analysis [19].

Figure 41 is a profile of a normal adult which includes methyl esters of the plasma free fatty acids and trimethylsilyl ethers of α-tocopherol and cholesterol. The column was programmed from 180°C at 2°C min^{-1}. The indicated components include: 14:0, methyl myristate; 16:0, methyl palmitate; 16:1, methyl palmitoleate; STD, methyl heptadecanoate; 18:0, methyl stearate; 18:1, methyl oleate (containing isomers); 18:2, methyl linoleate; α-TOC, α-tocopherol trimethylsilyl ether; CHOL, cholesterol trimethylsilyl ether.

In a representative profile for a stroke patient [19], a compositional change in the long-chain acid part of the profile was evident. The linoleic acid component was smaller and the oleic acid component, larger for the stroke patient. Lin and Horning [19] reported values for the linoleic acid/oleic acid ratio (18:2/18:1). The range was from 0.37 to 0.66 for the control group and from 0.11 to 0.36 for a group of 18 stroke patients. Considerably

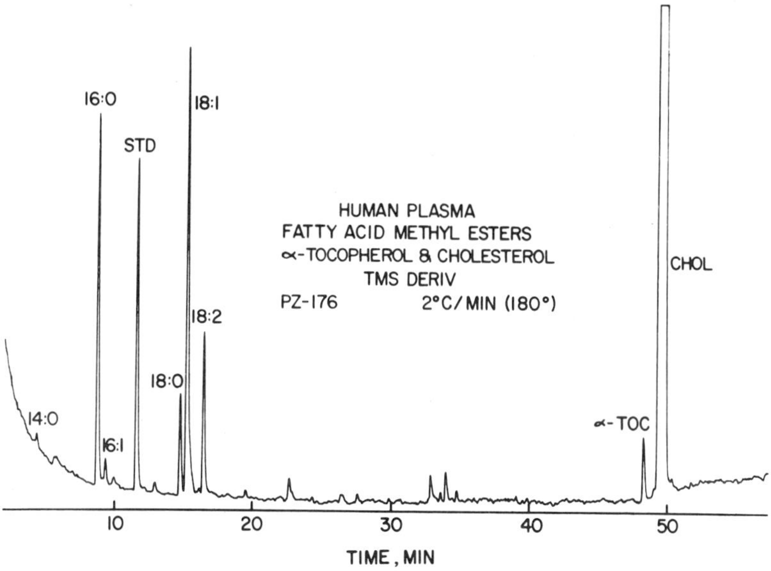

Figure 41. Separation of human plasma fatty acids as methyl esters and α-tocopherol and cholesterol as trimethylsilyl ethers on a 20-m PZ-176 (now Poly S-176) glass capillary column temperature programmed 2°C min^{-1} from 180°C. The chromatogram includes: 14:0—methyl myristate; 16:0—methyl palmitate; 16:1—methyl palmitoleate; STD—methyl heptadecanoate; 18:0—methyl stearate; 18:1—methyl oleate (containing isomers); 18:2—methyl linoleate; α-TOC—TMS ether of α-tocopherol; CHOL—TMS ether of cholesterol. (Source: Ref. 19. Reproduced courtesy of Elsevier Scientific Publishing Co., Amsterdam.)

lower plasma levels of α-tocopherol were found in the patient group, and low levels of plasma α-tocopherol and low linoleic anid/oleic acid ratios were almost invariably associated with the patient group.

Low plasma concentrations of α-tocopherol have been observed in association with a-β-lipoproteinemia [203], with disorders having as a common element a severe impairment of lipid absorption [203-205], and with cystic fibrosis [203,206,207]. The nature of the association in humans indicates that these disorders casue low plasma concentrations, rather than the reverse, and Lin and Horning [19] point out that their data should be used only to define associative, not causative relationships. Current views of α-tocopherol function in man have been summarized in a review by Olson [208], and a conference report [209] includes examples of most lines of investigation in this area, including reviews of current hypotheses of biological action.

The above results suggested that compositional studies should be carried out for plasma phosphatidylcholines and plasma cholesteryl esters for the two groups of subjects. Lin and Horning [19] expected that stroke patients would have depressed linoleic acid/oleic acid ratios, because they had found the depressed ratio for free fatty acids. Their view was in accordance with the results of Rosenlund et al. [210], who found that in children with cystic fibrosis both of these lipid classes showed depressed linoleic acid/oleic acid ratios. Underwood et al. [206] determined the concentration of α-tocopherol in children with cystic fibrosis to be about 0.2 mg dl^{-1}.

Lin and Horning [20] found that the stroke patient group showed uniformly depressed ratios of linoleic acid to oleic acid in plasma phosphatidylcholines, when compared with normal subjects. Many, but not all, patients showed a depressed ratio for cholesteryl esters. Some patients showed a marked elevation in arachidonic acid in plasma phosphatidylcholines. From their work they formulated this hypothesis:

> Thrombic strokes are a consequence of platelet aggregation, caused by endoperoxide formation resulting from an increased substrate supply of arachidonic acid, and possibly aided by an abnormal phospholipid composition in plasma and platelets. This situation results from a sequence of events starting with a marginal increase in the rate of metabolism over the rate of absorption of α-tocopherol, leading to α-tocopherol depletion; the later steps include a decrease in incorporation of linoleic acid into phosphatidylcholines and an increase in substrate availability of arachidonic acid. [20].

(It is not known if the effect of α-tocopherol depletion is specific or if another substance is involved.)

5. CONCLUDING REMARKS

The use of glass capillary gas chromatography in clinical medicine is still in its infancy. Although the technology and methodology for obtaining metabolic profiles have been worked out, many more applications of this approach are possible, particularly in studies of chronic diseases and subtle metabolic disorders. The powerful separating characteristics of glass capillary columns have brought capillary chromatographers almost to the point of being able to claim ability to separate any complex mixture of volatilizable compounds into all of its component parts. In the few instances when compound coelution occurs, a second column of slightly differing polarity will usually bring about the desired separation.

When glass capillary columns are used in combination with a mass spectrometer, which in turn is interfaced to a totally dedicated computer system, a truly sophisticated level of analytical capability is achieved [64,211-223]. Over the next decade, application of such analytical systems to studies of drug-drug interaction (synergistic and antagonistic effects) will come to the fore as well as investigations of drug-environmental agent interactions. Some of these considerations fall within the purview of biomedical applications, and the 1980s will witness the "coming of age" of glass capillary gas chromatographic applications in the clinical chemistry laboratory.

ACKNOWLEDGMENTS

Sincere appreciation is expressed to many colleagues in the fields of glass capillary gas chromatography and clinical medicine. Those who have been of particular assistance to me include C. J. W. Brooks, Ph.D., A. L. German, Ph.D., E. C. Horning, Ph.D., M. G. Horning, Ph.D., E. Jellum, Ph.D., S.-N. Lin, Ph.D., L. R. Malinak, MD., B. S. Middleditch, Ph.D., J. Szafranek, Ph.D., J.-P. Thenot, Ph.D., and A. Zlatkis, Ph.D.

REFERENCES

1. Zlatkis, A., *Advances in Chromatography, 1973*, University of Houston Publishers, Houston, Texas, 1973.
2. Zlatkis, A., and Ettre, L. S., *Advances in Chromatography, 1974*, Elsevier, New York, 1974.

3. Zlatkis, A., Bayer, E., Ettre, L. S., and Halász, I., *Advances in Chromatography, 1975*, Elsevier, New York, 1975.

4. Zlatkis, A., Shanfield, H., and Ettre, L. S., *Advances in Chromatography, 1976*, Elsevier, New York, 1976.

5. Zlatkis, A., Ettre, L. S., and Dijkstra, G., *Advances in Chromatography, 1977*, Elsevier, New York, 1978.

6. Zlatkis, A., Ettre, L. S., and Shanfield, H., *Advances in Chromatography, 1978*, Elsevier, New York, 1978.

7. Cram, S. P., and Juvet, R. S., Jr., *Anal. Chem.*, 48:411R (1976).

8. Cram, S. P., and Risby, T. H., *Anal. Chem.*, 50:213R (1978).

9. Szafranek, J., Pfaffenberger, C. D., and Horning, E. C., *J. Chromatogr.*, 88:149 (1974).

10. Szafranek, J., Pfaffenberger, C. D., and Horning, E. C., *Anal. Lett.*, 6:479 (1973).

11. Pfaffenberger, C. D., Szafranek, J., Horning, M. G., and Horning, E. C., *Anal. Biochem.*, 63:501 (1975).

12. Pfaffenberger, C. D., Szafranek, J., and Horning, E. C., *J. Chromatogr.*, 126:535 (1976).

13. Adam, S., and Jennings, W. G., *J. Chromatogr.*, 115:218 (1975).

14. Scholly, P. R., and Ettre, L. S., *Chromatogr. Newsl.*, 2:12 (1973).

15. Fedyanin, A. A., Prikhodko, A. A., and Kukushkina, N. A., *Zavod. Lab.*, 40:1066 (1974).

16. Apon, J. M. B., and Nicolaides, N., *J. Chromatogr. Sci.*, 13:467 (1975).

17. Smith, A., and Lough, A. K., *J. Chromatogr. Sci.*, 13:487 (1975).

18. Fenimore, D. C., *J. Chromatogr.*, 112:219 (1975).

19. Lin, S. N., and Horning, E. C., *J. Chromatogr.*, 112:465 (1975).

20. Lin, S. N., and Horning, E. C., *J. Chromatogr.*, 112:483 (1975).

21. Gloor, R., and Leidner, H., *Chromatographia*, 9:618 (1976).

22. Krupcik, J., Hrivnak, J., and Janak, J., *J. Chromatogr. Sci.*, 14:4 (1976).

23. Pullarkat, R. K., and Reha, H., *J. Chromatogr. Sci.*, 14:25 (1976).

24. Anderson, I., Norkrans, B., and Odham, G., *Anal. Biochem.*, 53:629 (1973).

25. Flanzy, J., Boudon, M., Leger, C., and Piket, J., *J. Chromatogr. Sci.*, 14:17 (1976).

26. Gough, T. A., and Sugden, K., *J. Chromatogr.*, 109:265 (1975).

27. Berthou, F. L., Picart, D., Bardou, L. G., and Flock, H. H., *J. Chromatogr. Sci.*, 12:662 (1974).

28. Delaforge, M., Maume, B. F., Bournot, P., Prost, M., and Padieu, P., *J. Chromatogr. Sci.*, 12:545 (1974).

29. Bailey, E., Fenoughty, M., and Chapman, J. R., *J. Chromatogr.*, 96:33 (1974).

30. Horning, E. C., Horning, M. G., Szafrank, J., Van Hout, P., German, A. L., Thenot, J. P., and Pfaffenberger, C. D., *J. Chromatogr.*, 91:367 (1974).

31. Leferink, J. G., and Leclercq, P. A., *J. Chromatogr.*, 91:385 (1974).

32. Pfaffenberger, C. D., and Horning, E. C., *J. Chromatogr.*, 112:581 (1975).

33. Sandra, P., Verzele, M., and Vanluchene, E., *Chromatographia*, 8:499 (1975).

34. Berthou, F., Picart, D., Bardou, L., and Flock, H. H., *J. Chromatogr.*, 118:135 (1976).

35. Deelder, R. S., Ramaekers, J. J. M., Van den Berg, J. H. M., and Wetzels, M. L., *J. Chromatogr.*, 119:99 (1976).

36. Pfaffenberger, C. D., and Horning, E. C., *Anal. Biochem.*, 80:329 (1977).

37. Edmonds, C. G., Smith, A. G., and Brooks, C. J. W., *J. Chromatogr.*, 133:372 (1977).

38. Matthiesen, U., and Staib, W., *Chromatographia*, 10:70 (1977).

39. Pfaffenberger, C. D., and Horning, E. C., *Anal. Biochem.*, 88:689 (1978).

40. Pfaffenberger, C. D., Malinak, L. R., and Horning, E. C., *J. Chromatogr.*, 158:313 (1978).

41. Cram, S. P., *High Resolution Gas Chromatography*, Academic Press, New York, 1979.

42. Novotny, M., Mc Connell, M. L., and Lee, M. L., *J. Agr. Food Chem.*, 22:765 (1974).

43. Blomberg, L., and Widmark, G., *J. Chromatogr.*, 106:59 (1975).

44. Goldberg, E. M., and Sandler, S., *Chromatographia*, 8:331 (1975).

45. Franken, J. J., and Trijbels, M. M. F., *J. Chromatogr.*, 91:425 (1974).

46. Eyem, J., and Sjoquist, J., *Anal. Biochem.*, 52:255 (1973).

47. Koenig, W. A., and Nicholson, G. J., *Anal. Chem.*, 47:951 (1975).

48. Adams, R. F., Vandemark, F. L., and Schmidt, G. T., *J. Chromatogr. Sci.*, 15:63 (1977).

49. Koenig, W. A., Rahn, W., and Eyem, J., *J. Chromatogr.*, 133:141 (1977).

50. Buser, H. R., *Anal. Chem.*, 48:1553 (1976).

51. Franken, J. J., and Vader, H. L., *Chromatographia*, 6:22 (1973).

52. Franken, J. J., and Luyten, B. J. M., *J. Assoc. Off. Anal. Chem.*, 59:1279 (1976).

53. Krijgsman, W., and Van De Kamp, C. G., *J. Chromatogr.*, 131:201 (1976).

54. Krijgsman, W., and Van De Kamp, C. G., *J. Chromatogr.*, 117:201 (1973).

55. Caddy, B., Fish, F., and Scott, D., *Chromatographia*, 6:251 (1973).

56. Caddy, B., Fish, F., and Scott, D., *Chromatographia*, 6:293 (1973).

57. Caddy, B., Fish, F., and Scott, D., *Chromatographia*, 6:335 (1973).

58. Bailey, E., Fenoughty, M., and Richardson, L., *J. Chromatogr.*, 131:347 (1977).

59. Wu, A., and Pearson, M. A., *Anal. Lett.*, 10:381 (1977).

60. Novotny, M., Lee, M. L., Low, C. E., and Raymond, A., *Anal. Chem.*, 48:24 (1976).

61. Novotny, M., Lee, M. L., Low, C. E., and Maskarinec, M. P., *Steroids*, 27:665 (1976).

62. Novotny, M., McConnell, M. L., Lee, M. L., and Farlow, R., *Clin. Chem.* (Winston-Salem, N. C.), 20:1105 (1974).

63. Horning, E. C., and Horning, M. G., *Methods Med. Res.*, 12:369 (1970).

64. Jellum, E., Storseth, P., Alexander, J., Helland, P., Stokke, O., and Teig, E., *J. Chromatogr.*, 126:487 (1976).

65. Kutsdy, R. J., *Handbook of Vitamins and Hormones*, Van Nostrand Reinhold, New York, 1973.

66. Jellum, E., *J. Chromatogr.*, 143:427 (1977).

67. Goodman, S. I., Helland, P., Stokke, O., Flatmark, A., and Jellum, E., *J. Chromatogr.*, 142:497 (1977).

68. Heininger, J., Munthe, E., Pakle, J., and Jellum, E., *J. Chromatogr.*, 158:297 (1978).

69. Stear, E. B., and Kadish, A. H., *Hormonal Control Systems*, Elsevier, New York, 1969.

70. Sawin, C. T., *The Hormones*, Little, Brown, Boston, 1969.

71. McKerns, K. W., *Steroid Hormones and Metabolism*, Appleton-Century-Crofts, New York, 1969.

72. Litwack, G., *Biochemical Actions*, Vols. 1 and 2, Academic Press, New York, 1970.

73. Briggs, M. H., and Brotherton, J., *Steroid Biochemistry and Pharmacology*, Academic Press, New York, 1970.

74. Briggs, M. H., *Advances in Steroid Biochemistry and Pharmacology*, Vols. 1 and 2, Academic Press, New York, 1970.

75. Frieden, E., and Lipner, H., *Biochemical Endocrinology of the Vertebrates*, Prentice-Hall, Englewood Cliffs, N.J., 1971.

76. Turner, C. D., and Bagnara, J. T., *Gerneral Endocrinology*, Saunders, Philadelphia, 1971.

77. Brooks, C. J. W., and Watson, J., in *Gas Chromatography, 1968* (C. L. A. Marbourn, ed.), Institute of Petroleum, London, 1969.

78. Brooks, C. J. W., and Harvey, D. J., *J. Chromatogr.*, 54:193 (1971).

79. Brooks, C. J. W., Middleditch, B. S., and Harvey, D. J., *Org. Mass Spectrom.*, 5:1429 (1971).

80. Baille, T. A., Brooks, C. J. W., and Middleditch, B. S., *Anal. Chem.*, 44:30 (1972).

81. Gaskell, S. J., Edmonds, C. G., and Brooks, C. J. W., *Anal. Lett.*, 9:325 (1976).

82. Brooks, C. J. W., Edmonds, C. G., and Gaskell, S. J., *Advan. Mass Spectrom.*, 7:1578 (1978).

83. Gaskell, S. J., and Brooks, C. J. W., *J. Chromatogr.*, 158:331 (1978).

84. German, A. L., and Horning, E. C., *J. Chromatogr. Sci.*, 11:76 (1973).

85. German, A. L., and Horning, E. C., *Anal. Lett.*, 5:619 (1972).

86. Van Hout, P., Szafranek, J., Pfaffenberger, C. D., and Horning, E. C., *J. Chromatogr.*, 99:103 (1974).

87. Horning, E. C., Horning, M. G., Ikekawa, N., Chambaz, E., Jaakonmaki, P. I., and Brooks, C. J. W., *J. Gas Chromatogr.*, 5:283 (1967).

88. VandenHeuvel, W. J. A., Gardiner, W. L., and Horning, E. C., *J. Chromatogr.*, 19:263 (1965).

89. Baillie, T. A., Brooks, C. J. W., and Horning, E. C., *Anal. Lett.*, 5:351 (1972).

90. Devaux, P. G., Horning, M. G., and Horning, E. C., *Anal. Lett.*, 4:151 (1971).

91. Brooks, C. J. W., and Harvey, D. J., *Steroids*, 15:283 (1970).

92. Chambaz, E. M., and Madam, C., 3rd Int. Cong. on Hormonal Steroids, Hamburg, G.F.R., Sept. 7-12, 1970.

93. Fales, H. M., and Luukkainen, T., *Anal. Chem.*, 37:955 (1965).

94. Horning, M. G., Moss, A. M., and Horning, E. C., *Anal. Biochem.*, 22:283 (1968).

95. Dray, F., and Weliky, I., *Anal. Biochem.*, 34:387 (1970).

96. Thenot, J. P., and Horning, E. C., *Anal. Lett.*, 5:21 (1972).

97. Gardiner, W. L., and Horning, E. C., *Biochim. Biophys. Acta*, 115:524 (1966).

98. Chamboz, E. M., and Horning, E. C., *Anal. Biochem.*, 30:7 (1969).

99. Sakauchi, N., and Horning, E. C., *Anal. Lett.*, 4:41 (1971).

100. Thenot, J.-P., and Horning, E. C., *Anal. Lett.*, 5:905 (1972).

101. Thenot, J.-P., and Horning, E. C., *Anal. Lett.*, 5:801 (1972).

102. Horning, M. G., Lau, S. S., Hung, A., Stillwell, W. G., and Hill, R. M., Abstract 279, 4th Int. Cong. on Hormonal Steroids, Mexico City, Sept. 2-7, 1974; *J. Steroid Biochem.*, 5:362 (1974).

103. Thenot, J.-P., and Horning, E. C., *Anal. Lett.*, 4:683 (1971).

104. Venning, E. N., *Endocrinology*, 39:203 (1946).

105. Schuller, E., *Acta Endocrinol.* (Copenhagen), 27:389 (1958).

106. Birke, G., Genzell, C. A., Plantin, L.-O., and Robbe, H., *Acta Endrocrinol.* (Copenhagen), 27:389 (1958).

107. Fishman, J., Brown, J. B., Hellman, L., Zumoff, B., and Gallager, T. F., *J. Biol. Chem.*, 237:1489 (1962).

108. Diczfalusy, E., and Troen, P., *Vitam. Horm.* (New York), 19:229 (1961).

109. Klopper, A., *Amer. Clin. Res.*, 2:289 (1970).

110. Green, J. W., Smith, K., Kyle, G. C., Touchstone, J. C., and Duhring, J. L., *Amer. J. Obstet. Gynecol.*, 91:684 (1956).

111. Dässler, C.-G., *Acta Endocrinol.* (Copenhagen), 56:333 (1967).

112. Rosner, J. M., Conte, N. F., Horita, S., and Forsham, P. H., *Amer. J. Med.*, 37:638 (1964).

113. Staple, E., Lynn, W. S., Jr., and Gurin, S., *J. Biol. Chem.*, 219:845 (1956).

114. Plantin, L.-O., Diczfalusy, E., and Birke, G., *Nature*, 179:421 (1957).

115. Dorfman, R. I., and Shipley, R. A., *Androgens: Biochemistry, Physiology, and Clinical Significance*, Wiley, New York, 1956.

116. Forchielli, E., Brown-Grant, K., and Dorfman, R. I., *Proc. Soc. Exp. Biol. Med.*, 99:594 (1958).

117. Maume, B. F., Bournot, P., Lhuguenot, J.-C., Baron, C., Barbier, F., Maume, G., Prost, M., and Padieu, P., *Anal. Chem.*, 45:1073 (1973).

118. Padieu, P., Barbier, F., Chessebeuf, M., Cordier, D., Gerique, M., Lallemant, C., and Olsson, A., 1st Int. Con. on Cell Differentiation, Nice, France, 1971.

119. Padieu, P., Lallemant, C., Barbier, F., and Chessebeuf, M., 7th Cong. Fed. Eur. Biochem. Soc., Varna, Bulgaria, 1971.

120. Begue, R.-J., Gustafsson, J.-A., and Gustafsson, S. A., *Eur. J. Biochem.*, 40:361 (1973).

121. Berg, A., Gustafsson, J.-A., *J. Biol. Chem.*, 248:6559 (1973).

122. DeMoor, P., and Denef, C., *Endrocrinology*, 82:480 (1968).

123. Denef, C., and DeMoor, P., *Endrocrinology*, 85:259 (1969).

124. Einarsson, K., Gustafsson, J.-A., and Stenberg, A., *J. Biol. Chem.*, 248:4978 (1973).

125. Gustafsson, J.-A., and Gustafsson, S. A., *Eur. J. Biochem.*, 44:225 (1974).

126. Gustafsson, J.-A., Gustafsson, S. A., Ingleman-Sundberg, Ponsette, A., Stenberg, A., and Wrang, O., *J. Steroid Biochem.*, 5:855 (1974).

127. Gustafsson, J.-A., and Ingleman-Sundberg, M., *J. Biol. Chem.*, 249:1940 (1974).

128. Gustafsson, J.-A., and Ponsetts, A., *Biochem.*, 13:875 (1974).

129. Gustafsson, J.-A., and Stenberg, A., *Endrocrinology*, 96:501 (1975).

130. Gustafsson, J.-A., and Stenberg, A., *J. Biol. Chem.*, 249:711 (1974).

131. Gustafsson, J.-A., and Stenberg, A., *J. Biol. Chem.*, 249:719 (1974).

132. Garris, G. W., *Endrocrinology*, 75:627 (1964).

133. DeKloet, R., Wallach, G., and McEwen, B. S., *Endrocrinology*, 96:598 (1975).

134. Dörner, G., *Endokrinologie*, 56:280 (1970).

135. Edwards, D. A., and Herndon, J., *Physiol. Behav.*, 5:993 (1970).

136. Gorski, R. A., and Wagner, J. W., *Endrocrinology*, 76:226 (1965).

137. Lieberburg, I., and McEwen, B. S., *Brain Res.*, 85:165 (1975).

138. Lieberburg, I., and McEwen, B. S., *Brain Res.*, 91:171 (1975).

139. Luine, V. N., Khylchevskaya, R. I., and McEwen, B. S., *J. Neurochem.*, 23:925 (1974).

140. McEwen, B. S., Pfaff, D. W., Chaptal, C., and Luine, V. N., *Brain Res.*, 86:155 (1975).

141. Reinisch, J. M., *Arch. Sex. Behav.*, 3:51 (1974).

142. Young, W. C., Goy, R. W., and Phoenix, C. H., *Science*, 143:212 (1964).

143. Begue, R.-J., Gustafsson, J.-A., and Gustafsson, S. A., *J. Steroid Biochem.*, 4:393 (1973).

144. Denef, C., *Endrocrinology*, 96:598 (1975).

145. Plapinger, L., McEwen, B. S., and Clemens, L. E., *Endrocrinology*, 93:1129 (1973).

146. Flores, F., Naftolin, F., and Ryan, K. J., *Neuroendocrinology*, 11:177 (1973).

147. Naftolin, F., Ryan, K. J., and Petro, Z., *Endocrinology*, 90:295 (1972).

148. Ehrhardt, A. A., Epstein, R., and Money, J., *Johns Hopkins Med. J.*, 122:160 (1968).

149. Ehrhardt, A. A., Evers, K., and Money, J., *Johns Hopkins Med. J.*, 123:115 (1968).

150. Money, J., and Meredith, T., *Pediat. Res.*, 1:59 (1967).

151. Ehrhardt, A. A., and Money, J., *J. Sex Res.*, 3:83 (1967).

152. Money, J., Ehrhardt, A. A., and Masica, D. N., *Johns Hopkins Med. J.*, 123:105 (1968).

153. Ehrhardt, A. A., Greenberg, N., and Money, J., *Johns Hopkins Med. J.*, 126:237 (1970).

154. Money, J., and Ehrhardt, A. A., in *Endrocrinology and Human Behavior* (R. P. Michall, ed.), Oxford University Press, London, 1968.

155. Masica, D. N., Money, J., Ehrhardt, A. A., and Lewis, V. G., *Johns Hopkins Med. J.*, 124:34 (1969).

156. Allen, B. J., Hayward, J. L., and Merivale, W. H. S., Lancet i, 496 (1957).

157. Bulbrook, R. D., Grenwood, P. C., and Hayward, J. L., Lancet i, 1154 (1960).

158. Bulbrook, R. D., Hayward, J. L., Spicer, C. C., and Thomas, B. S., Lancet ii, 1238 (1962).

159. Bondy, P. K., in *Diseases of Metabolism*

160. Bishop, C. T., *Adv. Carbohyd. Chem.*, 19:95 (1964).

161. Wolfrom, W. L., and Thompson, A., *J. Amer. Chem. Soc.*, 53:622 (1931).

162. Deulofeu, V., Cattaneo, P., and Mendivelza, G., *J. Chem. Soc.*, 147 (1934).

163. Dmitriev, B. A., Bachinowsky, L. V., Chizoo, O. S., Zolotarev, B. M., and Kochetkov, N. K., *Carb. Res.*, 19:432 (1971).

164. Unpublished observations.

165. Tesarik, K., *J. Chromatog.*, 65:295 (1972).

166. Morrison, I. M., and Perry, M. B., *Can. J. Biochem.*, 44:1115 (1966).

167. Sweeley, C. C., Bentley, R., Makita, M., and Wells, W. W., *J. Am. Chem. Soc.*, 85:2497 (1963).

168. Eklund, G., Josefsson, B., and Roos, C., *J. Chromatogr.*, 142:575 (1977).

169. DelGereco, F., and Scapellato, L., *Diabetes*, 2:457 (1953).

170. Isselbacher, K. J., in *The Metabolic Basis of Inherited Disease*, (J. B. Stanbury, J. B. Wyngaarden, and D. S. Fredrickson, eds.), 2nd ed. McGraw-Hill, New York, 1966.

171. Hiatt, H. H., in *The Metabolic Basis of Inherited Disease*, (J. B. Stanbury, J. B. Wyngaarden, and D. S. Fredrickson, eds.), 2nd ed. McGraw-Hill, New York, 1966.

172. Szafranek, J., Pfaffenberger, C. D., and Horning, E. C., *J. Chromatogr.*, 88:149 (1974).

173. Jones, T. M., and Albersheim, P., *Plant Physiol.*, 49:926 (1972).

174. Blake, J. D., and Richards, G. N., *Carbohyd. Res.*, 8:275 (1968).

175. Blake, J. D., and Richards, G. N., *Carbohyd. Res.*, 14:375 (1970).

176. Pirie, A., and van Heyningen, R., *Exp. Eye Res.*, 3:124 (1964).

177. Gitzelmann, R., Curtins, H.-C., and Schneller, I., *Exp. Eye Res.*, 6:1 (1967).

178. Kinoshita, J. H., *Invest. Ophthalmol.*, 4:713 (1974).

179. Heaf, D. J., and Galton, D. J., *Clin. Chim. Acta*, 63:41 (1975).

180. VanHeyninger, R., *Nature*, 184:194 (1959).

181. Cotlier, E., *Invest. Ophthalmol.*, 7:118 (1968).

182. Kinoshita, J. H., Dvornik, D., Kraml, M., and Gabbay, K. H., *Biochim. Biophys. Acta*, 158:472 (1968).

183. Kinoshita, J. H., Merola, L. O., and Tung, B., *Exp. Eye Res.*, 7:80 (1968).

184. Chylack, L. T., and Kinoshita, J. H., *Invest. Ophthalmol.*, 8:401 (1969).

185. Kinoshita, J. H., Barber, W. G., Merola, L. O., and Tung, B., *Invest. Ophthalmol.*, 8:625 (1969).

186. Dvornik, D., Simard-Duquesne, N., and Kraml, M., *Science*, 182:1146 (1973).

187. Obazawa, H., Merola, L. O., and Kinoshita, J. H., *Invest. Ophthalmol.*, 13:204 (1974).

188. Varma, S. D., and Kinoshita, J. H., *Biochim. Biophys. Acta*, 338:632 (1974).

189. Broekhuyse, R. M., *Biochim. Biophys. Acta*, 163:269 (1968).

190. Stewart, M. A., Kurien, M. M., Sherman, W. R., and Cotlier, E. V., *J. Neurochem.*, 15:941 (1968).

191. Pirie, A., *Invest. Ophthalmol.*, 4:629 (1965).

192. Sherman, W. R., Stewart, M. A., Kurien, M. M., and Goodwin, S. L., *Biochim. Biophys. Acta*, 158:197 (1968).

193. Sherman, W. R., Stewart, M. A., Simpson, P. C., and Goodwin, S. L., *Biochemistry*, 7:819 (1968).

194. Pitkänen, E., *Clin. Chim. Acta*, 48:159 (1973).

195. Caird, F. I., Hutchinson, M., and Pirie, A., *Brit. Med. J.*, 2:665 (1964).

196. Caird, F. I., Burditt, A. F., and Draper, G. J., *Diabetes*, 17:121 (1968).

197. Horning, M. G., Stratton, C., Nowlin, J., Harvey, D. J., and Hill, R. M., *Drug Metab. Dispos.*, 1:569 (1973).

198. German, A. L., Pfaffenberger, C. D., Thenot, J. P., Horning, M. G., and Horning, E. C., *Anal. Chem.*, 45:930 (1973).

199. Horning, E. C., and Horning, M. G., *Methods Med. Res.*, 12:387 (1970).

200. Lin, S.-N., Pfaffenberger, C. D., and Horning, E. C., *J. Chromatogr.*, 104:319 (1975).

201. Mathews, R. G., Schwartz, R. D., Pfaffenberger, C. D., Lin, S.-N., and Horning, E. C., *J. Chromatogr.*, 99: (1974).

202. Dole, V. P., and Meinertz, H., *J. Biol. Chem.*, 235:2595 (1960).

203. Muller, D. P. R., and Harris, J. T., *Biochem. J.*, 111:28P (1969).

204. Göransson, G., Norden, A., and Akesson, B., *Scand. J. Gastroenterol.*, 8:21 (1973).

205. Binder, H. J., Herting, D. C., Hurst, V., Finch, S. C., and Spiro, H. M., *New Eng. J. Med.*, 273:1289 (1965).

206. Underwood, B. A., Denning, C. R., and Navab, M., *Ann. N.Y. Acad. Sci.*, 203:237 (1972).

207. Bennett, M. J., and Medwadowski, B. F., *J. Clin. Nutr.*, 20:415 (1967).

208. Olson, R. E., *Circulation*, 48:179 (1973).

209. Nair, P. P., and Kayden, H. J., *Ann. N.Y. Acad. Sci.*, 203:1 (1972).

210. Rosenlund, M. L., Kim, H. K., and Kritchevsky, D., *Nature*, 251:719 (1974).

211. Burlingame, A. L., Kimble, B. J., and Derrick, P. J., *Anal. Chem.*, 48:368R (1976).

212. Eldjarn, L., Jellum, E., and Stokke, O., *J. Chromatogr.*, 91:353 (1974).

213. Mamer, D. A., Mitchell, E. J., and Scriver, C. R., in *Application of Gas Chromatography-Mass Spectrometry to the Investigation of Human Disease*, McGill University, Montreal Children's Hospital Research Institute, 1974.

214. Melville, R. S., and Dobson, V. F., DHEW Publication #(NIH) 75-762, 1974.

215. Brock, A., Heilskov, N. S. C., Keiding, R., Munck, O., and Nielsen, A. B., *Scand. J. Clin. Lab. Invest., Suppl.*, 35:143 (1975).

216. Burlingame, A. L., Shackleton, C. H. L., Howe, I., and Chizhov, D. S., *Anal. Chem.*, 50:347R (1978).

217. Abramson, F. P., *Microtechniques for the Clinical Laboratory: Concepts and Applications*, Wiley Biomedical, New York, 1976.

218. Stokke, O., *Biomed. Mass Spectrom.*, 3:97 (1976).

219. Johnstone, R. A. W., *Mass Spectrometry*, Vol. 4, Chemical Society, London, 1977.

220. Blau, K., and King, G. S., *Handbook of Derivatives for Chromatography*, Heyden, London, 1977.

221. Brooks, C. J. W., and Middleditch, B. S., *Mass Spectrometry*, Vol. 4, Chemical Society, London, 1977.

222. Jellum, E., *Adv. Mass Spectrom.*, 7:1651 (1978).

223. Frigerio, A., *Recent Developments in Mass Spectrometry in Biochemistry and Medicine*, Plenum Press, New York, 1978.

8

ANALYSIS OF STEROID HORMONES

ERIC VANLUCHENE

Academic Hospital, State University of Ghent
Ghent, Belgium

P. SANDRA

State University of Ghent
Ghent, Belgium

1. Introduction 289

2. Sample Preparation 291
   - 2.1 Hydrolysis 292
   - 2.2 Extraction of steroids 292
   - 2.3 Extract purification and fractionation 292
   - 2.4 Formation of derivatives 294
   - 2.5 Typical standard laboratory procedure for profiling urinary steroids 301

3. Instrumentation 302
   - 3.1 Columns 302
   - 3.2 Injection systems 306

4. Applications 307

5. Conclusion 325

References 327

1. INTRODUCTION

The secretion of a steroid reflects the (possibly deviated) action of an endocrine gland (adrenal, testis, ovary, and fetoplacental unit). The qualitative and quantitative analysis of steroids in biological fluids is therefore of considerable importance for the diagnosis of endocrine disorders.

In this chapter the possibilities of capillary gas chromatography for this type of analysis will be presented. Routine analysis of hormonal steroids can be done by non-selective detection, as in fluorescence and absorption spectroscopic measurements. For these methods the steroid isolation and purification is very complex and time-consuming. Gas chromatography gives the possibility of separation and determination at the same time. Radioimmunoassay (RIA) methods have reached a high degree of specificity and are ideally suited to the analysis of large series of samples. Because of this unique sensitivity, most steroid determinations today are performed with RIA. A drawback, however, is that for each steroid a specific RIA must be developed. Chromatographic determinations, on the contrary, are multicomponent analyses. They have the important advantage of detecting several steroids, even if the analysis did not primarily aim at their determination. Qualitative and quantitative information on each steroid present in the sample from only one determination or the "steroid-profiling technique" has become possible by the development of gas chromatography. The progress made in GC of steroid hormones coincides with the optimization and development of this separation technique. The first practical application of GC to the separation of steroids was described in 1960 [1]. The analyses were performed on packed columns filled with silanized diatomaceous earth of which a thin film of a polysiloxane stationary phase with high-temperature stability was coated. From then on, reports of the applicability of GC in routine analysis of hormonal steroids appeared regularly. Packed columns are still used for the assessment of estriol and pregnanediol in the fetoplacental unit, the analysis of 17-keto-steroids, etc. The complexity of the steroid profile, however, required GC columns of higher efficiency. The introduction of capillary columns by Golay [2] was welcome in this field. Steroid analyses on stainless steel capillary columns were reported in 1960 [3] and 1961 [4] but the separations obtained were poorer than on packed columns. The difficulties of early capillary GC remained unsolved for several years. The value of capillary columns for steroid profiling could only be developed fully by the optimization of glass capillary gas chromatography, or $(GC)^2$, about 10 years later. Other important points were the development of injection techniques, sensitive detectors, and suitable derivatives. Inter alia, pioneering work was done in the development of techniques for sample preparation by Curtius [5] and research workers of the Karolinska Institute, Sweden [6]; the formation of suitable derivatives by Horning and collaborators [7]; and the preparation of glass capillary columns by Grob [8], Verzele [9], Novotny [10], and the group of the Eindhoven University, the Netherlands [11].

Before 1975, $(GC)^2$ was largely used for research purposes. Now it is routinely applied in clinical laboratories in the diagnosis of disorders involving steroid biosynthesis and steroid metabolism. Routine analysis, however, implies that the procedures for steroid profiling are "simple" and "foolproof." In this respect, the estimation of urinary steroid hormones is relatively easy and achievable for routine analysis, whereas the analysis of blood or plasma samples is difficult and complicated. The concentration of steroids is indeed much higher in urine than in plasma samples, e.g., total corticoids 0.5 to 10 ppm in urine, 50 to 200 ppb in plasma.

The aim of this chapter is mainly to describe the procedures running routinely for urinary steroid profiling in the laboratory of one of the authors (c. 2000 analyses a year). This will be done in sufficient detail to allow the clinical chemists involved in endocrinology to take up the technology if they should wish to do so. Several applications also detailed in the chapter should illustrate the possibilities of $(GC)^2$ for hormonal analysis.

Because of the vast literature of steroid profiling by $(GC)^2$ considerable selection was necessary in preparing this contribution. This is per force subjective.

2. SAMPLE PREPARATION

The sample preparation for biological fluids depends on the steroid concentration and on the presence of other compounds which have to be removed prior to $(GC)^2$, as they disturb the qualitative and quantitative evaluation of the steroid profiles. Although a universal procedure cannot be presented, the following steps normally are performed:

1. Hydrolysis before or after the isolation of the steroids
2. Extraction of the steroids
3. Extract purification and fractionation
4. Derivative formation

2.1 Hydrolysis

In biological fluids, steroids occur free, associated with proteins, conjugated with glucuronic acid or with sulfuric acid (mono- and diesters), and as mixed conjugates. Because of low volatility and thermal instability the conjugates are not suitable for gas chromatographic analysis and have to be transformed to free steroids. The hydrolysis is carried out directly in the biological fluid or after isolation of the free steroid and steroid conjugates on XAD-2 resin [12]. The protein association is broken by denaturation during extraction or adsorption on XAD-2 resin at 64°C [13]. Hydrolysis of the steroid glucuronides is mostly performed enzymatically with β-glucuronidase. Most sulfate conjugates can be hydrolyzed with sulfatase except the conjugates of the 3α-hydroxyl group in 5α steroids and of some 17- and 20-hydroxyl groups [14]. Therefore, when reliable quantitative data are required the enzymatic hydrolysis is followed by a solvolysis [15].

2.2 Extraction of Steroids

The extraction of the liberated steroids can be carried out with all organic solvents having a favorable partition coefficient with regard to aqueous solutions. Ethyl acetate and diethylether are mostly used. The former is preferred because of the difficulty of peroxide elimination from diethyl ether. The drawback of solvent isolation is the dependency of extraction efficiency on the polarity of the steroids; e.g., up to 30% of corticoids is lost by a single extraction with ratio aqueous phase/extraction solvent 1:2.

A better alternative, especially for research purposes, is the isolation of steroids (free and/or conjugates) by column chromatography on Amberlite XAD-2 [12]. This resin exhibits adsorptive properties toward large molecules in aqueous solution. The steroids are eluted from the column with methanol or ethanol.

2.3 Extract Purification and Fractionation

Acidic impurities in urinary extracts are removed by washing with aqueous sodium bicarbonate (8%). We experienced that 90% of the urinary tetrahydrocortisone is lost when ether extracts are washed with 1 N sodium hydroxide. The chief urinary steroid in a neonatal profile, 3α,16α,18-trihydroxy-5-

androstene-17-one, is altered (95%) by solvolysis and alkali wash. These examples illustrate that care in purification is necessary.

Alternative procedures are ion exchange cleanup on the anion exchanger Amberlyst A-26 in the bicarbonate form [16] or on diethylamino-hydroxylpropyl Sephadex LH-20 [13]. Because of better recovery, DEAP-LH-20 is preferred.

Other biological fluids contain high quantities of proteins and lipids. Considering the low concentration of steroids, extract purification in this case is very important. Purification methods depend on the problem at hand and on the selectivity of the detection used. For single-ion detection (SID) solvent denaturation of the proteins and precipitation of the lipids at -20°C in a 70:30 methanol-water mixture may be sufficient (e.g., Ref. 17). Lipid impurities are best removed by reversed-phase chromatography on Lipidex 5000, the hydroxyalkoxypropyl derivative of Sephadex G-25 (Packard C. [13]). High-molecular-weight compounds are separated from the extracts by Sephadex G-25 gel [18]. Apolar impurities can be removed by gradient elution on small silicic acid columns [19].

Another approach for purification includes at the same time fractionation of liberated steroids. Fractionation of free steroids has been achieved by paper chromatography, thin-layer chromatography, and column chromatography on alumina or silica gel. For the last techniques Ref. 20 can be consulted. Drawbacks are the poor recovery and destruction during adsorption.

Sephadex LH-20 is frequently used for the separation of free steroids. Setchell [21] described a separation method on Sephadex LH-20 based on the different polarity of the steroids and using a solvent mixture of cyclohexane and ethanol. This method is of special interest for the estimation of urinary steroids from newborns [22]. As pointed out by Anderson [23] and Bègue [24], the hydroxyalkoxypropyl derivatives of Sephadex G-25 should give better fractionation.

Worth mentioning is the separation of 3-keto-4-ene steroids, based on the positive charge of their oximes, on the cation exchanger sulfoethyl LH-20 (SE-LH-20) [25].

For some applications it is interesting to analyze separately the different conjugates. Differential hydrolysis methods exist in which, prior to the enzymatic hydrolysis of glucuronides, the sulfates are solvolyzed and extracted from the steroid mixture [23]. A better approach is the separation of different conjugates with column chromatography on Sephadex

LH-20 [26]. The anion exchanger diethylaminohydroxypropyl Sephadex LH-20 should be better suited as stated by Setchell et al. [27].

2.4 Formation of Derivatives

Steroids are high-molecular-weight polyfunctional substances which, by their polar and thermolabile character, are difficult to analyze as such by gas chromatography. Some success can only be obtained on very short capillary columns [28]. An analysis of selected free steroids on a short capillary column coated with Superox-4 [29] is shown in Fig. 1. The numbers refer to steroids listed in Table 1.

For quantitative analysis the formation of derivatives is recommended. By derivatization the polar and active parts of the molecule are changed in apolar and inert parts. The intra- and intermolecular interactions are excluded, resulting in an increase of volatility and thermal stability. Another advantage of derivatization is the reduction of interaction between the substances and the column wall.

Several procedures are described for the derivatization of steroids, which contain primary, secondary, and tertiary hydroxyl groups together with keto groups. A prerequisite for steroid profiling is that for each steroid a single derivative is formed. This is a difficult task, as the reactivity

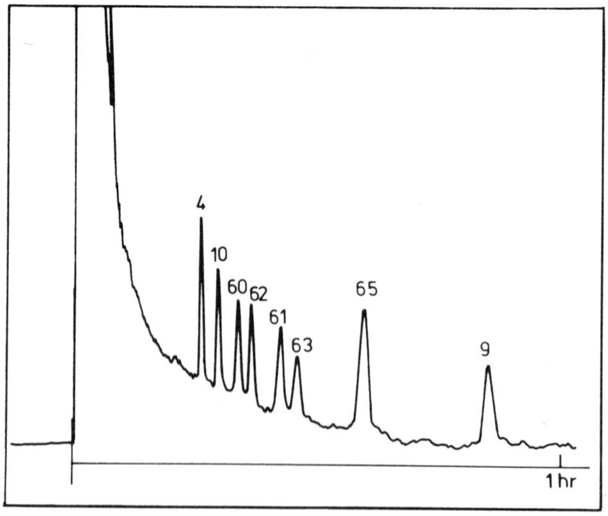

Figure 1. Gas chromatogram of free steroids on a 7.5-m open tubular column coated with Superox-4 (250°C). Numbers refer to Table 1.

of the hydroxyl groups and keto groups is strongly dependent on their nature and their steric environment.

For flame ionization detection, the hydroxyl groups are derivatized by the formation of trimethylsilyl ethers (TMS derivatives). The silyl donor ability of several silylating agents and silylating mixtures has been reviewed by Gleispach [30] and Poole [31]. The preferred reagents are trimethylsilylimidazole (TSIM) or the mixture trimethylsilylimidazole (TSIM), bistrimethylsilylacetamide (BSA), and trimethylsilylchlorosilane (TMCS) 3:3:2; both reacting with all hydroxyl groups (persilylation). Some steroids containing unprotected ketone groups form secondary products arising from the formation of enol-TMS ethers. To avoid enolization the keto groups in the 3, 7, 16, 17, and 20 positions can be protected by the formation of oxime derivatives prior to silylation. This is achieved quantitatively by reaction with methoxylamine (MO) hydrochloride [32], butoxylamine (BuO) hydrochloride, or benzyloxylamine (BO) hydrochloride [33]. The shielded 11-keto group only reacts under extremely hard conditions. Because of their high volatility, MO derivatives are preferred for the steroid-profiling technique. BuO and BO derivatives are used for the specific estimation of ketosteroids, as the derivatives shift to higher elution times. For 3-keto-4-ene and for 17-desoxy-pregnane-20-one steroids, the oximation reaction produces two products; the syn and anti isomers. The separation of the syn and anti isomers by $(GC)^2$, even on nonselective silicone phases, is not a hard obstacle for the steroid-profiling technique. The sensitivity for the estimation of the particular steroid is decreased, but when syn and anti peaks are observed, identification becomes easier and thus selectivity is increased. The ratio of the syn and anti isomer of 16-hydroxy-DHA, for instance, is dependent on the concentration and age of the methoxylamine hydrochloride solution. Only a freshly prepared saturated solution produces a constant ratio.

The two-step derivatization—methoxime (MO) and trimethysilyl (TMS) formation—as optimized by Thenot and Horning [7] has found wide application and is the method of choice for profiling steroids. A typical structure after hydrolysis and MO-TMS derivatization is given in Fig. 2.

Because of the low volatility and the harmful effect on the column performance, the silylating reagents, especially trimethylsilyl imidazole (TSIM), have to be removed before $(GC)^2$ analysis. A suitable method is purification of the sample on Lipidex 5000 according to Axelson and Sjövall

Table 1. List of Compound Names and Numbers

1. Dioctylphtalate
2. Androsterone
3. Etiocholanolone
4. Dehydroepiandrosterone
5. Epiandrosterone
6. 3β,17β-Dihydroxy-5-androstene
7. 11-Oxoandrosterone
8. 11-Oxoetiocholanolone
9. 11-Hydroxyandrosterone
10. 3β,17β-Dihydroxy-5α-androstane
11. 3α,17α-Dihydroxy-5β-pregnane-20-one
12. 3β-Hydroxy-5β-pregnane-20-one
13. 11-Hydroxyetiocholanolone
14. 3α-Hydroxy-5α-pregnane-20-one
15. 3α-Hydroxy-5β-pregnane-20-one
16. 3α,17β-Dihydroxy-5β-pregnane-20-one
17. 3β,16α-Dihydroxy-5-androstene-17-one
18. 3α,20α-Dihydroxy-5α-pregnane
19. 3α,20α-Dihydroxy-5β-pregnane
20. Squalene
21. 3β-Hydroxy-5α-pregnane-20-one
22. 3α,17α,20β-Trihydroxy-5 -pregnane
23. 3α,17α,20α-Trihydroxy-5α-pregnane
24. 3β,16α-Dihydroxy-5β-pregnane-20-one
25. 3β,17β-Dihydroxy-5-androstene-16-one
26. 3β,16α,17β-Trihydroxy-5-androstene
27. 3ξ,16ξ-Dihydroxy-5ξ-pregnane-20-one
28. 3α,17α,21-Trihydroxy-5β-pregnane-20-one
29. 3α,21-Dihydroxpregnane-20-one
30. 3β,16α,20α-Trihydroxy-5-pregnene
31. 3α,17α,20α-Trihydroxy-5β-pregnane-11-one
32. 3β,16α-Dihydroxy-5-pregnene-20-one
33. 3β,17α,20α-Trihydroxy-5-pregnene
34. 3α,17α,21-Trihydroxy-5β-pregnane-11,20-dione

Table 1 (cont'd)

35.	3α,21-Dihydroxy-5β-pregnane-11,20-dione
36.	3α,11β,21-Trihydroxy-5β-pregnane-20-one
37.	3α,11β,21-Trihydroxy-5α-pregnane-20-one
38.	3α,11β,17α,21-Tetrahydroxy-5β-pregnane-20-one
39.	3α,11β,17α,21-Tetrahydroxy-5α-pregnane-20-one
40.	3α,17α,20α,21-Tetrahydroxy-5β-pregnane-11-one
41.	3α,17α,20β,21-Tetrahydroxy-5β-pregnane-11-one
42.	3α,11β,17α,20β,21-Pentahydroxy-5β-pregnane
43.	Cholesterol
44.	3β-Hydroxy-5α-cholestane
45.	3α,11β,17α,20α,21-Pentahydroxy-5β-pregnane
46.	3β-Acetoxy-5-cholestene
47.	3β,20α-Dihydroxy-5-pregnene
48.	3β,21-Dihydroxy-5-pregnene-20-one
49.	3α,17α,21-Trihydroxy-5α-pregnane-20-one
50.	3α,17α,20α,21-Tetrahydroxy-5β-pregnane
51.	3-Hydroxy-1,3,5,(10)-estratriene-17-one
52.	3, 17β-Dihydroxy-1,3,5,(10)-estratiene
53.	3, 16α-Dihydroxy-1,3,5,(10)-estratriene-17-one
54.	3, 17β-Dihydroxy-1,3,5,(10)-estratriene-16-one
55.	3, 16α,17β-Trihydroxy-1,3,5,(10)-estratriene
56.	3, 16β,17β-Trihydroxy-1,3,5,(10)-estratriene
57.	3, 15α,16α,17β-Tetrahydroxy-1,3,5,(10)-estratriene
58.	3α-Hydroxy-5β-cholestane
59.	17β-Hydroxy-5α-androstane-3-one
60.	5-Androstene-3,17-dione
61.	Testosterone
62.	Pregnenolone
63.	Progesterone
64.	17α-Hydroxypregnenolone
65.	20α-Hydroxyprogesterone
66.	17α-Hydroxyprogesterone
67.	3α,17α,20β,21-Tetrahydroxy-5β-pregnane
68.	3β,16α,18-Trihydroxy-5-androstene-17-one

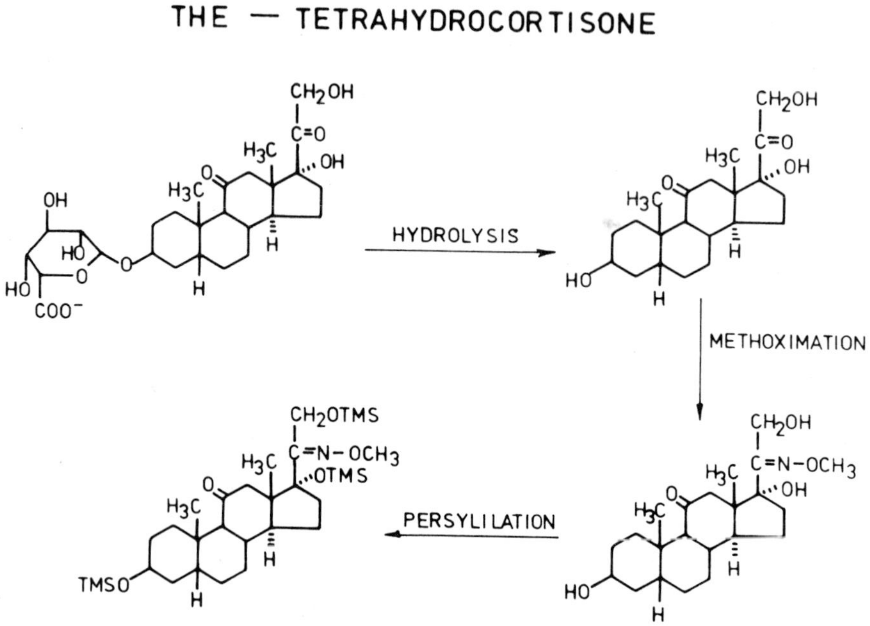

Figure 2. Hydrolysis and derivative formation (MO-TMS) of tetrahydrocortisone (THE).

[13]. Due to the different partition coefficients of derivatized steroids, solvent extraction to remove TSIM is dangerous. The method of De Jong [14], however, with the more polar dichloromethane as extractant and washing with sulfuric acid, looks promising.

Steroids are also derivatized into silyl ethers with N,O-bis(trimethylsilyl)trifluoroacetamide (BSTFA) or N-methyl-N-trimethylsilyl trifluoroacetamide (MSTFA) catalyzed with potassium acetate [34]. This reaction gives a quantitative yield of the TMS-enol-TMS derivatives. The applicability of this derivatization procedure is limited by incomplete reaction for the 17- and 3-keto-enol-TMS derivatives, while 3-keto-4-ene steroids form isomeric products [31]. Another drawback is the long elution time for the TMS-enol-TMS derivatives.

Derivatization of a sample by two different procedures can be very helpful for the qualitative and quantitative estimation of a steroid profile. This is illustrated in Fig. 3, showing the analysis of the TMS and MO-TMS urinary steroids of a woman with a hyperstimulated corpus luteum. Numbers above the peaks refer to Table 1. Steroids with a reactive keto group shift to longer elution times when methoximes are formed. Axelson

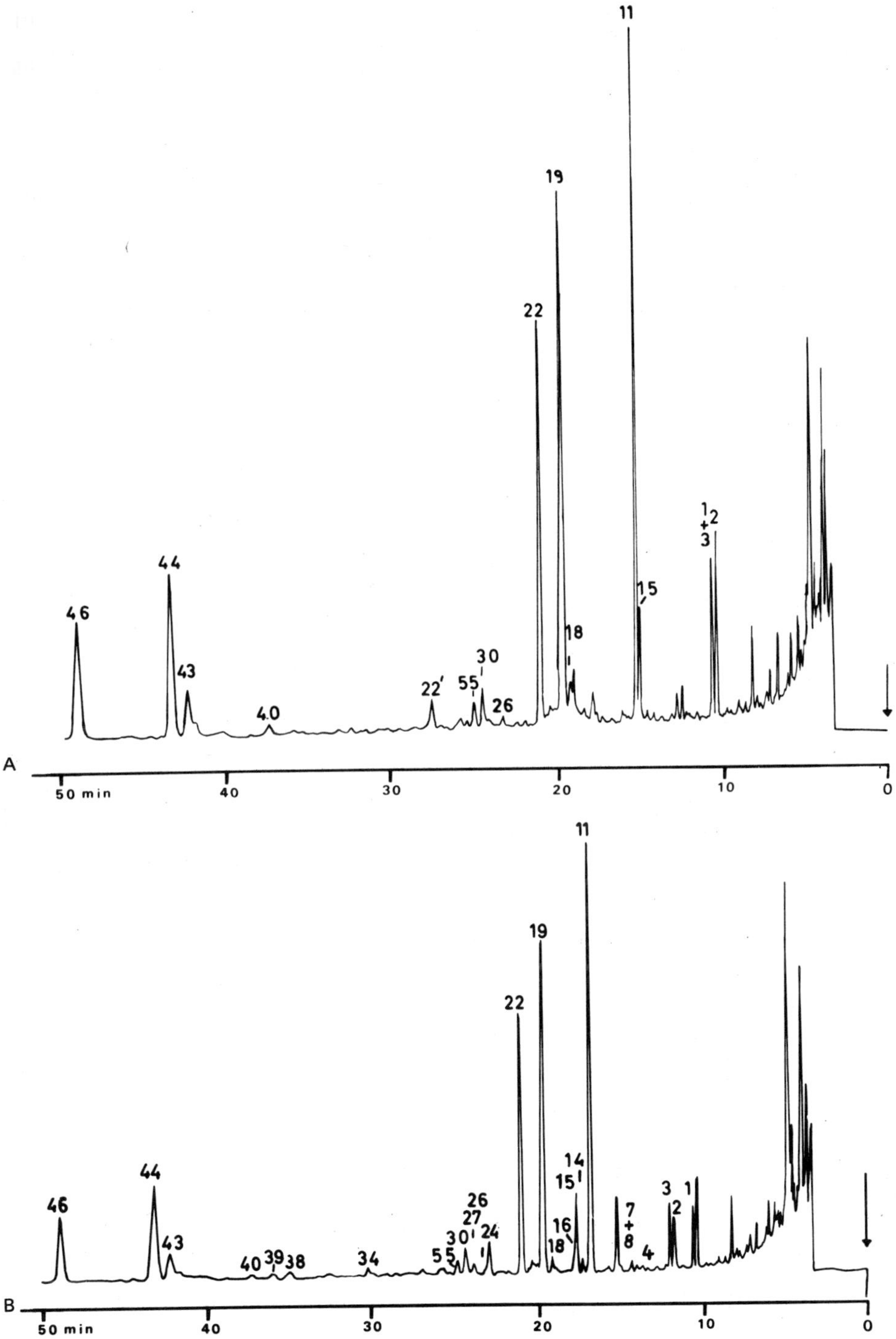

Figure 3. Chromatograms of a urine from a woman with hyperstimulated corpus luteum (molar pregnancy): (A) TMS ethers; (B) MO-TMS derivatives.

[35] has recently pointed out the possibility of converting steroid methyloximes to ethyl- or isopentyloximes. The conversion rate is a function of the position and environment of the methoxime function. The method allows the position determination of keto groups.

Steroids are detected by electron capture detection (ECD) after formation of the methoxime heptafluorobutyrate (MO-HFB) derivatives, through methoximation of the keto groups followed by conversion of the hydroxyl groups with heptafluorobutyryl anhydride or heptafluorobutyryl chloride. To avoid the formation of acidic side products decreasing the stability of the esters, perfluoroacetylated imidazoles can be used. Carbonyl groups are converted to electron capture derivatives by the formation of pentafluorobenzyloximes as described by Koshy [36] and Nambara [37]. Touchstone [38] analyzed DHA, 16-hydroxy-DHA, and estriol in plasma samples of pregnant women with ECD by converting the steroid sulfates directly to heptafluorobutyrates. This author gives a good review of earlier ECD derivatization of steroids.

Recently, Morgan and Poole [39,40] investigated the use of flophemsyldiethylamine and flophemesylchloride for the formation of electron capture derivatives of steroids. The good volatility, ability to respond to the electron capture detector, and also the useful mass spectral properties of these derivatives may find application in the future.

Dimethylthiophosphinic esters are new derivatives; picogram detection is possible with the alkali flame detector [41]. Mass spectral data of (M-)TMS derivatives are well documented in the literature [27]. However, steroids without methoxime functions yield poor mass spectra. To overcome this problem, t-butyldimethylsilyl ether and allyl-dimethylsilyl ether (or enol ether) derivatives are formed [42,43]. Recently, Quilliam and Westmore [44], extended the field by including other "crowded" silylation reagents as cyclotetramethylene isopropyl and cyclotetramethylene t-butylsilylchloride.

The stability of the derivatives varies greatly with the chemical nature of the steroid and the selected reagent. If persilylation is carried out and the $(GC)^2$ analysis is not performed immediately, derivatives are stored in the reagent or, after removal of the reagents by the Lipidex 5000 procedure, in a solvent mixture containing 98 parts cyclohexane, 1 part pyridine, 10 parts dimethoxy propane, and 2 parts hexamethyldisilazane. The tubes closed with a PTFE stopper are stored in a dessicator. For unpersilylated steroids, the reagents must be removed as further derivatization can occur. For systems where the steroids are introduced as solids with automatic injection, it is obvious that the derivatives must be stable in their solid state. Shackleton [22] and Fantl [45] indicated that a slight but gradual deterioration of the sample during storage was observed.

2.5 Typical Standard Laboratory Procedure for Profiling Urinary Steroids*

1. Before analysis, 24-hr urine collections are stored in plastic (PVC) bottles at -20°C without adding any preservative.
2. A fraction (1 ml for normal and newborn steroid profiling, 0.5 ml for pregnancy urines, 2 to 3 ml for infant urines) is transferred to a centrifuge tube and diluted to 8 ml with distilled water.
3. The pH is adjusted to 5.2 with a 20% acetic acid solution. One ml 2 M acetate buffer (pH 5.2) and 100 μl Helix Pomatia enzyme solution (1'Industrie Biologique Francaise) are added. The tube is placed in a thermostat at 37°C during an incubation period of at least 36 hr.
4. After the hydrolysis, 100 μl of a standard solution containing 5 μg 5α-cholestane-3β-ol and 5 μg cholesteryl acetate in ethanol is added. By means of a shaker the steroids are extracted with 21 ml ethyl acetate for 8 min after which the mixture is centrifuged at 3000 rpm for 2 min. The water layer is removed by suction with a water pump and the extract is washed with 5 ml of a saturated sodium bicarbonate solution. After centrifugation at 3500 rpm for 2 min, the water layer is sucked off and the extract is washed with 5 ml distilled water. The mixture is centrifuged for 6 min at 3500 pm. The water layer is removed and the extract is dried over 3 g of anhydrous sodium sulfate. The organic layer is decanted and the solvent is evaporated at 50°C with a dry nitrogen flow. If there are no emulsions, the centrifugation steps may be omitted.
5. The formation of methoxylamine (MO) derivatives is carried out by adding 100 μl of a saturated methoxylamine hydrochloride solution in pyridine. The mixture is kept at 62°C for 30 min. The pyridine is completely evaporated at 50°C under a dry nitrogen flow. The MO derivatives are persilylated by addition of 100 μl trimethylsilyl imidazole (TSIM) and by reaction for 2 hr at 100°C.

---

*This is the procedure employed by one of the authors (E.V.).

The mixture is allowed to cool and is diluted with 0.5 ml of a solution containing 98 parts cyclohexane, 1 part pyridine, 10 parts 2,2-dimethoxypropane, and 2 parts hexamethyldisilazane.

6. Reagents (methoxylamine hydrochloride and trimethylsilyl imidazole) are removed by column chromatography on a Lipidex 5000 column in the following way:
   Six cm of a Pasteur pipette (4.7 mm ID × 15 cm long) is filled with a suspension of Lipidex 5000 (0.3 g dry gel) in the solvent mixture mentioned before. The total sample (600 μl) is pipetted onto the column, and after the sample has passed through the column is washed with 1.5 ml of the solvent mixture.
   The eluate is collected in a pointed test tube, the solvent is evaporated by introducing a nitrogen flow, and the residue is dissolved in 100 μl of the cyclohexane reagent solution.
7. This step completes the sample preparation. Samples are transferred in 0.5-ml test tubes with polyethylene caps and stored in a vacuum dessicator filled with calcium chloride. The samples thus prepared are stable for several weeks.

3. INSTRUMENTATION

3.1 Columns

Of the materials currently available for the preparation of capillary columns, glass is the only one inert enough for the analysis of steroids. On metal surfaces, even derivatized steroids are irreversibly adsorbed or decomposed. Moreover, the surface of metal tubing cannot be deactivated by current procedures. In 1969, the first successful application of $(GC)^2$ for the analysis of steroid derivatives was described by Grob [8], showing that sufficiently inert capillary columns could be prepared. For clinical studies, Grob columns were used by Völlmin [46] and Ros [47]. Novotny [10] analyzed steroids on etched Pyrex glass and concluded that to avoid an undesirable interaction between the chromatographed solutes and the glass wall, silylation of the glass surface was necessary. The deactivation of the glass surface by silylation was also investigated by Rutten and Luyten [11]. They noted fast deterioration of the column performance if insufficiently silanized. Better deactivation was obtained by the surface active agent BTPPC. German and Horning [48] described the analysis of steroids on support-coated open tubular (SCOT) columns. The silanox layer increased the thermal stability and sample capacity of the capillary column.

The possibilities of wide-bore columns for steroid profiling in clinical routine work were described by Sandra et al. [49]. The columns were prepared from alkaline glass, which shows a considerably lower activity than borosilicate glass [50]. A deactivation seemed unnecessary if the columns were coated with a heavy film (0.3 to 0.6 μm). The advantage of wide-bore capillary columns is the high sample capacity which allows the analysis of unknown biological samples where steroids in high and low concentration must be estimated.

Madani et al. [51,52] prepared capillary columns of different polarities and with high-temperature stability by polymerization of silicon oil on the column wall. They showed the phase-glass wall interaction but did not investigate the catalytic activity of the modified wall.

Today, several methods of preparing glass capillary columns with high separation power and low catalytic activity are available. These methods can be found in the literature specialized in chromatography.

Concerning the choice of the stationary phase, by their high-temperature stability and nonselective properties, the methylsiloxane polymers (SE-30, OV-1, OV-101) are the most satisfactory now known for profiling steroids.

Especially the gum phases SE-30 and OV-1 yield a perfect homogeneous film on the glass surface. Moreover, the wettability of the glass wall need not to be improved by intermediate layers. They can also be used to decrease the activity. The columns are best coated according to the static coating procedure of Verzele [9,53]. A distinct advantage of this coating procedure over the dynamic method is the reproducibility of the film thickness, which is a must for routine work. Other stationary phases used in steroid work include OV-225 [54], AN-600 [55], methylphenylsiloxanes [52], Silar 5 CP [56], Superox-4 [57], and OV-17 [58].

Considering the high separation efficiency of capillary columns coated with OV-1 or SE-30, the selectivity of other stationary phases is seldom needed for steroid work.

A drawback of polydimethylsiloxane films is that they do not sufficiently suppress the activity of the column wall. In this context, the acidity or basicity of the glass wall is very important. In our experience, for routine work, the deactivation of the glass wall is not a critical problem if the columns are prepared from soda-lime glass and covered with a moderate film layer (>0.25 μm). For thinner films a wall modification (sodium chloride crystals [54,59], sodium chloride dendrites [60], $BaCO_3$ [61]) and a deactivation (Carbowax 20M [62,63], BTPPC [11]) must be included.

Using borosilicate glass, even a film thickness of 0.25 μm is insufficient to mask the active sites. Wall modification (HCl gas, HF gas, $BaCO_3$) and deactivation (C 20M, poly NIA [50], poly CHAZ [56]) is for all circumstances carried out.

It is well known, however, that whatever the deactivation procedure, the catalytic activity of a OV-1 or SE-30 column increases as a function of time and working temperature. After long use (weeks) at elevated temperatures (250°C), a well-deactivated column shows a certain activity for the polar substances (nonanol, DMA, DMP). This activity has little influence on the gas chromatographic process of most derivatized steroids. Normally, in routine analysis of urinary steroids, the diagnosis is based on large deviations from the normal steroid pathway and a small activity does not affect the estimation of the disorder.

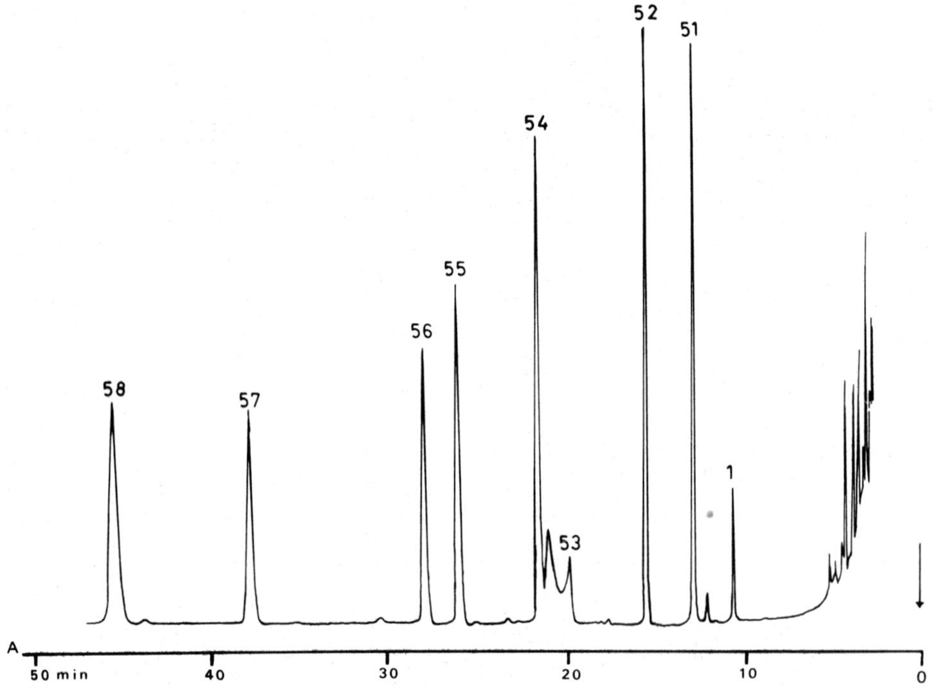

Figure 4A. Chromatogram of estratriene steroids on a soda lime glass capillary column.

Small quantitative deviations and some steroids however, are, only detected with reliable and well-deactivated columns. As an example, Fig. 4 shows the analysis of a test mixture containing eight TMS-estratriene steroids on a soda-lime glass capillary column and coated with a 0.3 μm SE-30 film. Except 3,16α-dihydroxy-1,3,5(10)-estratriene-17-one, all compounds elute quantitatively. On a Pyrex column coated with the same film thickness after surface modification by hydrochloric acid gas etching and deactivation with poly NIA [50], the $16e_1$ compound was quantitatively detected, whereas on a Pyrex column deactivated with BTPPC only 50% of the $16e_1$ quantity was detected. For research purposes and for profiling disorders which are not well established, one must pay attention to the quantitative aspect of the $(GC)^2$ analysis. High-quality inert capillary columns must be used. Also,

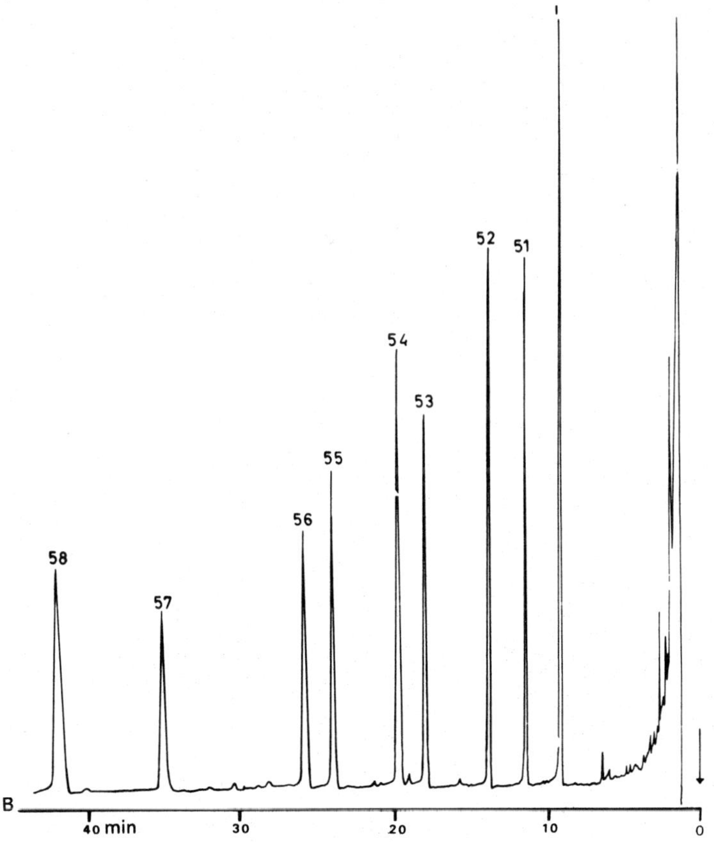

Figure 4B. Chromatogram of estratriene steroids on poly-NIA-deactivated borosilicate glass.

the quantitative aspect of the extraction procedure and sample preparation must be checked out very carefully.

Normally, the $(GC)^2$ analyses are performed on capillary columns 0.25 to 0.30 mm ID (narrow bore) and 20 to 30 m long. However, routine analysis of steroids as carried out by technicians is more reliable and much easier on wide-bore capillary columns with an ID of 0.5 mm and 40 to 50 m long. The columns are statically coated with a 0.2% OV-1 hexane solution. The lifetime of such capillary columns used isothermally at 245°C is approximately 300 analyses. As the urinary samples are not purified, after 100 analyses the first meter of the column is broken off.

3.2 Injection Systems

The requirements of a good injection system for high-boiling compounds, e.g., steroids mostly present in low concentration, can be summarized as follows:

The system may not influence the column performance.
The introduction of the reagent must be avoided.
A splitting device must be avoided if only a limited sample quantity is available.

Requirements for routine analysis are:

Simple and foolproof.
Can perform isothermally as it requires less time than for programmed runs.
Easy installation in different gas chromatographs.

Injection methods successfully used for steroid analysis are as follows.

Injection on a Cold Zone [8]. This injection system developed by Grob is a splitless injection with solvent bypass, in which the high-boiling compounds are trapped and separated from the solvent and reagent at the beginning of a cold column.

After elution of the solvent, the temperature is quickly raised to about 170°C and then temperature-programmed to about 250°C.

Application of Capillaries or Ampoules [22,45]. The samples are brought into open-ended capillary tubes. The solvent is evaporated and the capillary tube containing the solid steroids is dropped in a hot evaporation room. This system enables storage of samples and automation. The limiting factor of this system is the lability of the steroid derivatives in the solid state.

Injection on a Precolumn [64]. German and Horning described a glass injection system suitable for the quantitative analysis of derivatized biological samples. The system consists of a packed precolumn and a splitting device (1:5). This system had proved to be satisfactory and was routinely used by Pfaffenberger [65,66].

The All-glass Solid-state Injection Device. Introduced by Ros [67] and optimized by the Eindhoven group [68,69], the requirements for an injection technique for steroids are largely met in this system. In this technique, the steroids are injected on the tip of a needle, movable by means of a magnet. The solvent is evaporated via a bypass and the needle is introduced in the hot GC oven a few millimeters above the column connection. All analyses of our laboratory described in the application section have been carried out with this injection device.

On-column Injection. For quantitative work, on-column injection has been described by Sandra et al. [49]. Wide-bore capillaries were recommended. Recently, Grob [70] extended on-column injection for narrow-bore capillary columns, and preliminary results look very promising [71].

4. APPLICATIONS

In the first part of this section we will emphasize some published data which have in our opinion contributed to the present state of the art. A chronological sequence will be respected. The second part discusses some steroid profiles as routinely obtained in our laboratories.

The first $(GC)^2$ publications dealing with urinary steroid profiling originate from J. A. Völlmin and co-workers [46,72,73]. The low corticosteroid levels in their profiles undoubtly are caused by their washing procedures (ether extracts washed with 1 M sodium hydroxide). Improvement was obtained by extraction with ethyl acetate. The procedures described by Ros and Sommerville [47] included also an ether extraction at higher pH values, and the corticosteroids level was very low.

Routine steroid profiling was also described by Novotny [74] and Luyten [75]. Emphasis was placed on the reproducibility of the retention indices for qualitative estimation.

Pfaffenberger [66] analyzed urinary profiles in men and women employing the techniques developed by German [48] and Horning [76]. 5α and 5β reduction patterns of steroids were related to sex. The same technique was used for profiling urinary steroids in women with breast tumors [77].

Berthou et al. [78] analyzed pregnane- and androstanediols in urines. Sample purification was carried out by Girard separation of the keto steroids and chromatography on alumina. The purification was improved by the introduction of hydroxyalkylpropyl LH-20 gel chromatography [79]. The clinical importance of 3α,17β-dihydroxy-5α-androstanediol analysis in urines was pointed out (index of androgenicity). Three derivatization methods (TMS, HFB, and TFA) assured the high specificity for diol estimation. A chromatogram of 5α-androstanediols in male urine extract (TMS derivatives) is shown in Fig. 5.

The result with the column preparation of Madani et al. [51,52] is illustrated in Fig. 6. Benzyloxime-trimethylsilyl derivatives of the hormonal steroids of a 3-day-old newborn were analyzed at a column temperature starting from 190 to 310°C. Curtius et al. [80] describe the "routine" determination of urinary steroids with $(GC)^2$. Variation coefficients for

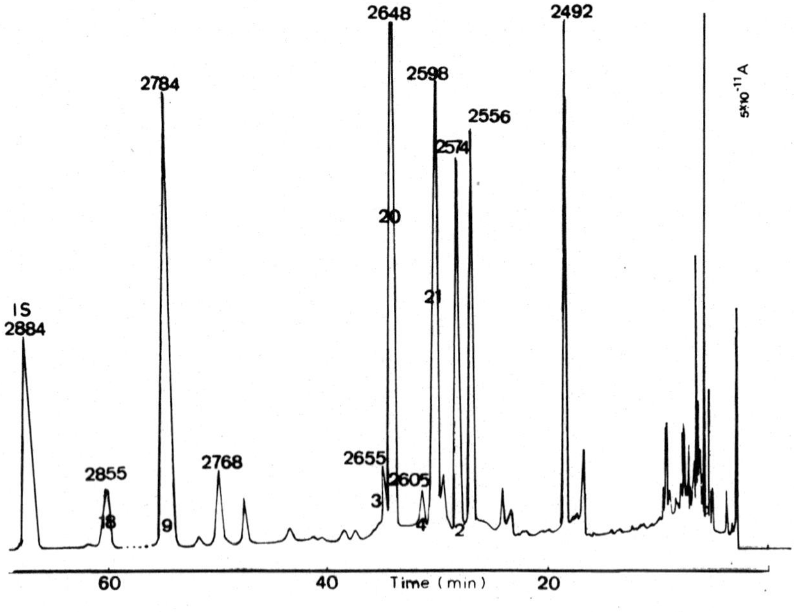

Figure 5. Gas chromatogram of male urine extract (5α-polarity, TMS derivatives); OV-101 column at 250°C: 2—5α-androstane-3α,17β-diol, 21—5-androstene-3β,17α-diol, 4—5α-androstane-3β,17α-diol, 20—5-androstene-3β,17β-diol, 3—5α-androstane-3β,17β-diol, 9—5α-pregnane-3β,20α-diol, 18—5-pregnene-3β,20α-diol, IS—5-androstene-3β,16β,17β-triol. (Source: Ref. 70. Reproduced with permission.)

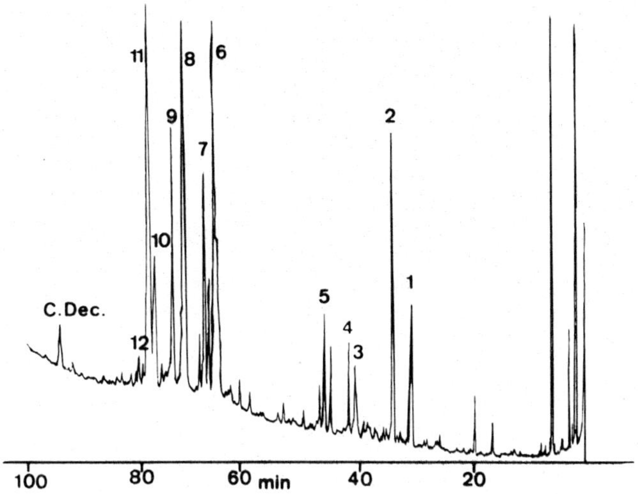

Figure 6. Methylsiloxane-coated capillary column separation of the steroids excreted as sulfates in a urine pool from 3-day-old human newborns. BO-TMS derivatives; temperature programming from 190 to 310°C at 2°C min^{-1}. Peaks: 1—androstenetriol, 2—androst-5-ene-3β,16α,17β-triol, 3—androstenetetrol, 4—androstenetetrol, 5—pregnanetriol, 6—3β,16α-dihydroxy-androst-5-ene-17-one, 7—androstenetriolone, 8—3β,17β-dihydroxyandrost-5-ene-16-one + pregnanediolone, 9—pregnenetriolone, 10—androstenetriolone, 11—3β,16α-dihydroxypregn-5-ene-20-one, 12—3β,21-dihydroxypregn-5-ene-20-one, C.Dec.—cholesteryldecylate (internal standard). (Source: Ref. 51. Reproduced with permission.)

the whole procedure were between 7 and 25%. Possibilities of loading tests with stable isotopes by $(GC)^2$/MS for metabolic studies were advanced.

Fantl and Gray [45] describe the routine determination of urinary steroids as trimethylsilyl enol ethers on commercial capillary columns.

In 1973, Setchell [21] demonstrated the possibilities of fractionation of urinary steroids on LH-20 Sephadex columns. Figure 7 represents the analysis of six fractions from a urine. The method is especially suited to the analysis of the complex profiles of newborn infant urines and was tested on reproducibility by Shackleton et al. [22]. In this work a study was undertaken [81] to detect, from the first moment of life, adrenal 21-hydroxylase defects in newborns. Figure 8 gives the profile change during the first month of life for a child with the defect. Because the cortisol production is impossible, excessive adrenal stimulation by ACTH (adrenocorticotropic hormone) leads to high secretion of cortisol precursors. These are pathognostic metabolites which appear in the urine, e.g., pregnanetriol, 11-ketopregnanetriol, and 17α-hydroxypregnanolones. The early recognition

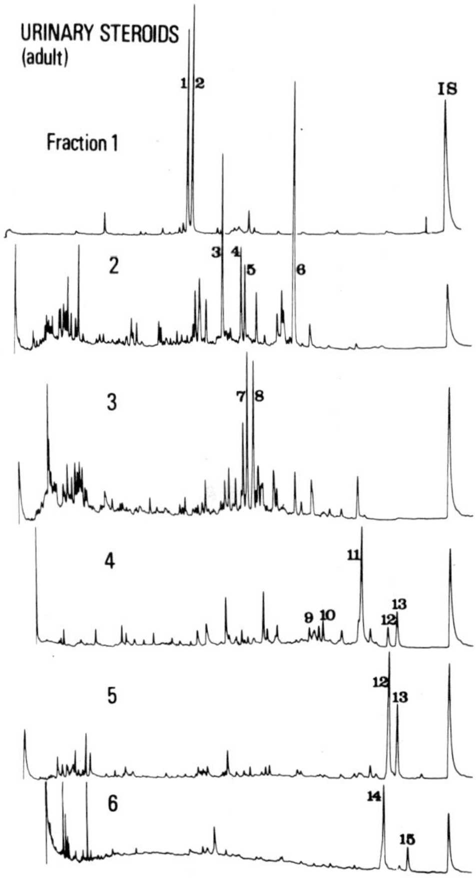

Figure 7. Fractionation of (adult) urinary steroids as methyloxime trimethylsilyl ethers on 25-m OV-101 capillary column. The following steroids are indicated: 1—androsterone, 2—aetiocholanonlone, 3—11-oxoaetiocholanolone, 4—11-hydroxyandrosterone, 5—11-hydroxyaetiocholanolone, 6—pregnanetriol, 7 and 8—androstanetriols, 9—tetrahydro-11-dehydrocorticosterone, 10—tetrahydrocorticosterone, 11—tetrahydrocortisone, 12—α-cortolone, 13—β-cortolone, 14—tetrahydrocortisol, 15—cortol, IS—cholesteryl-butyrate. (Source: Ref. 21. Reproduced with permission.)

of the symptom is extremely important in the clinical management of the patients. However, Shackleton concluded that for the first days of life only single-ion mass fragmentographic detection of the mentioned metabolites is competitive with RIA analysis. The recognition in older children and adults of steroid metabolizing defects, in contrast, is easily feasable with $(GC)^2$. Examples will be given later. Several authors involved in steroid profiling select these disorders to illustrate the possibilities of $(GC)^2$.

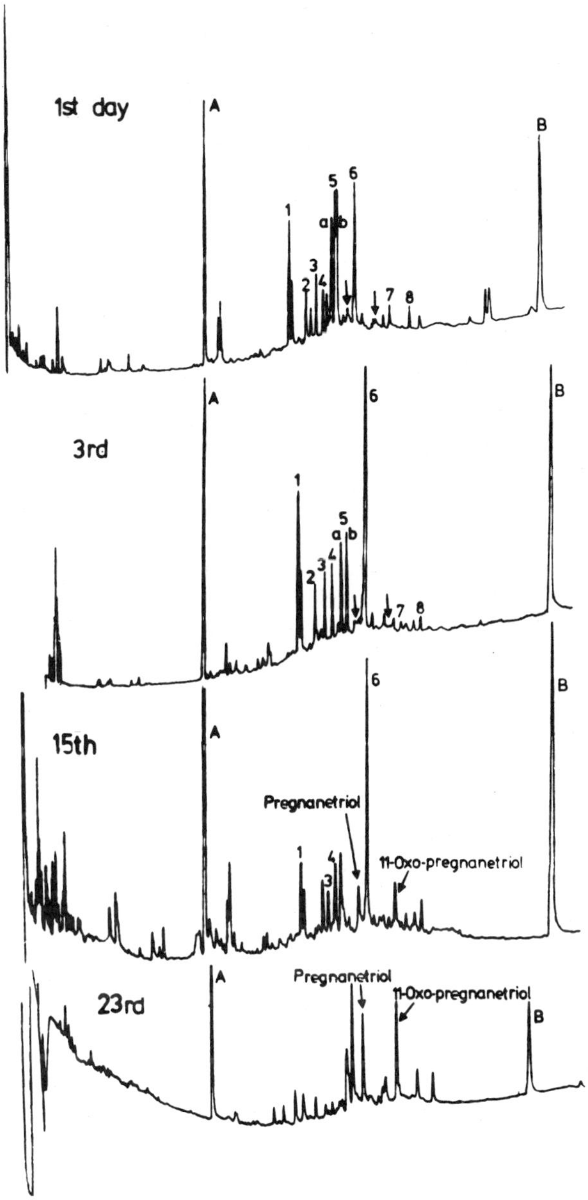

Figure 8. Multicomponent analysis of the major urinary steroids (as oxime-trimethylsilyl ethers) excreted in a case of 21-hydroxylase deficiency during first weeks of life. The following steroids are indicated: A and B internal standards 5α-androstane-3α,17α-diol and cholesteryl butyrate, 1—16α-hydroxy-DHA (doublet), 2—16β-hydroxy-DHA, 3—16-oxo-androstenediol, 4—5-androstene-3β,16α,17β-triol, 5—3β,16,18-trihydroxy-5-androstene-17-one (contaminated with oestriol on first day of life), 6—16α-hydroxy-pregnenolone, 7—21-hydroxypregnenolone, and 8—5-pregnene-3β,20α,21-triol. In the chromatograms illustrating steroid analysis on the first to fourth days of life arrows indicate the position where pregnanetriol and 11-oxopregnanetriol would appear if present. (Source: Ref. 81. Reproduced with permission.)

An excellent illustration of urinary sample prefractionation [27] is shown in Fig. 9 where the neutral and the different conjugated steroids were separated by ion exchange chromatography on the lipophilic gel diethylaminohydroxypropyl Sephadex LH-20 (DEAP LH-20).

Egger et al. [82] stated that in hirsute females the secretion rate of 5-ene steroids (DHA, androstenediol, 16-hydroxy DHA, androstenetriol, etc.) is increased. Moreover emotional stress caused a 20-fold increase of the DHA excretion. This phenomenon is not evident if one analyzes 24-hr urine samples and misses the diurnal rhythmic secretion.

Ludwig and Spiteller [83] determined the urinary profile of patients with uremia; 11 oxygenated C_{19} steroids are secreted at a very high level, whereas 5α-reduced steroids occur in low concentrations. Intensive sample purification due to the low steroid concentration prevents routine analysis of plasma steroid profiles. Sensitive and specific detectors (mass fragmentographic, ECD) are mostly employed.

Adlercreutz [84,85] described the possibilities of GC/MS for the analysis of estrogens and neutral steroids in plasma and other biological fluids. Prepurification proceeded with Sephadex G-25 (LH-20) for the fractionation of the different conjugates, Girard separations of the ketones, and column chromatography. Deuterated and tritiated steroids were used as internal standards.

Purification of unconjugated steroids in plasma samples was described by Axelson and Sjövall [13,25]. A strong-anion exchanger improved their method [86] and made it possible to determine also the unconjugated estrogens in plasma samples.

Breuer [87] analyzed plasma estradiol in pregnant women by simple extraction and derivatization after the addition of deuterated estradiol and selective ion monitoring. Analogous methods allow the determination of cortisol, aldosterone, testosterone, and dihydrotestosterone in plasma samples. Novotny [88] showed profiles from plasma steroids. Apart from conventional FID detection, the nitrogen selective detector was evaluated for

Figure 9 (Opposite). Gas chromatographic analyses comparing the steroid conjugate fractions isolated (by means of DEAP-LH 20 Sephadex) from the urine of a patient with Cushing's syndrome. Column: 25 m OV-1. Derivatives: MO-TMS. Column temperature was programmed at 1°C min^{-1} from 160 to 250°C. The following steroids are indicated: 1—3α-hydroxy-5α-androstan-17-one, 2—3α-hydroxy-5β-androstan-17-one, 3—3α,11β-dihydroxy-5α-androstan-17-one, 4—3α,11β-dihydroxy-5β-androstan-17-one, 5—3α,17α,21-trihydroxy-5β-pregnane-11

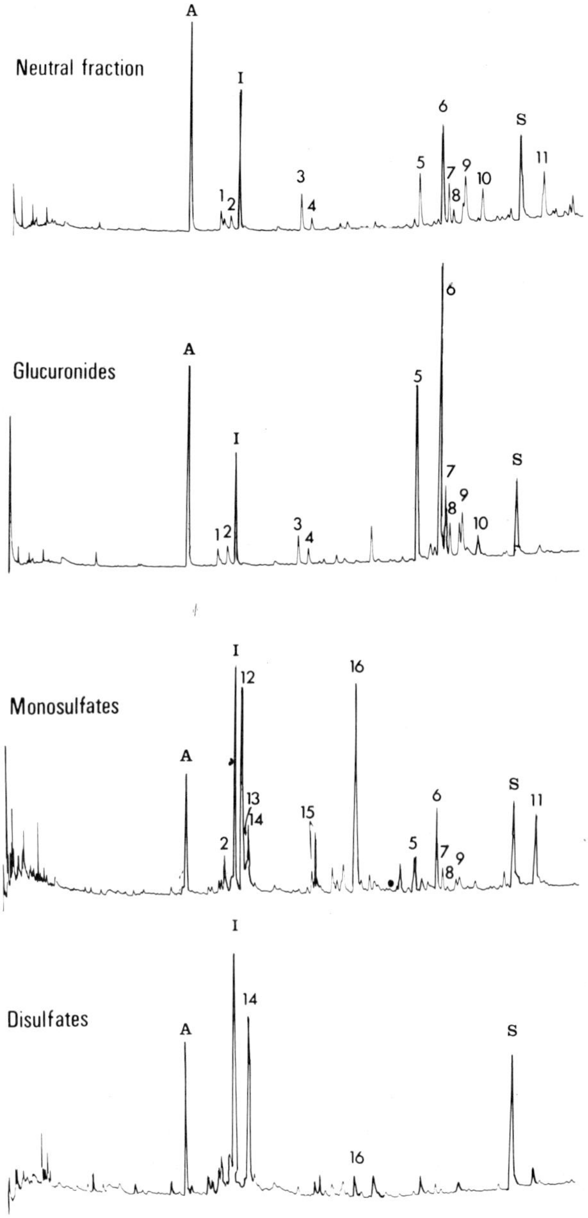

Figure 9 (continued). 20-dione, 6—3α,11β,17α,21-tetrahydroxy-5β-pregnan-20-one, 7—3α,11β,17α,21-tetrahydroxy-5α-pregnan-20-one, 8—3α,17α,20α,21-tetrahydroxy-5β-pregnan-11-one, 9—3α,17α,20β,21-tetrahydroxy-5β-pregnan-11-one and 5β-pregnane-3α,11β,17α,20β,21-pentol, 10—5β-pregnane-3α,11β,17α,20α,21-pentol, 11—cortisol, 12—dehydroepiandrosterone, 13—3β-hydroxy-5α-androstan-17-one, 14—5-androstene-3β,17β-diol, 15—3β,16α-dihydroxy-5-androsten-17-one (two peaks), 16—5-androstene-3β,16α,17β-triol, I—impurity in trimethylsilylimidazole. (Source: Ref. 27. Reproduced with permission.)

steroid analysis. The selectivity for ketosteroids after methoxime formation is illustrated in Fig. 10. Ludwig et al. [83] analyzed plasma and hemofiltrates from uremic patients. In comparison to normal plasma, high levels of androstenediol and low levels of dehydroepiondrosterone were found. Figure 11 shows the chromatogram of the sulfate-conjugated steroids from a plasma of a normal male. Hemofiltrates of uremic patients contain more 11-hydroxylated steroids.

Although most of the disorders are described in the literature, it is of interest to present the relevant chromatograms obtained in a unified way as this facilitates comparison of urinary profiles. The sample preparation, derivatization (persilylated MO-TMS derivatives), and gas chromatographic conditions were practically the same for all figures. $(GC)^2$ details are: column temperature, 245°C; injection and detection temperature, 250°C; falling needle injector [68]; carrier gas, helium; flowrate 2 ml min^{-1}. Glass capillary columns with internal diameters 0.4 to 0.5 mm were coated statically [9] with a 0.2% hexane solution of OV-1. Sample preparation was described previously.

A chromatogram of methoxime-persilylated derivatives of a series of urinary steroids is shown in Fig. 12. Each peak in the chromatogram represents 50 ng pure compound. Qualitative identification is carried out by self-established Kováts indices. Numbers in the chromatograms refer to Table 1. A chromatogram of MO-TMS derivatives of a series of 3-ketosteroids with a double bond in the 4 or 5 position is given in Fig. 13. Most 3-keto-4-ene steroids produce syn- and antiisomeric methoximes which are separated in our $(GC)^2$ conditions, e.g., testosterone (peak 61).

Despite the elementary sample preparation, the principal urinary metabolites can be analyzed quantitatively. By the high sample capacity of the wide-bore capillary columns symmetrical peaks are obtained in a broad concentration range (no leading peaks). Quantitative estimation for routine analysis can be done with peak height measurements using the internal standard method. In order to exclude interferences with our internal standard (dihydrocholesterol), a second one (cholesterylacetate) (respectively peak 44 and peak 46) is included and the height ratio of the two is checked. The reproducibility of the analysis procedure was determined by repeating 16 times the total analysis of one urine sample during 3 weeks. Variation coefficients (VC) and mean excretion values are given in Table 2. For a group of 16 normal females aged 18 to 30 years, the mean 24-hr excretion

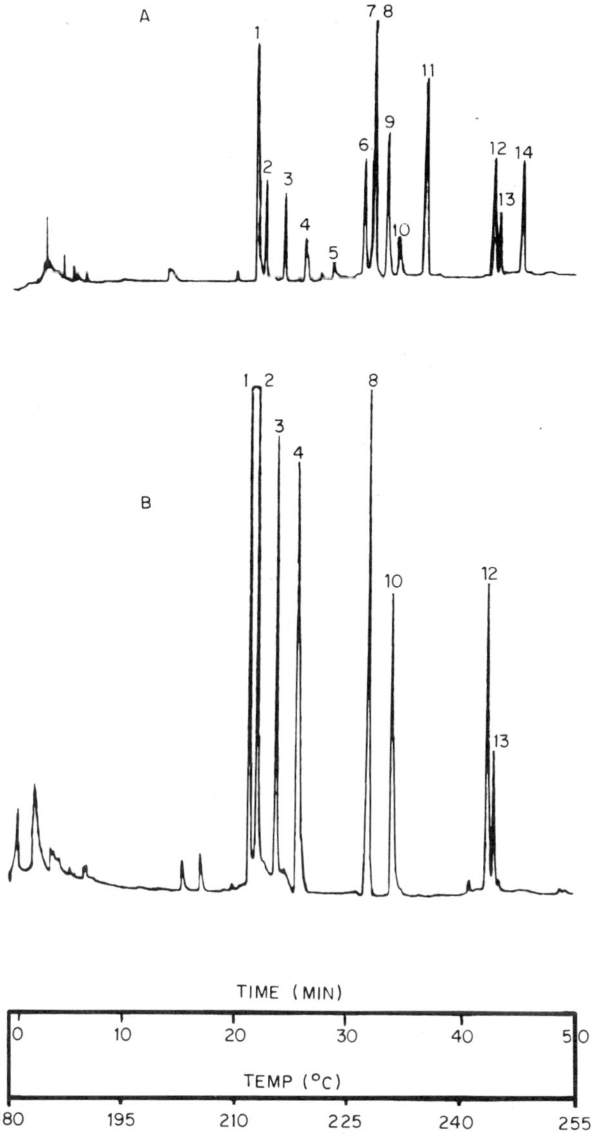

Figure 10. Chromatogram of a standard steroid mixture detected by (A) the flame ionization detector, and (B) nitrogen-sensitive (thermionic detector). Chromatographic conditions: 21 m × 0.26 mm ID glass capillary column coated with GE SE-30: 1—androsterone, 2—etiocholanolone, 3—dehydroepiandrosterone, 4—androstenedione, 5—androstanediol, 6—allopregnanediol, 7—pregnanediol, 8—pregnanolone, 9—pregnanetriol, 10—progesterone, 11—estriol, 12—tetrahydrocortisol, 13—allotetrahydrocortisol, 14—cholesterol. (Source: Ref. 88. Reproduced with permission.)

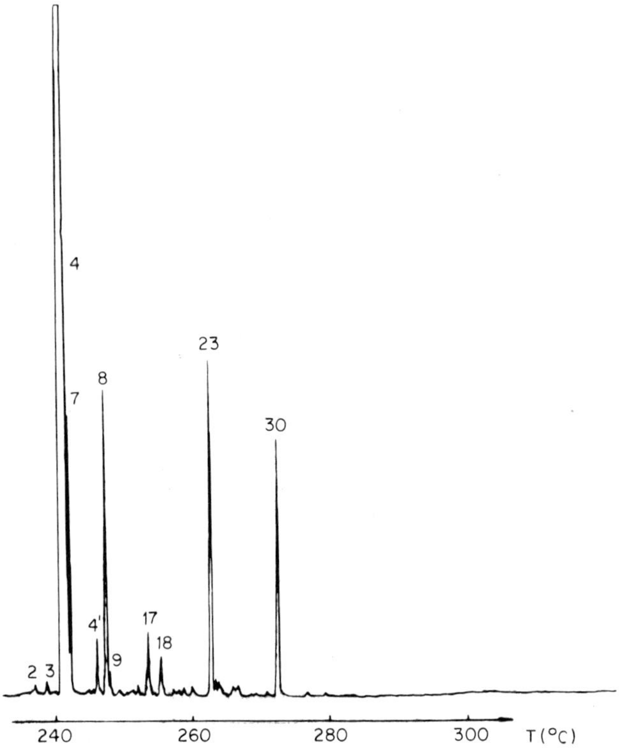

Figure 11. Gas chromatogram of steroidtrimethylsilylethers of the sulfate fraction from male plasma. Column: 30 m SE-30, programmed from 150° to 300°C, 2°C min^{-1}. 2—androsterone, 3—11-oxo-etiocholanolone, 4—DHA, 7—epiandrosterone, 8—androstenediol, 9—5α-androstane-3β,17β-diol, 17—pregnenolone, 18—16-hydroxy-DHA, 23—pregnenediol, 30—3β,17α,20α-pregnenetriol.

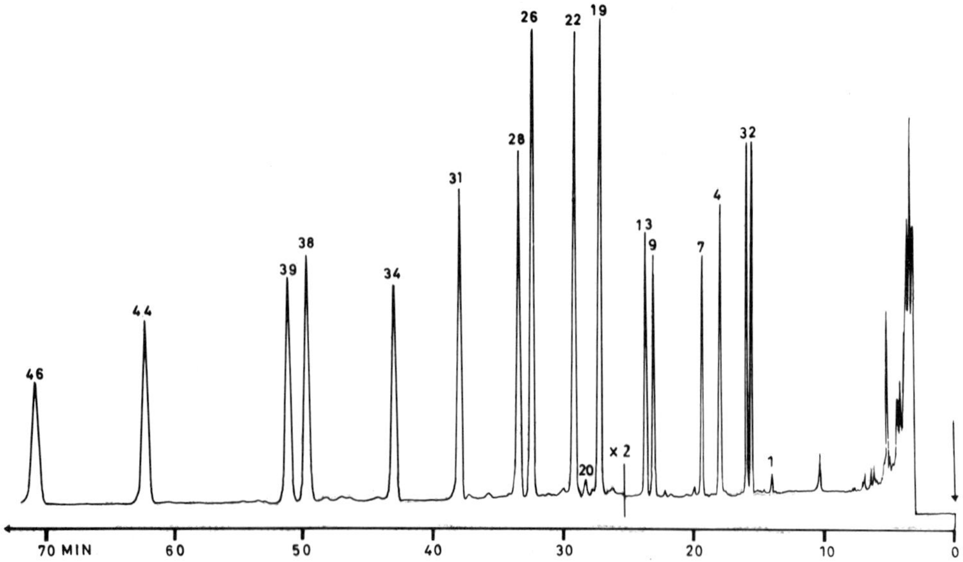

Figure 12. Chromatogram of MO-TMS derivatives of a number of steroids occurring in urinary extracts (50 ng per compound). OV-1 column, 245°C isothermally.

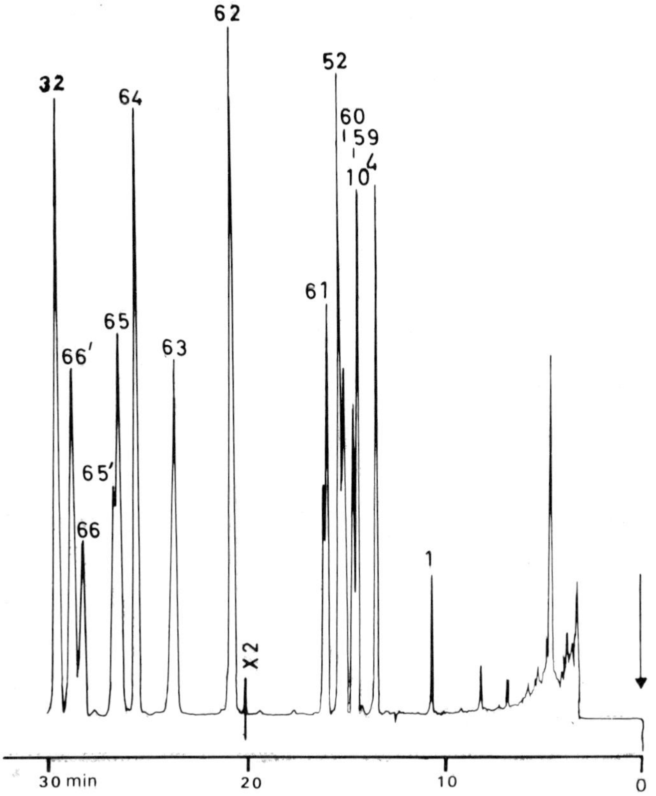

Figure 13. Chromatogram of MO-TMS derivatives of a series of steroids occurring in plasma samples.

Table 2. Reproducibility of Urinary Steroid Determination

Steroid	Mean mg 24 hr^{-1}	VC
Androsterone	1.26	4.8
Etiocholanolone	1.6	5.2
Dehydroepiandrosterone	0.05	24.0
Pregnanediol	4.20	5.1
Pregnanetriol	1.60	4.9
Tetrahydrocortisone	2.04	4.5
Tetrahydrocortisol	0.96	3.8
5α-Tetrahydrocortisol	0.54	6.6

and standard deviation (SD) have been determined (Table 3). These data agree with those published by others [45] except that values for corticosteroids are higher in our determinations. The persilylation of steroid methoximes (MO-TMS) makes an isothermal $(GC)^2$ analysis possible as the retention times are shortened. On our apolar capillary columns few of the naturally occurring steroids have equal retention times. An exception is the nearly simultaneous elution of 3α,17α-dihydroxy-5β-pregnane-20-one and 11-hydroxyandrosterone and the simultaneous elution of 3α-hydroxy-5α-pregnane-20-one with 11-hydroxyetiocholanolone. The two pregnanolones mentioned occur in high concentration in the luteal phase of the cycle (especially after ovarian stimulation) and in pregnancy. Differentiation can be made by peak shift techniques (see the above discussion of MO-TMS/TMS ether).

In Fig. 14, a urinary profile of a woman in the luteal phase of the cycle is represented. After ovulation, the ovary and especially the corpus luteum secretes progesterone and 17α-hydroxyprogesterone. The metabolites of both hormonal steroids are found in the urine profile. The pregnanolones with the 20-keto group unchanged and the pregnanediols originate from progesterone. From the four possible epimeric pregnanolones only 3α-hydroxy-5β-pregnane-20-one (peak 15) and its 3α-hydroxy-5α isomer (peak 14) are encountered. The diols 3α,20α-dihydroxy-5β-pregnane (peak 19) and 3α,20α-dihydroxy-5α-pregnane (peak 18) are the most prominent dihydroxy metabolites of progesterone. Peaks 11 and 22, respectively 3α,17α-dihydroxy-5β-pregnane-20-one and 3α,17α,20α-trihydroxy-5β-pregnane, originate from 17α-hydroxyprogesterone. Apart from the metabolites of progesterone and 17α-hydroxyprogesterone, the steroid profile of a woman in the luteal phase is similar to the profile in the proliferative (preovulation) phase or the profile of

Table 3. Mean Excretion Levels for a Group of 16 Normal Females Aged 18 to 30 Years

Steroid	Mean mg 24 hr^{-1}	SD
Androsterone	2.25	0.88
Etiocholanolone	2.37	0.71
Dehydroepiandrosterone	0.21	0.47
Tetrahydrocortisone	3.24	0.98
Tetrahydrocortisol	1.31	0.33
5α-Tetrahydrocortisol	0.97	0.38

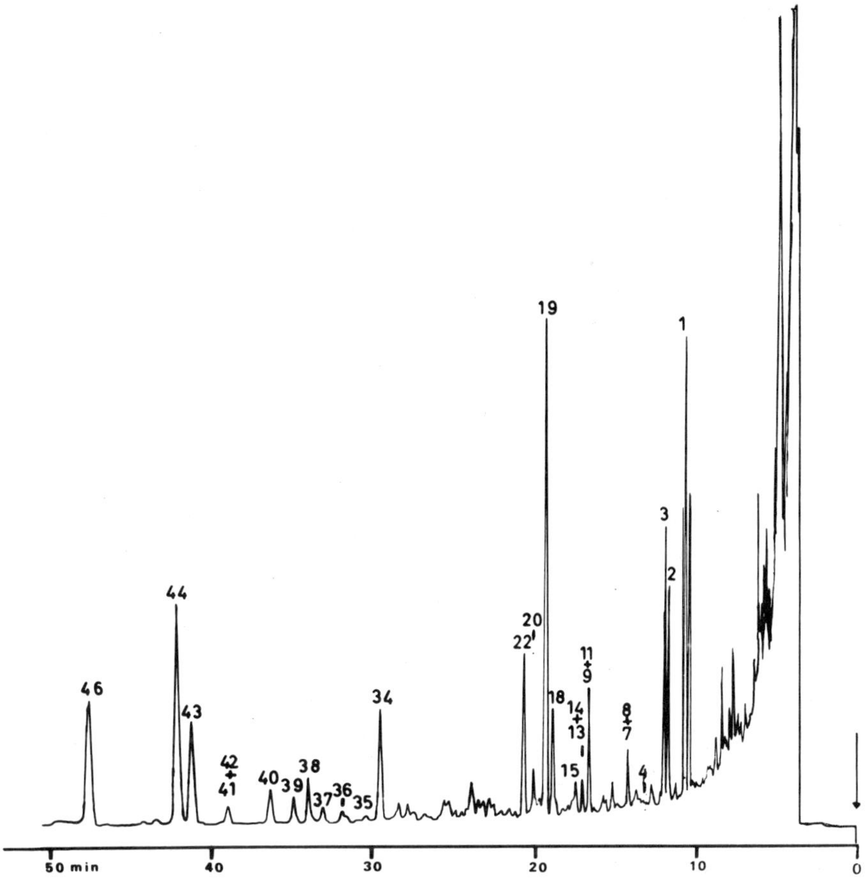

Figure 14. Urinary steroid profile for a woman in the luteal phase of the cycle.

a male urine. Androsterone (peak 2) and etiocholanolone (peak 3), their 11-oxo (peaks 7 + 8) and 11-hydroxy analogs (peaks 9 + 13), and the corticosteroids THE (peak 34), THF (peak 38), and allo-THF (peak 39) are the main steroids. 20α- or 20β-hydroxylation of the corticosteroids leads to cortolones (peak 40,41) and cortols (peak 42).

A profile of a urine of a pregnant woman is given in Fig. 15. In pregnancy high quantities of pregnenolone and progesterone are produced. The fetus converts these steroids to 16-hydroxylated structures. From 16-hydroxy-DHA, the placenta by its high aromatase capacity produces oestriol (peak 55). Also pregnanediol (3α,5β,20α) (peak 19) and the pregnanolones (peaks 11 and 15) appear as large peaks.

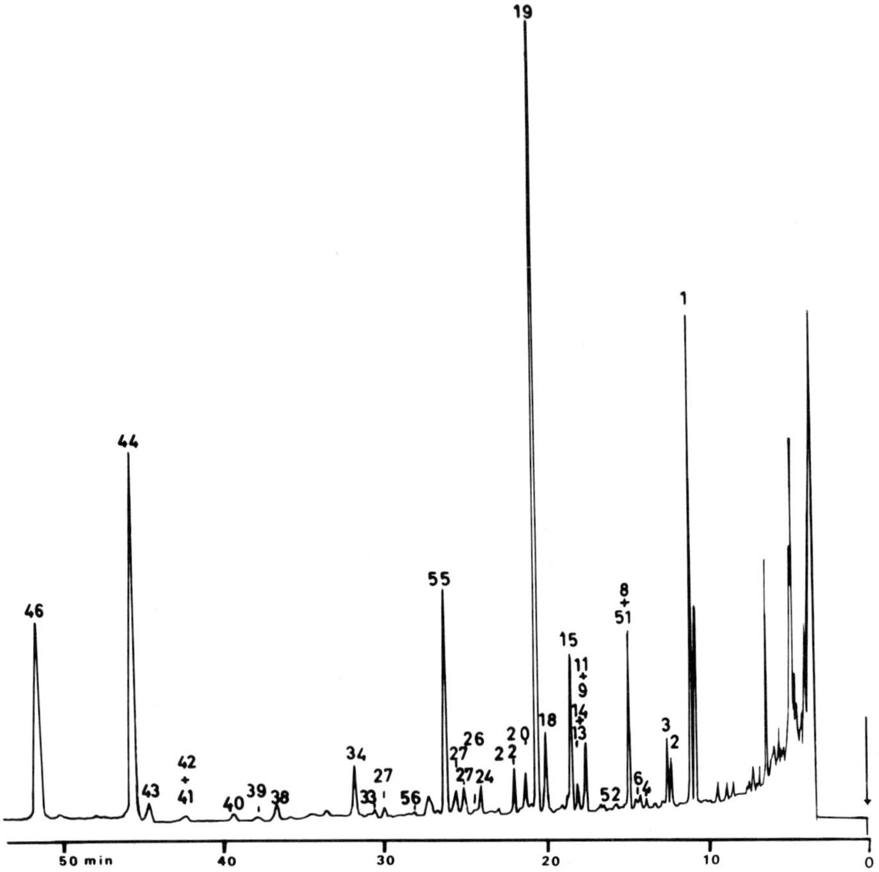

Figure 15. Urinary steroid profile of a pregnant woman.

In the first months of pregnancy, the corpus luteum graviditatis is the main producer of progesterone and by human chroionic gonadotrophin (HCG) stimulation high amounts of 17α-hydroxyprogesterone are formed. In such urines, more 17α-hydroxypregnanolone (peak 11) and pregnanetriol (peak 22) are found. Exogeneous stimulation [ovulation induction with clomiphene or HMG (human menopausal gonadotrophin) and HCG] leads to much higher 17α-hydroxylated metabolites. In endogeneous stimulation—as occurs in molar pregnancy, which is characterized by extremely high HCG blood levels—the preponderance of 17α-hydroxylated pregnanes (peaks 11 and 22) is obvious from Fig. 3 [89].

The overproduction of steroids with 5-ene structure in a patient with an adrenal carcinoma is illustrated in Fig. 16. Some of the cells of adrenal

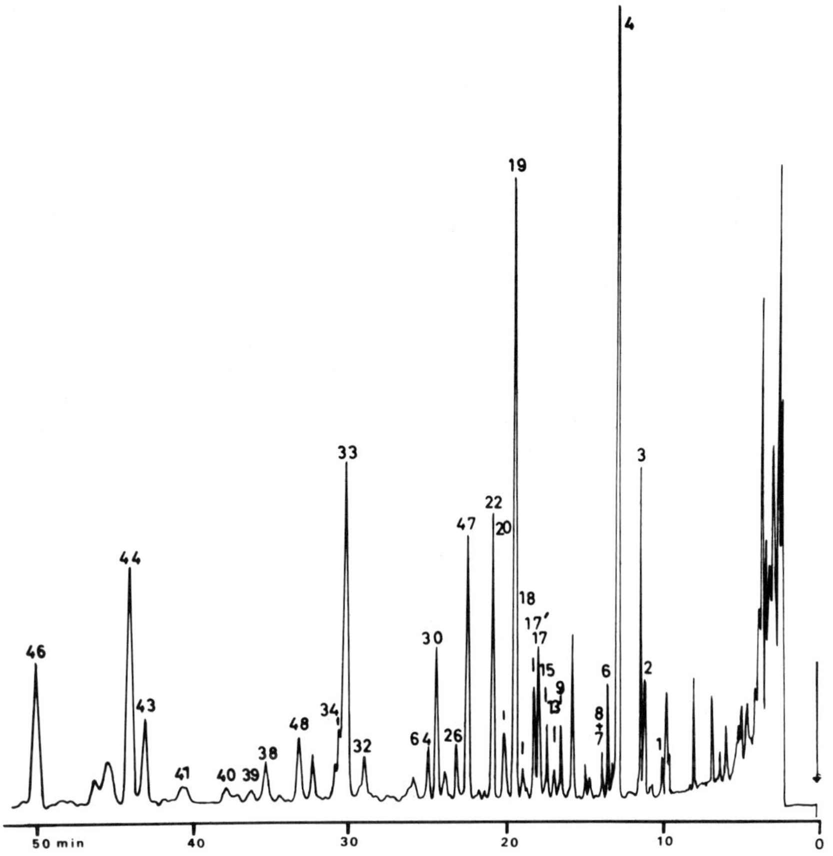

Figure 16. Urinary steroid profile of a patient with an adrenal carcinoma.

carcinoma deviate in growth and enzymatic system from normal cells. Although adrenal carcinoma with excessive Δ4 steroid production are known, mostly the Δ5 production dominates. The pathognomic steroid is in most cases DHA (peak 4). Regression of the tumor growth can be monitored by DHA measurements in blood or urine. Total urinary profiling also reveals high quantities of Δ5 pregnene structures (peaks 30, 33, 47, 48).

In Fig. 17, the 21-hydroxylase defect is illustrated. Cortisol is the principal stress hormone. Its production is regulated by a negative feedback mechanism; higher blood levels of cortisol depress the levels of the CRF (corticotropin-releasing factor) and of the ACTH (adrenocorticotropic hormone) and also decrease the adrenal steroid production. When 21-hydroxylation is totally or partially impossible, none or only minor quantities of

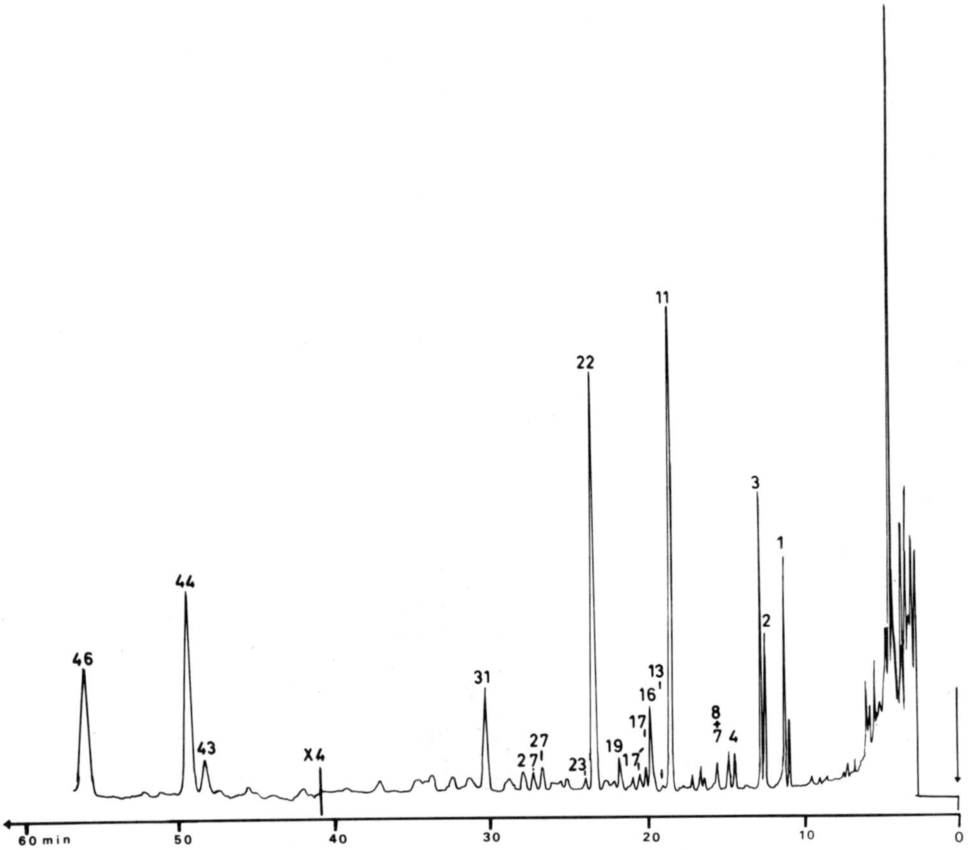

Figure 17. Urinary steroid profile of a patient with an adrenal carcinoma.

cortisol are produced. This results in a continuous stimulation of the adrenals which produce excessive amounts of androgens using precursors of cortisol synthesis (17-hydroxyprogesterone, for example). In females, this syndrome leads to androgenization. Female children who have the disease in their fetal life can have male differentiated sex organs. They are pseudohermaphrodites due to a congenital adrenal hyperplasia. In females, in which the disease only emerges at a more advanced age, hirsutism or sterility are the symptoms. As can be deduced from Fig. 17, high quantities of 21-desoxypregnanes, e.g., 17α-OH-pregnanolone (peak 11), 17β-OH-pregnanolone (peak 16), pregnanetriol (peak 22), and 11-ketopregnanetriol (peak 31) appear in the urinary profiles. Even in the luteal phase, the diagnosis is possible, because, apart from pregnanetriol (peak 22) and 17α-OH-pregnanolone (peak

11), which are normally present in luteal phase profiles, 11-ketopregnanetriol (peak 31) can be quantitized. The last compound cannot originate from the ovaries and is therefore the selective indicator of the disease.

Figure 18 represents a chromatogram for a patient with 11-hydroxylase deficiency. The precursor of cortisol, 11-desoxycortisol is accumulated in the blood circulation and excreted in the urine as 3α,17α,21-trihydroxy-5β-pregnane-20-one (peak 28). Other indications of the disease are the presence of two isomeric hexahydro metabolites (peaks 50, 67) and the high etiocholanolone (peak 3) levels.

A profile for a patient with lowered liver metabolisation (as occurs in annorexia nervosa) is given in Fig. 19. Especially the 5α reduction and the 11-hydroxysteroid dehydrogenation is decreased. Androsterone (peak

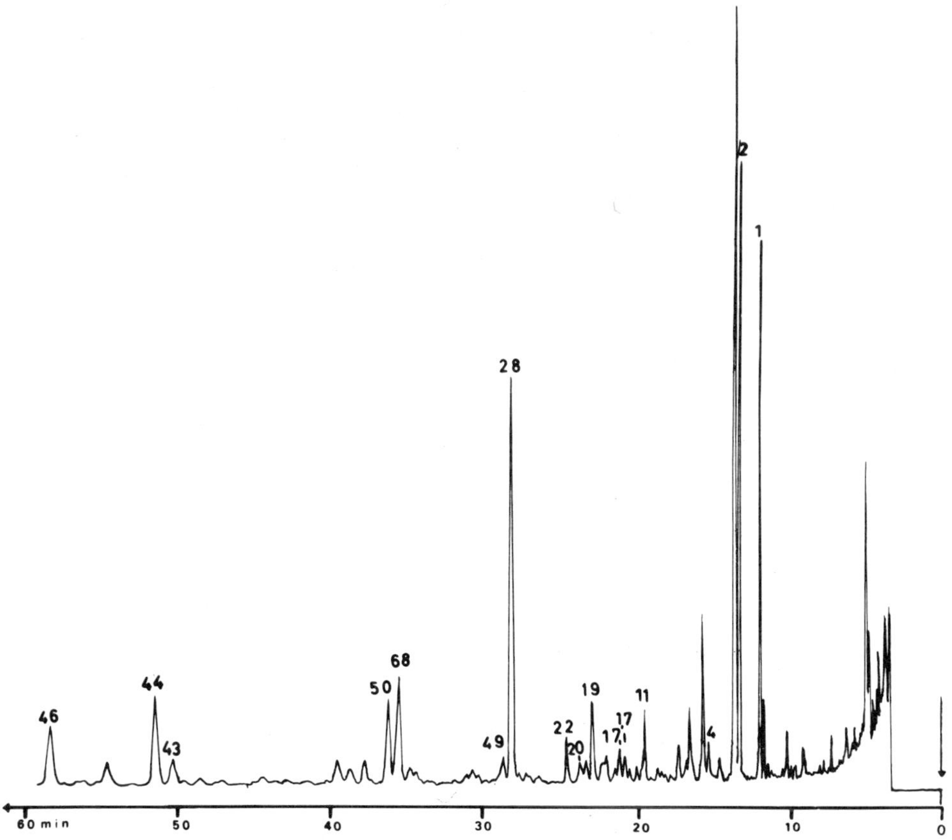

Figure 18. Urinary steroid profile of a patient with 11-hydroxylase deficiency.

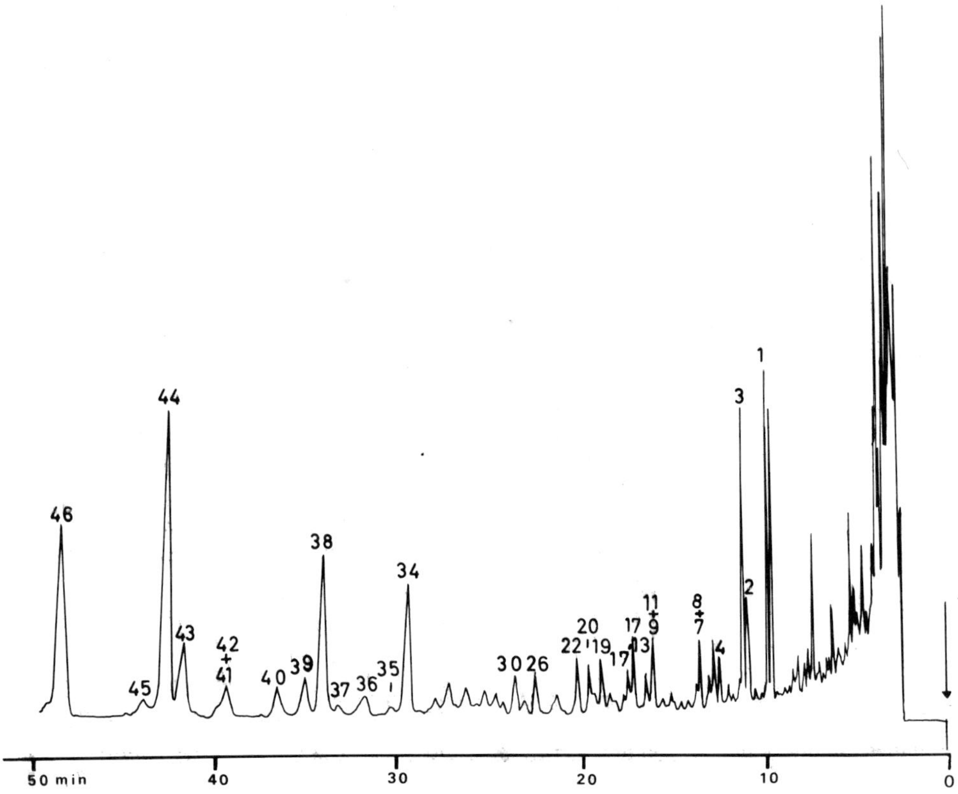

Figure 19. Urinary steroid profile of an anorexia nervosa patient.

2), 5α-tetrahydrocortisol (peak 39), and tetrahydrocortisone (peak 34) are in comparison to normal patients present in low concentration [90].

Liver metabolism is extremely important in steroid studies. The liver determines not only what is secreted in the urine, but also what is held in the blood. 5α and 5β reduction (or deactivation) of steroids are principally liver processes. The metabolic defects of the liver can be deduced from the urinary steroid metabolites. High quantities of pharmacologic substances can be liver-toxic and can change steroid metabolism [91].

In urines of newborns, most steroids have the 3β-hydroxy-5-ene group. Figure 20 is a profile of a urine of the first day of life, where maternal steroids [e.g., pregnanediol (peak 19), 16α-hydroxypregnanolones (peak 27), and oestriol (peak 55)] still occur. The profile for the third day of life is given in Fig. 21. The fetal adrenal zone, which involutes during the first months of life, is a very active steroid synthesyzing organ and pro-

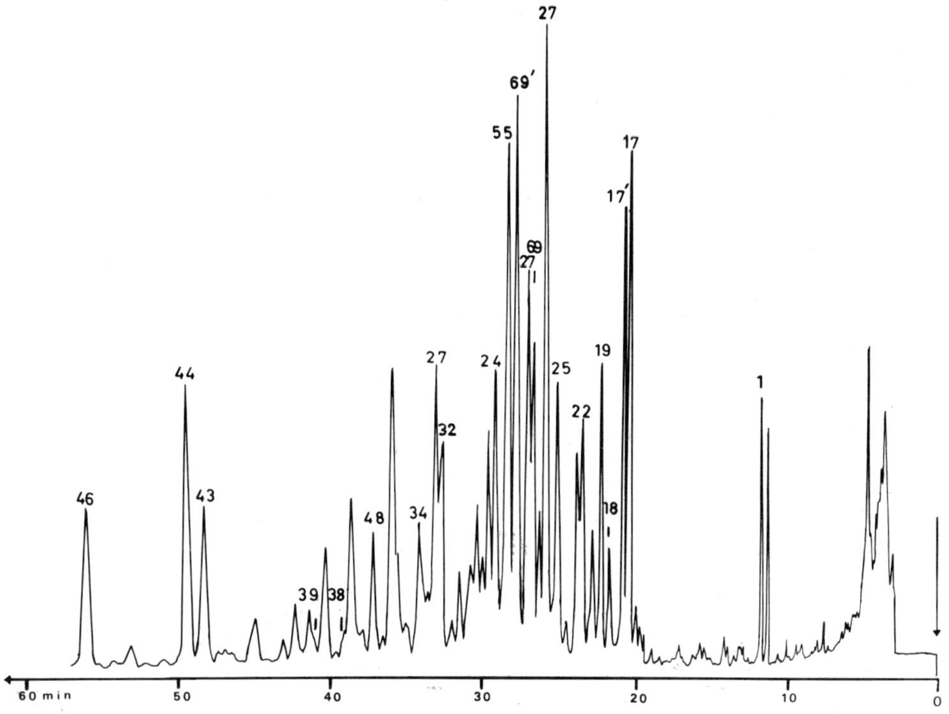

Figure 20. Urinary steroid profile of a newborn on the first day of life.

duces Δ5-pregenolone, 17-hydroxypregnenolone, and dehydroepiandrosterone. 16-hydroxylated metabolites of the former compounds are excreted in the urine (peak 17 syn- and anti-, peak 25, peak 69 syn- and anti-, peak 32, peak 33, and peak 48). The double bond at the 5 position persists by lack of reducing enzymes in sufficient quantities.

Even for these rather complicated profiles, the elementary sample preparation described above is thus sufficient.

5. CONCLUSION

The development of glass capillary gas chromatography has improved steroid analysis. Urinary steroid profiling is now carried out routinely. The advantage over the other techniques for urinary steroid analysis is that up to 50 steroids can be estimated in one analysis. Nowadays, the technology of $(GC)^2$ is sufficiently optimized and high-quality capillary columns are commercially available. A breakthrough of this technique in clinical

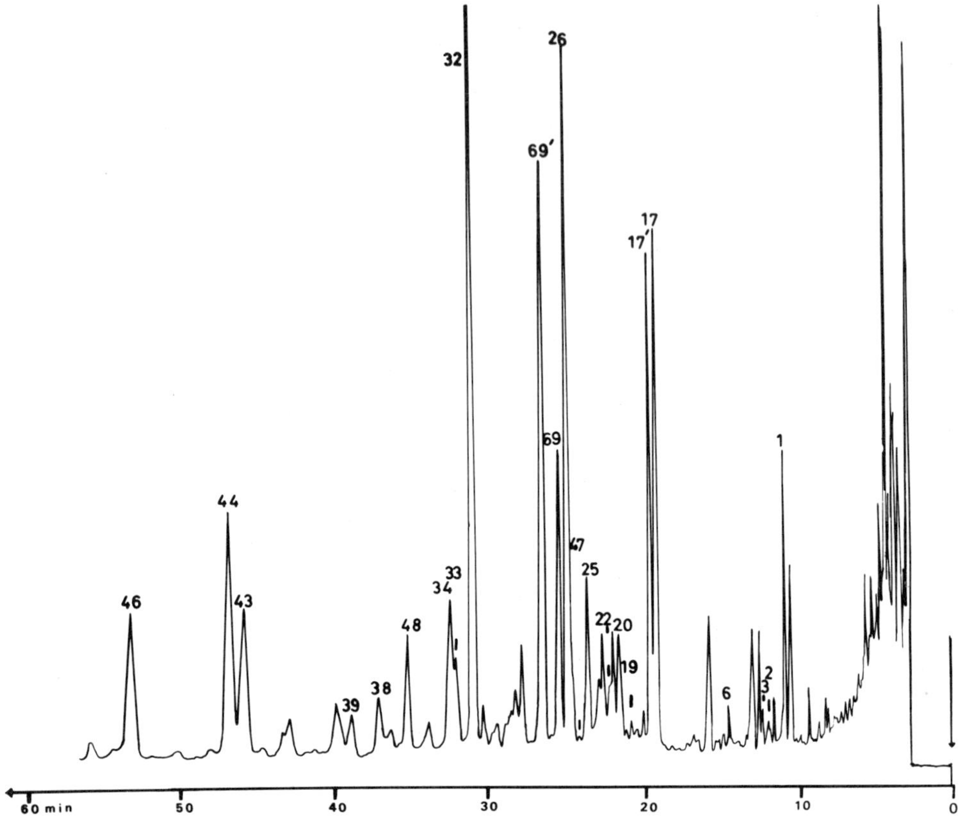

Figure 21. Urinary steroid profile of a newborn on the third day of life.

laboratories may therefore be expected in the near future. The results described in this chapter can be obtained in each clinical laboratory. On the other hand, due to the low concentration of the steroids and the different prepurification steps, the analysis of biologically active steroids in biological fluids such as blood and tissue extracts is complicated, and has thus become the subject of present research work in different laboratories. The work to be done in this field concerns not only the sample preparation. Injection techniques by which the whole biological sample can quantitatively be introduced in the capillary column must be developed. Selective and sensitive detectors (mass fragmentography, electron capture, and nitrogen detector) must be employed. The use of capillary columns with high selectivity and high temperature stability will further broaden the appliciability of $(GC)^2$ for steroid analysis.

ACKNOWLEDGMENTS

Thanks are due to Prof. Dr. D. Vandekerckhove, head of the Gynecological Clinic, and to Prof. Dr. M. Thiery, head of the Obstetrics Clinic, of the Academic Hospital, State University of Ghent for the facilities extended in preparing this contribution. We thank Prof. Dr. M. Verzele for his interest and for help with the manuscript.

REFERENCES

1. Vanden Heuvel, W. J. A., Sweeley, C. C., and Horning, E. C., *J. Amer. Chem. Soc.*, 82:3481 (1960).
2. Golay, M. J. E., in *Gas Chromatography* (V. J. Coates
3. Chen, C., and Lantz, C. A., *Biochem. Biophys. Res. Common*, 3 (1960).
4. Lipsky, S. R., and Landown, R. A., *Anal. Chem.*, 33:818 (1961).
5. Curtius, H. C., and Muller, H., *J. Chromatogr.*, 30:410 (1967).
6. Eriksson, H., and Gustafsson, J.-A., *Clin. Chim. Acta*, 41:79 (1972).
7. Thenot, J. P. T., and Horning, E. C., *Analyt. Letters*, 5:21 (1972).
8. Grob, K., and Grob, G., *J. Chromatogr. Sci.*, 7:584 (1969).
9. Bouche, J., and Verzele, M., *J. Chromatogr.*, 6:501 (1968).
10. Novotny, M., and Tessarik, K., *Chromatographia*, 1:332 (1968).
11. Rutten, G. A. F. M., and Luyten, J. A., *J. Chromatogr.*, 74:177 (1972).
12. Bradlow, H. L., *Steroids*, 11:265 (1972).
13. Axelson, M., and Sjövall, J., *J. Steroid Biochem.*, 5:733 (1974).
14. Leunissen, W. J. J., and Thyssen, J. H. H., *J. Chromatogr.*, 146:365 (1978).
15. Burstein, S., and Lieberman, S., *J. Biol. Chem.*, 233:331 (1958).
16. Shackleton, G. H. L., Sjövall, J., and Wisen, O., *Clin. Chim. Acta*, 27:354 (1970).
17. Millington, D., Jenner, D. A., Jones, T., and Griffiths, K., *Biochem. J.*, 139:473 (1974).
18. Adlercreutz, H., in *Methoden der Hormonbestimmung*, (H. Breuer, D. Hamel, and H. L. Krüskemper, eds.), G. Thieme, Stuttgart, 1975, p. 416
19. Laatikainen, T., and Vikko, R., *Eur. J. Biochem.*, 10:165 (1969).
20. Heftmann, E., ed., *Chromatography of Steroids*, Elsevier, Amsterdam, 1976.
21. Setchell, K. D. R., and Shackleton, C. H. L., *Clin. Chim. Acta*, 47:381 (1973).
22. Schackleton, C. H. L., and Honour, J. W., *Clin. Chim. Acta*, 69:267 (1976).

23. Anderson, R. A., Chambaz, E. M., Defaye, G., Madani, C., Baillie, T. A., and Brooks, C. J. W., *J. Chromatogr. Sci.*, 12:686 (1974).

24. Bèque, R. J., Desgrès, J., Padieu, P., and Gustafsson, J. A., *J. Chromatogr. Sci.*, 12:763 (1974).

25. Axelson, M., and Sjövall, J., *J. Chromatogr.*, 126:705 (1976).

26. Sjövall, J., *Steroids*, 11:703 (1968).

27. Setchell, K. D. R., Almè, B., Axelson, M., and Sjövall, J., *J. Steroid Biochem.*, 7:615 (1976).

28. Matthiesen, U., and Staib, W., *Chromatographia*, 10:70 (1977).

29. Verzele, M., and Sandra, P., *J. Chromatogr.*, 158:111 (1978).

30. Gleispach, H., *J. Chromatogr.*, 91:407 (1974).

31. Poole, C. F., in *Handbook of Derivatives for Chromatography* (C. Blau and G. King, eds.), Heyden, London, 1977.

32. Gardiner, W. L., and Horning, E. C., *Biochem. Biophys. Acta*, 115:524 (1968).

33. Devaux, P. G., Horning, M. G., Hill, R. M., and Horning, E. C., *Anal Biochem.*, 41:7 (1971).

34. Chambaz, E. M., Defaye, G., and Madani, C., in *Advances in Chromatography* (A. Zlatkis, ed.)

35. Axelson, M., *Anal. Biochem.*, 86:133 (1978).

36. Koshy, K. T., Kaiser, D. G., and Van Der Slik, A. L., *J. Chromatogr. Sci.*, 13:97 (1975).

37. Nambara, T., Kigasawa, K., Iwata, T., and Ibuki, M., *J. Chromatogr.*, 114:81 (1975).

38. Touchstone, J. C., and Dobbins, M. F., *J. Steroid Biochem.*, 6:1389 (1975).

39. Morgan, E. D., and Poole, C. F., *J. Chromatogr.*, 89:225 (1974).

40. Morgan, E. D., and Poole, C. F., *J. Chromatogr.*, 104:351 (1975).

41. Jacob, K., and Vogt, W., *J. Chromatogr.*, 150:339 (1978).

42. Blair, I. A., and Philipou, G., *J. Chromatogr. Sci.*, 16:201 (1978).

43. Blair, I. A., *J. Chromatogr. Sci.*, 15:478 (1977).

44. Quillian, M. A., and Westmore, J. B., *Anal. Chem.*, 50:59 (1978).

45. Fantl, V., and Gray, D. H., *Clin. Chim. Acta*, 79:237 (1977).

46. Völlmin, J. A., *Chromatographia*, 3:233 (1970).

47. Ros, A., and Sommerville, I. F., *J. Obst. Gyn. Brit. Comm.*, 78:1096 (1971).

48. German, A., and Horning, E. C., *J. Chromatogr. Sci.*, 11:76 (1973).

49. Sandra, P., Vanluchene, E., and Verzele, M., *Chromatographia*, 8:501 (1975).

50. Sandra, P., and Verzele, M., *Chromatographia*, 10:419 (1977).

51. Madani, C., Chambaz, E. M., Rigaud, M., Durand, J., and Chebroux, P., *J. Chromatogr.*, 126:161 (1976).

52. Madani, C., Chambaz, E. M., Rigaud, M., Chebroux, P., Breton, J. C., and Berthou, F., *Chromatographia*, 10:466 (1977).

53. Boogaerts, T., Verstappe, M., and Verzele, M., *J. Chromatogr. Sci.*, 10:217 (1972).

54. Alexander, G., and Rutten, G. A. F. M., *Chromatographia*, 6:231 (1973).

55. Cramers, C. A., Vermeer, E. A., and Franken, J. J., *Chromatographia*, 10:412 (1977).

56. Sandra, P., Verstappe, M., and Verzele, M., *J. High Res. Chromatogr. Chromatogr. Comm.*, 1:28 (1978).

57. Sandra, P., and Verzele, M., unpublished results.

58. Sandra, P., and Vanluchene, E.,

59. Schomburg, G., Husmann, H., and Weeke, F., *J. Chromatogr.*, 99:63 (1974).

60. Sandra, P., Verstappe, M., and Verzele, M., *Chromatographia*, 11:223 (1978).

61. Grob, K., and Grob, G., *J. Chromatogr.*, 125:471 (1976).

62. Cronin, D. A., *J. Chromatogr.*, 97:263 (1974).

63. Schomburg, G., Dielman, A., Husmann, H., and Weeke, F., *J. Chromatogr.*, 122:55 (1976).

64. German, A. L., and Horning, E. C., *Anal. Lett.*, 5(9):619 (1972).

65. Pfaffenberger, C. D., and Horning, E. C., *J. Chromatogr.*, 112:581 (1975).

66. Pfaffenberger, C. D., and Horning, E. C., *Anal. Biochem.*, 80:329 (1977).

67. Ros, A., *J. Chromatogr. Sci.*, 3:252 (1965).

68. Van den Berg, P. M. J., and Cox, T. P. H., *Chromatographia*, 5:301 (1972).

69. Cramers, C. A., and Vermeer, E. A., *Chromatographia*, 8:479 (1975).

70. Grob, K., and Grob, K., *J. Chromatogr.*, 151:311 (1978).

71. Documentation Carlo Erba.

72. Völlmin, J. A., *Clin. Chim. Acta*, 34:207 (1971).

73. Völlmin, J. A., and Curtius, H. C., *Z. Klin. Chem. Klin. Biochem.*, 9:43 (1971).

74. Novotny, M., and Zlatkis, A., *J. Chromatogr. Sci.*, 8:346 (1970).

75. Luyten, J. A., and Rutten, G. A. F. M., *J. Chromatogr.*, 91:393 (1974).

76. Horning, E. C., Horning, M. G., Szafranek, J., Van Hout, P., German, A. L., Thenot, J. P., and Pfaffenberger, C. D., *J. Chromatogr.*, 91:367 (1974).

77. Pfaffenberger, C. D., Malinak, L. R., and Horning, E. C., *J. Chromatogr.*, 158:313 (1978).

78. Berthou, F. L., Picart, D., Bardou, L. G., and Floch, H. H., *J. Chromatogr. Sci.*, 12:662 (1974).

79. Berthou, F., Picart, D., Bardou, L. G., and Floch, H. H., *J. Chromatogr.*, 118:135 (1976).

80. Curtius, H. C., Völlmin, J., Zagalak, M. J., and Zachman, M., *J. Steroid Biochem.*, 6:677 (1975).

81. Shackleton, C. H. L., *Clin. Chim. Acta*, 67:287 (1976).

82. Egger, H. J., Reiner, J., Spiteller, G., and Häffele, R., *J. Chromatogr.*, 145:359 (1978).

83. Ludwig, H., Spiteller, G., Matthaei, D., and Scheler, F., *J. Chromatogr.*, 146:381 (1978).

84. Adlercreutz, H., Tikkanen, M. J., and Hunneman, D. H., *J. Steroid Biochem.*, 5:211 (1974).

85. Adlercreutz, H., Martin, F., Wahlroos, O., and Soini, E., *J. Steroid Biochem.*, 6:247 (1975).

86. Axelson, M., and Sjövall, J., *J. Steroid Biochem.*, 8:683 (1977).

87. Breuer, H., and Siekmann, L., *J. Steroid Biochem.*, 6:685 (1975).

88. Novotny, M., Maskarinec, M. P., Steveinck, A. T. G., and Farlow, R., *Anal. Chem.*, 48:468 (1976).

89. Vanluchene, E., Dhont, M., Vandekerckhove, D., and Thiery, M., *Z. Geburtsh Perinarologie*, 181:294 (1977).

90. Vanluchene, E., Aertsens, W., and Vandekerkhove, D., *Acta Endocrinolo.*, 90:133 (1979).

91. Van Eyck, J., Dhont, M., Vandekerckhove, D., Vanluchene, E., Voorhoof, L., and Delbeke, L., in *Benelux Sym. Danatrol* (A. M. C. M. Schellen, ed.), European Press, Ghent, 1978, p. 51.

9

GLASS CAPILLARY GAS-LIQUID CHROMATOGRAPHY OF AMINO ACIDS

HALVOR JAEGER
HANS U. KLOER
HANS DITSCHUNEIT

University of Ulm
Ulm, Federal Republic of Germany

HARTMUT FRANK

University of Tübingen
Tübingen, Federal Republic of Germany

1. Introduction 331

2. Sample Preparation and Derivatization 334

3. Instrumentation 339
   3.1 Injection Ports 340
   3.2 Columns 340
   3.3 Detectors 342

4. Applications 342

5. Summary 357

References 361

1. INTRODUCTION

Amino acids, basic constituents of living matter, comprise 20 compounds with common structural features, but with diverse functionality and reactivity of their individual side chains. They represent the building blocks of proteins and, through their respective number and relative alignment, confer to these biopolymers their infinite variation in properties and functions. They also serve other purposes as metabolic intermediates.

For illumination of the biological role of amino acids and for determination of the primary structure of polypeptides, adequate methods for separation of the individual amino acids and their quantitation were needed. While 30 years ago the determination of the amino acid composition of a protein took months or years, the situation changed dramatically when Stein et al. [1] introduced an automatic amino acid analyzer based on separation of the individual amino acids by ion-exchange chromatography and detection by postcolumn reaction with ninhydrin.

Several hours were required for one analysis and sensitivity was in the micromole range. Therefore, soon after the advent of gas-liquid chromatography (GLC) the application of the new technique to analysis of amino acids was tried, in view of its greater speed and sensitivity and the lower cost and higher versatility of the required equipment. However, amino acids are zwitterionic compounds and are not amenable to GLC as such, due to their lack of volatility. One route to overcome this difficulty by blocking at least one of the two inoic groups was first envisaged by Bayer in 1957 [2]. He employed methylesters in this early attempt to separate amino acids by GLC. In one of the earliest reports on capillary GLC of amino acids, ethylesters were used [3].

In spite of their high volatility, amino acid methylesters exhibit undersirable properties like condensation to polymers and diketopiperazines, and marked tailing of the GLC peaks. One year later the same investigator reported that N-trifluoroacetyl amino acid methylesters [4] were well suited for GLC. This general type of derivative, N-acyl amino acid alkylester, has been widely used and investigated since then and nowadays represents virtually the only class of derivative suitable for quantitative work. The method used in the earliest reported separation of amino acids [5] utilized degradation of amino acids to volatile aldehydes. However, the methods of chemical degradation or transformation of amino acids to less polar compounds like amines, aldehydes and amino alcohols are not generally applicable. The reaction products of trifunctional amino acids retain very high polarity and are thus not amenable to GLC.

During the early 1960s the main work on GLC of amino acids centered around improvement of derivatization procedures, assessment of alternate methods of derivatization, and choice of suitable stationary liquid phases and inert diatomaceous supports. Later on, the quantitation of amino acids by GLC came into focus, necessitating a closer investigation of reproducibility and yields of derivatization procedures. An excellent review on derivatization of amino acid has been published by Husek and Macek [6].

Especially the trifunctional amino acids with their differently reactive carboxyl, amino, hydroxyl, and thiol groups proved to be the touchstone for evaluation of proposed methods. Gehrke's group [7-10] in a series of publications delivered an exhaustive treatment of all parameters affecting GLC of amino acids. Several others contributed alternate methods of chemical derivatization [11-13]. Most of these derivatives were intended for separation on packed columns, still the prevailing mode in GLC of amino acids, but are in principal equally applicable to capillary gas-liquid chromatography (CGLC).

The first successful attempts to separate amino acids on capillaries were published in the early 1960s [14-16]; at those times appropriate derivatization procedures were not yet fully developed and the capillaries used were rather long and of low efficiency compared to modern standards. Therefore, capillary GLC of amino acids during the 1960s was restricted to only a few examples [17-19], while at the same time the most active research on derivatization and quantitation of amino acids was done on packed columns.

The area where capillary GLC has been used almost exclusively at an early stage was in separation of optical isomers of amino acids. Here the high efficiencies of capillaries were needed in order to achieve resolution of diastereomers [20-23] or enantiomers on special chiral stationary phases [24-30] in spite of their small differences in solvation energy. In many cases rather long capillaries (up to 150 m) had been used, mostly made out of stainless steel.

It soon became evident that the catalytic and adsorptive activity of steel surfaces caused the breakdown of sensitive amino acid derivatives and the severe tailing often observed with polyfunctional compounds. Due to these reasons and the commercial availability of glass-drawing machines, glass capillaries became more popular. In spite of their greater fragility, glass capillaries are now almost exclusively employed. As the insight into properties and deactivation of glass surfaces extends, the art of building suitable capillaries for amino acid analyses has become common knowledge to many investigators.

A recent development in quantitative amino acid analysis involved highly stable chiral stationary phases of the polysiloxane type, such as Chirasil-Val [31-33]. By use of these phases many problems preventing the use of GLC for quantitative amino acid analysis are eliminated.

As all enantiomeric pairs of the amino acids can be separated on these phases, the optical antipode to each natural L-amino acid can be employed as chemically identical internal standard [34]. This method strongly recommends gas-liquid chromatography as a modern alternative to amino acid analysis by ion-exchange chromatography.

2. SAMPLE PREPARATION AND DERIVATIZATION

The majority of samples for which determination of amino acid content is required are of biological origin. The ultimate goal therefore is the applicability of CGLC to such "real" problems. Most reports published during the past 20 years concentrated on methodological aspects and certainly advanced the state of art. However, compared to the rapid development of the classical ion-exchange procedure, progress in application of gas-liquid chromatography to amino acid analysis has been rather slow. After 20 years of elaboration, gas-liquid chromatographic amino acid analysis is still restricted to a few special applications like separation of enantiomers, identification of phenylhydantoins of Edman degradation, and ultramicrodetermination if only small amounts of sample are available.

In any case, for performing the full train of proceedings from sample preparation to evaluation of the analytical data, highly skilled and experienced personnel are required. The results are generally believed to be of limited reliability, and in fact, if comparative studies are done, standard deviations for GLC analysis of amino acids usually are higher than for the ion-exchange procedure [35].

A high degree of sophistication and experience must be acquired for sample preparation and derivatization procedures. Applications of amino acid analysis to biological questions involves either determination of free amino acids in biological fluids or analysis of the amino acid composition of isolated proteins and protein tissue.

Free amino acids from biological fluids are separated from proteins by precipitation, centrifugation, or ultrafiltration and purified by intermediate adsorption on ion exchangers.

Recoveries usually range between 80 and 100% depending on the procedure, amino acids, and absolute amount of the compounds [10,35-37].

For analysis of proteins the first step is acid hydrolysis to the constituent amino acids, less frequently by alkaline hydrolysis or use of proteolytic enzymes. Careful exclusion of oxygen and the absence of foreign contaminants, traces of metals, and other impurities have been found to be essential. For GLC analysis of amino acids from proteins and tissues again the above-described ion-exchange cleanup is required in order to remove lipid and carbohydrate contaminants giving rise to a multitude of peaks in the GLC chromatogram.

As an example the method of Adams [37] is presented here: 0.1 ml of serum or 0.5 ml of urine, 20 μg of norleucine as internal standard, and 0.1 ml of a pH-adjusting solution were added to a tube. The last solution consisted of 50% acetic acid in deionized water and containing 5.0 mg of $SnCl_2$ 100 ml^{-1}.

The sample was mixed gently and subsequently checked for pH adjustment at 2 to 2.5. The mixture was transferred to a resin column containing 50 mg of Dowex 50-X8(H^+)100 to 120 mesh over glass wool and was allowed to pass through the column at a rate of about 1 drop 5 sec^{-1}. The column must not be allowed to run dry. After washing the column with deionized water, the sample was eluted with 2 ml of 2 N NH_4OH, evaporated to dryness in a water bath at 90°C, and derivatized.

The next series of steps involves transformation of amino acids into volatile compounds with a general goal of "one compound—one derivative—one peak." Research in this area has been most active in the last 20 years. Establishment of appropiate derivatization procedures is the most decisive aspect in success or failure of amino acid analysis by GLC. While most of the methods have been elaborated for packed columns they are equally applicable to GLC on capillary columns without major changes.

In the early times of GLC on capillary columns derivatives like simple ethylesters [14], 1,3-oxazolidinones [16], or N-alkyl amino acid esters have been used. These approaches unfortunately are restricted to simple amino acids, since the side chains of trifunctional amino acids are not blocked under these conditions. Their polarity in most cases is therefore too high.

Derivatization today is directed toward blocking all functional groups with appropriate reagents. Due to the diverse reactivity of the different functions present in protein amino acids not all methods proposed in the past achieve this goal.

Apart from carboxyl and amino groups, functional groups like aliphatic and aromatic hydroxyl groups, mercapto groups, and the nitrogen-containing guanidino, indol, and imidazol residues must be derivatized for optimal chromatographic properties. Naturally the reactions employed for these purposes involve reagents of high reactivity which in turn create the danger of formation of multiple derivatives and degradation of some of the more sensitive amino acids such as cysteine, methionine, tyrosine, and tryptophan.

In general, an ideal procedure for chemical derivatization must fulfill many requirements:

1. It is complete and reproducible over a wide range of concentration.
2. It generates only one derivative, which is chemically and thermally stable.
3. It does not interfere with other derivatization reactions.
4. It involves reagents which can easily be removed.
5. It blocks polyfunctional compounds in one step.
6. It involves nontoxic chemicals.
7. It does not react with impurities and accompanying substances.
8. It generates derivatives with high compound-specific response of the corresponding detector.

Maximal volatility is not always strived for, in order to avoid losses of the more volatile derivatives prior to chromatography.

For practical reasons the formation of fully blocked derivatives by a one-step procedure appears especially attractive. The closest approximation to this goal involves formation of N-(O,S)-trimethylsilyl amino acid trimethylsilylethers [40]. The main problems here are the large differences in reactivity of the different amino acids and formation of duplicate derivatives [41]. So far, conditions enabling the complete trimethylsilylation of the less reactive amino acids and at the same time generation of only one derivative for others could not be established. The most potent methods for derivatization of amino acids involves at least two steps, namely, esterification of carboxyl groups and acylation of amino, hydroxyl, and mercapto groups, and other nitrogen functions. This type of derivative has been proposed relatively early with the N-trifluoroacetyl amino acid esters [39] and applied to GLC on capillary columns of amino acid diastereomers [20,21].

Esterification and acylation are usually performed consecutively; however, simultaneous derivatization of amino and carboxyl groups has also been reported [42]. The sequence of the two steps is usually first esterification and then N-acylation. This may be converted, but increases the danger of formation of oxazolinones as side products during N-acylation [43,44]. The general procedure involves heating of the amino acid in the corresponding alcohol to about 100°C in the presence of an acid as catalyst, usually hydrochloric acid in concentrations between 1 and 8 N for times of 20 min to 1 hr. After evaporation of excess reagent the amino acid ester hydrochlorides are treated with the anhydride of the corresponding acid in methylene chloride or ethylacetate as solvent at various temperatures.

All alcohols from methyl to amyl have been used for esterification, but most frequently the isopropyl [20,28,31-34,45], n-propyl [46,47], isobutyl [19,21-25,48,49], and n-butylester [17,18,20,29,50,51] have been employed in gas-liquid chromatography. Concentrations of the catalyst are usually between 3 and 8 N. Limited solubility of some amino acids, especially the basic ones, is overcome by ultrasonication or transesterification of methylesters. The direct use of the latter is limited, due to the high volatility of these derivatives resulting in excessive loss during final evaporation of solvents and reagents. For blocking of amino groups, acetyl, trifluoroacetyl, pentafluoropropionyl, and heptafluorobutyryl groups have been employed most frequently, the latter three are preferred for their enhanced volatility and reactivity. Their high response factors with electron capture detectors make them especially suitable for microdeterminations of amino acids. The conditions for N-acylation are usually chosen so as to block all other functional groups at the same time. However, a few of these derivatives are very reactive and sensitive to moisture. Arginine and histidine are the two amino acids most difficult to derivatize with some esterifying reagents [46,51]. Arginine, in addition to its low solubility, requires high temperatures (150°C) for acylation of its highly polar guanidino group; with heptafluorobutyric anhydride, though, partial acylation of the guanidino group was achieved even at room temperature [48]. This compound is highly reactive and is easily destroyed by slightly basic glass surfaces, possibly by nucleophilic silanol groups of the glass surface of capillaries. The perfluoroacyl group attached to the imidazol ring of histidine is even less stable [10,52] and splits off on mere evaporation of the solvent and reagent, obviously by reaction with free acid which accumulates in the solution on evaporation. The monoacyl derivative of histidine is still sufficiently volatile, but due to the high basicity of the free imidazol it may be chromatographed with slight peak asymmetry only on capillaries with slightly alkaline glass surfaces. Therefore, the chromatography of the diacylated form is preferred, which is achieved by coinjection of a small quantity of the corresponding anhydride [51,52]. Such a treatment, however, shortens the life span of capillaries through decomposition of the stationary phases. For the chiral polysiloxanes [31] such a procedure is inadequate. On injection of a reactive anhydride the chiral moieties are partially racemized or split off. The alternative is a third derivatization step for histidine with diethyl pyrocarbonate, resulting in a stable ethyl oxycarbonyl derivative with good GLC properties [34,48,53].

Another possibility is acetylation. Acetyl derivatives generally show higher chemical stability, which also applies to the O-(S)-acetyl esters of serine, threonine, tyrosine, and cysteine. N(O,S)-acetyl amino acid propylesters have successfully been used for capillary GLC of amino acids from biological samples [47].

As mentioned above, capillaries were applied to GLC of amino acids relatively early for the separation of enantiomers. The choice of derivatives in these cases was directed mostly by the need for sufficient resolution of the two antipodes. Amino acids esterified with secondary alcohols and acylated with pentafluoroacetic or trifluoroacetic anhydride showed highest resolution factors [29a].

A large number of other derivatives remain unmentioned in this report. Many of them have been used for special purposes, some only for the sake of the publication. Actually the only class of derivatives employed in capillary GLC of amino acids are the N(O,S)-acyl amino acid alkylesters. The trimethylsilylated amino acids may be applicable in some special instances, but further optimization of derivatization is desirable.

The derivatives of choice for separation of amino acid enantiomers are the N(O,S)-trifluoroacetyl amino acid isopropylesters [28-30,45,54-59] or the N(O,S)-pentafluoropropionyl derivatives [29a,30-34,57].

For separation of enantiomers by transformation to diastereomers, either chiral ester groups [20-23,60-62] or N-acyl groups were employed [62-65]. This approach is less versatile due to the usually lower volatility of derivatives, the necessity of complete reaction in order to avoid unequal formation of the two diastereomers, and the high requirement for optical purity of the reagent.

As an example of enantiomer labeling, the method of Frank and co-workers [34] is described here:

As standard solution a mixture of 20 μmol of each D-amino acid is dissolved in 80 ml of water containing 1 drop of HCl and 0.01% azide and diluted to 100 ml.

The derivatization with esterification prior to acylation is performed as follows:

To the dried sample 1 ml of 2 N HCl in isopropanol is added. Air is displaced with nitrogen and 1 ml of mercaptoethane is added to prevent oxidative decomposition of tryptophane and to convert cystine into cysteine. The reaction mixture is heated at 110°C for 1 hr. Subsequently the solvent and excess reagent are removed under a gentle stream of nitrogen.

For acylation 250 μl of ethyl acetate and 50 μl of pentafluoropropionic anhydride are added to the dried sample and the reaction mixture is heated at 110°C for 10 min. Increasing of the temperature to 150°C must be performed if arginine is to be determined. Solvent and excess reagent are evaporated with a gentle stream of nitrogen. The residue is dissolved in methylene chloride and an aliquot in the range of 40 pmol is injected into the gas chromatograph. For derivatization of N^{im}-dethoxycarbonylhistidine 100 μl of benzene and 2 μl of diethyl pyrocarbonate are added to the dry residue of N-(O,S)-acyl amino acid isopropylesters.

Derivatization is performed in a heavy-walled screw-capped vial by heating at 150°C for 20 min. The solvent is evaporated and the residue is dissolved in methylene chloride and injected.

3. INSTRUMENTATION

Derivatization procedures for amino acids were originally developed for GLC on packed columns and the acquired proficiency is adaptable to capillary GLC without significant alterations. Far more decisive for the advancement of capillary GLC (CGLC) of amino acids was the development of appropriate equipment and instrumentation, as is true for other compound classes as well. The design of adequate inlet devices, columns, and stationary phases had the greatest bearing upon applicability of CGLC on amino acids.

In view of the potentially smaller amount of sample required for amino acid analysis by CGLC, minimization of sample preparation and derivatization is a logical consequence. Several studies on minimization in amino acid determination have been published [66-68]. Also, in order to develop CGLC to a routine method for amino acid analysis, the aspect of automatization should not be overlooked. A semiautomatic method for isolation of amino acids from biological sources has been devised [69], but still no complete automatization of the full train of manipulations from sample preparation to derivatization has been accomplished. Only through further development of standardized, reliable ways of manipulating biological samples may the full potential of CGLC of amino acids be exploited for routine determinations, e.g., for clinical use in detection of metabolic disorders of newborns.

3.1 Injection Ports

As for the gas chromatograph itself, the existing split or splitless inlet systems, ovens, temperature controls, pressure gauges, and capillary suspensions are satisfactory in most modern gas chromatographs intended for CGLC. Another aspect is the automatic introduction of samples into the gas chromatograph. Basically two designs are available, one working by conventional through-septum injection of an aliquot of the sample solution with an automatic syringe, the other by introduction of a small capillary containing the dried derivative into the hot injection chamber. Advantages and disadvantages of each design are known; each serves its purpose. (For further detail see Chap. 11 on fatty acids.)

In our view the latter device is preferable since it eliminates the "soaking" of capillaries with relatively large volumes of solvents. In addition, it is less prone to mechanical failure. However, since the dry samples are stored in open capillaries, the more volatile amino acids may be partially lost during extended storage. This problem can be solved by cooling the device to very low temperatures and by coating the sample capillaries with a silicone. Another possibility is the employment of so-called hot-pop capsules made from Teflon.

3.2 Columns

A most important contribution to CGLC of amino acids came from the general progress made in respect to the heart of such a system, the capillary itself. The parameters regulating the properties of "activity" of the inner glass surfaces of capillaries are fairly well controlled and understood by now. Especially for such delicate compounds as amino acids, a good deactivation of the glass surface is required, both to stabilize the film of stationary phase and to prevent the breakdown or adsorption of sensitive derivatives. Many alternate procedures have been described for this purpose. The common methods of etching the glass surface with aqueous hydrofluoric acid [45], gaseous hydrogen chloride [51,57], or pretreatment by silylation [57,70] all seem to produce satisfactory results, but no comparative studies have been made. Further deactivation with benzyltriphenylphosphonium chloride or triethanolamine have been proposed, but no evidence as to the beneficial effect was given. In some cases support-coated open tubular capillaries have been prepared by depositing colloidal silicic acid [31-34] or Silanox 101 [49] or by growing "whiskers" [51].

Most stationary phases employed now are the nonpolar polydimethylsiloxanes SE-30 and OV-101, or polydimethylsiloxanes prepared in situ and chemically bonded to the glass walls [46]. In one instance a highly polar mixture of a cyanopropylsiloxane (Silar 5 CP) and a polyoxyethylene (Carbowax 20M) has been employed [47]. Stationary phases used for separation of enantiomers of amino acids were mostly amion acid or dipeptide derivatives, the most efficient being N-lauroyl-valine t-butyl amide [70]. The latter phases exhibit larger separation factors than most others, but their generally high volatilities and/or low thermal stabilities restricted their use to temperatures below 160°C, in most cases even below 120°C.

A new development in the field of CGLC of amino acids is opened with the synthesis of a polysiloxane to which chiral groups are covalently bound; it is named Chirasil-Val. Its formula is

$$(CH_3)_3\text{-Si-}[O\text{-Si}(CH_3)_2]_n\text{-}[O\text{-Si}(CH_3)(CH_2)\text{-CONH-}\underset{\displaystyle CH\,(CH_3)_2}{\underset{|}{CH}}\text{-CONH-C}(CH_3)_3]_m\text{-O-Si}(CH_3)_3$$

These polysiloxanes are stable up to 250°C and enable the gas chromatographic separation of all 20 amino acids in a single run of 30 min. All enantiomers are sufficiently well separated in order to allow the use of the unnatural antipode as an ideal, chemically identical internal standard for each amino acid [34]. By this procedure most of the inherent shortcomings of the quantitative gas-liquid chromatographic amino acid analysis are eliminated, and even the problem of nonquantitative recovery of amino acids during work-up of biological fluids is thus eliminated. The exact determination of relative molar response factors, yields of derivative formation, or calculation of the injected aliquot become irrelevant. Even formation of duplicate derivatives does not affect quantitation as long as the new peaks do not overlap with other amino acids. The method of enantiomer labeling quite certainly will gain wide application in many areas, especially in the biomedical field. The potential of this procedure to biomedical application was shown in a recent study on amino acid levels in serum [71,72].

In a very short period of 1 week roughly 100 serum samples were analyzed with a reproducibility so far achieved only with the ion-exchange procedure.

3.3 Detectors

In keeping with this development the application of selective detectors like the thermoionic nitrogen detector and the electron capture detector opens up new horizons of sensitivity, accuracy, and precision not realized even by modern ion-exchange procedures. The enhancement of sensitivity of the thermoionic detector compared to the flame ionization detector is in the range of 80 to 180 depending on the amino acid [47], but of very low response toward compounds not containing nitrogen.

Finally, the construction of rapidly responding electronic integrators is no less important for the routine analyses of amino acids by CGLC. There are a few instruments capable of integrating chromatographic peaks down to a width of only 1 sec, which is sufficient in most cases. However, their computing capabilities need to be broadened especially in view of the new procedure of enantiomer labeling, where the amount of each amino acid is calculated by comparison of the ratios of each individual enantiomeric pair.

4. APPLICATIONS

A topic that has not been covered above is the gas-liquid chromatography of iodine-containing amino acids, the thyroid hormones. As we have not yet seen an application on glass capillary chromatography, some of the literature dealing with packed columns is cited here [73-78]. Jönsson (present name Sjödahl) and co-workers [46] investigated the chromatography of N-HFB-amino acid n-propylesters. The derivatization procedure was carried out according to the method of Moss and Lambert [11] with slight modifications.

The column used was a 6-m × 0.2-mm ID glass capillary column with dimethylsiloxane polymer chemically bound to the inner glass wall. The number of theoretical plates was measured at 24,000. (For equations, see Chap. 11, on fatty acids.) The partition ratios, which gives a measure of how long a solute spends in the stationary phase relative to the time it spends in the gas phase, was calculated for n-octadecane to be 13.4. The separation number TZ was found to be 13. The column temperature was initially held at 70°C for 4 min and subsequently programmed to 230°C at a rate of 5°C min^{-1}. Injection port and flame ionization detector were held at 230 C. Carrier gas was nitrogen at a velocity of 16 cm min^{-1} at a split ratio of 1:100.

Figure 1 shows a standard chromatogram of 19 amino acids as their corresponding N-HFB n-propylester derivatives. As can be seen from the figure, all derivatives were separated well enough for quantitative analysis. Analysis time was 35 min. The relative standard deviation for the quantitative results expressed as relative molar response ratios was found to be 5% in most cases and reached the 10% level for cystine.

Figure 2 shows the analysis of hydrolyzed insulin. The derivatives were found to be sufficiently stable for automatic injection, thus permitting the analysis of 30 samples in a 24-hr period.

Pearce [48] investigated amino acids as their corresponding N-heptafluorobutyryl isobutylesters. The derivatization procedure was based on that of MacKenzie and Tenaschuk [13]. The author compared a packed column with 3% SE-30 on Gas-Chrom Q 80 to 100 mesh with a support-coated open

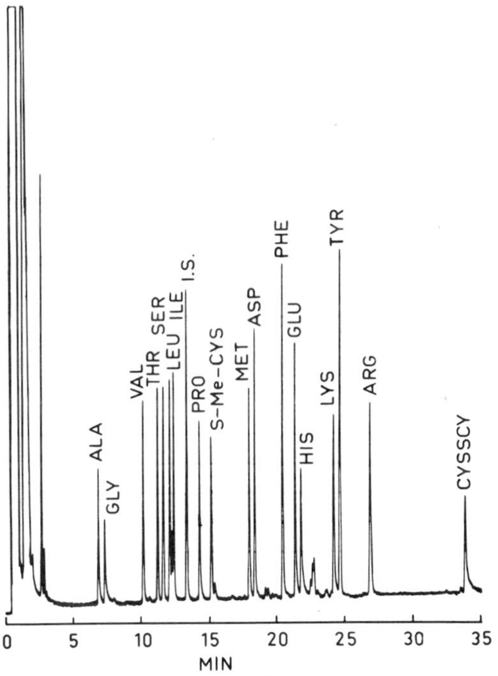

Figure 1. Separation of the N-HFB-amino acid n-propyl ester derivatives of 19 amino acids on a 5.85-m-long glass capillary column. Temperature program: 70°C for 4 min, then 5°C min^{-1}. Full-scale deflection on the recorder represents a detector current of 1.3×10^{11} A; 1/8 of the maximal sensitivity of the instrument was used. The peak between leucine and isoleucine probably represents allo-isoleucine. The peak behind S-methylcysteine represents cysteine (impurity from cystine). (Source: Ref. 46. Reproduced by permission of Academic Press, Inc., New York.)

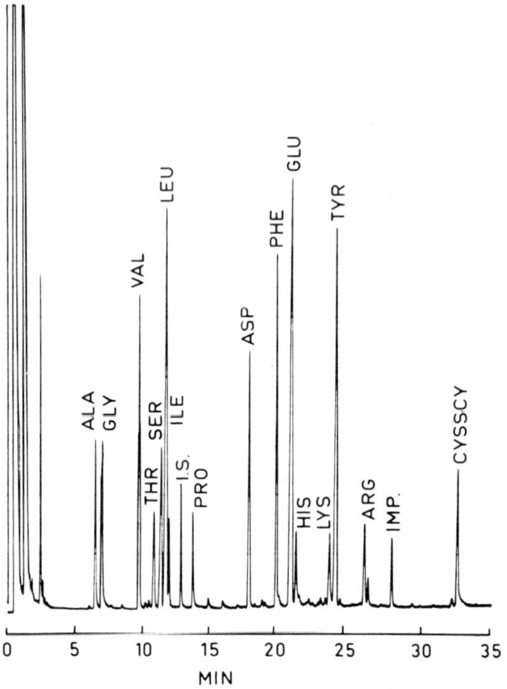

Figure 2. Analysis of hydrolyzed insulin. The amount taken for derivatization was 0.3 mg. The peaks represent 120 ng insulin. The impurity (IMP) is probably a hexosamine. Analysis conditions as in Fig. 1. (Source: Ref. 46. Reproduced by permission of Academic Press, Inc., New York.)

tubular (SCOT) capillary column. The latter was a 60-m × 0.5-mm ID glass capillary with SE-30 as stationary phase. The glass capillary column was temperature programmed from 100 to 250°C at 5°C min^{-1}. Injection port temperature was held at 260°C. A FID was used as detector at a temperature of 270°C. Carrier gas was nitrogen with a flowrate of 20 ml min^{-1}. The injected sample volume ranged from 0.1 to 0.5 μl.

Figure 3 shows a chromatogram of the amino acid calibration mixture on the above-described glass capillary column. As can be seen from the figure, all peaks were resolved by baseline separation. Two major peaks and one minor peak were produced as a result of the derivatization procedure. Glycine, proline, and methionine produced tailing peaks, due to minor unresolved peaks. Histidine showed severe tailing, due to polar interactions.

After 3 months of use the capillary column still exhibited satisfactory performance. The relative standard deviations remained satisfactory, except for cystine and ornithine, where the relative molar response ratios were found to be 4.64 ± 1.1 and 1.03 ± 0.15, respectively.

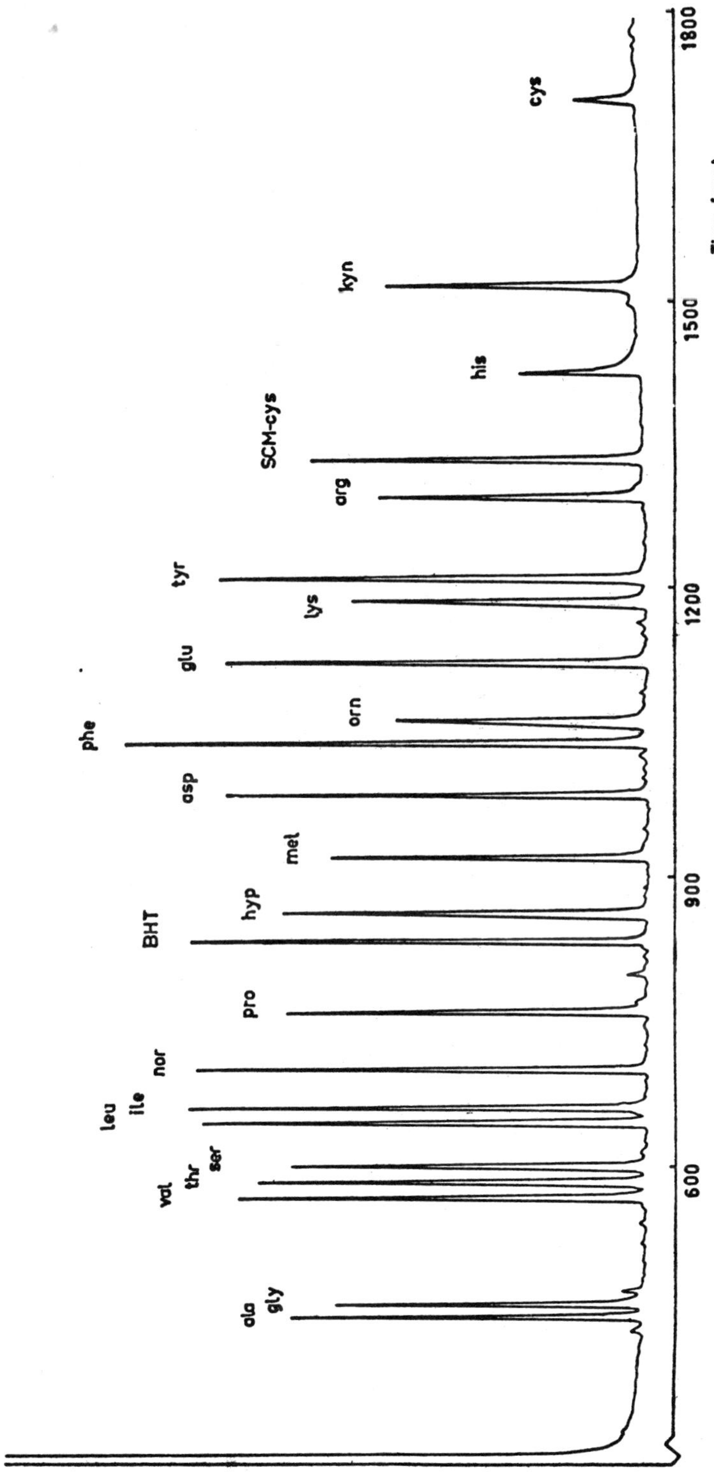

Figure 3. Standard chromatogram of the amino acid calibration mixture. Chromatographic conditions are given in the text. (Source: Ref. 48. Reproduced by permission of Elsevier Scientific Publishing Company, Amsterdam.)

Figure 4 shows a chromatogram of amino acids from a β-lactoglobuline hydrolysate. The values measured were in good agreement with those obtained with ion-exchange chromatography. The method was found to be most suitable for analysis of small amounts of amino acids from different biological sources.

Poole and Verzele [51] compared the elution properties of the major protein amino acids as their N(O)-acyl alkylester derivatives on columns coated with SE-30, OV-17, OV-210, and EGA, respectively. On the SE-30 column a single column baseline separation of the protein amino acids as their N(O)-heptafluorobutyryl n-butylester derivatives could be obtained in less than 35 min. The properties of the OV-210 column were found to be complementary to those of the SE-30 column, thus permitting better compound identification from retention time data. The columns were prepared as described previously [79]. Chromatographic conditions can be seen from the figures. Figure 5 shows a standard chromatogram of the N(O)-heptafluorobutyryl n-butylester derivatives of amino acids on the SE-30 glass capillary column. Figure 6 shows the separation of the same mixture on the OV-210 column. Figure 7 shows the amino acid composition of beer after ion-exchange column cleanup.

One of the earlier papers dealing with the resolution of racemic amino acids on optically active stationary phases is that of Parr and co-workers [30]. Derivatization involved esterification with 3 N HCl in isopropanol and subsequent trifluoroacetylation. A stainless steel column 400 ft × 0.02 in was coated with N-TFA-L-valyl-L-valine cyclohexylester. As the column was made of stainless steel, only one chromatogram is presented here. Figure 8 shows the separation of nine amino acids. Quantitation was possible only to a certain degree, due to overlapping problems. Chromatographic conditions are given in the figure.

In further studies Parr and Howard [80] investigated the structural effects of selected dipeptides as stationary phases for the chromatographic separation of enantiomeric amino acids. As stationary phases they used N-TFA-L-alanyl-L-alanine cyclohexylester, N-TFA-L-α-amino-n-butyryl-L-α-amino-n-butyric acid cyclohexylester, N-TFA-L-norvalyl-L-norvaline cyclohexylester, and N-TFA-L-norleucyl-L-norleucine cyclohexylester. N-TFA-L-α-amino-n-butyryl-L-α-amino butyric acid cyclohexylester was found to be suitable for complete resolution of different amino acids and their enantiomers.

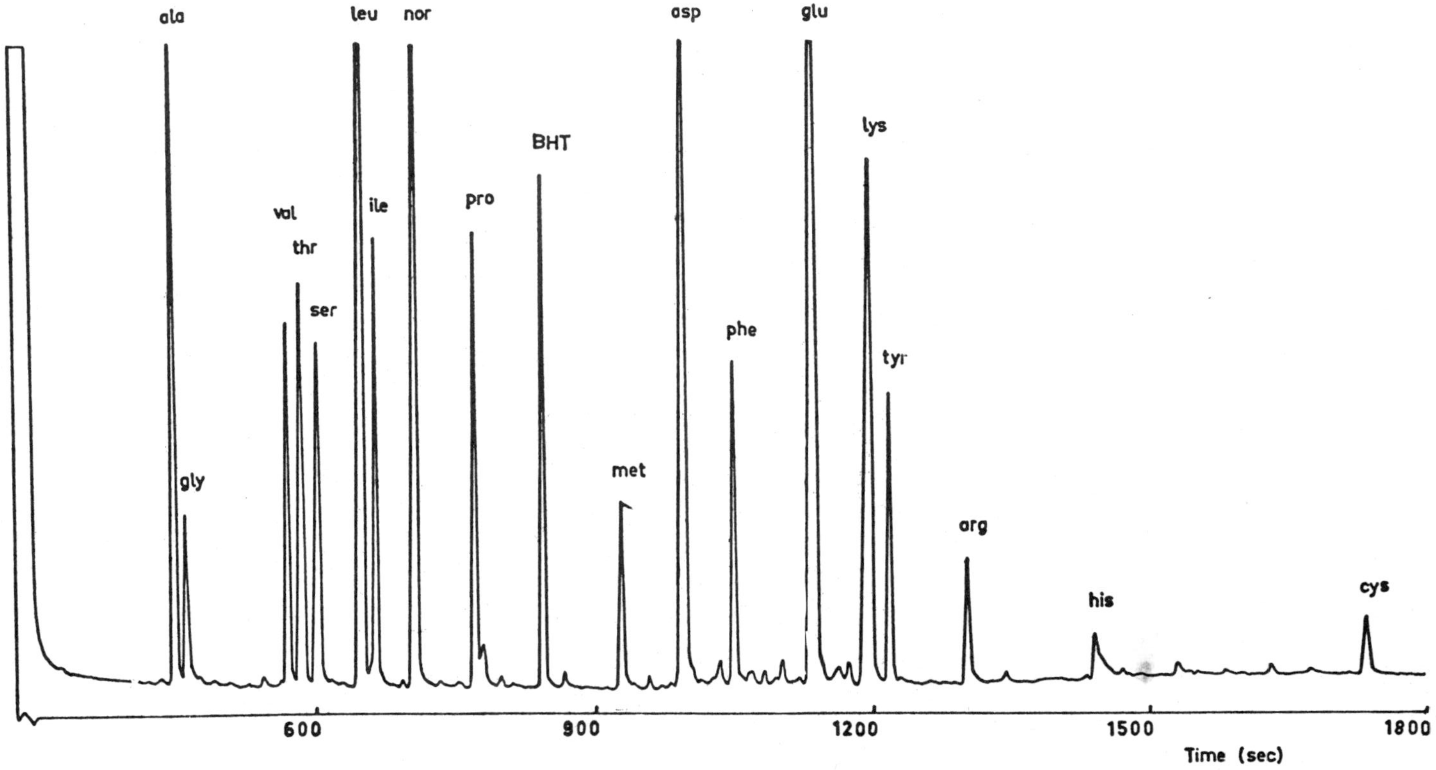

Figure 4. Chromatogram of amino acids from a β-lactoglobulin acid hydrolysate. Chromatographic conditions are given in the text. (Source: Ref. 46. Reproduced by permission of Elsevier Scientific Publishing Company, Amsterdam.)

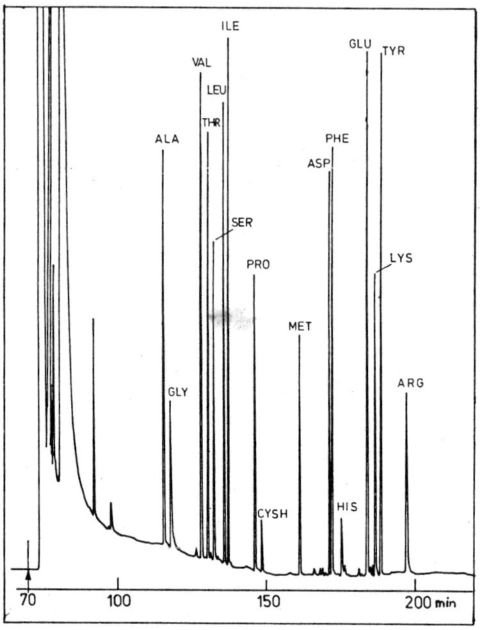

Figure 5. Separation of the N(O)-heptafluorobutyryl n-butyl ester derivatives of amino acid mixture H using a 30-m × 0.25-m ID glass capillary column coated with SE-30. Operating conditions: hydrogen carrier gas, flowrate, 3.3 ml min^{-1}; temperature program, 70 to 220°C at 4°C min^{-1}. (Source: Ref. 51. Reproduced by permission of Elsevier Scientific Publishing Company, Amsterdam.)

Figure 9 shows the complete resolution of different enantiomeric pairs (alanine, valine, glycine, threonine, leucine, isoleucine, serine, and proline) in a reasonable time. The chromatographic conditions are given in the figure. Less volatile amino acids, e.g., aspartic acid, phenylalanine, methionine, glutamic acid, tyrosine, arginine and tryptophan, could be resolved on a comparable column with a length of 100 ft. Thus even in further studies [81] complete resolution of the enantiomeric pairs of all protein amino acids could not be achieved on a single column.

In later studies of the same group [56] N-TFA-L-methionyl-L-methionine cyclohexylester, N-TFA-L-methionyl sulfoxide-L-methionine sulfoxide cyclohexylester, and N-TFA-methionyl sulfone-L-methionine sulfone cyclohexylester were used as stationary phases for the analyses of N-TFA-D,L-amino acid isopropylesters. These changes resulted in improved separations and shorter analysis times.

In a further study Brazell and co-workers [82] demonstrated the use of N-TFA-L-valyl-L-valine-O-carboranyl propylester as stationary phase for

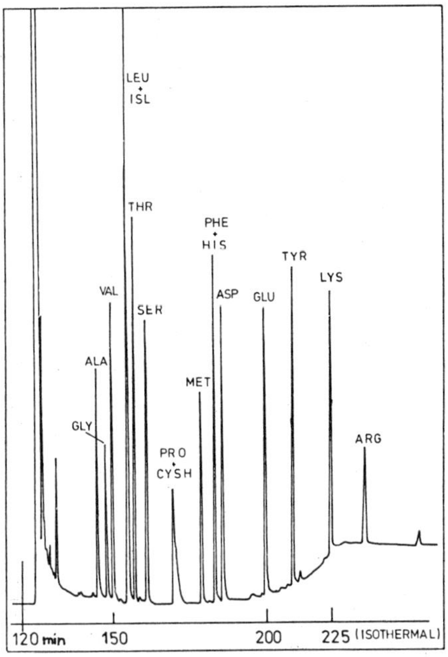

Figure 6. Separation of the N(O)-heptafluorobutyryl n-butyl ester derivatives of amino acid mixture H using a 25-m × 0.25-mm ID glass capillary column coated with OV-210. Operating conditions: hydrogen carrier gas, flowrate, 3.0 ml min^{-1}; temperature program, 120 to 225°C at 4°C min^{-1}. (Source: Ref. 51. Reproduced by permission of Elsevier Scientific Publishing Company, Amsterdam.)

the chromatography of N-TFA-D,L-amino acid isopropylesters. By using a 165-m × 0.5-mm ID stainless steel capillary column they obtained the separation of the enantiomeric pairs of alanine, threonine, valine, isoleucine, leucine, serine, aspartic acid, methionine, phenylalanine, and glutamic acid. As can be seen from Fig. 10, a separation of L-alanine from D-threonine, and D- from L-serine could not be achieved. D- and L-aspartic acid and D- and L-alanine show incomplete separation, thus leading to errors in integration. Chromatographic conditions are given in the figure. Thus the paper must be regarded as further step toward the complete separation of the enantiomeric pairs of protein amino acids on a single column.

In studies by König and Nicholson [57], glass capillary columns with their higher separation efficiency were used. Because of the lower catalytic activity of the glass surface in relation to stainless steel, no decomposition of cysteine, serine, and threonine was found. The new stationary phases tested were N-TFA-L-phenylalanyl-L-phenylalanine cyclohexylester (phe-phe), N-TFA-L-phenylalanyl-L-aspartic acid bis-(cyclohexyl)ester

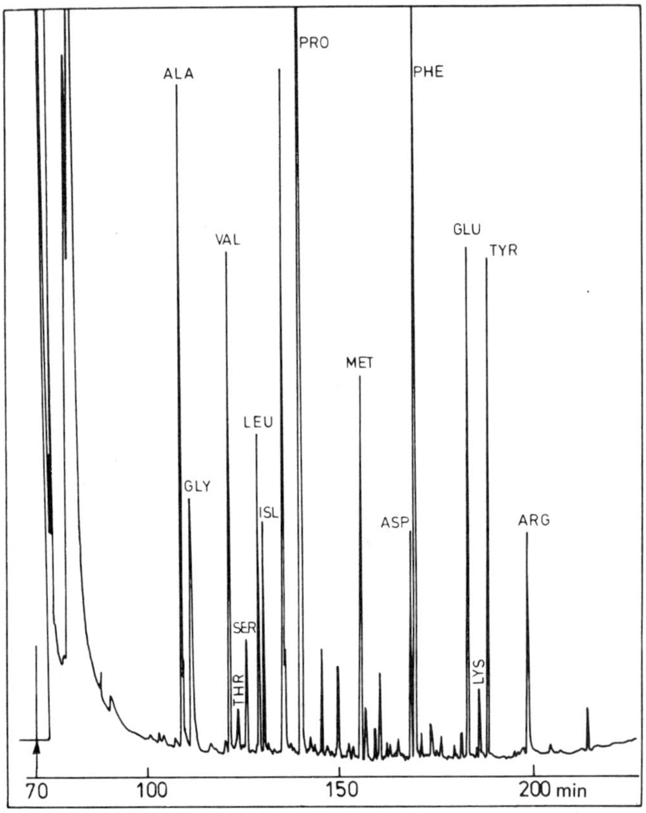

Figure 7. Determination of the amino acid composition of beer (Belgian lager) after ion-exchange column cleanup. Operating conditions as in Fig. 5. (Source: Ref. 51. Reproduced by permission of Elsevier Scientific Publishing Company, Amsterdam.)

(phe-asp), and N-TFA-L-valoyl-L-valine cyclohexylester (val-val). The columns were pretreated either with a solution of 1% 3-amino-propyltriethoxysilane in CH_2Cl_2 at a reaction temperature of 120 to 130°C or etched with dry HCl gas according to Novotny and Tešarik [83] with subsequent coating according to Bouche and Verzele [84]. The preparation of derivatives was performed according to the method of König and co-workers [28].

The glass capillary columns showed a much higher sensitivity than comparable stainless steel columns. A column coated with val-val yielded 100,000 theoretical plates, or 2570 meter^{-1} at a partition ratio of k = 23.3. (For equations, see Chap. 11 on fatty acids.)

The highest temperature stability of up to 165°C was found for the phe-asp phase, combined with a low melting point, thus enabling the use of a temperature program. The phases were further compared according to their

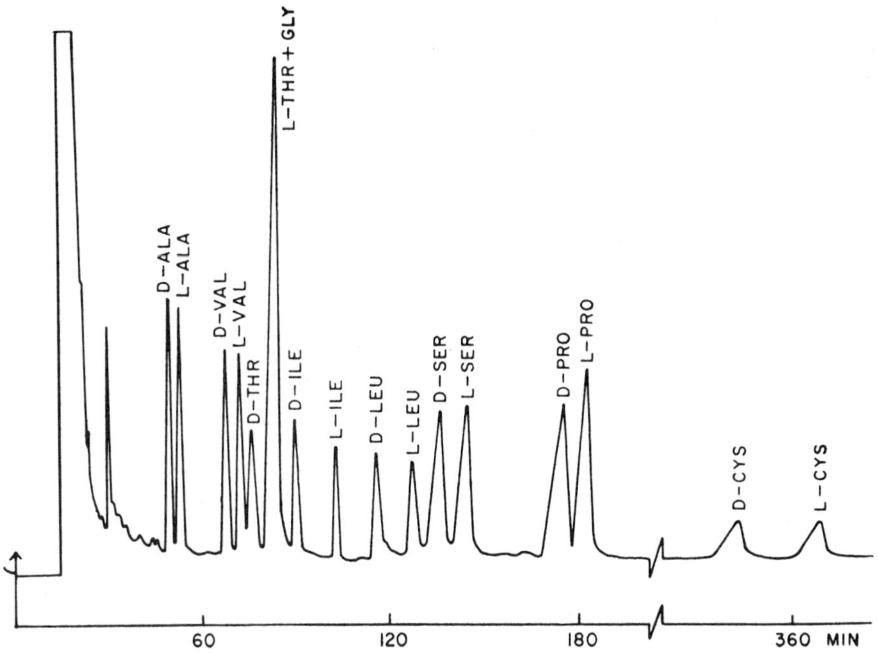

Figure 8. Gas chromatogram of PFP-amino acid isopropyl esters. Instrument: Varian 1200-1. Column: 400 ft, 0.02 in ID, stainless steel capillary coated with N-TFA-L-valyl-L-valine cyclohexyl ester, temperature 110°C isothermal; injector 180°C; detector FID 250°C; carrier gas He, pressure 10 lb in^{-2}. (Source: Ref. 30. Reproduced by permission of Preston Publications, Inc.)

separation factors between the D and L forms. The relative retention $\alpha_{D/L}$ expresses the ratio of the distribution coefficients of the enantiomeric pair of an amino acid. It is calculated by

$$\alpha_{D/L} = t'_{(L)}/t'_{(D)}$$

The requirements of a high $\alpha_{D/L}$ was met with phase val-val most adequately. Although these values could be improved with N-lauroyl-L-valine-tert-butylamide as stationary phase [70], a higher degree of overlapping was found between the amino acid species.

Figure 11 shows a separation of TFA-amino acid isopropylesters on a 39-m glass capillary column coated with val-val. Column temperature was held isothermally at 110°C.

The phe-phe phase could be used between 130 and 160°C. It was found to be more stable to temperature degradation and racemization, although its volatility is somewhat higher than that of phe-asp.

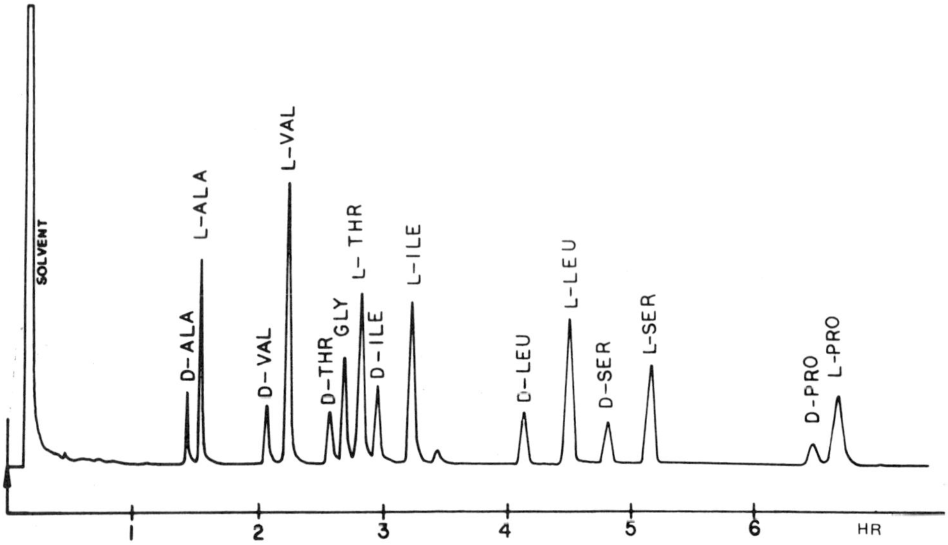

Figure 9. Chromatogram of N-TFA-D,L-amino acid isopropyl esters with N-TFA-L-α-amino-n-butyryl-L-α-amino-n-butric acid cyclohexyl ester as stationary phase. Chromatographic conditions: 400-ft × 0.02-in stainless steel capillary column; 110°C isothermal operation; injector 180°C; detector 275°C; carrier gas He at 12 lb in^{-2} gauge. (Source: Ref. 80. Reproduced by permission of *Analytical Chemistry*, copyright by the American Chemical Society.)

König and co-workers [61] investigated the use of pentafluoropropionyl amino acid (+)-3-methyl-2-butylesters for separations of diastereoisomers on glass capillary columns. By means of the derivatization a second asymmetric center was introduced into the amino acid molecule and diastereoisomers were formed, which could be separated on conventional, optically inactive phases. The second asymmetric center was introduced by esterification of the amino acids with (+)-3-methyl-2-butanol. The amino groups and all other susceptible functional groups were substituted by pentafluoropropionyl groups. The authors succeeded in separating the derivatives of all common amino acids, including histidine, arginine, and tryptophan.

Hence it must be accepted that quantitative esterification of amino acids with higher alcohols is difficult, because of the poor solubility of the chiral alcohols constitute sources of error.

The authors demonstrated that esterification was nearly quantitative with all amino acids except the basic ones (lysine, histidine, ornithine, diaminopropionic acid, and arginine). Interesterification improved the yield, but this method was not fully optimized.

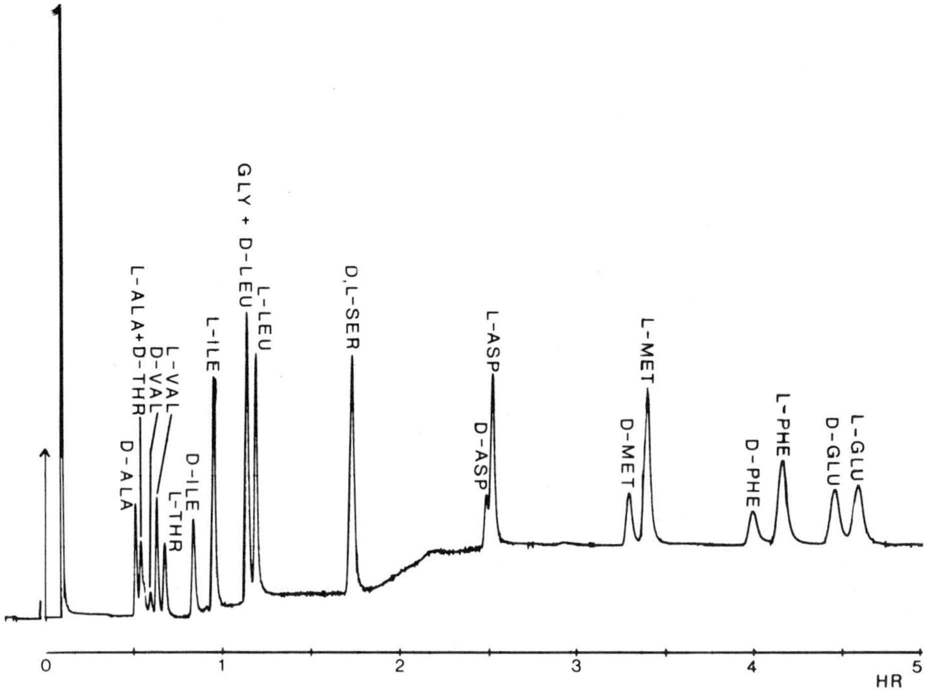

Figure 10. Chromatogram of N-TFA-D,L-amino acid isopropyl esters with N-TFA-L-valyl-L-valine-o-carboranyl propyl ester as the stationary phase. Chromatographic conditions: 165-m × 0.5-mm ID stainless steel capillary column, operation at 75°C isothermal for 30 min, then temperature-programmed with 0.5°C min^{-1} to 110°C; injector 180°C; detector 275°C; carrier gas He at 24 lb in^{-2}. (Source: Ref. 82. Reproduced by permission of the author.)

The column used was a 25-m glass capillary coated with SE-30. The number of theoretical plates was calculated to be 114,000 at 120°C for n-tetradecane (with nitrogen as carrier gas). Hydrogen was used as carrier gas. The derivatization was checked by GC/MS. Figure 12 shows a chromatogram of a standard mixture of diastereoisomeric pentafluoropropionyl amino acid (+)-3-methyl-2-butylesters on the above described column.

By comparison of this method with the following one, the two major ways of separation of D- and L-amino acids are described. In the above-described approach the diastereoisomers are formed by introduction of the second asymmetric center; in the method described in the next part of this section optically inactive derivatives are chromatographed on a chiral phase, thus leading to the separation of enantiomers by formation of diastereoisomeric molecule complexes between the D- and L-amino acid derivatives and the chiral liquid phase.

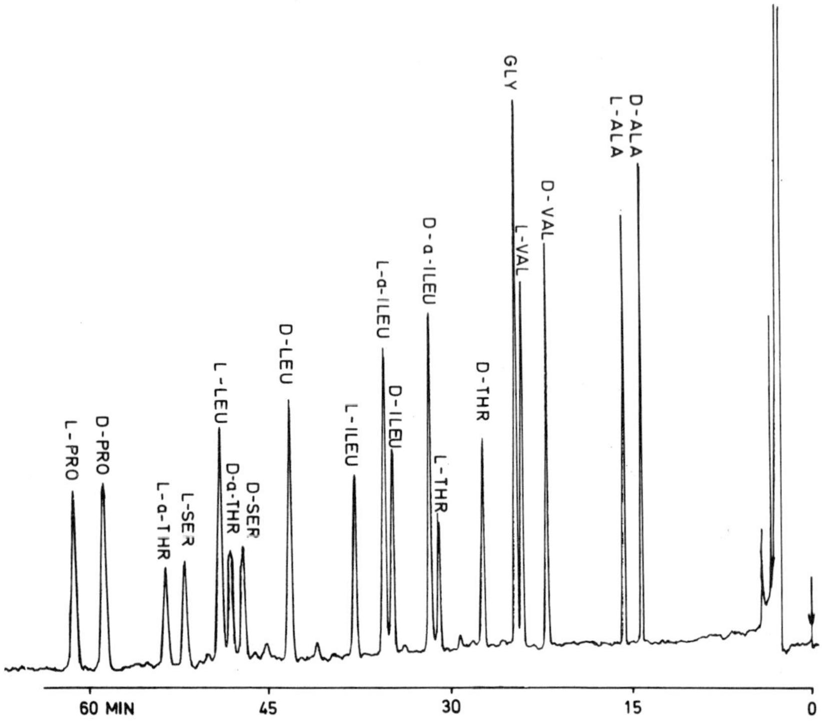

Figure 11. Separation of TFA-amino acid isopropyl esters on 39-m glass capillary coated with N-TFA-L-val-L-val cyclohexyl ester. Column temperature: 110°C. (Source: Ref. 57. Reproduced by permission of *Analytical Chemistry*, copyright by the American Chemical Society.)

The above-described method is not fully optimized, but must be seriously compared as an alternative to the method further below, which has the advantage of easy derivatization. Furthermore, the disadvantages of the method below, i.e., problems with the stationary phases, have been overcome now that columns are commercially available.

By application of the newly synthesized phase Chirasil-Val (for further detail see Sec. 3) Frank and co-workers [31-34] could separate the 20 protein amino acids and their unnatural enantiomers in a single run in about 30 min. The samples were derivatized as described in Sec. 2.

The glass capillaries used were 20 m × 0.3 mm ID coated with Chirasil-Val. The film thickness was about 0.1 μm. Hydrogen was used as carrier gas at an inlet pressure of approximately 0.35 bar. The column temperature was programmed as follows: 5 min isothermally at 87°C and from there with 4°C min^{-1} up to 200°C, and then maintained at this temperature.

As the unnatural D-amino acids are used as internal standards for each

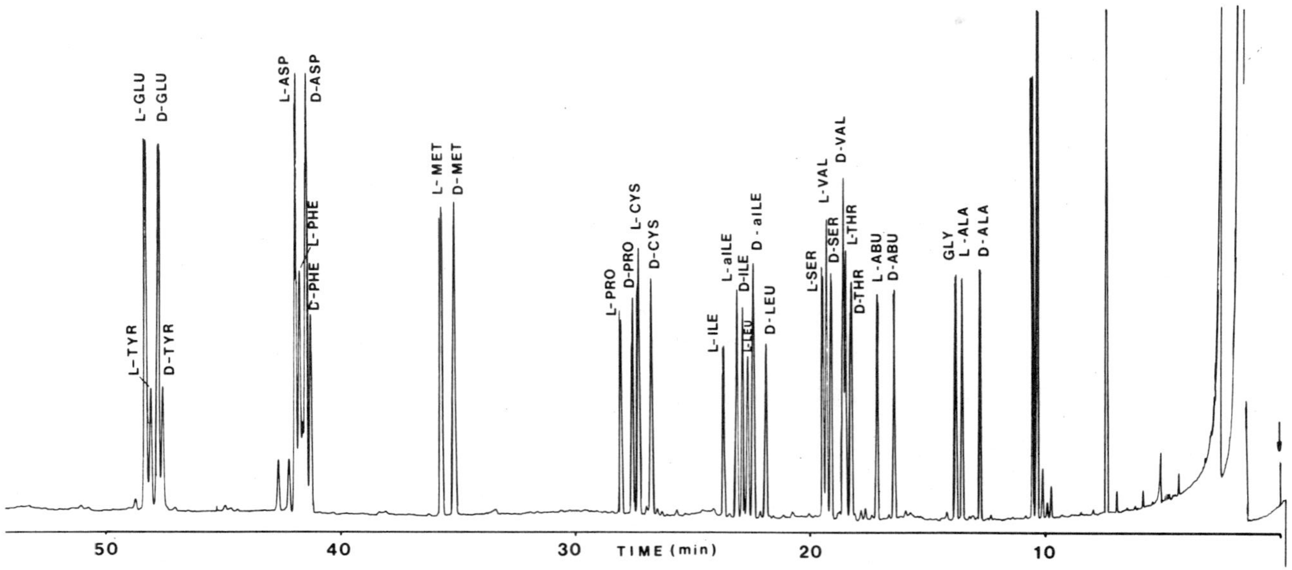

Figure 12. Gas chromatogram of diastereoisomeric pentafluoropropionylamino acid (+)-3-methyl-2-butyl esters on a 25-m glass capillary column (SE-30 AmAc, LKB 2101-210). Column temperature: 85°C; temperature program 2°C min^{-1}. Carrier gas: hydrogen. Split ratio: 1:30. (Source: Ref. 61. Reproduced by permission of Elsevier Scientific Publishing Company, Amsterdam.)

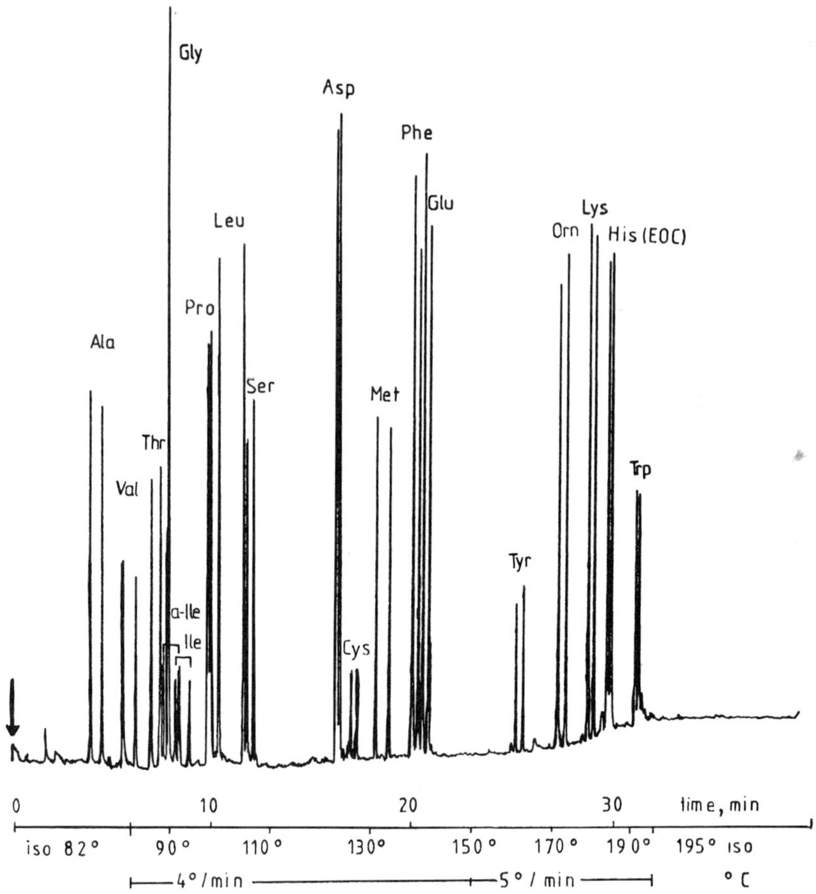

Figure 13. Gas chromatographic separation of a racemic standard mixture of amino acids on Chirasil-Val. Chromatographic conditions are given in the text.

respective natural L-enantiomer, the D forms have to be checked for absolute purity before use. The amount of the natural L-enantiomer can be calculated as the ratio of the peak areas of the L- and D-enantiomer multiplied by the amount of internal standard added and by application of correction factors, which can be calculated in a standard run.

After optimization of the method the standard deviations of absolute errors expressed as percentages of the nominal amount of amino acids in the sample were found to be between 0.5 and 3.6%.

For glycine, which is not optically active, D-alanine was chosen as the most suitable internal standard.

Figure 13 shows a standard chromatogram of a racemic standard mixture of amino acids on Chirasil-Val. The temperature program was varied compared to the above described and can be seen from the figure.

The method is of course applicable to the amino acid analysis from samples of biological origin [71,72].

Figure 14 shows a chromatogram of amino acids extracted from human serum, chromatographed as described above, but without the addition of the unnatural D-enantiomers, and Fig. 15 shows the same mixture with addition of the internal standards. As can be seen from the figures, the protein amino acids can be analyzed easily in a single run of 40 min.

Our own experiences with the glass capillary gas-liquid chromatography of amino acids were not very successful in the beginning [85]. Since then, we have undertaken the above-described method [86] and found it to be most suitable for routine analysis of clinical problems.

Figure 16 shows the gas-liquid chromatographic separation of a standard mixture of D- and L-amino acids on Chirasil-Val. Chromatographic conditions are the same as described above.

We regard this method as the most accurate and sensitive analytical approach for quantitative gas-liquid chromatographic analysis of amino acids. It is superior with regard to analysis time, and mostly superior or at least equal with regard to accuracy, compared to ion-exchange chromatography. Thus it will surely find widespread application in the field of amino acid analysis, especially when only small sample amounts are available.

5. SUMMARY

Investigations in the field of amino acid analysis have gained more and more interest in the last decades. Amino acid analysis for clinical investigations of metabolism, physiology, and pathophysiology of these acids is impossible without chromatographic techniques.

Although most of the scientific work is done by ion-exchange chromatography, gas-liquid chromatography of these substances has gained more and more interest because of lower costs, greater versatility, and shorter analysis times of the latter method.

It could be shown that extraction and derivatization of the different amino acid species yielded different results, thus making the application of a single internal standard impossible. Hence it was a logical consequence to add the unnatural D-enantiomers as internal standards for every natural acid in the L form.

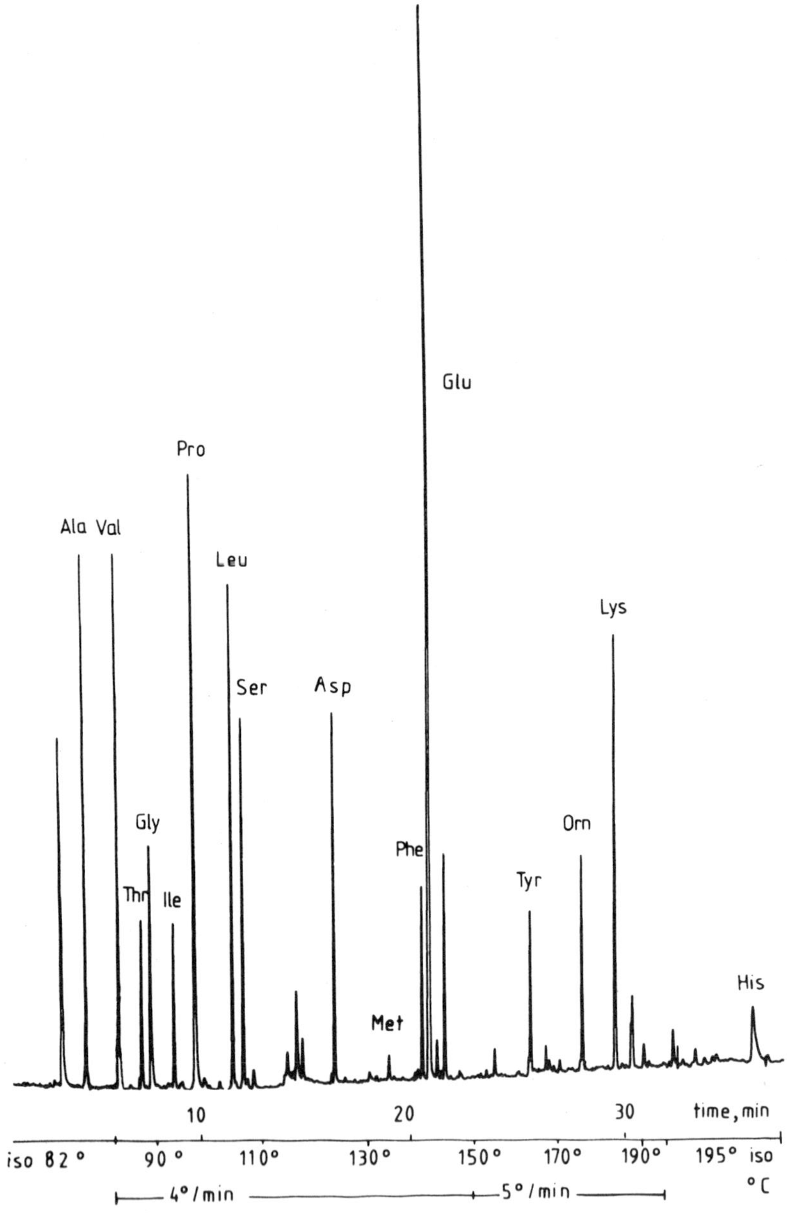

Figure 14. Gas chromatographic separation of amino acids from human serum as their N-pentafluoropropionyl isopropyl esters on Chirasil-Val *without* addition of the D-enantiomers. Chromatographic conditions are given in the text.

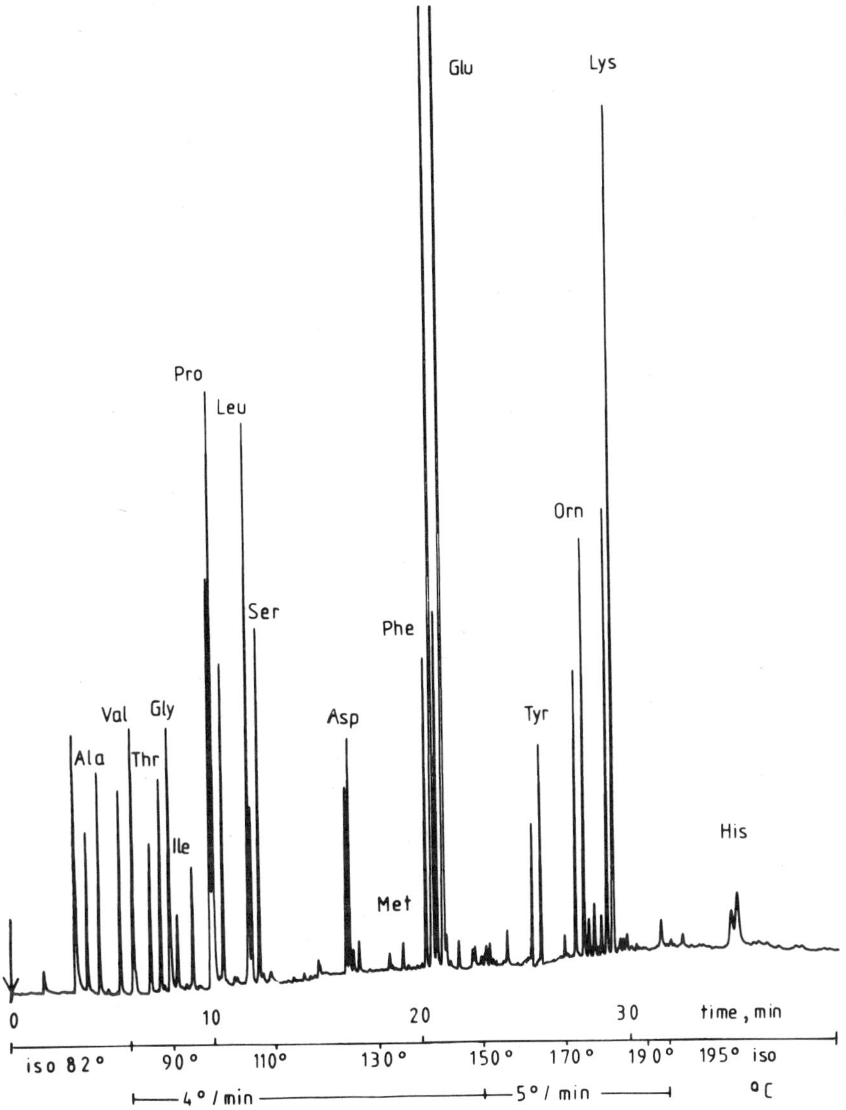

Figure 15. Gas chromatographic separation of amino acids from human serum as their N-pentafluoropropionyl isopropyl esters on Chirasil-Val *with* addition of the D-enantiomers. Chromatographic conditions are given in the text.

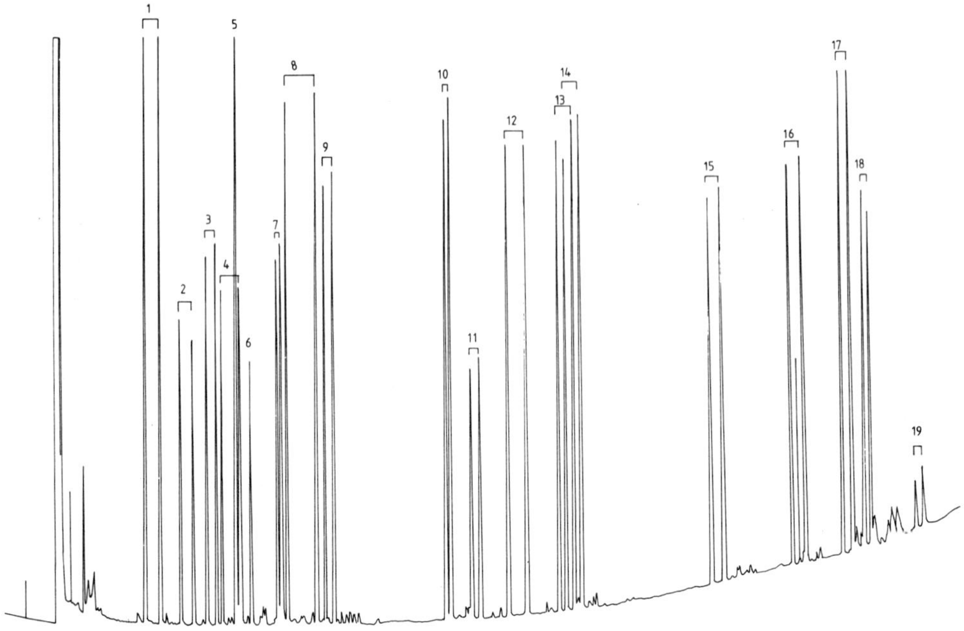

Figure 16. Gas chromatogram of a standard mixture of amino acid enantiomers as their N-pentafluoropropionyl isopropyl esters on Chirasil-Val. Glass capillary column: 20 m × 0.3 mm. Carrier gas: hydrogen with an inlet pressure of 0.35 bar. Temperature program: initially 90°C for 3 min, subsequently heated with 4°C min to 180°C. Injector: 225°C. Identification of peak numbers: 1—D/L-alanine, 2—D/L-valine, 3—D/L-threonine, 4—D/L-alloisoleucine, 5—D-isoleucine + glycine, 6—L-isoleucine, 7—D/L-proline, 8—D/L-leucine, 9—D/L-serine, 10—D/L-aspartic acid, 11—D/L-cysteine, 12—D/L-methionine, 13—D/L-phenylalanine, 14—D/L-glutamic acid, 15—D/L-tyrosine, 16—D/L-ornithine, 17—D/L-lysine, 18—D/L-histidine (N^{im}-EOC), 19—D/L-tryptophane.

For the separation of enantiomeric pairs of amino acids two separation methods are possible. The first approach was formation of diastereoisomers by introduction of a second asymmetric center in the molecule, leading to optically active derivatives, which can be chromatographed on an optically inactive liquid phase. The second approach was formation of optically inactive derivatives and subsequent chromatography on chiral phases.

Both methods have been applied with success, but the second has been more highly optimized.

At present the chromatography of amino acid enantiomers on chiral phases, e.g., Chirasil-Val, which is commercially available, can be regarded as superior in analysis time and money savings and at least equal in accuracy and precision when compared with ion-exchange chromatography.

REFERENCES

1. Stein, N. H., Moore, S., and Spackman, D. H., *Anal. Chem.*, 30:1119 (1958).
2. Bayer, E., *Angew. Chem.*, 69:640 (1951).
3. Andersson, C. O., Ryhage, E., Ställberg, S., and Stenhagen, E., *Ark. Kemi*, 19:405 (1962).
4. Bayer, E., in *Gas Chromatography, 1958* (D. H. Desty, ed.), Butterworths, London, 1958, p. 333.
5. Hunter, I. R., Dimick, K. P., and Corse, J. W., *Chem. Ind.*, London, 16:294 (1956).
6. Husek, P., and Macek, K., *J. Chromatogr.*, 113:139 (1975).
7. Gehrke, C. W., Lamkin, W. M., Stalling, D. L., and Schahrokki, F., *Biochem. Biophys. Res. Commun.*, 19:328 (1965).
8. Gehrke, C. W., Roach, D., Zumwalt, R. W., Stalling, D. L., and Wall, L. L., Quantitative Gas-Liquid Chromatography of Amino Acids in Proteins and Biological Substances, Report, Anal. Biochem. Lab. Inc. Columbia, Mo. 1968.
9. Roach, D., and Gehrke, C. W., *J. Chromatogr.*, 43:303 (1969).
10. Zumwalt, R. W., Roach, D., and Gehrke, C. W., *J. Chromatogr.*, 53:171 (1970).
11. Moss, C. W., and Lambert, M. A., *J. Chromatogr.*, 60:134 (1971).
12. Zanetta, D., and Vincedon, G., *J. Chromatogr.*, 76:91 (1973).
13. MacKenzie, S. L., and Tenaschuk, D., *J. Chromatogr.*, 97:19 (1974).
14. Andersson, C. O., Ryhage, E., Ställberg, S., and Stenhagen, E., *Ark. Kemi*, 19:405 (1962).
15. Charles, R., Fischer, G., and Gil-Av, E., *Israel J. Chem.*, 1:234 (1963).
16. Weygand, F., *Z. Anal. Chem.*, 205:406 (1964).
17. Pollock, G. E., *Anal. Chem.*, 39:1195 (1967).
18. Gelpi, E., König, W. A., Gibert, J., and Orό, J., *J. Chromatogr. Sci.*, 7:604 (1969).
19. Pollock, G. E., and Kawanchi, A. H., *Anal. Chem.*, 40:1357 (1968).
20. Gil-Av, E., Charles, R., and Fischer, G., *J. Chromatogr.*, 17:408 (1965).
21. Pollock, G. E., Oyama, V. I., and Johnson, R. D., *J. Gas Chromatogr.*, 3:174 (1965).
22. Gil-Av, E., Charles-Sigler, R., Fischer, G., and Nurok, D., *J. Gas Chromatogr.*, 4:51 (1966).
23. Pollock, G. E., and Oyama, V. I., *J. Gas Chromatogr.*, 4:126 (1966).
24. Gil-Av, E., Feibush, B., and Charles-Sigler, R., in *Gas Chromatography, 1966* (A. B. Littlewood, ed.), Institute of Petroleum, London, 1967, p. 227.
25. Gil-Av, E., Feibush, B., and Charles-Sigler, R., *Tetrahedron Lett.*, p. 1009 (1966).
26. Feibush, B., and Gil-Av, E., *J. Gas Chromatogr.*, 5:257 (1967).

27. Feibush, B., Gil-Av, E., and Tamari, T., *Israel J. Chem.*, 8:50 p (1970).

28. König, W. A., Parr, W., Lichtenstein, H. A., Bayer, E., and Oró, J., *J. Chromatogr. Sci.*, 8:183 (1970).

29. Nakaparksin, S., Birrell, P., Gil-Av, E., and Oró, J., *J. Chromatogr. Sci.*, 8:177 (1970).

29a. Parr, W., Yang, C., Bayer, E., and Gil-Av, E., *J. Chromatogr. Sci.*, 8:591 (1970).

30. Parr, W., Pleterski, J., Yang, C., and Bayer, E., *J. Chromatogr. Sci.*, 9:141 (1971).

31. Frank, H., Nicholson, G. J., and Bayer, E., *J. Chromatogr. Sci.*, 15:174 (1977).

32. Frank, H., Nicholson, G. J., and Bayer, E., *Angew. Chem. Intern. Ed.*, 17:363 (1978).

33. Frank, H., Nicholson, G. J., and Bayer, E., *J. Chromatogr.*, 146:197 (1978).

34. Frank, H., Nicholson, G. J., and Bayer, E., *J. Chromatogr.*, 167:181 (1978).

35. Pellizzari, E. D., Rising, C., Brown, J. H., Farmer, R. W., and Fabre, Jr., L. F., *Anal. Biochem.*, 44:312 (1971).

36. Kaiser, F. E., Gehrke, C. W., Zumwalt, R. W., and Kuo, K. C., *J. Chromatogr.*, 94:113 (1974).

37. Adams, R. F., *J. Chromatogr.*, 95:189 (1974).

38. Davis, J. W., and Fürst, A., *Anal. Chem.*, 40:1910 (1968).

39. Weygand, F., Geiger, R., and Swodenk, W., *Angew. Chem.*, 68:307 (1956).

40. Rühlmann, K., and Giesecke, W., *Angew. Chem.*, 73:113 (1961).

41. Gehrke, C. W., and Leimer, K. R., *J. Chromatogr.*, 57:219 (1971).

42. Brooks, J. B., Alley, C. C., and Liddle, J. A., *Anal. Chem.*, 46:1930 (1974).

43. Halász, I., and Bünnig, K., *Z. Anal. Chem.*, 211:1 (1965).

44. Grahl-Nielsen, O., and Solheim, E., *Chem. Commun.*, p. 1092 (1972).

45. Abe, I., Kohno, T., and Musha, S., *Chromatographia*, 11:393 (1978).

46. Jönsson,* J., Eyem, J., and Sjöquist, J., *Anal. Biochem.*, 51:204 (1973).

47. Adams, R. F., Vandemark, F. L., and Schmidt, G. J., *J. Chromatogr. Sci.*, 15:63 (1977).

48. Pearce, R. J., *J. Chromatogr.*, 136:113 (1977).

49. MacKenzie, S. L., and Hogge, L. R., *J. Chromatogr.*, 147:388 (1978).

50. Lawless, J. G., and Boynton, C. D., *Nature*, 243:405 (1973).

51. Poole, C., and Verzele, M., *J. Chromatogr.*, 150:439 (1978).

52. Roach, D., Gehrke, C. W., and Zumwalt, R. W., *J. Chromatogr.*, 43:311 (1969).

*Present name: Sjödahl.

53. Moodie, I. M., *J. Chromatogr.*, 99:495 (1975).

54. Parr, W., and Howard, P., *Chromatographia*, 5:18 (1972).

55. Grohmann, K., and Parr, W., *Chromatographia*, 5:18 (1972).

56. Andrawes, F., Brazell, R., Parr W., and Zlatkis, A., *J. Chromatogr.*, 112:197 (1975).

57. König, W. A., and Nicholson, G. J., *Anal. Chem.*, 47:951 (1975).

58. Beitler, U., and Feibush, B., *J. Chromatogr.*, 123:149 (1976).

59. Weinstein, S., Feibush, B., and Gil-Av, E., *J. Chromatogr.*, 126:97 (1976).

60. König, W. A., *Chromatographia*, 9:72 (1976).

61. König, W. A., Rahn, W., and Eyem, J., *J. Chromatogr.*, 133:141 (1977).

62. Bonner, W. A., *J. Chromatogr. Sci.*, 10:159 (1972).

63. Bonner, W. A., *J. Chromatogr. Sci.*, 11:101 (1973).

64. Florer, J. J., Bonner, W. A., and van Doot, M. A., *J. Chromatogr.*, 132:152 (1977).

65. König, W. A., Stoelting, K., and Kruse, K., *Chromatographia*, 10:444 (1977).

66. Gehrke, C. W., Zumwalt, R. W., and Wall, L. L., *J. Chromatogr.*, 37:398 (1968).

67. Gehrke, C. W., Zumwalt, R. W., Stalling, D. L., Roach, D., Aue, W. A., Ponnamperuma, C., and Kvenvolden, K. A., *J. Chromatogr.*, 59:305 (1971).

68. Frick, W., Chang, D., Folers, K., and Daves, G. D., *Anal. Chem.*, 49:1241 (1971).

69. Blanshard, K. C., Bradford, H. F., Dodd, P. R., and Thomas, A. J., *Anal. Biochem.*, 67:233 (1975).

70. Feibush, B., *Chem. Commun.*, p. 544 (1971).

71. Frank, H., Rettenmeyer, A., Weicker, H., Nicholson, G. J., and Bayer, E.,

72. Rettenmeyer, A., Frank, H., Ebert, W., and Weicker, H., Deutscher Sportärtze Kongress, Bad Nauheim, 1978.

73. Jaakonmaki, P. I., and Stouffer, J. E., *J. Gas Chromatogr.*, 1:303 (1967).

74. Sharoklu, F., and Gehrke, C. W., *Anal. Biochem.*, 24:281 (1968).

75. Alexander, N. M., and Scheig, R., *Anal. Biochem.*, 22:187 (1968).

76. Rall, J. E., *Clin. Chim. Acta*, 25:339 (1969).

77. Funakoski, K., and Calmmann, J. J., *Anal. Biochem.*, 27:150 (1969).

78. Petersen, B. A., Giese, R. W., Larsen, P. R., and Karger, B. L., *Clin. Chem.*, 23:1389 (1977).

79. Sandra, P., and Verzele, M., *Chromatographia*, 10:419 (1977).

80. Parr, W., and Howard, P. Y., *Anal. Chem.*, 45:711 (1973).

81. Haegele, K. D., Howard, P. Y., and Parr, W., *Origins of Life*, 6:195 (1975).

82. Brazell, R., Parr, W., Andrawes, F., and Zlatkis, A., *Chromatographia*, 9:57 (1976).

83. Novotny, M., and Tesarik, U., *Chromatographia*, 1:332 (1968).

84. Bouche, J., and Verzele, M., *J. Gas Chromatogr.*, 6:501 (1968).

85. Jaeger, H., Wagner, W., Homoki, J., Klör, H. U., and Ditschuneit, H., *Chromatographia*, 10:492 (1977).

86. Jaeger, H., Schlegel, E., Klör, H. U., Muck, A., and Ditschuneit, H., 2nd Can. Conf. on Chromatography, Toronto, 1979.

10

GLASS CAPILLARY GAS-LIQUID CHROMATOGRAPHY OF BILE ACIDS

HALVOR JAEGER
WILHELM NEBELUNG
HANS U. KLOER
HANS DITSCHUNEIT

University of Ulm
Ulm, Federal Republic of Germany

HARTMUT FRANK

University of Tübingen
Tübingen, Federal Republic of Germany

1. Introduction 366
   1.1 General Structure and Function 366
   1.2 Metabolism of Bile Acids 366
   1.3 Occurrence and Distribution of Individual Bile Acids 367
   1.4 Specific Properties of Different Bile Acids 368
   1.5 Application of Bile Acid Determination 369
2. Sample Preparation and Derivatization 370
   2.1 Extraction Procedures 371
   2.2 Hydrolysis and Solvolysis 377
   2.3 Derivatization 378
3. Columns and Detectors 380
4. Applications 380
5. Summary 387
   References 391

1. INTRODUCTION

Although the importance of bile acids has long been recognized [1], a widespread interest into their physiology and pathophysiology has arisen only in the last two decades, requiring more and more sophisticated analytical tools for their determination.

1.1 General Structure and Function

The main function of bile acids is their ability to operate as biological detergents, a property which is due to their chemical structure. The sterol backbone which is derived from cholesterol provides an apolar, hydrophobic character, while the modifying metabolic steps in the liver cell (dehydroxylation at the 3, 7, or 12 position, addition of a carboxyl group, conjugation) add a contralateral hydrophilic moiety to the molecule. This amphipathic structure enables the bile acids to "coat" lipids of low polarity such as lecithin, cholesterol, or glycerides to form lipid droplets called mixed micelles [2]. In this physicochemical arrangement, which serves to disperse dietary lipids in the gut lumen for hydrolysis and subsequent absorption, the hydrophobic side of the bile acid merges into the lipid core of the micelle, while the hydrophilic side faces the water phase. On their way from the liver to the gut the bile acids play an important role in solubilizing the other components of bile, i.e., lecithin and cholesterol. Thus, the concentration of bile acids in bile is crucial for the prevention of gallstone formation, which occurs when the sensitive balance between the three components of bile is shifting toward a relatively higher concentration of cholesterol with subsequent crystallization [3].

1.2 Metabolism of Bile Acids

The conversion of cholesterol to the different bile acid species takes place at different subcellular locations of the liver cell. While side chain oxidation is mediated by mitochondrial enzyme systems, all the modifications of the sterol ring are catalyzed by the action of enzymes in the smooth endoplasmatic reticulum. The rate-limiting enzyme of these processes seems to be cholesterol-7-α-hydroxylase [4]. The end products of this synthetic pathway in man are the primary bile acids cholic (3α,7α,12α-trihydroxy-5β-cholanic acid), chenodeoxycholic acid (3α,7α,12α-trihydroxy-5β-cholanic acid) and, probably, ursodeoxycholic acid (3α,7α-dihydroxy-5β-cholanic acid) [5]. Especially under conditions of impaired bile secretion, the formation of β-muricholic acid (3α,6β,7β-trihydroxy-5β-cholanic acid) has been observed in rat liver [6]. Both ursodeoxycholic acid and β-muricholic acid are formed from chenodeoxycholic acid in the liver.

The primary bile acids leave the liver cell in a conjugated form, i.e., the carboxyl group forms an amide linkage with the amino group of glycine or taurine. In the conjugated form the bile acids seem to have less cellular toxicity [7]. While most of the primary bile acids are reabsorbed without alteration in the ileum, some undergo modifications by bacterial enzymes, especially in the terminal ileum and colon. The modification consists of 7α-dehydroxylation leading to the formation of secondary bile acids, i.e., of deoxycholic (3α,12α-dehydroxy-5β-cholanic acid) from cholic and of lithocholic (3α-monohydroxy-5β-cholanic acid) from chenodeoxycholic acid. Whether ursodeoxycholate is also metabolized by bacteria is still controversial.

While no conversion to lithocholic acid was found in rhesus monkeys [8], a significant rise in fecal lithocholate content was observed after ursodeoxycholate administration in man [9]. This difference may be due to a higher specificity for the 7α position of the bacterial dehydroxylase present in the monkey, while the bacterial enzymes present in the human intestine may not be that selective. Intestinal bacteria are also capable of splitting the amide linkage and thereby deconjugating primary and secondary bile acids. Deconjugation occurs in the mid and terminal ileum and in the colon, where bacterial growth is most pronounced in man [10]. In some disease states such as inflammation of the ileum or in the "blind loop syndrome" resulting from gastric resection, the bacterial overgrowth may expand proximally, thus leading to a high degree of intestinal deconjugation [10].

Free and conjugated primary and secondary bile acids are transported back to the liver via the portal vein, where they are almost completely taken up by the liver during a single passage. This completes the so-called enterohepatic circulation of bile acids which prevents the need for continuous supplementation of fecal losses by de novo synthesis in the liver.

1.3 Occurrence and Distribution of Individual Bile Acids

The most important bile acids are cholate, chenodeoxycholate, and deoxycholate, which normally occur as glycyl conjugates. They comprise over 90% of the bile acids in bile, serum, and urine [5]. Less than 5% of the bile acids present in human bile have been identified as ursodeoxycholate and lithocholate conjugates [10]. Together with hyocholic and hyodeoxycholic acid they have recently been found in small amounts in serum and urine of patients with cholostasis [11]. Epilithocholic acid has been reported in trace amounts in normal urine [12].

1.4 Specific Properties of Different Bile Acids

While all bile acids seem to share the property of forming mixed micelles with lipids in bile and intestine, their intracellular detergent effects seem to differ somewhat. Cholic acid and, probably, β-muricholic acid are relatively weak detergents if tested as inhibitors of microsomal enzyme systems of the liver [6].

Dihydroxylated bile acids, especially chenodeoxycholate, seem to interfere much more with microsomal enzymes important for the metabolism of the bile acids, lipids, and drugs [13]. In higher concentrations this bile acid has a toxic effect on the liver of rhesus monkeys [8].

Another dihydroxylated bile acid, deoxycholate, has also been used to induce severe parenchymal liver damage in rats, due to its detergent effect on the membrane of subcellular organelles [14].

In spite of this, the excellent detergent properties of chenodeoxycholate have stimulated its application in the therapy of cholesterol gallstones [15]. By the administration of low, relatively nontoxic amounts it was possible to decrease sharply the lithogenic index of bile and thereby induce the dissolution of cholesterol gallstones. Recent reports favor the substitution of ursodeoxycholate for chenodeoxycholate because it seems to have almost no toxicity when given in amounts sufficient to dissolve biliary calculi in monkeys and man [8,9]. Both ursodeoxycholate and chenodeoxycholate are potent inhibitors of HMG-CoA-reductase, the key enzyme of cholesterol synthesis, while the trihydroxylated bile acids interfere only weakly with this enzyme [8]. Deoxycholate and chenodeoxycholate have another specific effect in common which other bile acids do not seem to have. They inhibit water and sodium absorption in the colon and thereby induce diarrhea [16].

Lithocholate appears to be the most liver-toxic bile acid, at least in most animal species and perhaps in man [17]. It induces severe microsomal dysfunction and, eventually, parenchymal necrosis. A potent means of detoxification of lithocholate seems to be sulfation, a mechanism which remains intact even in obstructive jaundice [18]. Sulfated lithocholate is poorly reabsorbed in the gut and the little returning to circulation is readily excreted with the urine.

As lithocholate is derived from chenodeoxycholate, considerable amounts of this toxic bile acid are expected to be formed when chenodeoxycholate is administered for gallstone dissolution. However, the capacity of the liver to sulfate lithocholate is so enhanced that only minor increases in free lithocholate are observed in bile and serum, while most of it is in the sulfated, nontoxic form [19].

1.5 Application of Bile Acid Determination

Although they share a common detergent effect in bile and gut, the various bile acids differ in many other properties. This has led to great interest in their differentiation by various chromatographic techniques, which have been increasingly applied in biochemical and clinical studies. Regarding the synthesis of the different bile acids, sophisticated techniques have been applied to study the contribution of different precursor pools such as newly synthesized cholesterol [20-22], lipoprotein cholesterol [23], and cholesterol ester [24].

The clearance of individual bile acids has also been determined by comparing levels in portal and systemic blood in order to learn how the liver is able to extract the different bile acids from the portal venous system [25].

The different bile acids have also been followed in bile, duodenal fluid, and feces to determine bile acid pool size and turnover in normals [26] and in patients with disturbed bile acid metabolism such as liver cirrhosis [27,28].

In an attempt to reduce the potential damage of bile acid reflux from the duodenum to the stomach, bile acids have been measured in the gastric juice and the binding properties of antacids for bile acids have been determined [29].

As gallstone frequency seems to increase as a result of some lipid-lowering treatments, bile acids were also measured in bile, serum, and feces in hyperlipidemia in order to clarify the mechanisms of this side effect [30-32].

As a severly disturbed bile acid metabolism is found in most patients with liver disease, much work has been devoted to the analysis of individual bile acids in these states. For example, significant alterations in individual bile acids of bile and serum have been observed in cirrhosis, where the proportion of chenodeoxycholate is increased, while that of cholate is decreased [33-36]. The serum concentration of deoxycholate is frequently reduced, sometimes to the point of disappearance [37-39].

In sophisticated studies using a combination of gas-liquid chromatography and mass spectrometry, traces of abnormal bile acids were detected in serum and urine of cholostatic patients [11].

Studies of cholostatic patients also showed that conjugation pathways are activated which are not used normally. Thus, large amounts of sulfated monohydroxy bile acids are excreted in the urine in cholostasis [40]. Also, large amounts of bile acid glucuronides are found in the urine, which are synthesized in the cholostatic liver [41].

2. SAMPLE PREPARATION AND DERIVATIZATION

The first scientists to apply the tool of gas-liquid chromatography (GLC) to bile acid analysis were VandenHeuvel and co-workers [42], who chromatographed the bile acids as their methylesters. These have the disadvantage of high boiling points, and degradation has been observed. The same author [43] later demonstrated that additional derivatization leading to trifluoroacetylesters of the bile acid methylesters decreased their boiling points and improved their volatilities which made analysis at lower temperatures possible. These difficulties in the 1960s have now been overcome by modern column technology and application of glass capillary GLC.

To overcome the problems merely of chromatographing isolated bile acids took a rather short time. Problems arose when it came to determining steroid acids in body fluids. The only exceptions were bile acids from gallbladder bile, where the content of the acids is so high that their extraction can be easily performed. Their concentrations in feces and gallstones is also high, but extraction from serum or urine presented many more difficulties. The small amounts of bile acids present in these body fluids necessitates more sophisticated techniques for their extraction. Besides that, the bile acids occur mostly as taurine and glycine conjugates, which require hydrolyzation prior to derivatization. The restricted amounts of sample material require optimal standardized conditions and high-purity solvents for the processes of extraction, hydrolysis, and derivatization. On top of this, the number of different bile acids is so high that it is impossible to separate them on one column. A table of the more common natural and derived bile acids has been given by Kuksis [44]. Consequently, efforts were concentrated primarily on the analysis of the major bile acids present in serum, urine, and bile.

2.1 Extraction Procedures

Sjövall [45] has made an attempt to summarize the extraction procedures from various natural sources presented by numerous authors. In the present chapter only very few of these can be discussed. As there is very little literature on glass capillary gas-liquid chromatography, most of these papers deal with packed columns. Nevertheless, the extraction procedures might be transferred to analysis on glass capillary columns as well.

Extraction from Serum

In the beginning most of the extraction procedures were performed by the so-called hot ethanol method, which is based upon the work of Josephson [46]. This method has been modified extensively. As one example we have chosen to review the extraction procedure of Norman [47]:

Plasma is mixed with 10 volumes of ethanol and heated to boiling. The precipitate is centrifuged off and the supernatant is diluted with water to about 70% ethanol. The neutral lipids and cholesterol are removed by subsequent extraction with petroleum ether. After removal of ethanol by evaporation, the bile acids and their conjugates are redissolved in water acidified to pH 1. They can now be extracted with butanol.

These methods were replaced in most laboratories by extraction procedures based on adsorption chromatography, which makes easier the removal of interfering substances, e.g., neutral lipids and cholesterol. One of the first papers describing the application of adsorbent columns for the extraction of bile acids was published by Danielson and co-workers in 1963 [48].

Kuksis [49] reported in 1965 the application of Dowex columns. An extraction with hot ethanol was followed by a subsequent continuous extraction with diethyl ether over 24 hr.

After application of a Dowex-50 cation-exchange column for further purification and desalting, the glycine and taurine conjugates could be recovered separately from a Dowex-1 anion-exchange column by application of a gradient of HCl in aqueous alcohol or acetone as described by Gorden and co-workers [50].

In the same year more papers appeared referring to bile acid extraction by application of adsorbent columns [51,52]. The latter group (Sandberg and co-workers [52]) used an Amberlyst XN-1006 anion-exchange resin in the chloride form, which was washed with aqueous organic solvent prior to use. The ion exchanger was first converted into the bicarbonate form and then into the hydroxide form. A serum sample of 5 to 10 ml was diluted 1:1 with distilled water. The ion exchanger in the bicarbonate form was pipetted into a chromatographic column and subsequently washed with 50 ml of 1 N NaOH, followed by 50 ml of 1 N NaOH in 80% ethanol and finally with distilled water until the effluent was neutral. The diluted serum sample was adjusted to pH 11 with 1 N NaOH and then transferred to the anion-exchange column. The column was washed with distilled water until the effluent was neutral, and subsequent washings were performed with 20 ml of 95% ethanol, 20 ml of ethanol/ethylene chloride 1:1 (v/v), and 20 ml of 80% ethanol. The bile acids and their conjugates could then be eluted with 150 ml of 0.2 M $(NH_4)_2CO_3$ in 80% ethanol as suggested by Kuron and Tennent [53]. Overall recovery of added ^{14}C-labeled deoxycholic, chenodeoxycholic, cholic, and taurocholic acid was between 82 and 96%.

In 1972 Makino and Sjövall [54] presented the extraction procedure most widely used at present. They were trying to reduce the solvent evaporation steps to a minimum. The work was based on a liquid-solid extraction procedure applying Amberlite XAD-2 resin, as had been suggested before by Bradlow [55]. The resin is washed before use with water, methanol, acetone, and water, after which it can be stored in water.

For extraction of bile acids and their conjugates, 2 ml of serum or plasma are diluted with 18 ml of 0.1 M sodium hydroxide in saline. The sample is placed in an ultrasonic bath for 15 min and then transferred to a 1-g Amberlite XAD-2 column. It is allowed to pass through the column at a rate of about 0.1 ml min^{-1}. After washing of the column with water until neutral, the bile acids are eluted with 3 ml of ethanol. After subsequent alkaline hydrolysis, the hydrolysate is acidified with 4 N HCl and diluted to 30 ml with water and the sample is passed through an XAD-2 column again. After washing to neutrality, excess water is removed from the column with 5 ml of hexane and the adsorbed free bile acids are eluted with 3 ml of a mixture of hexane/chloroform/methanol, 1:1:1 (v/v/v). This gave recoveries of 85 to 95% of added lithocholic acid. Recoveries of glycocholic, chenodeoxycholic, and cholic acid were 80, 103, and 93% respectively. The recoveries were about the same as described by Sandberg and co-workers [52].

Makino and co-workers [56] demonstrated that the sulfated bile acids were extracted together with the free and conjugated acids by application of the method of Makino and Sjövall [54]. The separation of these compounds could be achieved as follows:

After evaporation to dryness of the eluate from the Amberlite XAD-2 column the residue was dissolved in 2 ml of chloroform/methanol 1:1 (v/v) containing 0.01 M sodium chloride. The sample was allowed to pass through a column containing 4 g of Sephadex LH-20. The nonsulfated fraction (containing free acids and conjugates) was eluted with 70 ml of the solvent and the sulfated fraction with 50 ml of methanol.

In our own laboratory we use the method of Makino and Sjövall [54] with slight modifications. The XAD-2 column is washed before use not only with the solvents described in the original paper but also with 20 ml of 0.1 M sodium hydroxide in saline. As the pores of the resin differ in width according to the type of solvent passed through the column, it is necessary to perform the last cleaning step with the solvent of the sample [57]. For this same reason, it is necessary to follow the sample through the column with the saline solution, after which the column is washed with distilled water until the eluent is neutral. The bile acids and their conjugates are then eluted with two 5 ml portions of methanol. We found recovery rates of 92% after the first and an additional 4% after the second elution with methanol, as determined with radio-labeled glycocholic and taurocholic acid. The recovery rate of free cholic acid was measured at 98% after the first elution step. The eluates are pooled and evaporated to dryness. This is followed by enzymatic hydrolysis (see Sec. 2.2). After acidification of the hydrolysate with HCl the free bile acids are extracted with three 5 ml portions of diethyl ether. The first extraction gave 39.6%, the second gave an additional 40.1%, and the third an additional 6% rate of recovery. Further extraction steps do not yield significantly higher recovery rates. The combined ethereal extracts must be dried with 5 g of Na_2SO_4 to remove water and water-soluble compounds. Without this step we were not able to achieve a chromatogram on a glass capillary column with correct peak area measurement by the computing integrator. (For a summary, see the flowsheet in Fig. 1.)

Extraction from Bile

In gallbladder bile the amount of bile acid present is so high that there are no requirements for a high-grade purification of the sample.

Klaasen [58] has described a simple method based on the previous work of Sandberg and co-workers [52] and Okishio and co-workers [59]. To the diluted bile 1 ml of 0.2 M PO_4 buffer (pH 5.8) is added. After enzymatic hydrolysis the internal standard is added, the solution acidified with 1 ml of 6 N HCl, and the free bile acids are extracted with two 5 ml portions of diethyl ether. In general, the method of Makino and Sjovall [54] can be applied to gallbladder bile without further modifications.

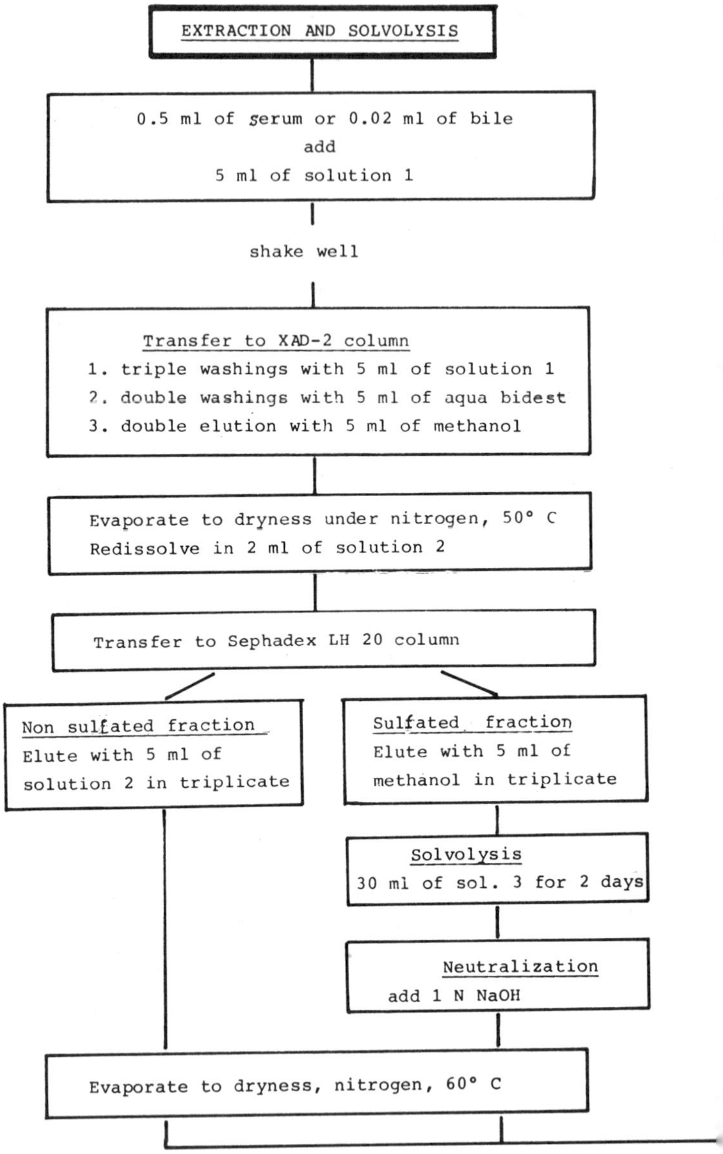

Figure 1. Flowsheet diagram of bile acid extraction and derivatization. Solutions: 1—0.1 M sodium hydroxide in saline; 2—chloroform/methanol 1:1 (v/v), containing 0.01 M sodium chloride; 3—ethanol/acetone 1:9 (v/v), acidified with 2 N HCl to pH 1; 4—5 ml of aqua bidest/0.25 ml of EDTA buffer/0.5 ml of acetate buffer pH 5.6/50 μl cholyl-glycyl hydrolase/200 μl trifluoroacetic anhydride.

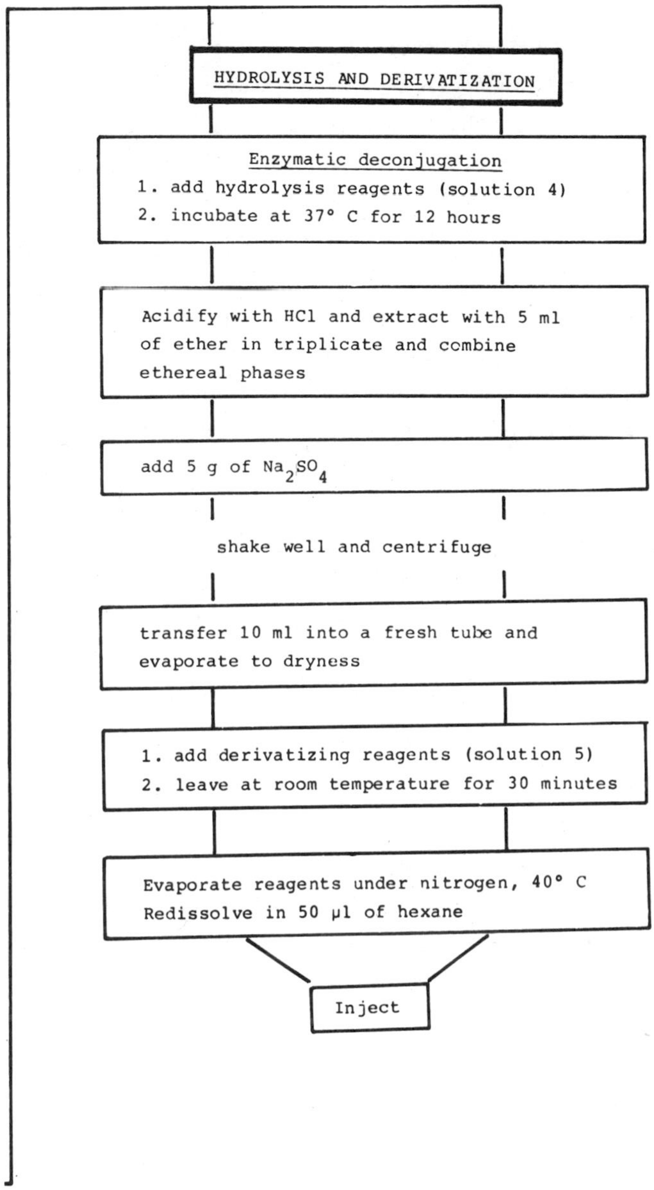

Figure 1 (cont'd).

Extraction from Feces or Duodenal Contents

The bile acids are excreted in feces as a mixture of primary and secondary bile acids [60]. Numerous authors have reported the occurence of individual bile acids [61-63]. Eneroth and co-workers [64] made an attempt at a more complete characterization of bile acids present in feces. The feces sample was homogenized in 1.5 liters of chloroform/methanol 1:1 (v/v), and the homogenate extracted for 48 hr in a Soxhlet apparatus with hot chloroform/methanol 1:1 (v/v). After hydrolysis the acids were extracted with diethyl ether for 16 hr. The extract was washed several times and the washings were reextracted with diethyl ether. After methylation the bile acids were preseparated by thin-layer chromatography according to Eneroth [65]. Several studies documented the composition of human fecal bile acids [66,67]. Evrard and Janssen [68] presented a method where the freeze-dried feces was heated together with glacial acetic acid containing the internal standard (1 hr at 120°C). An acetic acid/toluene extract was prepared and evaporated to dryness. After alkaline hydrolysis the unsaponifiable matter was extracted with petroleum ether. The bile acids were extracted from the hydrolysate after acidification with diethyl ether and subsequently methylated. The recovery rate was found to be about 80%.

Extraction from Urine

For extraction from urine, the modification of the original method by Makino and Sjövall [54,56] is usually employed. Thirty ml of urine are percolated directly through a 10-g Amberlite XAD-2 column and eluted with 60 ml of methanol. The separation of the sulfated from the nonsulfated fraction was performed as described in the earlier section "Extraction from Serum."

Extraction from Tissue Cells

For extraction from tissue cells, Greim and co-workers [69] presented a method where cylinders of liver tissue were solubilized after addition of 50 μl of 3.4 N NaOH in a boiling water bath. After neutralization and enzymatic hydrolysis the deconjugated bile acids were extracted with diethyl ether. After evaporation to dryness the residue was reextracted and the acids were derivatized.

Okishio and co-workers [70] freeze-dried a tissue homogenate and extracted it twice with 100 ml of 95% ethanol containing 0.1 ml of aqueous NH_3. The residue was washed with an additional 50 ml of solvent. After pooling and concentration the extracts were chromatographed on Amberlite A-26.

The eluates were hydrolyzed enzymatically and extracted with diethyl ether after acidification.

Extraction from Gallstones

Nakayama [71] powdered gallstones and subsequently extracted the acids with acidified chloroform/methanol 2:1 (v/v). Another method was presented by Schoenfield and co-workers [72].

2.2 Hydrolysis and Solvolysis

Bile acids occur mainly in biological sources as the taurine or glycine conjugates. The two types of conjugates can be separated from each other by ion-exchange chromatography [50,73] or differential extraction [74], where the glycine conjugates can be recovered by extraction with ethyl acetate and the taurine conjugates with butanol.

Both types of conjugates are usually hydrolyzed together by alkaline hydrolysis. Destruction of the acids by heating with hot alkali has been reported [52,75,76]. This was found to be due to their adsorption on silicates released from the glass during hydrolysis. This could be overcome by substitution of nickel or stainless steel bombs [44]. Best results were obtained with 1 M NaOH at 115°C for 6 hr or by autoclaving in 1.25 M NaOH for 3 hr at 15 lb pressure. Even then some ketocholanic acids are destroyed [44].

Therefore, an enzymatic method is preferred. The first to report such a method were Nair and co-workers [77]. But as was demonstrated in their work, the conjugates were not hydrolyzed at the same rate or to the same extent when working with a cell-free extract of *Clostridium perfringens*. Kelly and Doisy [78] found that an extract of *Bacillus sphericus* preferentially hydrolyzed taurocholate. Further attempts were made to obtain purer enzyme preparations [79-81], because the enzymatic hydrolysis yielded improved recovery rates and avoided the alkali induced destruction of the acids [81]. The enzymatic hydrolysis with cholyl-glycyl hydrolase has therefore become the most widely used method for the hydrolysis of bile acids, although other authors have described enzymatic methods where the enzyme is obtained from other bacterial sources [82].

The method of Nair and co-workers [77] as used in our laboratory is performed as follows:

After elution from the XAD-2 column with methanol the solvent is evaporated under a stream of nitrogen in a water bath at 45°C. The bile acids are dissolved in a mixture of 2 ml of 0.1 M acetate buffer (pH 5.6), 25 μl of the cholyl-glycyl hydrolase preparation (specific activity not less than 100 μmol mg^{-1} protein), 0.5 ml of 0.2 M EDTA, and 0.2 ml of 0.2 M β-mercaptoethanol. In contrast to the original method, we incubate the mixture at 37°C in a water bath overnight. After subsequent acidification with 6 N HCl to pH 1 the free bile acids can be extracted into diethyl ether as described in the section "Extraction from Serum."

For separate determination of the sulfated and nonsulfated fractions the two fractions are separated on Sephadex LH-20, as described under "Extraction from Serum," according to the method of Makino and co-workers [56].

The sulfated fraction is solvolyzed in 30 ml of ethanol/acetone 1:9 (v/v), acidified with a few drops of 2 N HCl to pH 1 and the mixture is left standing for 2 days at room temperature. After neutralization by addition of 1 N NaOH the solvent was evaporated as originally suggested by Palmer [83]. The solvolysis is followed by an alkaline hydrolysis and subsequent extraction of the free bile acids into diethyl ether.

2.3 Derivatization

Bile acids as polyfunctional steroids cannot be chromatographed in the original form. Early gas-liquid chromatography of bile acids used their methyl ester derivatives [42]. Because of their higher polarity, the methylesters tend to adsorb on the column [84]; hence a double derivatization, methylation followed by trimethylsilylation or trifluoroacetylation, is usually employed.

Ester Derivatization

Alkylesters. Most frequently prepared are the methylesters. Usually diazomethane is employed as the methylating agent according to the method of Schlenk and Gellerman [85], which was originally developed for methylation of fatty acids:

The acids are dissolved in methanol and an ice-cold solution of freshly distilled ethereal diazomethane is added until a yellow color persists, which indicates that the reaction is complete. The diazomethane can be generated from N-methyl-N-nitroso-p-toluenesulfonamide [86]. This method has some great disadvantages; diazomethane is highly explosive and has been reported to be carcinogenic. The hazards associated with diazomethane are discussed by Gutsche [87].

Alternatively, methylation can be achieved by treatment with 5% HCl in methanol [51], but this results in dehydration of the steroid nucleus and eventual formation of colored complexes [88].

Esters of other alcohols have also been prepared. Hofman and Mosbach [89] derivatized bile acids to their propylesters by esterification with n-propanol/sulfuric acid 99:1 (v/v).

Because of the hazards involved in the derivatization with diazomethane and the disadvantages of other methods, we have chosen in our laboratory the method of Imai and co-workers [90], which gave excellent results.

Hexafluoroisopropanol together with trifluoroacetic anhydride were previously employed for the derivatization of acidic metabolites of catecholamines [91]. The derivatization is performed as follows:

To the bile acids 100 μl of hexafluoroisopropanol and 200 μl of trifluoroacetic anhydride are added and the mixture is kept for 30 min in a water bath at 37°C. After evaporation of the solvent and resolution in acetonitrile, the mixture is injected into the gas chromatograph. The authors found by GC/MS that the carboxyl groups of the bile acids were selectively esterfied and the hydroxyl groups acetylated (see also the flowsheet, Fig. 1). We have tested the stability of the derivatives and found no alteration in the chromatogram after 5 days refrigerated storage. The sample, containing 11 bile acids and cholesterol, was chromatogramed 10 times, and the mean of coefficent variation was found to be 5 to 7% of the individual bile acids. During the 5 days of the test, no alteration could be seen in the corrected relative peak areas. Hence we feel quite astonished that the diazomethane methylation with all its hazards and disadvantages is still used so frequently.

Acylesters. VandenHeuvel and co-workers [43] have demonstrated the use of trifluoroacetic anhydride in the derivatization of the hydroxy groups of bile acids. Sjovall [92] prepared the acylesters as follows:

The bile acids are dissolved in 0.2 to 0.5 ml of trifluoroacetic anhydride and heated to 30 to 35°C for 15 min. Excess reagent is evaporated and the residue is dissolved in acetone and injected into the gas chromatograph.

A definitively complete trifluoroacetylation of all hydroxyl groups can be obtained when the sample is derivatized in the presence of dry pyridine [93].

Although there have been reports that the trifluoroacetates may undergo thermal decomposition [94], one must emphasize that these derivatives have yielded the most complete resolutions of natural bile acids by gas-liquid chromatography and have proven to be the derivatives of choice for combined GC/MS analysis [94]. Excellent agreement between the relative retention times in different laboratories [95] and accurate comparisons with reference compounds has made this derivatization the method of choice.

Donike [96] has described the acylation of steroids with N-methyl-bis-(trifluoroacetamide) (MBTFA) and bis-(trifluoroacetamide) (BTFA). We have applied this method successfully to the derivatization of bile acid methyl- or propylesters. As no difference in derivatization compared with trifluoroacetic acid could be found, the older method was maintained.

Ethers. Grundy and co-workers [51] derivatized bile acid methylesters to their corresponding trimethylsilyl ethers by reaction with a solution of dry pyridine/hexamethyldisilazane/trimethylchlorosilane 9:3:1 (v/v/v). The reaction was allowed to proceed for 1 hr or longer.

Supina and co-workers [97] found variable amounts of secondary products in the presence of keto acids. This could be avoided by substitution of dimethylformamide for pyridine.

By comparison of the trimethylsilyl, trifluoroacetyl, and heptafluorobutyrate derivatives [98] it could be shown that under some conditions the trifluoroacetyl derivatives yielded slightly improved chromatographic behavior. Most steroids exhibit their highest response factor with trimethylsilyl derivatives when flame ionization detectors are used, but the trifluoroacetyl and heptafluorobutyrate derivatives make application of the more sensitive electron capture detector possible.

3. COLUMNS AND DETECTORS

On packed columns best results could be obtained with QF-1 [98] because this phase gives better separations of a hydroxy and the corresponding keto compound. As only a few papers have been published on glass capillary gas-liquid chromatography of bile acids, the columns and the respective liquid phases thereon are discussed in the next section. With reference to detectors, one must recall the well-known fact that halogenated derivatives of bile acids permit the use of the more sensitive electron capture detector, but in most cases flame ionization detection is used.

4. APPLICATIONS

To our knowledge and at the time of this writing, only three groups have chromatographed bile acids on glass capillary columns. The first scientists to apply this modern tool of gas-liquid chromatography were Laatikainen and Hesso [99].

For extraction they used the original method of Makino and Sjövall [54], with subsequent enzymatic hydrolysis according to Nair and Garcia [81], or alternatively alkaline hydrolysis. From the hydrolysate, free bile acids were extracted with ethylacetate. Before methylation with diazomethane, 7-ketodeoxycholic acid was added as internal standard. The methylesters were subsequently purified on aluminum oxide and converted to their corresponding trimethylsilyl ether derivatives prior to GLC. Fifty-m glass capillaries with an inner diameter of 0.3 mm were coated with OV-101 as described by Grob [100,101]. Carrier gas was nitrogen with a flow of 2 ml min^{-1}. Split ratio was 1:10. Nitrogen was used as well as makeup gas with a flow of 40 ml min^{-1}. The column was programmed from 150 to 255°C with 8°C min^{-1}. Temperatures of the injection port and the flame ionization detector are not given. Analysis time was 40 min. As stated, correction factors were taken into account and measurement of the individual bile acids utilized the internal standard method.

Figure 2 shows chromatograms of a standard bile acid mixture (A) and bile acids extracted from serum (B) and from gallbladder bile (C). Methylester trimethylsilyl ethers of cholic, chenodeoxycholic, deoxycholic, and lithocholic acid were completely separated. Overall recovery rates of glycocholic and taurocholic acid with enzymatic hydrolysis were 68 to 80% (mean 81%) and 87 to 100% (mean 97%), respectively.

No changes were found in the properties of the column after 300 bile acid analyses. The bile acid levels were in close agreement with the data given by Sandberg and co-workers [52].

A disadvantage of the method is that the internal standard is not added prior to extraction, and thus the loss during extraction is not taken into account. Furthermore a separation between lithocholic acid and cholesterol could not be achieved. Even time-consuming purification steps presented here cannot totally remove cholesterol from the samples. Thus an accurate measurement of the important lithocholic acid is not possible. Moreover, the use of split injection in quantitative glass capillary gas-liquid chromatography does not seem advisable.

Karlaganis and Paumgartner [102,103] presented a method in which many of these disadvantages were overcome:

After addition of hyodeoxycholic acid as internal standard 2 ml of

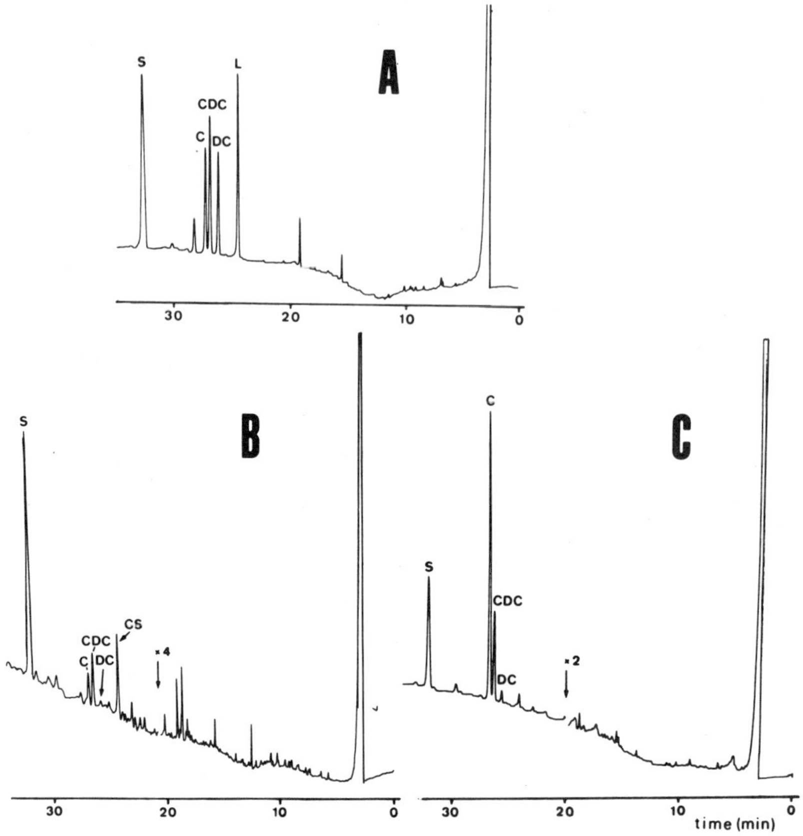

Figure 2. Glass capillary gas chromatograms of trimethylsilyl ethers of bile acid methylesters. A 50-m OV-101 column programmed with 8°C min^{-1} from 150 to 255°C. A: Reference cholic (C), chenodeoxycholic (CDC), deoxycholic (DC), and lithocholic (LC) acid, and the internal standard (S). B: A serum sample from a healthy female. C: A serum sample from a patient with cholostasis of pregnancy; CS—cholesterol. (Source: Ref. 99. Reproduced by permission of Elsevier/North Holland Biomedical Press, Amsterdam.)

serum of 20 µl of bile were extracted according to Makino and Sjövall [54] as modified by Schwarz et al. [104], including solvolysis of the sulfated fraction. After alkaline hydrolysis and extraction with ethylacetate the free bile acids were methylated with diazomethane [85] and subsequently converted to their corresponding trimethylsilyl ether derivatives according to Makita and Wells [105]. The analysis was performed on a Carlo Erba gas chromatograph with an injection port system as described by Grob and Grob [106] and equipped with a flame ionization detector. The glass capillary column had an inner diameter of 0.3 mm and a length of 20 m. The internal surface was covered with a crystal layer of barium carbonate at room temperature according to Grob and co-workers [107]. The layer was coated with a 0.2% solution of polyethylene glycol 20M in methylene chloride by the method of static coating [108].

The injection port temperature was held at 300°C and the column temperature was 230°C. Hydrogen was used as carrier gas with a flow of 4.4 ml min^{-1}. The split ratio was 1:5. The column provides an outstanding separation of the major bile acids and cholesterol. As can be seen from Fig. 3, baseline separation of cholesterol and cholic, deoxycholic, chenodeoxycholic, hydodeoxycholic, lithocholic, ursodeoxycholic, and 3β-hydroxy-5-cholenoic acids was achieved. Time of analysis was about 15 min and therefore extraordinarily short for a separation of this quality, although of course shorter separations have been reported on packed column [109]. Standard curves are absolutely linear. The variation of the method was rather low. Figures 4 and 5 show the applicability of the method to the analysis of bile acids from serum and gallbladder bile, respectively.

Even this outstanding method has some disadvantages. As Fig. 4 shows, almost all peaks are superimposed on the broad tail of the solvent peak, which makes correct automatic integration nearly impossible. This broad solvent peak may be due in part to the silylation and in part to insufficient purification of the sample. Furthermore, as discussed in the section "Alkylesters," we regard the use of diazomethane as a methylating agent as hazardous. Moreover, the lifetime of the column is limited to 5 months, which is too short for routine analyses.

The same authors [110] have recently described the same method with application of the new direct injection systme described by Grob and Grob [111]. They found improvements in reproducibility and sensitivity resulted from use of this on-column injector.

Our own laboratory's first results in glass capillary gas-liquid chromatography of bile acids were published in 1977 [112]. At that time we used the extraction procedure of Makino and Sjövall [54] and the methylation was performed with diazomethane. As Figure 6 shows, we were able to separate cholanic, cholic, lithocholic, chenodeoxycholic, hyodeoxycholic, 3,12-diketocholic, 7,12-diketolithocholic, and dihydrocholic acid. Chromatography was performed on a 20-m glass capillary column with an inner diameter of 0.32 mm coated with SE-54. Carrier gas was hydrogen with a flow of 3.2 ml min^{-1}. Temperature of the injection port was maintained at 275°C. The temperature program ranged from 180 to 250°C with 1.2°C min^{-1}. Separation was unsatisfactory, and as the derivatization with diazomethane was

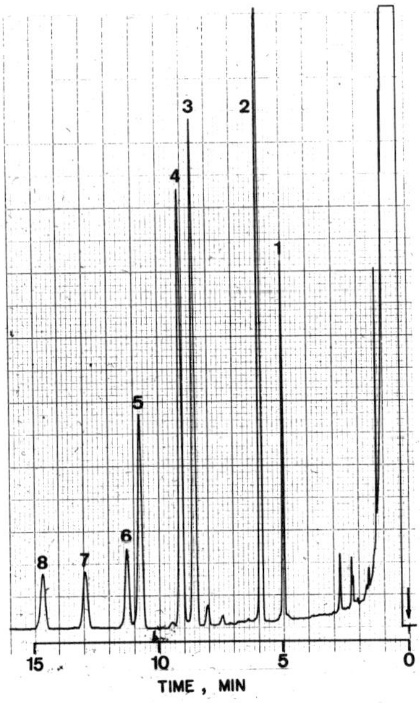

Figure 3. Capillary gas chromatogram of TMS ethers of methylesters of a standard bile acid mixture containing cholesterol (1), cholic (2), deoxycholic (3), chenodeoxycholic (4), hyodeoxycholic (internal standard, 250 pmol per injection) (5), lithocholic (6), ursodeoxycholic (7), and 3-β-hydroxy-5-cholenoic acid (8). A 20-m PG 20 M glass capillary column was used at 230°C. (Source: Ref. 102. Reproduced by permission of *Journal of Lipid Research*.)

regarded as hazardous, we developed our present method. The extraction and derivatization procedure can be seen from the flowsheet diagram in Fig. 1.

We use now a 50-m glass capillary column with an inner diameter of 0.30 mm coated with OV-1, which is commercially available (H. and G. Jaeggi, Trogen, Switzerland).

For evaluation of column efficiency the separation number TZ was measured and determined to be 57.1. (For equations, see Chap. 11 on fatty acids.) This shows an optimal TZ m^{-1}, as good capillaries have a TZ m^{-1} greater than 1. In this case it was TZ $m^{-1} = 1.14$. From the separation number the number of theoretical plates could be calculated according to the equation

$$n = 100(TZ + 1)^2$$

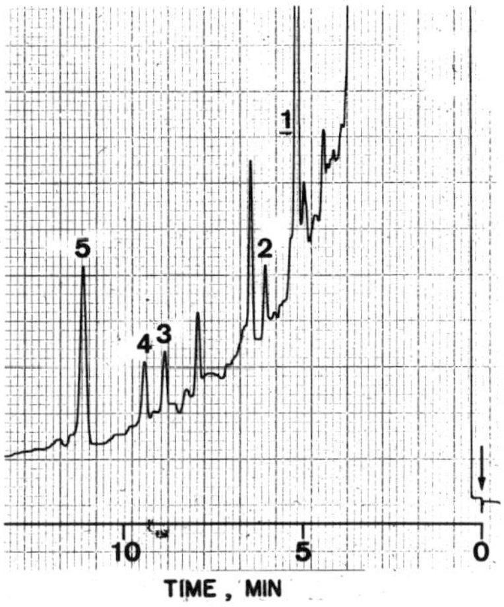

Figure 4. Capillary gas chromatogram of TMS ethers of methylesters of bile acids extracted with XAD-2 from 2 ml of normal human serum containing 6.36 nmol hyodeoxycholic acid as internal standard. Peak numbers are the same as in Fig. 3. A 20-m PG 20 M glass capillary column was used at 230°C. (Source: Ref. 102. Reproduced by permission of *Journal of Lipid Research*.)

and was thus found to be 337,561 or expressed as n/m 6751. We usually choose to calculate the separation number only, as the correct calculation of the number of theoretical plates is limited to specified running conditions.

The capacity ratio k was 18.9. The capacity C of the column for cholanic acid was found to be 9.7×10^{-6} g, following the equation

$$C = 0.05Md^3(1 + k)$$

where M = molecular weight of cholanic acid, being 360.56

d = inner diameter

Carrier gas was nitrogen, as we intended to use an electron capture detector, but as we have found that concentration of bile acids in serum is sufficiently high a flame ionization detector for the use of (FID), we chose to use hydrogen as carrier gas, thus shortening the time of analysis considerably. Carrier gas flow was 4.2 ml min^{-1}, and flows of hydrogen and air in the FID were 14 ml min^{-1} and 160 ml min^{-1}, respectively. The flows were adjusted by optimizing the FID.

The injection port temperature was held at 275°C. The temperature program was as follows: injection at 70°C, initial temperature 180°C, programmed at 1°C min^{-1} to 250°C. Total analysis time was 70 min and could be reduced considerably by using hydrogen as carrier gas. An automatic injection system as described in Sec. 3 could be used.

As can be seen from Fig. 7 deoxycholic and cholic acid were separated

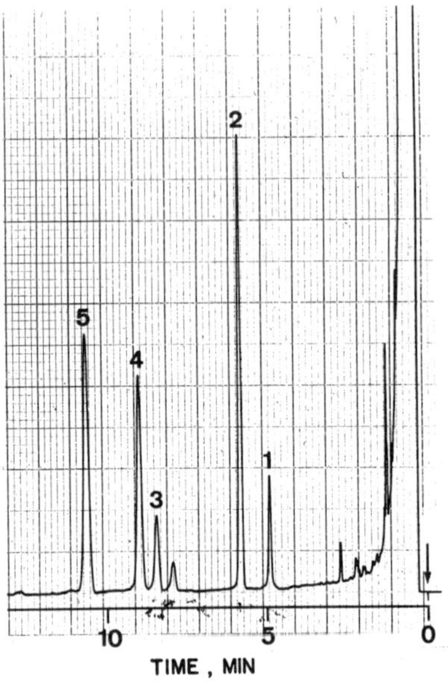

Figure 5. Capillary gas chromatogram of TMS ethers of methylesters of bile acids extracted from 20 μl of human bile containing 127 nmol of hyodeoxycholic acid as internal standard. Peak numbers are the same as in Fig. 3. a 20-m PG 20 M glass capillary column was used at 230°C. (Source: Ref. 102. Reproduced by permission of *Journal of Lipid Research*.)

by 4.60σ separation, as well as lithocholic, chenodeoxycholic, and Δ5-cholenic acid. All other peaks were separated by baseline separation.

Figure 8 shows a chromatogram of bile acids as their hexafluoroisopropylester trifluoroacetates extracted from serum and Fig. 9 a chromatogram of bile acids from human gallbladder bile. The method has proven to be applicable to routine analysis in clinical problems [113].

At present, only two comparable methods are available. The main disadvantage of our own method is the relatively long analysis time of 70 min. Karlaganis and Paumgartner [102,103] could reduce their time to 15 min, but as our samples are quite stable, an automatic solid sampler can be used. Furthermore, the lifetime of the column is increased by use of the solid sampler system because of the absence of a solvent peak. Our present column has now been in use for 1 year, running over 23 hr day^{-1} and shows no significant reduction of its properties.

Thus the decision as to which of these methods to use depends on the possibilities and inclinations of the laboratories in which they may be used. Both of the methods are superior in column efficiency, accuracy, and precision to packed columns and to thin-layer or high-performance liquid

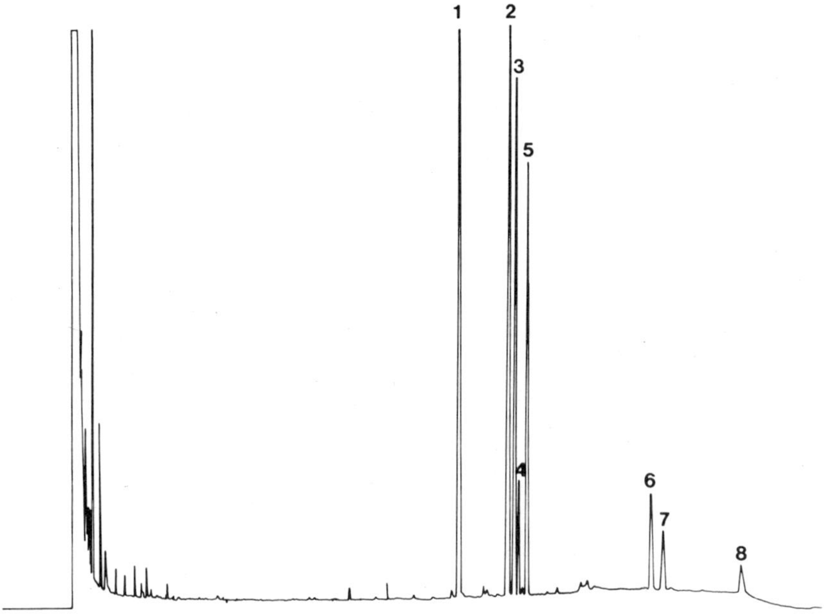

Figure 6. Chromatogram of a standard mixture of bile acid methylester trifluoroacetates. Column: 20 m × 0.32 mm ID, coated with SE-54. Injection port: 275°C. Program: 180 to 250°C with 1.2°C min^{-1}. Peak numbers: 1—cholanic acid, 2—cholic acid, 3—lithocholic acid, 4—chenodeoxycholic acid, 5—hyodeoxycholic acid, 6—3,12-diketocholic acid, 7—7,12-diketolithocholic acid, 8—dihydrocholic acid.

chromatography, tool should open up new possibilities in the investigation of bile acid metabolism.

5. SUMMARY

The rising interest in the physiology and pathophysiology of bile acids requires more and more sophisticated techniques. Thus the development of methods for their analysis by glass capillary gas-liquid chromatography was a logical consequence.

The most widespread methods for extraction of the conjugated and free bile acids are performed on resins, such as XAD-2. After hydrolysis of the conjugated bile acids and solvolysis of the sulfated fraction, the steroid acids are subjected to derivatization. Usually a double derivatization procedure is performed, applying first an esterification (still mostly with diazomethane, despite its hazards) followed by acylation with trifluoro-

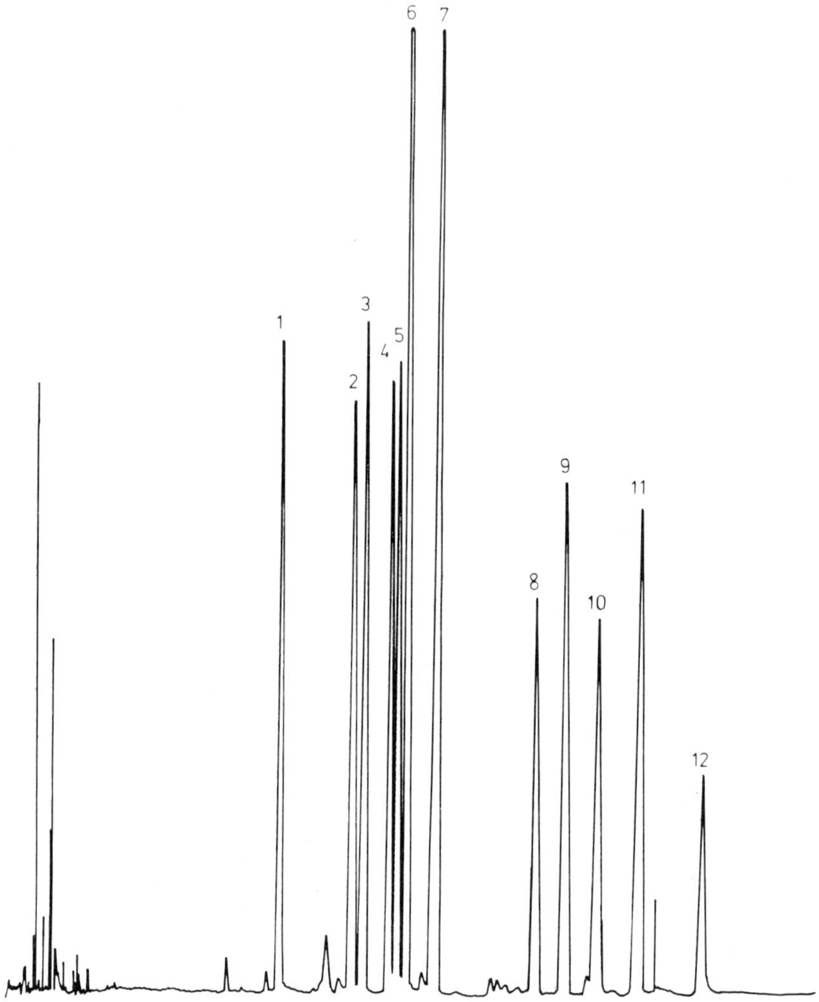

Figure 7. Chromatogram of a standard mixture of bile acid hexafluoroisopropylester trifluoroacetates. Column: 50 m × 0.33 ID, coated with OV-1. Injection port: 275°C. Program: 180 to 250°C. with 1.5°C min^{-1}. Peak numbers: 1—cholanic acid, 2—deoxycholic acid, 3—cholic acid, 4—lithocholic acid, 5—chenodeoxycholic acid, 6—Δ5-cholanic acid, 7—ursodeoxycholic acid, 8—cholesterol, 9—7-ketodeoxycholic acid (internal standard), 10—3,12-diketocholanic acid, 11—7,12-diketolithocholic acid, 12—dihydrocholic acid.

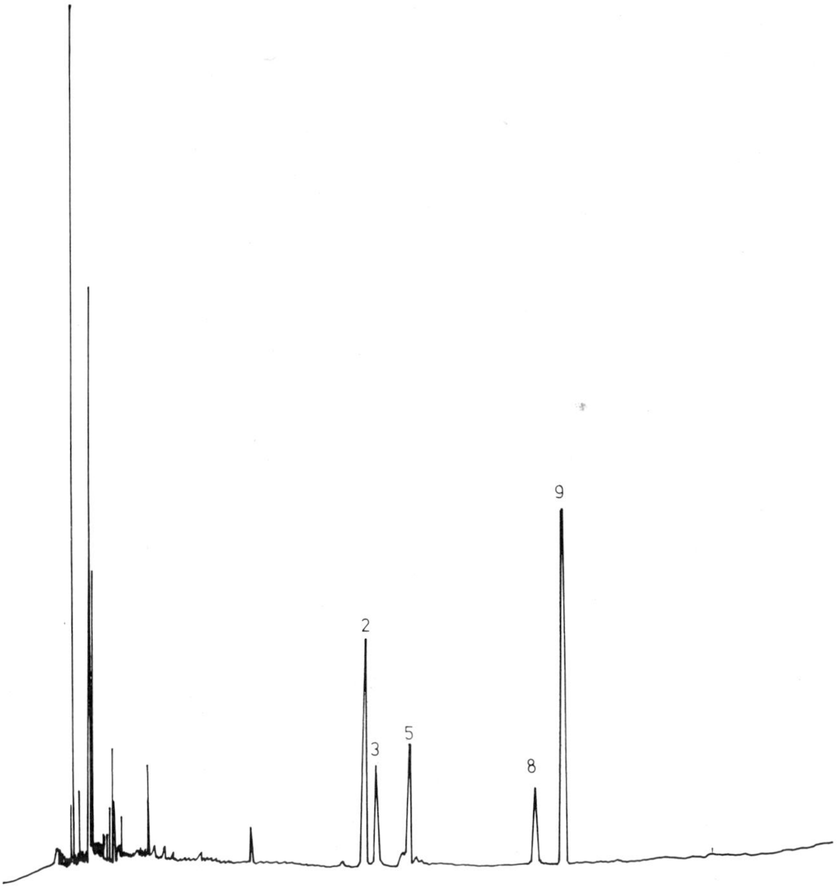

Figure 8. Chromatogram of bile acid hexafluoroisopropylester trifluoroacetates extracted from human serum. GLC conditions and peak numbers as in Fig. 7.

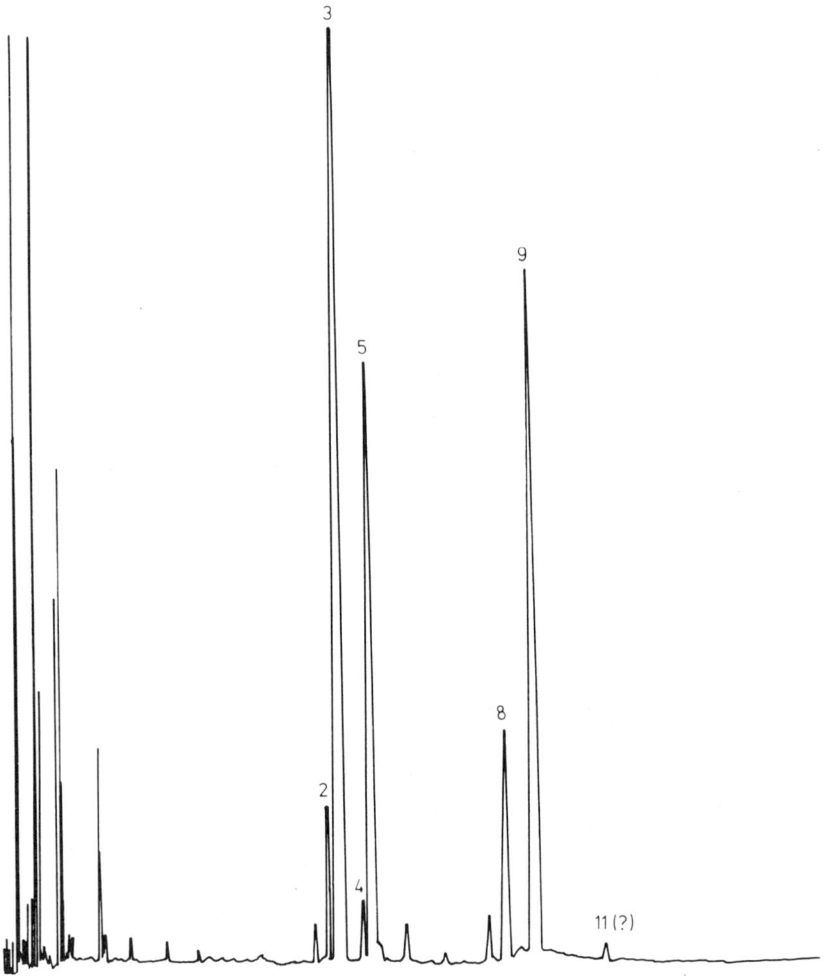

Figure 9. Chromatogram of bile acid hexafluoroisopropylester trifluoroacetates extracted from human bile. GLC conditions and peak numbers as in Fig. 7.

acetic acid anhydride or etherification with silylating reagents. No special columns or detectors are needed. For trimethylsilyl ether derivatives of bile acid methylesters, Carbowax 20M was applied with great success. For the hexafluoroisopropylester trifluoroacetates an OV-1 column can be used. Both methods have minor disadvantages but serve their purpose well. Glass capillary gas-liquid chromatography of bile acids can be regarded as the most sophisticated technique in the analysis of these substances. Further attempts must be made to increase the separation efficiencies as many new bile acids and metabolites are described.

REFERENCES

1. Josephson, B., *Physiol. Rev.*, 21:463 (1941).
2. Carey, M. C., and Small, D. M., *Arch. Intern. Med.*, 130:506 (1972).
3. Hofmann, A. F., and Small, D. M., *Amer. Rev. Med.*, 18:433 (1967).
4. Nair, P. P., and Kritchevsky, D., *The Bile Acids: Chemistry, Physiology, and Metabolism*, Plenum Press, New York, 1971.
5. Sjövall, J., *Clin. Chim. Acta*, 5:33 (1960).
6. Greim, H., Trülzsch, D., Roboz, J., Dressler, K., Czygan, P., Hutterer, F., Schaffner, F., and Popper, H., *Gastroenterology*, 63:837 (1972).
7. Stiehl, A., *Europ. J. Clin. Invest.*, 4:59 (1974).
8. Fedorowski, T., Salen, G., Zaki, F. G., Shefar, S., and Mosbach, E. H., *Gastroenterology*, 74:75 (1978).
9. Makino, J., and Nakagawa, S., *J. Lipid Res.*, 19:723 (1978).
10. Hofmann, A. F., in *Advances in Internal Medicine* (G. H. Stollermann, ed.), Yearbook Publishers, Chicago, 1976, pp. 501-534.
11. Summerfield, J. A., Billing, B. H., and Shackleton, C. H. L., *Biochem. J.*, 154:507 (1976).
12. Makino, I., Sjövall, J., Norman, A., and Strandwick, B., *FEBS Lett.*, 15:161 (1971).
13. Hutterer, F., Bacchin, P. G., and Denk, H., *Life Sci.*, 9:1159 (1970).
14. Schriewer, H., von Bassewitz, D. B., and Raven, H. M., *Klin Wochenschr.*, 51:39 (1973).
15. Hofmann, A. F., and Paumgartner, G., eds., *Chenodeoxycholic Acid Therapy of Gallstones: Progress Report*, Schattauer, Stuttgart and New York, 1975.
16. Mekligian, H. S., Phillips, S. F., and Hofmann, A. F., *J. Clin. Invest.*, 50:1569 (1971).
17. Palmer, R. H., *Arch. Intern. Med.*, 130:606 (1972).
18. Stiehl, A., Earnest, D. L., and Admirand, W. H., *Gastroenterology*, 68:534 (1975).
19. Stiehl, A., Raedsoh, R., and Kommerell, B., *Digestion*, 12:105 (1975).
20. Kim, D. N., Lee, K. T., Reiner, J. M., and Thomas, W. A., *Exp. Mol. Path.*, 22:284 (1975).
21. Ogura, M., Ayaki, Y., and Goto, M., *J. Biochem.*, 80:537 (1976).
22. Mitropoulos, K. A., Myout, N. B., Gibbons, G. F., and Balasubramainam, B. E. H., *J. Biol. Chem.*, 249:6052 (1974).
23. Schwartz, C. C., Vlahcevic, Z. R., Halloran, L. G., Nisman, R., and Swell, L., *Proc. Soc. Exp. Biol. Med.*, 156:261 (1977).
24. Mathé, D., d'Hollander, F., and Chevallier, F., *Biochimie*, 54:1479 (1972).
25. Ahlberg, J., Angelin, B., Björkhem, J., and Einarsson, K., *Gastroenterology*, 73:1377 (1977).
26. Mok, H. J., von Bergmann, K., and Grundy, S. M., *Gastroenterology*, 73:684 (1977).

27. Knodell, R. G., Kinsey, M. D., Boedeker, E. C., and Collin, D. P., *Gastroenterology*, 71:196 (1976).

28. Mehta, S. J., Struthers, J. E., Kaye, M. D., and Naylor, J. L., *Gastroenterology*, 67:674 (1974).

29. Caspary, W. F., and Graf, S., *Dtsch. Med. Wochenschr.*, 103:825 (1978).

30. Grundy, S. M., *J. Clin. Path.*, 55:269 (1975).

31. Pennington, C. R., Ross, D. E., Bateson, M. C., and Bouchier, J. A. D., *J. Clin. Path.*, 31:58 (1978).

32. Einarsson, K., and Hellström, H., *Eur. J. Clin. Invest.*, 2:225 (1972).

33. Osborn, E. C., Wootton, I. D. P., da Silva, L. C., and Sherlock, S., *Lancet*, II:1049 (1959).

34. Carey, J. B., *J. Clin. Invest.*, 37:1494 (1958).

35. Vlahcevic, Z. R., Juttijudata, D., Bell, C. C., and Swell, L., *Gastroenterology*, 62:1174 (1972).

36. McCormick, W. C., Bell, C. C., Swell, L,, and Vlahcevic, Z. R., *Gut*, 14:895

37. Turnberg, L. A., and Grahame, G., *Gut*, 11:126 (1970).

38. Vlahcevic, Z. R., Buhac, J., Bell, C. C., and Swell, C., *Gut*, 11:420 (1970).

39. Struthers, J. E., Mehta, S. J., Kaye, M. D., and Naylor, J. L., *Amer. J. Dig. Dis.*, 22:861 (1977).

40. Back, P., *Z. Gastroenterologie*, 11:477 (1973).

41. Fröhling, W., and Stiehl, A., *Europ. J. Clin. Invest.*, 6:67 (1975).

42. VandenHeuvel, W. J. A., Sweeley, C. C., and Horning, E. C., *Biochem. Biophys. Res. Commun.*, 3:33 (1960).

43. VandenHeuvel, W. J. A., Sjövall, J., and Horning, E. C., *Biochim. Biophys. Acta*, 48:596 (1961).

44. Kuksis, A., in *Lipid Chromatographic Analysis*, Vol. 2 (G. V. Marinetti, ed.), Marcel Dekker, New York, 1969, p. 230.

45. Sjövall, J., in *Methods of Biochemical Analysis* (D. Glick, ed.), Wiley-Interscience, New York, 1964, p. 97.

46. Josephson, B., *Biochem. J.*, 29:1519 (1935).

47. Norman, A., *Acta Physiol. Scand.*, 32:1 (1954).

48. Danielson, H., Eneroth, P., Hellstrom, U., Lindstedt, S., and Sjövall, J., *J. Biol. Chem.*, 238:2299 (1963).

49. Kuksis, A., *J. Amer. Oil Chem. Soc.*, 42:276 (1965).

50. Gordon, B. A., Kuksis, A., and Beveridge, J. M. R., *Can. J. Biochem. Physiol.*, 41:77 (1963).

51. Grundy, S. M., Ahrens, Jr., E. H., and Miettinen, T. A., *J. Lipid Res.*, 6:397 (1965).

52. Sandberg, D., Sjövall, K., Sjövall, J., and Turner, D. A., *J. Lipid Res.*, 6:182 (1965).

53. Kuron, G. W., and Tennent, D. M., *Fed. Proc.*, 20:268 (1961).

54. Makino, I., and Sjövall, J., *Anal. Lett.*, 5(6):341 (1972).

55. Bradlow, H., *Steroids*, 11:265 (1968).

56. Makino, I., Shinozaki, K., Nakogawa, S., and Mashimo, K., *J. Lipid Res.*, 15:132 (1974).

57. Blome, J., personal communication.

58. Klaassen, C. D., *Clin. Chim. Acta*, 35:225 (1971).

59. Okishio, T., Nair, P. P., and Gordon, M., *Biochem. J.*, 102:654 (1967).

60. Danielson, H., *Advan. Lipid Res.*, 1:335 (1963).

61. Jenke, M., and Bandow, F., *Z. Physiol. Chem.*, 249:16 (1937).

62. Carey, J. B., and Watson, C. J., *J. Biol. Chem.*, 216:847 (1963).

63. Norman, A., and Palmer, R. H., *J. Lab. Clin. Med.*, 63:986 (1964).

64. Eneroth, P., Gordon, B., Ryhage, R., and Sjövall, J., *J. Lipid Res.*, 7:511 (1966).

65. Eneroth, P., *J. Lipid Res.*, 4:11 (1963).

66. Eneroth, P., Gordon, B., and Sjövall, J., *J. Lipid Res.*, 7:524 (1966).

67. Eneroth, P., Hellström, K., and Sjovall, J., *Acta Chem. Scand.*, 22:1729 (1968).

68. Evrard, E., and Janssen, G., *J. Lipid Res.*, 9:226 (1968).

69. Greim, H., Czygan, P., Schaffner, F., and Popper, H., *Biochem. Med.*, 8:280 (1973).

70. Okishio, T., Nair, P. P., and Gordon, M., *Biochem. J.*, 102:654 (1967).

71. Nakayama, F., *J. Lab. Clin. Med.*, 72:602 (1968).

72. Schoenfield, L., Sjövall, J., and Sjövall, K., *J. Lab. Clin. Med.*, 68:186 (1966).

73. Wiggins, H. S., and Wootton, I. D. P., *Biochem. J.*, 70:349 (1958).

74. Levin, S. J., Johnston, G., and Boyle, A. J., *Anal. Chem.*, 33:1407 (1961).

75. Mosbach, E. M., Kalinsky, H. J., Halpern, E., and Kendall, F. E., *Arch. Biochem. Biophys.*, 51:402 (1954).

76. Levin, S. J., Irvin, J. L., and Johnston, C. G., *Anal. Chem.*, 33:856 (1961).

77. Nair, P. P., Gordon, M., Gordon, S., Reback, J., and Mendeloff, A. I., *Life Sci.*, 4:1887 (1965).

78. Kelly, R. L., and Doisy, Jr., E. A., *Fed. Proc.*, 26:849 (1967).

79. Nair, P. P., Gordon, M., and Reback, J., *J. Biol. Chem.*, 242:7 (1967).

80. Nair, P. P., in *Bile Salt Metabolism* (L. Schiff, J. B. Carey, and J. H. Dietschy, eds.), Thomas, Springfield, Ill., 1969, p. 172.

81. Nair, P. P., and Garcia, C., *Anal. Biochem.*, 29:164 (1969).

82. Roovers, J., Evrard, E., Vanderhaeghe, J., *Clin. Chim. Acta*, 19:449 (1968).

83. Palmer, R. H., *Proc. Nat. Acad. Sci. U.S.*, 58:1047 (1967).

84. Bloomfield, D. K., *Anal. Chem.*, 34:737 (1962).

85. Schlenk, H., and Gellerman, J. L., *Anal. Chem.*, 32:1412 (1960).

86. de Boer, T. J., and Backer, H. S., *Rec. Trav. Chim.*, 73:229 (1954).

87. Gutsche, C. D., *Org. Reactions*, 8:391 (1954).

88. Hammarsten, O., *Z. Physiol. Chem.*, 61:459 (1909).

89. Hofmann, A. F., and Mosbach, E. H., *J. Biol. Chem.*, 239:2813 (1964).

90. Imai, K., Tamura, Z., Mashige, F., and Osuga, T., *J. Chromatogr.*, 120:181 (1976).

91. Dziedzic, S. W., Dziedzic, L. B., and Githow, S. E., *J. Lab. Clin. Med.*, 82:829 (1973).

92. Sjövall, J., *Acta Chem. Scand.*, 16:1761 (1962).

93. Kuksis, A., and Gordon, B. A., *Can. J. Biochem. Physiol.*, 41:1355 (1963).

94. Jung, G., Pauschmann, J., Voelter, W., Breitmaier, E., and Bayer, E., *Chromatographia*, 3:26 (1970).

95. Sjövall, J., in *Biomedical Applications of Gas Chromatography* (H. A. Szymanski, ed.), Plenum Press, New York, 1964, p. 151.

96. Donike, M., *J. Chromatogr.*, 78:273 (1973).

97. Supina, W. R., Kruppa, R. F., and Henly, R. S., *J. Amer. Oil Chem. Soc.*, 43:138 (1966).

98. VandenHeuvel, W. J. A., Haahti, E. O. A., and Horning, E. C., *J. Amer. Chem. Soc.*, 83:1513 (1961).

99. Laatikainen, T., and Hesso, A., *Clin. Chim. Acta*, 64:63 (1975).

100. Grob, K., *Helv. Chim. Acta*, 48:1362 (1965).

101. Grob, K., *Helv. Chim. Acta*, 51:718 (1968).

102. Karlaganis, G., and Paumgartner, G., *J. Lipid Res.*, 19:771 (1978).

103. Karlaganis, G., and Paumgartner, G., *Clin. Chim. Acta*

104. Schwarz, H. P., von Bergmann, K., and Paumgartner, G., *Clin. Chim. Acta*, 50:197 (1974).

105. Makita, M., and Wells, W. W., *Anal. Biochem.*, 5:1 (1972).

106. Grob, K., and Grob, Jr., K., *J. Chromatogr.*, 94:53 (1974).

107. Grob, K., Grob, G., and Grob, Jr., K., *Chromatographia*, 10:181 (1977).

108. Grob, K., Grob, G., and Grob, Jr., K., *J. Chromatogr.*, 125:471 (1976).

109. Jones, D. C., Melchior, G. W., and Watkins Reeves, M. J., *J. Lipid Res.*, 17:273 (1976).

110. Karlaganis, G., and Paumgartner, G., *HRC & CC*, 1:54 (1978).

111. Grob, K., and Grob, Jr., K., *J. Chromatogr.*, 151:311 (1978).

112. Jaeger, H., Wagner, W., Homoki, J., Klör, H. U., and Ditschuneit, H., *Chromatographia*, 10:492 (1977).

113. Nebelung, W., Jaeger, H., Klör, H. U., and Ditschuneit, H., 2nd Can. Conf. on Chromatography, Toronto, 1979.

11

GLASS CAPILLARY GAS-LIQUID CHROMATOGRAPHY OF FATTY ACIDS

HALVOR JAEGER
HANS U. KLOER
HANS DITSCHUNEIT

University of Ulm
Ulm, Federal Republic of Germany

HARTMUT FRANK

University of Tübingen
Tübingen, Federal Republic of Germany

1. Introduction 396
   1.1 Structure and Nomenclature 396
   1.2 Metabolism and Function 397
   1.3 Biochemical Applications of Fatty Acid Analysis 398
   1.4 Clinical Application 400
   1.5 Fatty Acid Analysis in Biochemical and Clinical Problems of Nutrition 400
   1.6 History 401
2. Extraction And Derivatization 402
   2.1 Extraction 402
   2.2 Derivatization 404
3. Instrumentation 409
   3.1 Inlets 409
   3.2 Columns 411
   3.3 Detectors 413
4. Applications 414
5. Summary 442
References 445

1. INTRODUCTION

1.1 Structure and Nomenclature

Fatty acids consist of an alkyl chain of varying length with a terminal carboxyl group. The kind of C-C bonding in the alkyl chain determines the three-dimensional behavior. Thus, C-C double bonds which occur in unsaturated fatty acids will cause the chain to deviate by an angle of about 40°. The unsaturated fatty acids will thereby require more space in the lipid structure. The space-requiring property is most pronounced in fatty acids with several double bonds, so-called polyunsaturated fatty acids (PUFAs) which play an important regulatory role in cell membranes of animal species because they fit into hydrophobic parts of functional membrane proteins and determine membrane fluidity [1].

Most of the 300 naturally occurring fatty acids belong to families with unbranched straight chains. Traces of branched-chain fatty acids may also occur in mammalian tissue and body fluids, while they are abundant in plant tissues.

Fatty acids usually contain an even number of carbon atoms. In human physiology and pathology only a few branched-chain fatty acids have been found, the most important of which is phytanic acid, occurring in body fluids and tissues of patients with Refsum's syndrome (see Sec. 1.4). According to their different metabolic fate in mammals the fatty acids are divided into short-chain (C_2-C_8), medium-chain (C_8-C_{14}), and long-chain (C_{14}-C_{24}) fatty acids. The systematic chemical nomenclature refers to the alkyl chain (e.g., hexadecanoic acid) while the trivial name which is mostly used often refers to the source the fatty acid has been isolated from (e.g., palmitic acid from palm oil). As an abbreviation fatty acids may be designated by chain length and number of double bonds, e.g., C 16:0 or singly 16:0 for palmitic acid, a long-chain fatty acid with no double bond. Palmitic acid (C 16:0) and stearic acid (C 18:0) are encountered in almost all natural lipid mixtures of animal tissues. Fatty acids with a chain length of more than 20 carbon atoms are found in phospholipids of plasma and nervous tissue.

The double bonds of PUFAs may be arranged in a cumulated, conjugated, or isolated fashion. Most of the biologically important PUFAs contain isolated double bonds arranged in the so called divinyl methane rhythm, i.e., with a methyl group interposed between two double bonds. In the systematic nomenclature PUFAs are designated according to the position of double bonds when counting from the carboxyl group. However, this pure chemical nomenclature does not take into account the biochemical relationship between the different unsaturated fatty acids. Therefore, another biochemical nomenclature is more widely used which is based on determining the number of C atoms between the methyl end and the first double bond of the respective fatty acid. For example, the double bond of oleic acid is at carbon atom 9 when counting from the terminal methyl group. Accordingly, fatty acids metabolically related to oleic acid and having the first double bond at carbon atom 9 when counting from the methyl end belong to the "oleic acid" or "ω9" family. According to the chemical nomenclature, a fatty acid with 20 carbon atoms called 5,8,11-eicosatrienoic acid would not be supposed to be related to oleic acid (9-octadecenoic acid). The alternate designation of this C 20 fatty acid in the biochemical nomenclature (20:3ω9) reveals the metabolic relationship with oleic acid (C 18:1ω9). While C 20:3ω9 is a biochemical indicator of essential fatty acid deficiency, its chemically but not biochemically related counterpart 8,11,14-eicosatrienoic acid belongs to the "linoleic acid family" (ω6). It is therefore designated C 20:3ω6 in the biochemical nomenclature and is not at all related to essential fatty acid deficiency. Thus, it is evident that the designation of PUFAs according to a "family" is useful and that analytical differentiation between positional isomers of otherwise identical PUFAs may be important in metabolic studies on mammals, as will be discussed later.

1.2 Metabolism and Function

Fatty acids are found in intra- and extravascular spaces in the "free" form (free fatty acids, FFAs) or integrated into the molecular structure of different lipids. The pool of FFA in plasma and tissue is subjected to a constant and rapid turnover which serves to deliver energy to the various tissues and to provide structual components for building cellular and intracellular membranes throughout the body. The pool size of plasma FFA is relatively small, but a very short half-life (2 to 3 min in man) in connection with a high turnover rate (up to 40% min^{-1}) supplies the necessary amounts to the organs and peripheral tissues where they are needed [2].

About half of the plasma FFA are transported as an albumin complex, while the rest is distributed among the different plasma lipoproteins where it is supposed to reside within the membrane [3].

In the fasting state most of the plasma FFA is derived from hydrolysis of adipose tissue triglyceride, while in a postprandial situation FFA are generated by the action of lipoprotein lipase. This enzyme, which resides on the surface of endothelial cells in capillaries of adipose and muscle tissue [4], hydrolyzes the triglyceride core of triglyceride-rich lipoproteins (chylomicrons, very-low-density lipoproteins) and thus releases FFAs which remain in plasma complexed to albumin or which cross the capillary wall and enter the underlying tissues. Thus, the FFA spectrum of plasma which can be determined by gas-liquid chromatography may reveal the source from which the fatty acid has been released and indicates the metabolic state of the subject [5]. Besides being utilized for energy production by oxidation, fatty acids that have once entered liver, gut, and other peripheral tissues are incorporated into lipids for local or systemic need. Locally, mainly phospholipids for inner and outer cell membranes are synthesized from intracellular fatty acids, while the lipoprotein-synthesizing organs (liver and gut) incorporate them into triglycerides, cholesterol esters, and phospholipids which are released from the synthesizing cells together with free cholesterol and protein in the form of lipoproteins.

1.3 Biochemical Applications of Fatty Acid Analysis

The gas chromatographic analysis of fatty acids has been used for a variety of problems regarding structure, function, and metabolism of lipids in animal and plant tissues. For example, the fatty acid components at the different ester bonds of naturally occurring tryglycerides (1, 2, and 3 position) have been identified by a very elegant method combining hydrolysis of the 1 and 3 position of the native triglycerides by pancreatic lipase, subsequent isomerization, synthesis of artificial phospholipids as an intermediate product of analysis, and finally determination of the fatty acid species located at the three different positions of the triglyceride molecule [6]. Using this method, it has been found that the adipose tissue triglycerides of swine and other animals contain a different fatty acid spectrum at the 1, 2, or 3 position of the glycerol backbone [7]. Structural analysis of phospholipids can be performed in a similar way using, for example, phospholipase A_2 from snake venom (*Naja naja*) which selectively splits the ester bond at the 2 position of lecithin or other phosphoglycerides. The fatty acids released by this enzymatic cleavage can be compared to the fatty acid spectrum of the remaining 1-lysophosphatide. Analyses of naturally occurring lecithins (erythrocytes, plasma, etc.) with this method showed that the 2 position is esterified mostly with an unsaturated fatty acid, while the 1 position is occupied by a saturated fatty acid [8,9]. Metabolic studies concerning metabolism of fatty acids have often applied the combined use of radiolabeled FFA or precursors and of gas-liquid chromatography (GLC) analysis of the fatty acid spectrum. The follow-up of specific activities of single fatty acids has permitted them to predict the metabolic fate of the respective fatty acid [10].

By feeding a type of fatty acid like linolenic acid (18:3ω3) which is only found in traces in body fluids and not broken down, the metabolic fate of exogenous fatty acids can be followed without using radiolabeled compounds [11]. This so-called cold labeling technique has also been used to study the relative incorporation of linoleic and other PUFAs into the different lipids of plasma and organs [12]. It was also attempted to use the fatty acid spectrum of different lipids as "cold markers" for determining the metabolic handling of fatty acid containing lipids of the different plasma lipoproteins [13,14]. However, the results of these studies indicated that the differences among the fatty acid composition of lipids of lipoprotein classes were not great enough to permit their use as endogenous markers. In order to determine the metabolic conversions of fatty acids in body fluids and tissues without using radiolabeled compounds the determination of positional fatty acid isomers may be required. Using this kind of analysis it was noted, for example, that only normal liver cells incorporated the isomeric C 18:1 fatty acids oleic acid (ω9) and vaccenic acid (ω11) in fixed proportions into different lipids, while hepatoma cells had lost this specificity and produced lipids containing the isomers at random [15].

In another study with hepatoma cells there was also a definite difference in metabolism of the two C 18:1 isomers which would not have been detectable without using very sophisticated fatty acid analysis [16]. Tumor cells (Ehrlich ascites tumor) were also used as a model system to study intracellular metabolism and incorporation into cellular cholesterol ester of fatty acids being contained in cell culture media. By comparison of fatty acid spectra of the medium and after extraction of the cells it was possible to detect metabolic changes and selectivity for the incorporation of monoenoic fatty acids into tissue cholesterol ester, a selectivity which is also important for atherogenesis [17].

In another study regarding the influence of dietary fat on atherosclerosis, the fatty acid composition of aortal lipids was compared to that of plasma lipoproteins in swine on different lipid-elevating and lipid-lowering diets. There were distinct differences in fatty acid patterns of tissue and plasma and it was concluded that atherogenic proteins were not retained unchanged [18].

1.4 Clinical Application

Concerning the lipid-lowering action of PUFA, studies have been performed on the fate of dietary linoleate, i.e., on its incorporation into various plasma lipids [19,20]. When analyzing the fatty acid spectrum of various lipid classes an almost selective accumulation of dietary linoleate was observed in cholesterol ester and phospholipids of hypercholesterolemic patients treated with linoleate-enriched lecithin (EPL substance) [21].

Analysis of the plasma free fatty acid spectrum has also been applied to study the problem of cardiac arrythmias being induced by elevated free fatty acid levels in myocardial infarction [22,23]. To answer the question of the contribution of endogenous and exogenous fatty acids the fatty acid content of lipids in plasma and liver in alcoholism has been studied. It was found that due to the stimulation of lipolysis the fatty acid spectrum in body fluids and tissues was dominated by adipose tissue fatty acids and their metabolic derivatives [24,25].

The fatty acid compostion of liver and adipose tissue lipids was found to be so typical for alcoholics versus nonalcoholics that a postmortem diagnosis of alcoholism appeared to be possible [26].

The analysis of short-chain fatty acids in breath can be used as a diagnostic tool in assessing liver function of patients with compensated and decompensated cirrhosis of the liver [27].

Another clinical application concerns a rare neurological disorder called Refsum's syndrome, the true nature of which could only be elucidated by fatty acid analysis of plasma and tissue. It was found that this disease is caused by an abnormal accumulation of a branched C 20 fatty acid called phytanic acid which cannot be broken down as usual by the liver and other cells of these patients [28,29].

In plasma of patients with Refsum's syndrome, phytanic acid may be found in amounts corresponding to 5 to 30% of total fatty acids [30].

1.5 Fatty Acid Analysis in Biochemical and Clinical Problems of Nutrition

Biochemical and clinical problems of nutrition constitute probably the one field of research where fatty acid analysis has been most widely used. Important practical differences in fatty acid pattern have been established between human and cow's milk, indicating that breast milk contains more of the PUFA necessary for the infant to thrive [31]. Fatty acid analysis of stool has been used to prove that substitution of medium-chain triglycerides for normal dietary fat circumvents the problem of providing enough calories for growth in children with severe malabsorption due to a variety of diseases [32].

Analysis of positional fatty acid isomers in commercially available hydrogenated fats such as margarines has become important as it was shown that polyunsaturated trans-fatty acids which are generated by vigorous hydrogenation may be atherogenic, while their naturally occurring cis-analogs would be antiatherogenic [33].

Finally, qualitative and quantitative determination of individual fatty acids has become most important on the detection of essential fatty acid (EFA) deficiency, a syndrome that has been observed first in animals on a fat-free diet and which is due to a lack of linoleic acid (C 18:2ω6) and its derivatives [34]. While EFA deficiency is due to a lack of the ω6 family in most mammals and man, linolenic acid (C 18:3ω3) has been discovered to be essential for simian primates [35].

EFA deficiency is not easily detectable in adult man without fatty acid analysis, while newborn children, who have no PUFA stores, show clinical signs rather rapidly on a deficient diet [36].

In adult patients EFA deficiency with clinical symptoms such as skin rashes developed only after prolonged, fat-free intravenous nutrition [37, 38]. However, biochemical signs of EFA deficiency, i.e., the appearance of 5,8,11-eicosatrienoic acid (C 20:3ω9) and a significant decrease of linoleic and arachidonic acid (C 20:4ω6) could be detected in normals only after 2 weeks on fat-free intravenous nutrition [39]. This observation seems to make fatty acid analysis in patients on long-term parenteral nutrition almost mandatory. Without appropriate substitution fatty acid deficiency with either biochemical or even clinical manifestation also develops in patients with malabsorption due to surgery, inflammatory bowel disease, or liver disease [40]

1.6 History

In 1952 the first separation of fatty acids by gas-liquid chromatography was carried out by James and Martin [41]. On a packed column coated with a mixture of silicone oil DC 550 and stearic acid (9:1) on Celite 545 they chromatographed underivatized fatty acids from C_1 to C_{12}. This classical paper opened up a new area of scientific research in the field of lipid chemistry.

Since then, gas-liquid chromatography of fatty acids and glycerides has been used as the tool of choice in investigations of their physiology and pathophysiology.

2. EXTRACTION AND DERIVATIZATION

As described in Sec. 1, fatty acids occur in most natural sources to a high degree as glycerides; one to three fatty acids can be esterified to glycerol. In oils and butterfat lipids occur mostly as triglycerides. In serum of man and animals lipids are bound to protein as lipoprotein. They can be separated according to density into three major classes: very-low-density lipoproteins (VLDL), low-density lipoproteins (LDL), and high-density lipoproteins (HDL) [42]. The lipoproteins differ also in their electrophoretic mobility, so that a separation by lipoprotein electrophoresis is also possible [43]. The major lipid fractions bound to lipoproteins are, in order of decreasing polarity: phospholipids, monoglycerides, diglycerides, triglycerides, and cholesterol esters. A free fatty acid fraction also occurs. The phospholipids are subdivided in glycerophospholipids and sphingolipids. The first subgroup contains important fractions such as phosphatidyl choline and phosphatidyl serine as well as lysophosphatides and phosphatidyl inositol. In the latter subgroup sphingomyelins, cerebrosides, and gangliosides can be found. The only major lipid group that does not contain fatty acids is cholesterol.

2.1 Extraction

The extraction procedures do not vary widely according to the source; most can be applied to serum, stool, urine, gallbladder bile, cell membranes, oil, butterfat, or any other natural source after appropriate handling. With most of the more usual methods the neutral lipids can be extracted very easily. Problems occur with the extraction of the more polar compounds.

One of the first-described methods was that of McFarlane [44], who performed successive extractions with ether at a temperature range of -20 to -25°C. The method did not achieve complete delipidation, for about 35% of the lipid remained with the protein. The method was modified by Scanu and co-workers [45,46] by the use of a mixture of ethanol/ether 3:1 (v/v) for the extraction of lipids from lipoprotein fractions. The method was again modified for the extraction from serum [47,48]. Disadvantages were long extraction times, the need to work at low temperatures, and the utilization of large quantities of organic solvents so that the method has not found widespread application.

One of the earlier methods, that of Folch et al. [49], is still widely used today, as modified by Sperry [50]. It is performed as follows:

To plasma a solution of chloroform/methanol 2:1 (v/v) is added in an amount of 1:20. After precipitation of the protein the sample is centrifuged or the protein material is filtered off. The filtrate is washed with 4 ml of a 0.02% solution of $CaCl_2$. After phase separation the upper phase is discarded and the chloroform phase is evaporated to dryness on a water bath at 40°C under a gentle stream of nitrogen.

In recovery experiments in our own laboratory we found rather low recovery rates for this method, as indicated by radiolabeled compounds. For phospholipids, (phosphatidyl choline) the recovery rate was 60% after filtration and only 41% after washing with the $CaCl_2$ solution. For free fatty acids the recovery rates were 82 and 62% respectively. For triglycerides we found 83 and 82% respectively. Best results were obtained with cholesterol esters. Here we found recovery rates of 86% after filtration and 84% after the subsequent washing. Hence the method of Folch and co-workers in the described modification can be used for the analysis of triglycerides and cholesterol esters. For phospholipid analysis other methods should be applied.

Cham and Knowles [51] presented a method utilizing butanol-diisopropyl ether (DIPE), leading to a miscible single organic phase, thus avoiding resolubilization of protein. The method is described as follows:

Five-ml serum or plasma aliquots, each containing 0.5 mg of ethylenediamine tetraacetate (EDTA), were added to 10 ml of organic phase consisting of DIPE with butanol. The tubes were subjected to end-over-end rotation for varying periods of time with subsequent centrifugation at 2000 rpm for 2 min to separate the aqueous and the organic phase. The aqueous phase was removed and subjected to quantitative lipid analysis.

As demonstrated, the authors found a delipidization of 100% for triglycerides, cholesterol, phospholipid, and unesterified fatty acids by application of butanol/DIPE 25:75 (v/v) over a period of 3 hr. When trying to reproduce these data we found the same results for the non-lipid-containing aqueous phase. Recovery rates in the organic phase were not as expected. Measured with radio-labeled compounds we found recovery rates of 41 to 47% for phospholipids, 63% for free fatty acids, and 56% for triglycerides. Hence it must be concluded that much of the lipid material remains with the protein.

Best results were obtained with a modification of the original Folch method, as reported by Nelson [52,53]. We further modified the method slightly as follows:

First 0.2 ml of serum is added dropwise to 3 ml of methanol with subsequent mixing. Then 4 ml of chloroform is added and after subsequent mixing an additional 4 ml of the same solvent is added and the mixture is shaken again. The sample is filtered and nonlipid material removed by the addition of 0.2 g of Sephadex G-25. The solvent is evaporated on a water bath and the residue is dissolved in 4 ml of chloroform, subjected to evaporation, redissolved in another 4 ml of chloroform, and acidified with a few drops of sulfuric acid. The sample is pipetted into a small column and the lipids are eluted with an additional 26 ml of chloroform. This yields recovery rates of 99, 98, 97, 95, and 87% for triglycerides, cholesterol esters, phosphatidyl choline, phosphatidyl serine, and phosphatidyl inositol, respectively. The use of Sephadex G-25 for the purification of the lipid extract has also been demonstrated by Wells and Dittmer [54].

Other methods [55] also yielded unsatisfactory recovery rates for phospholipids; the most satisfactory method in terms of overall recovery is the above-described procedure.

For most extractions, including those described here, the use of an antioxidant like BHT (4-methyl-2,6-di-tert-butylphenol) is of utmost importance. Wren and Szezepanowska [56] described the necessity of an antioxidant, especially when dealing with phospholipids, and from our own experience we support these findings in full.

After extraction the lipids from natural sources have to be separated into the lipid classes described above. It cannot be described here in full detail how this can be achieved, but adsorption chromatography is applied in most cases. For further details see Refs. [57-61].

2.2 Derivatization

Esters

Prior to esterification the lipids with ester-bound fatty acids have to be saponified. Usually a two-step procedure is avoided by application of a "transesterification." This term refers to the exchange of acyl groups among different esters. Most of the methods dealing with biological samples have the ability to achieve transesterification of the common lipids.

Therefore, we shall describe only one esterification method here, which esterifies only free fatty acids, but which because of its simplicity is detailed below.

Thenot and co-workers [62] reported on this method, based on the findings of Brechbühler and co-workers [63], in which amide acetals were used to prepare esters from carboxylic acids in good yield. Any kind of esters—methyl-, ethyl-, propyl-, or n-butylesters—could be prepared. The method is performed as follows:

Fatty acids dissolved in pyridine are placed in a reaction vessel and 200 μl of N,N-dimethyl formamide dimethylacetal (DMF-DMA) are added and the mixture is injected directly into the gas chromatograph. A "sandwich" injection, in which the sample in pyridine was contained in the same syringe with the DMF-DMA, separated by an air bubble, yielded good results as well. Pyridine is interchangeable with many other organic solvents.

The methylation with diazomethane as described by Schlenk and Gellerman [64] is hazardous for several reasons [65] (see also Chap. 10 on bile acids). Although a reaction gas chromatographic method has been described [66] which decreases the risks by using a precolumn reactor, we still feel that methylation with diazomethane cannot be recommended.

The methylesters are suitable for gas chromatography of higher fatty acids, whereas methylesters of lower fatty acids may undergo losses during derivatization, due to their high volatility.

Although bases as well as acids are effective catalysts for esterification reactions, most scientists use acid catalysis. This may be due to possible isomerization of methylene-interrupted double bonds in the presence of excessive base [67,68]. The most commonly used acids are hydrochloric or sulfuric acid. The concentration of acid for effective catalysis is not critical.

The method of Stoffel and co-workers [69] serves as an example for the transesterification with hydrochloric acid in methanol:

The fatty acid sample is boiled under reflux with a 20-fold amount of 5% hydrochloric acid in methanol for 60 to 90 min. The methylesters in pure form are isolated from the reaction products by sublimation in vacuo. Sheppard and Ford [70] have redesigned the sublimators and reported excellent recovery rates.

Other methods also employ hydrochloric acid as catalyst [71-77] and utilize acid concentrations from 2.5 to 5% HCl (w/v) at temperature ranges of 60 to 100°C and time intervals of 1 to 17 hr. Esterification procedures using HCl-methanol for the transesterification of lipids and the esterification of free fatty acids have been outlined by James [78].

The use of sulfuric acid as catalyst has also found widespread application. The method is quantitative for methylation of both cis and trans acids [79,80]. The same results were achieved by the Special Task Group for Preparation of Methyl Esters of the American Oil Chemists' Society [81]. Published methods vary in acid concentration from 1 to 20% (w/v), in temperature range from 65 to 100°C, and in time for reaction from 1 to 16 hr [89].

Mason and co-workers [90-92] used HCl as catalyst and also added 2,2-dimethoxypropane to the reaction mixture, thus avoiding ester losses occurring during refluxing and formation of polymers.

Shapiro and Kritchevsky [93] reported that sulfuric acid and 2,2-dimethoxypropane yielded good recoveries for methylation at 70°C after 4 hr. The dimethoxypropane also acted as a drying agent.

As early as 1931 boron trifluoride was regarded as a good catalyst for the esterification of organic acids [94,95]. Metcalfe and Schmitz [96] were the first to use a 12.5% solution of BF_3-methanol for the preparation of fatty acid methylesters. The results were in good agreement with derivatizations carried out with diazomethane and HCl-methanol.

With a 50% BF_3-methanol solution Lough [97] reported that considerable amounts of methoxy-substituted fatty acids were produced as artifacts and found that the problem could be overcome by reduction of the percentage of boron trifluoride [98].

Metcalfe et al. [99] revised the procedure by incorporation of the rapid saponification procedure of Ast [68]. This method replaced the H_2SO_4-methanol derivatization of the Association of Official Analytical Chemists (AOAC) in 1968 [100].

This change was due to a collaborative study of the American Oil Chemists' Society [101], which adopted the method [102]. To our knowledge the most excellent reference for transmethylation of fatty acids of various lipids has been given by Morrison and Smith [103]. We have tested this method and found it to be excellent for the determination of fatty acids from various lipids of human and animal origin. We recommend the consultation of this paper especially when dealing with more uncommon lipids. The method is performed as follows:

To an aliquot of lipid solution, which is evaporated to dryness,

BF_3-methanol reagent is added under nitrogen in the proportions of 1 ml reagent per 4 to 16 mg of lipid. The tube with Teflon-lined screwcap is then heated for a requisite time at 100°C. After cooling the esters are extracted by addition of 2 volumes of pentane and 1 volume of water, shaking and centrifuging until both layers are clear. To recover dimethylacetals the method was modified. From table 3 of that paper it can be seen that triglycerides and sterol esters were derivatized with a mixture of 35% BF_3-methanol (14% w/v). The heating times ranged from 2 min for free fatty acids to 90 min for sphingomyelins. When lipids of different types are present, they are treated like sphingomyelins. Transmethylation of fatty acid esters may also be readily accomplished by application of sodium methoxide [104].

For the derivatization of acids with lower carbon numbers the use of higher alcohols is recommended. Dummel and Kum [105] prepared ethylesters. These as well as butylesters are prepared in a similar manner.

Butanol and 2% of sulfuric acid was used by Bezard and Bugant [106]. Aue and co-workers [107,108] and McCone and Hance [109] used butanol saturated with hydrogen chloride.

Furthermore, benzylesters [110,111] and p-substituted benzylesters [112] have been applied to the derivatization of short-chain fatty acids.

Of course, fatty acids with lower carbon numbers can be chromatographed without derivatization [113,114], but these techniques have not been perfected for routine analysis. A review has been given by Ref. 115.

Chromatography of the lipids, e.g., triglycerides, that are separated according to the carbon numbers of their fatty acids has been performed by Kuksis and co-workers [116]. The high temperatures required for the elution have limited so far the applicability of glass capillary columns to this problem. Perhaps new columns for use in higher temperature [117] and the new on-column injection system described by Grob and Grob [118] will bring further improvements in this area.

The use of electron capture detection can be exploited by derivatization with halogenated compounds, thus permitting an ultratrace analysis.

Kawahara [119] has derivatized organic acids to their corresponding pentafluorobenzylesters by application of acetone, α-bromo-2,3,4,5,6-pentafluorotoluene and potassium carbonate. The mixture was refluxed for 3 hr, washed, and dried over anhydrous sodium sulfate. He found recovery rates from 81 to 88%.

Silyl Ethers

As demonstrated by Pierce [120], silylation agents react with the carboxyl group and with other groups present in the molecule of a carboxylic acid. This is of special interest when fatty acids have to be analyzed in the presence of other substances.

Tallent and Kleinman [121] silylated the hydrolysate from lipids with BSA, thus converting the fatty acids as well as the glycerol to the corresponding silyl ethers. This yields quantitative results not only for the fatty acids but for the glycerol as well, both being part of the lipids.

Previous work on the trimethylsilylation of fatty acids had been done by Kuksis and co-workers [116,122], Dalgiesh and co-workers [123], and Birkhofer and Donike [124]. Donike [125] even used trimethylsilyl ethers of fatty acids for the determination of the "actual activity" of columns, which represents an expression for the quality of liquid phases, support materials, and columns.

Other Derivatives

Other derivatization procedures are of minor interest. Methylation of fatty acids by pyrolysis of their tetramethylammonium salts in the gas chromatograph has been described by Downing and Greene [126].

Direct hydrogenation of unsaturated fatty acids with palladium catalyst on the column support was described by Koman [127].

Pohl and co-workers [128] analyzed polyenoic fatty acids on a microscale. The methylesters of fatty acids are analyzed after separation by thin-layer chromatography of their mercury adducts and oxidatively cleaved into mono- and dicarboxylic acids by potassium permanganate in glacial acetic acid. Identification of the methylated cleavage products is achieved by programmed gas chromatography.

Comparison of Derivatization Procedures

Excellent reviews on derivatization have been given by Fontell and co-workers [129], Kuksis [130], Sheppard and Iverson [131], Drozd [132], Ahuja [133], and Ackman [134]. Over the years several comparative studies about the different derivatization procedures have appeared. In 1959 Craig and Murty [135] compared diazomethane and sodium methoxide as catalysts and found almost identical results.

In 1960 Luddy and co-workers [136] compared sodium or potassium methoxide as catalysts with direct esterification with diazomethane, HCl-methanol, and H_2SO_4-methanol. The latter methods yielded quantitative conversion, whereas even large excesses of sodium or potassium methoxide produced "substantially quantitative" esterification.

In 1961 Vorbeck and co-workers [137] compared the methods of Roper and Ma [138] using diazomethane, Stoffel and co-workers [69] using HCl-methanol with sublimation, Hornstein and co-workers [139] using HCl-methanol on ion-exchange resins, and Metcalfe and Schmitz [96] using BF_3-methanol. According to the results achieved with fatty acids of higher molecular weight the authors demonstrated a slightly better standard deviation for the diazomethane method with $\pm$0.46, $\pm$0.52, and 0.77%, respectively. For the shorter-chain fatty acids great losses occurred with the HCl method with sublimation. Best results were obtained with the diazomethane method with a standard deviation of $\pm$0.32%, whereas the BF_3-methanol method and the HCl method on resin had higher standard deviations with $\pm$13.8 and $\pm$26.5% respectively.

In 1965 Jamieson and Reid [139a] found similar results for six different techniques: sodium methoxide with a reaction time of 40 min [140], KOH [141], sodium methoxide with a reaction time of 4 hr [142], saponification with 7.8% KOH in ethanol, saponification with 0.5 N KOH in ethanol [143], and cold saponification with 50% aqueous KOH [144].

In 1967 de Man [145] compared the methods of Hornstein and co-workers [139] using HCl-methanol on ion-exchange resin, that of Morrison and Smith [103] using BF_3-methanol, and that of de Man [146] called the "sealed-tube method." He reported similar results for the higher fatty acids and losses of the lower-molecular-weight fatty acids with the first two methods.

In summary, one may state that most of the methods described here give comparable results and should be checked for their suitability to a specific application. In our laboratory we prefer the BF_3-methanol method as described by Morrison and Smith [103] because of its wide applicability to derivatization of fatty acids from almost all natural sources.

3. INSTRUMENTATION

No need could be found for any special equipment, so this section only gives a brief description of commonly used inlets, columns, and detectors.

3.1 Inlets

As capillary columns posess a much lower sample capacity compared to packed columns, the sample introduction is somewhat more complicated. The sample load of a capillary column is in the ng range, whereas that of packed columns is about 200-fold higher. This led to the introduction of splitters, where most of the sample is vented and only a small portion of it is brought onto the column. Standard splitter designs have been described [147,148]. Normal split ratios are between 1:20 and 1:1000. One of the difficulties occurring with this system is the maintenance of the exact split ratio during long periods of routine analysis. Thus injection of the complete sample without splitting was introduced. Grob and Grob [149] have described a splitless injection technique. The straightened end of the capillary is inserted into the vaporization tube. In order to achieve a full transmittance of the vaporized sample from the vaporization tube with its large volume onto the front part of the capillary one has to perform what Grob and Grob [150] called "cold trapping," which means injection of the sample at a column temperature that is at least 30 to 40°C below the separation temperature. By performing this procedure the solvent is removed before the sample and the sample is condensed on the beginning of the column, leading to well-separated peaks.

The same authors have also demonstrated the influence of the septum. When injecting the sample the septum will sometimes absorb part of it, and this slowly migrates into the column, leading to tailing peaks. This is due to the slow carrier gas flow of the capillary, compared to the packed column, where this problem does not occur. It can be overcome by septum flushing.

In summary, the handling of the injection port system as described by Grob and Grob [149,150] is as follows: The sample is injected at a column temperature 30°C below the separation temperature with the splitter outlet closed. After an appropriate time (0.5 to 3 min) the split is opened and functions as a septum flush so that any part of the sample associated with the septum is vented. The cold trapping procedure ends by heating the column oven to the initial separating temperature, and instituting the normal temperature program. We have had good experiences with this type of sample injection. Other authors [151] regard it as undesirable to first vaporize the sample in the injection port, then to transfer it to a cold column inlet with subsequent condensation and afterward vaporizing it again by starting the temperature program. They prefer to use a double-column system with intermediate trapping.

These differences could be overcome by the introduction of an on-column injection system as described by Grob and Grob [118]. The sample, up to 5 µl, is injected directly onto the column, thus avoiding disturbances by the septum and the difference in size between injection port and column inlet. Such an injection port is now commercially available (Brechbühler A. G., Urdorf, Switzerland); other types of on-column injectors have been described before [114,118a].

Routine analysis needs automatic injection. Most of the systems available on the market are based on the normal performance of an injection with a syringe. The syringe of the automatic injector sucks the sample in and injects it into the gas chromatograph.

We have in use a so-called solid sampler system (Carlo Erba, S.p.A., Milan, Italy) [152]. An aliquot of the sample is sucked into a container capillary, which is coated with a silicone liquid phase (e.g., SE-30). Before coating with a 0.05% solution of SE-30 in methylene chloride (w/v), the container capillary has to be treated with a 4% solution of trimethylchlorosilane in chloroform (v/v) in order to obtain an even film of liquid on the inner column wall, because chemical modification of the surface decreases the contact angle between the liquid and the surface.

Subsequently, the solvent is evaporated from the container capillaries with a gentle stream of nitrogen in a cooled system at 4°C, so that only the sample and no solvent remains in the container capillaries. These are placed in the automatic injector, permitting up to 22 analyses to be performed.

Of course, this type of automatic injector cannot be used for gases, and in our analyses we found poor reproducibility for fatty acids with less than 14 carbon atoms, even when cooling the system to 4°C. The system combines the advantages of high reproducibility, injection of the entire sample of 2.3 μl, absence of a solvent peak (thus increasing the lifetime of the column), lack of any disturbances by the septum, low cost, and high stability.

3.2 Columns

Orr and Callen [153] have introduced synthetic polyesters from short-chain dicarbonic acids and lower-molecular-weight diols. Craig and Murty [154] and James [78] have recommended the synthesis of polyesters with ethylene glycol, diethylene glycol, or 1,4-butanediol.

Separations on different polar packed columns have been described by numerous authors [155-158]. A review is given by Haken [158a]. For the complete mixtures of naturally occurring fatty acids the low separation factors of packed columns have proven to be insufficient. Thus capillary columns are columns of choice for the determination of fatty acid patterns. It is beyond the scope of this chapter to describe the methods for preparing such columns. Acceptable procedures have been given by numerous authors [148-170]. For complex mixtures, where cis/trans isomerism occurs, polar phases must be employed.

Nelson [171] has tested the elution characteristics of fatty acid methylesters mainly on diethylene glycol succinate or butanediol succinate polyester support-coated open tubular columns.

A comparison between butanediol succinate (BDS), Silar-5CP, and Apiezon-L liquid phases has been carried out by Ackman and Hooper [172]. Apiezon-L was insufficient for the separation of complex mixtures, whereas Silar-5CP reduced elution time sufficiently. The stability of the phase was reported to be good.

Golovnya and co-workers [173] have compared SE-30, OV-225, Silar-5CP, and Silar-10C by means of the elution characteristics of fatty acid methylesters. The authors did not favor one type of liquid phase but thought that the wide range of polarity covered by these phases should minimize possible overlapping, thus leading to better identification of complex fatty acid mixtures.

Prevot and Mordret [174] regarded Carbowax 20M as sufficient for the separation of unsaturated fatty acids but did not demonstrate separation of cis and trans isomers.

Van Hout and co-workers [175] have compared as liquid phases PZ-176, a polyphenyl ether sulfone, with HI-EFF-8 BP, a cyclohexanedimethanol succinate, Silar 5CP, a cyanoalkylphenyl silicone, OV-17, a methylphenyl silicone, and SP-2401, a trifluoropropylmethyl silicone. Best results according to separation were accomplished with the Silar phase, but the stability of the columns was very poor, lasting only up to 2 weeks. This is in good correlation with the findings of Jaeggi [117], who was unsuccessful in producing stable and reproducible columns with Silar phases, due to high variability between different batches.

We have had the best experiences with FFAP (free fatty acid phase), which is a reaction product between Carbowax 20M and 2-nitroterephthalic acid (see also Sec. 3.3), a phase which was described as well by Goretti and Liberti [176].

A comparison between the cyanopropyl polysiloxanes like Silar-10C, SP-2340, and OV-275, and the polyethylene glycol has been given by van Vleet and Quinn [177].

Some excellent chromatograms of extraordinarily good separations of cis and trans isomers on commercially available glass capillary columns have been published [178], but to our knowledge no scientist has been able so far to work with these columns. They should resolve all problems of separation of cis and trans isomers, but unfortunately we have not been able to buy such a column.

Grob and co-workers [179] compared the elution characteristics of cis/trans isomers on Silar-10C columns and FFAP columns. They concluded that on Silar-type phases an error is made by giving too small values for the acids present in the trans form, whereas on FFAP phases a smaller error is made by giving too small values for the cis isomers. It is our feeling that the separation efficiency was not high enough for either phase, as it is possible to separate the ω-9, ω-7 and ω-5 isomers of oleic acid on an optimal FFAP column [180]. Furthermore, this column demonstrated remarkable stability, over a period of 2 years, even under continuous use with automated injections over 24-hour days.

In summary, we do think that the Silar phases should separate cis and trans isomers much better than does FFAP, but the chromatograms we have seen so far do not show superior properties as compared to the FFAP column.

3.3 Detectors

For routine analysis of fatty acid methylesters detection with flame ionization detectors (FIDs) is sufficient. Problems occur with the use of an electron capture detector (ECD). Its effective detector volume is much larger than in the FID, which decreases separation factors. These problems have been partially overcome by construction of micro-ECDs specially designed for glass capillary chromatography [181]. A micro-ECD is also commercially available from Carlo Erba S.p.A., Milan, Italy.

The application of radio-gas chromatography, in which a radioactivity-monitoring system is substituted for the conventional detector has been described by several authors [182-185]. The technique offers no advantage in accuracy and has been used by only a few authors; hence a description of the method is beyond the scope of this manuscript.

The use of mass spectrometry for detection of fatty acid methylesters is limited by the fact that isomers differing in the position of unsaturation give similar spectra, regardless of the double bond position. Some attempts have been made to overcome this problem [186-189].

Automatic integration is certainly advisable for the routine analysis of fatty acid mixtures producing large numbers of peaks. The data should be transferred, by means of punched tapes, e.g., to computers, as it is far too time-consuming to compare such a large amount of information without data processing. The computers can also be used as an integrator. For a review of data processing in gas chromatography, see Ref. 190.

4. APPLICATIONS

The study of lipids that contain a wide and complicated spectrum of fatty acids is nearly impossible with the low separation factors delivered by packed columns. Furthermore, collaborative studies with packed columns [191,192] have revealed that the packed column results vary to such a degree that accurate measurements cannot be obtained.

Glass capillary chromatography of free fatty acids has been performed by Novotny and co-workers [193], using short (12-m) glass capillary columns coated with Dexsil 300 GC. This phase possesses a high temperature stability, but the peaks tailed considerably, and quantification was not optimal.

Verzele and co-workers [194] showed good separations of free fatty acids with on-column injection, thus avoiding splitting errors. They regarded columns with an ID of 0.5 mm and more as better suited for this injection technique. To avoid glass breakage the authors used gold capillaries as an "interface," as this material has been found to be even less catalytically active than glass [194]. The injection port described was adaptable to most commercially available gas chromatographs. Advantages of the system are that the sample does not contact catalytically active material and the dead volume in the injection port is limited to a minimum. They achieved values of less than 1 mm for the height equivalent to a theoretical plate (HETP, or h), and recommended high carrier gas flows, with hydrogen favored as the carrier gas because of the balance between axial and radial diffusion [195].

Figure 1, showing the separation of the free fatty acids up to linolenic acid (C 18:3) on a 10-m glass capillary coated with FFAP, is an example of this work. Details are given in the legend to the figure. Elution times are rather long, but separation is excellent in view of the fact that the acids are not derivatized. In further developments of these columns, the authors showed separations on wall-coated (WCOT) and support-coated (SCOT) open tubular columns coated with FFAP.

Figure 2 shows almost the same sample except for the fatty acid C 10:0, analyzed isothermally at 200°C. As reported, quantitative analysis was better with on-column injection than with any of the splitters used.

Hrivnák and co-workers [113] presented a separation of the short-chain free fatty acids. Baseline separation was achieved on sixteen saturated

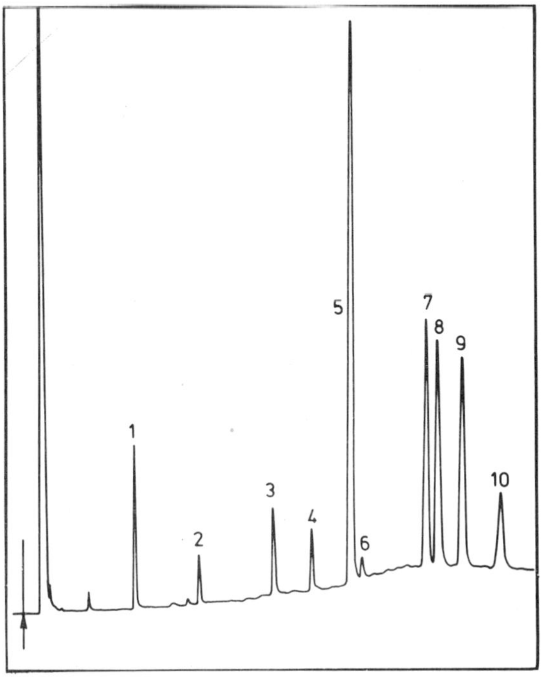

Figure 1. Analysis of long-chain free fatty acids. Column: 10 m × 0.5 mm; FFAP; WCOT; temperature programmed 140 to 204°C at 2°C min^{-1}; detector and injector 250°C; carrier gas hydrogen, flowrate 4 ml min^{-1}; elution time C 18:3: 75 min. Peaks: 1—C 10:0, 2—C 12:0, 3—C 14:0, 4—C 15:0, 5—C 16:0, 6—C 16:1, 7—C 18:0, 8—C 18:1, 9—C 18:2, 10—C 18:3. (Reproduced by permission of Prof. M. Verzele, University of Ghent, Belgium.)

fatty acids in the carbon number range from C_2 to C_6, using stainless steel capillary columns with an ID of 0.25 mm and a length of 45 m. The best of the phases tested was Ucon LB-550-X with the addition of phosphoric acid, operated at 125°C. At this temperature the minimum plate number "n req." necessary for the baseline separation of all pairs of acids was found to be less than 100,000 theoretical plates. Here "n" is calculated from $n = 5.54[t_R/w_{.5}]^2$. The method could be applied to the analysis of lower fatty acids in biological materials and foodstuffs.

Ettre and March [196] compared efficiency, resolution, and speed on open tubular columns and packed columns. Three types of columns were used: a packed column with a length of 8 ft and an ID of 2.16 mm with a liquid phase loading of 10 wt % on Chromosorb W 80 to 100 mesh, a stainless steel WCOT column with a length of 150 ft and an ID of 0.25 mm, and a stainless steel SCOT column with a length of 50 ft and an ID of 0.5 mm with Aerosil plus Minusil as support. Diethylene glycol succinate (DEGS) was used as

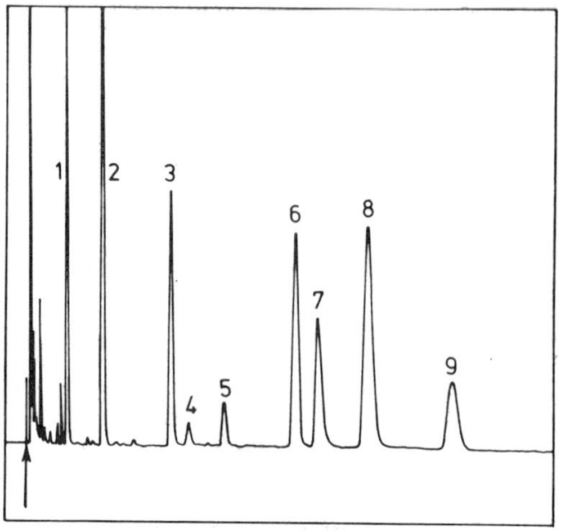

Figure 2. Analysis of long-chain free fatty acids. Column: 28 m × 0.6 mm ID; FFAP; SCOT; temperature 200°C; detector and injector 300°C; carrier gas hydrogen, flowrate ±10 ml min^{-1}; elution time C 18:3: 60 min. Peaks: 1—C 12:0, 2—C 14:0, 3—C 16:0, 4—C 16:1, 5—C 17:0, 6—C 18:0, 7—C 18:1, 8—C 18:2, 9—C 18:3. (Reproduced by permission of Prof. M. Verzele, University of Ghent, Belgium.)

liquid phase on all three columns. The phase ratios were 10.3, 120, and 54, respectively. The phase ratio β was calculated from $\beta = V_G/V_L$, thus describing the volume of the column occupied by the gas phase relative to the volume occupied by the stationary phase. The mean values of the partition ratios were 58.6, 5.1, and 11.2 for the packed, the WCOT, and the SCOT column, respectively. The partition ratio was calculated from $k = t'_R/t'_M$, thus describing how long a solute spends in the stationary phase relative to the time it spends in the gas phase. From β and k the partition coefficient of methyloleate on DEGS at 180°C was calculated and found to be 597.4, 612.0, and 604.8 for the packed, the WCOT, and the SCOT column, respectively. When comparing the three columns at their optimum average linear gas velocities the resolution of the methylstearate/methyloleate peaks was found to be 1.22, 5.85, and 3.85 for the packed, the WCOT, and the SCOT column, respectively.

Figure 3 shows comparable sections of the respective chromatograms obtained by calculating from the plate numbers and retention times the standard deviations of the respective Gaussian peaks, and reconstructing that particular part of the chromatogram from those values.

The true performance of the three columns could be tested in three different ways: by comparison of retention times corresponding to the same resolution, or by comparison of the resolution corresponding to the same retention time, or by preparation of resolution vs. retention time plots.

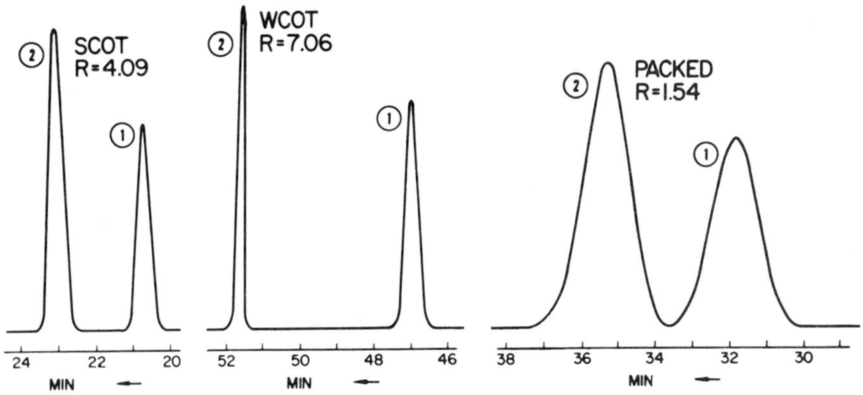

Figure 3. Reconstructed chromatograms showing the methylstearate/methyloleate peak pair on the three columns, at optimum average linear gas velocity. Peaks: 1—methylstearate, 2—methyloleate. (Source: Ref. 196. Reproduced by permission of Elsevier Scientific Publishing Company, Amsterdam.)

When comparing the resolution corresponding to the same retention time the resolutions of the methylstearate/methyloleate peaks were found to be 1.22, 5.85, and 3.85 for the packed, WCOT, and SCOT column, respectively, as mentioned above. HETP values calculated from the column length divided by the theoretical plate number, were found to be 1.61, 1.46, and 1.13 mm, respectively.

Comparative chromatograms at the same retention time for methyloleate on the packed column and the two open tubular columns are given in Figs. 4, 5, and 6 [197]. Of course, one would expect better resolutions on glass capillary columns. These same authors also chromatographed a sample of fatty acid methylesters on a 100-m WCOT glass capillary column with an ID of 0.27 mm coated with DEGS [198]. In this latter case the resolution between the methylstearate and methyloleate peaks was measured at 12.8. The chromatogram is shown in Fig. 7.

Ackman and Hooper [199] investigated the so-called retarding load effect, which is exerted by excessive proportions of some components on adjacent later-eluting isomers [200]. The results indicated that structurally dissimilar materials such as trans acids in a mixture of cis isomers do not cause this effect.

Figure 8 shows that as early as 1969 remarkable separations of cis and trans isomers could be obtained on stainless steel columns. A 150-ft × 0.25-mm ID column coated with BDS was used at a temperature of 170°C, with helium carrier gas at 50 lb in^{-2} gauge.

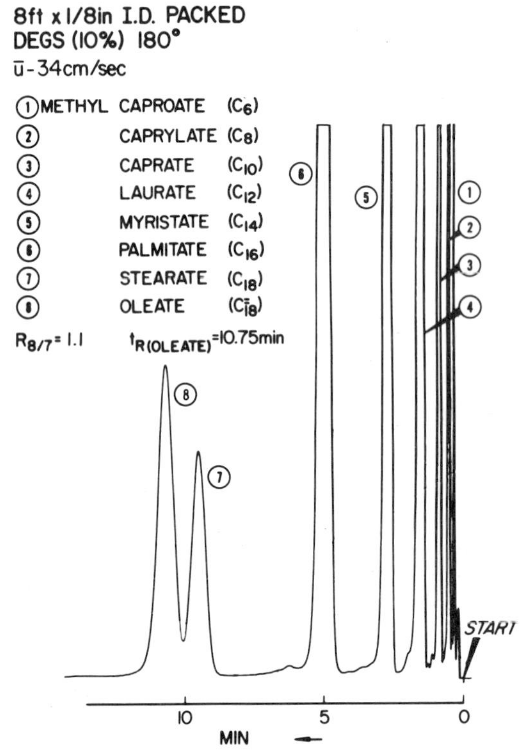

Figure 4. Chromatogram of a fatty acid methylester mixture obtained on a 8-ft × 1/8-in ID packed column coated with DEGS stationary phase. Temperature 180°C; average linear carrier gas (helium) velocity 34 cm sec^{-1}; resolution of the methylstearate/oleate peaks 1.1. (Source: Ref. 197. Reproduced by permission of Friedr. Vieweg und Sohn Verlagsgesellschaft, Wiesbaden.)

Pollock [201] investigated two Dexsil 400-GC phases, end groups uncapped and capped. The columns were 150 ft × 0.25 mm ID, temperature-programmed from 100 to 300°C. The uncapped phase was found to have lower bleed and greater stability. The tailing of esters C_6 through C_{11} was also much worse on the capped phase. As can be seen from Fig. 9, separation of saturated from unsaturated fatty acids was achieved, but not the separation of cis and trans isomers.

Egge and co-workers [202] investigated minor constituents of human milk. The branched-chain fatty acids of human milk fat were enriched 100- to 500-fold by urea fractionation, following catalytic hydrogenation of the unsaturated fatty acids. Methylesters were obtained by transesterification with alkaline methanolysis. The authors compared packed, WCOT, and SCOT

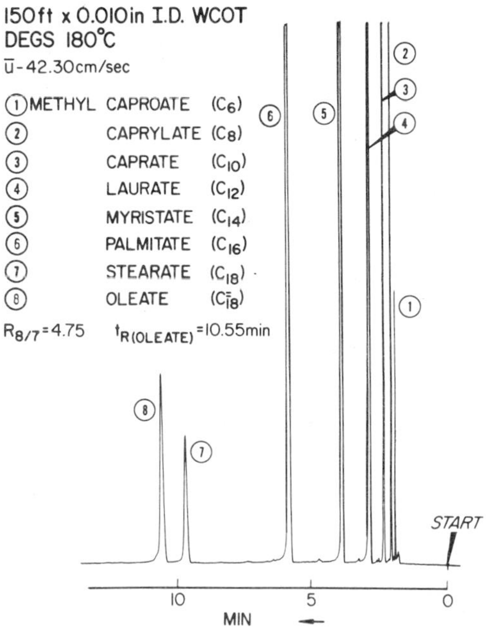

Figure 5. Chromatogram of a fatty acid methylester mixture obtained on a 150-ft × 0.010-in ID WCOT column coated with DEGS stationary phase. Temperature 180°C; average linear carrier gas (helium) velocity 42.3 cm sec^{-1}; resolution of the methylstearate/oleate peaks 4.75. (Source: Ref. 197. Reproduced by permission of Friedr. Vieweg und Sohn Verlagsgellschaft, Wiesbaden, and Elsevier Scientific Publishing Company, Amsterdam.)

capillary columns by using a combination of gas chromatography/mass spectrometry. Figure 10 shows a chromatogram of the branched-chain fatty acids from the lipids of human milk. Used was a 100-ft × 0.25-mm ID open tubular column coated with DEGA (diethylene glycoladipate polyester). The chromatographic conditions are given in the legend to the figure.

Krupcik and co-workers [203] studied the influence of temperature on the separation of fatty acid methylesters on columns with three different phases. Apiezon-L, butane-1,4-diolsuccinate (BDS), and DEGS. The values of the equivalent carbon chain length (ECL) were determined with a reproducibility of better than ±0.01 carbon atom.

Figure 11 shows a chromatogram of fatty acid methylesters from butterfat on BDS. As can be seen from the figure, the methylesters of the unsaturated acids are eluted after the corresponding saturates, which is due to a polar interaction of the polyester phase with the methylesters of unsaturated acids. The chromatogram shows a good separation of cis 16:1. A disadvantage was the undefined polarity of the stationary phase, which

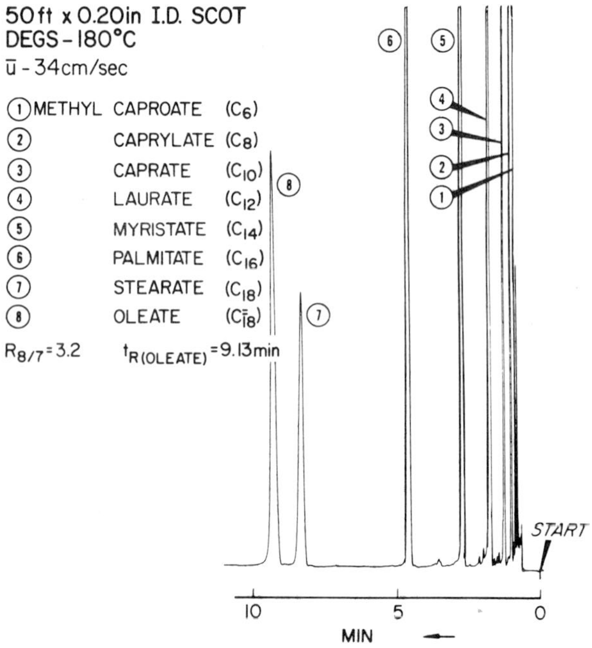

Figure 6. Chromatogram of a fatty acid methylester mixture obtained on a 50-ft × 0.020-in ID SCOT column prepared with DEGS stationary phase. Temperature 180°C; average linear carrier gas (helium) velocity 34.0 cm sec^{-1}; resolution of the methylstearate/oleate peaks 3.2. For peak identification, see the legend to Fig. 5. (Source: Ref. 197. Reproduced by permission of Friedr. Vieweg und Sohn Verlagsgesellschaft, Wiesbaden, and Elsevier Scientific Publishing Company, Amsterdam.)

changed with aging of the column [204]. The chromatogram in the figure was performed on a 100-m × 0.25-mm ID capillary column, coated with BDS and operated at a temperature of 160 to 180°C. The column had 190,000 theoretical plates. Nitrogen was used as carrier gas with a pressure of 2.1 kp cm^{-2} at a split ratio of 1:250.

In one study [205] Apiezon-L and BDS were compared as stationary phases on capillary columns according to their separations of cis/trans isomers. The separation of cis/trans isomers was found to be poorer on BDS, but this phase makes the separation of positional isomers possible.

Figure 12 shows a chromatogram on BDS at 180°C, on columns possessing some 100,000 theoretical plates. The other chromatographic conditions were as described above.

From the same group [206] came another comparison of the BDS, DEGS, and Apiezon-L phases. The authors concluded that optimal separation of all

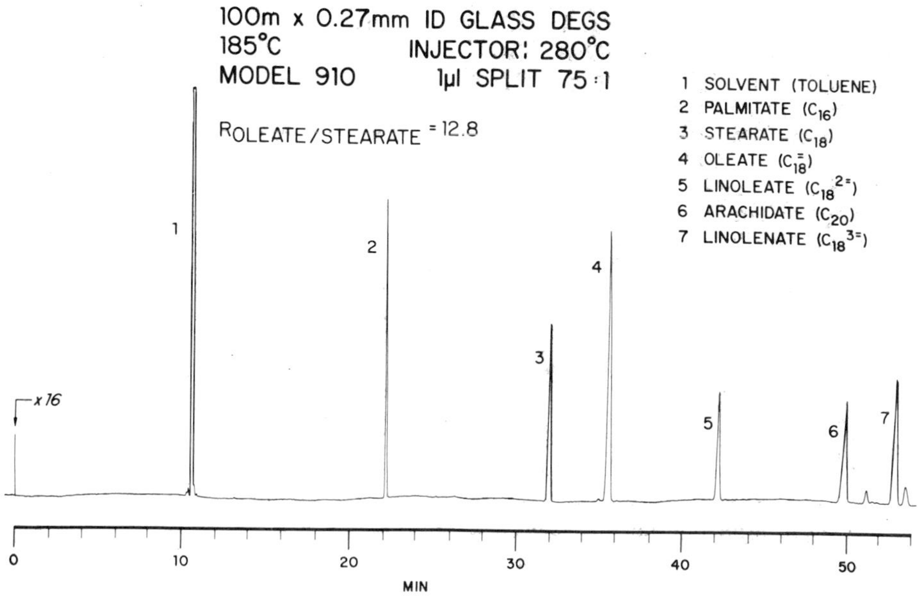

Figure 7. Analysis of fatty acid methylester sample. Column: 100 m × 0.27 mm ID open tubular coated with DEGS liquid phase; glass tubing. Column temperature: isothermal at 185°C. Resolution (R) between methylstearate and methyloleate peaks is 12.8. (Source: Ref. 198. Reproduced by permission of Perkin Elmer, Inc.)

fatty acids isolated from butterfat could only be obtained by application of several columns with different polarities.

Figure 13 shows a chromatogram of fatty acid methylesters from butterfat on Apiezon-L, Fig. 14 a chromatogram of the same mixture on DEGS, and Fig. 15 a chromatogram of the same mixture on BDS for comparison of the separation achieved by the different liquid phases. An FID was used in all three cases. Column temperatures were 200, 170, and 180°C and the carrier gas pressures 2.1, 1.1, and 2.1 kp cm^{-2} for the Apiezon-L, the DEGS, and the BDS columns, respectively. The columns were 45 m × 0.25 mm ID, 45 m × 0.25 mm ID, and 100 m × 0.25 mm ID with theoretical numbers of 40,000, 60,000, and 190,000 for the Apiezon-L, the DEGS, and the BDS columns, respectively.

Lin and co-workers [207] described a method of coating glass capillary columns with PZ-176. The basic difficulty of breakup of the liquid film on the glass surface [208] was overcome by combining very fine particles of silanized silicic acid (Silanox 101) with the liquid phase on the inner wall of a smooth glass capillary column, thus providing a coating that does not break up on repeated heating and cooling. The procedure of German and

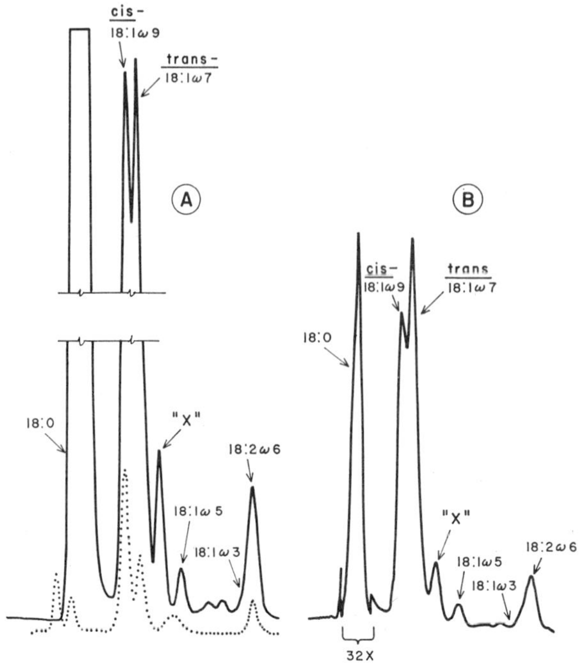

Figure 8. Parts of the chromatograms of C-18 monoethylenic and adjacent esters from two ruminant fats. In A the normal separation of cis-18:1ω9, ω7 and ω5 in nonruminant fat (fish oil) is shown by the dotted curve at bottom. Attenuation of B was 8× except for 32× for duration of 18:0 component. Attenuation of A was 16× for ruminant material, 16× for fish material. (Source: Ref. 199. Reproduced by permission of Preston Publications, Inc.)

Horning [209], involving a two-step coating process was used. The liquid phase was a phenyl ether-phenyl sulfone polymer which can be used in the temperature range of 180 to 320°C. The columns showed efficiencies of 1700 to 1800 theoretical plates m^{-1}.

Similar columns were used by Lin and Horning [210] for the analysis of fatty acid methylesters from human plasma phosphatidyl cholines and cholesterol esters. Lipids were extracted according to the method of Folch and co-workers [49], with subsequent separation by thin-layer chromatography. Transesterification was carried out by addition of methanol and sulfuric acid, followed by heating at 60°C for 3 hr. The methylesters were extracted into isooctane.

Figure 16 shows a chromatogram of the fatty acid methylesters from phosphatidyl cholines of a stroke patient. The column used was a 32 m × 0.25 mm ID glass capillary coated with PZ-176. Injector temperature programming was set at 2°C min^{-1}.

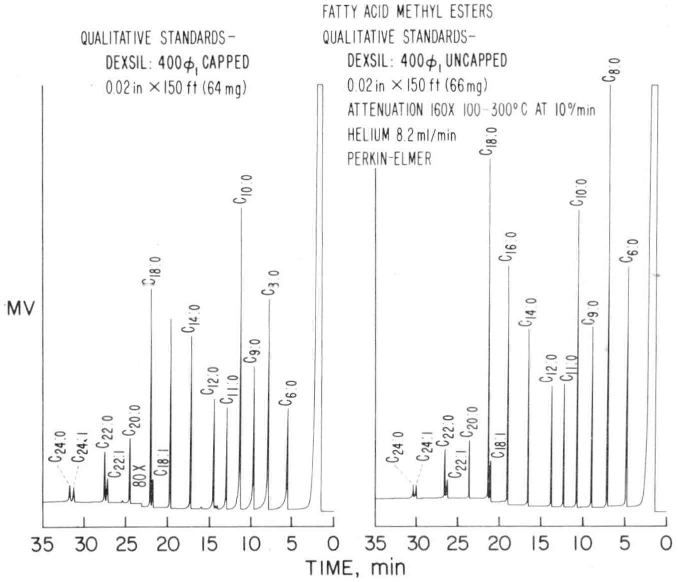

Figure 9. Separation of saturated and unsaturated fatty acid methylesters on capped and uncapped Dexsil 400 GC stationary phase. (Source: Ref. 201. Reproduced by permission of the American Chemical Society.)

Schomburg and Husmann [159] compared the separation of a mixture of fatty acid methylesters on two different stationary phases, the first being Carbowax 20M and the second being DEGS, providing a higher polarity. The number of carbon atoms of the fatty acids tested ranged from C 14 to C 18. The Carbowax column proved to be more stable at temperatures up to 260°C. A separation of the esters of oleic and elaidic acid could not be accomplished. This could be overcome on the DEGS column, which showed lesser stability than the Carbowax column. Figures 17 and 18 show chromatograms of the same sample on Carbowax 20M and DEGS, respectively. Chromatographic conditions are given in the legends to the figures.

Flanzy and co-workers [211] used a Carbowax 20M column for the analysis of nutritionally important fats and oils. The authors stated that Carbowax 20M combines the advantages of the nonpolar but very efficient Apiezon phases with those of the BDS phase, which is less efficient but exhibits a higher-resolution capability than do nonpolar columns. Hence the advantages combined in the Carbowax 20M phase are high efficiency, high resolution, and high stability. The capillaries were of stainless steel but are included here because of the good separations shown.

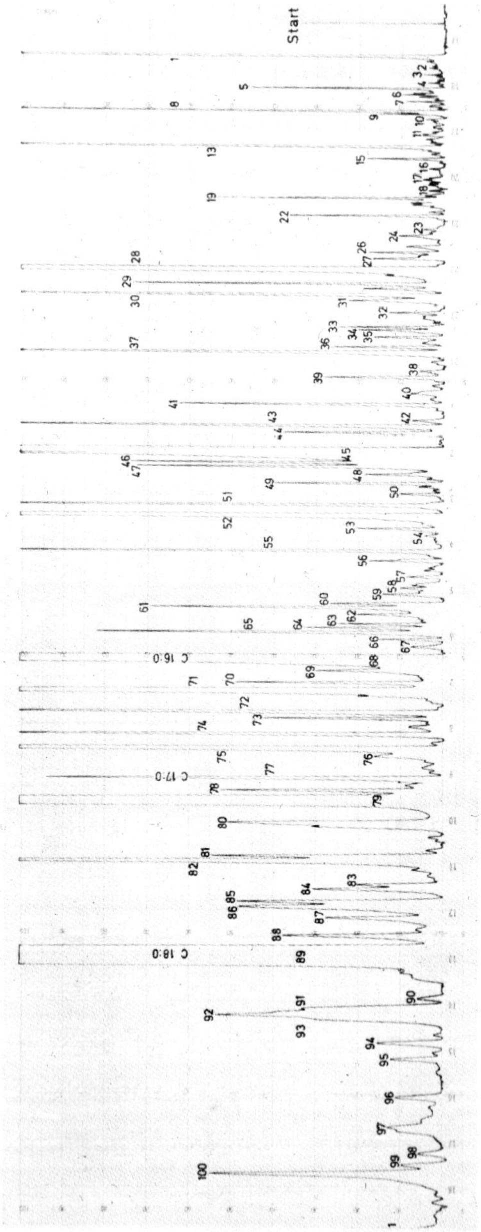

Figure 10. Gas chromatogram of methylesters of branched-chain fatty acids derived from the lipids of human milk. Open tubular column 0.25 mm × 50 m coated with polyphenylether (OS-138, Perkin Elmer); temperature program 160 to 220°C, 1°C min^{-1}; carrier gas He, 1 ml min^{-1}; injection block 280°C; molecule separator 200°C; sample size 20 nl injected without splitting; chart speed 0.5 cm min^{-1}. Peaks: 1—methyl-n-decanoate, 2—methyl-3 and 4-methyldecanoate, 3—ethyldecanoate and methyl-7- and 5-methyldecanoate, 4—methyl-9-methyldecanoate, 5—methyl-8-methyldecanoate, 6—methyl-n-undecanoate, 7—methyl-n-undecenoate, 8—methyl-3-methylundecanoate, 9—methyl-4- and 5-methylundecanoate, 10—methyl-10-methylundecanoate, 11—methyl-n-dodecanoate, 12—methyl-3-methyldodecanoate, 13—methyl-5- and 6-methyldodecanoate, 14—methyl-4-methyldodecanoate, 15—methyl-2-, 7-, and 8-methyldodecanoate, 16—methyl-11-methyldodecanoate, 17—methyl-10-methyldodecanoate, 18—2, 6-di-tert-butyl-p-cresol, 19—methyl-n-tridecanoate, 20—methyl-n-tridecenoate, 21—methyl-3,7,11-trimethyldodecanoate, 22—methyl-5-methyltridecanoate, 23—methyl-4-methyltridecanoate, 24—methyl-12-methyltridecanoate, 25—methyl-11-methyltridecanoate, 26 and 27—isomers of peak 18, 28—methyl-n-tetradecanoate, 29—methyl-7- and 8-methyltetradecanoate, 30 and 31—isomers of peak 18, 32—methyl-5- and 6-methyltetradecanoate, 33—methyl-4,8,12-trimethyltridecanoate, 34—methyl-4- and 2-methyltetradecanoate, 35—methyl-13-methyltetradecanoate, 36—methyl-12-methyltetradecanoate, 37—solvent impurity, 38—methyl-n-pentadecanoate, 39—methyl-n-pentadecanoate, 40—methylpentadecanoate, 41—methyl-5,9,13-trimethyltetradecanoate, 42—methyl-4-methylpentadecanoate, 43—methyl-14-methylpentadecanoate, 44—methyl-13-methylpentadecanoate, 45—solvent impurity, 46—methyl-n-hexadecanoate, 47—methyl-9-cyclohexylnonanoate, 48—solvent impurity, 49—o-dimethyl-phthalate, 50—methyl-2,6,10,14-tetramethylpentadecanoate, 51—methyl-7- and 8-methylhaxadecanoate, 52—methyl-4-methylhexadecanoate, 53—methyl-12-methylhexadecanoate, 54—methyl-15-methylhexadecanoate, 55—methyl-14-methylhexadecanoate, 56—methyl-n-heptadecanoate, 57, 58, and 59—solvent impurities, 60—methyl-9-,5-, and 10-methylheptadecanoate, 61—methyl-3,7,11,15-tetramethylhexadecanoate, phytanic acid methylester, 62—methyl-11- and 12-methylheptadecanoate, 63—methyl-16-methylheptadecanoate, 64—solvent impurity, 65—methyl-n-octadecanoate, 66—methyl-11-cyclohexylundecanoate, 67—oleic acid methylester, 68—linoleic acid methylester, 69—methyl-10- and 11-methyloctadecanoate, 70—methyl-n-nonadecanoate, 71—di-isobutyl-phthalate, 72—not identified, 73—n,i-butyl-phthalate. (Source: Ref. 202. Reproduced by permission of North Holland Publishing Company, Amsterdam.)

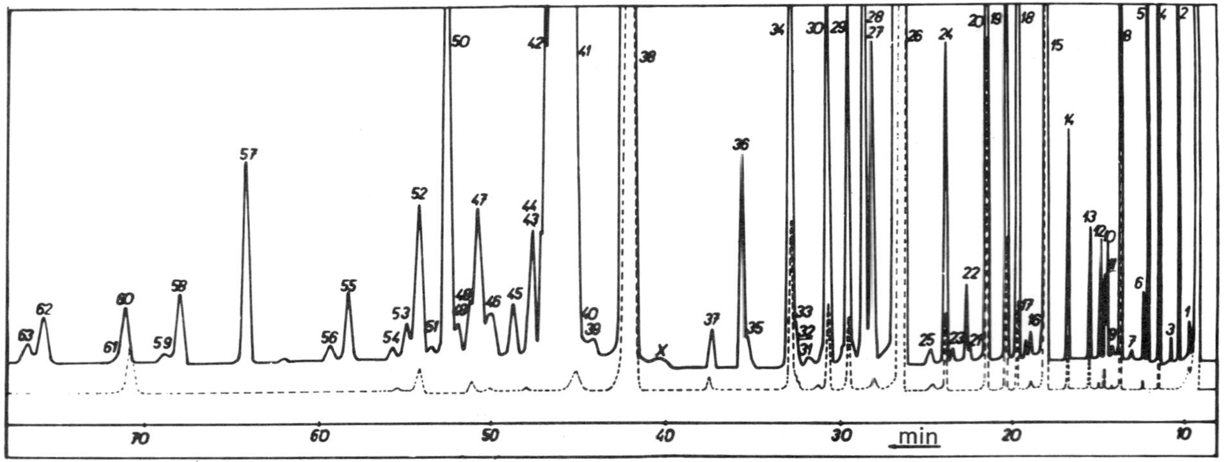

Figure 11. Separation of the methylesters of butterfat acids on BDS at 180°C. Solid line—original sample; dotted line—sample after hydrogenation. Peaks: 1—7:0, 2—8:0, 3—9:0, 4—10:0, 5—10:1, 6—11:0, 7—iso-12:0, 8—12:0, 10—12:1, 11—12:1 + iso-13:0, 12—12:1, 13—13:0, 14—iso-14:0, 15—14:0, 18—14:0 + iso-15:0, 19—anteiso-15:0, 20—15:0, 22—15:1, 24—iso-16:0, 26—16:0, 27—trans-16:1, 28—cis-16:1, 29—iso-17:0, 31—anteiso-17:0, 34—17:0, 36—17:1, 37—iso-18:0, 38—18:0, 41—cis-18:1 (9), 50—cis,cis-18:2, 52—19:0, 57—cis,cis,cis-18:3 (9,12,15), 60—20:0, 62—20:1. (Source: Ref. 203. Reproduced by permission of Elsevier Scientific Publishing Company, Amsterdam.)

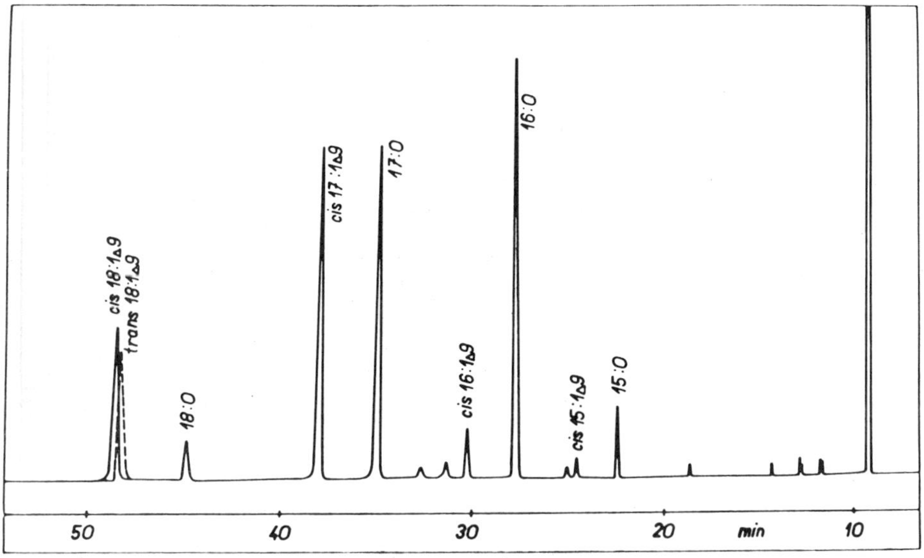

Figure 12. Separation of the methylesters of fatty acids isolated from biological material by capillary gas chromatography on BDS at 180°C. Methylelaidate (trans-18:1 9) (dotted peak) was added to the mixture. (Source: Ref. 205. Reproduced by permission of the American Oil Chemists' Society.)

Figure 19 shows a chromatogram of fatty acid methylesters from human milk fat. Figure 20 shows a chromatogram of fatty acid methylesters from pig leaf fat and from sheep peritoneal fat on column of 100 m × 0.5 mm ID with 100,000 theoretical plates. Temperature of the column and the FID was maintained at 190°C, and the injection port temperature was 210°C. The split ratio was 1:100. Carrier gas was nitrogen at a flow of 2 ml min^{-1}, corresponding to a pressure of 0.6 bar. The method could be used for routine analyses of usually well-defined raw materials.

Sandra and co-workers [212] have experimented with "whisker" surfaces in glass capillary columns, representing yet another roughening technique, which allows efficient coating of the column with any stationary phase. The procedure was introduced by Schieke and co-workers [213,214] and modified by Sandra and Verzele [215]. Advantages and disadvantages of whisker surfaces are described.

Figure 21 shows a chromatogram of a standard fatty acid methylester sample with Silar-5CP as liquid phase on a whiskered surface. The separation of cis and trans 18:1 (oleic and elaidic acid) is quite good. The separation powers for additional isomers of oleic acid should be investigated.

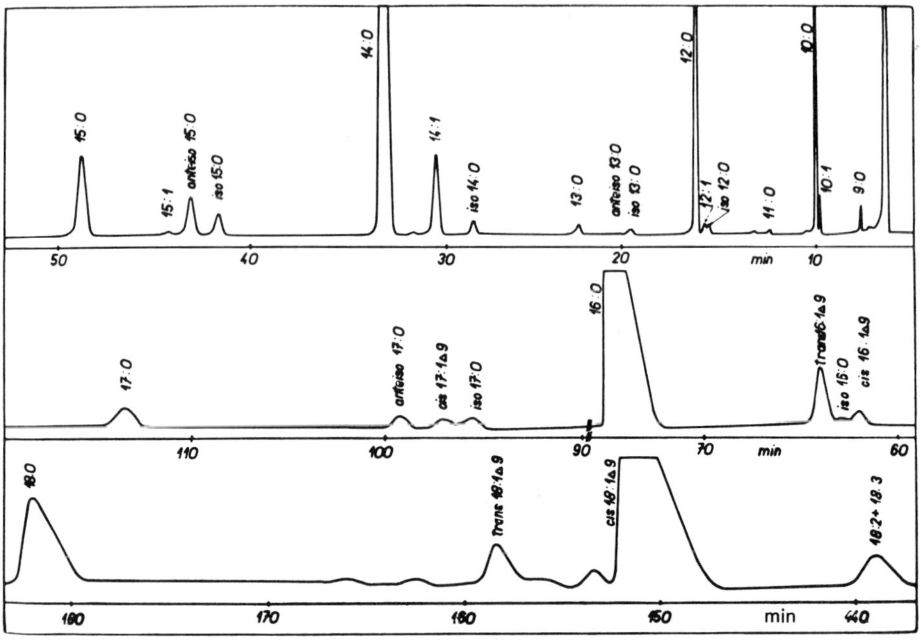

Figure 13. Chromatogram of fatty acid methylesters from butterfat analyzed on Apiezon-L. For further details see text. (Source: Ref. 206. Reproduced by permission of Volkswirtschaftlicher Verlag GmbH, Kempten, West Germany.)

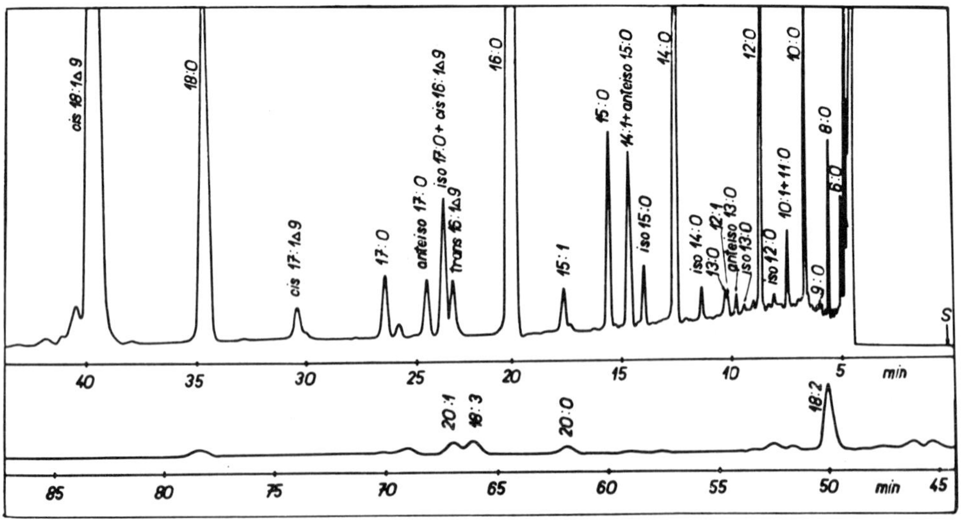

Figure 14. Chromatogram of the same sample as in Fig. 13 analyzed on DEGS. (Source: Ref. 206. Reproduced by permission of Volkswirtschaftlicher Verlag GmbH, Kempten, West Germany.)

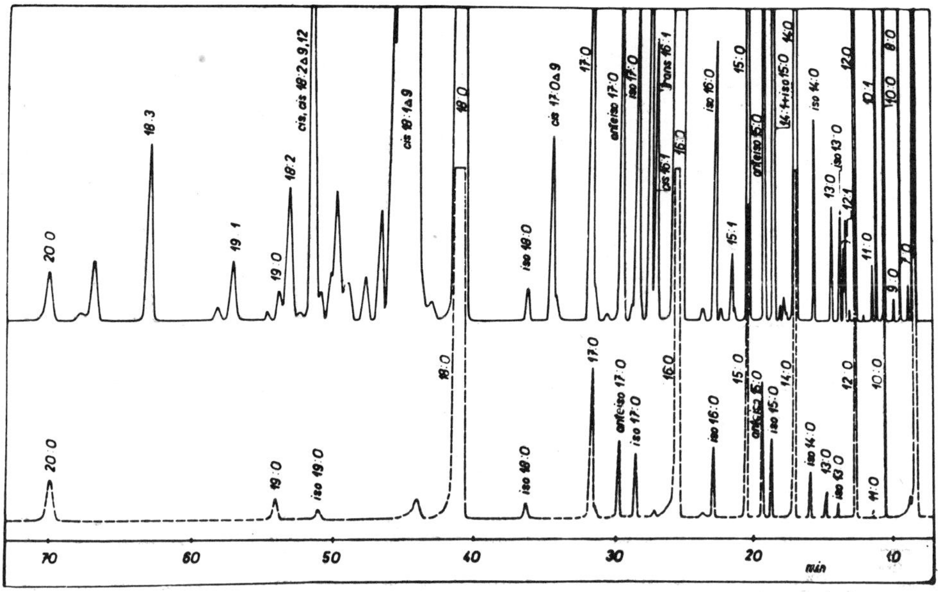

Figure 15. Chromatogram of the same sample as in Fig. 13 analyzed on BDS. (Source: Ref. 206. Reproduced by permission of Volkswirtschaftlicher Verlag GmbH, Kempten, West Germany.)

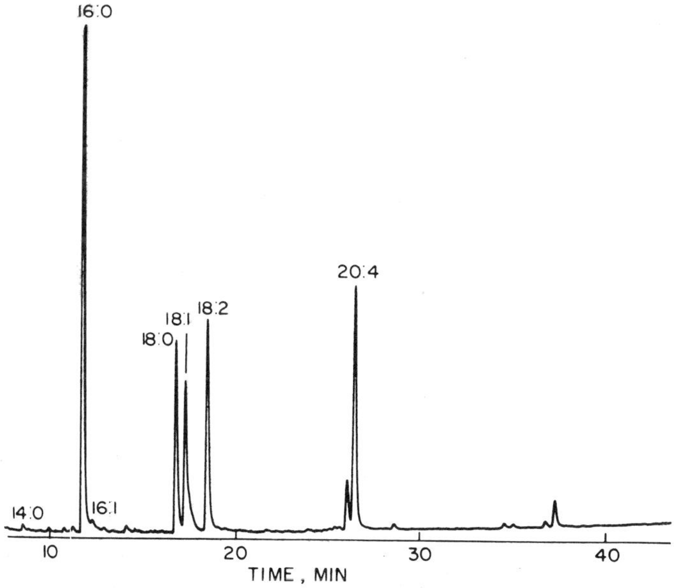

Figure 16. GC analysis of long-chain methylesters from plasma phosphatidyl cholines of a stroke patient. (Source: Ref. 210. Reproduced by permission of Elsevier Scientific Publishing Company, Amsterdam.)

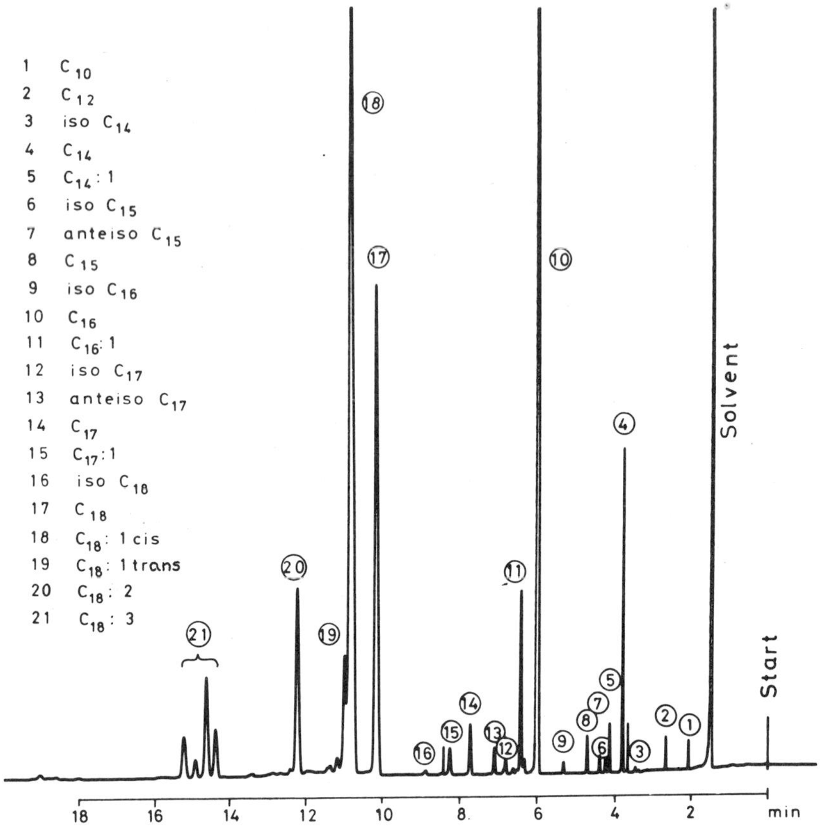

Figure 17. Fatty acid methylesters (sample courtesy of Henkel and Cie., Düsseldorf). Column: 50 m × 0.25 mm CW 20M; glass, 200°C, 2.0 bar H_2. (Source: Ref. 159. Reproduced by permission.)

The highly polar cyanopropyl siloxanes Silar-5CP and Silar-10CP are of utmost interest for the separation of cis/trans as well as positional isomers.

Verzele and Sandra [216] have tested Superox-4, a polyethylene glycol of molecular weight 4×10^6. The phase has the advantages of high-temperature stability and good adherence to glass. Its separation efficiency is comparable to Carbowax 20M. In one example, the authors showed a chromatogram of a separation of fatty acid methylesters up to linolenic acid (C 18:3) in 138 sec. For routine analysis, where the separation of cis/trans isomers or positional isomers is not needed, this separation offers a remarkable advantage by shortening the analysis time to a minimum.

Figure 22 shows the above-described separation. Chromatographic conditions are given in the legend to the figure.

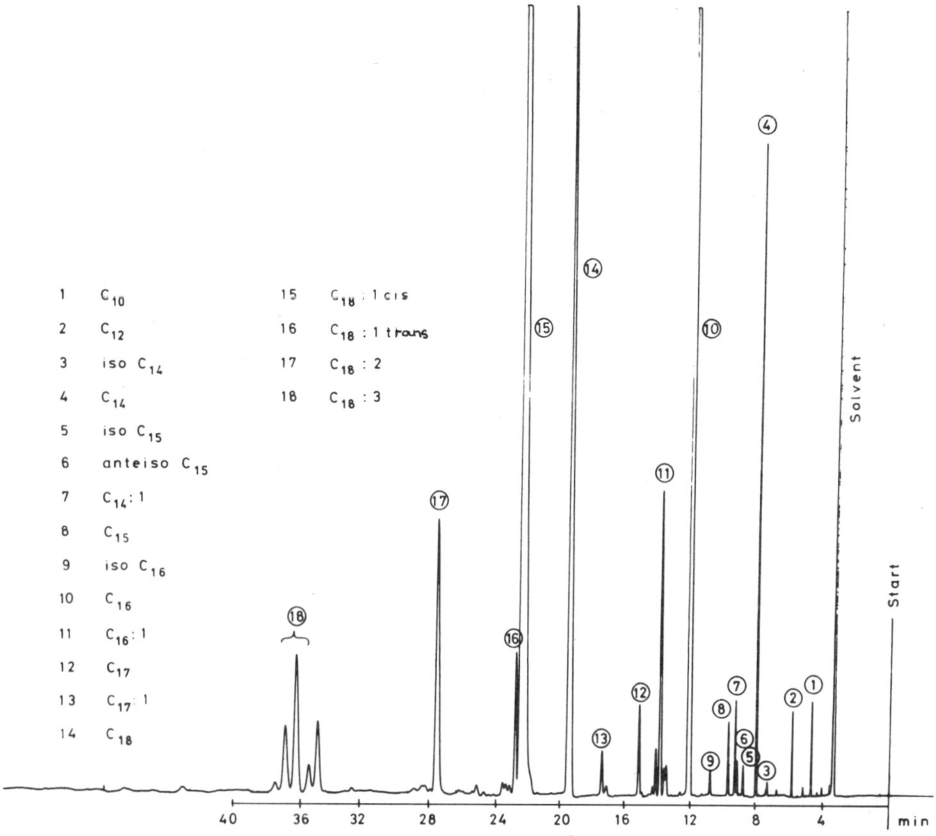

Figure 18. Fatty acid methylesters (sample courtesy of Henkel and Cie., Düsseldorf). Column: 50 m × 0.25 mm DEGS; glass, 180°C, 0.5 bar N_2. (Source: Ref. 159. Reproduced by permission.)

Heckers and co-workers [217] compared the separation efficiencies of the liquid phases Silar-10C, Silar-9CP, SP-2340, and OV-275, for fatty acid methylesters on packed columns. The resolution for cis/trans isomeric pairs of monounsaturated fatty acids decreased both with increasing oven temperature and increasing nitrogen flowrate on all four phases. All of the packings were found suitable for the separation of fatty acid isomers, but different lots of Silar-10C possessed different chromatographic properties, and thus could not be recommended by the authors. From the same group [218] came a report on glass capillary chromatography of fatty acid methylesters on SP-2340. The fatty acid composition of phospholipids from human red cell ghosts was investigated. The column used was a 30 m × 0.3 mm ID glass capillary, coated with the cyanopropylsiloxane SP-2340 with an average molecular weight of 2800. The temperature program entailed heating for 5 min at a

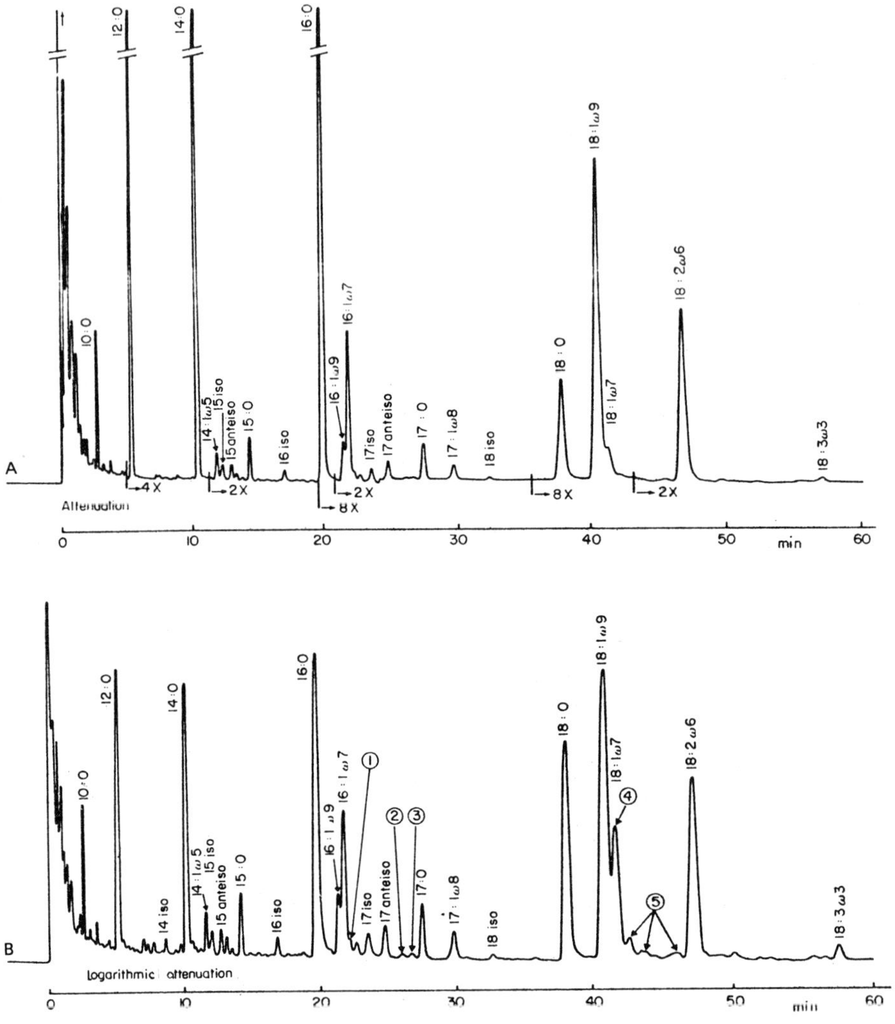

Figure 19. Comparison of standard linear recorder response (A) and logarithmic recorder response (B) for same sample of methylesters of fatty acids from human milk. Note manual attenuations in linear recording mode and the fact that all peaks are on scale in the logarithmic mode of recording. (Source: Ref. 211. Reproduced by permission of Preston Publications, Inc.)

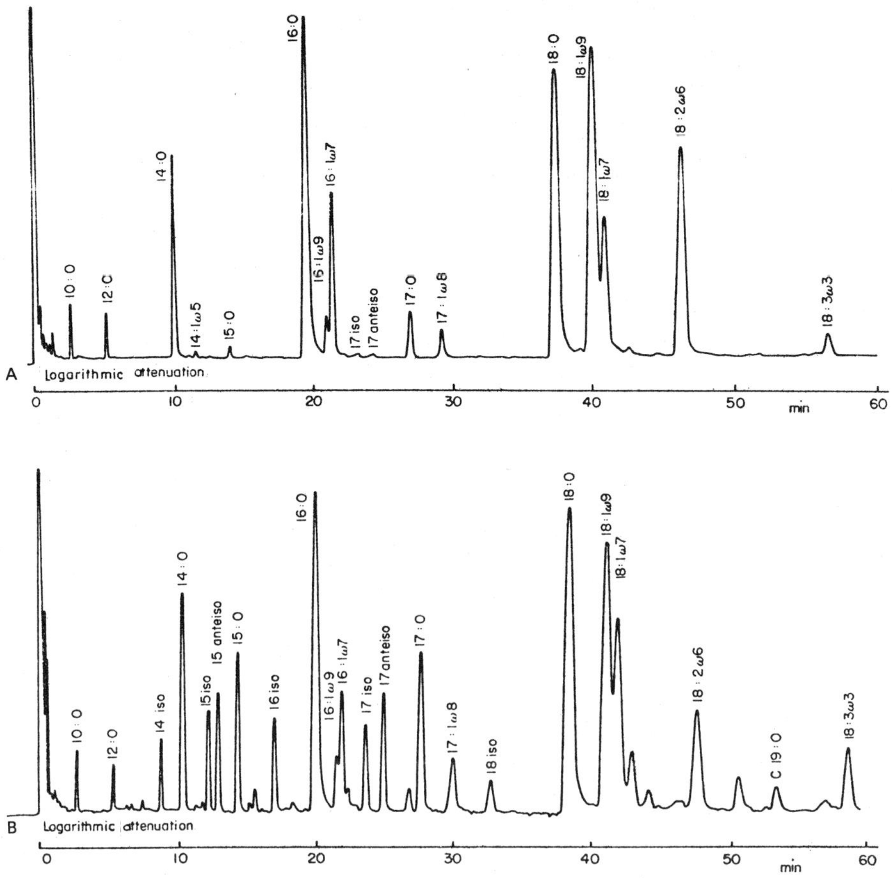

Figure 20. Comparison of methylesters of fatty acid of pig leaf fat (A) and sheep peritoneal fat (B), both recorded in logarithmic mode. The additional complexity of the ruminant (sheep) fat is apparent. (Source: Ref. 211. Reproduced by permission of Preston Publications, Inc.)

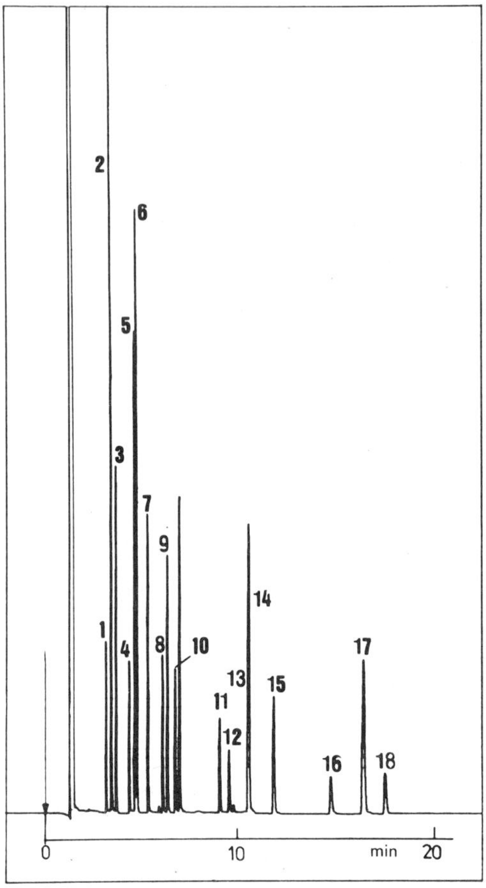

Figure 21. Gas chromatogram of a fatty acid methylester standard mixture. Column: 50 m × 0.25 mm ID deactivated with *N*-cyclohexyl-3-azetidinal (CHAZ) and coated with Silar-5CP. Column temperature 240°C, carrier gas flow ∿2 ml min^{-1} hydrogen. Peaks: 1—C16:0, 2—C16:1, 3—C17:0, 4—C18:0, 5—C18:1 (trans), 6—C18:1 (cis), 7—C18:2, 8—C18:3, 9—C20:0, 10—C20:1, 11—C20:4, 12—C22:0, 13—C20:5, 14—C22:1, 15—C23:0, 16—C24:0, 17—C24:1, 18—C22:6. (Source: Ref. 212. Reproduced by permission of Dr. Alfred Hüthig Verlag GmbH, Heidelberg, West Germany.)

rate of 5°C min^{-1}, followed by heating at 0.5°C min^{-1} to 35.5 min, and finally at 1°C min^{-1} to the maximum temperature of 190°C, which was maintained for an additional 10 min. Analysis time was 95 min. Carrier gas was nitrogen at a flowrate of 2.5 ml min^{-1} at a split ratio of 1:70. The injection port and detector were maintained at 240°C. The separation number TZ, a measure of column efficiency, based on the degree to which two members of a homologous series are separated, which takes into account the sharpness of the peaks and their relative retentions, was found to be 7.76. It was calculated from the equation

$$TZ = \frac{t_{R(B)} - t_{R(A)}}{W_{0.5(A)} + W_{0.5(B)}} - 1$$

where t_R is the retention time of substances A and B, and $W_{0.5}$ is their width at half-height. Calculated per meter column length TZ/m was 0.26. The number of theoretical plates n was found to be 7674 or 256 n/m. The loadability of the column for decanoic acid methylester was found to be 1.80×10^{-6} g.

In spite of the relatively low separation number (for a capillary column), the separation efficiency for geometric and positional isomers of fatty acid methylesters was very high.

All five pairs of the geometric isomers of the monoethylenic fatty acids, and the all-cis and all-trans components of octadecadiethylenic and octadecatriethylenic acids exhibited separations better than 4.6σ. Although the authors emphasize that not all technical problems of coating these capillaries have been solved, the cyanopropylsiloxane phases such as SP-2340 and Silar-10C are regarded as future phases of choice for the separation of fatty acid methylesters from complex mixtures.

Figures 23 and 24 show the applicability of SP-2340 capillary columns to the separation of complex fatty acid mixtures. These were generously supplied by Dr. Heckers and have not been published previously. Further details are given in the legends to the figures.

The above-described method has been successfully applied by the same authors to the analysis of trans-isomeric fatty acids in human myocardicum, jejunum, and aorta in studies of their relationship to different degrees of atherosclerosis [219] as well as to the analysis of trans-isomeric acids present in West German margarines, shortenings, and frying and cooking fats [220].

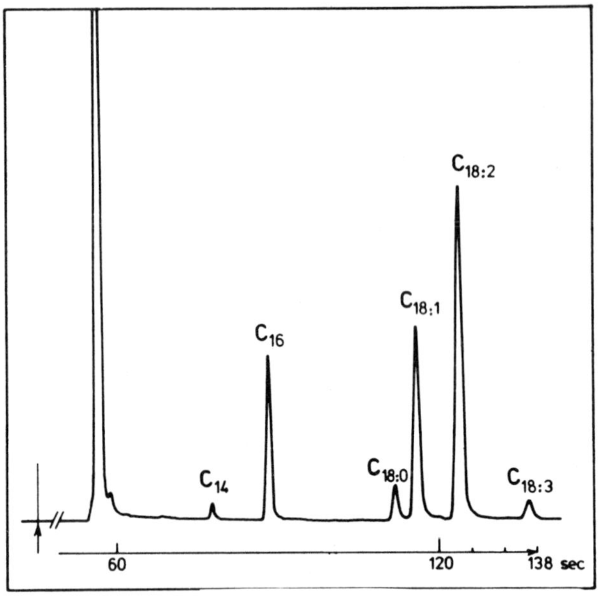

Figure 22. FAME analysis on a 25-m × 0.5-mm ID NaCl dendrite column coated with Superox-4. Temperatures: column, injector, and detector, 250°C. Carrier gas (hydrogen) flowrate: 5 ml min^{-1}. (Source: Ref. 216. Reproduced by permission of Elsevier Scientific Publishing Company, Amsterdam.)

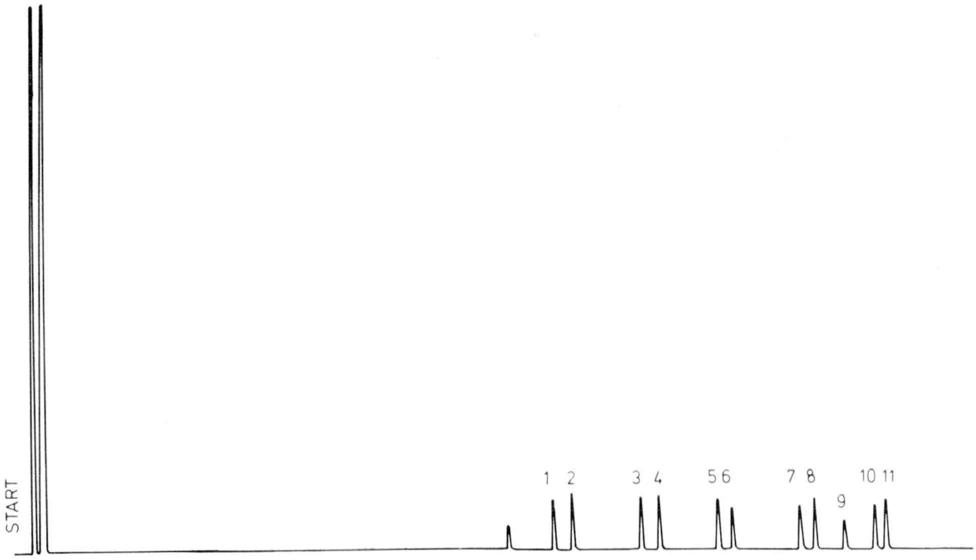

Figure 23. Chromatogram of a standard mixture of cis/trans fatty acid methylesters on SP-2340. For chromatographic conditions, see text. Peaks: 1—C 14:1 trans, 2—C 14:1 cis, 3—C 15:1 trans, 4—C 15:1 cis, 5—C 16:1 trans, 6—C 16:1 cis, 7—C 17:1 trans, 8—C 17:1 cis, 9—C 18:0, 10—C 18:1 trans, 11—C 18:1 cis. (Reproduced courtesy of Dr. Heckers, University of Giessen, West Germany.)

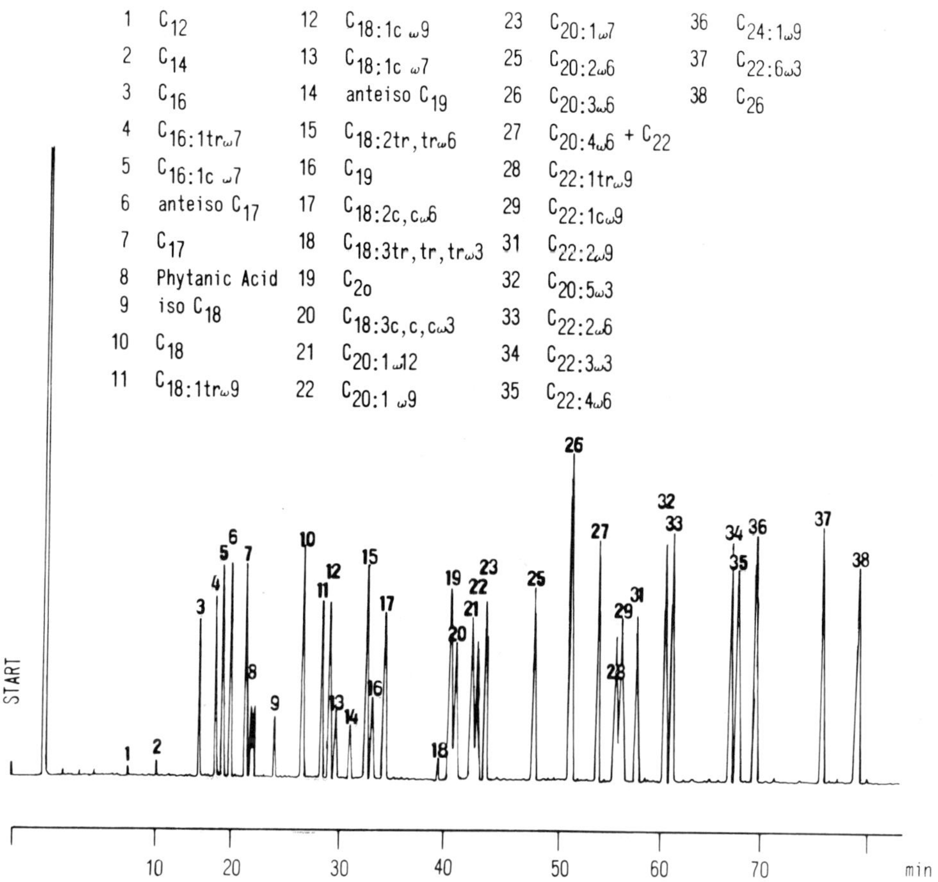

Figure 24. Chromatogram of a standard mixture of fatty acid methylesters including positional and geometrical isomers on SP-2340. (Reproduced courtesy of Dr. Heckers, University of Giessen, West Germany.)

Our first experiences with glass capillary chromatography of fatty acid methylesters were reported at Hindelang I in 1975 [180]. In investigations of fatty acid metabolism we extracted lipids from human serum according to the method of Folch and co-workers [49]. Transmethylation was performed according to the method of Morrison and Smith [103], as described in Section 2.2. Purified hydrogen was used as carrier gas with an inlet pressure of 0.9 atm, equivalent to a flow of 4 ml min^{-1} at room temperature. Flows of hydrogen and air in the FID were 12 and 150 ml min^{-1}, respectively.

The injection port temperature was held at 250°C. Injection was performed with an automatic solid sampler as described in Sec. 3.1 at a temperature of 40°C with subsequent heating to 95°C, according to the "cold trapping" method described by Grob and Grob [150]. From the initial temperature the column was programmed to 198°C at 1.2°C min^{-1}. The upper temperature was held for 7 min before cooling.

The 50 m × 0.33 mm ID glass capillary column was purchased from H. and G. Jaeggi, Trogen, Switzerland. It was coated with FFAP (free fatty acid phase), a modified polyether phase. The separation number TZ was calculated at 47.65, or TZ/m = 0.93. Good nonpolar columns values reach above 1.0, but for a polar column this represents an outstanding separation number. The number of theoretical plates was 227052 or n/m = 4541. The partition ratio k, which represents a measure of how long a solute spends in the stationary phase relative to the time it spends in the gas phase, was calculated by $k = t'_R/t'_M$ where t'_R is the adjusted retention time. The gas holdup time t'_M was calculated under isothermal conditions at 90°C and a recorder chart speed of 80 in hr^{-1} with octanoic (a), nonanoic (b), and decanoic (c) acid by using the equation

$$t_M = \frac{t_{M+R}a\, t_{M+R}c - (t_{M+R}b)^2}{t_{M+R}a + t_{M+R}c - 2(t_{M+R}b)}$$

where t_{M+R} is the total retention time of the indicated solute. The partition ratio of decanoic acid was found to be 16.04. By using these data in the equation

$$C = 0.05Md^3(1 + k) \times 10^{-6} \text{ g}$$

where M is the molecular weight of decanoic acid and d is the inner diameter of the column, we calculated the capacity (C) of the column to be 5.28×10^{-6} g.

The standard deviations of relative retention times do not exceed 0.26% and those of the weight percent values do not exceed 1.53% of the mean values for 10 chromatograms.

The correction factors ranged from 965 to 992 for saturated esters and from 1010 to 1100 for unsaturated fatty acid esters. Correction factors changed by about 10% after 1 year of column use, and column to column variations were similar.

Figure 25 shows a chromatogram of a standard mixture of fatty acid methylesters. A 4.6σ separation of cis- and trans-palmitoleic acid (C 16:1ω7) was achieved, meaning that peak overlap did not exceed more than 1 to

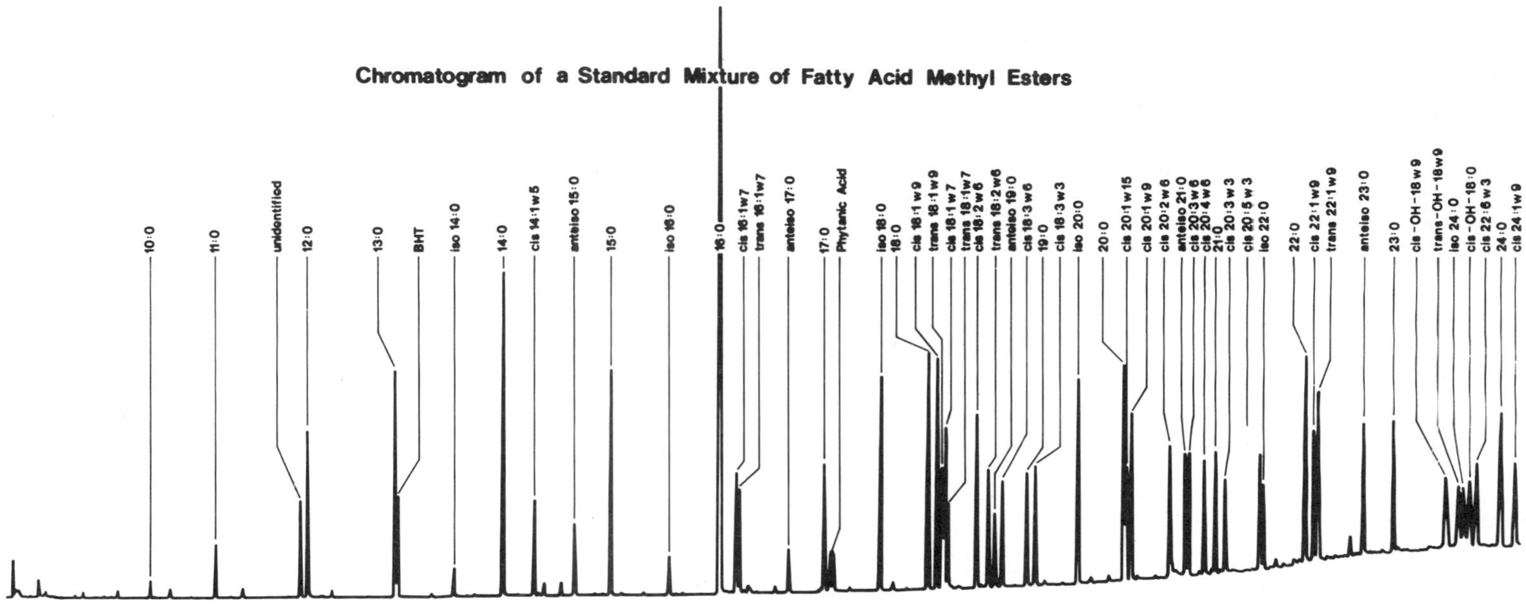

Figure 25. Chromatogram of a standard mixture of fatty acid methylesters including positional and geometrical isomers on FFAP. Chrcmatographic conditions are given in the text.

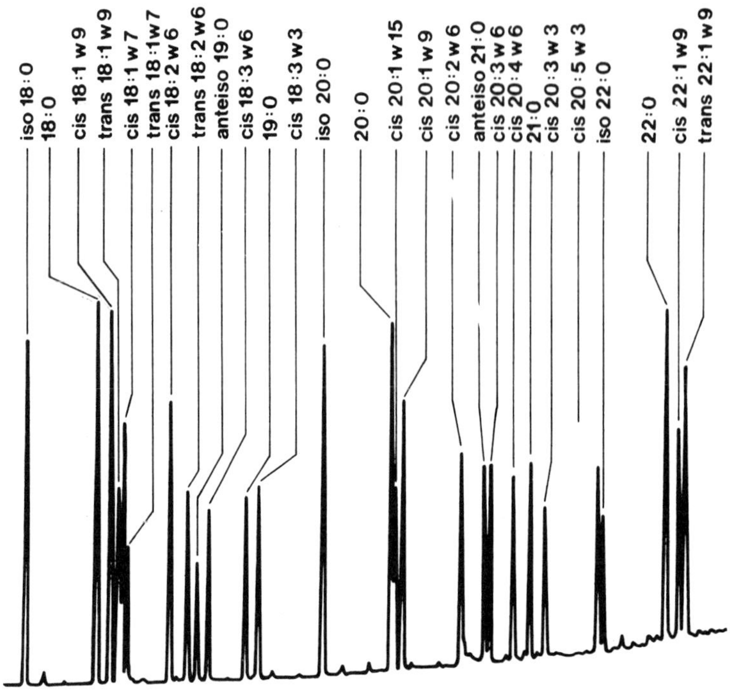

Figure 26. Part of Fig. 25 in higher magnification.

2% of the peak area value, thus providing a precision of about ±0.3% with electronic integration. On close examination of the chromatogram (Fig. 26) one can see that the important separations of cis- and trans-linoleic acid (C 18:2ω6), of α- and λ-linolenic acid (cis-18:3ω3, cis-18:3ω6), of 8,11, 14- and 11,14,17-eicosatrienoic acid (cis-20:3ω6, cis-20:3ω3), and of 5- and 11-eicosaenoic acid (cis-20:1ω15, cis-20:1ω9) are all complete. Four isomers of C 18:1 were separated: oleic (cis-18:1ω9), elaidic (trans-18:1 ω9), cis-vaccenic (cis-18:1ω7), and trans-vaccenic (trans-18:1ω7). The separation of cis-18:1ω6 and trans-18:1ω6 acid esters is possible as well.

The separation of other cis and trans isomers, e.g., cis- and trans-13-docosenoic acid (erucic and brassidic acid; 22:1ω9) and cis- and trans-ricinoleic acid (OH-18:1ω9) were all either baseline or 4,6σ separations. In general, the positional isomers were separated better from each other than were the geometrical cis/trans isomers.

Figure 27 shows a chromatogram of the fatty acid methylesters from human plasma phospholipids, revealing that most of the fatty acids from the standard mixture occur in plasma, too.

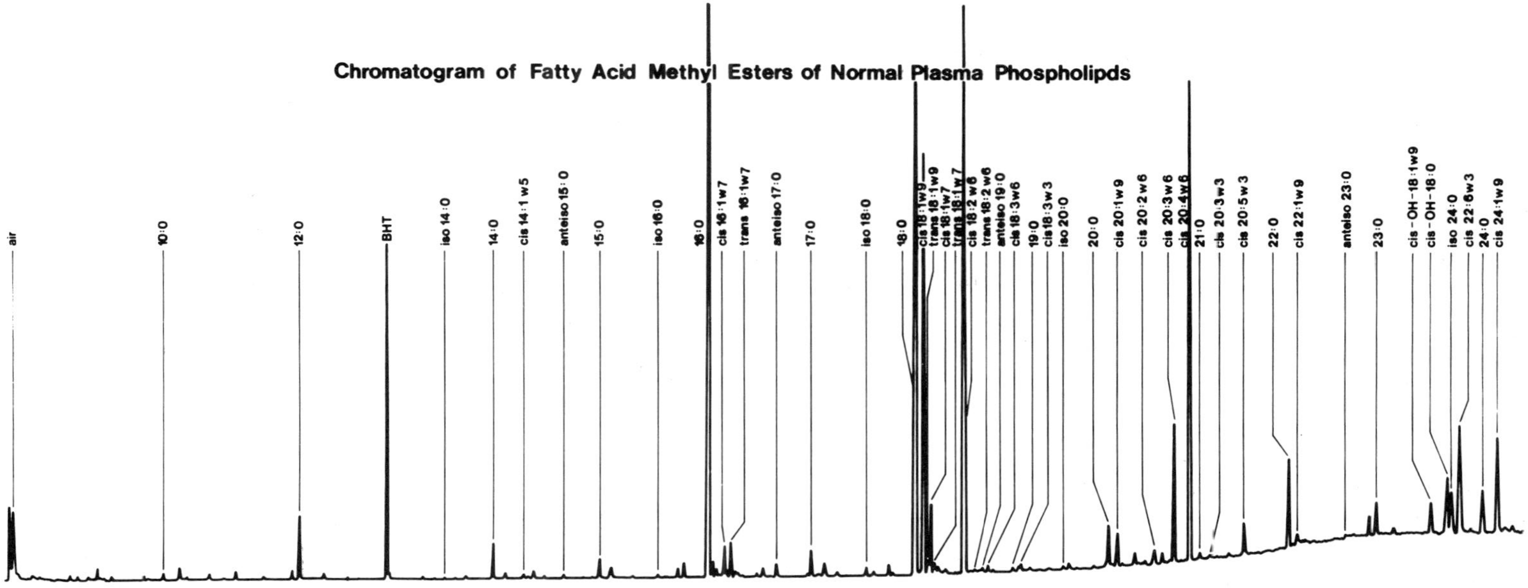

Figure 27. Chromatogram of fatty acid methylesters from normal human phospholipids on FFAP. Chromatographic conditions are given in the text.

Figure 28 shows a chromatogram of normal plasma triglycerides and Fig. 29 a chromatogram of normal plasma cholesterol esters. The method can also be applied to the analysis of fatty acids in lipid fractions from other sources in the human body (e.g., Ref. 221) and has found acceptance in medical research [222].

5. SUMMARY

A number of extraction procedures have been suggested; the investigator must choose from among acid catalysis with methanolic sulfuric acid, hydrochloric acid, or BF3-methanol, whereas alkaline catalysis seems to give poor reproducibility. With this exception, most of the methods available are suitable for routine analysis.

Regarding injection techniques we have shown advantages and disadvantages of the widely used methods, and conclude that further work must be accomplished in the on-column injection technique. Further work must also be done on automatization of chromatography and data processing.

Regarding the columns one must consider the analytical problem: for routine analysis, where separation of positional or geometrical isomers is not needed, columns capable of shorter analysis times have been introduced. For the resolution of isomers it was shown that polyether phases with high separation numbers give separations equal or superior to cyanopropylsiloxane phases with smaller separation numbers. As soon as the technical problems occurring with the latter phases are solved, analyses superior to those demonstrated here should be possible. In general, it is apparent that the methods of analysis of fatty acids--either underivatized or as esters--have improved considerably during recent years, especially since 1973. Detection limits are much lower and separation efficiencies have been continually improved. Glass capillary chromatography of fatty acids has provided the lipid chemist with a powerful tool.

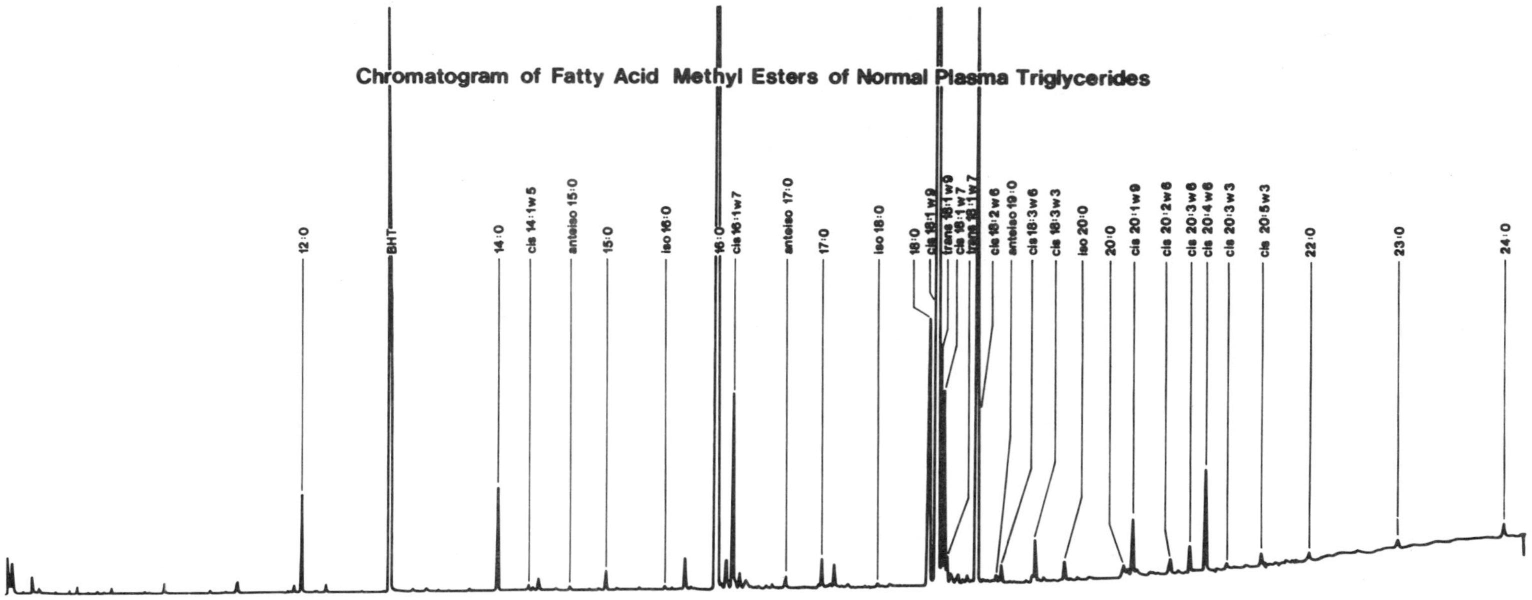

Figure 28. Chromatogram of fatty acid methylesters from normal human triglycerides on FFAP. Chromatographic conditions are given in the text.

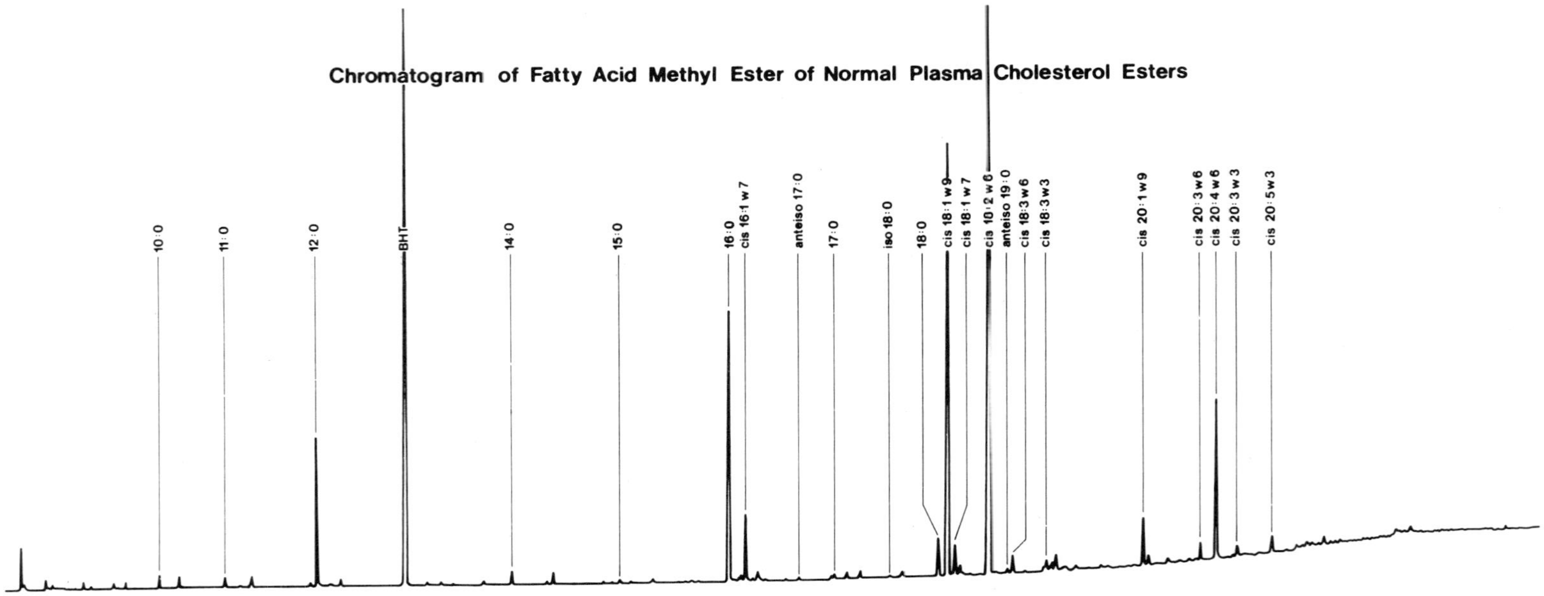

Figure 29. Chromatogram of fatty acid methylesters from normal human cholesterol esters on FFAP. Chromatographic conditions are given in the text.

REFERENCES

1. van Deenen, L. L. M., *Pure Appl. Chem.*, 25:25 (1971).
2. Dole, V. P., and Rizack, M. A., *J. Lipid Res.*, 2:90 (1961).
3. Skipski, V. P., Barclay, M., Barclay, R. K., Fetzer, V. A., Good, J. J., and Archivald, F. M., *Biochem. J.*, 104:340 (1967).
4. Olivecrona, T., Bengtsson, G., Höck, M., and Lindahl, U., in *Lipoprotein Metabolism* (h. Greten, ed.), Springer, Berlin, 1976, p. 13.
5. Cornwell, D. G., Kruger, F. A., Hanwi, G. J., and Brown, J. B., *Metabolism*, 11:840 (1962).
6. Christie, W. W., and Moore, J. H., *Biochim. Biophys. Acta*, 176:445 (1969).
7. Christie, W. W., and Moore, J. H., *Biochim. Biophys. Acta*, 210:46 (1970).
8. van Golde, L. M. G., Thomasi, V., and van Deenen, L. L. M., *Chem. Phys. Lipids*, 1:282 (1967).
9. Marai, L., and Kuksis, A., *J. Lipid Res.*, 10:141 (1969).
10. Hagenfeld, L., and Wahren, J., *Metabolism*, 24:1299 (1975).
11. Oette, K., Frese, W., and Phlippen, R., *Z. Ges. Exp. Med.*, 154:208 (1971).
12. Nestel, P. J., and Couzens, E. A., *J. Lipid Res.*, 7:487 (1966).
13. Lindgren, F. T., and Nichols, A. V., *Ann. N.Y. Acad. Sci.*, 94:55 (1961).
14. Goodman, D. S., and Shiratori, T., *J. Lipid Res.*, 5:30 (1964).
15. Wood, R., and Wiegand, R. D., *Lipids*, 10:746 (1975).
16. Wamerstrom, D. E., and Jenkin, H. M., *Biochim. Biophys. Acta*, 431:469 (1976).
17. Brenneman, D. E., McGee, R., Liepkalnis, V. A., and Spector, A. A., *Biochim. Biophys. Acta*, 388:301 (1975).
18. Goldsmith, D. P. J., and Jacobi, H. P., *Lipids*, 13:174 (1978).
19. Chait, A., Onitiri, A., Nicole, A., Rabaya, E., Davies, J., and Lewis, B., *Atherosclerosis*, 20:347 (1974).
20. Nestel, P. J., Havenstein, N., Scott, T. W., and Cook, L. J., *Aust. N.Z. J. Med.*, 4:497 (1974).
21. Peeters, J., Bloton, V., Soetewey, F., Declercq, B., and Vandamme, D., *Münch. Med. Wochenschr.*, 115:1358 (1973).
22. Grünert, A., Olbermann, M., and Grünert-Fuchs, M., *Med. Welt*, 27:1304 (1976).
23. Hagenfeldt, L., and Wester, P. O., *Acta Med. Scand.*, 194:357 (1973).
24. Alling, C., Dencker, S. J., Svennerholm, L., and Tichy, J., *Acta Med. Scand.*, 185:99 (1969).
25. Jaillard, J., Sezille, G., Scherpereel, P., and Fruchert, J. C., *Nutr. Metabol.*, 13:114 (1971).

26. Berghaus, G., Kurz, A., and Dotzauer, G., *Dtsch. Med. Wochenschr.*, 100:1233 (1975).

27. Chen, S., Mahadevan, V., and Zieve, L., *J. Lab. Clin. Med.*, 75:622 (1970).

28. Klenk, E., and Kahlke, W., *Hoppe-Seylers Z. Physiol. Chem.*, 333:133 (1963).

29. Herndon, J. A., Steinberg, D., Uhlendorf, B. W., and Fales, H. M., *J. Clin. Invest.*, 48:1017 (1969).

30. Kahlke, W., *Klin. Wochenschr.*, 42:1011 (1964).

31. Droese, W., Pape, E., and Stolley, H., *Durop. J. Pediat.*, 122:57 (1976).

32. Leyland, F. C., Fosbrooke, A. S., Lloyd, J. K., Segale, M. M., Tamir, J., Tomkins, R., and Wolff, O. H., *Arch. Dis. Child.*, 44:170 (1969).

33. Sgoutas, D., and Kummerow, F. A., *Amer. J. Clin. Nutr.*, 23:1111 (1970).

34. Holman, R. T., in *Progress in the Chemistry of Fats and other Lipids*, Vol. 9, Pergamon Press, Oxford, 1968, pp. 279ff.

35. Fiennes, R. N., Sinclair, A. J., and Crawford, M. A., *J. Med. Prim.*, 2:155 (1973).

36. Paulsrud, J. R., Witten, C. F., Stewart, S. E., and Holman, R. T., *Amer. J. Clin. Nutr.*, 25:897 (1972).

37. Collins, F. D., Sinclair, A. J., Royle, J. P., Coats, D. A., Maynard, A. T., and Leonard, R. F., *Nutr. Metabol.*, 13:150 (1971).

38. Richardson, T. J., and Sgoutas, D., *Amer. J. Clin. Nutr.*, 28:258 (1975).

39. Wene, J. D., Connar, W. E., and den Besten, L., *J. Clin. Invest.*, 56:127 (1975).

40. Shimoyama, T., Kikuchi, H., Press, M., and Thompson, G. R., *Gut*, 14:716, 1973.

41. James, A. T., and Martin, A. J. P., *Biochem. J.*, 50:679 (1952).

42. Bragdon, J. H., Eder, H. A., Gould, G., and Havel, R. J., *Circul. Res.*, 4:129 (1956).

43. Fredrickson, D. S., Levy, R. I., and Lees, R. S., *New. Engl. J. Med.*, 276:273 (1967).

44. McFarlane, A. S., *Nature*, 149:439 (1942).

45. Scanu, A., Lewis, L. A., and Bumpus, F. M., *Arch. Biochem. Biophys.*, 74:390 (1958).

46. Scanu, A., *J. Lipid Res.*, 7:295 (1966).

47. Scanu, A., and Schiano, N., *Riv. lst. Sieroter., Italy*, 29:276 (1954).

48. Scanu, A., and Schiano, N., *Riv. lst. Sieroter, Italy*, 29:457 (1954).

49. Folch, J., Lees, M., and Sloane Stanley, G. H., *J. Biol. Chem.*, 226:497 (1957).

50. Sperry, W. M., in *Methods of Biochemical Analysis*, Vol. 2, Interscience, New York and London, 1955, p. 83.

51. Cham, B. E., and Knowles, B. R., *J. Lipid Res.*, 17:176 (1976).

52. Nelson, G. J., *J. Lipid Res.*, 3:71 (1962).

53. Nelson, G. J., *J. Lipid Res.*, 8:374 (1967).

54. Wells, M. A., and Dittmer, J. C., *Biochemistry*, 2:1259 (1963).

55. Sigrist, J., Sigrist-Nelson, K., and Gitler, C., *Biochem. Biophys. Res. Commun.*, 74:178 (1977).

56. Wren, J. J., and Szezepanowska, *J. Chromatogr.*, 14:405 (1964).

57. Borgström, B., *Acta Physiol. Scand.*, 25:111 (1952).

58. Hirsch, J., and Ahrens, Jr., E. J., *J. Biol. Chem.*, 233:311 (1958).

59. Wren, J. J., *J. Chromatogr.*, 4:173 (1960).

60. Marinetti, G. V., ed., *Lipid Chromatographic Analysis*, Vol. 1, Marcel Dekker, New York, 1967.

61. Nelson, G. J., *J. Amer. Oil. Chem. Soc.*, 44:86 (1967).

62. Thenot, J.-P., Horning, E. C., Stafford, M., and Horning, M. G., *Anal. Lett.*, 5(4):217 (1972).

63. Brechbühler, H., Büchi, H., Hatz, E., Schreiber, J., and Eschenmoser, A., *Angew. Chem. Internat. Edit.*, 2:211 (1963).

64. Schlenk, H., and Gellerman, J. L., *Anal. Chem.*, 32:1412 (1960).

65. Middleditch, B. S., and Desiderio, D. M., *Anal. Lett.*, 5:605 (1972).

66. de Oliveira, D. B., and Harris, W. E., *Anal. Lett.*, 6(12):1107 (1973).

67. Farquhar, J. W., Insull, Jr., W., Rosen, P., Stoffel, W., and Ahrens, Jr., E. H., *Nutr. Rev. Suppl.*, 17:1 (1959).

68. Ast, H. T., *Anal. Chem.*, 35:1539 (1963).

69. Stoffel, W., Chu, F., and Ahrens, Jr., E. H., *Anal. Chem.*, 31:307 (1959).

70. Sheppard, A. J., and Ford, L. A., *J. Assoc. Offic. Anal. Chem.*, 46:947 (1963).

71. Nelson, G. J., and Freeman, N. K., *J. Biol. Chem.*, 235:578 (1960).

72. Heath, T. J., Adams, E. P., and Morris, B., *Biochem. J.*, 92:511 (1964).

73. Spencer, W. A., and Schaffrin, R., *Can. J. Biochem.*, 42:1659 (1964).

74. Kates, M., *J. Lipid Res.*, 5:132 (1964).

75. Toote, J. L., Allan, R. J., and Agranoff, B. W., *J. Lipid Res.*, 6:518 (1965).

76. Kishimoto, Y., and Radin, N. S., *J. Lipid Res.*, 6:435 (1965).

77. O'Brien, J. S., Fillerup, D. L., and Mead, J. F., *J. Lipid Res.*, 5:329 (1964).

78. James, A. T., in *Methods of Biochemical Analysis*, Vol. 8 (D. Glick, ed.), Interscience, New York and London, 1960, pp. 18-20.

79. American Oil Chemists' Society, *Official and Tentative Methods of the American Oil Chemists' Society, 3rd ed;* Method Ce 2-66, Champaign, Ill., 1966.

80. American Oil Chemists' Society, *Official Methods of Analysis, 10th ed.*, Washington, D.C., 1965, sec. 26.052, p. 429.

81. American Oil Chemists' Society, *J. Amer. Oil. Chem. Soc.*, 41:158 (1964).

82. Davidoff, F., and Korn, E. D., *J. Biol. Chem.*, 238:3199 (1963).

83. Ways, P., Reed, C. F., and Hanahan, D. J., *J. Clin. Invest.*, 42:1248 (1963).

84. Bowyer, D. E., Leat, W. M. F., Howard, A. N., and Gresham, G. A., *Biochim. Biophys. Acta*, 70:423 (1963).

85. Chalvardjian, A. M., *Biochem. J.*, 90:518 (1964).

86. Feldman, G. L., and Ronser, G., *J. Amer. Oil Chemists' Soc.*, 42:290 (1965).

87. Fosbrooke, A. S., and Tamir, I., *Clin. Chim. Acta*, 20:517 (1968).

88. Rogozinski, M., *J. Gas Chromatogr.*, 2:136 (1964).

89. Rogozinski, M., *J. Gas Chromatogr.*, 2:328 (1964).

90. Mason, M. E., and Waller, G. R., *Anal. Chem.*, 36:583 (1964).

91. Mason, M. E., Eager, M. E., and Waller, G. R., *Anal. Chem.*, 36:587 (1964).

92. Mason, M. E., Johnson, B., and Hamming, M. C., *Anal. Chem.*, 37:769 (1965).

93. Shapiro, I. L., and Kritchevsky, D., *J. Chromatogr.*, 18:599 (1965).

94. Bowlers, H., and Nieuwland, J. A., *J. Amer. Chem. Soc.*, 53:3835 (1931).

95. Hinton, J. D., and Nieuwland, J. A., *J. Amer. Chem. Soc.*, 54:2017 (1932).

96. Metcalfe, L. D., and Schmitz, A. A., *Anal. Chem.*, 32:363 (1961).

97. Lough, A. K., *Biochem. J.*, 90:40 (1964).

98. Lough, A. K., *Nature*, 202:795 (1964).

99. Metcalfe, L. D., Schmitz, A. A., and Pelka, J. R., *Anal. Chem.*, 38:514 (1966).

100. Firestone, D., *J. Assoc. Off. Anal. Chem.*, 52:254 (1969).

101. O'Conner, R. T., Allen, R. R., Chipault, J. R., Herb, S. F., and Hoerr, C. W., *J. Amer. Oil Chem. Soc.*, 45:103 (1968).

102. American Oil Chemists' Society, *Official and Tentative Methods of the American Oil Chemists' Society*, Vols. 1 and 2, 3rd ed. Champaign, Ill., 1964.

103. Morrison, W. R., and Smith, L. M., *J. Lipid Res.*, 5:600 (1964).

104. Storry, J. E., and Rook, J. A. F., *Biochem. J.*, 98:210 (1965).

105. Dummel, R. J., and Kun, E., *J. Chromatogr.*, 54:130 (1971).

106. Bezard, J., and Bugant, M., *J. Chromatogr. Sci.*, 7:639 (1969).

107. Aue, W. A., Hastings, C. R., Gerhardt, K. O., Pierce, J. O., Hill, H. H., and Moseman, R. F., *J. Chromatogr.*, 72:259 (1972).

108. Gerhardt, K. D., and Aue, W. A., *J. Chromatogr.*, 82:382 (1973).

109. McCone, G. E., and Hance, R. J., *J. Chromatogr.*, 69:204 (1972).

110. Corina, D. L., and Dunstan, P. M., *Anal. Biochem.*, 53:571 (1973).

111. Klemm, H. P., Hintze, U., and Gercken, G., *J. Chromatogr.*, 75:19 (1973).

112. Watson, J. R., and Crescuolo, P., *J. Chromatogr.*, 52:63 (1970).

113. Hrivnák, J., Soják, L., Beska, E., and Janak, J., *J. Chromatogr.*, 68:55 (1972).

114. Verzele, M., Verstappe, M., Sandra, P., Van Luchene, E., and Vuge, A., *J. Chromatogr. Sci.*, 10:688 (1972).

115. Cochrane, G. C., *J. Chromatogr. Sci.*, 13:440 (1975).

116. Kuksis, A., Stachnyk, O., and Holub, B. J., *J. Lipid Res.*, 10:660 (1969).

117. Jaeggi, J., personal communication.

118. Grob, K., and Grob, Jr., K., *J. Chromatogr.*, 151:311 (1978).

118a. Karasek, F. W., *Anal. Chem.*, 43:1982 (1971).

119. Kawahara, F. K., *Anal. Chem.*, 40:2073 (1968).

120. Pierce, A. E., *Silylation of Organic Compounds*, Pierce Chemical Co., Rockford, Ill., 1968.

121. Tallent, W. H., and Kleiman, R., *J. Lipid Res.*, 9:146 (1968).

122. Kuksis, A., Marai, L., and Gornall, D. A., *J. Lipid Res.*, 8:352 (1967).

123. Dalgliesh, C. E., Horning, C. E., Horning, M. G., Knox, K. L., and Yarger, K., *Biochem. J.*, 101:792 (1966).

124. Birkofer, L., and Donike, M., *J. Chromatogr.*, 26:270 (1967).

125. Donike, M., *Chromatographia*, 6:190 (1973).

126. Downing, D. T., and Greene, R. S., *Anal. Chem.*, 40:827 (1968).

127. Koman, V., *J. Chromatogr.*, 45:311 (1969).

128. Pohl, P., Glasl, H., and Wagner, H., *J. Chromatogr.*, 42:75 (1969).

129. Fontell, K., Holman, R. T., and Lambertsen, G., *J. Lipid Res.*, 1:391 (1960).

130. Kuksis, A., *Fette, Seifen, Anstrichm.*, 73(2):130 (1971).

131. Sheppard, A. J., and Iverson, J. L., *J. Chromatogr. Sci.*, 13:448 (1975).

132. Drozd, J., *J. Chromatogr.*, 113:303 (1975).

133. Ahuja, S., *J. Pharm. Sci.*, 65:163 (1976).

134. Ackman, R. G., in *Methods in Enzymology*, Vol. 14 (J. M. Lowenstein, ed.), Academic Press, New York and London, 1969, pp. 354-357.

135. Craig, B. M., and Murty, N. L., *J. Amer. Oil Chem. Soc.*, 36:549 (1959).

136. Luddy, F. E., Barford, R. A., and Riemenschneider, R. W., *J. Amer. Oil Chem. Soc.*, 37:447 (1960).

137. Vorbeck, M. L., Mattick, L. R., Lee, F. A., and Pederson, C. S., *Anal. Chem.*, 33:1512 (1961).

138. Roper, D. M., and Ma, T. S., *Microchem. J.*, 1:245 (1957).

139. Hornstein, I., Alford, J. A., Elliot, L. E., and Crowe, P. F., *Anal. Chem.*, 32:540 (1960).

139a. Jamieson, G. R., and Reid, E. H., *J. Chromatogr.*, 17:230 (1965).

140. Hert, S. F., Magidman, P., Luddy, F. E., and Tiemenschneider, R. W., *J. Amer. Oil Chem. Soc.*, 39:142 (1962).

141. Wolff, J. P., *Am. Fals. Expert Chim.*, 53:310 (1960).

142. Anselmi, S., Boniforti, L., and Monacelli, R., *Boll. Lab. Chim. Prov.*, 2:257 (1960).

143. Burchfield, H. P., and Storrs, E. E., *Biochemical Applications of Gas Chromatography*, Academic Press, New York, 1962, p. 528.

144. Paeck, K., and Tracey, M. W., eds., *Modern Methods of Plant Analysis*, Vol. 2, Springer, Berlin, 1955, p. 626.

145. de Man, J. M., *Lab. Pract.*, 16:150 (1967).

146. de Man, J. M., *J. Dairy Sci.*, 47:546 (1964).

147. Halasz, I., and Schneider, W., *Anal. Chem.*, 33:978 (1961).

148. Schomburg, G., Husmann, H., and Weecke, F., *J. Chromatogr.*, 99:63 (1974).

149. Grob, K., and Grob, G., *Chromatographia*, 5:1 (1972).

150. Grob, U., and Grob, Jr., K., *J. Chromatogr.*, 94:53 (1974).

151. Schomburg, G., Dickmann, R., Husmann, H., and Weeke, F., *J. Chromatogr.*, 122:55 (1976).

152. Jaeger, H., Klör, H. U., and Ditschuneit, H., *J. Lipid Res.*, 17:185 (1976).

153. Orr, C. H., and Callen, J. E., *J. Amer. Chem. Soc.*, 80:246 (1958).

154. Craig, B. M., and Murty, N. L., *J. Amer. Oil Chem. Soc.*, 36:549 (1959).

155. Lipsky, S. R., Landowne, R. A., and Godet, M. R., *Biochim. Biophys. Acta*, 31:336 (1959).

156. Hornstein, I., Elliott, L. E., and Crowe, P. F., *Nature*, 184:1710 (1959).

157. Böttcher, C. J. F., Woodford, F. P., Boelsma-van Houte, E., and van Glut, E. M., *Rec. Trav. Chim. Pays-Bas*, 78:794 (1959).

158. Craig, B. M., *Acta Chem. Scand.*, 14:1442 (1960).

158a. Haken, J. K., *J. Chromatogr. Sci.*, 13:430 (1975).

159. Schomburg, G., and Husmann, H., *Chromatographia*, 8:517 (1975).

160. Alexander, G., and Rutten, G. A. F. M., *J. Chromatogr.*, 99:81 (1974).

161. Desty, D. H., Haresmape, I. M., and Whyman, B. H. F., *Anal. Chem.*, 32:302 (1960).

162. Novotny, M., and Tesarik, *Chromatographia*, 1:332 (1968).

163. Novotny, M., and Bartle, U. D., *Chromatographia*, 7:122 (1974).
164. van Rijswick, M. H. J., and Tesarik, K., *Chromatographia*, 7:135 (1974).
165. Szafranek, J., Pfaffenberger, C. D., and Horning, E. C., *J. Chromatogr.*, 88:149 (1974).
166. Grob, K., *Helv. Chim. Acta*, 48:1362 (1965).
167. Grob, K., *Helv. Chim. Acta*, 51:718 (1968).
168. Grob, K., and Jaeggi, H. J., *Chromatographia*, 5:382 (1972).
169. Grob, K., and Grob, G., *J. Chromatogr.*, 125:471 (1976).
170. Grob, K., Grob, G., and Grob, Jr., K., *Chromatographia*, 10:181 (1977).
171. Nelson, G. J., *Lipids*, 9:254 (1974).
172. Ackman, R. G., and Hooper, S. N., *J. Chromatogr. Sci.*, 12:131 (1974).
173. Golovnya, R. V., Kralets, V. P., and Kuzmenko, T. E., *J. Chromatogr.*, 121:118 (1976).
174. Prevot, A. F., and Mordret, F. X., *Rev. Franc. Corps Gras*, 23:409 (1976).
175. Van Hout, P., Szafranek, J., Pfaffenberger, C. D., and Horning, E. C., *J. Chromatogr.*, 99:103 (1974).
176. Goretti, G., and Liberti, A., *J. Chromatogr.*, 61:334 (1971).
177. van Vleet, E. S., and Quinn, J. G., *J. Chromatogr.*, 151:396 (1978).
178. *Gas-Chromatogr. Newsletter* (Applied Science Lab., State College, Pa.), 19 (1978).
179. Grob, Jr., K., Neukom, H. P., Fröhlich, D., and Battaglia, R., *HRC & CC*, 1:94 (1978).
180. Jaeger, H., Klör, H. U., Blos, G., and Ditschuneit, H., *Chromatographia*, 8:507 (1975).
181. Brechbühler, B., Gay, L., and Jaeger, H., *Chromatographia*, 10: (1977).
182. Karmen, A., in *Methods in Enzymology*, Vol. 14 (J. M. Lowenstein, ed.), Academic Press, New York, 1969, p. 465.
183. Swell, L., *Proc. Soc. Exp. Biol. Med.*, 121:1290 (1966).
184. Breckenridge, W. C., and Kuksis, A., *Lipids*, 5:342 (1970).
185. Kolb, B., and Wiedeking, E., *Z. Anal. Chem.*, 243:129 (1968).
186. Capella, P., and Zorzut, C. M., *Anal. Chem.*, 40:1458 (1968).
187. Eglinton, G., and Hunnemann, D. H., *Phytochemistry*, 7:313 (1968).
188. Argoudelis, D. J., and Perkins, E. G., *Lipids*, 3:379 (1968).
189. Perkins, E. G., and Argoudelis, D. J., *Lipids*, 4:619 (1969).
190. Schomburg, G., Weeke, F., Weimann, B., and Ziegler, E., *Angew. Chemie Internat. Edit.*, 11:366 (1972).
191. Horning, E. C., Ahrens, Jr., E. H., Lipsky, S. R., Mattson, F. H., Mead, J. F., Turner, D. A., and Goldwater, W. H., *J. Lipid Res.*, 5:20 (1964).

192. De la Court, F. H., Van Cassel, N. J. P., and van der Falk, J. A. M., *J. Chromatogr.*, 72:249 (1972).

193. Novotny, M., Segura, R., and Zlatkis, A., *Anal. Chem.*, 44:9 (1972).

194. Verzele, M., Van Cauwenberghe, K., and Bouche, J., *J. Gas Chromatogr.*, 5:114 (1967).

195. Kaiser, R. E., *Chromatographie in der Gasphase*, Bibliographisches Institut, Mannheim, 1961, p. 20.

196. Ettre, L. S., and March, E. W., *J. Chromatogr.*, 91:5 (1974).

197. Ettre, L. S., *Kapillarsäuren in der Gaschromatographie: Eine Einführung*, Vieweg, Braunschweig, 1976.

198. Averill, W., and March, E. W., *Chromatography Newsletter*, 4:20 (1976).

199. Ackman, R. G., and Hooper, S. N., *J. Chromatogr. Sci.*, 7:549 (1969).

200. Ackman, R. G., and Castell, J. D., *J. Gas Chromatogr.*, 5:489 (1967).

201. Pollock, G. E., *Anal. Chem.*, 44:634 (1972).

202. Egge, H., Murawski, U., Ryhage, R., György, R., Chatranon, W., and Zilliken, F., *Chem. Phys. Lipids*, 8:42 (1972).

203. Krupcik, J., Hrivnák, J., Barnoky, L., and Janak, J., *J. Chromatogr.*, 65:323 (1972).

204. Prevot, A., in *Chromatography and Methods of Immediate Separation*, (G. Parissatiis, ed.), Union of Greek Chemists, Athens, 1966, p. 141.

205. Hrivnák, J., Soják, L., Krupcik, J., and Duchesne, Y. P., *J. Amer. Oil Chem. Soc.*, 49:68 (1972).

206. Hrivnák, J., Palo, V., and Krupcik, J., *Milchwissenschaft*, 28:11 (1973).

207. Lin, S.-N., Pfaffenberger, C. D., and Horning, E. C., *J. Chromatogr.*, 104:319 (1975).

208. Merle d'Aubigne, J., Landault, C., Guichon, G., *Chromatographia*, 4:309 (1971).

209. German, A. L., and Horning, E. C., *J. Chromatogr. Sci.*, 11:76 (1973).

210. Lin, S.-N., and Horning, E. C., *J. Chromatogr.*, 112:483 (1975).

211. Flanzy, J., Boudon, M., Leger, C., and Pihet, J., *J. Chromatogr. Sci.*, 14:17 (1976).

212. Sandra, P., Verstappe, M., and Verzele, M., *J. HRC & CC*, 1:28 (1978).

213. Schieke, J. D., Comins, N. R., and Pretorius, V., *J. Chromatogr.*, 112:97 (1975).

214. Schieke, J. D., Comins, N. R., and Pretorius, V., *Chromatographia*, 8:354 (1975).

215. Sandra, P., and Verzele, M., *Chromatographia*, 10:419 (1977).

216. Verzele, M., and Sandra, P., *J. Chromatogr.*, 158:111 (1978).

217. Heckers, H., Dittmar, K., Melcher, F. W., and Kalinowski, H. O., *J. Chromatogr.*, 135:93 (1977).

218. Heckers, H., Melcher, F. W., and Schloeder, U., *J. Chromatogr.*, 136:311 (1977).

219. Heckers, J., Korner, M., Tuschen, T. W. L., and Melcher, F. W., *Atherosclerosis*, 28:389 (1977).

220. Heckers, H., and Melcher, F. W., *Amer. J. Clin. Nutr.*, 31:1041 (1978).

221. Jaeger, H., and Ditschuneit, H., *Verh. Dtsch. Ges. Inn. Med.*, 81:1411 (1975).

222. Jaeger, H., Wagner, W., Homoki, J., Klör, H. U., and Ditschuneit, H., *Chromatographia*, 10:492 (1977).

12

ANALYSIS OF ESSENTIAL OILS

TAKAYUKI SHIBAMOTO

University of California
Davis, California

1. Introduction 455

2. Preparation of Essential Oils 456
   - 2.1 Historical Aspects 456
   - 2.2 Industrial Methods 457
   - 2.3 Laboratory Methods 458

3. Instrumentation 466

4. Developments in Gas Chromatographic Analysis of Essential Oils 469
   - 4.1 Historical Aspects 469
   - 4.2 Columns and Analytical Techniques 472
   - 4.3 Retention Index (Kováts Index, I) 475

5. Applications 476
   - 5.1 Examples 476
   - 5.2 Perfumery 492

6. Future Developments 498

References 506

1. INTRODUCTION

An essential oil is defined as a volatile material derived by physical processes from the odorous material of a single botanical form and/or species with which it agrees in name and odor [1]. There are over 200 different kinds of essential oils, and they are widely used in cosmetic and pharmaceutical products and as flavor ingredients for food products. They generally give the characteristic fragrance to the plant in which they are found.

Essential oils are classified into three groups [2]. The first group is the monoterpenes, which have been known for hundreds of years as components of essential oils. Much work has been done on this group of chemicals, and the complexity of their structures has been elucidated with increasing clarity since the development of instrumental analysis, including gas chromatography.

The second group is composed of sesquiterpenes, diterpenes, and a residue group containing high-molecular-weight fatty acids, polymers, etc. Terpene chemists have studied the structures of sesquiterpenes and diterpenes intensively. The most difficult aspect of investigation into these compounds is that they are, for the most part, quite similar in their physical and chemical natures. At present, only high-resolution gas chromatographic column technology can separate certain of these compounds.

The third group is of most interest and consists of a variety of oxygenated compounds ranging from alcohols, ketones, aldehyde, esters, and ethers to complicated macrocyclic lactones. These compounds possess characteristic odors and are the ones most often used as flavor and fragrance ingredients.

Since essential oils contain important aroma-emitting chemicals, many studies have been made to identify their chemical constituents. To date, over 300 chemicals have been identified in more than 200 different kinds of essential oils. Essential oil analysis requires refined microanalytical techniques and high-resolution separation systems. Glass capillary column techniques have played and will continue to play an important role in shedding light on the composition of essential oils.

2. PREPARATION OF ESSENTIAL OILS

2.1 Historical Aspects

The history of essential oil use stretches back centuries. There were many production plants existing in Alexandria as early as 7 B.C. The variety of oils was very limited, and it is not known exactly how the oils were used.

We do know, however, that they were first developed as pharmaceutical materials rather than as perfumery products. By the seventeenth century 175 different essential oils had been produced. Following the development of modern chemistry in the nineteenth century, methods for the production of essential oils were also improved. It was in the middle of the nineteenth century that the great German chemist Kekulé introduced the word *terpene* [3]. In addition to physical methods (extraction, distillation, etc.), chemical methods were introduced [4].

2.2 Industrial Methods

Expression

Expression is used mainly for citrus peel oils. Peel is pressed to give a mixture of oil and juice, and the oil layer is then separated by centrifugation. Because citrus oils are very susceptible to heat degradation, the process of pressing is carried out at low temperatures. The oils obtained by this method are the "cold press oils" and are of a higher grade than the oils obtained by steam distillation.

Extraction

The essential oils from certain flowers (jasmin, violet, mimosa, etc.) are very unstable at high temperature. Steam distillation causes decomposition and polymerization of the constituents of these oils. For these oils, therefore, extraction methods are usually applied, both volatile and nonvolatile solvents being used.

Nonvolatile solvent extraction or "enfleurage" has been used for many years and is still used on a small scale, primarily in southern France. It may be used with flowers that are too unstable when subjected to heat or steam. The flower petals are placed on the surface of a fatty or greasy base on a plate. The trays or stacks of greased plates, with the flowers adhering to the fat, are left for 24 hours. After the fat has absorbed the oils from the flowers, the essential oils are recovered by extraction with volatile solvents.

Volatile solvent extraction was first used by Robiquet in 1835 [5]. It is widely used to obtain high-quality concentrated flower oils from jasmin, tuberose, jonquil, mimosa, and violet. The fresh flowers are extracted with volatile solvents (e.g., petroleum ether, benzene, ethyl alcohol) at room temperature. The oil is then recovered after the solvent is removed by distillation under reduced pressure. This extraction method is costly and requires a high degree of skill in comparison with the distillation method and has therefore been used only for high-quality flower oils.

Distillation

The principle of distillation is the one most commonly used to obtain oils on an industrial scale. There are three methods of distillation: steam distillation, water distillation, and water and steam distillation.

In steam distillation, steam produced in a boiler is blown through a container into which the plant material has been placed. The mixture of water and oils is collected after the resulting vapor is condensed and separated as they float to the surface of the water.

In water distillation, the plant material makes contact with the boiling water during the distillation process. This method is simpler than steam distillation and is used for flowers like roses, which coagulate upon contact with steam. The collection and separation of oil are the same as in steam distillation.

Water and steam distillation is a primitive form of steam distillation. The plant material is placed on a grid or on a stack of trays located immediately over boiling water, and the collection and separation of the oil are the same as in the other methods of distillation.

2.3 Laboratory Methods

General Considerations

Samples for gas chromatographic analysis have been prepared in many ways. When preparing essential oil samples for instrumental analysis, one usually thinks of ways to obtain samples sufficiently large for infrared analysis (IR), nuclear magnetic resonance (NMR), mass spectrometry (MS), etc. Gas chromatographic (GC) analysis does not, however, require a large amount of sample. Often a 5- to 10-μl sample is sufficient for routine GC analysis. In particular, with the development of the high-resolution glass capillary column, the retention indices of compounds have become much more dependable and reproducible. Many analyses can be performed using gas chromatography/mass spectrometry (GC/MS) and GC retention indices. In such cases, the sample size can be less than 10 μl; accordingly, methods of preparing small samples have been developed. It is, of course, still necessary to use IR, NMR, UV and other instrumental analytical methods to confirm the structure of some compounds.

Two excellent texts in which sample preparation of flavor materials for GC analysis is described are those by Teranishi et al. [6] and Jennings [7]. One can readily apply the methods described therein to the essential oils. The most important aspect of sample preparation is that the oils in the sample isolated should be, as nearly as possible, identical to their natural form in the plant with respect to constituents and aromas. Some changes will occur since the oils must undergo certain physical processes (e.g., distillation and extraction) which involve heat treatment and mixing with other chemicals. To reduce distortion to the lowest possible level, however, the method of preparation should utilize low temperatures, inert atmospheres, and the absence of contaminating elements.

Headspace Collection

Samples obtained directly from headspace resemble most closely the natural forms because no physical treatments have been applied. Headspace collection is widely used, therefore, in perfumery studies, since the headspace constituents are those which one actually smells in the natural plant. The disadvantages of these methods are that the sample obtained is very small and some problems of reproducibility can occur. The detection of trace volatiles by this technique is also somewhat limited by the large volume of headspace gas required to detect such trace components.

Fleming et al. [8] developed a direct headspace analysis technique for fermented olives using gas chromatography. They placed a diced olive sample into a 125-ml Erlenmeyer flask with brine water. Using butanol as the internal standard, headspace vapor was collected with a 6-ml disposable syringe. This headspace vapor in the syringe was then injected directly into the GC. The authors detected five major compounds by gas chromatography, three of which were identified as acetaldehyde, methyl sulfide, and ethanol.

Only major volatile constituents can be detected by this method. It is therefore desirable to concentrate the headspace vapor in some way.

Liebich et al. [9] prepared headspace volatiles of rum in the following way. A sample of rum (1.5 liters) was placed in a 2-liter flask, and the flask was connected to a 20-ml flask cooled in a dry ice/acetone bath. After 36 hr, 0.3 ml of highly volatile components was collected in the 20-ml flask. The collected material was then transferred from the flask to a gas chromatographic column for analysis. Samples obtained in this way were free of artifacts and contaminants, and the investigators were able to identify 22 highly volatile compounds, including ethylformate, methanol, acetaldehyde, and acetone. They pointed out, however, that the gas chromatographic analysis of the sample was complicated by its high ethanol content.

In many cases, headspace samples prepared from essential oils contain large quantities of water or ethanol which complicate gas chromatographic analysis. Entraining headspace vapor on materials such as silica gel, active alumina, charcoal [10], or porous polymers eliminates the water or ethanol and enriches the samples. Porous polymers have been used successfully by Yabumoto et al. [11], Murray [12], and Jennings and Filsoof [13]. The great advantage of using porous polymers is that water and low-molecular-weight alcohols show very short retentions compared with other low-molecular-weight organic compounds [7]. It is possible, therefore, to remove most of the water or ethanol from the samples using porous polymers without losing any other compounds from the samples. Yabumoto [14] developed the apparatus shown in Fig. 1 to collect melon volatiles on a porous polymer (Porapak Q). After trapping the volatiles on Porapak Q, a tube containing the polymer was heated to 100°C while purified nitrogen gas was passed through it in the direction opposite to the direction of trapping. The volatile compounds were recovered by condensation on the chilled glass wall (Fig. 2). In this study, Yabumoto identified 46 volatile compounds using porous polymer trapping followed by glass capillary GC. A typical gas chromatogram will be shown in the next section.

The Porapak trap condensation method is, however, time-consuming and has some limitations. The precolumn method, in which a precolumn is directly connected to the inlet port of the gas chromatograph, has been studied by several investigators as a possible solution to these problems [15,16]. In this method the headspace samples are trapped on porous polymers, charcoal, or a similar material at room temperature or below. The trapped samples are then transferred into the gas chromatographic injector by heating the trap while introducing a nitrogen carrier. This method is less time-consuming, but requires special apparatus, that is, the precolumn and the joint at the inlet port of the GC. Due to the complex inlet system, satisfactory gas chromatograms have not yet been obtained by this method.

Distillation and Extraction

The classic way to obtain essential oil samples is by distillation, and this method is still widely used. Distillation is, however, a very

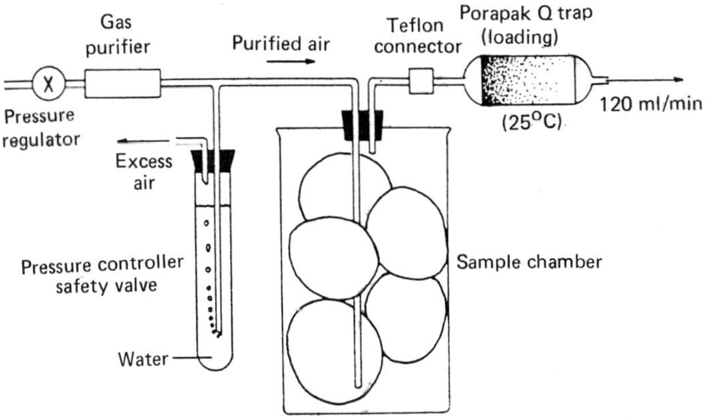

Figure 1. Apparatus of headspace trapping for melon volatiles. (Source: Ref. 14. Reproduced by permission of Dr. K. Yabumoto.)

tedious procedure that often produces artifacts or promotes rearrangements of sample constituents.

Steam distillation has been used most widely in industry, and has also been used to obtain essential oil volatiles on an analytical scale [17-20]. The yield of oil is usually less than 0.1% relative to the plant used. An additional step required to obtain oils from the steam distillate is organic solvent extraction with dichloromethane, ethyl ether, pentane, etc.. Shibamoto and Jennings [21] used a liquid-liquid continuous extractor to recover juniper leaf oil from a steam distillate (Fig. 3).

A unique apparatus conducts water distillation and organic solvent extraction simultaneously [22]. Figure 4 shows a prototype apparatus used

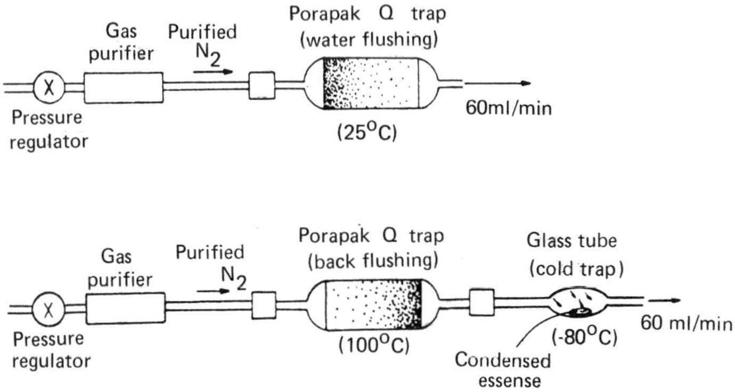

Figure 2. Method of recovering oil from porous polymer trap. (Source: Ref. 14. Reproduced by permission of Dr. K. Yabumoto.)

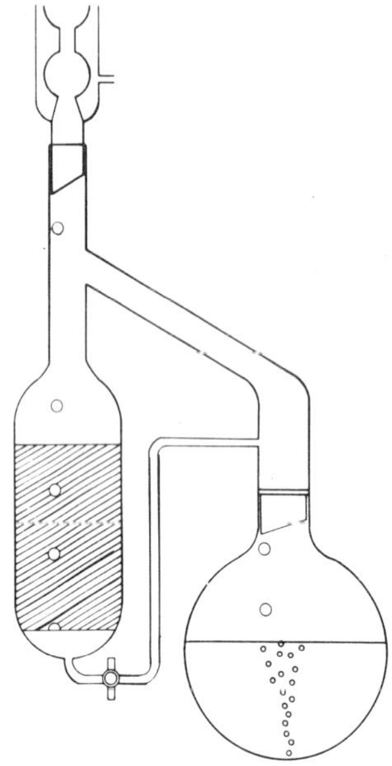

Figure 3. Liquid-liquid continuous extractor used to recover juniper leaf oil from a steam distillate with methylene chloride. (Source: Ref. 22. Reproduced by permission of Drs. S. T. Likens and G. B. Nickerson.)

to isolate hop oil from beer. Since this apparatus was first developed, many researchers have applied the technique and the apparatus has been improved [23-26]. Figure 5 shows a modified distillation-extraction apparatus developed by Schultz et al. [27]. The disadvantage of this technique is that water distillation is not as efficient as steam distillation in the recovery of the less volatile constituents of natural plants. Shibamoto developed an apparatus which can first conduct steam distillation and subsequently water distillation-extraction (Figs. 6,7).

Fractional distillation under reduced pressure has been applied to samples which can be obtained in large quantities, such as commercial citrus oils and strawberry essence [28-32]. This technique may produce artifacts, however, while the samples pass through the fractionating columns. If conditions of high vacuum are maintained at low temperatures, however, the problem may be avoided. The type of fractionation column used is another critical point in this method. Table 1 summarizes the types of fractionating columns used for essential oil analysis described by Lawrence [33].

Thin-layer and Column Chromatography

Thin-layer and column chromatography have also been used to fractionate

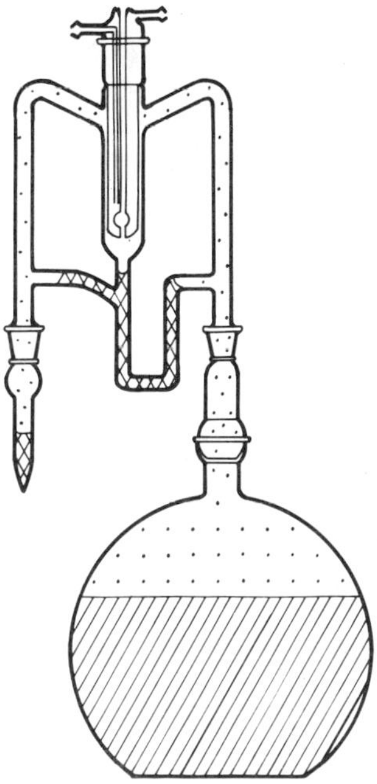

Figure 4. Simultaneous distillation and extraction used to isolate hop oil from beer. (Source: Ref. 22. Reproduced by permission of Drs. S. T. Likens and G. B. Nickerson.)

samples. Pattabhiraman et al. [34] reviewed the use of thin-layer chromatography (TLC) in essential oil analysis in 1968. A compound cannot generally be identified by this technique, but it is useful in separating complex samples into groups. In particular, silver-nitrate-impregnated silica gel and other adsorbents have been used to separate unsaturated compounds according to the position and number of double bonds [35-39].

Column chromatography has also been used to fractionate essential oils. Table 2 shows the columns and solvents used for various essential oil analyses.

Preparative Gas Chromatography

Since the development of the high-resolution gas chromatographic column, in combination with mass spectrometric analysis, extensive pretreatment of samples is required less frequently. In recent years, most essential oils have been analyzed using GC/MS and gas chromatographic retention indices. The agreement of an MS pattern and a GC retention index of an unknown with those of an authentic sample is generally sufficient to identify the unknown.

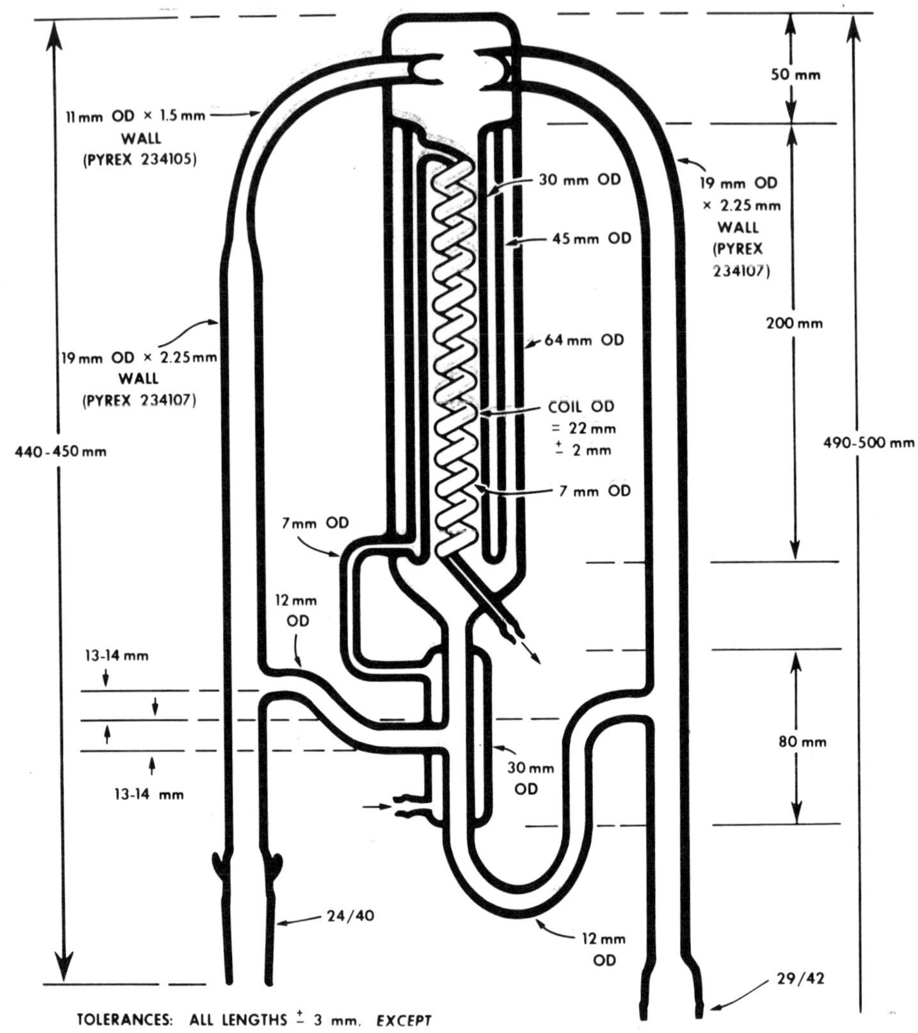

Figure 5. Modified simultaneous distillation and extraction apparatus. (Source: Ref. 14. Reproduced courtesy of Dr. Schultz, Western Regional Research Laboratory, Agriculture Research Service, USDA, Berkeley, California.)

If authentic samples are not available, preparative gas chromatographic techniques can be applied to isolate the target compound for further instrumental analysis. Table 3 summarizes the types of columns used for preparative gas chromatography in various essential oil analyses.

Table 1. Types of Fractionating Columns

Column	Types	General features	Information
1. Empty	Straight or smooth spirals, increased surface area, concentric tube	Small pressure drop and low hold-up	Concentric tube columns; very efficient—used for micro-, high-temperature, and vacuum distillation
2. Packed	Straight smooth tubes	High pressure drop and hold-up; in general packing size ~1/10 diameter of column	Packing such as Raschig rings, spirals, helices, spheres, saddles, etc.—random packing not very efficient
3. Plate	Bubble plate, bubble cap and sieve plate	High pressure drop and large liquid hold-up	Special-purpose columns for repetitive distillations where distillate high purity is required; Oldershaw column of sieve plate type is favored
4. Stationary element	Continuous glass or wire gauze, spiral, wire helix or inclined film	Moderate pressure drop and hold-up	Heli-Pak a favored packing because of extremely low HETP[a] even at high loads
5. Rotating element	Spinning band, rotating concentric tube, or condenser conical segment or wire spiral	Moderately small pressure drop and hold-up	Spinning band columns useful for microvacuum distillation.

[a]Height equivalent to one theoretical plate

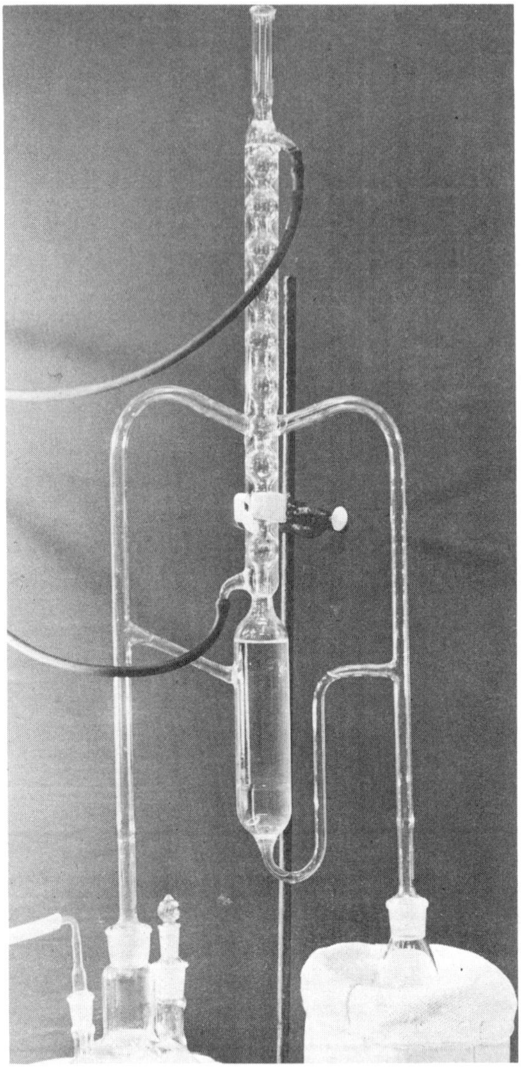

Figure 6. Photograph of simultaneous distillation and extraction apparatus modified for steam distillation.

3. INSTRUMENTATION

As noted previously, the analytical system should be inert and clean. Terpenes, which are the main constituents of essential oils, are particularly susceptible to interconversion reactions. Hunter and Brogden [57], for example, postulated a series of interconversions involving limonene, terpinolene, γ-terpinene, α-terpinene, α-phellandrene, and β-phellandrene. Wrolstad and Jennings [58] demonstrated the conversion of sabinene into α-thujene,

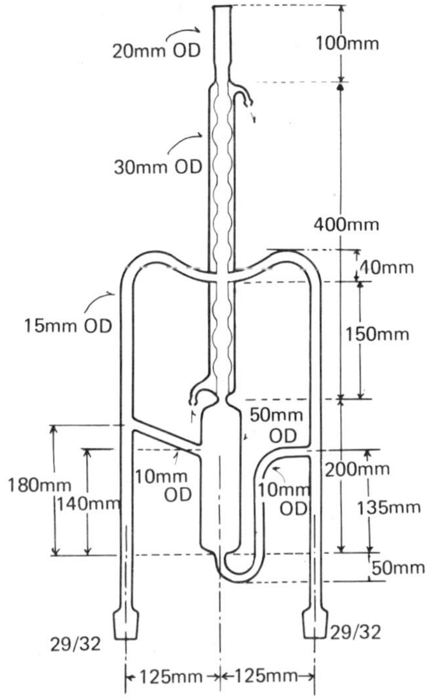

Figure 7. Details of Fig. 6.

α-terpinene, γ-terpinene, terpinolene, limonene, and β-phellandrene. These reactions occur quite readily even under anaerobic conditions, and are encouraged by acidic residues which may accumulate in the injector and by metallic surfaces of stainless steel columns. Glass columns and injectors are recommended to prevent samples from making contact with metallic surfaces; the glass capillary column exhibits not only high resolution but also minimizes the production of artifacts.

One must be very careful to use a clean injector, since repeated injections of samples cause accumulation of acidic residues in the injector. Cleaning an injector, however, is a very time-consuming process. Shibamoto developed a disposable glass inlet system (Figs. 8,9). The combination of this inlet system and a glass capillary column forms a system in which no metal components are in contact with the volatiles.

The proper selection of a GC stationary liquid phase is, of course, critical for essential oil analysis. At least two different polarity columns should be used for confirming the identities of compounds. A comparison of the retention indices of unknowns with those of authentic samples on a polar and nonpolar liquid phase is generally used in the qualitative analysis of essential oils at the present time.

Table 2. Column Chromatographs Used for Essential Oil Analyses

Essential oil	Packing material	Column length	Solvents (used successively)	Reference
Thyme	Florisil	7 m × 3 cm	Petroleum ether (500 ml); petroleum ether/benzene/ 9:1 (500 ml), 1:9 (500 ml); benzene/ethyl ether/95:5 (500 ml), 90:10 (500 ml), 80:20 (500 ml); ethyl ether (500 ml)	40
Marsh grass (*Juncus roemerianns*)	Florisil	0.25 m × 2 cm	Pentane, 2% ethyl ether in pentane, 5% ethyl ether in pentane, 10% ethyl ether in pentane, 20% ethyl ether in pentane, ethyl ether (100-ml portion)	19
Marsh grass (*Scirpus Americanns*)	Florisil	0.25 m × 2 cm	Pentane (100 ml), 10% ethyl ether in pentane (100 ml) ethyl ether (100 ml)	20
Black pepper	Silica gel	a	Isooctane (2500 ml), methanol (1000 ml)	41
Lettuce	Alumina	a	8 Portions of petroleum ether (total 2000 ml), benzene (500 ml)	42
Opoponax	Alumina	14in × 1in	25 Portions of 20-ml hexane	43
Olive	Silica gel	1 m × 5 mm	10 ml each of ethanol/ether 7:3	44
Japanese peppermint	Silica gel + selite 545 (1:1)	43.5 cm × 10.5 cm	n-Hexane, n-hexane/benzene, benzene, benzene/ethyl acetate, ethyl acetate	30
Orange essence	Florisil	0.6 m × 5 cm	Hexane, ether, ethanol (1 to 2 liters)	45
Grapefruit essence	Florisil	77.5 cm × 5 cm	Hexane (1100 ml), 5% ethyl ether in hexane, 10% ethyl ether in hexane (7 portions of 100 ml, total 2100 ml), 5% ethanol in ethyl ether (100 ml), ethanol (100 ml)	46
Meyer lemon	Florisil	17in × 1in	Hexane (400 ml), hexane/ethyl ether 1:1 (400 ml), ethanol (300 ml)	47
Olive	Silica gel	40 cm × 1.5 cm	n-Pentane, pentane/ethyl ether	25
Japanese pepper (leaf oil)	Silica gel	a	Petroleum ether, 2 to 5% of ethyl ether in petroleum ether	48

[a] Not specified in reference.

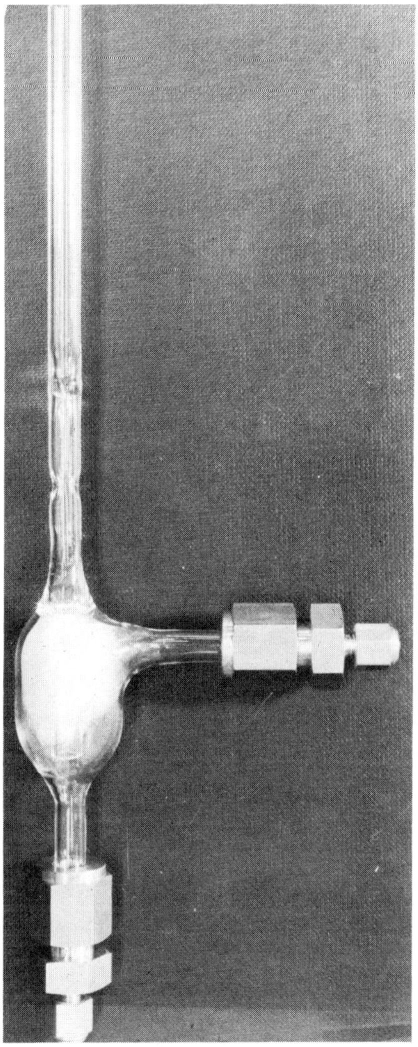

Figure 8. Photograph of disposable combined glass injector and splitter.

4. DEVELOPMENTS IN GAS CHROMATOGRAPHIC ANALYSIS OF ESSENTIAL OILS

4.1 Historical Aspects

It is well known that many essential oils contain over 300 compounds. The systematic study of essential oils began after Lavoisier (1743-1794) discovered the law of mass conservation [59]. Some compounds found in essential oils were identified before the development of chromatographic techniques. However, most of the work had been done to obtain physical and chemical

Table 3. Columns Used for Preparative Gas Chromatography

Liquid phase	Solid support	Column size	Type of column	Essencial oil	Reference
FFAP (4%)	Chromosorb G, AW DMCS (80/100)	4 m × 7 mm	Glass	Thyme	40
Apiezon-L (4%)	Chromosorb G, AW DMCS (80/100)	4 m × 7 mm	Glass	Thyme	40
OV-101 (4%)	Chromosorb G, AW DMCS (80/100)	4 m × 7 mm	Glass	Thyme	40
Carbowax 20M (10%)	Chromosorb G (30/40)	12 m × 11 mm	Glass	Black pepper	49
OV-101 (5%)	Chromosorb G, AW DMCS (70/80)	3 m × 4 mm	Glass	Angelica root	50
LAC-2R-446 (5%)	Chromosorb G, AW DMCS (70/80)	3 m × 4 mm	Glass	Angelica root	50
FFAP (10%)	Chromosorb W, AW (60/80)	3 m × 6 mm	Stainless steel	Angelica root	50
Apiezon-L (15%)	Chromosorb W, AW (60/80)	3 m × 6 mm	Stainless steel	Angelica root	50
Carbowax 20M (10%)	Chromosorb W, NAW (60/80)	15 ft × 0.25 in	Aluminum	*Mentha piperita* L.	32
SF-96 (10%)	Chromosorb W, NAW (60/80)	15 ft × 0.25 in	Aluminum	*Mentha piperita* L.	32
Apiezon-L (20%)	Chromosorb W, NAW (60/80)	15 ft × 0.25 in	Aluminum	*Mentha piperita* L.	32
Carbowax 20M	Chromosorb G, AW DMCS (70/80)	3.5 m × 3 mm	Glass	Cantaloupe	51
Triton X 305 (20%)	Chromosorb W (60/80)	25 ft × 1/4 in	Stainless steel	Japanese pepper	52
Carbowax 20M (20%)	Chromosorb W (60/80)	10 ft × 3/8 in	Aluminum	Japanese pepper	52

SF-96 (20%)	Chromosorb W (60/80)	10 ft × 3/8 in	Aluminum	Japanese pepper	52
FFAP (20%)	Chromosorb W, AW DMCS	20 ft × 3/8 in	Aluminum	*Achillea millefolium* L.	53
Carbowax 20M (20%)	Gas Chrom P (60/80)	20 ft × 1/4 in	Stainless steel	Lemon	47
UCW-98 (10%)	Gas Chrom P (60/80)	20 ft × 1/4 in	Stainless steel	Lemon	47
DEGS (10%)	Anakrom, AWS (80/90)	6 ft × 0.128 in	Stainless steel	Passion fruit	54
FFAP (10%)	Chromosorb W, AW (60/80)	3 m × 6 mm	Stainless steel	Sweet Marjoram	55
SE-30 (5%)	Chromosorb G (60/80)	3 m × 6 mm	Stainless steel	Sweet Marjoram	55
Apiezon-L (15%)	Chromosorb W, AW (60/80)	3 m × 6 mm	Stainless steel	Sweet Marjoram	55
LAC-2R-446 (5%)	Chromosorb G, AW DMCS (70/80)	3 m × 6 mm	Stainless steel	Sweet marjoram	55
Carbowax 20M (15%)	a	16 ft × 1/4 in	Glass	Nutmeg	56

[a]Not specified in reference.

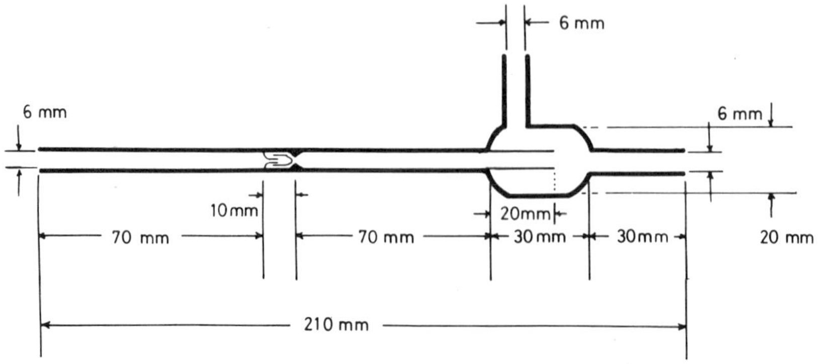

Figure 9. Details of Fig. 8.

information about the essential oils themselves (e.g., their solubility, boiling points, specific gravities, refractive indices, optical rotations, acid values, ester values, ester values after acetylation, alcohol content, aldehyde and ketone content), rather than information about the constituent compounds themselves. This was due to the different purposes to which essential oils were put at that time. Guenther [60] and Gildemeister and Hoffman [61] are the two standard texts in this field.

Since James and Martin succeeded in making a prototype gas chromatograph in 1952, it has become the most powerful instrument for essential oil analysis. Table 4 summarizes the pioneer work in essential oil analysis by gas chromatography. Following the development of gas chromatography, the number of essential oil constituents identified has increased dramatically. Comprehensive analysis of essential oils had to wait, however, for the development of high-resolution columns.

4.2 Columns and Analytical Techniques

The column is one of the most important and critical parts of the gas chromatograph. Analyses performed on the most modern gas chromatograph-mass spectrometer equipped with a computer data system will be unsatisfactory unless a high-resolution column is used, since it is the column which performs the separation.

Figure 10 shows a gas chromatogram of peppermint oil obtained on an early type of packed column. (Detailed descriptions of columns will not be provided here, as they are readily available elsewhere.) In the 1960s most essential oil analyses were conducted using packed columns. Kugler and Kováts [69] identified 48 components of mandarin orange oil using a

Table 4. Pioneer Works of Essential Oil Analyses by Gas Chromatograph

Essential oil	Liquid phase	Column size	Reference
Onion	Carbowax 1540	a × 4.6 mm	62
Eucalyptus	Silicon E 301	1.7 m × 8 mm	63
Lemon	Carbowax 600	11.5 ft × 6 mm	64
Thyme	Reoplex 100	2 m × 4.8 mm	65
Bois de rose	Silicon DC-550	2.25 m × 3 m × a	66
Lavender	"	"	66
Lavandin	"	"	66
Bergamot	"	"	66
Lemon	"	"	66
Orange	"	"	66
Orange flower	"	"	66
Rose	"	"	66
Timo	"	2 and 4 m × a	67
Acorus calamus L.	Ucon polar	10 ft × 1/4 in	68

[a]Not specified in the reference.

packed column (2.2 m × 1.5 cm ID glass column, 2% Apiezon-L or Emulphor-O on celite, 40/60 mesh). Even though the oil was pretreated by fractional distillation and solvent extraction, 93 compounds in the oil remained unidentified.

The capillary column was invented by Golay in 1958, and is still called the Golay column. It is difficult to obtain a packed column with more than 10,000 theoretical plates, but a capillary column can easily provide over 50,000 plates. This represents a significant advance in column technology. During the late 1960s and early 1970s, the stainless steel capillary column was used extensively for essential oil analysis. Figure 11 shows the gas chromatogram of the same peppermint oil (as in Fig. 10) obtained with an early type of stainless steel capillary column. This type of column is still used widely, not only in essential oil analysis, but also in various other fields. Lawrence et al. [31] identified 83 constituents in the oil of *Mentha piperita* L. using a stainless steel capillary column (100 ft × 0.01 in ID, coated with either Carbowax 6000 or SF 96) with a combination of alkali extraction, fractional distillation, column chromatography, and preparative gas chromatography.

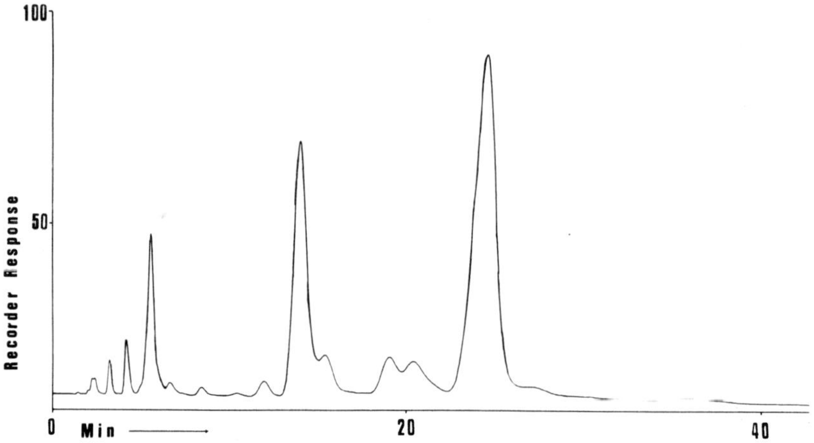

Figure 10. Chromatogram of peppermint oil: 2-m × 3-mm ID stainless steel column packed with 5% Carbowax 20M on Chromosorb W (80/100); oven temperature 165°C isothermal; helium carrier gas flowrate: 0.7 kg cm^{-2}.

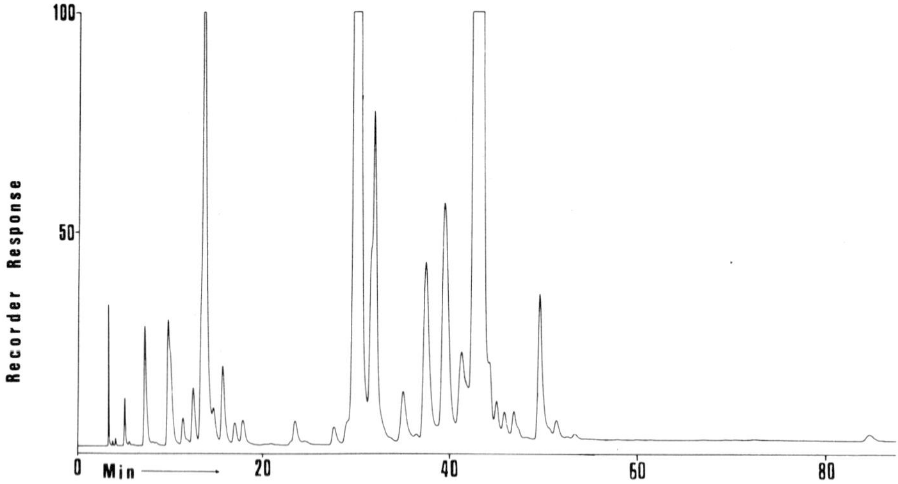

Figure 11. Chromatogram of peppermint oil: 45-m × 0.5-mm ID stainless steel capillary column coated with Carbowax 20M; oven temperature 145°C isothermal; nitrogen carrier gas flowrate 1.2 kg cm^{-2}.

Since the early 1970s, the technique of combined gas chromatography/mass spectrometry has been used extensively for essential oil analysis. Detailed descriptions of this technique will not be provided here since they are readily available elsewhere [70,71]. Flath et al. [25] reported the identities of 77 components of olive oil using GC/MS. Identification of constituents was made by comparison of their mass spectra and retention indices with those of authentic compounds. Since the development of high-resolution columns, this method has begun to be accepted as a means of identifying trace elements in essential oils, pesticides, air pollutants, and food flavors. The use of a second liquid phase for additional retention index confirmation makes the results more reliable. Figure 12 shows the gas chromatogram of the same oil as analyzed in Figs. 10 and 11 obtained with a modern glass capillary column.

The glass capillary column was created by Desty in 1960 [72]. Coating the internal surface of the glass capillary column was very difficult, but these difficulties have been overcome by several researchers [73-76]. Today high-quality glass capillary columns are readily available commercially. Although this type of column can offer a much more inert system than a stainless steel column, there are surprisingly few examples of glass capillary columns used in essential oil analyses.

4.3 Retention Index (Kováts Index, I)

The glass capillary column has given new significance to gas chromatographic retention data [7]. Recently improved high-resolution glass capillary systems give highly reproducible retention indices [77]. Jennings [7] reported that by determining retentions on two high-resolution columns of different polarity, the level of confidence of compound identity can be greatly increased. Retention systems have been studied by many investigators [78-83].

The Kováts retention index (I) proposed by Dr. E. Kováts [84], has been the most widely used in essential oils analysis [19-21,26,54]. This index expresses the retention behavior of a chemical compound along a uniform scale determined by a series of closely related standard n-paraffins. The linearity of the relationship between elution temperature and n-paraffin carbon number depends upon the isothermal retention volumes. Figures 13 and 14 show the gas chromatogram of peppermint oil containing added n-parraffin standards, on two different columns under isothermal conditions

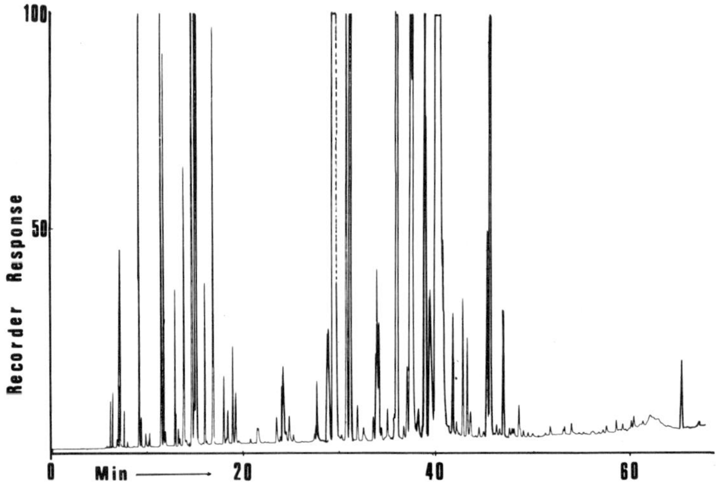

Figure 12. Chromatogram of peppermint oil: 48 m × 0.28-mm ID glass capillary column coated with Carbowax 20M and programmed from 80 to 200°C at 2°C min^{-1}; nitrogen carrier gas flowrate 15 cm sec^{-1}.

(Carbowax 20M and OV-101). The peak numbers in the C series indicate the number of carbon atoms in the n-paraffins. Guichon [85] reported experimental data showing that the programmed temperature retention index can be defined by interpolation of elution temperature. Figures 15 and 16 are gas chromatograms of the same peppermint oil as in Figs. 13 and 14 again with n-paraffin standards added and run under programmed temperature conditions. Peak identification and retention indices of the main constituents are listed in Table 5.

Theoretically, it is necessary to use isothermal conditions to obtain the Kovâts Index. In the case of programmed plot, however, a small section may be viewed as if it were a straight line (Fig. 17). Within these small sections, one can use the programmed plot as a linear relationship between elution temperature and n-paraffin carbon number. Practically, the programmed plot is more applicable, because peaks of higher-boiling-point compounds appear in ideal shape, which is essential in obtaining accurate retention values.

5. APPLICATIONS

5.1 Examples

One of the earliest studies of essential oil analysis using a glass capillary column was reported by Torline and Ballschmieter [86]. They identified 38

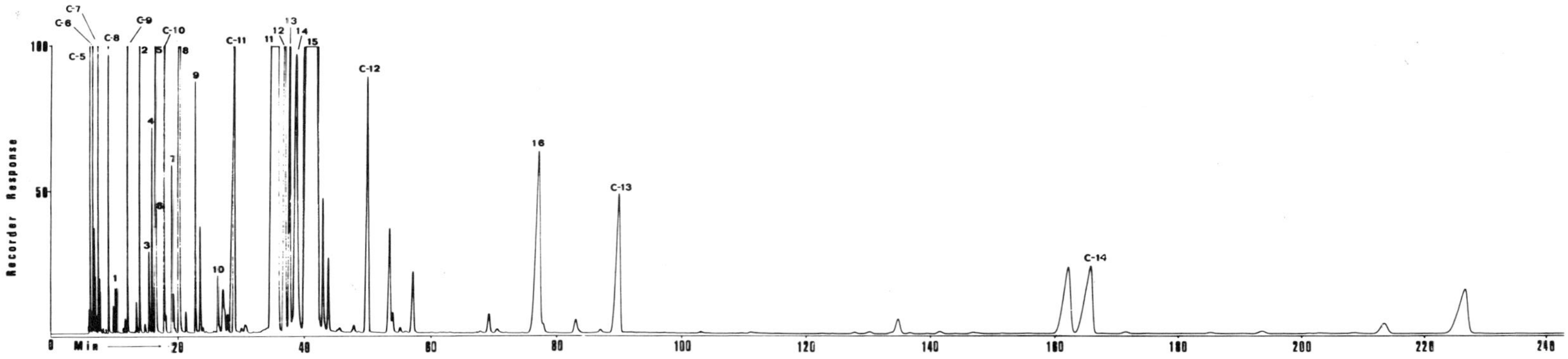

Figure 13. Chromatogram of peppermint oil with n-paraffin standards: 50-m × 28-mm ID glass capillary column coated with OV-101, isothermal at 100°C.

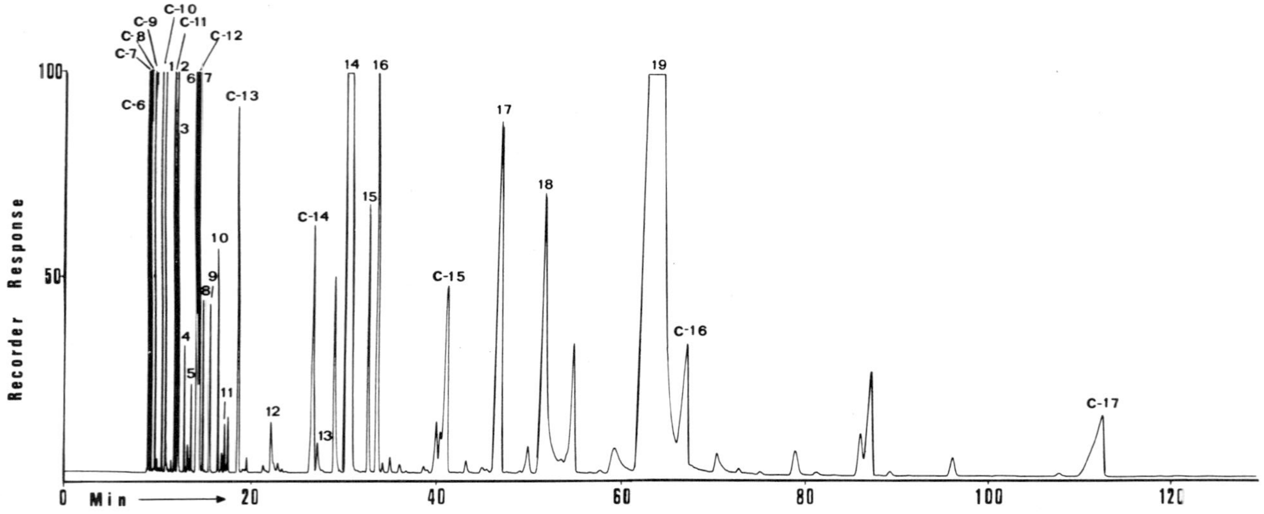

Figure 14. Chromatogram of peppermint oil with n-paraffin standards: 50-m × 0.28-mm ID glass capillary column coated with Carbowax 20M, isothermal at 100°C.

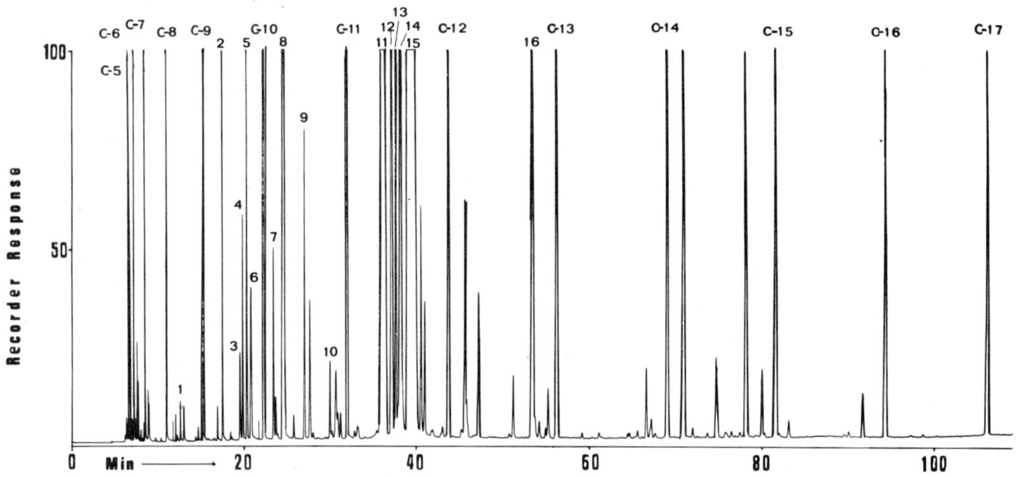

Figure 15. Chromatogram of peppermint oil with n-paraffin standards. The same column as in Fig. 13, and programmed from 80 to 200°C at 2°C min^{-1}.

compounds in the essential oil of freeze-dried guava puree using the GC/MS technique. A glass capillary column (110 m × 0.3 mm ID) coated with free fatty acid phase (FFAP) was used in this analysis. Figure 18 showing their capillary chromatogram indicates that this oil contains well over 100 components.

A combined GC/MS and GC retention indices were used in the analysis of components of marsh grass [20]. Gas chromatography utilized a glass capillary column (74 m × 0.75 mm ID) coated with OV-17, which was also used to determine the retention indices. Forty compounds, including 15 aliphatic and aromatic hydrocarbons, 9 aldehydes, 2 ketones, 12 alcohols, phenols, and esters, and benzothiazole were proposed as components of this oil.

Debrauwere and Verzele identified 51 substances in the oxygenated fraction of pepper oil using two kinds of glass capillary columns [41] after the oil was fractionated into five groups: free acids, esters, carbonyl compounds, alcohols, and residual oxygenated compounds. A glass open tubular column (70 m × 0.9 mm ID) coated with SE-30 was used for the isolation of free acids, such as methylesters (Fig. 19). The other oxygenated fractions were isolated by a glass open tubular column (75 m × 0.5 mm ID) coated with Carbowax 20M (Figs. 20 through 23).

The same authors also identified 26 hydrocarbons in black pepper oil [49]. After the separation of the hydrocarbon fraction, the oil was

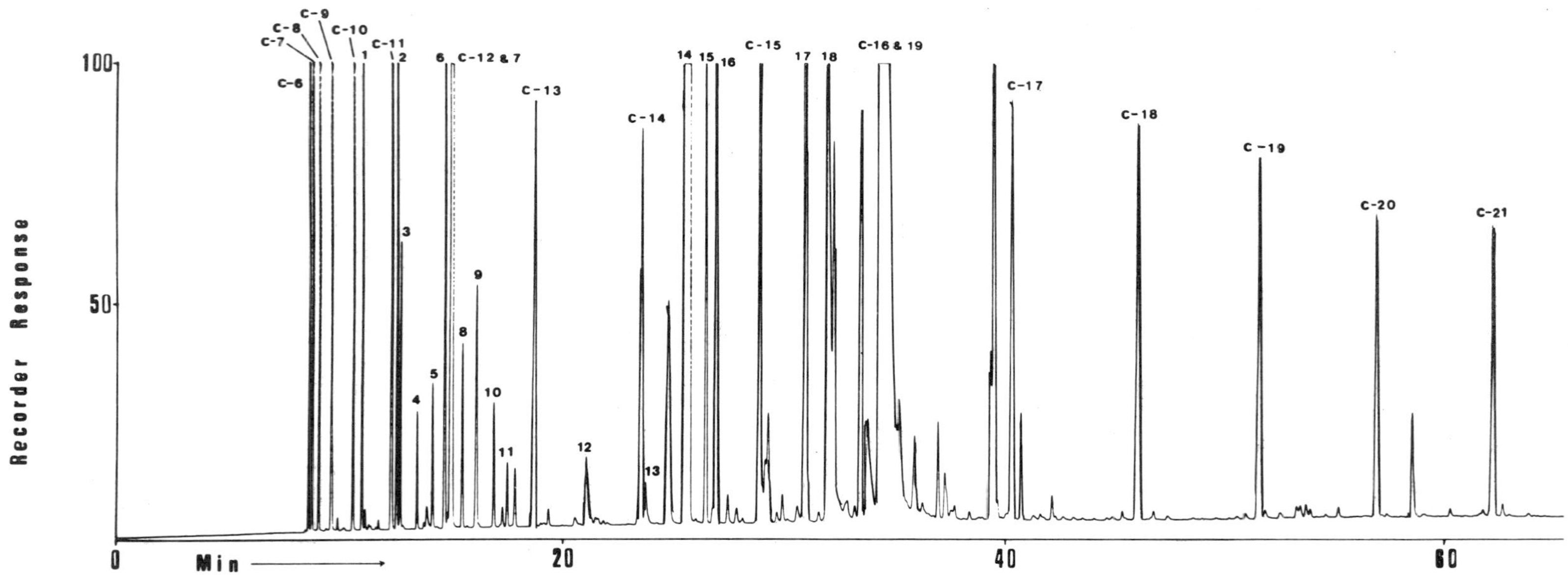

Figure 16. Chromatogram of peppermint oil with n-paraffin standard. The same column as in Fig. 14, and programmed from 80 to 200°C at 2°C min^{-1}.

Table 5. Kováts Index of Peppermint Constituents

Carbowax 20M				OV-101			
Peak no.	Compound	Isothermal[a]	Programmed[b]	Peak no.	Compound	Isothermal[c]	Programmed[d]
1	α-Pinene	1054	1029	1	cis-3-Hexene-1-ol	837	847
2	β-Pinene	1126	1111	2	α-Pinene	935	940
3	Sabinene	1126	1118	3	3-Octanol	963	971
4	Δ^3-Carene	1156	1149	4	Sabinene	968	975
5	Myrcene	1181	1174	5	β-Pinene	976	980
6	d-Limonene	1200	1191	6	Myrcene	982	986
7	1,8-Cineol	1209	1200	7	p-Cymene	1011	1015
8	β-Phellandrene	1221	1215	8	1,8-Cineol	1025	1028
9	γ-Terpinene	1260	1257	9	γ-Terpinene	1050	1055
10	p-Cymene	1260	1257	10	Terpinolene	1080	1083
11	Terpinolene	1275	1273	11	Menthone	1138	1142
12	cis-3-Hexene-1-ol	1354	1354	12	Isomenthone	1145	1151
13	3-Octanol	1406	1405	13	Menthofuran	1149	1155
14	Menthone	1436	1444	14	Neomenthol	1154	1159
15	Menthofuran	1450	1460	15	Menthol	1167	1171
16	Isomenthone	1457	1468	16	Menthyl acetate	1276	1281
17	Menthyl acetate	1529	1541				
18	Neomenthol	1548	1559				
19	Menthol	1590	1600				

[a]Refer to Fig. 13.
[b]Refer to Fig. 15.
[c]Refer to Fig. 14.
[d]Refer to Fig. 16.

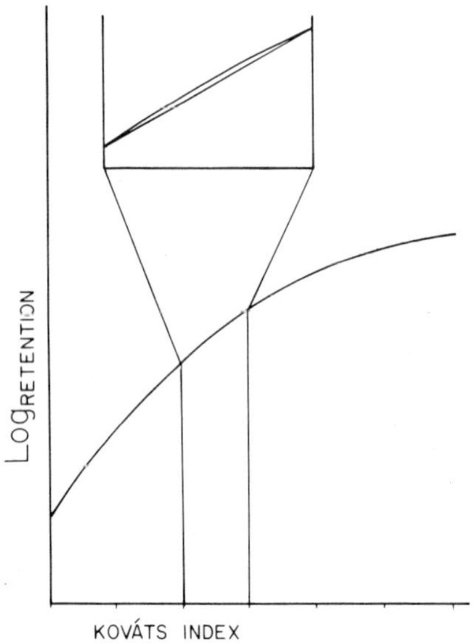

Figure 17. Approximate linear relationship between $\log_{\text{retention}}$ and Kováts index in small section of programmed plot.

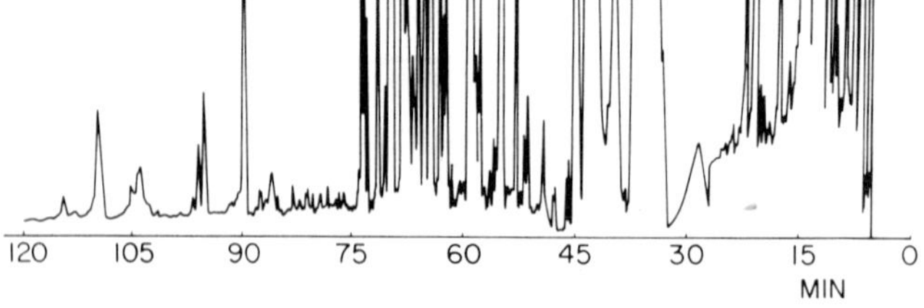

Figure 18. Chromatogram of the freon-12 extract of partially freeze-dried guava puree: 110-m × 0.3-mm ID glass capillary column coated with free fatty acid phase (FFAP) and programmed from 30 to 180°C at 2°C min^{-1}. (Reproduced courtesy of Dr. P. Torline, National Food Research Institute, Pretoria, South Africa.)

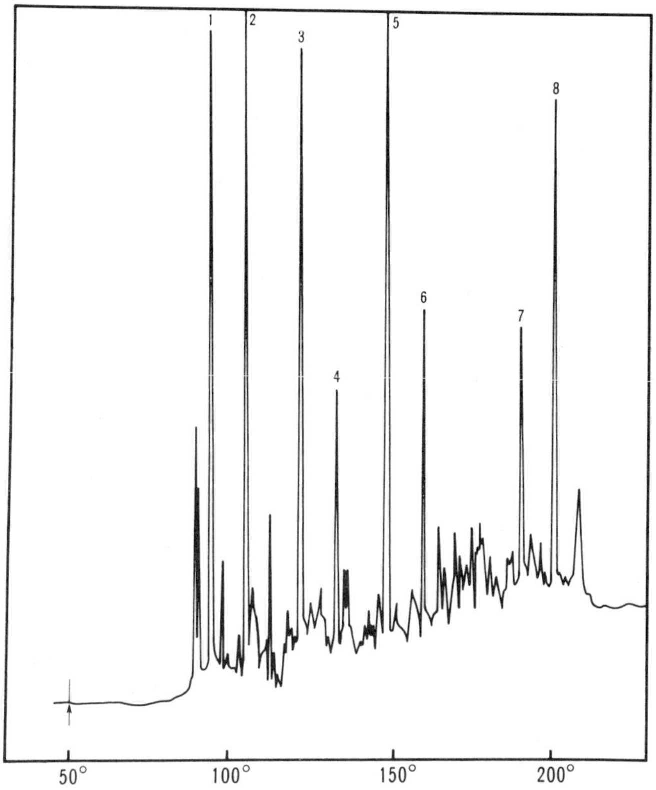

Figure 19. Monitogram of free acids (as methylesters) from black pepper oil: 70-m × 0.9-mm ID glass capillary column coated with SE-30 and programmed from 50 to 200°C at 4°C min^{-1}. (Reproduced courtesy of Dr. M. Verzele, Laboratory of Organic Chemistry, State University of Ghent, Krijgslaan, Ghent, Belgium.)

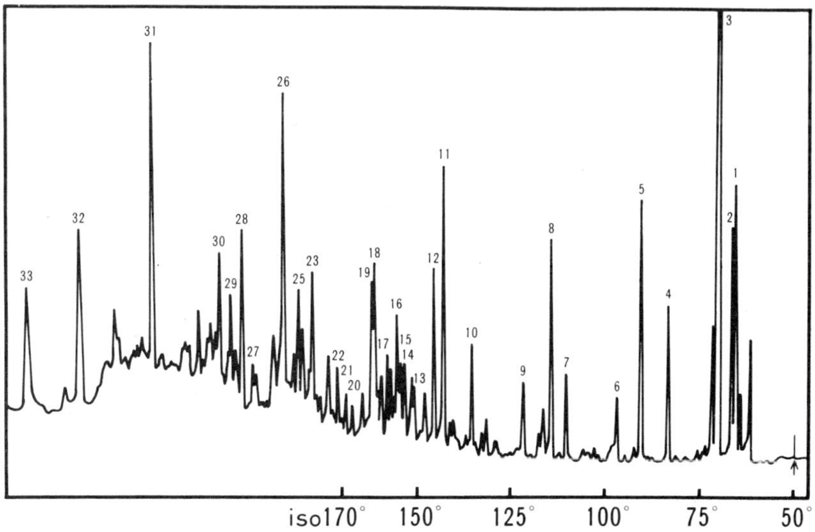

Figure 20. Monitogram of methylesters from black pepper oil: 75-m × 0.5-mm ID glass capillary column coated with Carbowax 20M and programmed from 50 to 170°C at 2°C min^{-1}. (Reproduced courtesy of Dr. M. Verzele, Laboratory of Organic Chemistry, State University of Ghent, Krijgslaan, Ghent, Belgium.)

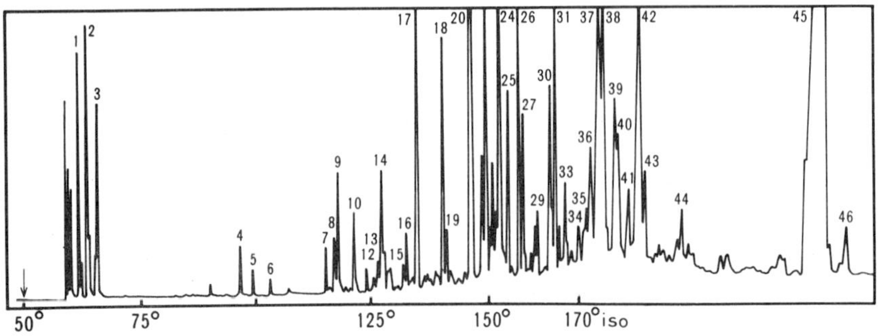

Figure 21. Chromatogram of carbonyl compounds from black pepper oil: 75-m × 0.5-mm ID glass capillary column coated with Carbowax 20M and programmed from 50 to 170°C at 2°C min^{-1}. (Reproduced courtesy of Dr. M. Verzele, Laboratory of Organic Chemistry, State University of Ghent, Krijgslaan, Ghent, Belgium.)

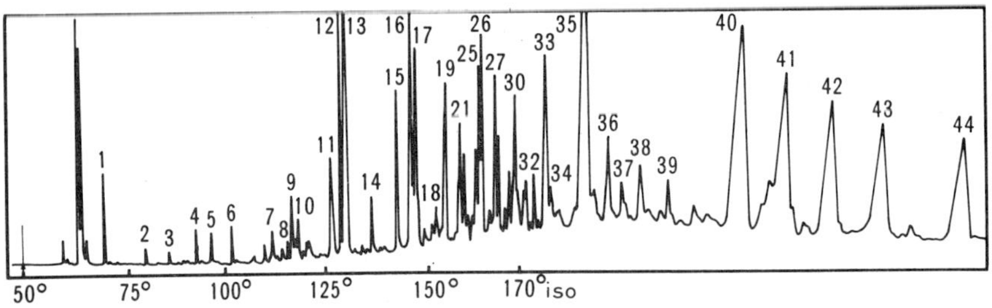

Figure 22. Chromatogram of alcohols from black pepper oil. Same circumstances as in Fig. 21. (Reproduced courtesy of Dr. M. Verzele, Laboratory of Organic Chemistry, State University of Ghent, Krijgslaan, Ghent, Belgium.)

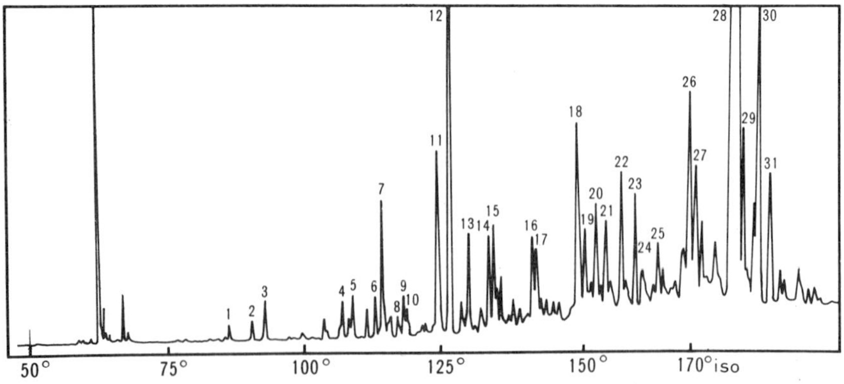

Figure 23. Chromatogram of the residual oxygenated fraction from black pepper oil. Same circumstances as in Fig. 21. (Reproduced courtesy of Dr. M. Verzele, Laboratory of Organic Chemistry, State University of Ghent, Krijgslaan, Ghent, Belgium.)

subjected to column chromatography using isooctane as the developing solvent. Sixteen sesquiterpene hydrocarbons were effectively separated by Carbowax 20M (glass open tubular column, 40 m × 0.4 mm ID). It is common knowledge that some sesquiterpenes are structually closely related and possess similar chemical and physical properties. It is very difficult to separate and detect such compounds, especially when their concentrations are so low that even though their separation may be achieved by TLC, fractional distillation, etc., they successfully resist subsequent detection. A high-resolution gas chromatographic column may offer the only technique by which these compounds can be separated at our present state of the art.

Taskinen and Nykanen [50] identified 70 volatile constituents of angelica root oil, which was separated by extraction with ether-pentane followed by steam distillation, using two dissimilar capillary columns (25 m × 0.28 mm ID, SP-1000 and OV-101). A typical gas chromatogram of this sample is shown in Fig. 24. The authors also introduced a glass inlet splitter which assured an inert system throughout the analysis.

A quantitative study of Moroccan thyme oil was conducted with a glass capillary column (86 m × 0.5 mm ID) coated with Carbowax 20M; 21 constituents were characterized on the basis of their infrared spectra [40].

Shibamoto and Jennings [21] reported the presence of 97 components in the essential oil of California juniper. Forty-two compounds were identified on the basis of their mass spectra and gas chromatographic retention indices (Kováts Index). Two types of WCOT glass capillary columns (100 m × 0.25 mm ID) were used in this investigation: Carbowax 20M or SE-30 admixed with 7% Igepal CO-880. These columns were used on many of the analytical determinations and also for the GC/MS studies. Columns (100 m × 0.25 mm ID) coated with these same liquid phases and with FFAP and OV-17 were also used in comparative analyses. The compounds identified included 12 monoterpene hydrocarbons, 9 monoterpene alcohols, 5 monoterpene aldehydes and ketones, 4 monoterpene acetates, and 12 sesquiterpene hydrocarbons. In addition to the above, 17 monoterpenes and 38 sesquiterpenes were recognized from their characteristic MS fragmentation patterns.

Figure 25 shows a gas chromatogram of the essential oil of juniper leaf obtained with an SE-30 column. Referring to this figure, the area from peaks 4 to 23 is dominated by monoterpene hydrocarbons, that from peaks 24 to 42 by monoterpene alcohols, that from peaks 43 to 50 by a series of mono-

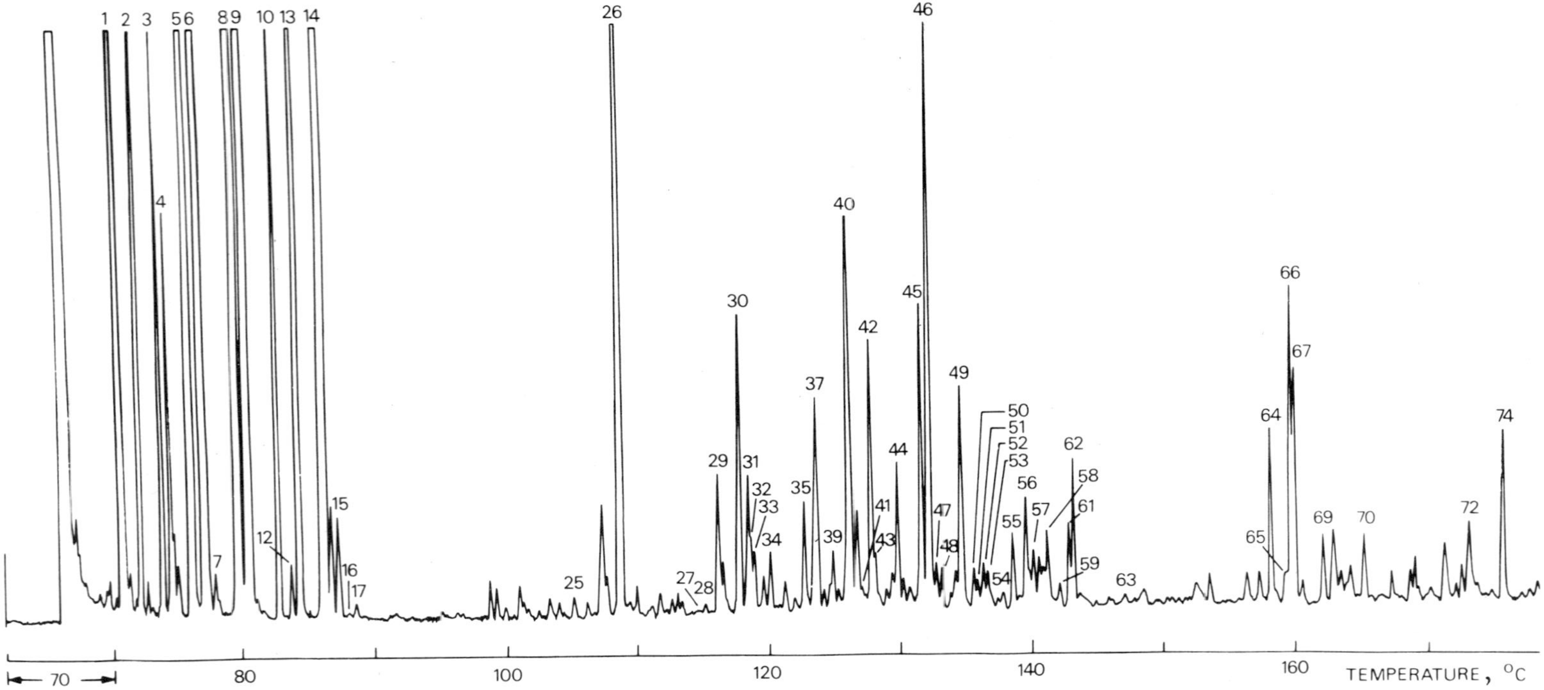

Figure 24. Chromatogram of angelica root oil on SP-1000 glass capillary column (25 m × 0.28 mm ID). Temperature programmed 4 min at 70°C, 2°C min^{-1} 70 to 215°C. (Reproduced courtesy of Drs. J. Taskinen and Nykanen, Research Laboratories of the State Alcohol Monopoly, Helsinki, Finland.)

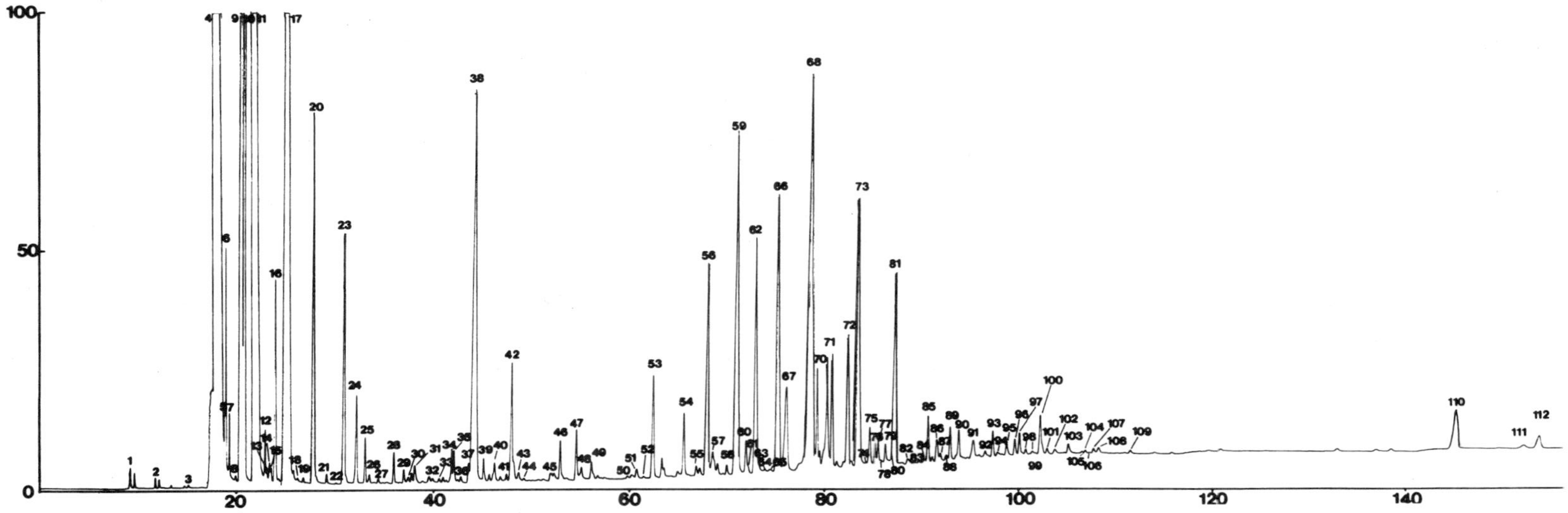

Figure 25. Chromatogram of California juniper leaf oil, split ~1:100, on a 100-m × 0.25-mm ID wall-coated open tubular (WCOT) glass capillary column coated with methyl silicone SE-30. Temperature-programmed from 70 to 170°C at 1°C min^{-1} and held. For details, see text.

terpene ketones, and that from peaks 51 to 55 by the acetate esters of monoterpene alcohols. In addition, peaks 56 to 88 represent sesquiterpene hydrocarbons, followed by sesquiterpene ketones and alcohols. Figure 26 shows the chromatogram obtained from a Carbowax 20M column. The numerical assignments on this chromatogram are those used on the former chromatogram (Fig. 25). Some sesquiterpenes were difficult to separate on the Carbowax 20M column, but the more volatile compounds were separated more efficiently by Carbowax 20M than by SE-30. Figure 27 shows a chromatogram from the FFAP column. Since polarity of this liquid phase is similar to that of Carbowax 20M, there is not too much difference between the Carbowax 20M and FFAP chromatograms.

Yabumoto et al. [51] investigated the volatile constituents of cantaloupe using three sampling techniques: (1) direct sampling of melon cavity gases via a septum-sealed glass probe cemented to the melon; (2) entrapment of large volumes of headspace gas on porous polymer traps; and (3) steam distillation-extraction of pieces of melon fruit tissue. Essences were subjected to gas chromatographic analysis on high-resolution WCOT glass capillary columns. Compound identifications were based on mass spectra, gas chromatographic retentions (Kováts Index), and occasionally infrared spectra. Two kinds of 80-m × 0.25-mm ID columns were used in these experiments; one was coated with SE-30 admixed with 7% Igepal CO-880, the other was coated with Carbowax 20M admixed with 5% benzyltriphenyl phosphonium chloride (BTPPC) and 1% Ionox 330. Figures 28, 29, and 30 show the chromatograms of the three samples obtained by the methods listed above. The authors pointed out that the chromatogram obtained from direct headspace sampling of the fruit cavity gas (Fig. 30) exhibited larger amounts of lower-boiling-point volatiles than the chromatogram of the Porapak essence (Fig. 28). The chromatogram of the steam distillation-extraction essence (Fig. 29) was reasonably comparable to the other two chromatograms, except for the early portion, in which the solvent peaks obliterated other peaks. Forty-six compounds, mainly esters, were identified in this essence. One of the most interesting compounds isolated was dimethyl disulfide. That some sulfur-containing compounds are present in fruits and vegetables is recognized, although very few have been reported. Sulfur-containing compounds are known to be very susceptible to metallic surfaces and rarely survive analysis in equipment utilizing metal columns or injectors. It may be possible to find more sulfur-containing compounds in essential oils using these more inert glass capillary columns.

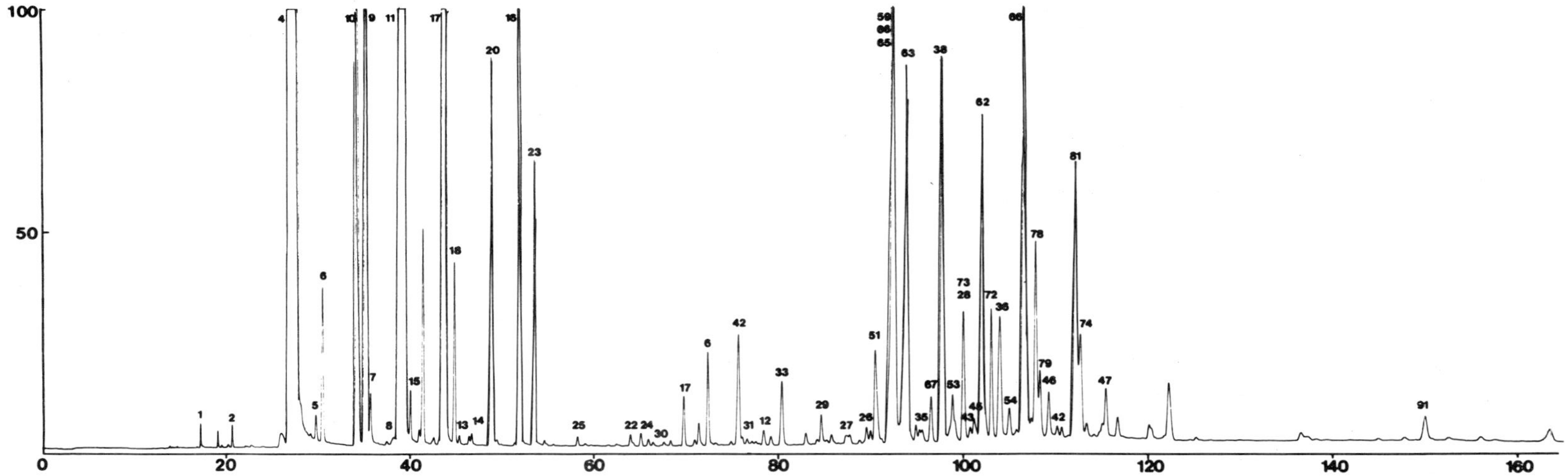

Figure 26. Chromatogram of California juniper leaf oil on a 100-m × 0.25-mm ID WCOT glass capillary column coated with PEG Carbowax 20M. Temperature programmed from 70 to 170°C at 1°C min^{-1} and held. Peaks as in Fig. 25.

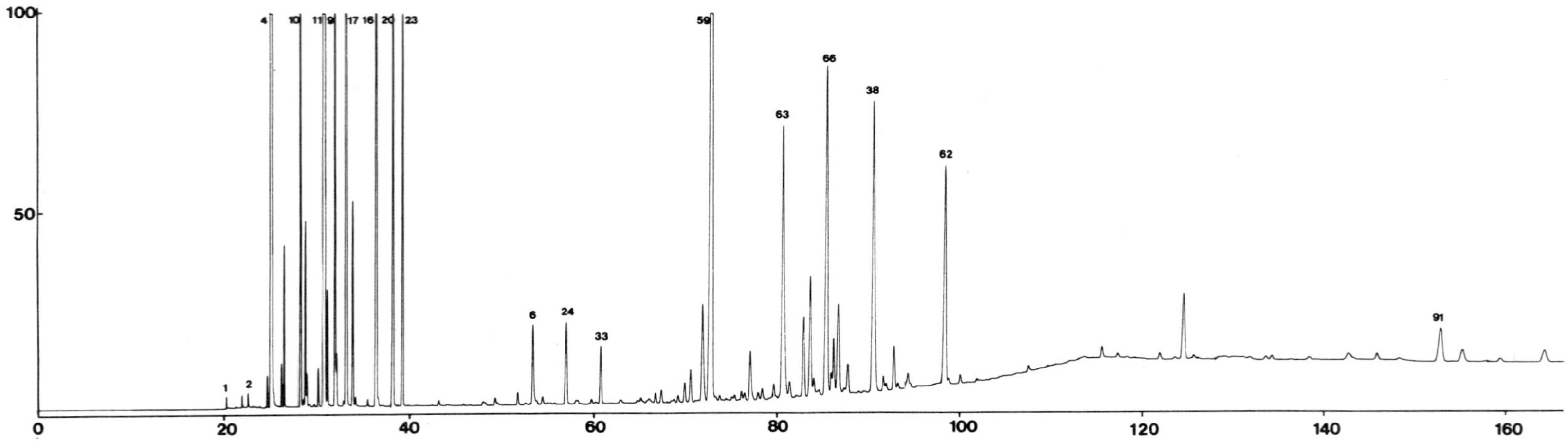

Figure 27. Chromatogram of California juniper leaf oil on a 100-m × 0.25-mm ID WCOT glass capillary column coated with the terphthalic-acid-terminated polyethylene glycol FFAP. Temperature-programmed from 70 to 170°C at 1°C min^{-1} and held.

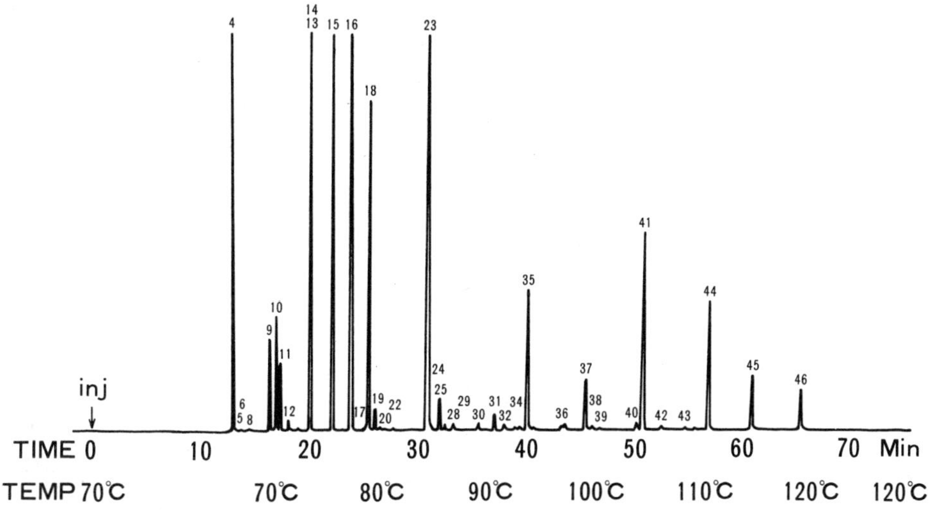

Figure 28. Chromatogram of cantaloupe Porapak essence on WCOT glass capillary column coated with Carbowax 20M. Column held at 70°C for 16 min, programmed at 1°C min^{-1} to 120°C and held. (Source: Ref. 60. Reproduced by permission of the copyright owner.)

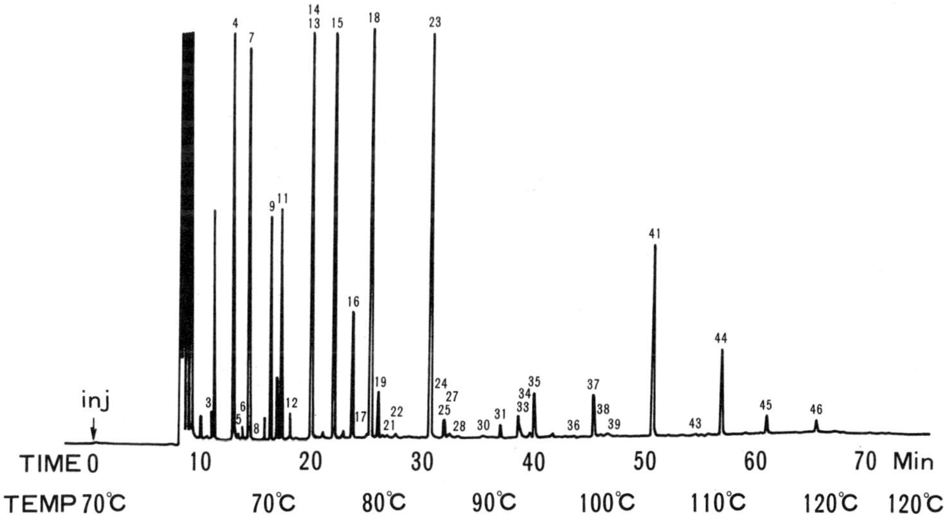

Figure 29. Chromatogram of a cantaloupe steam distillation-extraction essence on WCOT glass capillary column coated with Carbowax 20M. Same circumstances as in Fig. 28. (Source: Ref. 60. Reproduced by permission of the copyright owner.)

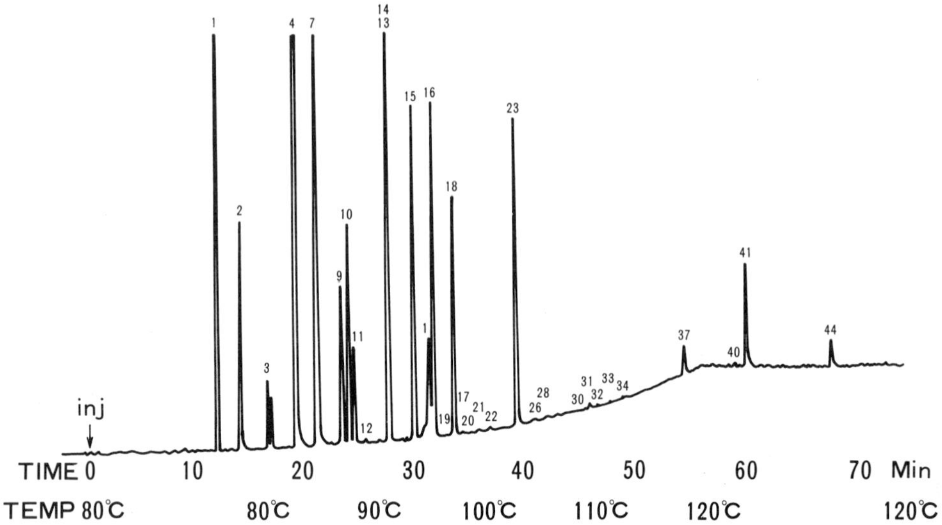

Figure 30. Chromatogram of a cantaloupe fruit cavity gas on WCOT glass capillary column coated with Carbowax 20M. Same circumstances as in Fig. 28. (Source: Ref. 60. Prproduced by permission of the copyright owner.)

Fifty-six compounds, including 11 monoterpenes and 4 sesquiterpenes, were identified in the essential oil of the castanopsis flower, which grows wild in central and southern Japan [26]. Figures 31 and 32 show typical chromatograms of the oil, obtained on a polar column (Carbowax 20M) and a nonpolar column (OV-101).

The glass capillary columns used in essential oil analyses are summarized in Table 6.

5.2 Perfumery

Glass capillary columns have not been used widely in the perfumery field to date. Since perfumers are equipped with the most sensitive existing detector for odorous chemicals (i.e., the human nose, which is 1000 times more sensitive than the flame ionization detector with regard to some compounds), the perfumer can more carefully assess the effect of new means of studying perfumery materials. The glass capillary column is highly recommended for perfumery studies because of its inertness and high resolution. Since there is no way to express odors in terms of physical measurements, a gas chromatographic pattern is one of the most convenient and useful ways to depict these odorous mixtures. Figures from 33 to 38 show gas chromatograms of the essential oils most commonly used for perfumery products (cosmetic products, toiletry products, etc.).

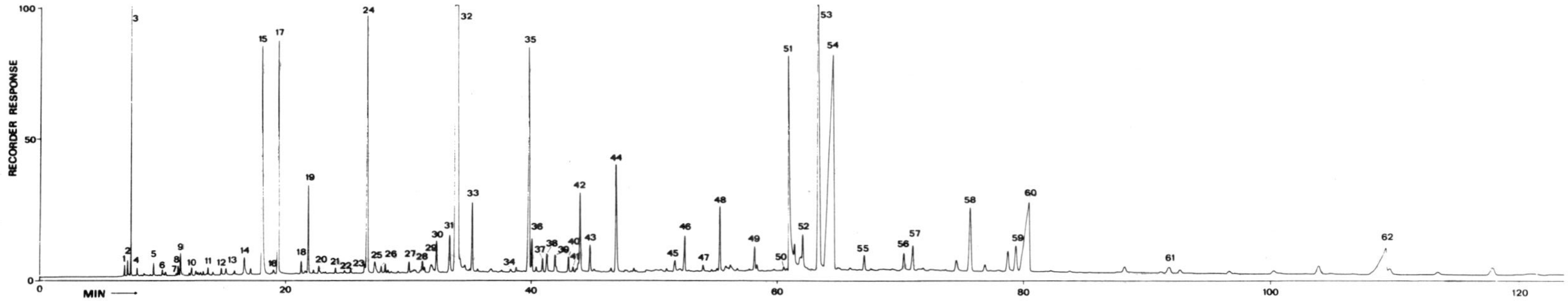

Figure 31. Chromatogram of an oil of castanopsis flower on a 50-m × 0.28-mm ID WCOT glass capillary column coated with Carbowax 20M.

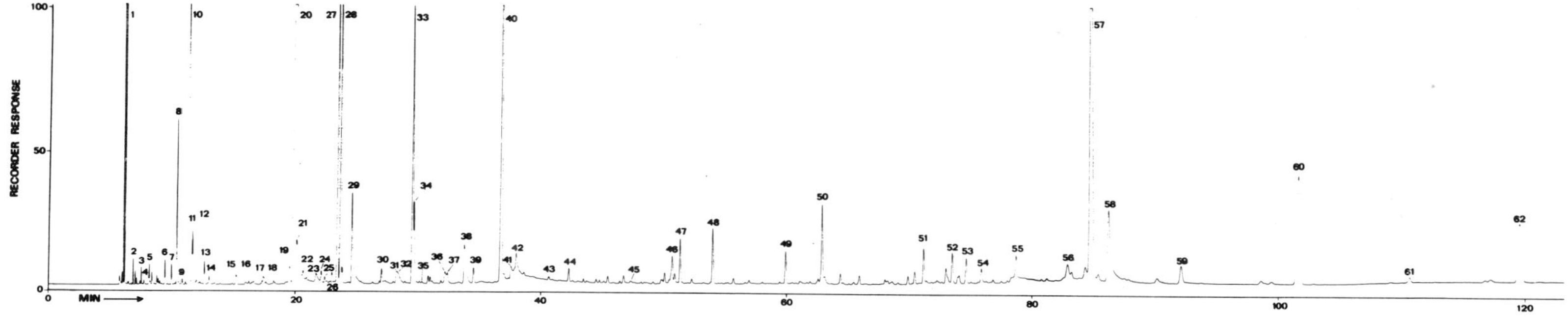

Figure 32. Chromatogram of an oil of castanopsis flower on 70-m × 0.25-mm ID WCOT glass capillary column coated with methyl silicone OV-101.

Table 6. WCOT Glass Capillary Columns Used for Essential Oil Analysis

Liquid phase	Column Size (m × mm)	Oil	Reference
FFAP	100 × 0.3	Guava	86
Polypropylene glycol (PPG)	50 × 0.3	Lemon oil	87
Carbowax 20M	86 × 0.5	Thyme	40
Carbowax 20M	40 × 0.4	Black pepper	49
Emulphor ON-870	40 × 0.2	Sage leaf	16
OV-17	74 × 0.75	Marsh grass	20
SE-30	70 × 0.9	Black pepper	41
Carbowax 20M	75 × 0.5	Black pepper	41
SP-1000	25 × 0.28	Angelica root	50
Carbowax 20M	48.6 × 0.24	Peppermint	38
SE-30	100 × 0.25	Cantaloupe	51
Carbowax 20M	100 × 0.25	Cantaloupe	51
Carbowax 20M	100 × 0.25	Juniper leaf	21
SE-30	100 × 0.25	Juniper leaf	21
FFAP	50 × 0.25	Juniper leaf	21
Carbowax 20M	50 × 0.28	Castanopsis flower	26
OV-101	70 × 0.28	Castanopsis flower	26

Both the horticultural and crop-to-crop variations influence the composition of an essential oil and the relative proportions of individual constituents. This is true even when dealing with exactly the same species; consequently, quality control of perfumery materials is very important. Gas chromatographic quality control is not only useful for analysis of perfumery ingredients, but also for checking the perfumes and flavor bases which comprise the final products. Figures from 39 to 42 show gas chromatograms of two famous perfumes. Both (polar and nonpolar) columns are widely used in industrial quality control, and one must choose a liquid phase suitable to the mixture. Generally, nonpolar columns are used for perfumes and polar columns for flavor essences.

A long column is not practical for industrial quality control, since the analysis time required per sample would be prohibitive. In some plants over 300 perfumery materials (each usually consisting of a mixture of 50

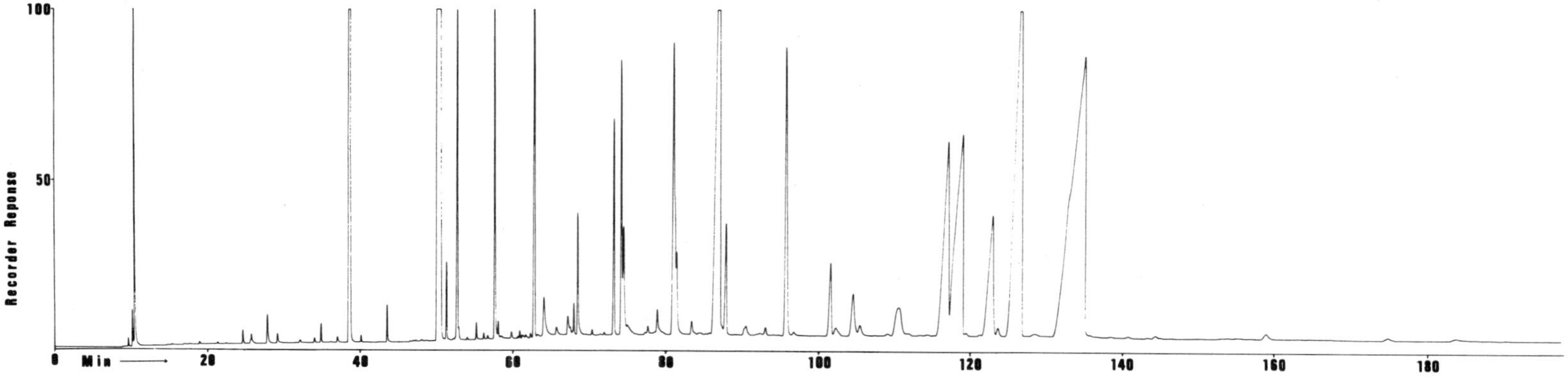

Figure 33. Chromatogram of Jasmin oil from Grasse, France: 50-m × 0.28-mm ID glass capillary column coated with Carbowax 20M and programmed from 80 to 200°C at 2°C min^{-1}.

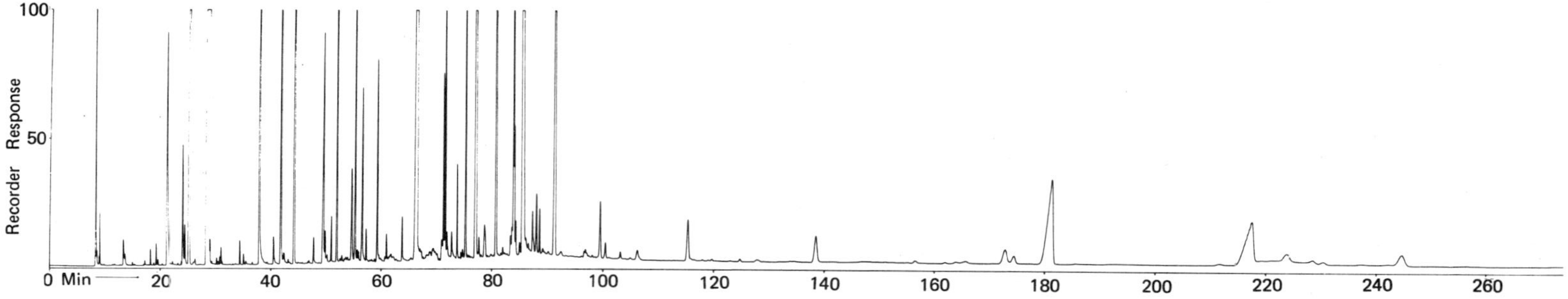

Figure 34. Chromatogram of the same oil as in Fig. 33: 50-m × 0.28-mm ID glass capillary column coated with OV-101 and programmed from 80 to 200°C at 2°C min^{-1}.

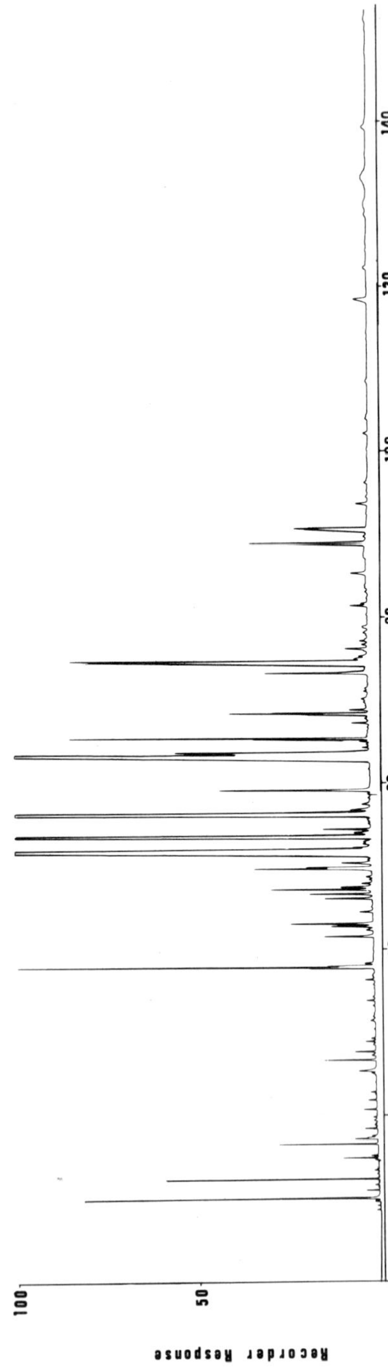

Figure 35. Chromatogram of rose oil from Bulgaria. Same circumstances as in Fig. 33.

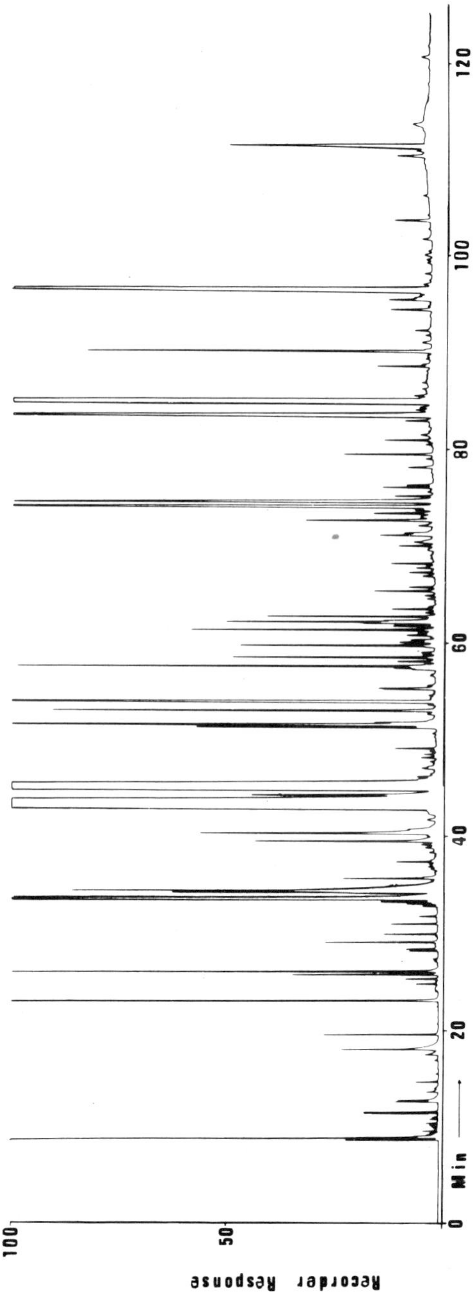

Figure 36. Chromatogram of the same oil as in Fig. 35. Same circumstances as in Fig. 34.

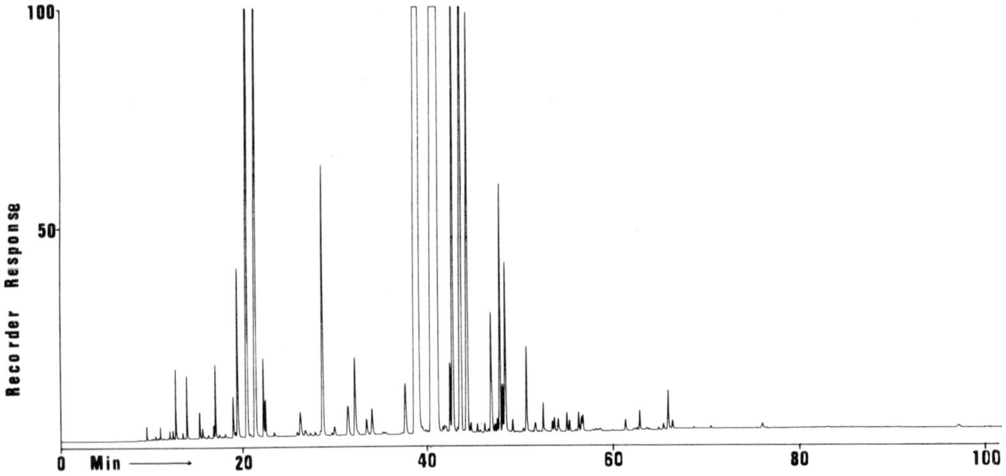

Figure 37. Chromatogram of lavender oil. Same circumstances as in Fig. 33.

to 500 chemicals) must be checked on a daily basis. Increasing the number of gas chromatographs would be one way to solve this problem, but it is much more economical to shorten the analysis time. The ideal time for one sample would be less than 30 min. Since it takes 2 to 3 hr to run an essential oil sample on a 50-m × 0.25-mm ID glass capillary column coated with Carbowax 20M, the analysis time may be decreased either by shortening the column or by applying a thinner coating. Figures 43 and 44 show chromatograms of the same peppermint oil shown in Fig. 12, obtained on columns which were modified for quality control. Adjusting the length for ideal analytical time is the easiest technique, and this is generally feasible only with the glass capillary column.

6. FUTURE DEVELOPMENTS

The use of glass capillary columns in essential oil analysis is becoming increasingly more practical. A wide range of commercial columns are available today. Although the handling of these columns is still somewhat more difficult than handling stainless steel columns, methods of column mounting which should alleviate the difficulty have been developed and are being improved upon. Better coating methods are also being studied. Columns approaching 100% coating efficiency may become available on a routine basis in the near future.

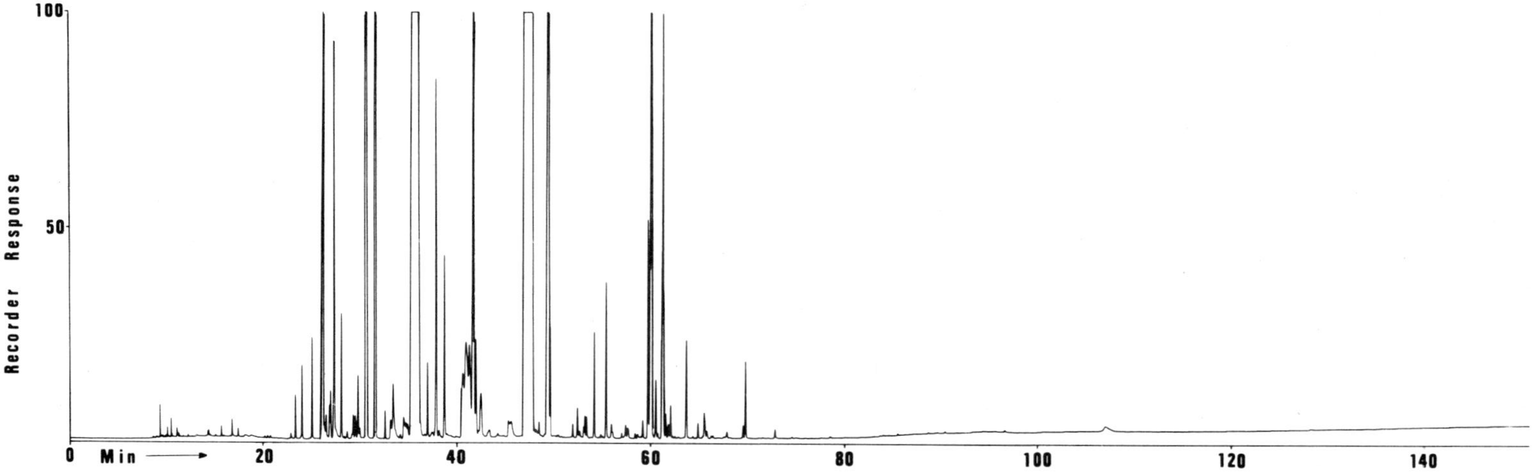

Figure 38. Chromatogram of the same oil as in Fig. 37. Same circumstances as in Fig. 34.

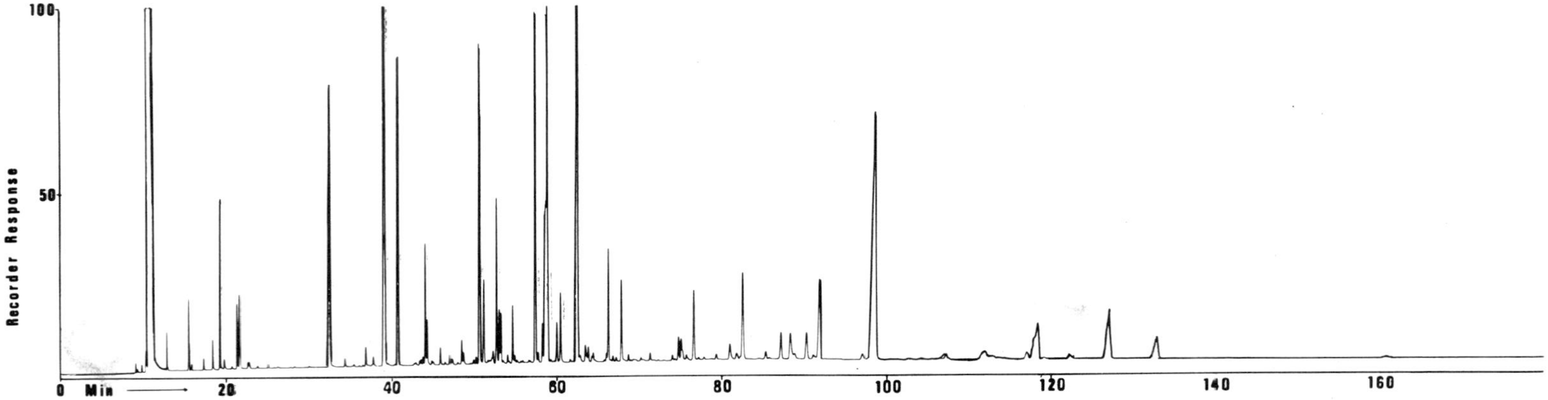

Figure 39. Chromatogram of a perfume, Chanel No. 5. Same circumstances as in Fig. 33.

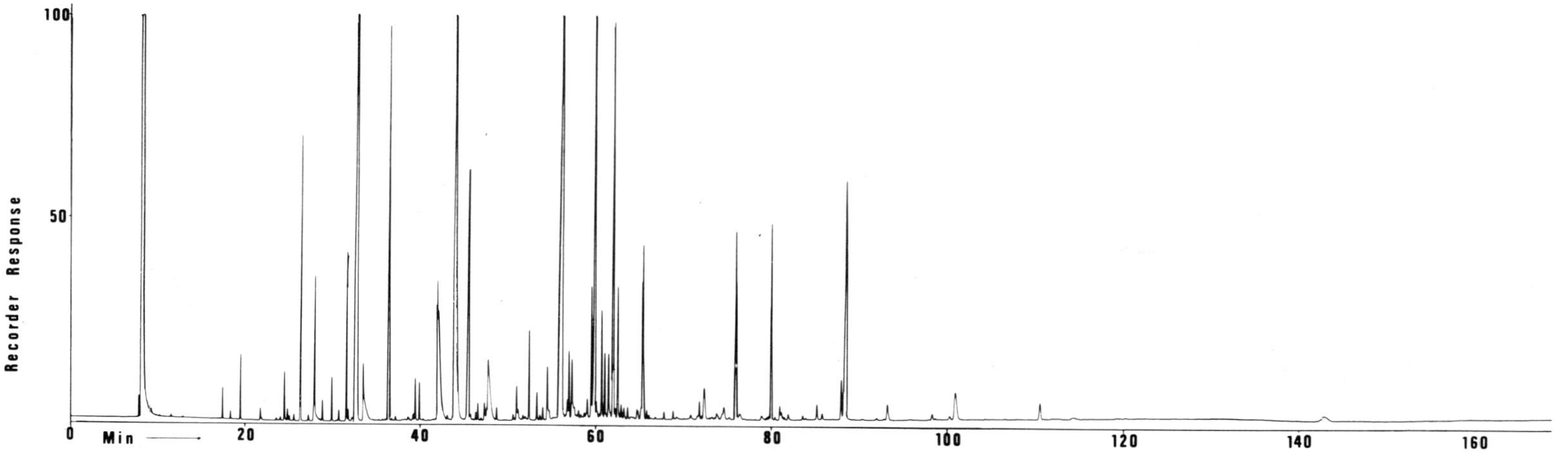

Figure 40. Chromatogram of the same perfume as in Fig. 39. Same circumstances as in Fig. 34.

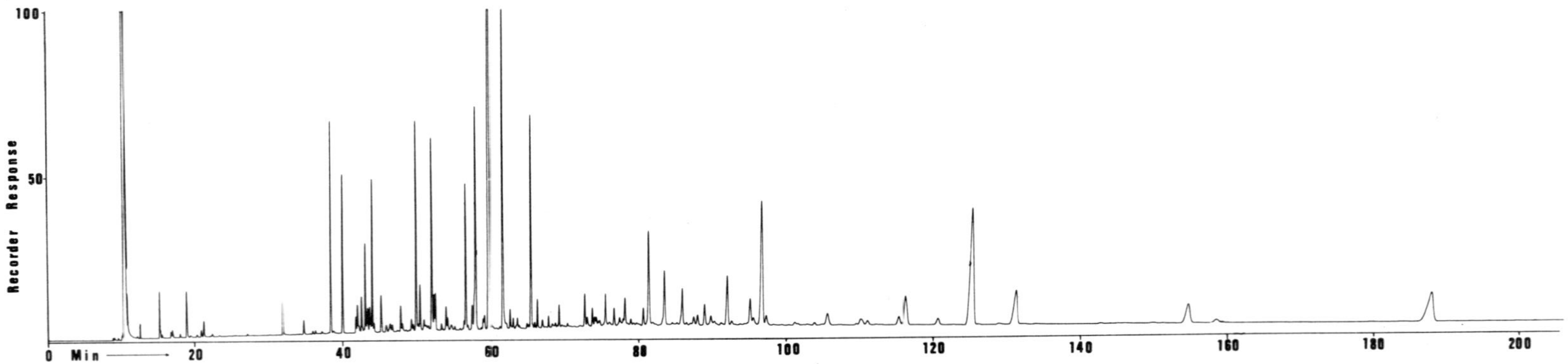

Figure 41. Chromatogram of a perfume, Arpége. Same circumstances as in Fig. 33.

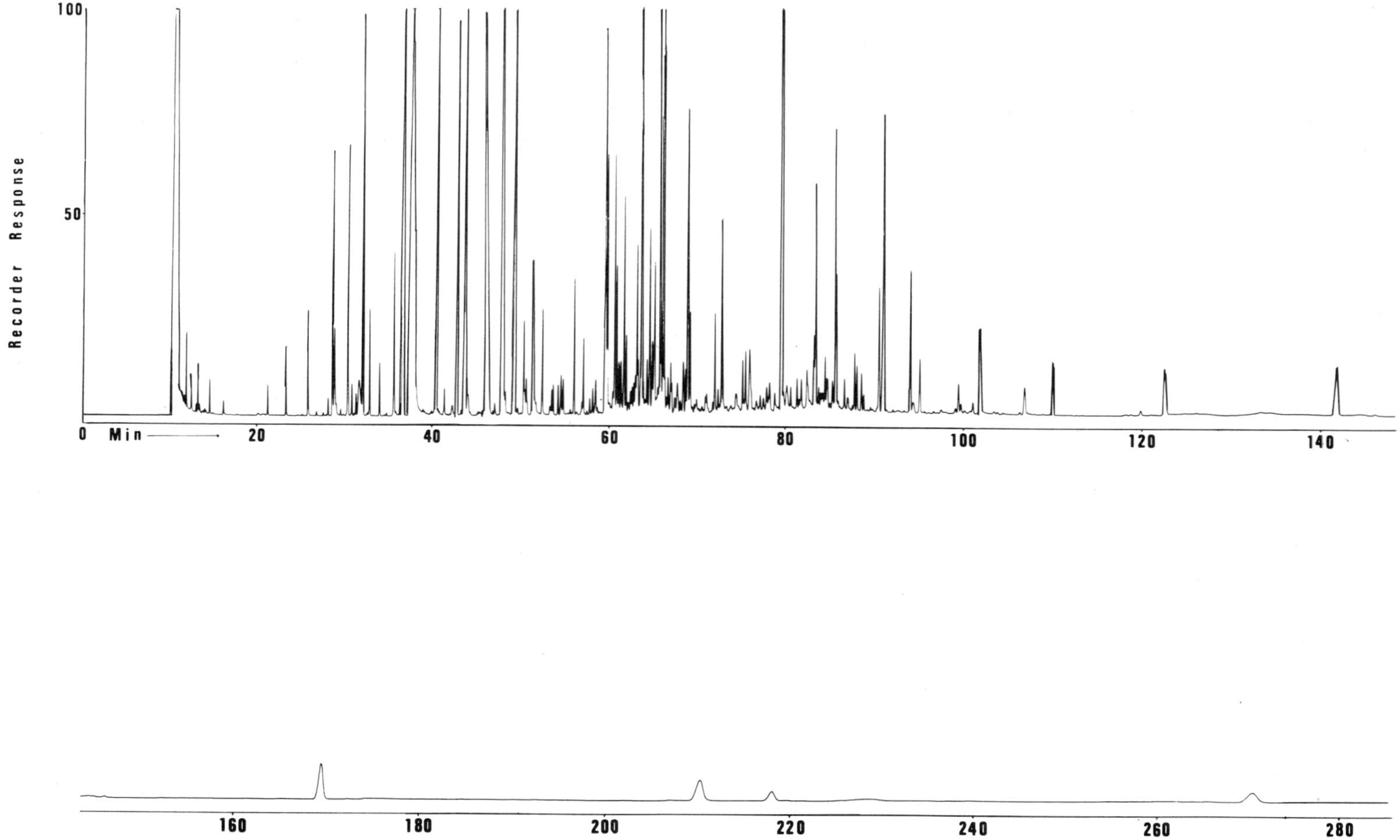

Figure 42. Chromatogram of the same perfume as in Fig. 41. Same circumstances as in Fig. 34.

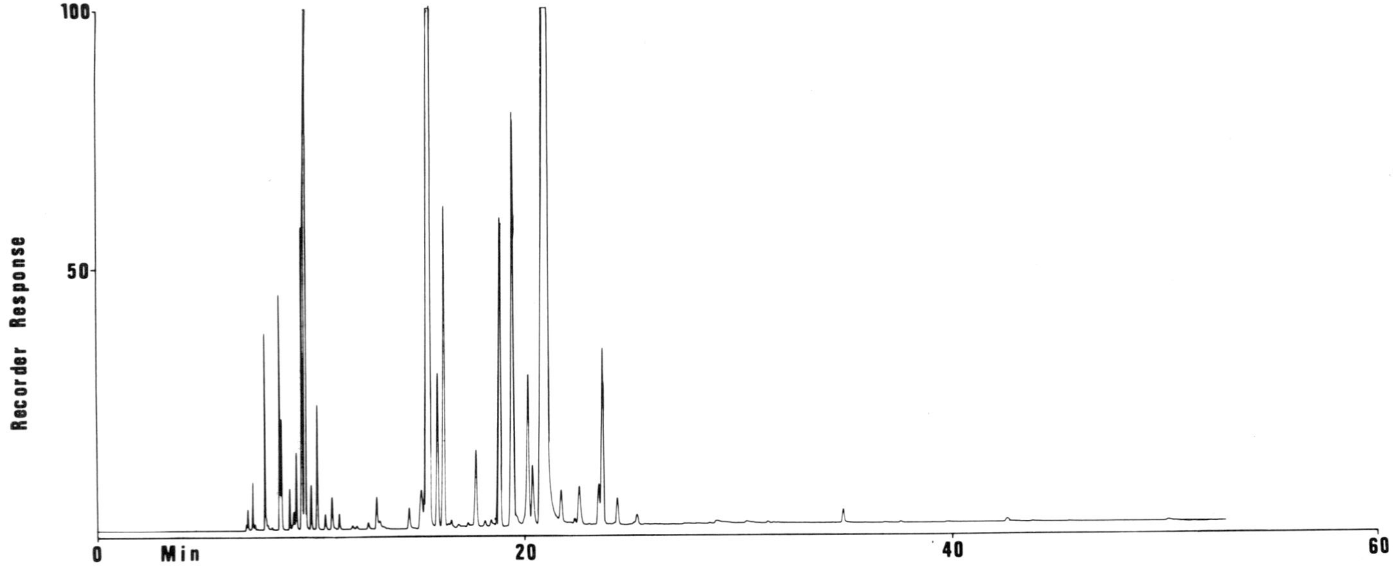

Figure 43. Chromatogram of the same peppermint oil as in Fig. 12: 50-m × 0.28-mm ID glass capillary column coated with Carbowax 20M (coated with dilute concentration of liquid phase, ~0.01%, by static method). Same circumstances as in Fig. 12.

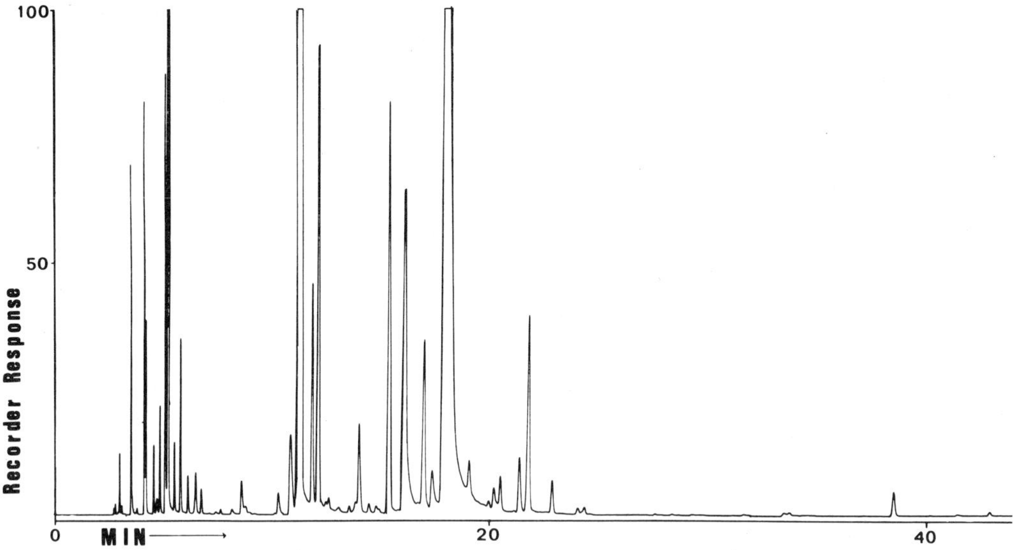

Figure 44. Chromatogram of the same peppermint oil as in Fig. 12: 16-m × 0.28-mm ID glass capillary column coated with Carbowax 20M (coated with 0.5% liquid phase in methylene chloride-pentane solution by static method). Same circumstances as in Fig. 12.

Analyses involving sulfur or halogen compounds are better conducted with inert glass capillary columns than with stainless steel columns having active metallic surfaces. The isolation of sulfur-containing compounds in essential oils may lead to new areas of essential oil research.

Improvements in reproducibility of retention data may also suggest new methods of qualitative analysis of microconstituents of essential oils. Identification of unknowns based on the comparison of their mass spectra and retention indices with those of authenticated samples is becoming more reliable. It may be interesting to reexamine common essential oils, which have been investigated by other methods, using the glass capillary column.

REFERENCES

1. Archtander, S., Perfume and Flavor Materials of Natural Origin, private publication, 1969.
2. Nicoll, B., *J. Chem. Proc. Eng.*, 49:99 (1968).
3. Kekulé, A., *Lehrb. Org.*, 2:437 (1866).
4. Kindt, F., and Trommsdorffs, J. J., *Pharmacie*, 11:132 (1802).
5. Robiquet, A., *J. Pharm.*, 21:335 (1835).
6. Teranishi, R., Hornstein, I., Issenberg, P., and Wick, E. L., *Flavor Research: Principles and Techniques*, Marcel Dekker, New York, 1971.
7. Jennings, W. G., *Gas Chromatography with Glass Capillary Columns*, 2nd ed., Academic Press, New York, 1980.
8. Fleming, G. P., Etchells, J. L., Bell, T. A., *J. Food Sci.*, 34:419 (1969).
9. Liebich, H. M., Koening, W. A., and Bayer, E., *J. Chromatogr. Sci.*, 8:527 (1970).
10. Klimes, I., and Lamparsky, D., *Int. Flavor and Food Additives*, Nov./Dec., p. 272 (1976).
11. Yabumoto, K., Jennings, W. G., and Pangborn, R. M., *J. Food Sci.*, 40:105 (1975).
12. Murray, K. E., *J. Chromatogr.*, 135:49 (1977).
13. Jennings, W. G., and Filsoof, M., *J. Agr. Food Chem.*, 25:440 (1977).
14. Yabumoto, K., Investigation of Certain Volatile Constituents of Ripening Cantaloups, *Cucumis melo*, Ph.D. thesis, University of California, Davis, 1976.
15. Dupuy, H. P., Fore, S. P., and Goldblatt, L. A., *J. Amer. Oil Chem. Soc.*, 50:340 (1973).
16. Novotny, M., Lee, M. L., and Bartle, K. D., *Chromatographia*, 7:333 (1974).
17. Buttery, R. G., McFadden, W. H., Lundin, R. E., and Kealy, M. P., *J. Inst. Brew.*, 70:396 (1964).
18. Sakai, T., Yoshihara, K., and Hirose, Y., *Bull. Chem. Soc. Jap.*, 41:1945 (1968).
19. Miles, D. H., Mody, N. V., and Minyard, J. P., *Phytochemistry*, 12:1399 (1973).
20. Mody, N. V., Parish, E. J., Bhattacharyya, J., Miles, D. H., and Hedin, P. A., *Phytochem.*, 13:2027 (1974).
21. Shibamoto, T., and Jennings, W. G., Proceedings 7th Int. Cong. of Essential Oils, 1977, Kyoto, Abstr. No. K-2, page 414.
22. Likens, S. T., and Nickerson, G. B., *J. Chromatogr.*, 21:1 (1966).
23. Buttery, R. G., Seifert, R. M., Guadagni, D. G., Black, D. R., and Ling, L. C., *J. Agr. Food Chem.*, 16:1009 (1968).
24. Cronin, D. A., and Ward, M. K., *J. Sci. Food Agr.*, 22:477 (1971).

25. Flath, R. A., Forrey, R. R., and Guadagni, D. G., *J. Agr. Food Chem.*, 21:948 (1973).

26. Yamaguchi, K., and Shibamoto, T., *J. Agr. Food Chem.*, 27:847 (1979).

27. Schultz, T. H., Flath, R. A., Mon, T. R., Eggling, S. B., and Teranishi, R., *J. Agr. Food Chem.*, 25:446 (1977).

28. Lawrence, B. M., Hogg, J. W., and Terhune, S. J., *J. Chromatogr.*, 50:59 (1970).

29. Lawrence, B. M., Terhune, S. J., and Hogg, J. W., *Amer. Perf. Cos.*, 85:53 (1970).

30. Sakata, I., and Hashizume, T., *Agr. Boil. Chem.*, 35:128 (1971).

31. Lawrence, B. M., Hogg, J. W., and Terhune, S. J., *Flav. Ind.*, 3:467 (1972).

32. Bauman, S., and Gierschner, K., *Riechstoffe Aromen Korperpflegemittel*, 24:92 (1974).

33. Lawrence, B. M., *Perf. Essen. Oil Rec.*, 59:421 (1968).

34. Pattabhiraman, T. R., Sastry, L. V. L., and Abraham, K. O., *Perfum. Essent. Oil Rec.*, 60:233 (1969).

35. Rijkens, F., *Parf. Cosm. Sav.*, 12:72 (1969).

36. Karawya, M. S., Balbaa, S. I., and Hifnawy, M. S., *Amer. Perf. Cosmet.*, 85:29 (1970).

37. Motl, O., and Trka, A., *J. Soc. Cosmet. Chem.*, 24:747 (1973).

38. Armstead, D. E. F., *Sch. Sci. Rev.*, 54:468 (1973).

39. Lawrence, B. M., *Can. Inst. Food Technol. J.*, 4:A44 (1971).

40. Migel, J. D., Richard, H. M. J., and Sandret, F. G., *J. Agr. Food Chem.*, 24:833 (1976).

41. Debrauwere, J., and Verzele, M., *J. Sci. Food Agr.*, 26:1887 (1975).

42. Knapp, F., Aexel, R., and Nicholas, H. J., *J. Food Sci.*, 33:159 (1968).

43. Wenninger, J. A., and Yates, R. L., *J. Assoc. Official Anal. Chem.*, 52:1155 (1969).

44. Fedeli, E., and Jacini, G., *Chim. Ind.*, Milan, 52:161 (1970).

45. Coleman, R. L., and Shaw, P. E., *J. Agr. Food Chem.*, 19:520 (1971).

46. Coleman, R. L., Lund, E. D., and Shaw, P. E., *J. Agr. Food Chem.*, 20:100 (1972).

47. Moshonas, M. G., Shaw, P. E., and Veldhuis, M. K., *J. Agr. Food Chem.*, 20:751 (1972).

48. Kusumoto, S., Ohsuka, A., and Kotake, M., *Bull. Chem. Soc. Jap.*, 41:1950 (1968).

49. Debrauwere, J., and Verzele, M., *J. Chromatogr. Sci.*, 14:296 (1976).

50. Taskinen, J., and Nykanen, L., *Acta Chem. Scand.*, B29:757 (1975).

51. Yabumoto, K., Jennings, W. G., and Yamaguchi, M., *J. Food Sci.*, 42:32 (1977).

52. Sakai, T., Yoshihara, K., and Hirose, Y., *Bull. Chem. Soc. Jap.*, 41:1945 (1968).

53. Falk, A. J., Bauer, L., and Bell, C. L., *Achillea Millefolium*, 39:598 (1974).

54. Parliment, T. H., *J. Agric. Food Chem.*, 20:1043 (1972).

55. Taskinen, J., *Acta Chem. Scand.*, B28:1121 (1974).

56. Baldry, J., Dougan, J., Matthews, W. S., Nabney, J., Pickering, G. R., and Robinson, F. V., *Int. Flav. Food Addit.*, 7:28 (1976).

57. Hunter, G. L. K., and Brogden, Jr., W. B., *J. Org. Chem.*, 28:1679 (1963).

58. Wrolstad, R. E., and Jennings, W. G., *J. Chromatogr.*, 18:318 (1965).

59. Levi, L., Thomson, J. L., Evans, J. C., Bernstein, H., Forman, S. A., and Miles, N. M., *Perf. Essent. Oil Rec.*, 49:715 (1958).

60. Guenther, E., *The Essential Oils*, Vols. 1-6, Van Nostrand, New York, 1952.

61. Gildemeister, H. P., and Hoffmann, F., *Die Atheriscen Ole*, Vols. 1-7, Akademie, Berlin, 1961.

62. Jacobs, M. B., *Amer. Perf. Aromat.*, 70:53 (1957).

63. Domange, L., and Longuevalle, S., *Ann. Pharm. Franc.*, 16:557 (1958).

64. Bernhard, R. A., *Food Res.*, 23:213 (1958).

65. Naves, Y.-R., *France et ses Parfums*, 2:23 (1959).

66. Irie, M., Takagi, H., and Takasago, T., *Koryo*, 54:33 (1959).

67. Runti, C., and Bruni, G., *Boll. Chim. Farm.*, 99:435 (1960).

68. Baxter, R. M., Dandiya, P. C., Kandel, S. I., Okany, A., and Walker, G. C., *Nature*, 185:466 (1960).

69. von Kugler, E., and Kováts, E., *Helv. Chim. Acta*, 46:1480 (1963).

70. Hamming, M. C., and Foster, N. G., *Interpretation of Mass Spectra of Organic Compounds*, Academic Press, New York, 1972.

71. McFadden, W., *Techniques of Combined Gas Chromatography/Mass Spectrometry*, Wiley-Interscience, New York, 1973.

72. Desty, D. H., *Anal. Chem.*, 32:302 (1960).

73. Grob, K., *Helv. Chim. Acta*, 51:718 (1968).

74. Novotny, M., and Zlatkis, A., *Chromatogr. Rev.*, 14:1 (1971).

75. Bouche, J., and Verzele, M., *J. Gas Chromatogr.*, 6:501 (1971).

76. Jennings, W. G., Yabumoto, K., and Wohleb, R. H., *J. Chromatogr. Sci.*, 12:344 (1974).

77. Rijks, J. A., Ph.D. Thesis, Technical University of Eindhoven, The Netherlands, 1973.

78. Miwa, T. K., Micolajczak, K. L., Earle, F. R., and Wolff, I. A., *Anal. Chem.*, 32:1739 (1960).

79. Smith, J. F., *Chem. Ind.*, p. 1024 (1960).

80. van den Dool, H., and Kratz, P. D., *J. Chromatogr.*, 11:463 (1963).

81. Kováts, E., *Advan. Chromatogr.*, 1:229 (1965).

82. Ackman, R. G., *J. Chromatogr. Sci.*, 10:535 (1972).

83. Novak, J., and Ruzickova, J., *J. Chromatogr.*, 91:79 (1974).

84. Kováts, E., *Helv. Chim. Acta*, 41:1915 (1958).

85. Guichon, G., *Anal. Chem.*, 36:661 (1964).

86. Torline, P., and Ballschmieter, H. M. B., *Lebensm-Wiss. Technol.*, 6:32, (1973).

87. Kugler, E., Langlais, R., Halang, W., and Hufschmidt, M., *Chromatographia*, 8:468 (1975).

13

ANALYSIS OF FOOD AND FLAVORS

WALTER G. JENNINGS

University of California
Davis, California

1. Introduction 511
2. Sample Preparation 512
3. Instrumentation 513
4. Applications 515
   4.1 Flavor and Aroma Volatiles 515
5. Summary 531
   References 531

1. INTRODUCTION

Gas chromatography (GC) is widely used in many areas of food analysis, and the range of volatiles embraced by those samples that are of interest to the food analyst is extremely diverse. Some of these compounds are relatively unstable, and others may contain substances that are prone to react with components of the analytical system, e.g., nitrogen or sulfur groups. Still others of these samples are quite complex, and their separation may require extreme powers of resolution. Occasionally a relatively simple mixture is encountered and the analyst is more concerned with short analysis times. To all of these purposes, glass capillary columns are especially well suited. Certainly the system can possess a high degree of inertness, and it can be used for the analysis of compounds that suffer heavy attrition or may entirely fail to negotiate packed columns or stainless steel or nickel capillary columns. The higher efficiencies obtainable with glass capillary columns are widely recognized, but in addition, shorter analysis times and higher sensitivities are also attainable. These advantages have been explored elsewhere [1].

GC has been widely applied not only to the identification of specific flavor compounds, but also to attempts to correlate the presence, absence, or relative ratios of specific volatiles with positive or negative flavor notes. It has been used to analyze for food contaminants such as pesticide residues and "reaction products" such as nitrosamines or histamine. The analyses of food constituents—amino acids, fatty acids, saccharides, and even some vitamins—have also enjoyed considerable progress that can be largely attributed to advances in GC. In the following pages, we will explore the application of glass capillary GC to these various areas.

2. SAMPLE PREPARATION

Because food products generally contain large amounts of water and nonvolatile materials, some method of isolating the volatile constituents, i.e., sample preparation, may be required. The simplest method of obtaining a sample is frequently the direct injection of the headspace gas, and this has yielded very satisfactory results in some cases [2,3]. The concentration of an individual component in that headspace is influenced by a number of factors, including its concentration in the product and its volatility; the headspace concentrations of the less volatile components are usually low. At the same time, the volume of the headspace sample that can be injected on these small-capacity columns is seriously limited, and unless this restricted sample size contains a detectable amount of a particular volatile, no indication of its presence will be obtained. As a crude rule of thumb, compounds with a volatility lower than that of an ester containing nine to ten carbon atoms [2] are not usually detected by simple direct headspace injection. The volatiles from much larger volumes of headspace gas can be concentrated by slowly injecting these more massive amounts through a chilled U-tube; alternatively, the front portion of the column itself can be utilized as the trap (cold trapping [1]). With many food samples, water is the major volatile collected, and while it is possible to purchase glass capillary columns that have a high tolerance for direct water injection [3-5], water has a devastating effect on others.

A great deal of attention is still being directed toward suitable methods of sample preparation; many of these employ steps involving distillation, extraction, and/or adsorption. Such treatments usually cause quantitative and can engender qualitative changes in the sample that is finally injected. A number of methods for the concentration of larger volumes of headspace gas have been explored, and this topic was the subject of a recent symposium [6]. A method using liquid carbon dioxide to prepare "solvent-free" sample extracts (or eluates of porous polymers or other adsorbants) was recently described [3,7]. Volatiles entrained in vapor streams (or air) have also been isolated by cocondensation with refluxing dichlorodifluoromethane (Freon 12) [7].

Amino-, hydroxyl-, or carboxyl-containing compounds (i.e., those with an active hydrogen atom, including fatty acids, saccharides, amino acids, etc.) exhibit a depressed volatility because of their tendency to interact with similar molecules, or with components of the analytical system (e.g., the liquid phase). Reactions such as methylation and trimethyl silylation are widely used to prepare derivatives with improved GC behavior. Appropriate reactions have been described in detail [1,8,9].

It must be emphasized that no step in the entire analytical process is more critical than that of sample preparation. All too frequently the investigator irretrievably compromises his results even before beginning the analysis. Unfortunately, neither an advanced level of instrumental excellence nor a high degree of sophistication in data processing can compensate for compositional changes engendered by errors committed in these early steps in the analysis.

3. INSTRUMENTATION

Most aspects of the special instrumental requirements accompanying the use of glass capillary columns have been discussed separately in an early chapter. One area that perhaps should be explored here is that of the role of column efficiency.

Because many of the samples with which he deals are of unknown composition, the analyst working on foods and natural products is never sure what degree of success he has achieved in any given separation. A sample may exhibit 200 separate peaks on a particular analytical system, but unless the mass spectrometer indicates that each of these peaks is composed of a single pure compound, the investigator has no assurance that a more efficient system might not further separate some of these individual peaks into two or more fractions. One or more of the unresolved components may represent a toxin or contaminant of interest, or may be related to positive or negative flavor notes. If they have not been resolved, measurement of their concentrations may not be possible, and any "correlations" would be meaningless. Hence the very highest separating power should be utilized for initial separation of samples of unknown composition. A decision can then be made as to what degree of separation is required for a particular analysis; this in turn permits the selection of a column and operating conditions [10] that will achieve, in the minimum time, the required separation of only those compounds that are of concern in that particular analysis. In some cases the application of preliminary separations or ancillary reactions may help to produce a starting material on which it is possible to achieve increased separation, or from which the GC analysis may be able to provide more specific information [11].

The variety of column efficiency measurements currently utilized by the different suppliers of glass capillary columns has left the average user in a state of some confusion; indeed, it sometimes seems that this is the supplier's goal. Claims of "up to 5000 theoretical plates per meter" are of questionable value from two points of view. First, such statements are meaningless unless the test conditions, including the partition ratio of the test compound, are specified [1]. Second, the consumer is assured only that none of the columns from that supplier will exceed the given figures; how many of them attain it is unstated and no minimum theoretical plate content is specified for the column a buyer will recieve. A more precise claim would guarantee that each column possess a certain minimum coating efficiency. This specification takes into account manufacturing variables such as the diameter of the column and testing variables such as the partition ratio of the test compound. Hence it is a value with little or no elasticity, which may account for the fact that it is not widely used by column suppliers. Coating efficiencies can be calculated by the prospective buyer from any test data that include the column diameter, the partition ratio of the test compound, and performance data that can be reduced to the height equivalent to a theoretical plate [1]. Short columns offer several advantages; their efficiencies per unit analysis time are higher and increase dramatically as the column is made shorter and shorter [10,12-14], and their lower pressure drops lead to fewer operational problems. Their disadvantage is that the number of theoretical plates that can be delivered by a short column is obviously limited. Hence, it is usually wise to use columns with the largest number of plates per unit length (allowing for differences in column diameters), i.e., columns of the highest coating efficiencies. Recent developments in recycling in short high-efficiency columns show great promise, not only in offering the highest degree of separation per unit time, but in converting theoretical plate numbers and column length from design parameters to operational parameters [10,12-14].

4. APPLICATIONS

4.1 Flavor and Aroma Volatiles

Yabumoto et al. used a variety of sampling techniques, including direct headspace injection, simultaneous steam distillation-extraction and porous polymer trapping to generate samples of muskmelon (cantaloupe) essence, which were then analyzed by glass capillary GC and GC/mass spectrometry (GC/MS) [15]. In later work concerned with extending aspects of this work and establishing the routes involved in the biogenesis of some of these compounds, Ingraham (see Jennings, Ref. 10) established that by injecting a given compound into the fruit cavity, it was possible to follow changes in the concentration of that precursor and its product(s) by direct headspace injections onto glass capillary columns (Fig. 1). While this offers an elegantly simple experimental approach as compared to the use of radiolabeled precursors, it is (as detailed earlier) limited to compounds whose volatility is reasonably high.

Winter et al. reported the isolation and mass spectral characterization of a number of compounds from a commercially prepared juice of the yellow passion fruit [16]. The glass capillary system utilized was sufficiently inert that they were able to identify two sulfur-containing compounds, 3-methylthiohexanol and cis- and trans-2-methyl-4-propyl-1,3-oxathiane as constituents of this fruit. It would not be surprising to find that the contribution of sulfur compounds to food flavors is more widespread than has been previously recognized partly because many such compounds possess

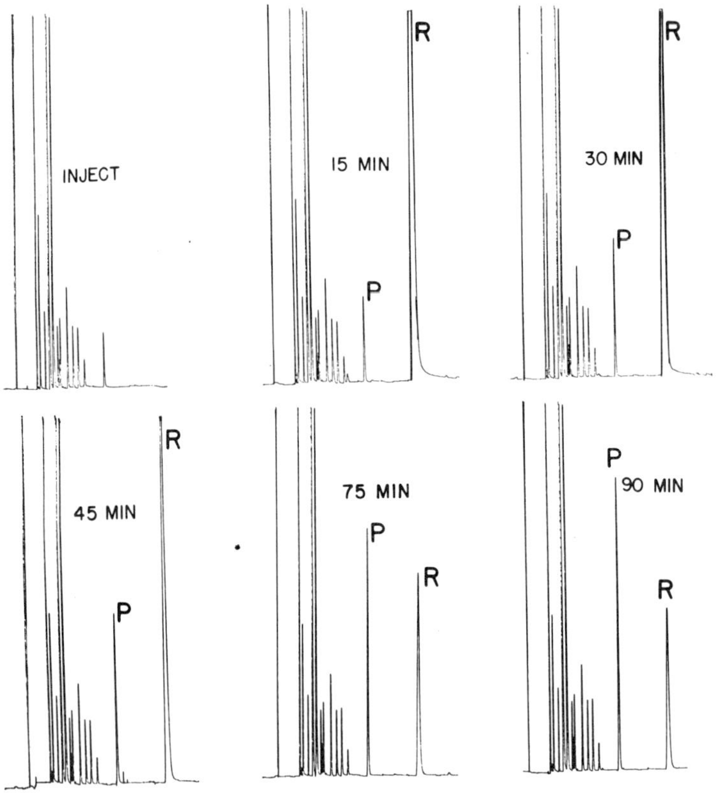

Figure 1. Sequential injections of 3-ml aliquots of cantaloupe cavity (headspace) gas, split ∿1:100. Column: J&W 20 m × 0.25 mm Carbowax 20M programmed from 70 to 170°C at 4°C min^{-1}. A 1.0-μl aliquot of 2-methyl butanol (R) was injected at time 0; the peak labeled P is 2-methyl butyl acetate. (Source: Ref. 10.)

very low thresholds, and partly because the survival of sulfur compounds in normal packed or metal-containing GC equipment is usually poor.

Figure 2 shows chromatograms of several varieties of figs [17]; the striking differences between these samples are very probably due to variations in fruit fermentation products resulting from the action of wild yeasts that insects, penetrating through the ostiole, carry into the ripening fruit. One method for concentrating the volatiles from large volumes of vapor is illustrated in Fig. 3. A massive amount of headspace gas from a closed chamber containing ripening bananas was swept through a tube containing precleaned porous polymer Porapak Q. The polymer was then placed in a Soxhlet extraction thimble and subjected to extraction with liquid carbon dioxide at 25 to 30°C and ∿75 atm pressure. The carbon dioxide was

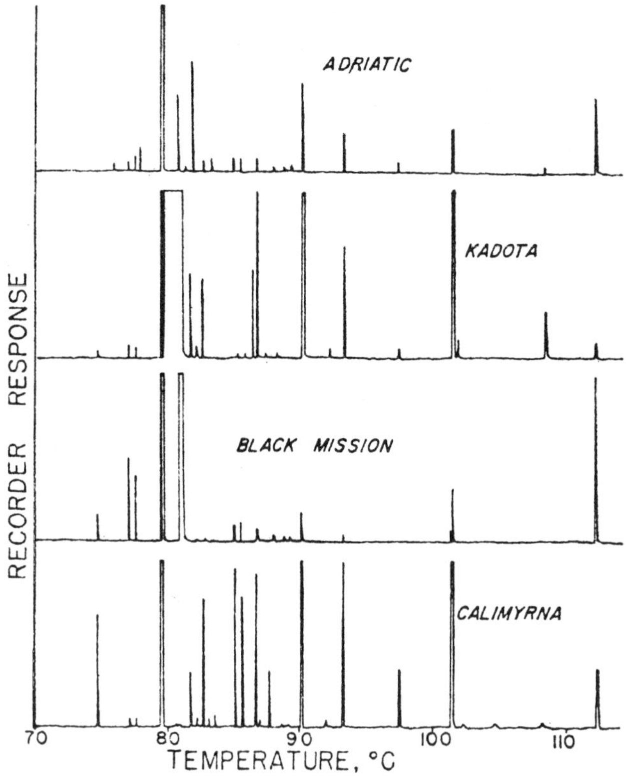

Figure 2. Chromatograms of Porapak concentrates of headspace volatiles from full ripe figs. Column: 0.25 mm × 60 m Carbowax 20M programmed from 70 to 120°C at 1°C min^{-1}. (Source: Ref. 17.)

slowly discharged at subzero temperatures to yield a solvent-free extract that produced the chromatogram shown in Fig. 3. No areas of this chromatogram are obliterated; all peaks are sample peaks. The carbon dioxide extraction apparatus, which has been described elsewhere [7], is useful for a wide variety of substances. Filter paper strips placed in the extraction flask become impregnated with the extract, and provide a convenient method of offering samples for aroma panel assessment or pheremone assay. Figure 4 shows a glass capillary chromatogram of a Bartlett pear essence, and Figure 5 a sample from Jalapeña pepper fruit tissues; both of these samples were produced by simultaneous steam distillation-extraction [1]. Buttery et al. [18] used a large-bore (0.75-mm) glass capillary column to examine the constituents of an atmospheric steam distillate from artichokes, and were able to identify a number of those constituents (Fig. 6). Kallio [19]

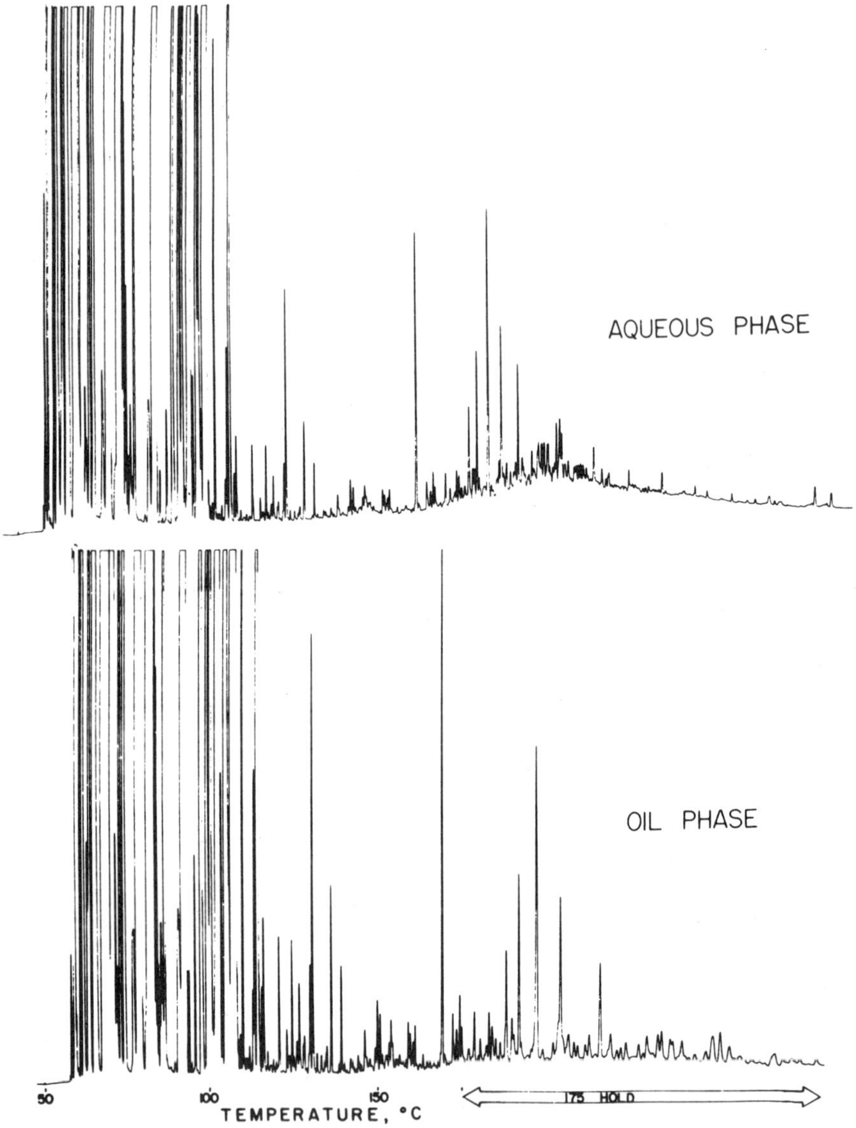

Figure 3. Chromatograms of liquid carbon dioxide extract of Porapak used in concentrating headspace vapors from ripening banana. Column: 0.25 m × 50 m SP-2100 programmed from 50 to 175°C at 2°C min^{-1} and held. (Source: Ref. 7.)

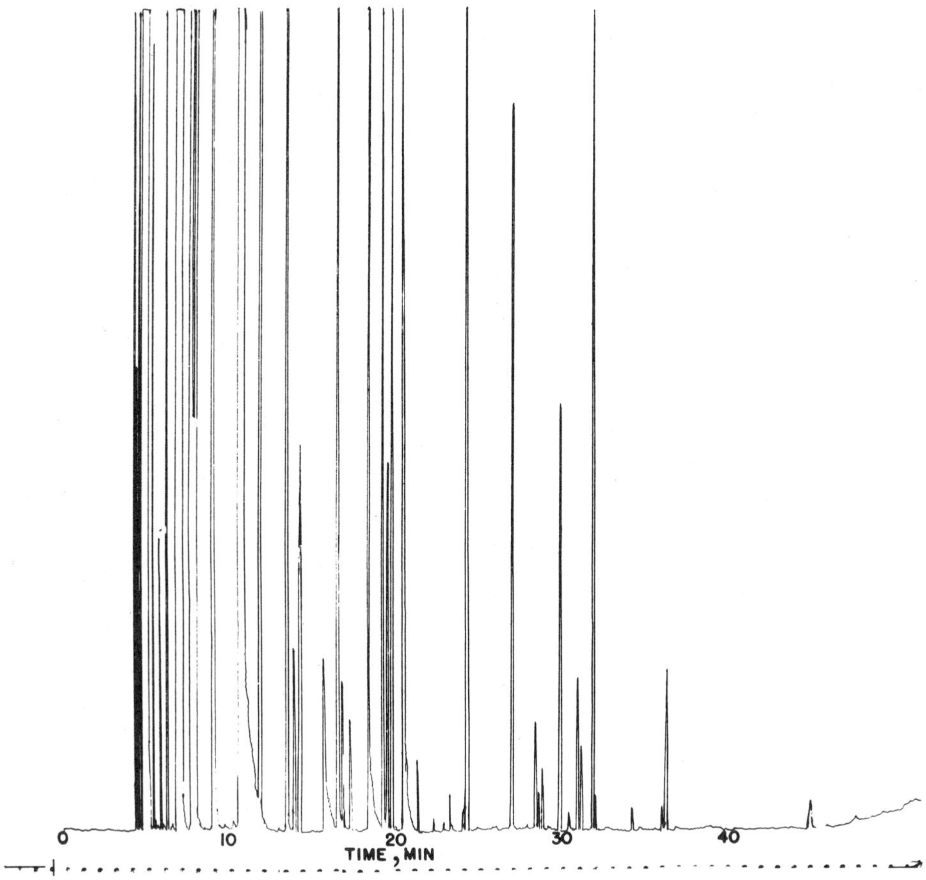

Figure 4. Chromatogram of a Bartlett pear essence. Column: 50 m × 0.25 mm SP-2100 programmed from 70 to 280°C at 2°C min^{-1}. Note that previous high-temperature use has not affected column efficiency, but deactivation has suffered as demonstrated by tailing of alcohols. This is a remedial defect.

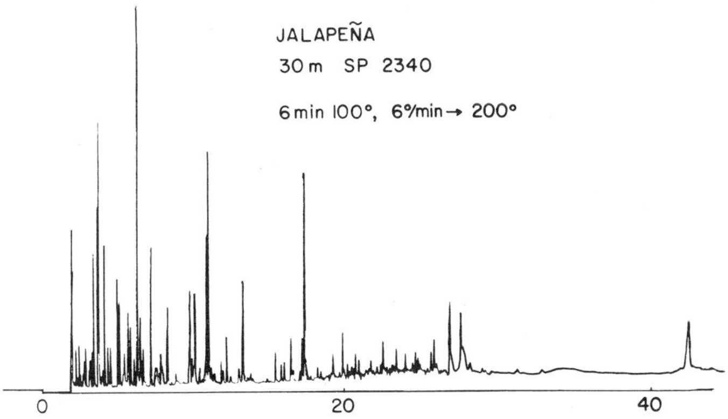

Figure 5. Steam distillation extraction essence of Jalapeña peppers. Column: 20 m × 0.25 mm coated with SE-52; 6 min 100°C, 6°C min^{-1} to 200°C.

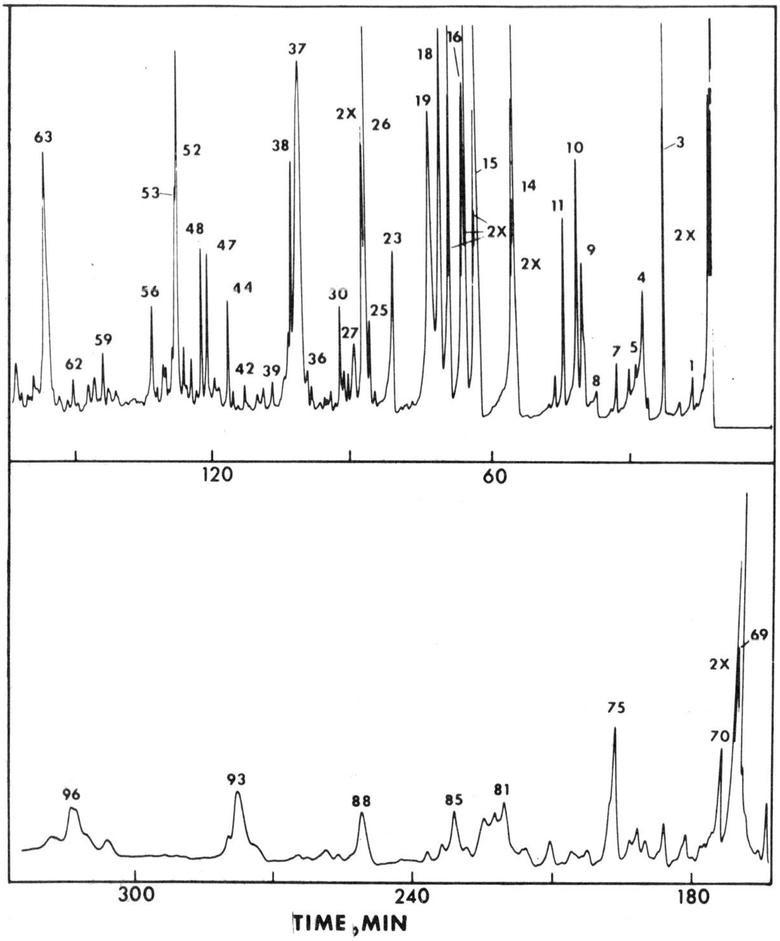

Figure 6. Large-bore glass capillary analysis of the vacuum steam distillate of artichoke leaves. Column: 0.75 mm × 150 m coated with Tween 20; 50°C 12 min, programmed to 170°C at 1°C min^{-1} and held. (Source: Ref. 18.)

subjected the extract of a vacuum steam distillate of an arctic bramble press juice to glass capillary GC/MS, and identified an impressive array of furanones, alcohols, esters, lactones, and terpenoids (Fig. 7). The whole field of food constituent analysis has been a major beneficiary of advances in GC technology, and that field now seems poised for even more spectacular advances as modern developments in glass capillary technology are applied to these very complex systems.

In recent years, more and more knowledgeable experts have become increasingly pessimistic regarding an ever-growing imbalance between population growth, particularly in the developing countries, and world food supplies. One promising approach to the world food problem involves the manufacture of foods from unconventional sources; synthetic meats from retextured soy protein are an example. Unless these products are palatable, however, consumer acceptance is poor. This has led to an intense interest in synthetic meat flavors, which in turn must be based on a comprehensive knowledge of natural meat flavors; many of the latter are sulfur-containing compounds. Because of these interrelationships, meat flavor analysis offers an especially fertile field for glass capillary GC.

Figure 8 shows a chromatogram of Tenax-entrained volatiles derived from the headspace of a heated pork product [20]. Alcohols, several cyclic products, and a number of sulfur-containing compounds dominate the mixture. A chromatogram of dichloromethane extract of heat reaction products from a model system of glucose, ammonia, and hydrogen sulfide [21] is shown in Fig. 9. Most of these, identified by GC/MS, are thiophenes, thiazoles, pyrazines, or furans. Buttery et al. [22] used wide-bore glass capillary columns to isolate a number of constituents in a roasted lamb fat. Pyridines, pyrazines, a thiazole, and several thiazolines were reported among the compounds identified.

During a recent symposium concerned with the application of headspace techniques to food analysis, several authors reported some interesting applications and analyses that utilized glass capillary columns. Figure 10 shows glass capillary chromatograms of headspace volatiles of hydrolyzed vegetable protein (also of interest in meat flavor studies), collected on three different porous polymers [23]. The fact that the nature of the trapping material used has apparently exercised a profound effect on the relative amounts of the various volatiles represented by these chromatograms is a familiar phenomenon, and points out again the difficulty of relating the composition of the beginning sample to the end data presented by the chromatogram. Volatile emanations of the vanilla bean trapped from recirculating vapors were examined by glass capillary chromatography by Klimes and Lamparsky [24]; Figure 11 shows a chromatogram reproduced from their work.

The aromas of tea and coffee have intrigued man for many years, and both of these beverages—as well as the leaf or bean—can produce very complex chromatograms. Figure 12 compares chromatograms from tea beverage

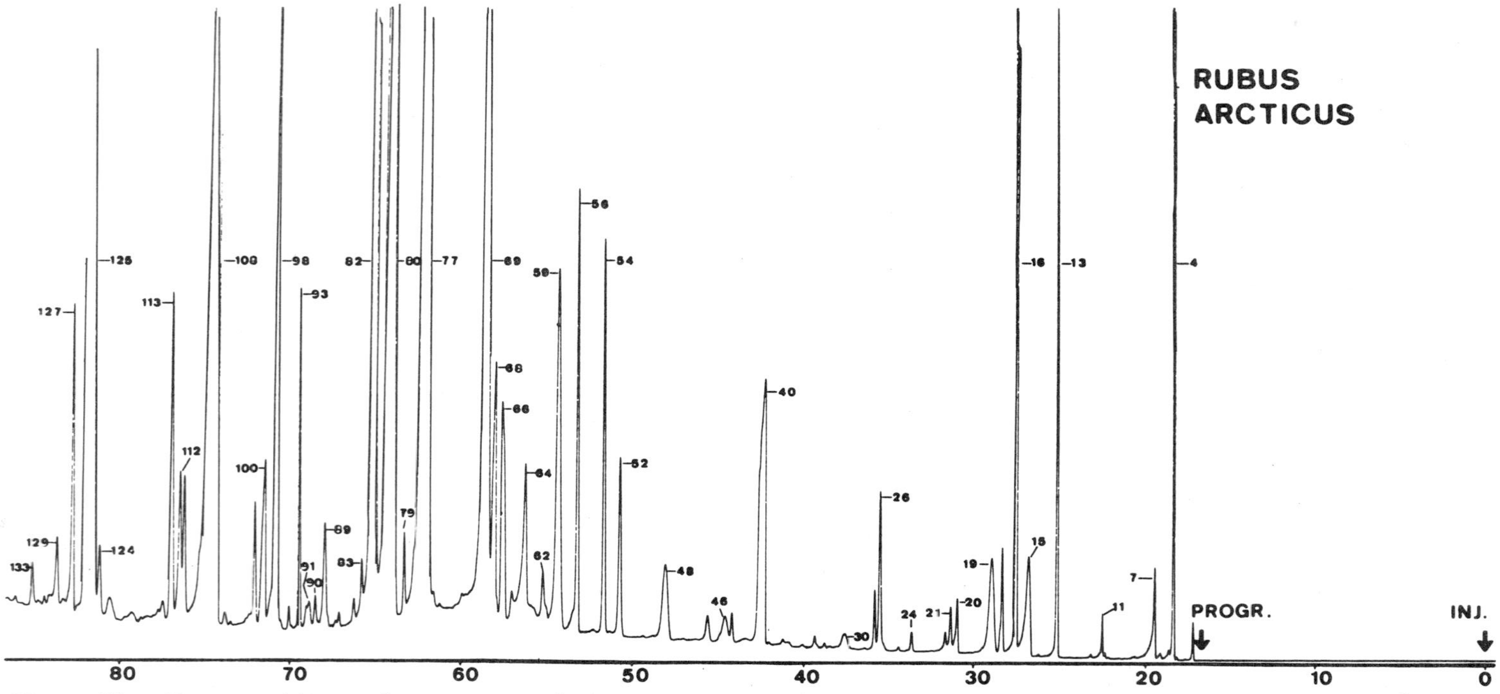

Figure 7A. Glass capillary chromatogram of the press juice of an arctic bramble. Programmed from 60 to 230°C at 2°C min^{-1}. (Source: Ref. 19.) (0 to 85 minute portion; chromatogram continues in Fig. 7B).

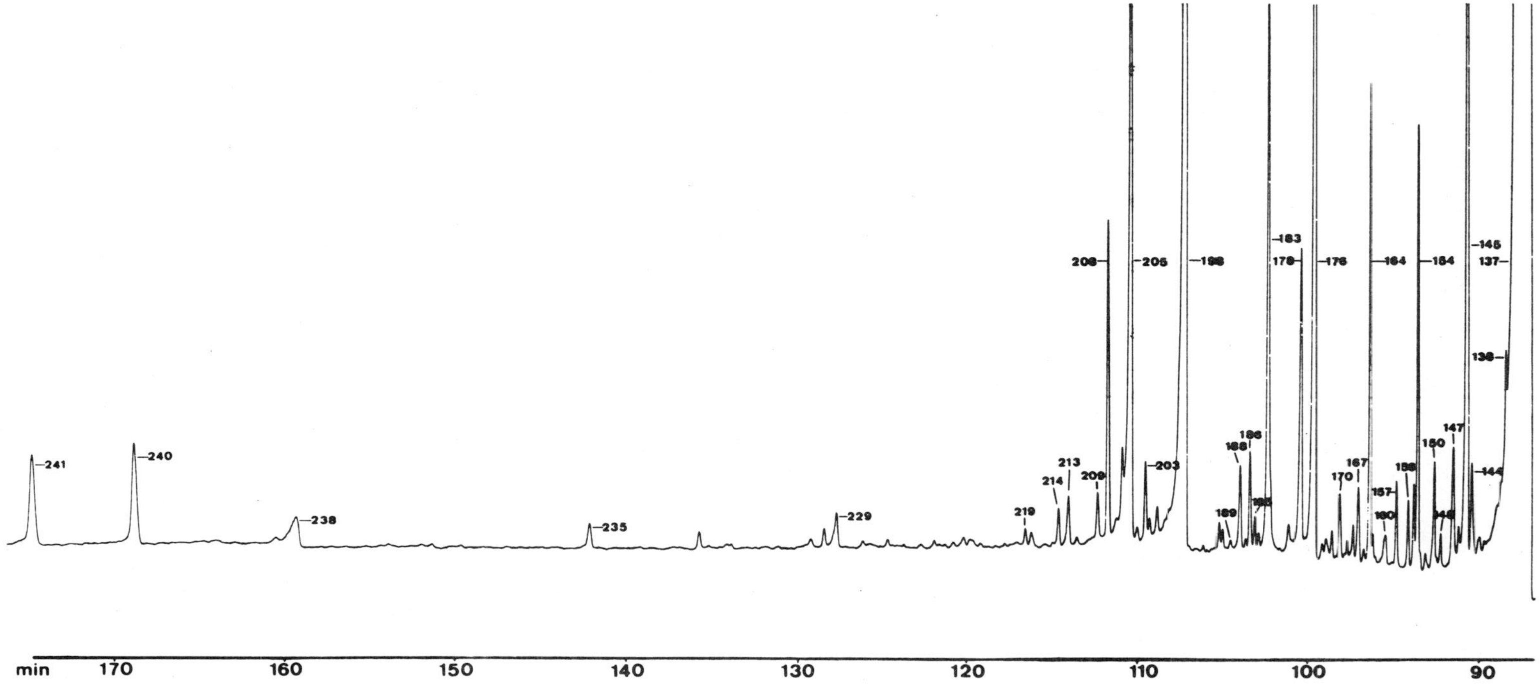

Figure 7B. Glass capillary chromatogram of the press juice of an arctic bramble. Programmed from 60 to 230°C at 2°C min^{-1}. (Source: Ref. 19.) (85 to 175 minute portion).

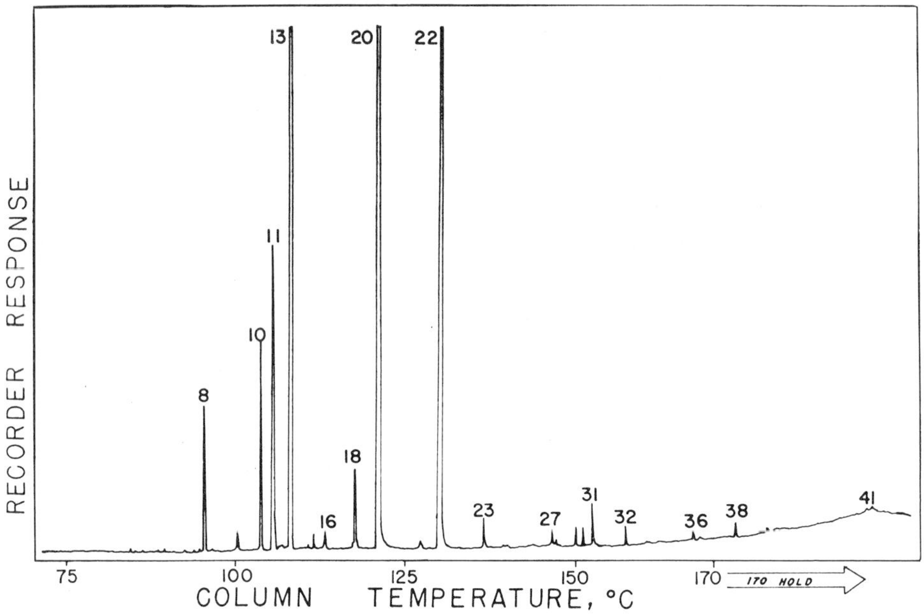

Figure 8. Volatiles from heated pork, trapped on Tenax GC. Column: 0.25 mm × 45 m, coated with Carbowax 20M, programmed from 50 to 170°C at 2°C min^{-1}. (Source: Ref. 20.)

headspace with a steam distillate of tea leaves, and Fig. 13 compares the headspace of ground coffee beans with a steam distillate of those beans [25].

GC is also widely used in studies of food constituent analysis, but the specific areas of fatty acid separations and amino acid analysis have been covered more completely in other chapters of this book, and the interested reader is referred there. Some attention has also been given to saccharides. Szafranek et al. [26] demonstrated the separation of monosaccharide derivatives on glass capillaries coated with methyl silicone and stabilized with a fumed silicone dioxide. The separation of compounds produced by irradiation of glucose was reported by Schomburg and Husmann [27] (Fig. 14), and Adam and Jennings [28] were able to separate most of the components of a mixture of disaccharide derivatives (Fig. 15).

Increasing concern with environmental contamination, reflected in several other chapters of this book, has led to more critical examination of our food materials as more of these contaminants enter the food chain. GC, coupled with specific detection modes, offers a fast, convenient method for determining levels of a number of pesticides, herbicides, or related

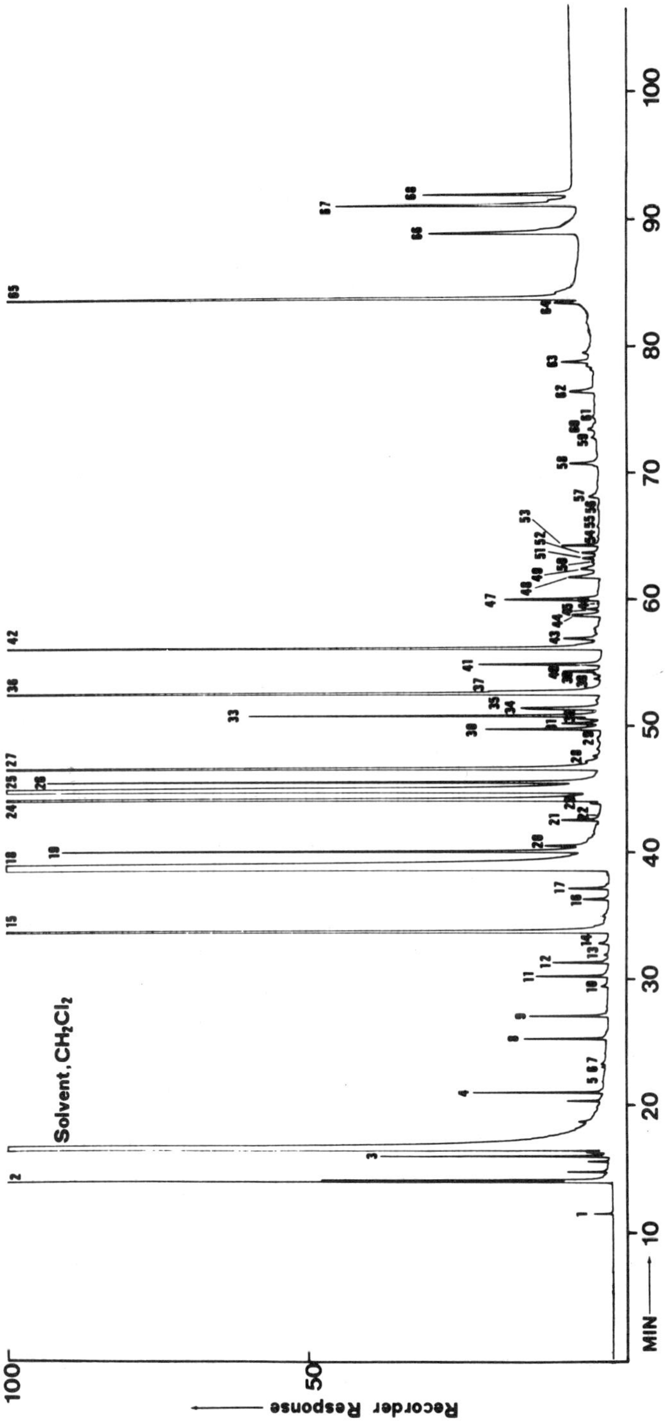

Figure 9. Chromatogram of the reaction products from a glucose-ammonia-hydrogen sulfide model system. (Source: Ref. 21.)

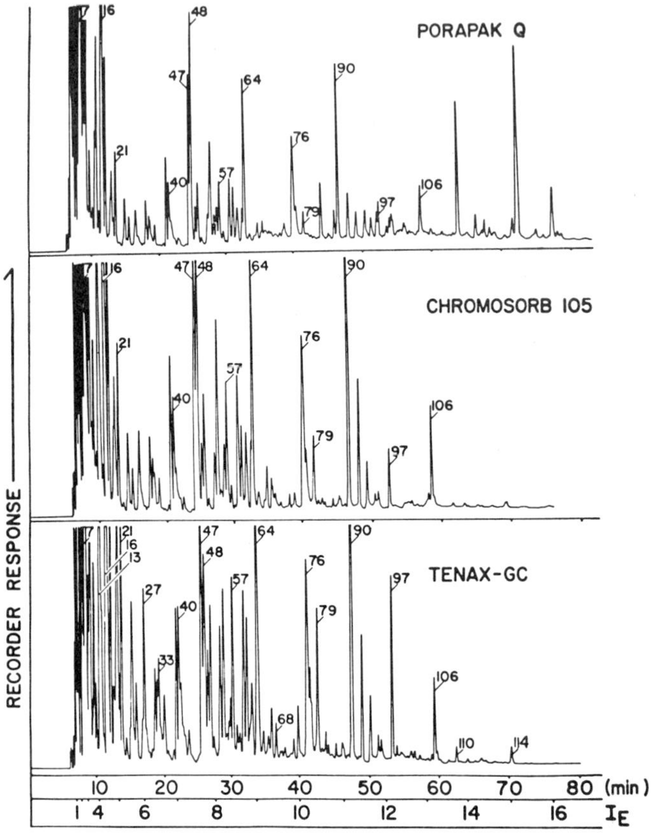

Figure 10. Chromatograms from hydrolyzed vegetable protein headspace collected on various porous polymers. (Source: Ref. 23.)

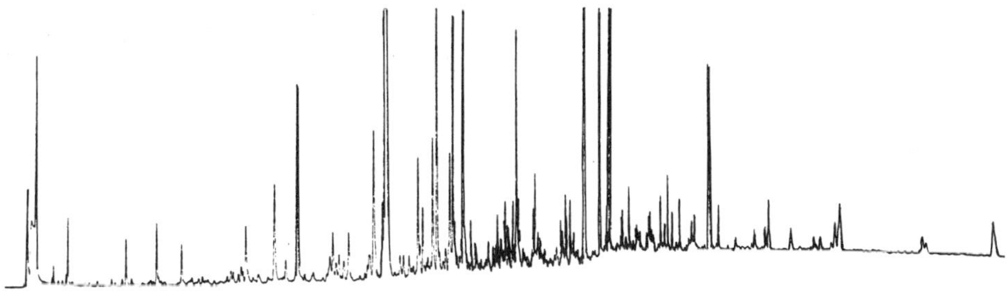

Figure 11. Chromatogram of vanilla bean headspace. (Source: Ref. 24.)

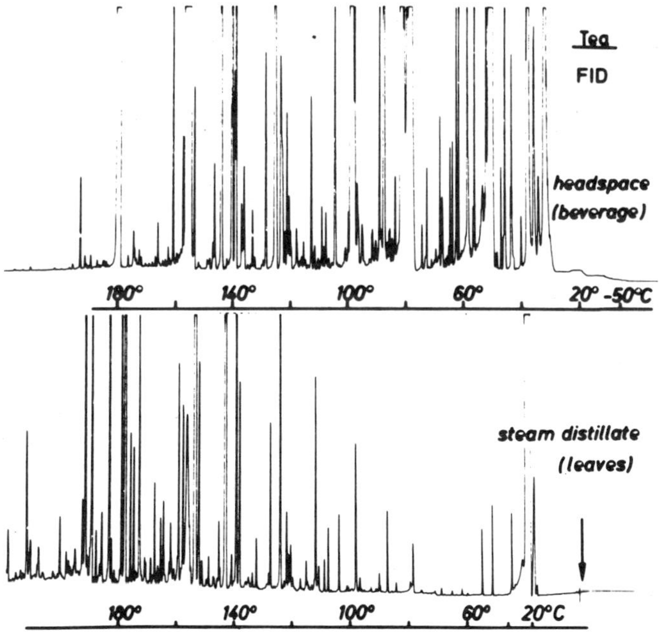

Figure 12. Chromatograms of (top) tea beverage headspace and (bottom) tea leaf steam distillate. (Source: Ref. 25.)

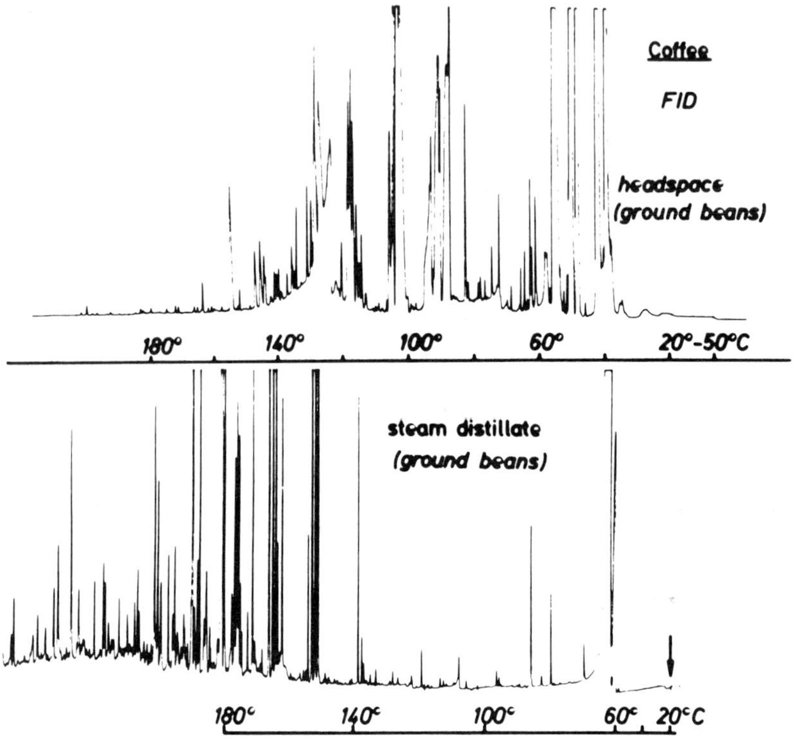

Figure 13. Chromatograms of (top) coffee bean headspace and (bottom) coffee bean steam distillate. (Source: After Ref. 25.)

compounds in foods. Work in our own laboratory is aimed at extending our capability for detecting volatile derivatives of certain metals, including selenium. An example of the isolation and characterization of specific natural toxins is offered by the work of Pyysalo and Niskanen [29], who identified nine N-methyl-N-formyl hydrazones in raw tissues of the fresh morrel, and were able to monitor the heat-induced destruction of these toxins (Fig. 16).

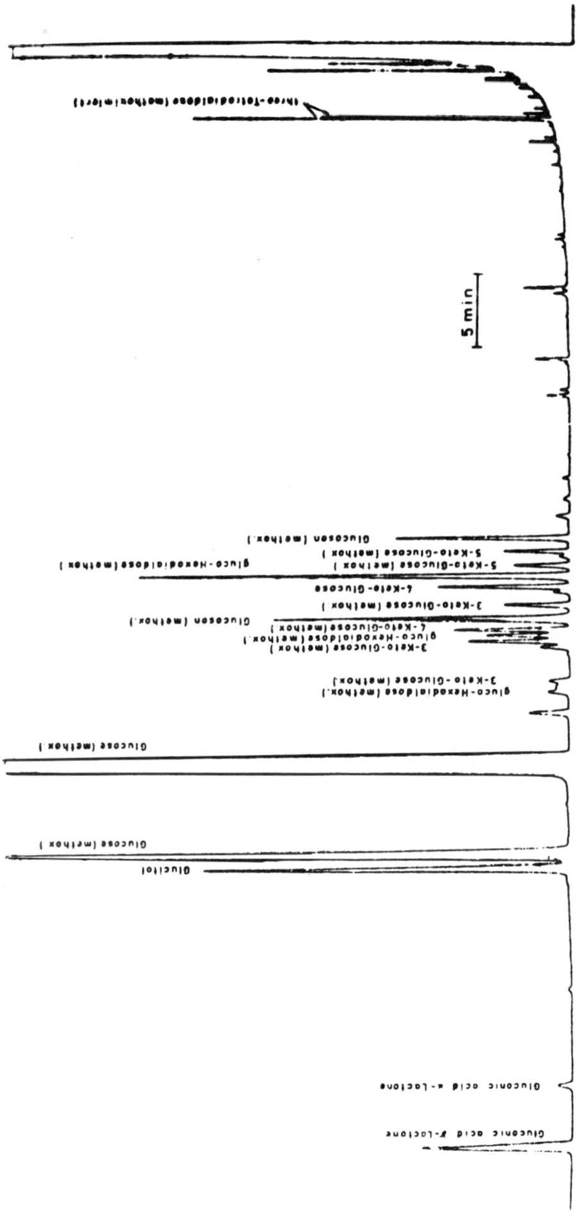

Figure 14. Chromatogram of the products of irradiated glucose. (Source: Ref. 27.)

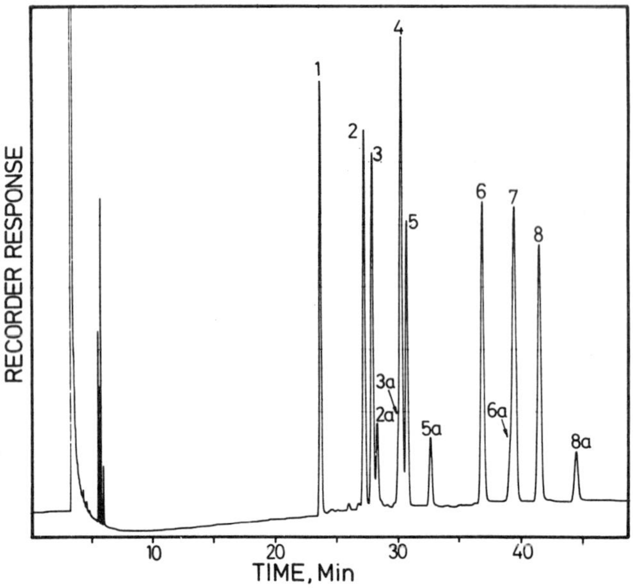

Figure 15. Chromatographic separation of disaccharides. Peak assignments: 1—sucrose TMS, 2—lactose MO-TMS, major, 3—cellobiose MO-TMS, major, 2a—lactose MO-TMS, minor, 3a—cellobiose, minor, 4—trehalose TMS, 5—maltose MO-TMS, major, 5a—maltose MO-TMS, minor, 6—gentiobiose MO-TMS, major, 6a—gentiobiose MO-TMS, minor, 7—palatinose MO-TMS, 8—melibiose MO-TMS, major, 82—melibiose MO-TMS, minor. (Source: After Ref. 28.)

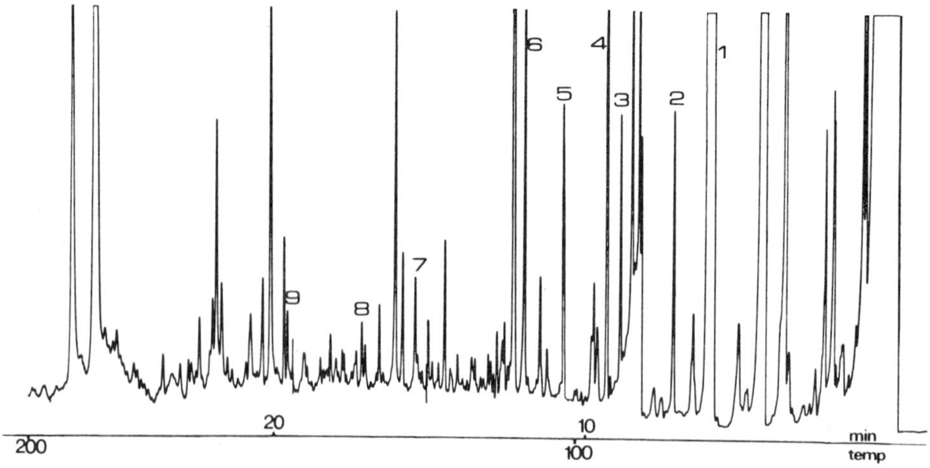

Figure 16. Chromatograms from tissues of the false morrel. (Source: After Ref. 29.)

5. SUMMARY

Because of their many advantages (improved component survivals, increased resolution, shorter analysis times, higher sensitivity), glass capillary columns offer great promise to the field of food analysis. Their one limitation is a restricted sample capacity, which poses problems for the worker interested in sniffing or collecting individual peaks from the effluent. Resolution (which is related to the potential for shorter analysis times) and sensitivity vary directly, while sample capacity varies inversely, with the diameter of the capillary. Those investigators wishing to utilize conventional nondestructive detectors to collect individual peaks or to subject the effluent stream to sniff-testing are usually forced to utilize less efficient larger-bore columns. On the other hand, those interested in optimizing sensitivity, resolution, and analysis time must usually forego sample collection, sniff-testing, and many of the classical methods for structural elucidation and confirmation.

There are two approaches that may help resolve this conflict. Several workers are exploring the feasibility of high-aspect ratio columns that possess both increased sample capacity and high resolution [30-34], but results have not been encouraging to date. Another approach would be the development of alternative methods of identification, including precise retention indices on different liquid phases [11] or the use of recycle systems permitting the liquid phase polarity to be optimized for a specific separation or characterization [10]. In some cases it may be feasible to extend the sensitivities of other methods of analysis to a point where they become compatible with the low sample capacities of the higher-resolution systems, but such developments will require much more attention than they are now receiving.

REFERENCES

1. Jennings, W. G., *Gas Chromatography with Glass Capillary Columns*, Academic Press, New York and London, 1978.
2. Yabumoto, K., Jennings, W. G., and Yamaguchi, M., *J. Food Sci.*, 42:32 (1977).
3. Jennings, W., Pittsburgh Conf. on Analytical Chemistry and Applied Spectroscopy, Cleveland, March 5-9, Paper #251, 1979.
4. Snyder, W. D., Application Note AN 228-8, Hewlett Packard Co., Avondale.

5. Rooney, T. A., *Ind. Res. Dev.*, 20:143 (1978).

6. Charalambous, G., ed., *Analysis of Foods and Beverages: Headspace Techniques*, Academic Press, New York and London, 1978.

7. Jennings, W., *HRC&CC*, 2:221 (1979).

8. Pierce, A. E., *Silylation of Organic Compounds*, Pierce Chemical Co., Rockford, Ill., 1968.

9. Pierce, A. E., *Handbook of Silylation*, Pierce Chemical Co., Rockford, Ill., 1977.

10. Jennings, W. G., *J. Chromatogr. Sci.*, 17:636 (1979).

11. Jennings, W., and Shibamoto, T., *Qualitative Analysis of Flavor and Volatiles with Glass Capillary Columns*, Academic Press, New York and London, 1980.

12. Yabumoto, K., and VandenHeuvel, W. J. A., *J. Chromatogr.*, 140:197 (1977).

13. Jennings, W., Settlage, J. A., and Miller, R. J., *HRC&CC*, 2:441 (1979).

14. Jennings, W., Settlage, J. A., Miller, R. J., and Raabe, O. G., *J. Chromatogr.*, 186:189 (1979).

15. Yabumoto, K., Jennings, W. G., and Yamaguchi, M. J., *Food Sci.*, 42:32 (1977).

16. Winter, M., Furrer, B., Willhalm, B., and Thommen, W., *Helv. Chim. Acta*, 59:1613 (1976).

17. Jennings, W. G., *J. Food Chem.*, 2:185 (1977).

18. Buttery, R. G., Gaudagni, D. G., and Ling, L. C., *J. Agr. Food Chem.*, 26:791 (1978).

19. Kallio, H., *J. Food Sci.*, 41:555 (1976).

20. Uchman, W., and Jennings, W., *J. Food Sci.*, 2:135 (1977).

21. Shibamoto, T., and Russell, G. F., *J. Agr. Food Chem.*, 25:109 (1977).

22. Buttery, R. G., Ling, L. C., Teranishi, R., and Mon, T. R., *J. Agr. Food Chem.*, 25:1227 (1977).

23. Withycombe, D. A., Mookherjee, B. D., and Hruza, A., in *Analysis of Foods and Beverages: Headspace Techniques*, (G. Charalambous, ed.), Academic Press, New York, 1978, p. 81.

24. Klimes, I., and Lamparsky, D., in *Analysis of Foods and Beverages: Headspace Techniques*, (G. Charalambous, ed.), Academic Press, New York, 1978.

25. Vitzthum, O. G., and Werkhoff, P., in *Analysis of Foods and Beverages: Headspace Techniques* (G. Charalambous, ed.), Academic Press, New York, 1978.

26. Szafranek, J., Pfaffenberger, C. D., and Horning, E. C., *J. Chromatogr. Sci.*, 88:149 (1974).

27. Schomburg, G., and Husmann, H., *Chromatographia*, 8:517 (1975).

28. Adam, S., and Jennings, W. G., *J. Chromatogr.*, 115:218 (1975).

29. Pyysalo, H., and Niskanen, A., *J. Agr. Food Chem.*, 25:644 (1977).

30. Schieke, J. D., Comins, N. R., and Pretorius, V., *J. Chromatogr.*, 112:97 (1975).

31. Pappendick, H. D., and Baudisch, J., *J. Chromatogr.*, 122:443 (1976).

32. Duplessis, G., Torline, P., and Kozma, N., *Chromatographia*, 10:624 (1977).

33. Desty, D. H., and Douglas, A. A., in *Advances in Chromatography* (Proc. 13th Int. Symp., St. Louis) (A. Zlatkis, ed.), 1978, p. 73.

34. Pierce, Jr., H. D., Unrau, A. M., and Oenschlager, A. C., *J. Chromatogr. Sci.*, 17:297 (1979).

14

ANALYSIS OF BEER AND HOPS

M. VERZELE AND P. SANDRA

State University of Ghent
Ghent, Belgium

1. Introduction 535

2. Instrumentation 536

    2.1 Wide-bore Versus Narrow-bore Capillary Columns 536
    2.2 Odor Evaluation 538
    2.3 Glass Capillary Gas Chromatography/Mass Spectrometry 540

3. Hop Analysis 540

    3.1 Aroma of Hops 540
    3.2 Essential Oil Analysis of Hops 541

4. Beer Analysis 552

    4.1 Analysis of Volatiles 552
    4.2 Contribution of Hop-derived Compounds to the Flavor of Beer 562
    4.3 Estimation of Fatty Acids in Beer 569
    4.4 Amino Acids 570
    4.5 $(GC)^2$ Analysis of Bitter Acids 570

5. Conclusion 574

References 575

1. INTRODUCTION

Gas chromatography (GC) has been applied from the beginning to the problems of the brewing industry. In the present chapter only capillary GC aspects connected to hops, hop extracts, isomerized hop extracts, wort, and beer will be discussed. Within the space allotted for this, it was not possible to develop most of the problems in depth. Specialized readers may therefore feel that particular points have not received sufficient attention. This is for example the case for hop variety characterization through essential oil fingerprinting. This chapter, however, is, as stated, about *capillary* GC and therefore many interesting packed column GC resutls on hops and beer are not even mentioned. The chapter should make clear that capillary GC has already contributed much to brewing science.

2. INSTRUMENTATION

The general instrumentation and techniques of capillary GC are amply described in other chapters of this book. We sometimes use technology which is not used so consistently elsewhere. We feel, for example, that wide-bore glass capillary GC should get more attention. This aspect of glass capillary gas chromatography or $(GC)^2$, as we call it, will be emphasized in the first part of this chapter. We also believe our approach to on-line smell evaluation and to $(GC)^2$/mass spectrometry [$(GC)^2$/MS] deserves to be described in some detail.

2.1 Wide-bore Versus Narrow-bore Capillary Columns

Most of the analyses presented in the following applications are performed on wide-bore capillary columns. We have favored [1,2] wide-bore capillary columns because in some aspects they have advantages over narrow-bore capillary columns.

The efficiency per meter for a wide-bore capillary column is less than for a narrow-bore column; the loss of resolution, however, can easily be compensated by greater column length. Wide-bore columns have practically no flow resistance, and by using a longer column the overall performance increases. As illustrated in Fig. 1, the $H\bar{u}$ plots and $TZ\bar{u}$ plots of a wide-bore and narrow-bore capillary column are compared. The capillary columns were coated statically [3] with SE-30 using a 0.3% solution (same β value for both columns). The columns were operated isothermally at 100°C with helium as carrier gas. Via an all-glass inlet splitter 20 ng of hydrocarbons (C_9 to C_{14}) were introduced. The $H\bar{u}$ curves were determined for C_{14} ($k = 6.7$), the $TZ\bar{u}$ plots for C_{13} to C_{14} ($k = 3.6$; 6.7). Dead time values

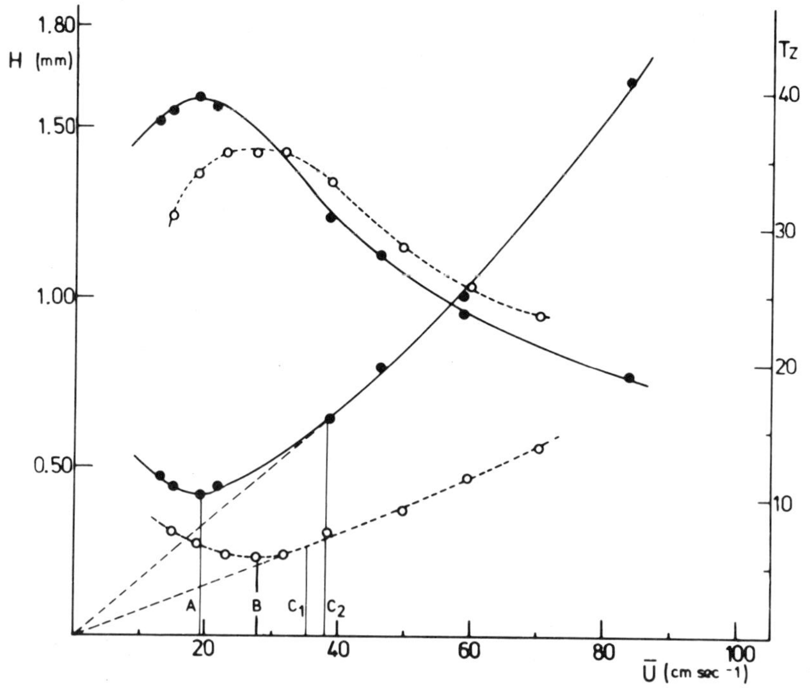

Figure 1. TZ-ū plots (top) and H-ū plots (bottom) for capillary columns. Solid circles, wide-bore column, 0.54 mm ID × 63 m. Open circles, narrow-bore column, 0.26 mm ID × 28 m.

were calculated following the method of Peterson-Hirsch for at least two series of hydrocarbons. From these plots the following conclusions can be drawn:

The resolution per meter of a wide-bore capillary column (TZ/m = 0.62) is half the resolution of a narrow-bore capillary column (TZ/m = 1.28). By doubling the column length, the same resolution is obtained. For the wide-bore capillary column the height-equivalent-to-a-theoretical-plate (HETP) value is minimum (0.41 mm) at a gas velocity of 19 cm sec^{-1} (A), whereas for the narrow-bore capillary column HETP min is obtained at a gas velocity of 27.5 cm sec^{-1} (B). This means that if the columns are used at the highest number of theoretical plates the analysis time on the wide-bore column (63 m) is 3.3 times the analysis time on the narrow-bore column (28 m).

The optimum practical gas velocity (OPGV) is for both columns practically equal--e.g., 35 cm sec^{-1} for the 0.26-mm column (C_1), 37.5 cm sec^{-1} for the 0.54-mm column (C_2). At the OPGV value the TZ values are respectively 35 and 31.75 and the time difference is reduced to a factor 2. If speed is all-important, narrow-bore columns must be favored.

On the other hand, wide-bore open tubular columns have a much larger sample capacity and therefore direct sample introduction can be carried out more easily. They can manage large solvent samples without unduly large solvent peaks. Conventional splitting can be omitted and much improved quantitative results are obtained. The requirements on instrument construction are less stringent (dead volumes, makeup gas) and conventional GC instruments can be used. Moreover, in our hands, more reliable data are obtained with wide-bore columns for smell analysis and $(GC)^2$/MS.

Figure 2, comparing the analysis of the oxygenated fraction of hop essential oil on a wide-bore and narrow-bore column, shows the value of wide-bore columns, which is, in our opinion, underestimated. Note that the sample was directly [2] introduced for the wide-bore column whereas an all-glass splitting device was employed for the narrow-bore capillary column. Both columns were statically coated [3] with Superox-4 [4].

Every chromatographer, however, has to find out which column is best suited for his problem. With regard to the great number of possibilities from which one can make a choice, this is a difficult task. It indeed is not easy to choose the optimum characteristics out of a full range of column radii (0.2 to 1 mm), column lengths (5 to 500 m), and stationary phases, apart from the selection of the carrier gas, the flowrate and the analysis temperature. In $(GC)^2$ the selectivity of the stationary phase is not the guiding factor and more important characteristics like performance, temperature stability, activity, and mass spectrometric compatibility must be favored [5]. In our laboratory, dealing with a wide range of applications, 90% of the $(GC)^2$ analyses are performed on OV-1 or SE-30 as nonpolar gum phase and on the Superoxes [4,6] as polar gum phases. As for the use of wide-bore (0.5-mm) and narrow-bore (0.25-mm) glass capillaries, this is for the moment about 50/50 in our laboratory.

2.2 Odor Evaluation

In the early development of $(GC)^2$, flavor analysis tried to identify all the compounds of an essential oil. Chromatograms with hundreds of identified compounds are published in the literature.

Today total identification is no longer the main goal of essential oil analysis. The flavor chemist is now more interested in the organoleptic properties of the compounds.

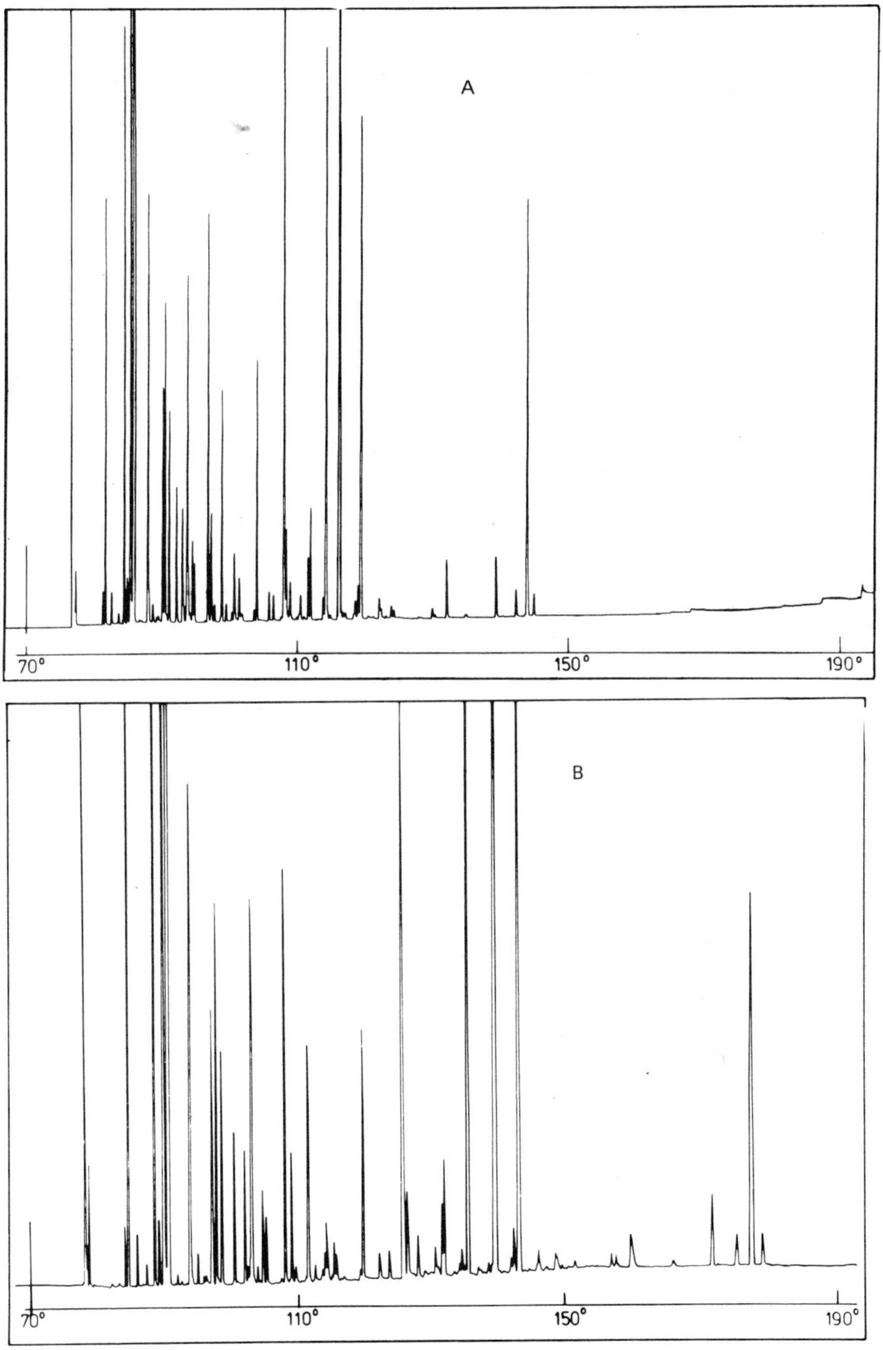

Figure 2. Chromatogram of the oxygenated fraction of hop essential oil on a narrow-bore (A) and wide-bore (B) capillary. Chromatographic conditions: (A) column 25 m × 0.25 mm ID Superox-4; temperature-programmed 70 to 190°C at 2°C min^{-1}; injection via an all-glass inlet splitter (1:20). (B) column 50 m × 0.5 mm ID Superox-4; temperature-programmed 70 to 190°C at 2°C min^{-1}; on column injection.

The human nose, as a detector of unique sensitivity and selectivity, can be used in combination with $(GC)^2$, and this technique gives organoleptic information of the eluted compounds. A simple device, used in combination with wide-bore capillary columns, which allows odor evaluation and flame ionization detection is shown in Fig. 3. The column outlet is split; one arm is connected to the flame ionization detector (FID) and the other end is led to the nose via Teflon tubing.

An analogous system is used for off-line heart cutting. The column is split in a ratio 1:5 (FID/side arm) and the side arm is connected to a microtrap [7].

2.3 Glass Capillary Gas Chromatography/Mass Spectrometry

In $(GC)^2$/MS, separators are more and more replaced by the direct inlet and the open split system. Choosing for full sample transfer, simplicity of construction and the absence of metal parts, we modified a Finnigan 3200 CI quadrupole mass spectrometer for direct connection via an all-glass heated inlet system (Fig. 4). The capillary (0.35 mm ID) is an extension of the open tubular column into the ion source. For chemical ionization the reagent gas is introduced in the ion source via a coaxial inlet with needle valve [8].

Even under these extremes of pressure drop at the column end, about 75% of the plate number of a 0.25-mm column remained. For most purposes this is more than enough.

3. HOP ANALYSIS

Hop is a perennial climbing vine, cultivated in central European countries, North America, Australia, and Japan. The cones of the female plant of the species *Humulus lupulus* contain the lupulin glands which include the resins and the essential oil. Hop is used exclusively for brewing purposes. Its brewing value is mainly based on its bittering power and characteristic aromatic odor.

3.1 Aroma of Hops

The odor of hops is not very pleasant. This is mainly due to the high concentration of myrcene in the headspace above fresh hops.

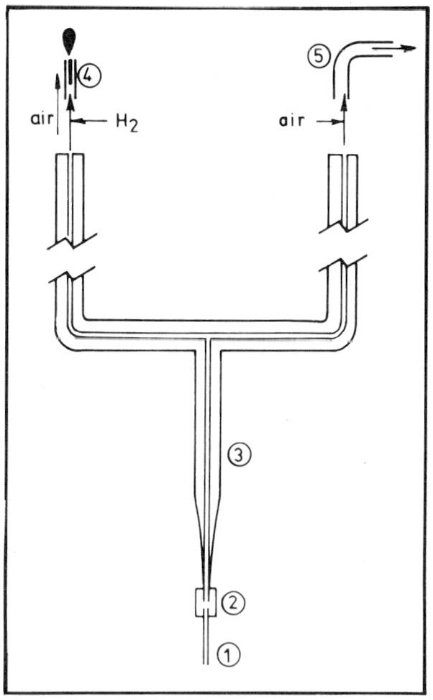

Figure 3. Splitting device for odor detection: 1—capillary column, 2—Teflon shrinkable tubing, 3—all-glass split device (6 mm OD × 0.5 mm ID), 4—FID detector, 5—Teflon tubing to nose.

This is illustrated in Fig. 5, showing the analysis of 2 ml headspace of hop pellets which were thermostated at 50°C [9]. The major compound is myrcene (peak 5); other interesting compounds are peak 1—isoprene, 2—acetone, 3—isobutylmethylketone, and 4—dimethylvinylcarbinol. The headspace sampling method is limited to the analysis of very volatile compounds present in high concentration. Therefore, the aroma of hops is normally evaluated on the basis of the amount of the essential oil and its composition.

3.2 Essential Oil Analysis of Hops

Hop oil is conventionally obtained by steam distillation. This technique can introduce artifacts, and therefore a milder technique—namely, vacuum steam distillation—was developed [10]. Hop essential oil has been extensively studied. Gas liquid chromatography was first used by Howard [11] as early as 1956. R. G. Buttery et al. [12] studied the constituents of hop oil by temperature-programmed capillary column gas chromatography combined with a Time-of-Flight mass spectrometer. The capillary column was 100 ft

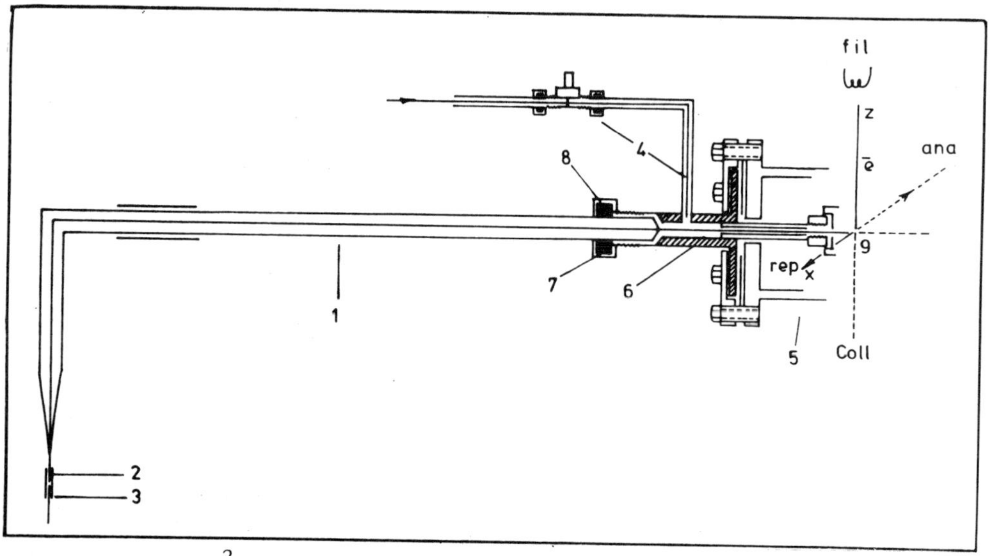

Figure 4. $(GC)^2$/MS interface: 1—glass capillary (0.35 mm ID), 2—Teflon shrinkable tubing, 3—capillary column, 4—coaxial reagent gas inlet with needle valve, 5—manifold Finnigan 3200, 6—modified transfer line with schematic view of ion source, 7—Teflon O-ring, 8—swagelock. (Source: Ref. 8.)

long with 0.01 in as internal diameter and was coated with Tween 20. The same technique was employed for the analysis of the volatile oxygenated constituents of hop oil [13]. Eighty-seven peaks were recorded (Fig. 6); most of them have been identified or a structure was suggested (Table 1). Since this pioneering work, the composition of hop oil has been investigated by many workers using capillary gas chromatography and more than 250 compounds have unequivocally been identified. An elegant and rapid method for the analysis of the volatile components of hops was described by Naya and Kotake [14]. A solid sampler containing 5 mg of lupulin is introduced in an injection block heated at 180°C for 1.5 min. Through a nitrogen stream, the volatilized constituents are swept onto a trapping column cooled by dry ice-isopropanol. The whole column system is heated by temperature programming from 30 to 150°C at a rate of 7.5°C min^{-1}. The capillary columns used for the analyses were 45 m long, 0.25 mm and 0.5 mm internal diameter, made of stainless steel, and coated with HB 2000. Seventy-six compounds were identified by the combination high-resolution chromatography/mass spectrometry and the qualitative and quantitative composition of the hop oil of five varieties was reported.

Figure 7 shows a chromatogram at two different intensities of hop essential oil obtained by vacuum steam distillation on a 40-m × 0.26-mm ID

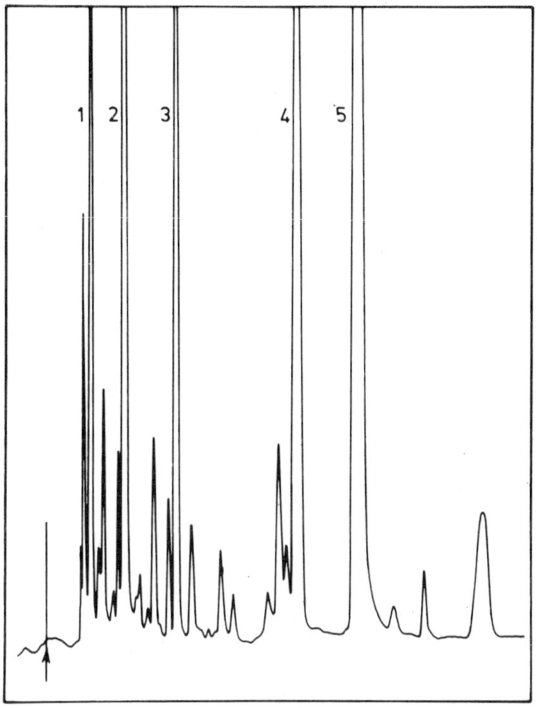

Figure 5. Headspace analysis of hop pellets. Column: 100 m × 0.5 mm ID coated with Carbowax 400. Injection: 2-ml headspace directly at 150°C. Temperature column: 70°C. Carrier gas: hydrogen, flowrate 10 ml min^{-1}. Peak identification: see text.

glass capillary column coated with Superox-4 [4]. The most important compounds are the hydrocarbons myrcene (peak 1), caryophyllene (peak 2), and humulene (peak 3), which constitute 70 to 80% of the whole oil.

As can be seen from Fig. 7(B), the chromatogram of hop oil is very complex, and in order to improve resolution the essential oil is usually first separated into a hydrocarbon and an oxygenated fraction by chromatography on silica gel. A $(GC)^2$ analysis of the oxygenated fraction of hop essential oil has been shown previously (Fig. 2). For better identification, the oxygenated constituents are fractionated further on the basis of their functional groups. Up to now, 62 hydrocarbons, 50 alcohols, 23 aldehydes, 37 ketones, 66 esters, 11 sulfur-containing compounds and some miscellaneous constituents have been identified. A drawback of several fractionation techniques, however, is that artifacts are introduced, e.g., terpenes are known to isomerize on silica gel [15,16]. For the distinction of varieties, a total analysis is preferred. The analysis of the essential oil of Saaz hops (1974) obtained by vacuum steam distillation is represented in Fig. 8.

Table 1. Identification of Constituents in Oxygenated Fraction of Brewers Gold Hop Oil Using Capillary GLC and Mass Spectral Analysis (see Fig. 6)

Peak number	Identification confirmed by comparison of mass spectrum and GLC retention time with those of authentic sample	Mass spectral predictions where authentic sample was not available
4	2-Methylpropyl isobutyrate[a]	
5	Methyl hexanoate	
6		Methyl thio-2-methylbutyrate
7	Butyl isobutyrate	
8	2-Methylbutyl propionate[a]	
9	Methyl 5-methylhexanoate[a]	
10		
11	2-Methylpropyl 2-methylbutyrate	
12	3-Methylbutyl isobutyrate	
13	2-Methylbutyl isobutyrate[a]	
14	Methyl heptanoate[a] and methyl 4-methylhex-2-enoate[a]	
15		
16	Pentyl isobutyrate	
17		2-Methylpropyl branched pentenoate and methyl thiohexanoate (branched)
18		Pentenyl isobutyrate
19		Methyl 2,5-dimethylhexanoate
20	Methyl thiohexanoate[a]	and methyl 6-methylheptanoate[b]
20a	Ethyl heptanoate	
21	2-Nonanone	and 2-methyl-5-pentenylfuran[b]
22	Hexyl propionate[a]	
23	2-Methylbutyl isovalerate	
24	2-Methylbutyl 2-methylbutyrate[a] and linalool[a]	
25	Methyl octanoate[a]	
26	Hexyl isobutyrate	
26c		Methyl thioisoheptanoate
27		Oxygenated terpene
28	Methyl thioheptanoate[a] and 2-decanone	
29		Methyl nonanoate (branched)
30	Heptyl propionate[a]	

Peak number	Identification confirmed by comparison of mass spectrum and GLC retention time with those of authentic sample	Mass spectral predictions where authentic sample was not available
31	Octyl acetate[a]	and methyl nonenoate[b]
32	Methyl nonanoate[a]	
33	Heptyl isobutyrate	
34	2-Methylbutyl hexanoate	
35		Branched 2-undecanone
35a		Methyl 2-methylnonanoate
36		Mol. wt. 204
37		
38		Methyl 8-methylnonanoate
39	2-Undecanone[a]	
40		
41	Methyl dec-4-enoate[c]	
42	Methyl deca-4,8-dienoate[c]	
43	Methyl geranate[a]	
44	Methyl decanoate[a]	
45		Methyl decenoate
46	Octyl isobutyrate	
47		2-Methylbutyl heptanoate
48		2-Dodecanone (branched)
48a	Neryl acetate	
48b		Branched nonanyl isobutyrate
48c		Methyl undecanoate (branched)
49	Geranyl acetate[a]	
50		Methyl 9-methyldecanoate
51		
52		Methyl undecenoate
53		Methyl undecadienoate
54		Methyl undecenoate
55	Methyl undecanoate	
56		Methyl undecenoate
57	Neryl propionate	
58		2-Tridecanone (branched)
59		

Table 1 (continued)

Peak number	Identification confirmed by comparison of mass spectrum and GLC retention time with those of authentic sample	Mass spectral predictions where authentic sample was not available
60	Geranyl propionate[a]	
61		
62	Neryl isobutyrate	and methyl dodecanoate (branched)
63	2-Tridecanone[a]	
64 65		Methyl dodec-8-enoate[b,d] and dodecadienoate Mol. wt. 208
66		Linalyl propionate
67	Geranyl isobutyrate[a]	
68		
69	Methyl dodecanoate	
70		Methyl dodecenoate
71		Branched 2-tetradecanone
72		Tetradec-9-en-2-one[b,d]
73		Sesquiterpene, mol. wt. 204
74		Mol. wt. 204
75		Terpene ester
76		
77	2-Tetradecanone	
78		Methyl tridecenoate
79		Methyl tridecenoate
80		Sesquiterpenoid, Mol. wt. 220
81		Sesquiterpenoid, Mol. wt. 222
82		
83		
84		Pentadeca-6,9-dien-2-one[b,d]
85		Mol. wt. 220
86		Mol. wt. 222
87	2-Pentadecanone	

[a]Infrared absorption spectrum identical with that of authentic sample.

[b]Identification supported by infrared absorption spectrum.

[c]Identity previously proved by NMR and ozonolysis.

[d]Identification supported by ozonolysis.

Source: Ref. 13.

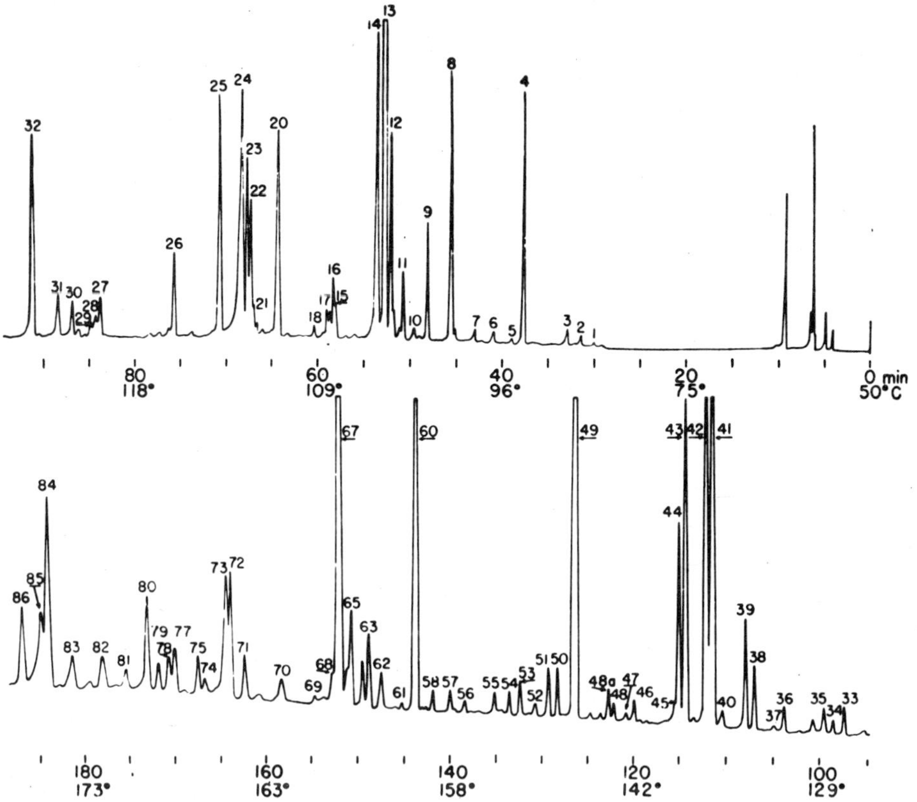

Figure 6. Temperature-programmed capillary column GLC analysis of the oxygenated fraction of hop oil (Brewer's Gold hops). Column 192 ft long × 0.01 in ID, coated with SF-96 plus tail reducers; helium carrier gas velocity 24.5 cm sec^{-1}. (Source: Ref. 13. Reproduced courtesy of Dr. R. G. Buttery.)

The essential oil was analyzed by $(GC)^2$ on a 90-m × 0.5-mm ID SE-30 column in combination with mass spectrometry (electron impact and chemical ionization). The results are given in Table 2, and the compounds typical for hop oil are indicated. As shown by smell analysis, the two regions A and B in the chromatogram are very important for the aroma of hops. They contain potent flavoring compounds with very low threshold values.

Isomerized hop extracts have a clear hoppy aroma, although the preparation of the extract involves two-phase extraction steps which should have removed the oil. In order to discover the origin of this aroma (essential oil or degradation products) isomerized hop preparations were extracted with isooctane. The organic layer was analyzed by $(GC)^2$ in

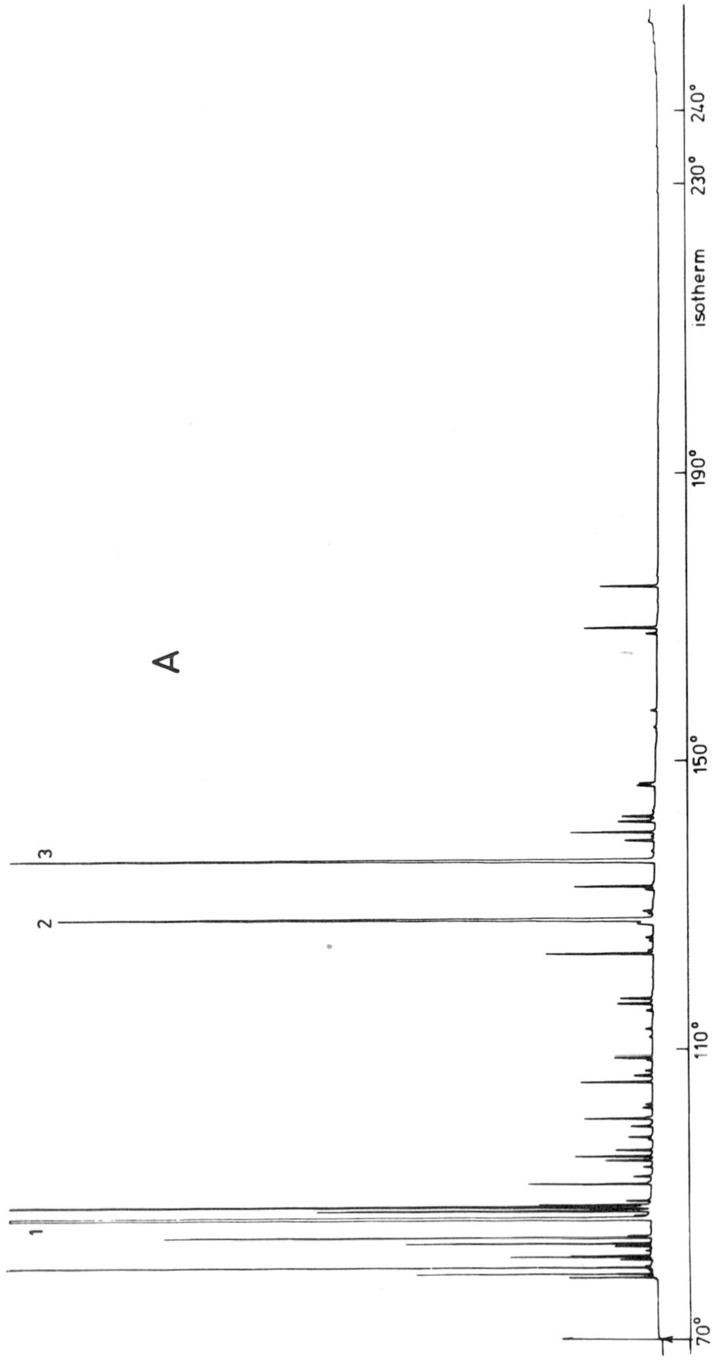
A
1
2
3
70°
110°
150°
190°
isotherm
230°
240°

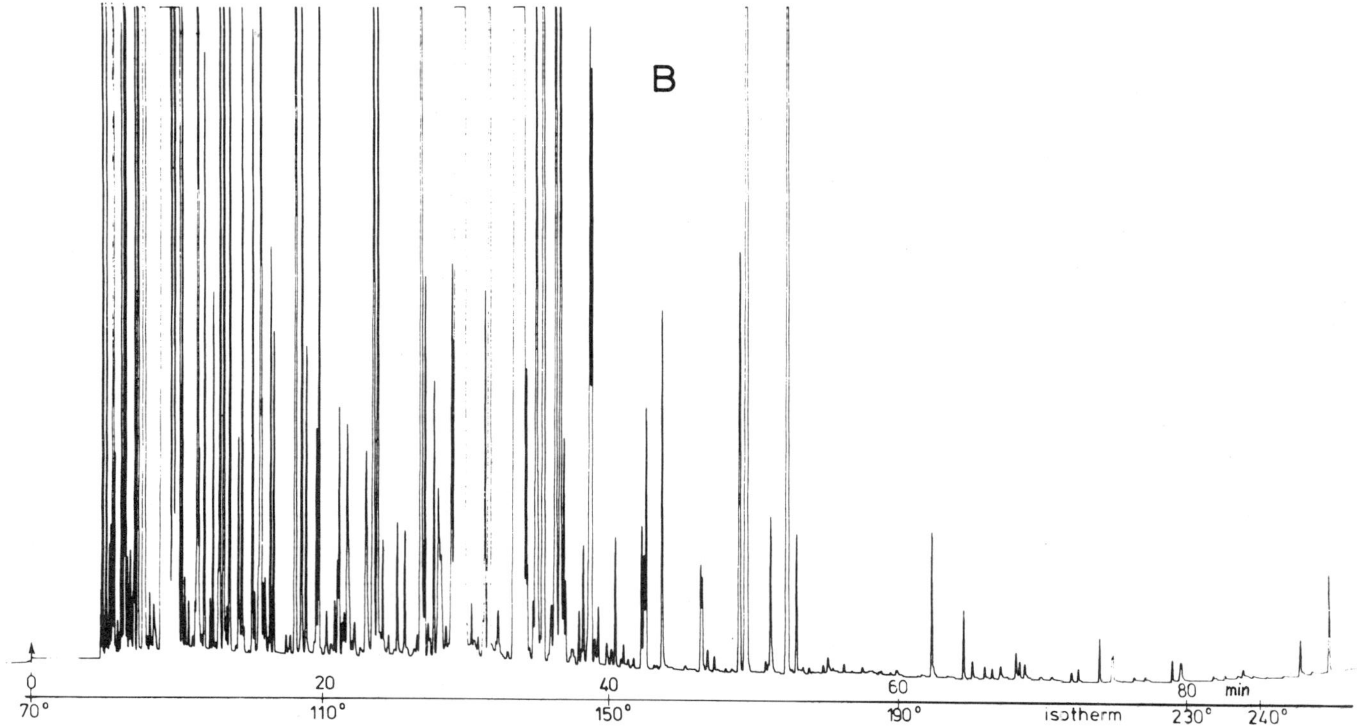

Figure 7. Chromatogram at two intensities of hop oil. Capillary column 40 m × 0.25 mm ID static-coated 0.2% Superox-4. Temperatures: detector and injector 250°C. Column programmed from 70 to 240°C at 2°C min^{-1}. Flowrate 1.5 ml min^{-1} (H_2). (A) Sample size 0.2 μl, split ratio 1:50. Sensitivity setting 8×10^{-11}. Peaks: 1--myrene, 2--caryophyllene, 3--humulene. (B) Sample size 0.7 μl, split ratio 1:50. Sensitivity setting 1×10^{-11}. (Source: Ref. 6.)

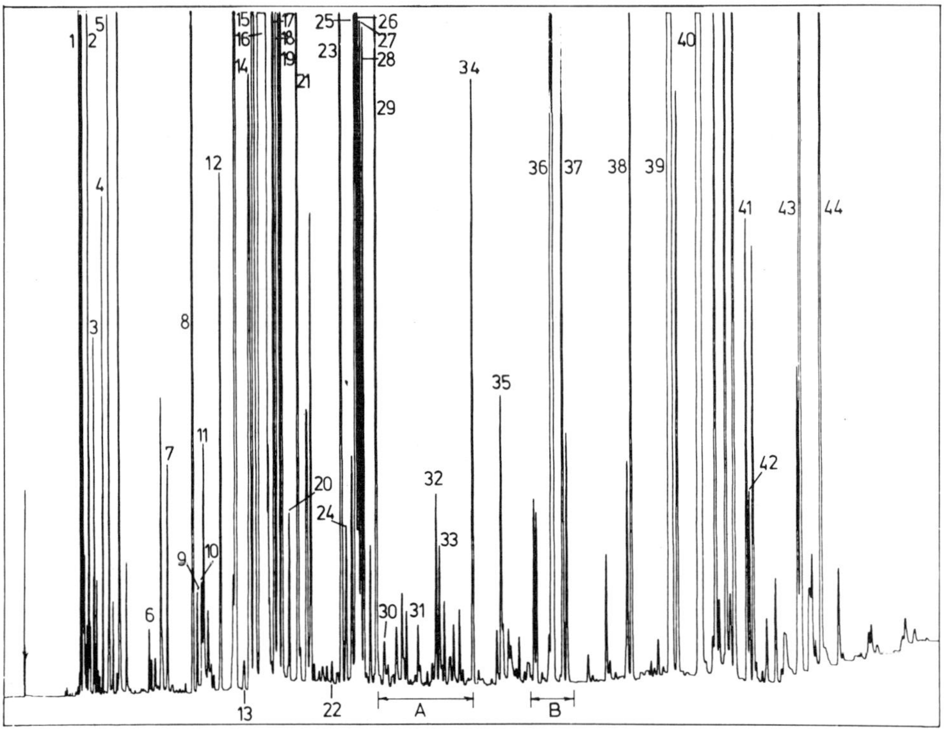

Figure 8. Vacuum-steam-distilled 1974 Saaz hop oil. Column: 90 m × 0.5 mm ID coated statically with SE-30. Column temperature-programmed from 80 to 200°C at 2°C min^{-1}. Flowrate 6 ml min^{-1} (H_2). Sample size 0.2 µl on column.

combination with smell analysis. In the aromagram (Fig. 9), the compounds with an important smell are indicated with an asterisk and their structures were determined by chemical ionization and electron impact mass spectrometry. Several compounds of the volatile fraction originate from hop essential oil, e.g., 4-methylpent-2-enoic acid lactone (peak 1), 6-methyl-5-hepten-2-one (peak 2), myrcene (peak 3), cis-linalooloxide (peak 4), nonanone-2 (peak 5), trans-linalooloxide (peak 6), linalool (peak 7), undecanone-2 (peak 8), methyl-dec-4-enoate (peak 9), methyl-decadi-4,8-enoate (peak 10), β-caryophyllene (peak 11), humulene (peak 12), and humuladienone (peak 13). These are all clearly hop essential oil constituents. The conclusion must be that alkaline aqueous solutions of hop bitter acids have remarkable dissolving power for the essential oil since they extract these neutral compounds from organic solvents.

The preparation of hop extract involves removal of a solvent by vacuum distillation. The recovered solvents have a clear hoppy smell, and the question of which are the responsible compounds was investigated.

Table 2. Composition of the Vacuum-distilled Essential Oil of Saaz Hops (1974) (see Fig. 8)

1. Acetone
2. Dimethylvinylcarbinol[a]
3. 3-Methylbutan-2-one
4. 3-n-Propoxy-but-1-ene
5. 4-Methylpentan-2-one
6. 4-Mercapto-3-methyl-butan-2-one (tentative?)
7. Isoamylacetate
8. 2-Methylbutyl-2-methylpropanoate
9. Methyl hexanoate
10. 4-Hydroxy-4-methylpent-2-enoic acid lactone[a]
11. Monoterpene
12. α-pinene
13. 6-Me-5-hepten-2-one
14. Methyl heptanoate (branched)
15. β-Pinene
16. Myrcene[a]
17. Isoamyl-2-methylpropanoate
18. Pentyl-2-methylpropanoate
19. Methyl heptanoate
20. p-Cymene
21. Limonene
22. Cis-linalooloxide[a]
23. Methyl octanoate (branched)
24. Trans-linalooloxide[a]
25. Linalool[a]
26. Perilene
27. Aliphatic esters
28. Aliphatic esters
29. Methyl octanoate
30. γ-decalactone[a]
31. Borneol
32. Terpineol-4
33. Methyl nonanoate (branched)
34. Methyl nonanoate
35. Geraniol
36. Methyl-4,8-decadienoate[a]

Table 2 (continued)

37.	Methyl-4-decenoate[a]
38.	α-Copaene
39.	β-Caryophyllene[a]
40.	Humulene[a]
41.	Germacrene D
42.	Calamenene
43.	Oxygenated sesquiterpenoid
44.	Humuladienone[a]

[a]Typical for hop oil.

This $(GC)^2$ analysis of an enriched distillate is shown in Fig. 10. The smell of the eluted compounds was evaluated; components with a hoplike or hop-reminding smell were marked and identified by $(GC)^2$/MS: isobutyraldehyde (3), pentenols (4, 5, 6), 3-methylbutyraldehyde (7), 2-methylbutyraldehyde (9), and methyl isobutanoate (12).

4. BEER ANALYSIS

In recent years, gas chromatography has found increasing application in the brewery. Capillary columns have been used for several problems. Buttery et al. [17] analyzed the essential oil of hopped and unhopped beer by capillary gas chromatography. A $(GC)^2$ analysis of beer volatiles has also been published by Jennings et al. [18]. For quality control, the Heineken group described the analysis of the low-volatile compounds of beer by headspace sampling and injection on a wide-bore capillary column coated with Carbowax 400 [19]. We used glass capillary columns for the analysis of the volatiles [20,21], the fatty acids [20,22], the bitter acids [23,24], and the amino acids [25] of beer. Other applications, which have not been yet published, will be introduced in this chapter.

4.1 Analysis of Volatiles

Beer volatiles can be analyzed by different procedures. Direct injection, as successfully applied for some alcoholic beverages [26], is, because of the high content of water and ethanol, not very useful for the estimation

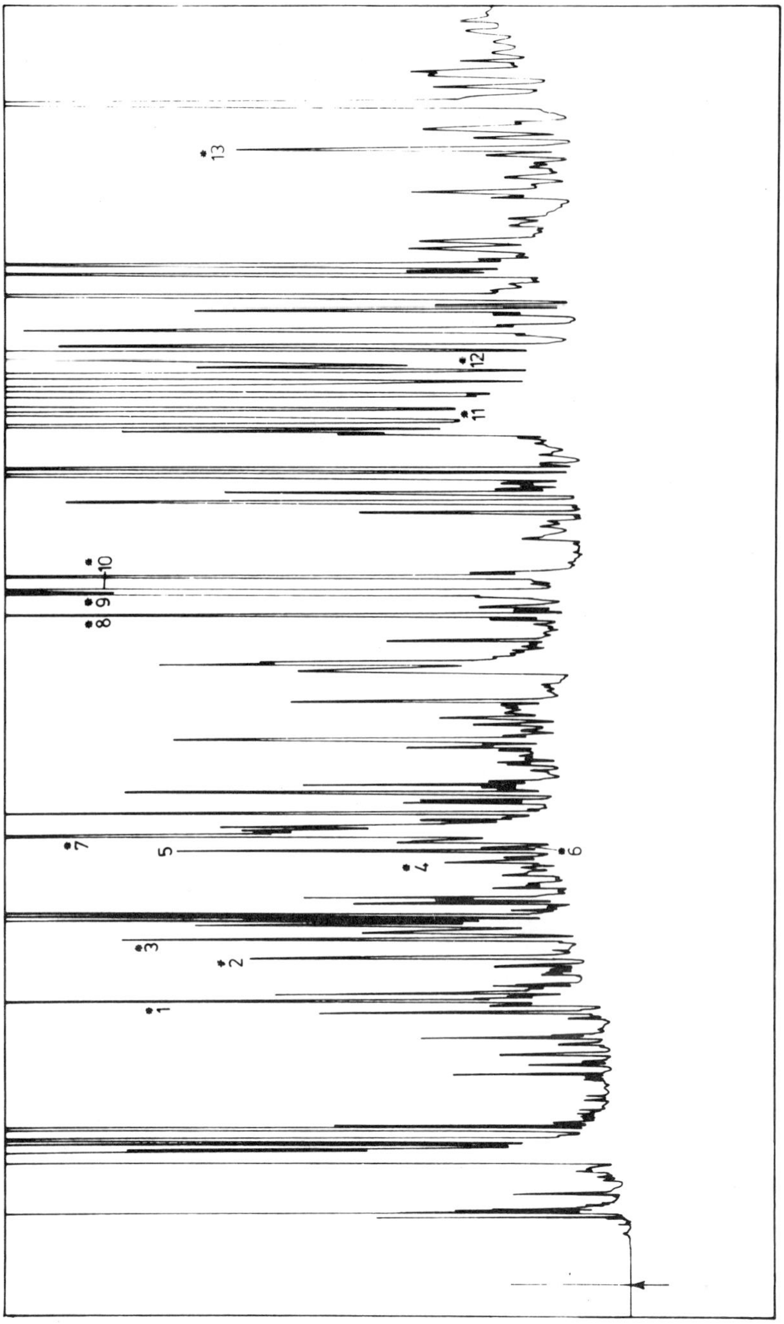

Figure 9. Volatile components in isomerized hop extracts. Column: 70 m × 0.6 mm ID coated with OV-1. Column temperature-programmed from 70 to 220°C at 2°C min^{-1}. Flowrate 5 ml min^{-1} (H_2).

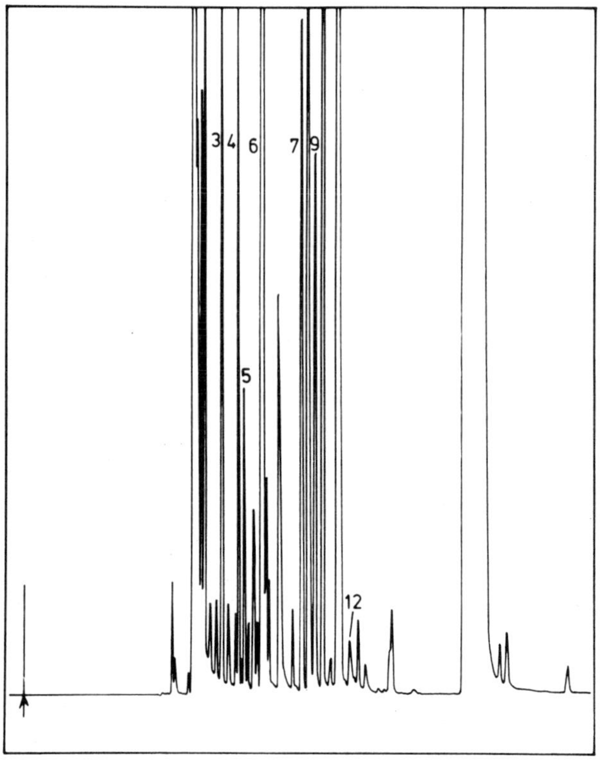

Figure 10. Volatile fraction collected from the residual solvent of the isomerized hop extract preparation. Column: 100 m × 0.5 mm ID coated with OV-1. Column temperature 40°C, flowrate 1.4 ml min^{-1} (H_2).

of beer volatiles. The methods most successful to date for routine assessments and research purposes are:

For low-molecular-weight volatiles: the direct headspace sampling technique and the headspace entrainment concentration technique employing porous polymers.

For medium-molecular-weight volatiles: the carbon disulfide extraction procedure.

For high-molecular-weight volatiles: the continuous liquid-liquid extraction technique

Direct Headspace Sampling Technique

The very volatile components in beer are present in high concentrations in the headspace of bottled beer. By injecting an aliquot of the headspace on a Carbowax 400 capillary column a picture of the beer flavor is obtained which gives useful information on the fermentation process [19]. A typi-

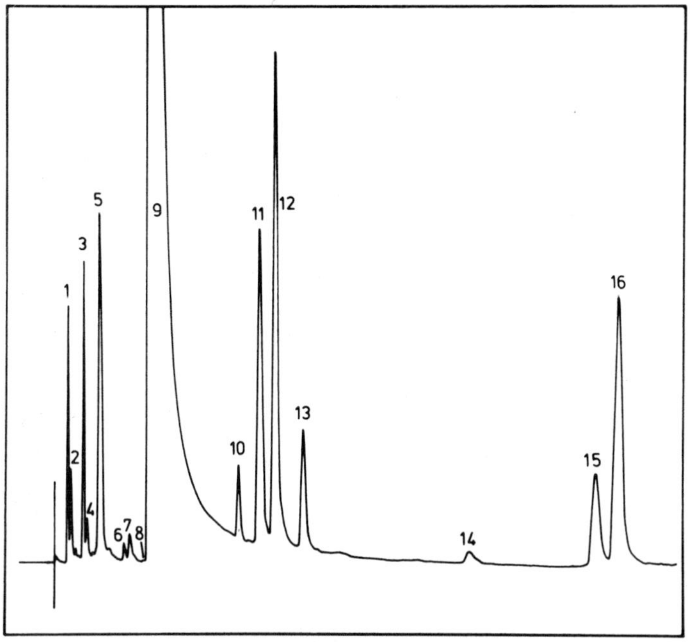

Figure 11. Headspace analysis of beer. Column: 100 m × 0.5 mm ID capillary column whiskered and coated with Carbowax 400. Detection: flame ionization (FID) at 150°C. Injection: 0.5 ml headspace directly at 150°C. Temperature column: 65.5°C. Carrier gas: nitrogen. Inlet pressure: 1.6 atm. Peaks: 1—acetaldehyde, 2—dimethylsulfide, 3—ethyl formate, 4—acetone, 5—ethyl acetate, 6—ethyl propionate, 7—methanol, 82—propanol, 9—ethanol, 10—n-propanol, 11—isoamyl acetate, 12—heptanone-4 (565 ppm internal standard), 13—2-methylpropanol, 14—ethyl caproate, 15—opt. active amyl alcohol, 16—isoamyl alcohol. (Reproduced courtesy of Dr. J. Strating, Heineken Breweries.)

cal chromatogram is presented in Fig. 11. Acetaldehyde (peak 1) is an indicator of the fermentation quality; higher content is less desirable. The concentration of the most important esters (ethyl acetate and isoamyl acetate) is influenced by the fermentation method. High-gravity fermentation and high temperatures produce beers with an estery character. Less estery beers are obtained with fermentations carried out under pressure.

The fermentation conditions are also responsible for the production of the amyl alcohols, and their concentrations are normally proportional to the ester concentrations.

The headspace technique in combination with electron capture detection allows the estimation of the butter flavor in beer. The vicinal diketones, diacetyl, and 2,3-pentanedione are responsible for this off-flavor in beer. Both ketones are by-products of the fermentation and their production is strongly influenced by the fermentation conditions. Their concentration

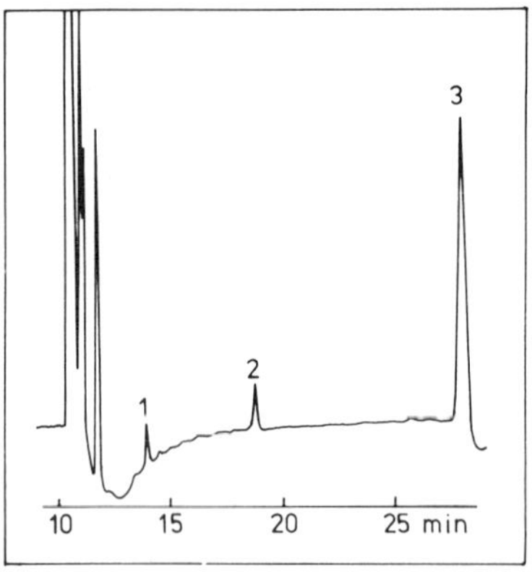

Figure 12. Estimation of the butter flavor in beer. Column: 165 m × 0.75 mm ID capillary column coated with didecylphthalate. Detection: electron capture (ECD) at 130°C. Injection: 0.5 ml headspace directly at 110°C. Temperature column: 60°C. Peaks: 1—diacetyl, 2—2,3-pentanedione, 3—2,3-hexanedione, (internal standard). (Source: Ref. 19. Reproduced by permission of the copyright owner.)

in beer should be kept as low as possible. Figure 12 shows a chromatogram of 0.5 ml headspace of a commercial beer injected on a stainless steel capillary column coated with didecylphthalate. In order to measure both diketones, 2,3-hexanedione is added as internal standard [19]. Both examples illustrate the usefulness of capillary gas chromatography for quality control in the brewery.

Headspace Entrainment Concentration Technique

A versatile technique for concentration of the low volatiles of beer has been developed by Jennings and co-workers [27,28] and was used for several purposes in our laboratory [29]. In this technique the flavor volatiles are trapped on porous polymers (Porapak Q, Tenax) while water and ethanol, exhibiting low retention volumes on these materials, are removed. The system (Fig. 13) consists of a glass vessel, containing 250 ml beer and 8 ppm octanol as internal standard and antifoam agent. The vessel is thermostated at 30°C. Purified nitrogen is bubbled through the sample at a rate of 37.5 ml min^{-1} for 1/2 hr. The gas flow, containing the headspace volatiles, is conducted in the entrapment system which consists of a 20-cm × 3-mm ID glass tube packed with 80-100 mesh Porapak Q (80-100 mesh) held at 38°C. After sampling the vessel is disconnected and residual water is removed by passing nitrogen through the column for 10 min. The Porapak Q trap is then reversed and the beer volatiles are back-flushed into a U-formed glass tube cooled in liquid air. The microtrap is broken in a pointed

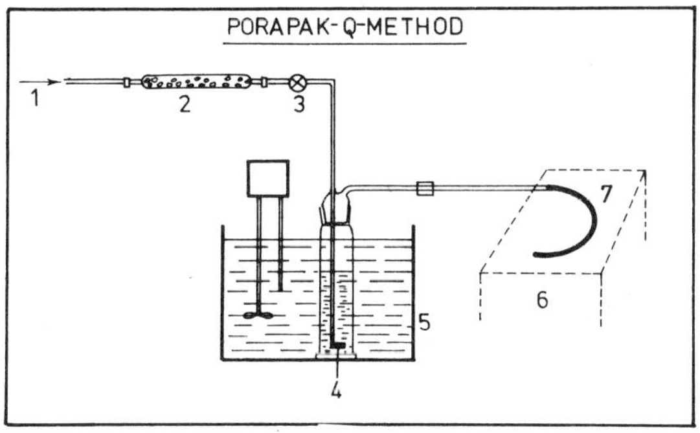

Figure 13. Porapak Q method: 1—lecture bottle, 2—drierite trap, 3—pressure regulator, 4—glass fiber filter, 5—thermostat, 6—oven, 7—Porapak Q column.

test tube which is centrifuged at 4000 rpm. Approximately 2 μl of beer volatiles are obtained; 0.5 μl is directly injected on a wide-bore capillary column statically coated with Carbowax 20M. The chromatogram of an essence obtained in this way from hopped beer is shown in Fig. 14. This technique supplements the headspace technique as the higher-boiling components of the volatile fraction are readily detected. Some interesting compounds are: 1—acetone, 2—ethyl acetate, 3—ethanol, 4—ethyl butyrate, 5—n-propanol, 6—isobutanol, 7—isoamyl acetate, 8—isoamyl alcohol, 9—ethyl hexanoate, 10—ethyl octanoate, S—octanol (internal standard), 11--ethyl decanoate. The reproducibility is very good and the method can be employed for the quantitative analysis of some beer flavor compounds.

The sunstruck flavor of beer has been ascribed by Kuroiwa and Hashimoto to the formation of 3-methyl-2-butene-1-thiol by photolysis of the iso-α acids in the presence of sulfur-containing amino acids [30]. The thiol could be formed by illumination of beer in a concentration of maximum 1 ppb. The isolation of the thiol was achieved by the Porapak Q trapping technique and its presence confirmed by $(GC)^2$ with flame photometric detection after cochromatography with synthetic thiol. Further proof was obtained by the combination $(GC)^2$/mass fragmentography selecting the typical ions m/e 102 (M^+), 69, 68, and 53. The flame photometric detector (FPD) chromatogram of illuminated and, for comparison, of unilluminated beer is shown in Fig. 15. The peak of the thiol is a major one. The other larger peaks are alcohols which are FPD-active in high concentrations.

Extraction Procedures

Carbon Disulfide Extraction. The extraction of beer volatiles with carbon disulfide for subsequent (GC) analysis was first reported by Powell

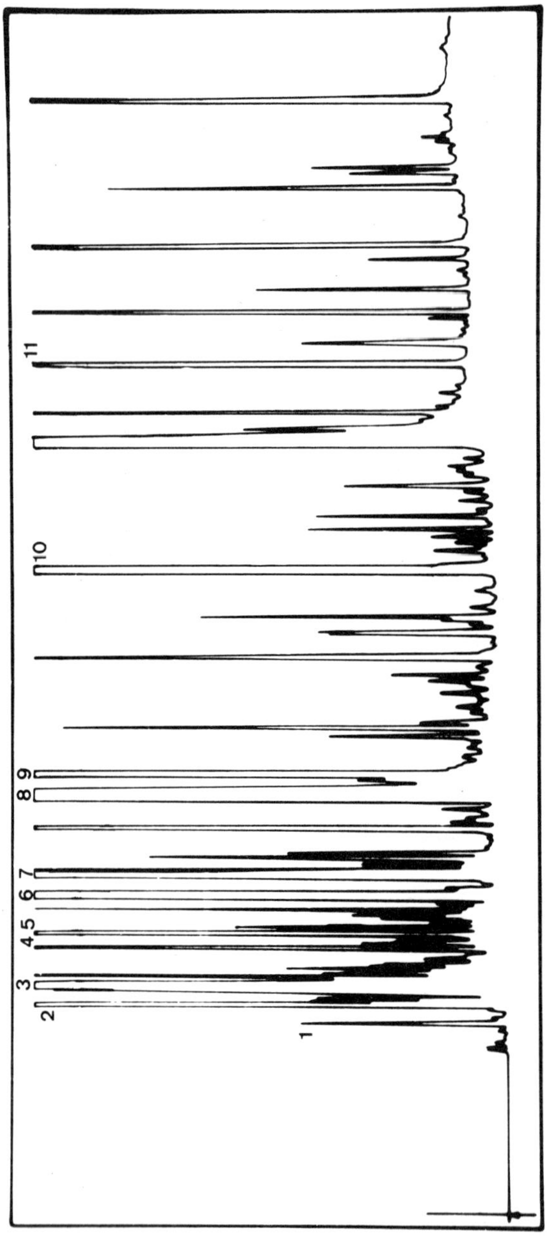

Figure 14. Chromatogram of the headspace volatiles of beer using the Porapak Q technique. Column: 85 m × 0.5 mm ID coated with Carbowax 20M. Column programmed from 50 to 180°C at 2°C min-1; detector and injector temperature 220°C. Carrier gas: hydrogen, flowrate 3 ml min-1. Sample: 0.5 µl essence, on-column injection via a glass insert. Peaks: see text.

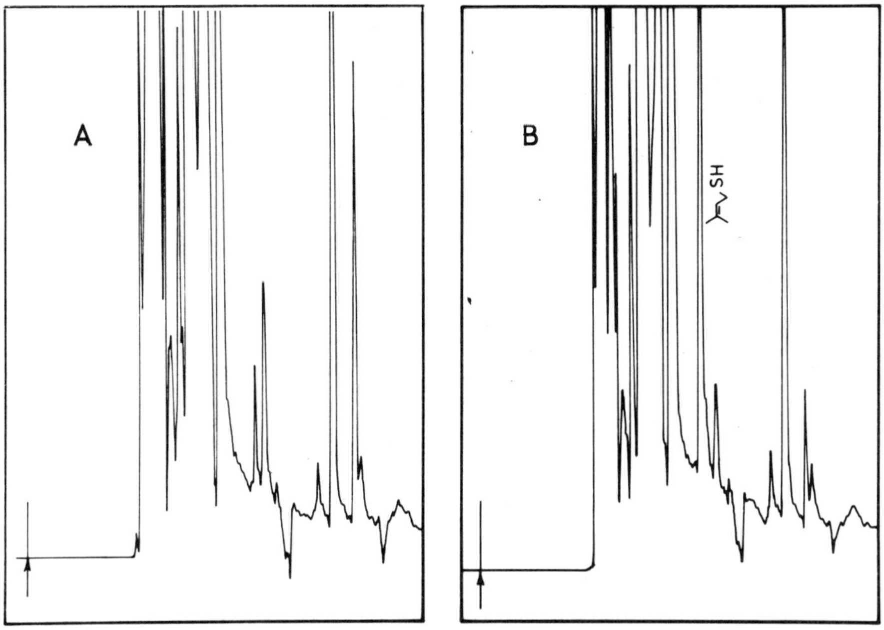

Figure 15. FPD chromatogram of unilluminated (A) and illuminated (B) beer. Column: 80 m × 0.6 mm ID coated with SE-30. Temperature: column programmed from 50 to 150°C at 2°C min^{-1}. Detection: FPD at 160°C. Oxygen flowrate: 40 ml min^{-1}. Hydrogen flowrate: 120 ml min^{-1}. Nitrogen flowrate: 80 ml min^{-1}. Injection: directly. Carrier gas: hydrogen flowrate 4 ml min^{-1}. Sample preparation: headspace concentration on Porapak Q. (Source: Ref. 31.)

and Brown [32]. Difficulties were encountered with reproducibility and very large variations between laboratories were found. By modification [33-36] and standardizing the conditions [37] reliable quantitative data were obtained. This concentration technique provides the quantitative estimation of a number of higher-molecular-weight substances. 1-Octanol is normally added as internal standard. A modified version of the CS_2 extraction procedure was used by Sandra and Verzele [20] for the investigation of the hop aroma of beer.

The enriched CS_2 extracts were analyzed on a wide-bore capillary column coated with FFAP. The $(GC)^2$ analyses of hopped and unhopped beer are shown in Fig. 16.

The similarity of both chromatograms is conspicuous. Noticeable differences are in the intensity of peaks 5 and 11. Both compounds are oxidation products of the bitter acids and were present in all hopped beers investigated. A typical Belgian beer (gueuze) obtained by spontaneous fermentation of wort was characterized by high concentrations of ethyl lactate,

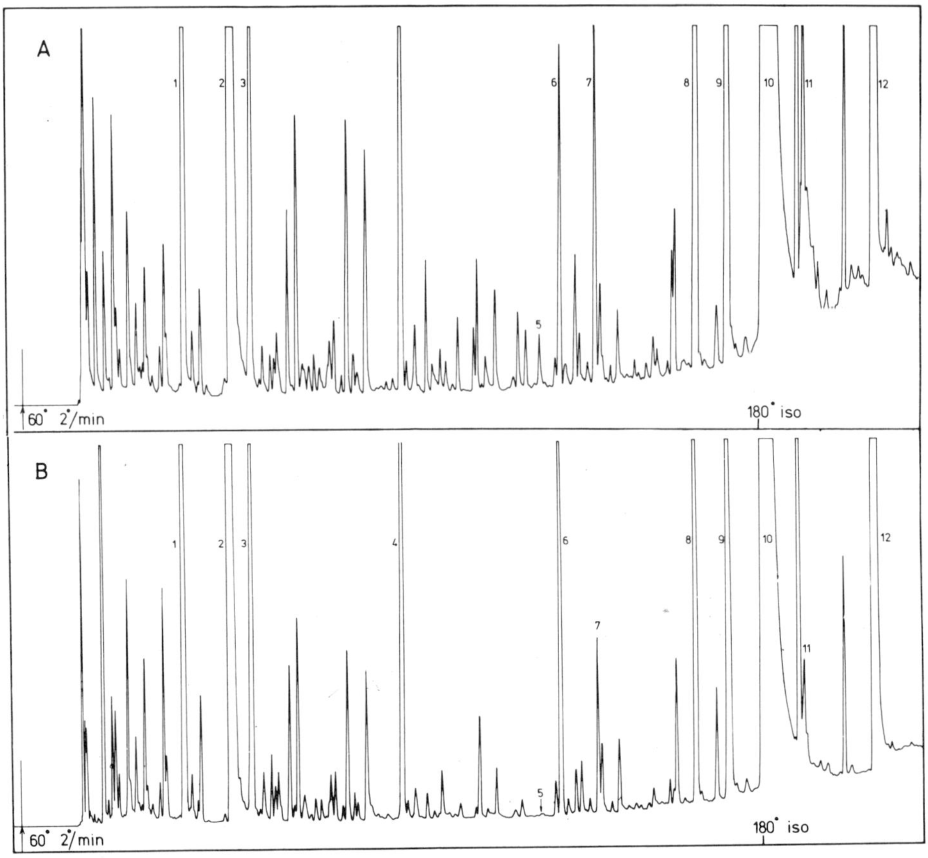

Figure 16. Chromatogram for hopped (A) and unhopped (B) beer—CS_2 extracts. Glass open tubular column 50 m × 0.5 mm ID, dynamically coated with 15% FFAP. Temperature of detector and injector: 250°C, column programmed from 60 to 200°C at 2°C min^{-1}. Flowrate: 3.2 ml min^{-1}. H_2 column, 17 ml min^{-1} H_2 and 250 ml min^{-1} air to FID. Peaks: 1—isoamyl acetate, 2—isoamyl alcohol, 3—ethyl n-hexanoate, 4—ethyl n-octanoate, 5—4-hydroxy-4-methyl-2-pentenoic acid lactone, 6—ethyl n-decanoate, 7—furfuryl alcohol, 8—phenyl ethyl acetate, 9—hexanoic acid, 10—phenyl ethyl alcohol, 11—4-methyl-3-pentenoic acid, 12—octanoic acid. (Source: Ref. 20.)

α-p-methoxyphenylethanol, and p-ethylphenol. The two last compounds were identified only by their mass spectral data.

Continuous Liquid-Liquid Extraction. Continuous liquid-liquid extraction of beer with ether, pentane, and the solvent mixture dichloromethane-pentane-ether 6:3:1 has been employed. The applicability of this method is somewhat limited for quality control. The method is mostly used for research purposes, e.g., trace analysis and the analysis of the higher-boiling components of beer.

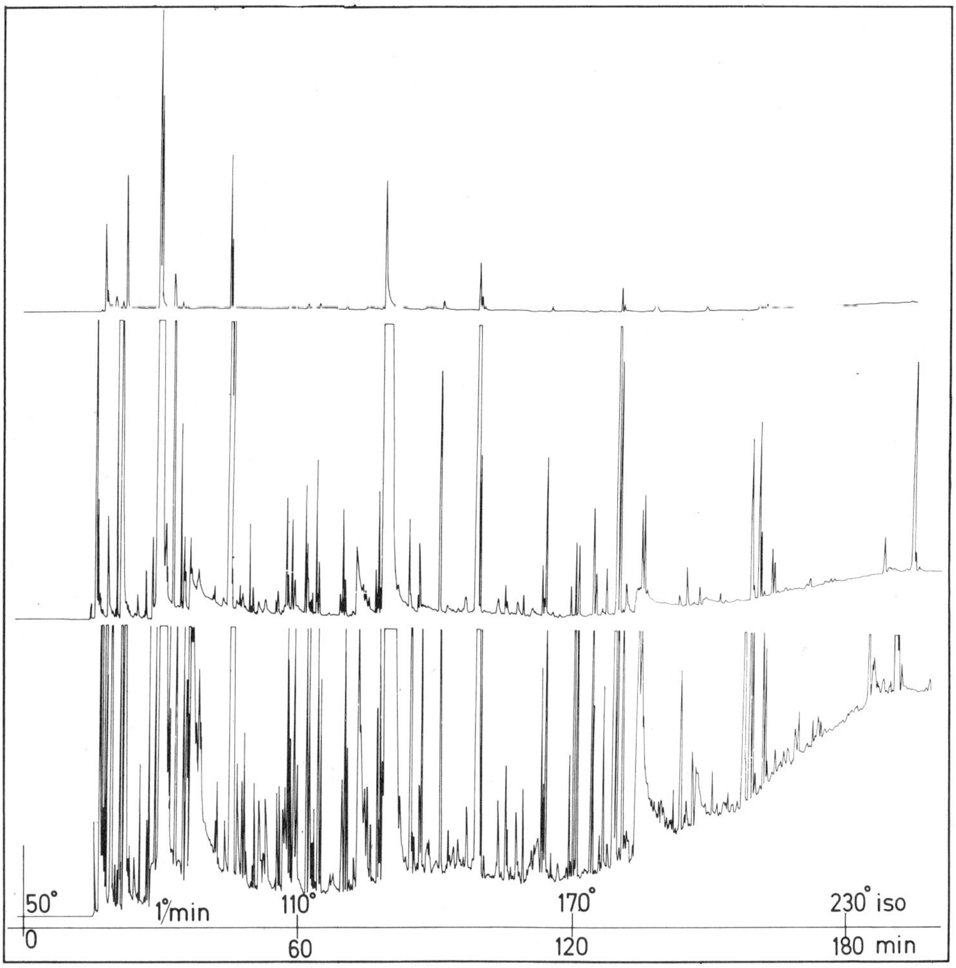

Figure 17. Chromatogram at different intensities of commercial beer—extraction method II. Glass open tubular column 400 m × 0.5 mm ID, static-coated 5 mg ml^{-1} SE-30. Temperature: detector and injector 250°C. Column programmed from 50 to 230°C at 1°C min^{-1}. Flowrates 5 ml min^{-1} H_2 column, 17 ml min^{-1} H_2 and 220 ml min^{-1} air to FID. Sample size: 1 μl undiluted mixture. (Source: Ref. 20.)

Figure 17 shows a chromatogram at different intensities of beer essential oil obtained by this method. The complexity is obvious and up to 400 different peaks can be counted in the last chromatogram without considering peak shoulders. In this way, minor constituents can be detected, organoleptically evaluated, and even identified by combined gas chromatography/mass spectrometry or gas chromatography/mass fragmentography. Till now, more than 300 compounds were identified in beer essential oil. These techniques have also been used to analyze the hoppy character of beer [17,20].

4.2 Contribution of Hop-derived Compounds to the Flavor of Beer

The questions if hops and how hops impart a characteristic hoppy aroma to beer have not yet been settled. Brewers still consider that hops with a pleasant aroma are superior in quality, and without further ado it is assumed that this finer aroma or smell is transferred to beer with such hops. The hoppy aroma of lager beer, however, has not been settled unequivocally by purely organoleptic methods.

Nevertheless, if there is a smell difference between hopped and unhopped beer, then there is a chemical difference which we can detect and analyze. Considering the complexity of beer essential oil, the high resolution of capillary columns is required for such analyses. Studies by several investigators have clearly shown that most of the components of the essential oil are by their volatility in steam and low solubility in water, lost during boiling [17,38-40] of the hopped wort. Remaining traces are removed by filtering the wort [38,41], during the fermentation [38,42] or the lagering of young beer [38]. Figure 18 shows the analysis of a hopped wort (200 g ml^{-1}) on a 80-m × 0.5-mm ID capillary column coated with SE-30. The wort extract, obtained by continuous liquid-liquid extraction, was derivatized with diazomethane. A wort extract is mainly composed of highly water-soluble products (acids, dicarboxylic acids, benzoic acids, and cinnamic acids), and without derivatization a reproducible analysis was not possible. In this chromatogram, the only product related with hops is peak 25, identified as the lactone of 4-hydroxy-4-methyl-2-pentenoic acid (oxidation product of the bitter acids). For the other compounds see Table 3.

In the analysis of beer essential oil [20] special attention was paid to the presence of the major hop oil compounds, e.g., the hydrocarbons myrcene, humulene, and caryophyllene, representing 70 to 80% of hop oil, and linalol, 4-methyldecenoate and 4,8-methyldecadienoate. Mass fragmentography (MF), which guaranteed by selection of the four m/e values 80, 93, 136, 208, sufficient sensitivity and selectivity for the detection of 1 ppb of hop oil hydrocarbons, was unsuccessfully applied in this respect. This is illustrated in Fig. 19, representing the relevant region of the $(GC)^2$/MS trace of normal commercial beer and one for the same beer to which 10 ppb humulene (peaks with asterisks) was added. By the same technique, linalool in a concentration of 10 ppb was detected in one investigated beer. This concentration, however, is too low to impart a characteristic aroma in comparison with the flavor threshold in beer (100 ppb). The methylesters of

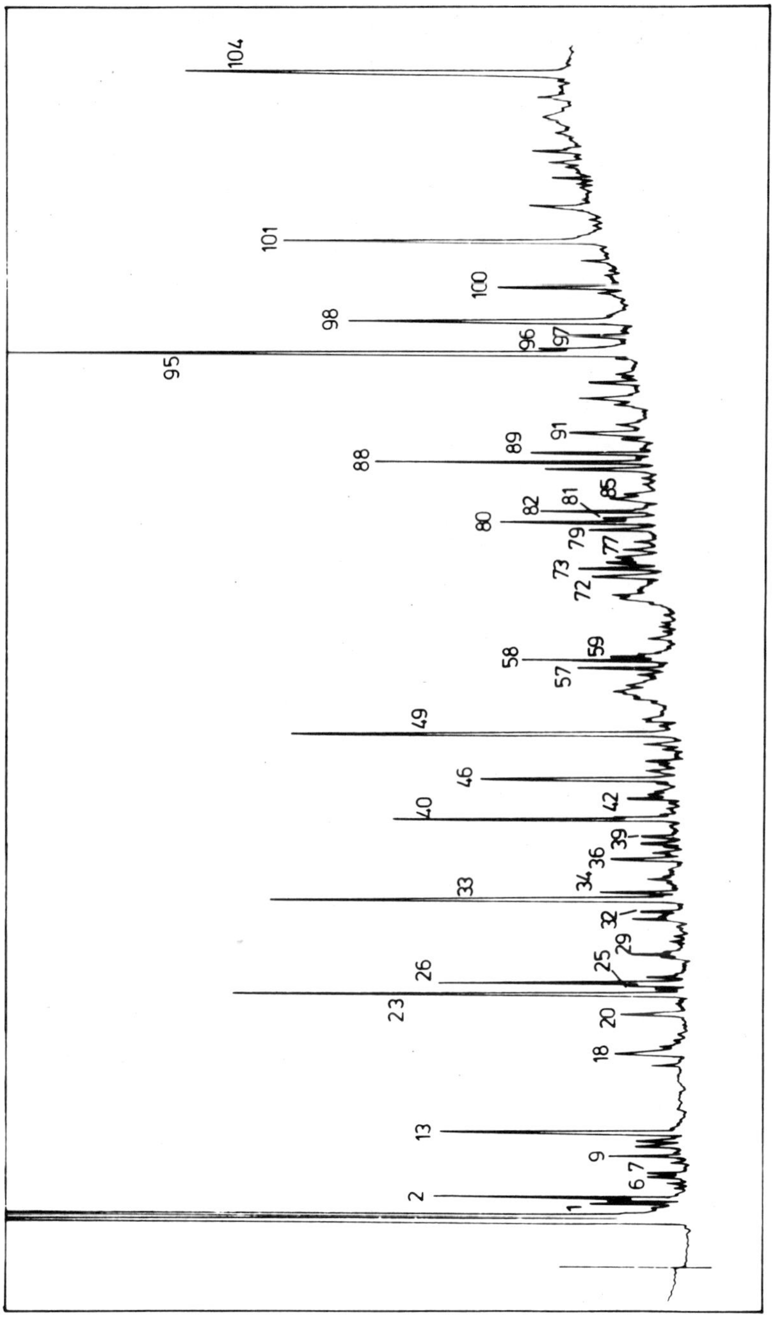

Figure 18. Analysis of the extract of hopped wort, after derivatization with diazomethane. CI monitogram $(GC)^2$/MS quadrupole Finnigan 3200. Column temperature-programmed 80 to 200°C at 2°C min^{-1}. Carrier gas helium, flowrate 4 ml min^{-1}.

Table 3. Compounds Identified in Hopped Wort (after Derivatization with Diazomethane) (See Fig. 18)

1	Butanone
2	Ethyl acetate Dimethylvinylcarbinol
6	Methyl isobutyrate
7	Ethyl methylcarbonate
8	Methyl n-butyrate
9	Acetaldehyde diethylacetal
13	Methyl isoamylate
18	Furfurylalcohol
23	Methyl hexanoate (branched)
25	4-Methyl-4-hydroxy-2-pentenoic acid lactone
26	Methyl hexanoate
29	Methyl furfurate
32	Ethyl hexanoate
33	Dimethyl succinate
34	Methyl heptanoate
36	Dimethyl glutarate
39	1-Octanol (standard)
40	Methyl benzoate
42	β-Phenylethanol
44	Phenylpropanone
46	β-Carbomethoxy pyridine + methyl octanoate
49	Methylphenyl acetate
57	Methyl nonanoate
59	(o,p) Methoxybenzaldehyde
73	o-Methoxymethyl benzoate
77	(p,m) Methoxyacetophenone
78	Methyl nonenoate
80	(p,m) Methoxymethyl benzoate
81	Methoxyphenylpropanone
82	Methyl cinnamate
85	Methyl decanoate
88	Methyl(methoxyphenyl)acetate
89	Dimethyl sebacate (tentative)
91	Dimethoxybenzaldehyde
96	Dimethoxyacetophenone

Table 3 (continued)

97	Dimethoxyphenylpropane
98	Dimethoxymethyl benzoate
101	Methoxymethyl cinnamate
103	Trimethoxymethyl benzoate
104	Dimethoxymethyl cinnamate

dec-4-enoic acid and deca-4,8-dienoic acid survive the boiling process but they are converted to the ethylesters during fermentation [17]. The concentration of ethyl-dec-4-enoate is 3 times higher in hopped than in unhopped beer [43]. To investigate their contribution to the hop aroma of beer, a fraction of hop oil containing both methylesters was collected by preparative gas chromatography. The $(GC)^2$ analysis of this fraction is shown in Fig. 20. Organoleptic experiments indicated that this fraction has a weak hop smell. By transesterification into the ethylesters, however, the hoppy smell was lost [44]. The Porapak Q technique has also been used in trying to isolate and identify volatiles from hop oil in beer [29]. Commercial beers, hopped and unhopped reference beers produced from the same malt and beers to which 1 ppm hop oil and 10 ppb myrcene was added, were investigated. Although 10 ppb myrcene could easily be enriched and detected, in none of the beers ppb amounts of the main compounds of hop oil could be recovered. Eight beers, brewed with seven different hops and one with an isomerized hop extract, were analyzed by the Porapak Q procedure, and it was not possible to indicate from the aromagrams which hop variety was used [45]. This is illustrated by the comparison of the aromagrams (Fig. 21) of beer brewed with Saaz hops (A) and isomerized hop extract (B). Moreover, no clear difference was made by an organoleptic panel, and the beer brewed with hops with the finest aroma was not selected as the best one.

In conclusion, no major compounds of the essential oil of hops can be shown to be present in normal lager beer, despite the availability and use of powerful analytical techniques. For these reasons, to impart a desirable hop character to some beers, e.g., English ale beers, a number of processes are employed, including dry hopping [46], addition of hop oil or adding a portion of hops to the kettle toward the end of boiling [47,48], or addition to beer in the conditioning tanks of an aqueous emulsion of hop oil obtained by vacuum steam distillation [49]. Volatile water-soluble compounds, formed when hops are boiled in wort or formed by oxidation of the bitter acids and

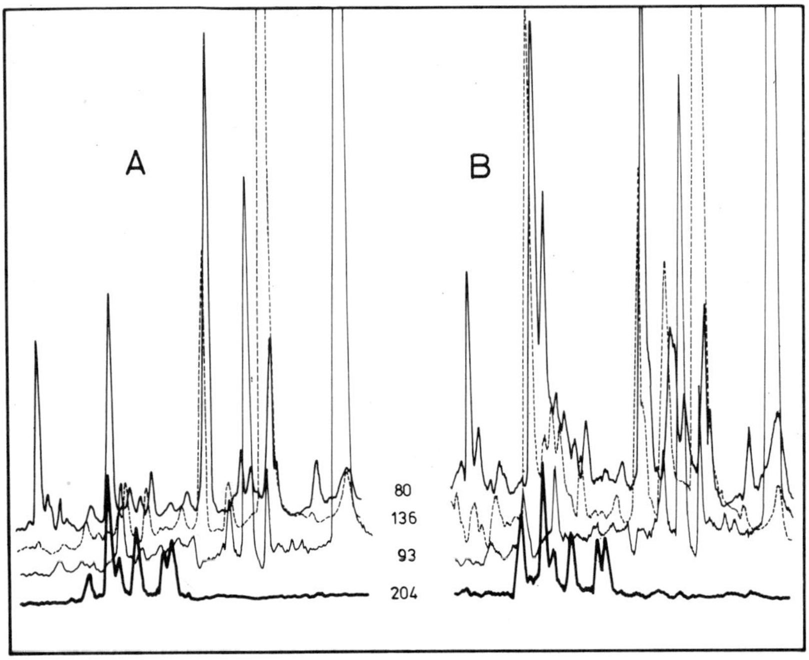

Figure 19. Mass fragmentogram in the relevant region of the chromatogram for commercial beer (A) and commercial beer + 10 ppb humulene (B). (Source: Ref. 20.)

terpene hydrocarbons, have been suspected to impart a "hoppy note" to beer [38,50]. Some of them, e.g., 3-methylbutan-2-one, 4-methylpentan-2-one, 2-methyl-3-buten-2-ol, 2,4-hexadien-1-al, and humuladienone, are considered to give a mild hoplike aroma to unhopped beer [38,51]. We investigated chemically and organoleptically oxidation profiles of beer bitter acids [52] and the terpene hydrocarbons [56].

The detection of 4-hydroxy-4-methyl-2-pentenoic acid lactone (∿10 to 50 ppb) and 4-methyl-3-pentenoic acid (50 to 100 ppb) in beer (Fig. 16) confirms the presence of oxidation compounds of the bitter acids [18] in hopped beer. This contribution was also demonstrated by the quantitative comparison of the lower fatty acids of hopped and unhopped beer. Acetic acid (a secondary oxidation product), 2-methylpropionic acid (oxidation of the acyl side chain of the cohomologs of the bitter acids), 3-methylbutyric acid (oxidation of humulone and lupulone), and 2-methylbutyric acid (oxidation of the ad-homologs) are present in higher concentration (100 to 500 ppb) in hopped than in unhopped beer.

Considering also the findings of Palamand [53,54] and Hashimoto [38, 55], we conclude that the hop-derived compounds summarized in Table 4 contribute to the flavor of beer. They do not contribute to the hoppy char-

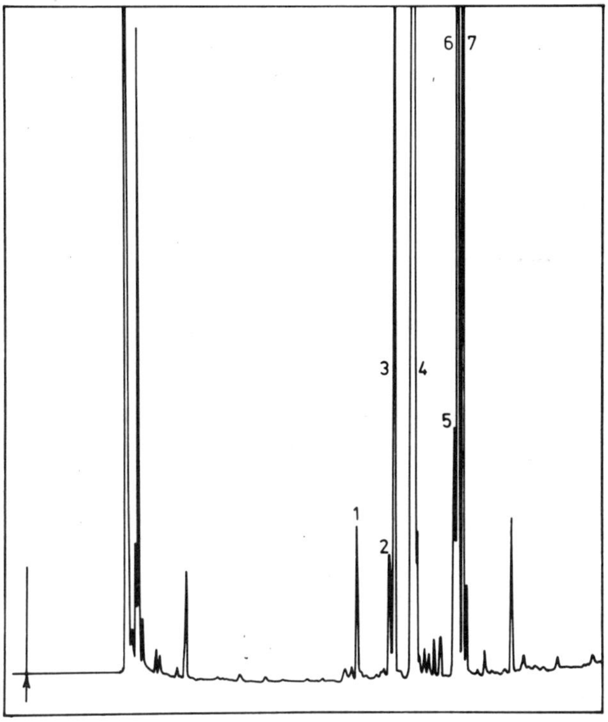

Figure 20. Fraction with a weak hop aroma of the oxygenated compounds of hop essential oil. Column: 50 m × 0.5 mm ID FFAP. Temperature column: programmed from 100 to 160°C at 2°C min^{-1}. Carrier gas: hydrogen, flowrate 3 ml min^{-1}. Peaks: 1—methyl isodecanoate, 2—methyl decanoate, 3—methylnonylketone, 4—methyl cis-4-decenoate, 5—isohexenoic acid, 6—methyl-4, 8-decadienoate, 7—benzyl butyrate.

acter of this flavor, as none of them has a smell that is even remotely reminiscent of hops.

Oxidation experiments on purified hydrocarbons of hops (myrcene, humulene, and caryophyllene) indicated that about 90% of the hydrocarbons polymerize during the oxidation process at 100°C (temperature of wort boiling). The volatile fractions, especially for myrcene and caryophyllene, are very complex mixtures. However, neither the oxidation mixture nor the individual compounds (smell analysis) possessed a hop character [56]. The oxidation profile of humulene is represented in Fig. 22. The major compound humuladienone was isolated by preparative capillary gas chromatography on a column of 0.9 mm ID with microkatharometer detection. The mass fragmentograms, by selecting the key ions 67, 96, 109, and 138, of commercial beer and commercial beer after addition of 30 ppb humuladienone were compared and the presence of this compound in European lager beers could not be confirmed [20,51].

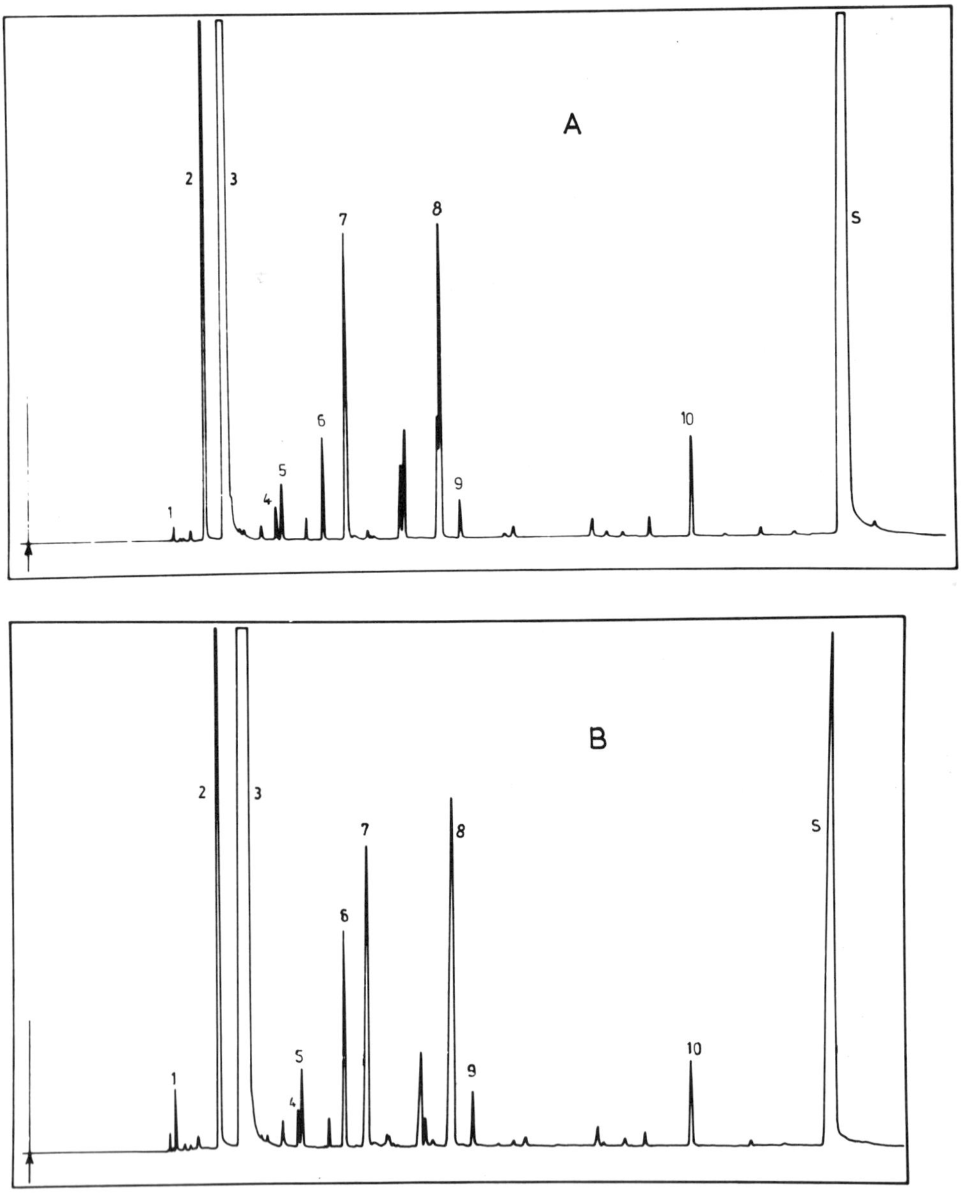

Figure 21. Aromagrams of beers brewed with Saaz hops (A) and isomerized hop extracts (B). Column: 60 m × 0.5 mm ID coated with Carbowax 20M. Column temperature-programmed from 50 to 150°C at 2°C min^{-1}. Carrier gas flowrate: 2.1 ml min^{-1}. Peak numbers as for Fig. 14.

In conclusion, till now no analytical evidence of the presence of the essential oil of hops in lager beer could be obtained. Oxidation products of the bitter acids are present in beer essential oil; they contribute to beer aroma but do not impart a hoppy note. If there is a hoppy aroma in

Table 4. Oxidation Products of Hop Resins Present in Beer

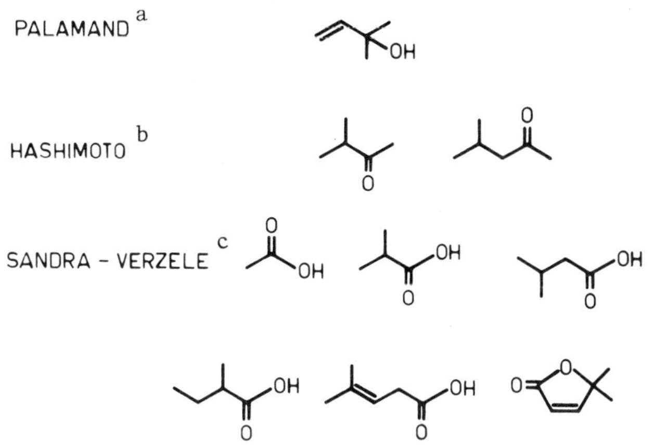

[a]Palamand: see Ref. 54.
[b]Hashimoto: see Ref. 38.
[c]Sandra-Verzele: see Ref. 20.

beer, then it must be due to a mass collaboration effect of hop essential oil components present below the detection limits of current techniques (∿1 ppb). In this respect, present research in our laboratories is directed at the isolation and identification of compounds with a pronounced hop aroma in the oxygenated fraction was found to be a more potent flavoring agent than the hydrocarbon fraction. If such a compound or compounds can be isolated, their presence in beer essential oil can be analyzed by mass fragmentography. Preliminary results indicate that such compounds are present in hop oil at the ppb level with very low threshold values [9].

4.3 Estimation of Fatty Acids in Beer

The lower fatty acids (C_2 to C_6) of lager beer were investigated with $(GC)^2$ by Sandra and Verzele [20]. Beer essential oil obtained by continuous liquid-liquid extraction is washed out with a saturated sodium bicarbonate solution. The water layer is acidified to pH 1 and extracted with ether. An aliquot is directly injected on a glass capillary column coated with FFAP, which is suitable for the analysis of nonderivatized fatty acids (Fig. 23). The higher fatty acids (C_6 to C_{10}) can be quantitatively estimated by the carbon disulfide extraction procedure as previously described (Fig. 16). The $(GC)^2$ analysis of long-chain fatty acids (C_{12} to $C_{18:3}$) was de-

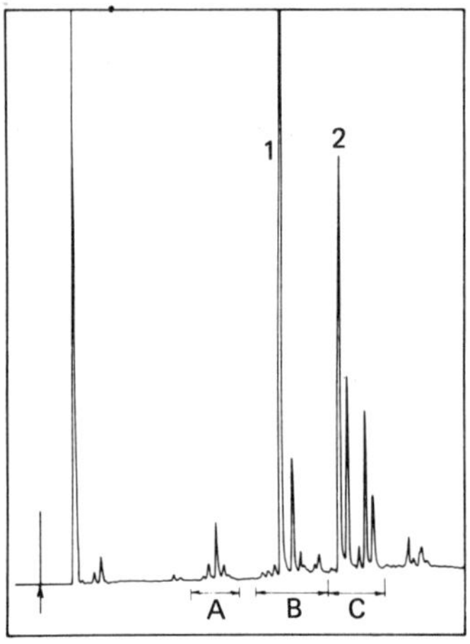

Figure 22. Oxidation profile of humulene. Column: 40 m × 0.5 mm ID coated with FFAP, temperature-programmed from 130 to 180°C at 2°C min^{-1}. Carrier gas flow: 2.8 ml min^{-1}. Peak regions: A—$C_{15}H_{24}$, B—$C_{15}H_{24}O$, C—$C_{15}H_{24}O_2$. Peaks: 1—humuladienone, 2—humulenediepoxide.

scribed by Sandra and Verzele [22]. The underivatized long-chain fatty acids were analyzed on a short (10-m × 0.5-mm) capillary column coated with FFAP. The analysis of the fatty acid methylester derivatives was reported by Spaepen et al. [57].

4.4 Amino Acids

The production of a consistent beer requires careful control of the amino acid composition. All of the protein amino acids are present as well as some other related acidic substances [58].

For quality control there is a need for a simple and rapid method of amino acid analysis that does not require expensive or delicate equipment. The analysis of the protein amino acids of a Belgian lager beer is shown in Fig. 24.

The amino acids of beer are purified by ion-exchange chromatography on a 5-cm × 1-cm column of Amberlite CG-120 (100-120 mesh) according to the procedure of Otter and Taylor [58]. For $(GC)^2$ analysis the amino acids are transformed into the N(O)-heptafluorobutyryl butylester derivatives. An optimized procedure for the preparation of these derivatives has been described by Poole and Verzele [25].

4.5 $(GC)^2$ Analysis of Bitter Acids

The α acids (I in Fig. 25) are the important bitter acids. These substances do not occur in beer, but by boiling with wort they are isomerized into the

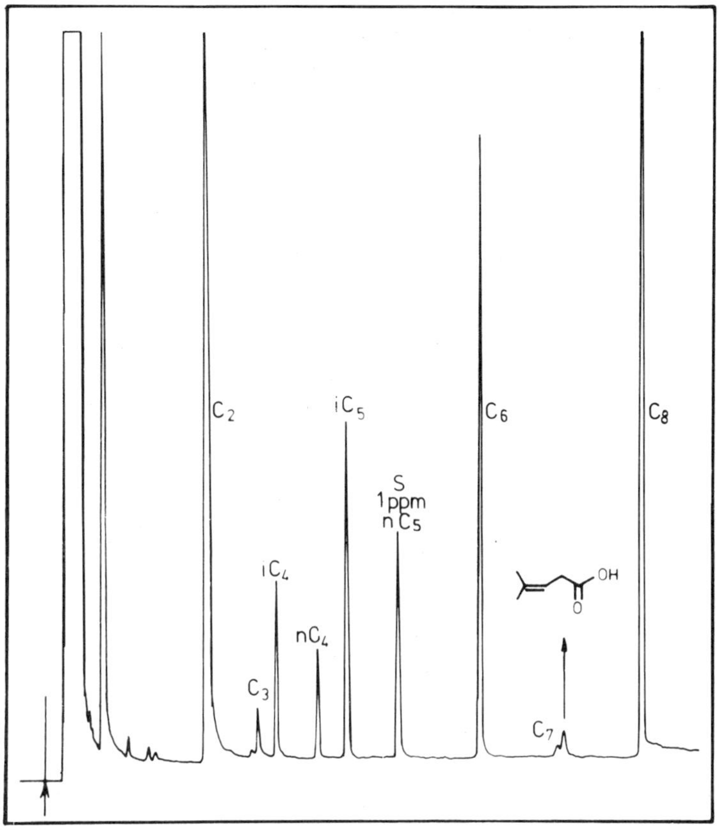

Figure 23. Analysis of lower fatty acids in beer. Column: 20 m × 0.5 mm coated with FFAP. Temperature, detector, and injector 250°C. Column temperature-programmed from 80 to 160°C at 2°C min^{-1}. Carrier gas: hydrogen, flowrate 3 ml min^{-1}. Peaks: 1—acetic acid, 2—propionic acid, 3—isobutyric acid, 4—n-butyric acid, 5—2-methylbutyric acid + 3-methylbutyric acid, 6—n-valeric acid, 7—hexanoic acid, 8—heptanoic acid, 9—isohexenoic acid, 10—octanoic acid. (Source: Ref. 20.)

beer-soluble cis- and trans-iso-α acids (IIIa and IIIb). Since there are three major α acids (humulone, cohumulone, and adhumulone), there are six major iso-α acids representing about 80% of the beer bitter substances. The other bitter beer substances are further transformation products of the iso-α acids (IV to VII) and oxidation products of the β acids (II), the hulupones (VIII). The qualitative analysis of these substances in wort, beer, and isomerized hop extracts can be carried out by $(GC)^2$. A quantitative analysis, however, is hindered by incomplete derivatization or by thermolability of the derivatives. Hop bitter acids contain vinylogous acid, enolic and alcoholic hydroxyl groups which must be derivatized to produce thermostable derivatives with sufficient GC volatility. Several derivatization methods, e.g., diazomethane, methoxylamine, trifluoroacetic

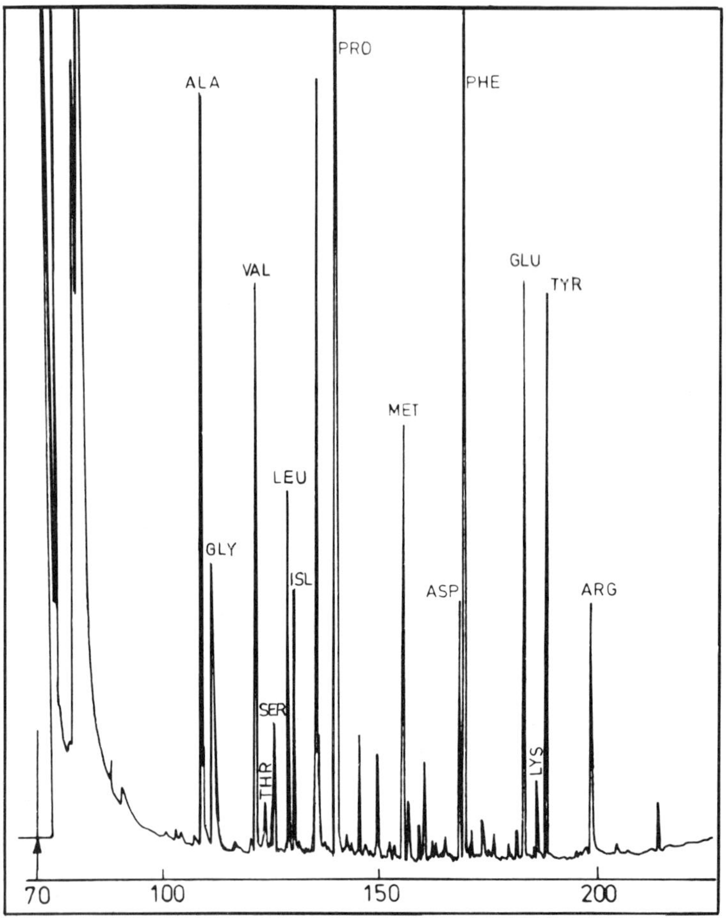

Figure 24. The amino acid composition of Belgian lager beer N(O)heptafluorobutyryl n-butyl ester derivatives. Column: 30 m × 0.25 mm ID glass capillary column coated with SE-30. Detection: FID at 250°C. Injection: all-glass inlet splitter at 225°C. Temperature column: programmed from 70 to 220°C at 4°C min^{-1}. Carrier gas: hydrogen, flowrate 3.3 ml min^{-1}. (Source: Ref. 25.)

acid anhydride, butyl- and phenylboronic acid, were tried, but none of them gave useful results.

Derivatization with different trimethylsilyl donors in a wide range of solvents was studied. The best results were obtained with HMDS in dimethoxypropane (DMP) in a ratio 2:3. The derivatives of some hop bitter acids (e.g., humulone) are unstable and slowly decompose on the column. Complete quantitative analysis of all hop bitter acids and their transformation products is therefore at present not possible with gas chromatography. More-

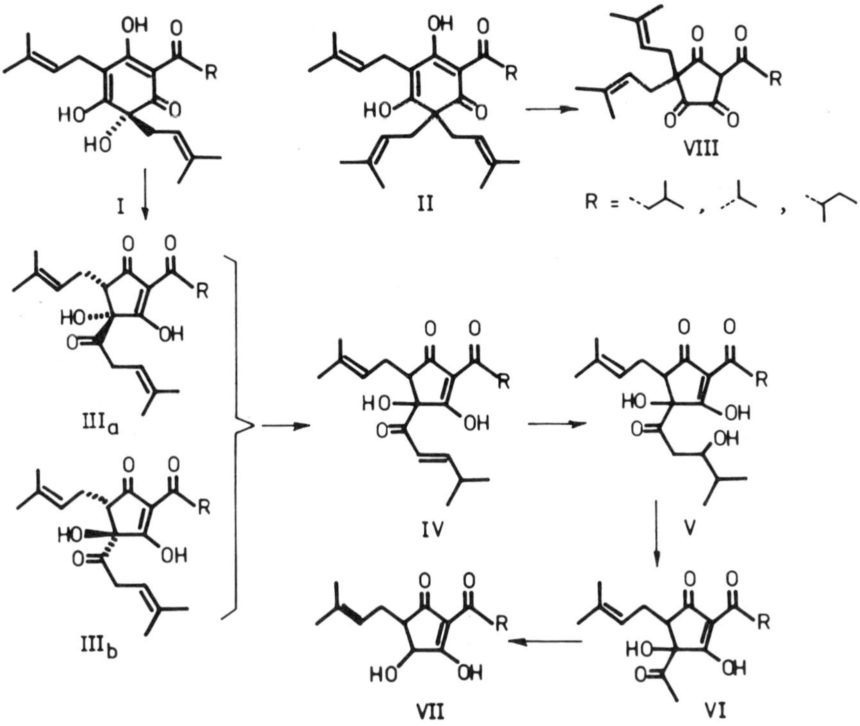

Figure 25. Important hop bitter acids: I_A—humulone, I_B—cohumulone, I_C—adhumulone, II_A—lupulone, II_B—colupulone, II_C—adlupulone, III_a—cis-iso-α-acid, III_b—trans-iso-α-acid, IV—allo-iso-α-acids, V—hydrated allo-iso-α-acids, VI—acetylhumulinic acid, VII—humulinic acid, VIII—hulupones.

over, we believe it is not worthwhile to do more $(GC)^2$ research in this respect.

Nevertheless, the $(GC)^2$ procedure allows the study of some problems about the bitter acids in beer and isomerized hop extracts. Some examples illustrate the possibilities. Figure 26 compares chromatograms of hopped and unhopped beer. The unhopped beer is prepared with yeast which has never been in contact with bitter acids. The peaks in this chromatogram come from the beer matrix. In the hopped beer, the major compounds are trans-iso-cohumulone (1), cis-iso-cohumulone (2), trans-iso-adhumulone (3), trans-iso-humulone (4), cis-iso-adhumulone (5), and cis-isohumulone (6). Minor peaks are cohulupone (7), hulupone (8), and humulinic acids (9 and 10). Peaks with an arrow are probably oxidation products of the bitter acids.

The fingerprint of an isomerized hop extract is presented in Fig. 27. The similarity with the beer bitter acids is obvious. For research purposes

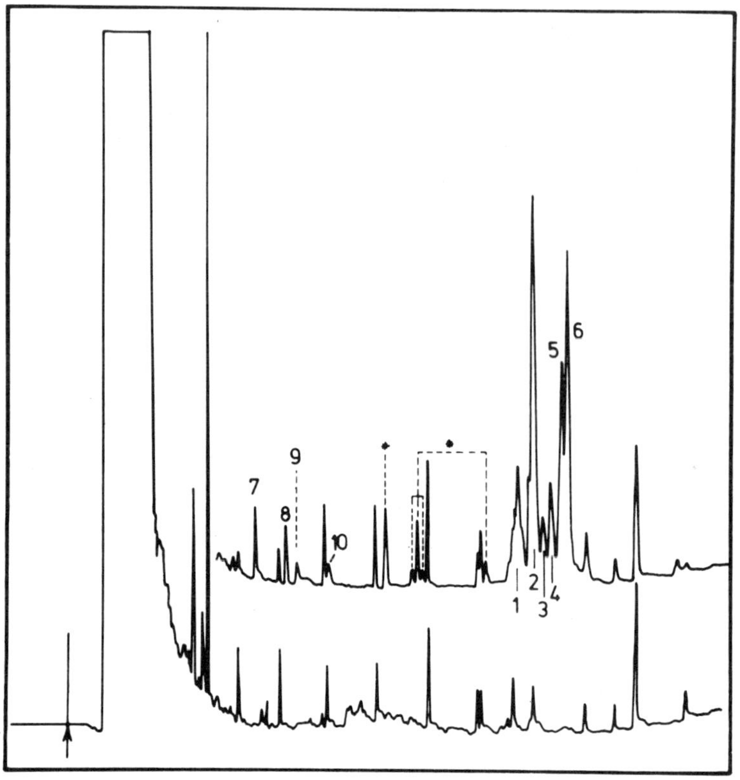

Figure 26. $(GC)^2$ analysis of hop bitter acids. Lower chromatogram: unhopped beer. Upper chromatogram: hopped beer. Column: 45 m × 0.5 mm ID coated with SE-30. Temperature: column programmed from 150 to 225°C at 2°C min^{-1}, detector and injector 250°C. Injection technique: on-column via a gold capillary interphase. Carrier gas: hydrogen, flowrate 3.5 ml min^{-1}. Peak identification: see text. Sample preparation: 250 ml beer is acidified with hydrochloric acid (pH 2) and extracted with 1,1-iso-octane. 750-ml iso-octane extract is evaporated to 10 ml. The residue is transferred to a pear-shaped flask and evaporated in vacuum. The residue is treated with 150 μl dimethyl phosphate and 100 μl HMDS. The mixture is heated at 50°C for 3 hr and 0.5 to 1 μl is injected.

another application of the technique is the purity and identity check for individual bitter acids obtained by countercurrent distribution.

5. CONCLUSION

Capillary GC has been applied to almost all aspects of brewing science. The results summarized in this chapter come practically exclusively from fundamental research. It can be expected that these will become the basis for more and more applied and routine analyses. Capillary GC and more specifically $(GC)^2$ therefore deserve a prominent place in the quality control

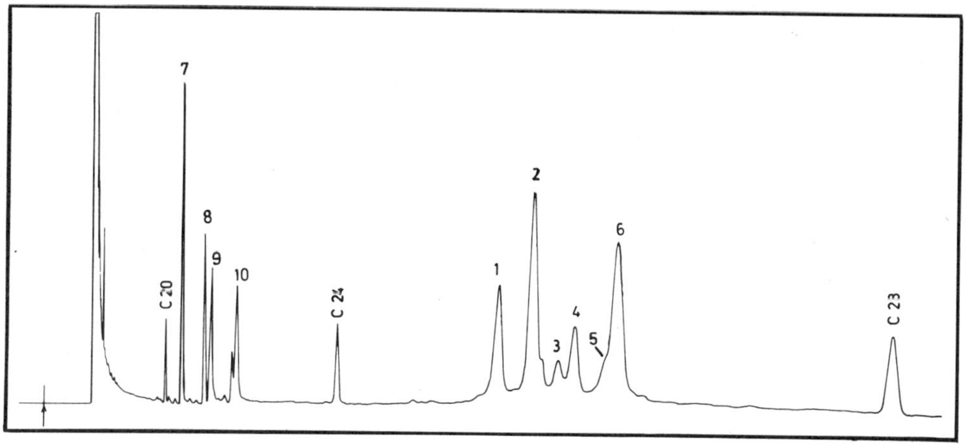

Figure 27. Chromatogram of purified isomerized hop extract. Column: 60 m × 0.8 mm ID coated with OV-1. Temperature: column 230°C isothermally, detector and injector 250°C. Carrier gas: hydrogen, flowrate 10 ml min^{-1}. Peak identification and injection technique as in Fig. 26. Sample preparation: 150-mg extract is weighed in a 30-ml test tube. Methanol (5 ml) and isooctane (10 ml) are added. Hydrochloric acid 10 ml 0.2 N is added stepwise with intermittent shaking. One ml of the isooctane layer is pipetted in a derivatization tube. The solvent is removed by blowing nitrogen over it and derivatization is carried out as for the beer sample. (Source: Ref. 23.)

laboratories of the brewing industry. The most important analyses in this respect seem to be those on the headspace of beer, estimating the off-flavors. The diketone buttery flavor is already routinely analyzed by $(GC)^2$ and the tracing like in Fig. 11 will certainly become important. For the near future analysis of the cardboard flavor (trans-2-nonenal) by $(GC)^2$ may be expected. Since these headspace analyses are most easily carried out on wide-bore capillary column, this again emphasizes the importance of this column type, as mentioned in the chapter.

ACKNOWLEDGMENTS

We thank the Ministerie voor Wetenschapsbeleid and the Nationaal Fonds voor Wetenschappelijk Onderzoek for financial help to the laboratory. We thank Heineken Breweries—Zoeterwoude for sponsoring our hop research.

REFERENCES

1. Verzele, M., Verstappe, M., Sandra, P., Vanluchene, E., and Vuye, A., *J. Chromatogr. Sci.*, 10:688 (1972).

2. Sandra, P., Verzele, M., and Vanluchene, E., *Chromatographia*, 8:499 (1975).

3. Bouche, J., and Verzele, M., *J. Gas Chromatogr.*, 6:501 (1968).

4. Verzele, M., and Sandra, P., *J. Chromatogr.*, 158:111 (1978).

5. Verzele, M., *HRC&CC*, 1:288 (1978).

6. Sandra, P., Verzele, M., Verstappe, M., and Verzele, J., *HRC&CC*, 2:288 (1979).

7. Sandra, P., Redant, G., Saeed, T., and Verzele, M., *HRC&CC*, submitted for publication.

8. Vangaever, F., Sandra, P., and Verzele, M., *Chromatographia*, 12:153 (1979).

9. Redant, G., Ph.D. Thesis, University of Ghent. 1979.

10. Pickett, J. A., Coates, J., and Sharpe, F. R., Proc. Eur. Brew. Conv., Nice, 1975, p. 123.

11. Howard, G. A., *J. Inst. Brew.*, 62:158 (1956).

12. Buttery, R. G., McFadden, W. H., Teranishi, D. R., Kealy, M. P., and Mon, T. R., *Nature*, 200:435 (1963).

13. Buttery, R. G., Black, D. R., and Kealy, M. P., *J. Chromatogr.*, 18:399 (1965).

14. Naya, Y., and Kotake, M., *Bull. Chem. Soc. Jap.*, 45:2887 (1972).

15. Scheffer, J. J. C., Koedam, A., and Baerheim Svendsen, A., *Chromatographia*, 9:425 (1976).

16. Saeed, T., Sandra, P. J., and Verzele, M. J. E., *Phytochemistry*, 17:1433 (1978).

17. Buttery, R. G., Black, D. R., Lewis, M. J., and Ling, L., *J. Food Sci.*, 32:414 (1967).

18. Jennings, W. G., Yabumoto, K., and Wohleb, R. H., *J. Chromatogr. Sci.*, 12:344 (1974).

19. Slotema, F. P., *Brewers Digest*, 8:48 (1978).

20. Sandra, P., and Verzele, M., Proc. Eur. Brew. Conv., Nice, 1975, p. 107.

21. Sandra, P., *Cerevisiae*, 1:37 (1976).

22. Sandra, P., and Verzele, M., *J. Inst. Brew.*, 81:302 (1975).

23. Verzele, M., Vanluchene, E., and Van Dyck, J., *Anal. Chem.*, 45:1549 (1973).

24. Vanluchene, E., and Verzele, M., *Chromatographia*, 6:519 (1973).

25. Poole, C. F., and Verzele, M., *J. Chromatogr.*, 150:439 (1978).

26. Grob, Jr., K., Neukom, H. P., and Kaderli, H., *HRC&CC*, 1:98 (1978).

27. Jennings, W. G., Wohleb, R. H., and Lewis, M. J., *J. Food Sci.*, 37:69 (1972).

28. Wohleb, R. H., Jennings, W. G., and Lewis, M. J., Proc. Amer. Soc. Brew. Chem. Proc., 1972, p. 1.

29. Gunst, F., Ph.D. Thesis, University of Ghent, 1977.

30. Kuroiwa, Y., and Hashimoto, N., Proc. Amer. Soc. Brew. Chem., 1961, p. 28; 1963, p. 181.

31. Gunst, F., and Verzele, M., *J. Inst. Brew.*, 84:291 (1978).

32. Powell, A. D. G., and Brown, I. H., Proc. Inst. Brew., Austral. Sect., 1966, p. 257.

33. Trachman, H., and Saletan, L. T., *Wallerstein Lab. Commun.*, 32:19 (1969).

34. Engan, S., *Brygmesteren*, 28:191 (1971).

35. Meilgaard, M. C., Elizondo, A., and MacKinney, A., *Wallerstein Lab. Commun.*, 34:95 (1971).

36. Zenz, H., Klaushofer, H., and Szep, H., *Mitt. Versuchssta. Gaerungsgewerte Wien*, 27:100 (1973).

37. Stenroos, L. E., Siebert, K. J., and Meilgaard, M. G., *J. Amer. Soc. Brew. Chem.*, 34(1):4 (1976).

38. Hashimoto, N., *Rep. Res. Lab. Kirin Brew. Co.*, 13:1 (1970).

39. Krüger, E., and Neuman, L., *Monatsschr. Brau.*, 23:117,269 (1970).

40. Silbereisen, K., Krüger, E., and Baron, G., *Monatsschr. Brau*, 23:93 (1970).

41. Neuman, L., Ph.D. Thesis, University of Berlin, 1973.

42. Maule, D. R., *J. Inst. Brew.*, 73:351 (1967).

43. Sandra, P., Ph.D. Thesis, University of Ghent, 1975.

44. Coppens, B., Sandra, P., and Verzele, M., unpublished results.

45. Sandra, P., Gunst, F., and Verzele, M., unpublished results.

46. Rabak, F., *Wallerstein Lab. Commun.*, 7:173 (1944).

47. Harris, J. O., *J. Inst. Brew.*, 82:27 (1976).

48. Hudson, J. R., *J. Inst. Brew.*, 71:482 (1965).

49. Laws, D. R. J., Peppard, T. L., Sharpe, F. R., and Pickett, J. A., *J. Amer. Soc. Brew. Chem.*, 36(2):69 (1978).

50. Buttery, R. G., and Ling, L. C., *Brewers Digest*, 41(8):71 (1966).

51. Shimazu, T., Hashimoto, N., and Kuroiwa, Y., Proc. Amer. Soc. Brew. Chem., 33(1):7, 1975.

52. Biot, J. M., and Verzele, M., *Bull. Soc. Chim. Belges*, 86:41 (1977).

53. Palamand, S. R., Hardwick, W. A., and Cole, P. W., Proc. Amer. Soc. Brew. Chem., 1969, p. 78.

54. Palamand, S. R., Markl, K. S., and Hardwick, W. A., Proc. Amer. Soc. Brew. Chem., 1971, p. 211.

55. Kuroiwa, J., Kukubo, E., and Hashimoto, N., Proc. 4th Int. Fermentation Symp., March 1972, Kyoto, Japan.

56. Sandra, P., Coppens, B., and Verzele, M., unpublished results.

57. Spaepen, M., Van Oevelen, D., and Verachtert, H., *J. Inst. Brew.*, 84:278 (1978).

58. Otter, G. E., and Taylor, L., *J. Inst. Brew.*, 82:264 (1976).

15

ANALYSIS OF GRAPES, WINES, AND BRANDIES

A. RAPP

Federal Research Institute of Viticulture
Siebeldingen, Federal Republic of Germany

1. Introduction 579

2. Enrichment of Aroma Substances 581

3. Gas Chromatographic Separation of the Aroma Concentrates 589

4. Applications 590
   - 4.1 Grape Berries 590
   - 4.2 Wine 603
   - 4.3 Wine Distillates 609

5. Summary 613

References 615

1. INTRODUCTION

Aroma substances are decisively important to foods, and aroma compounds make a major contribution to the selection, nature, and quality of our foods and beverages. As is the case with most foods, the characteristic aroma of grapes and wines is usually formed by the combined contributions of several hundred chemically quite different compounds. The concentrations of the individual components can range from 10^{-1} to 10^{-10} g kg^{-1} [1-33]. Since our sense organs are extremely sensitive to many aromatic substances, minute concentrations of those with low threshold values (i.e., 10^{-4} to 10^{-12} gl^{-1}) can be of greater sensory importance than those present in relatively large amounts. Together with the fact that these complex mixtures are composed of many hundreds of individual components, this complicates investigations of grape and wine aromas, especially with respect to an objective evaluation. Thus, successful quality evaluations and varietal characterizations require both the detection of as many aroma substances as possible and complete and reproducible separations, together with quantitative determination.

Initial investigations on volatile aroma substances of wine were conducted by Hennig and Villforth [39]. They extracted larger quantities of wine and identified therefrom a few compounds using the classical methods of detection. Advances in chemical methodology have permitted further separation and detection of the individual components associated with aroma substances.

In the late 1950s, Bayer et al. [40,41] used gas chromatography for the detection of aroma substances in wine. Besides the higher alcohols (2-methyl-propanol-1 and fermentation amyl alcohol), they also detected a few esters and found differences in the content of these components in the vine varieties investigated (Riesling, Morio-Muskat, Portugieser). Later on, many authors dealt with the aroma substances in grape berries, wine, and wine distillates; progress was limited because only packed columns were available, with their relatively limited separation potential. On the one hand, a few volatile fermentation products were determined quantitatively using direct injections of wine or wine distillates [3,4,12,13,15,18,21,23, 24,42-53]. It was also found that the formation of these substances was influenced by the composition of the fermentation substrate [9,48,54-59]. In some cases, clues to their biosynthesis were reported [21,57,60-62]. Much attention was directed toward attempts to separate aroma concentrates of grape berries, wine, and wine distillates into their individual components [1,3,9,17,21,30,32,33,44,49,51,63-72]. Because of the limited separations achieved by these packed columns, only a small number of peaks were obtained in most cases. The initial capillary column separations (steel capillary columns; 50-ft Apiezon-L, 50-ft Ucon LB 550X, 500-ft GESF-96) resulted in considerably more peaks, (40 to 60 components) [19,73-80], but a satisfactory detection of the aroma components of grape berries and wine was not yet possible. The more recent introduction of high-resolution capillary columns, combined with improved and more selective techniques of enrichment, finally supplied the necessary conditions for successful investigations on the aroma composition of grape berries, wine, and beverages made from wine.

2. ENRICHMENT OF AROMA SUBSTANCES

In order to analyze even those aroma components present in minute concentrations, enrichment of the constituents is necessary. Several physical methods are available for this purpose (e.g., salting out, extracting, distilling, freezing out). Each procedure causes some changes, and only an approximate quantitative analysis of the aroma substances can be made. In addition, the original relative proportions of the aroma substances can be changed by these analyses. Further, the enrichment of aroma substances has to be conducted as carefully as possible in order to avoid artifact development; otherwise an assessment of the authentic aroma composition is impossible.

For the enrichment of aroma substances from plant products (fruits, vegetables, etc.) several procedures that require decomposition of the cell structure have been suggested. These lead to the formation of considerable amounts of aldehydes and alcohols--mainly hexanal, trans-2-hexen-al-1, and cis-3-hexenol-1--probably due to enzymic oxidations in the presence of oxygen [81-84]. In addition, enzymatic hydrolytic processes can also lead to the degradation of some fruit esters. Because a comparison of the original aromagrams of the grape berries of different vine cultivars is already sufficiently complicated, it is much more desirable to investigate the primary or original aroma substances, without the complications offered by these artifacts.

To inhibit the fruit enzymes, numerous compounds are available (heavy metal salts, cyanides, sulfites, alcohols, etc.), and the individual enzyme response is influenced differently by these various substances. For instance, the formation of hexanal is not inhibited by HCN, whereas ascorbic acid is only a moderate inhibiting factor. Trichloroacetic acid and H_2SO_3, on the other hand, prevent the formation of C_6 compounds to a large extent. According to our own investigations, methanol also proved to be an effective inhibitor [83]. As can be seen from Table 1, the aroma composition is changed considerably when preparing the grape berries from a system with active enzymes. If enzymes are inhibited even 10 min after the decomposition of the cell structure starts, the hexanol content increases 77-fold compared with immediate inactivation; cis-3-hexenal-1 increases 102-fold [83]. Hence these samples must be processed under conditions where as few changes as possible occur in the original aroma constituents, and consequently the analytically identifiable differences represent the different varietal character.

The problem of analyzing aroma substances from wine and other alcoholic beverages is to enrich these trace components from solutions that contain large quantities of water and ethanol. When selecting the method, it should result in ethanol-free aroma concentrates, if possible. Samples prepared by freezing out, salting out, or distilling also achieve ethanol enrichment and are consequently hardly suited for the enrichment of trace components from these products.

Enrichment of volatile aroma substances from grape berries, grape juices, wine, and beverages made from wine is mainly conducted by liquid-liquid extraction [9,20,21,28,32,33,49,56,71,77,78,80,83,85-89] or by gas extraction [21,90-95].

The most satisfactory enrichment, avoiding solvent dilution, should be possible by gas extraction, the aroma substances being condensed in traps or adsorbed on solids (activated charcoal, Tenax phases, etc.). Simple freezing out of the gas flow (N_2 [90,91,94], propane/CO_2 [21]), however, usually results in cocondensing water and, in the case of alcoholic beverages, ethanol. Several authors adsorb the aroma substances from the sweeping gas by polymers (Tenax [96-98], Chromosorb [98], Porapak Q [96,99]) or by activated charcoal [100], which leads to the subsequent problem of desorption. Depending on temperature and time of desorption, significant quantitative differences in the aroma composition of the same initial samples can arise [101]. Owing to the catalytic character of some adsorptive agents, artifact formation has also been observed during the desorption of the aroma substances adsorbed by active charcoal [102]. Jennings et al. [99] initially eluted water and ethanol, coadsorbed on Porapak Q during the gas extraction of aroma substances from alcoholic beverages, with N_2 from the carrier phase.

Subsequently, the aroma substances were passed into the glass capillary (GC) column by heating the Porapak column [99]. Williams et al. [103] first enriched the aroma substances on Chromosorb 105. In a second trap level, these were passed via a short column with 10% SF-96 on Chromosorb A (20-30 mesh) and then to the GC column by reverse-flushing. The relative standard deviation for the individual components ranged between 2% (2-phenylethanol) and 10% (ethyl-hexanoate). We have developed a new technique [95] by which a moderate and sufficient concentration of the trace components from the steam phase is achieved above aqueous and aqueous-alcoholic solutions. The method produces aroma components that are water- and ethanol-free; enrichment is accomplished by introducing the gas flow (N_2) into Freon-11.

Table 1. Formation of C_6 Compounds When Processing Grape Berries cv. Silvaner—Inhibition of Enzymes by 65% Methanol

C_6 Compound	Inhibition of Enzymes				
	Immediately—peak high (mm)	After 10 min—peak high (mm)	Ratio 10 min: immed.	After 120 min—peak high	Ratio 120 min: immed.
Hexanal-1	194	14,938	77	8,750	45
Hexanol-1	100	2,563	27	11,350	114
Trans-2-Hexenal-1	3,163	15,750	5	20,250	6
Trans-2-Hexenol-1	143	4,500	31	15,500	108
Cis-3-Hexenal-1	55	5,625	102	1,500	27
Cis-3-Hexenol-1	14	191	31	320	21

Source: Ref. 83.

As a basis for quantitative investigations, liquid-liquid extraction, e.g., with pentane/methylene chloride [9,21,28,49,86,98], pentane [29,67, 87,104-106], isopentane [77,78], and methylene chloride [65,69,71,107], have been used.

Freon 13 [108], ether/pentane [20], and carbon disulfide [15], are well suited for the reproducible enrichment of aroma substances. By selecting an appropriate solvent with a high degree of purity and a low boiling point, a moderate, artifact-free extract of the aroma substances can be achieved. When extracting aroma substances from alcoholic samples, it is imperative that the method does not also achieve enrichment of the ethanol. A solvent largely satisfying these criteria (i.e., degree of purity, low boiling point, minute solubility of ethanol), and which has been used with advantage for the enrichment of aroma substances from aqueous and aqueous alcoholic solutions, is trichlorofluormethane (Freon II) [56,80,83,104,105,109-112]. The extractability of some alcohols (C_4 to C_{10}) by Freon II from aqueous alcohol is shown in Table 2 [111]. After 4 hr, the limiting value is reached for the extraction of alcohols with a chain length of greater than six carbon atoms (99%). For alcohols with less than six carbon atoms, longer extraction times are necessary for efficient recovery, i.e., 10 hr for pentane-1-ol. Esters with increasing carbon chain length require less time for efficient extraction: for the ethylesters of caproic acid, 99% of the limiting values are attained after 4 hr; for the ethylesters of caprylic acid, 84% after 4 hr and 96% after 20 hr [111]. Enrichment of the aroma substances with Freon II is conducted by the equipment shown in Fig. 1 [83]. To make a comparison, the extractability of alcohols by pentane/dichloromethane (2:1) is also shown in Table 2. This also exhibits an increase in yield with increasing chain length. Compared with Freon II, a longer period of extraction is necessary: 99% of hexanol-1 is extracted after 4 hr when using Freon-11; this result is achieved only after 15 hr by using pentane/dichloromethane. Cobb et al. [113] compared the extractability of some alcohols and esters in sugar-free and sugar-containing aqueous ethanolic solutions by various solvents (Freon-11, dichloromethane, ether, isopentane). Using the method of shaking, Freon-11 also gave the highest extraction yield of the compounds investigated. The smallest yield was obtained with isopentane. Our investigations on the enrichment of aroma substances from homogenates of grape berries, grape juices, and wine demonstrated that an extraction time of 20 hr was sufficient and that repetitive analyses were quite reproducible (Fig. 2). As can also be seen

Table 2. Extractability of Alcohols by Freon-II and Pentane/Dichloromethane from Aqueous Alcoholic Solutions (% of Available Final Value)

	Time (hr)						
Alcohol	2	4	6	10	15	20	25
Freon-II							
C_4	44	65	76	85	91	94	97
C_5	80	96	98	99			
C_6	93	99					
C_7	94	99					
C_8	95	99					
C_9	96	99					
C_{10}	96	99					
Pentane/Dichlormethan (2:1)							
C_4	55	71	83	95	97	98	98
C_5	61	80	93	98	98	98	99
C_6	61	80	93	98	99		
C_7	65	81	93	98	99		
C_8	68	83	95	99			
C_9	68	83	95	99			
C_{10}	68	83	95	99			

Source: Ref. 83.

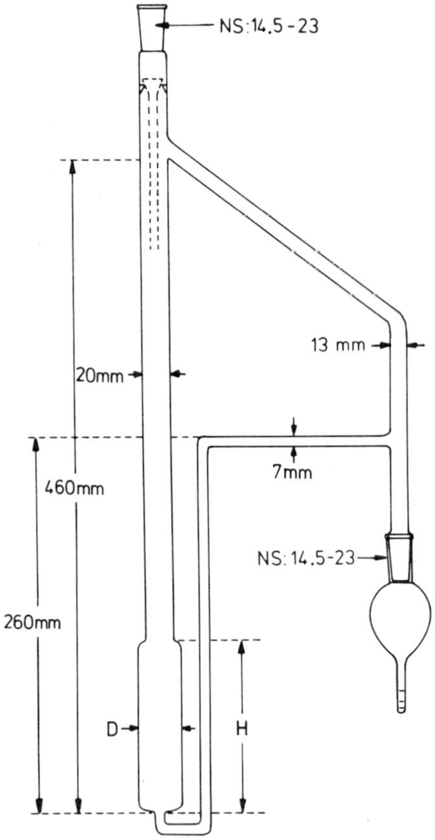

Figure 1. Extraction equipment (without cooler) for enriching the volatile aroma compounds from must and wine with Freon-11: D—32 mm, H = 120 mm (equipment for 100 ml), D—70 mm, H—260 mm (equipment for 1000 ml). (Source: Ref. 83.)

from Fig. 4, the size of the drops of the solvent (Freon-11) during the extraction period apparently does not influence the extraction yield.

Salting out is an easily applied method of enrichment, especially in routine investigations of alcoholic beverages. In this method, phase separation is used in the ternary system ethanol-water-salt. The method of salting out with NaH_2PO_4 and $(NH_4)_2SO_4$, developed by Bertrand [114], has been used by Woidrich et al. [115] for the quantitative determination of higher esters and 2-phenyl ethanol in alcoholic beverages. Our own investigations showed that this method, when used for the enrichment of trace components, is inferior to the continuous extraction with Freon-11. Although

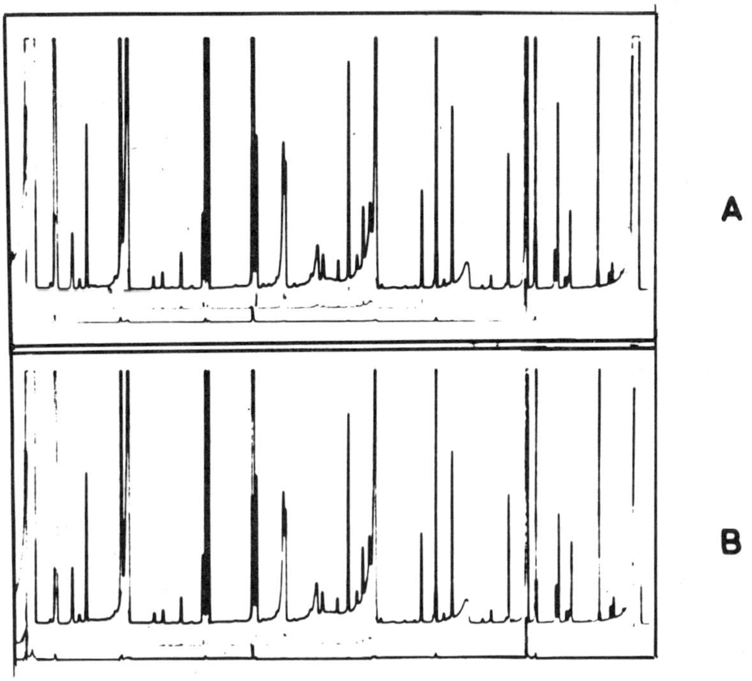

Figure 2. Extractability of volatiles from wines by Freon-11: duration of extraction 20 h; A—drip diameter ∿1 mm; B—drop diameter ∿5 mm.

the Freon technique is moderate and avoids development of artifacts with added components (salts), this does not apply to the ethanolic phase. In this case, formation of acetal and thus occurrence of nonoriginal aroma components is possible.

In Fig. 3, the aroma composition of orange peels is illustrated during three different procedures of enrichment (distillation-extraction, liquid-liquid extraction [Freon-11] [83], headspace extraction [95]). It reveals that there is a distinct difference in the distribution of the individual aroma components, depending on the method applied. For example, the content of the main component limonene (L) is significantly increased during the gas extraction (H) compared with that of the other components. In case of distillative enrichment (D), the content of components with longer retention times is increased. There is little difference between the various techniques of enrichment for some components.

Jennings et al. [96], Wyllie et al. [101], and Beernaert et al. [116] demonstrated the influence of various extraction methods of aroma concentrates on the composition of the aroma patterns. These results indicate quite clearly that the method of enrichment to be used should be examined

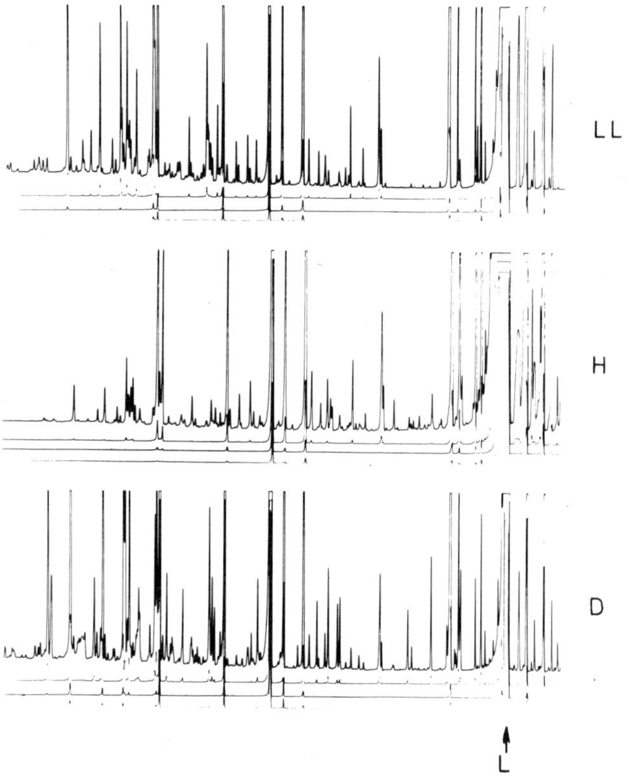

Figure 3. Volatiles of orange peel. Methods for extraction: LL—liquid-liquid-extraction (with Freon-11) (100), H—headspace (new technique) (107), D—distillation, L—limonene.

in each case with respect to the product and the planned aroma analysis. Aromagram comparisons, e.g., for varietal differentiation of grapevines, would require the preparation of aroma concentrates produced by the same method and under the same conditions.

A single internal standard to be used in quantitation may well be changed or even lost during an extraction. Therefore, the use of several standards, added in precisely defined amounts and ratios during the preparation of the samples, is preferable. This permits evaluation of any errors caused by enrichment, concentration, or gas chromatographic analysis. Errors in individual components detected by repetitive analyses has been found to range from 5 to 15%. More exact determinations of individual components is possible only when the investigator determines the extraction yield and the response factor of all of the individual components.

3. GAS CHROMATOGRAPHIC SEPARATION OF THE AROMA CONCENTRATES

As already mentioned, the enrichment of these aroma substances leads to aroma concentrates that are composed of hundreds of individual components and that contain compounds of several different classes of constituents. Separation of such complex mixtures composed of many components of different polarities (alcohols, esters, acids, hydrocarbons, etc.), different boiling points (50 to 350°C) and concentrations that range over several orders of magnitude (10^{-2} to 10^{-10} g $liter^{-1}$) demands columns of very high efficiency. Such separations have been made possible only by using capillary chromatography, particularly through the availability of suitable high-resolution glass capillary columns [83,85,117-122].

In many cases, the aroma concentrates are introduced directly into the capillary columns, without further pretreatment. In other instances [10, 98,123] the acids are first separated from the aroma concentrates with 5% Na_2CO_3 solution.

These then are esterified with diazomethane and analyzed with the neutral components in two separate analyses. Aroma concentrates may also be subjected to preliminary separations into several fractions of differing polarity by column chromatography (silica gel), using different elution agents (pentane, pentane/diethylester, diethylether/pentane, diethylether, methanol); subsequently, each fraction is characterized by gas chromatography on an appropriate column [26,86,98,124,125]. In quantitative investigations, this method possesses distinct disadvantages compared with the direct injection of aroma concentrates, although it is well suited to subsequent identification of the individual components using coupled gas chromatography/mass spectrometry. However, the possibility of artifact development or impurities contributed by the large quantities of the elution agents applied should be considered when column chromatographic separations are utilized. Using the multistage technique [118,126-130], especially the trapping method [127,130-133], preseparations of these many-component mixtures can be conducted much more easily and are very useful for mass spectrometric investigations. Thus, it is possible to recover the desired components in one trap, and even to effect their enrichment by multiple injections, and finally to separate them to an optimal degree from secondary substances by subsequent rechromatography.

Using appropriate columns and taking into consideration optimal conditions (split, flow, temperature), the aroma concentrate can be introduced without any pretreatment. Figure 4 shows a 60-min section (20th min to 80th min) from a gas chromatogram of an aroma concentrate of grape berries of the cultivar (cv.) Müller-Thurgau, using a 90-m Reoplex 400 column (inside diameter 0.4 mm). Under the applied conditions (p = constant; all-glass system; carrier gas H_2: 2.9 ml min^{-1} yielding an optimum flow velocity of 25 to 27 cm sec^{-1}: temperature program: 1.5°C min^{-1}), these aroma concentrates of grape berries and wines are separated into 350 to 400 peaks with a total analysis time of 180 to 200 min. This number of separated components should be sufficient for evaluating variety and quality of berries and wines of different grapevine cultivars.

In Fig. 5, sections of aromagrams (16th to 80th min; Reoplex 400) of aroma concentrates (Freon extracts) of different fruit juices are compared. There are significant differences in the content of some aroma components of the investigated fruit varieties, which can be used for the differentiation of the fruit juices. In further investigations attention will be directed to the question of whether these significant differences determined for several varieties are also present to the same degree between the samples of the different vine cultivars, and whether this will provide an analytical characterization of the varieties.

4. APPLICATIONS

4.1 Grape Berries

Comparison of Aromagrams of Different Grape Varieties

A comparison of the aromagrams of grape juices from different varieties (Fig. 6, aromagram section: 30th to 80th min; column: Reoplex 400) reveals distinct differences in the aroma composition of the individual grape varieties ("fingerprint pattern"). Although there are only quantitative differences in the fingerprint print patterns of the varieties, these differences are so distinct as to provide a basis for characterizing a variety [111,112,121]. The aromagram of the varieties investigated shows distinct differences in the quantity of the components ("key substances"). The calculation of these "key substances"—for Morio-Muskat, the variety-specific components have been marked in Fig. 6—results in relative values characterizing the individual varieties [134,135].

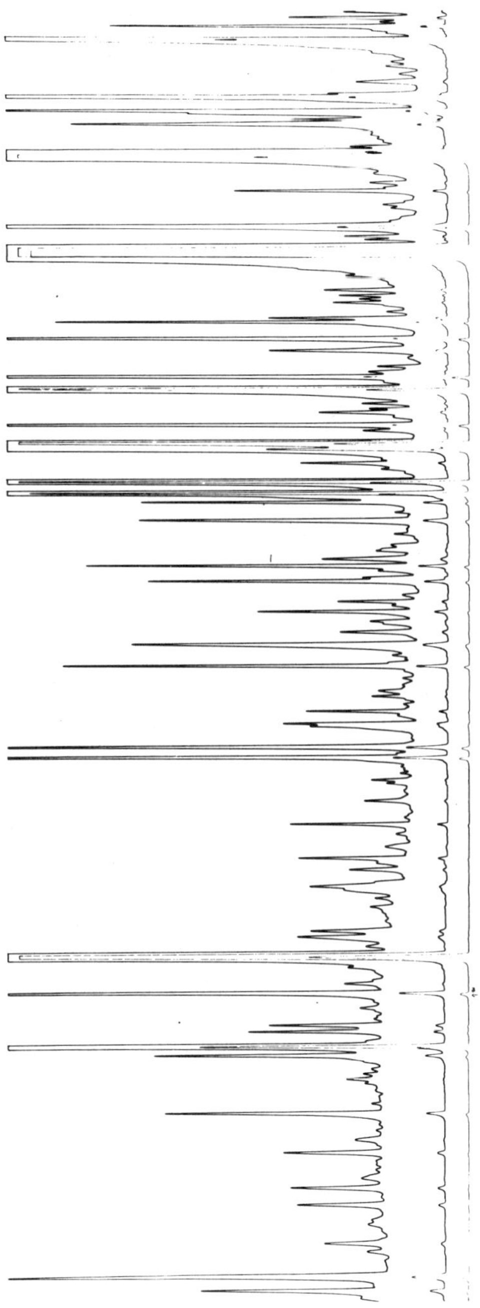

Figure 4. Aromagram of grape berries of the cv. Müller-Thurgau. Separation on a 90-m Reoplex glass capillary column. Section from the 20th to 80th min. Temperature-programmed 50 to 160°C, rate 1.5°C min-1.

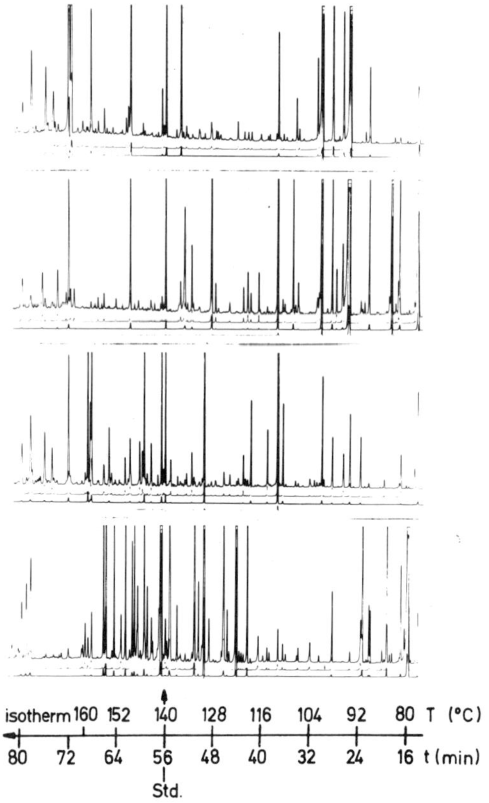

Figure 5. Aromagrams of different fruit juices; 90-m Reoplex glass capillary column. In order from top to bottom: grape juice, apple juice, black currant juice, orange juice.

Influence of Grape Maturity and the Location on the Composition of Aroma Components

As can be seen from the aromagrams shown in Fig. 6, volatile constituents of healthy berries at similar maturities and prepared under the same conditions were compared. To decide whether the variety-specific differences are a function of the stage of maturity, some varieties have been investigated at different phases of maturity. In Fig. 7 aromagram sections of Riesling samples at varying degrees of maturity are compared (90m LAC-3R-728: 12th to 82th min). It is apparent that numerous aroma substances are already present in the grape berries at the first date of vintage (immediately after sugar accumulation has set in) and that some components (key substances) predominate, which are also the main components at later dates of maturity. In spite of a difference in the degree of maturity ranging

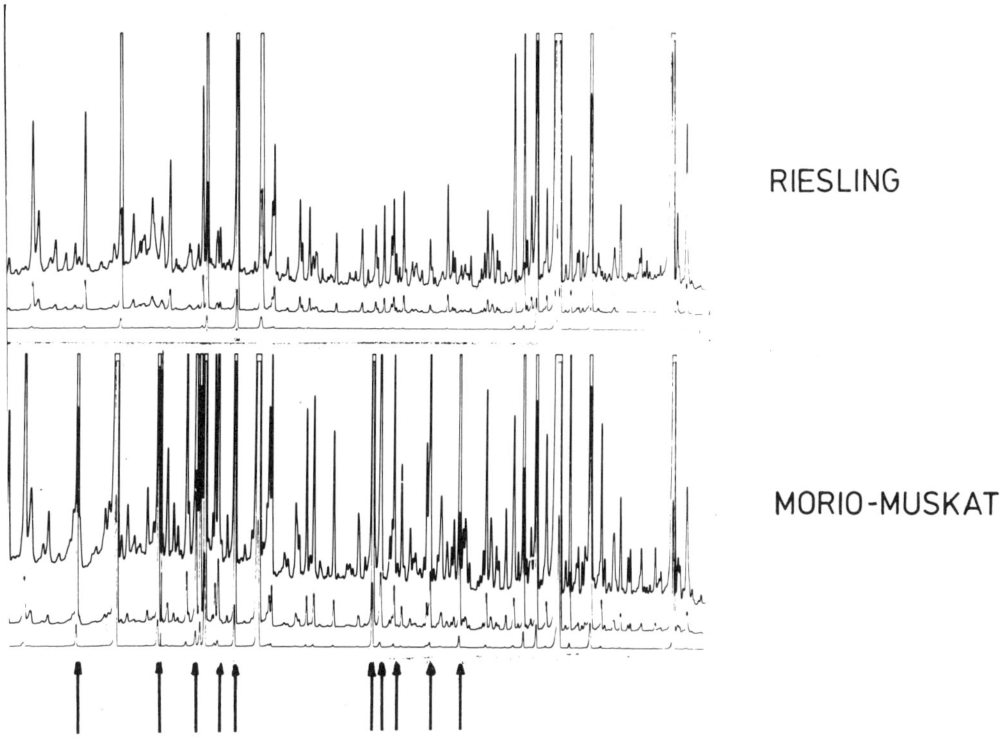

Figure 6. Aromagrams of grape berries of the cultivars Riesling and Morio-Muskat: ⟶ = prominent key substances of Morio-Muskat.

over 60° Oechsle,* the fingerprint pattern of the cv. Riesling is clearly recognizable from all aromagrams. In addition, a comparison of the aromagrams reveals that the content of some aroma components changes markedly with changes in the degree of maturity. Figure 8 shows changes in the content of some components in the course of berry ripening. For example, linalool (retention time 35) increases with the sugar accumulation in the cv. Riesling, whereas the geraniol content (retention time 58) decreases with increasing sugar content. Further investigations in this direction might permit an evaluation of the "aroma maturity" or of the optimum aroma quality of a vine variety.

---

*The approximate sugar content on the Oechsle scale is given by dividing degrees Oechsle by 4 and subtracting 2.5 from the result. Thus for a must of 80° Oechsle: 80/4 - 2.5 = 17.5. The approximate prospective percentage alcohol (by volume) is given by 0.125 × degrees Oechsle.

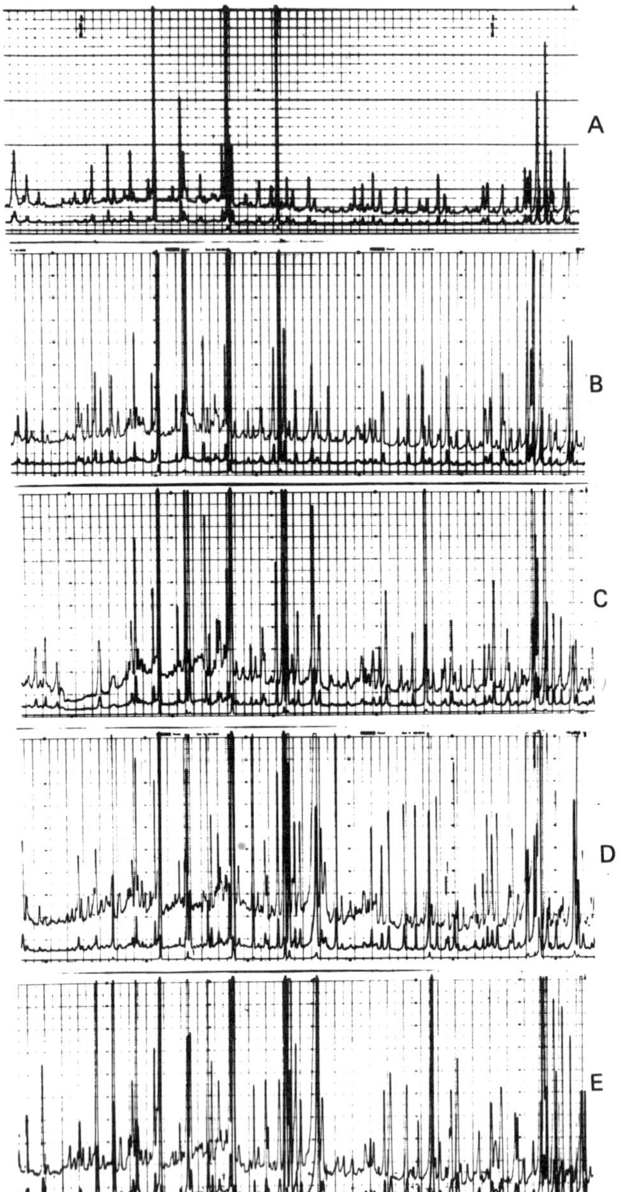

Figure 7. Influence of maturity on the aroma composition of Rieslin grapes; 75-m LAC-3R-728 glass capillary column. All samples from the Geilweilerhof vineyards. Dates and data: A (8/12/76)—32.5° Oechsle (SG ∿1.03181), 33.1% acid; B (8/19/76)—59.6° Oechsle (SG ∿1.05929); C (9/2/76)—67.4° Oechsle; (SG ∿1.06742); D (9/10/76)—74.3° Oechsle (SG ∿1.07430); E (9/27/76)—88° Oechsle (SG ∿1.08823).

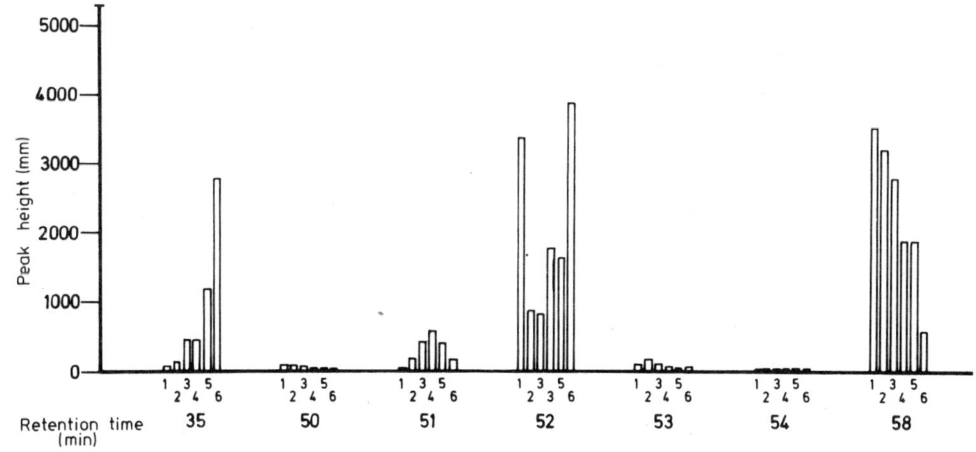

Figure 8. Influence of maturity on the content of some aroma compounds of the cultivar Riesling. Compound with retention time 35—linalool. Dates and data: 1 (8/19/76)—33° Oechsle (SG ∿1.03276), 2 (8/26/76)—60° Oechsle (SG ∿1.05995), 3 (9/2/76)—67° Oechsle (SG ∿1.06708), 4 (9/10/76)—74° Oechsle (SG ∿1.07404), 5 (9/27/76)—88° Oechsle (SG ∿1.08823), 6 (10/19/76)—87° Oechsle (SG ∿1.08733).

This influence of the berry maturity on the aroma composition, as shown for the cv. Riesling (Figs. 7 and 8), is also recognizable for the other varieties investigated which have markedly different fingerprint patterns.

Several authors [1,30,56,87] have reported on the influence of berry maturity on the content of some terpenoid compounds (inter alia linalool, nerol, geraniol, citronellol) in Muscat varieties (Muscat of Alexandria, Muscat Hamburg, Muscat Frontignan, etc.). Hardy [56] determined a continuous increase in the linalool content with increasing sugar accumulation for Muscat of Alexandria. Terrier et al. [30] and Bayonove et al. [90] also detected an increase in the linalool content depending on maturity for several Muscat cultivars, which, however, decreased with increasing "overmaturity." A relation between degree of maturity and ratio linalool/geraniol was established by Wagner et al. [87] for some Muscat crossings (progenies of the crossings Riesling × Muscat Ottonel and Muscat Ottonel × Auxerrois). Our investigations [136] indicate no significant influence of berry maturity on the linalool content or on the ratio linalool/citronellol and geraniol/citronellol for Morio-Muskat (Silvaner × Weisser Burgunder); for Riesling, however, there was a marked dependence on the degree of maturity (Table 3).

To allow successful characterization of a variety, the "fingerprint pattern" of the individual varieties with the specific key substances must

Table 3. Influence of Ripening and Oecological Factors on Varietal Specific Relations

Variety and location	Linalool/citronellol 40 to 100° Oechsle	Geraniol/citronellol 40 to 100° Oechsle
Morio-Muskat (Geilweilerhof)	4.4-5.5	11.0-10.9
Riesling (Geilweilerhof)	3.3-39.2	31.8-38.6
Riesling (Trier)	0.4-33.8	13.9-32.3
Riesling (Geisenheim)	2.0-30.8	13.7-23.7

differ significantly and must also be independent of the location. In Figs. 9 and 10 aromagram sections of samples of the cvs. Riesling and Morio-Muskat from different vine-growing areas of Germany are compared. Although the samples have been taken from different sites and soils, good agreement can be observed [134,136]. A peak pattern can be used for characterizing a variety only if the exogenous influences, e.g., maturity, location, year of vintage, etc., do not exceed certain limits. A comparison of the aromagrams of different varieties illustrates to what extent the patterns of an individual variety may vary and still be clearly recognizable. The fingerprint pattern of Morio-Muskat or Riesling, also shown in Figs. 9 and 10, differ markedly from those of the various samples of Riesling or Morio-Muskat. Some typical key substances predominate in Morio-Muskat and serve to differentiate this variety from the various Riesling samples. Similarly, citrus oils from fruits of different origin (Australia, South Africa, Argentina, Arizona, Israel, Italy) showed no significant qualitative change in their aroma patterns [137,138].

Characterizing Different Varieties by Some Key Substances

In Fig. 11 the contents (peak heights) of some key substances typical of Riesling samples taken at different locations (Winnenden, Trier, Geisenheim, Würzburg and Geilweilerhof) and varying degrees of maturity are compared with those of Morio-Muskat samples (location Geilweilerhof) [134]. The black columns represent the peak heights of the selected components at the beginning of berry maturity. The change in peak height during berry maturation is shown by the light column. With increasing berry maturity, a marked increase in the linalool content (component 35) is recognizable for the cv. Riesling. The components 53 and 58 (58 = geraniol) of Morio-Muskat show also an increase, whereas the components of all Riesling samples de-

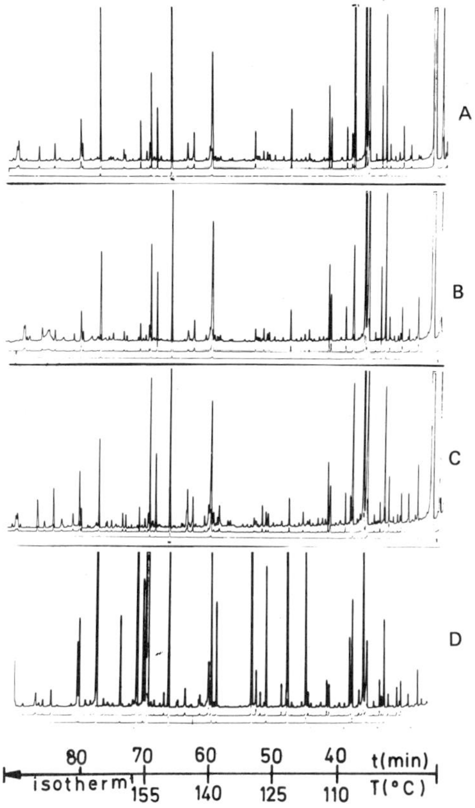

Figure 9. Comparison of fingerprint patterns of Riesling samples from different wine-growing areas of Germany and Morio-Muskat from Geilweilerhof: A—Geisenheim area (8/10/77), 19° Oechsle (SG ∿1.01885), 35% acid; B—Winnenden area (8/22/77), 16° Oechsle (SG ∿1.01647), 33% acid; C—Blankenhornsberg area (8/23/77), 15° Oechsle (SG ∿1.01567), 33% acid; D—Morio-Muskat, Geilweilerhof area (8/18/77), 16° Oechsle (SG ∿1.01647), 33% acid.

crease with increasing degree of maturity. It is apparent that, with regard to the peak height of these selected aroma components (Fig. 11: columns 1 to 5), all of the Riesling samples are in close agreement. When compared, the cvs. Riesling and Morio-Muskat (Fig. 11: column 6) show differences in the selected key substances which are so pronounced that it is possible to differentiate the varieties. In addition to those components studied, the total aromagram includes a number of other peaks which might also be useful for characterizing a variety. Some of these characteristic key substances are terpene compounds (inter alia linalool, nerol, citronellol, ho-trienol). Several authors have described the occurrence of these terpene compounds as aroma constituents of grapes, particularly as characteristic components of

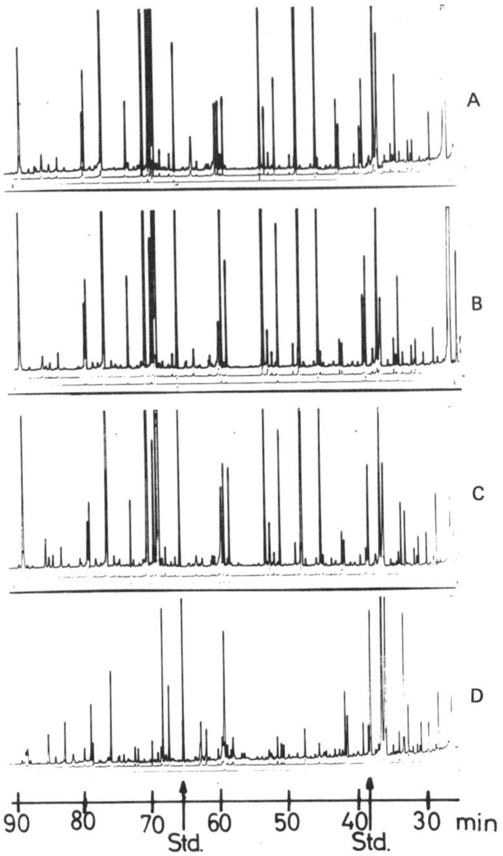

Figure 10. Comparison of fingerprint patterns of Morio-Muskat samples from different wine-growing areas of Germany with Riesling from Geilweilerhof: A—Morio-Muskat, Hohenheim area (8/22/77) 14.6° Oechsle (SG ∿1.01514), 29.1% acid; B—Morio-Muskat, Geilweilerhof area (8/18/77), 16.2° Oechsle (SG ∿1.01662), 32.8% acid; C—Morio-Muskat, Freiburg area (8/23/77), 22° Oechsle (SG ∿1.02165), 33.6% acid; D—Riesling, Blankenhornsberg area (8/23/77), 15.4° Oechsle (SG ∿1.01575), 32.8% acid.

muscat varieties [1,7,28,30,31,56,71,78,86,87,89,135,139,140]. For example, 80 to 1500 μg liter^{-1} linalool, 100 to 1000 μg liter^{-1} geraniol and 10 to 400 μg liter^{-1} nerol were found in muscat varieties [30]. There are distinct differences in the content of some terpene alcohols among other varieties [28]; thus, "Scheurebe" (Silvaner × Riesling) contains much more ho-trienol and linalool than Traminer or Riesling. No variety-specific differences were determined in the components of sesquiterpene hydrocarbons by Schreier et al. [86].

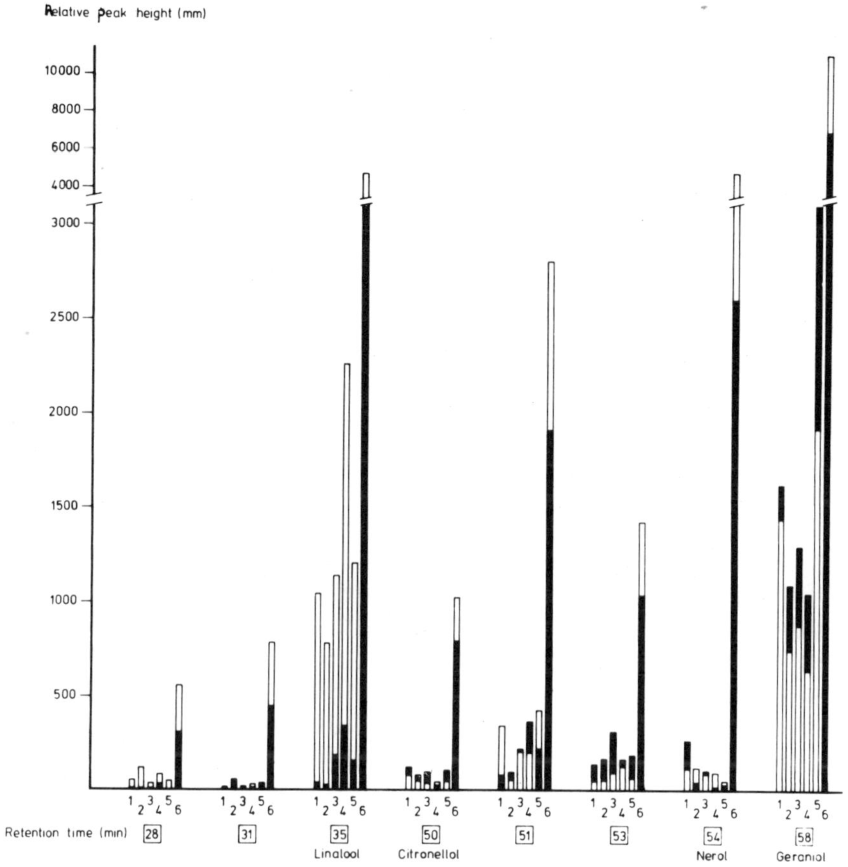

Figure 11. Content of some aroma compounds of the cultivars Riesling and Morio-Muskat: 1—Riesling, Winnenden area: □ 76° Oechsle (SG ∿1.07580), ■ 41° Oechsle (SG 1.04102); 2— Riesling, Trier area: □ 73° Oechsle (SG ∿1.07208), ■ 47° Oechsle (SG ∿1.04727); 3—Riesling, Geisenheim area: □ 100° Oechsle (SG ∿1.1000), ■ 64° Oechsle (SG ∿1.06427); 4—Riesling, Wurtzburg area: □ 88° Oechsle (SG ∿1.08823), ■ 61° Oechsle (SG ∿1.06104); 5—Riesling, Geilweilerhof: □ 88° Oechsle (SG ∿1.08823), ■ 59° Oechsle (SG ∿1.05886); 6—Morio-Muskat, Geilweilerhof: □ 75° Oechsle (SG ∿1.07456), ■ 59° Oechsle (SG ∿1.05886).

Wagner et al. [87,141] found that it was possible to differentiate the progenies of the crossing Muscat ottonel × Auxerrois into "Muscat type" and "neutral type" from their linalool and geraniol contents.

By multiple discriminate analysis (calculation program BMD—07M), 27 key components of Riesling and Morio-Muskat were tested to determine whether these were suitable for characterizing these varieties. The relative amount of each of the components analyzed was estimated by visual comparison of the aromagrams of the two varieties. Figure 12 shows the separation of Riesling and Morio-Muskat by these few key substances selected from several hundred components. The result of the multiple discriminate analysis shows that after five steps of calculation a highly significant separation between the cultivars Riesling and Morio-Muskat is apparent. This separation was obtained for samples of the two varieties from German vineyard regions and for berry maturities ranging from 40 to 100° Oechsle [111,134].

Differentiation of the cvs. Riesling, Silvaner, B-6-18 [(Vitis Riparia × Gamay F_2 × Fosters White Seedling)] is also possible by these 27 selected components. Differences within this group were not as great as is the case for the aromatic cv. Morio-Muskat. Excluding the values of Morio-Muskat, differentiation of these varieties can be achieved after five steps, and after fifteen steps this differentiation becomes highly significant (Fig. 13) [111,134].

The next question to arise is to what extent these varietal differences, which have previously been determined, are recognizable in new grape varieties from similar families. Figure 14 shows comparative aromagram sections (on Reoplex 400) of the cvs. Kö-50-100 [(Silvaner × Müller-Thurgau) × Müller-Thurgau], Bacchus [(Silvaner × Riesling) × Müller-Thurgau], B-6-18

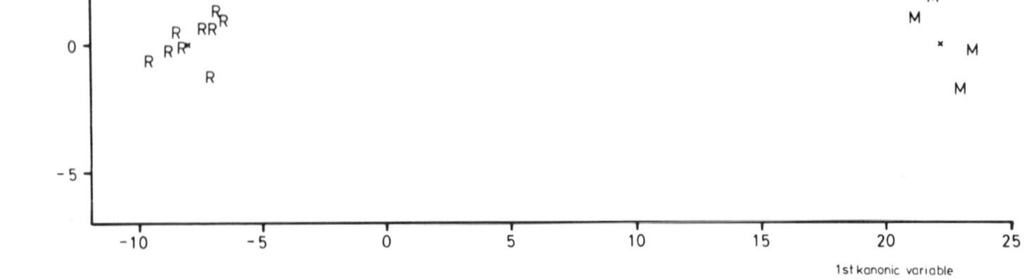

Figure 12. Differentiation between Riesling (R) and Morio-Muskat (M) by the aid of multiple discriminate analysis.

[(Vitis riparia × Gamay) F_2 × Fosters White Seedling], and Oberlin 595 (*Vitis riparia* × Gamay). It can be seen that there are significant differences in the fingerprint patterns between the new cvs. Kö-50-100 and Bacchus on the one hand and Oberlin 595 and B-6-18 on the other hand [111,112]. Furthermore, it was found that the aroma patterns of related varieties show such marked quantitative differences that multiple discriminate analysis of the major key substances allows clear differentiation of the varieties. There are also distinct differences in the aromagrams of the cvs. Müller-Thurgau (Riesling × Silvaner), V. Labrusca, and Taylor (Labrusca × Riparia) (Fig. 15), among several characteristic key substances of the individual varieties. These investigations show that separation of aroma concentrates on glass capillary columns and statistical evaluations of the different key substances can be used to form the basis of an analytical varietal characterization. An essential part of this work involves the selection of key substances suitable for this application. Up to now, these key substances have been determined by visual comparison of corresponding aromagram sections. However, using this technique those components can also be selected which may contribute little to the overall aroma. A more selective method might involve the selection of key substances on the basis of their importance to the aroma. A simple and rapid selection of those constituents

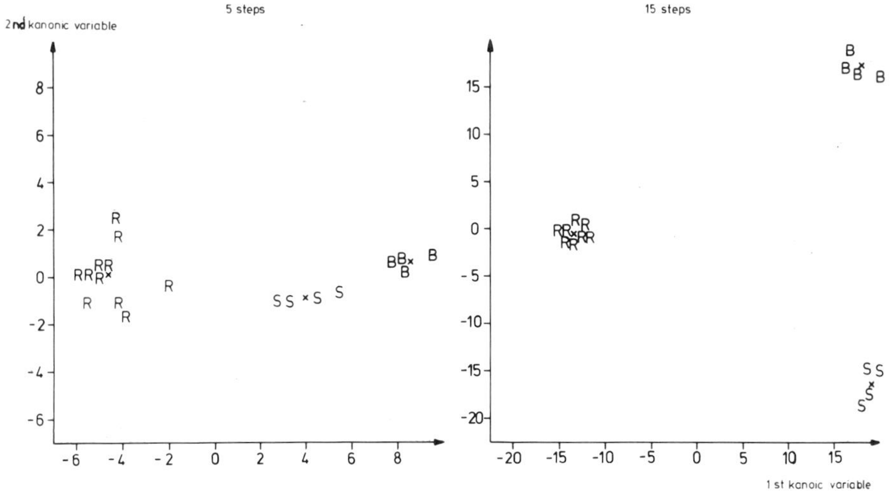

Figure 13. Differentiation between Riesling (R), Silvaner (S), and B-6-18 (B) by the aid of multiple discriminate analysis.

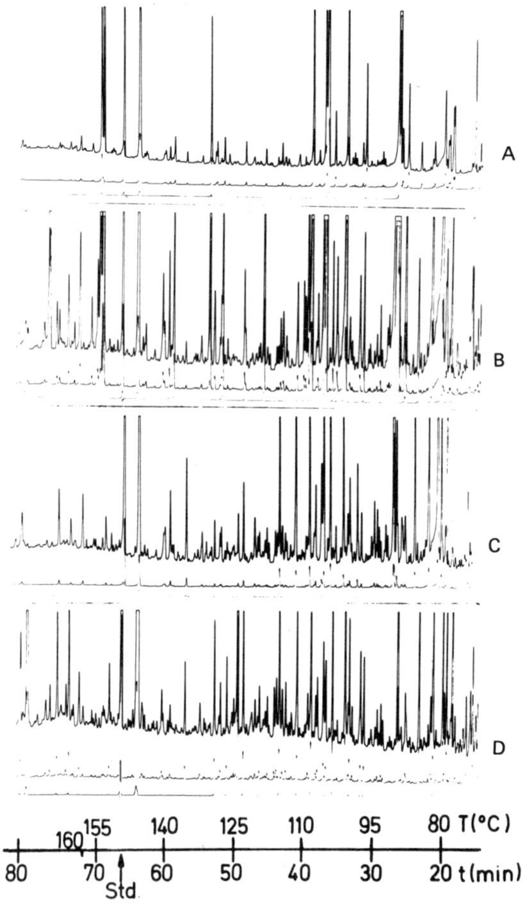

Figure 14. Comparison of fingerprint patterns of different vine varieties. Data (with dates): A—Kö-50-100 (10/10/77), 81° Oechsle (SG ∿1.08110), 11% acid; B—Bacchus (10/12/77), 72° Oechsle (SG ∿1.07318), 10% acid; C—B-6-18 (10/26/77), 87° Oechsle (SG ∿1.08733), 12% acid; D—Oberlin (11/8/77), 90° Oechsle (SG ∿1.09003), 19% acid.

which determine variety and quality can be effected by means of the sniffing technique. By this method, type and intensity of the odor of every individual component arriving at the end of the glass capillary column can be determined by means of a "sniffing detector," in paralled with the instrument detector [85,97,111,112,118,131,142]. Another method is to test the flavor of the analyzed components by a panel analysis, trapping them in neutral liquids [143]. By a combination of these approaches, variety-specific key substances may be more easily recognizable. This may also allow for a correlation between analytical patterns and organoleptic significance of these individual components.

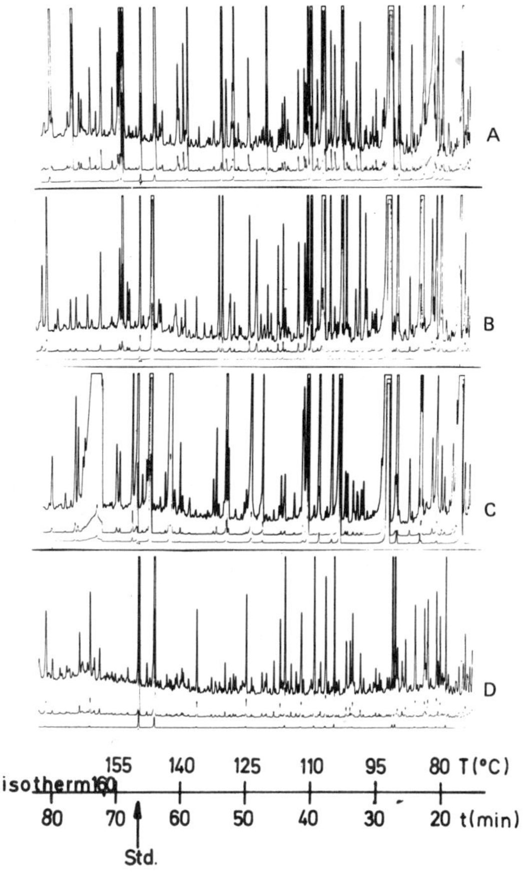

Figure 15. Comparison of fingerprint patterns of different vine varieties. Data (with dates): A—Bacchus (10/12/77), 72° Oechsle (SG ∿1.07318), 10% acid; B—Müller-Thurgau (10/31/77), 57° Oechsle (SG ∿1.05667), 10% acid; C—Labrusca (11/15/77), 59° Oechsle (SG ∿1.05886), 10% acid; D—Taylor (11/8/77), 72° Oechsle (SG ∿1.07318), 16% acid.

4.2 Wine

Wine aroma is composed of several hundred individual components. The total content of all of these aroma substances in wine ranges from approximately 0.8 to 1.0 g liter^{-1} (∿1% of the ethanol content). About one-half of this consists of the fermentation products 2-methyl-propanol-1, 2-methyl-butanol-1, 3-methyl-butanol-1, and 2-phenyl-ethanol, while the other half is distributed over 400 to 600 other aromas components. This complicates investigations on the aroma composition of wine. Besides the serious problems accompanying the enrichment of aroma components from this aqueous ethanolic solution, it is extremely difficult to separate these small concentrations of aroma compounds from the dominant alcohols. Recent developments allow selection of suitable techniques for extensive enrichment of these aroma substances. These include not only liquid-liquid extraction using Freon-11 [56,80,83,105,109-112] but also gas extraction [21,28,91-95,98] to produce ethanol-free aroma concentrates from wine. The steel capillary columns used earlier [19,73-77,80] were unable to resolve these aroma concentrations into individual components. As already mentioned, aromagrams of only 40 to 60 peaks could be achieved. In spite of the improved separation capabilities of glass capillary columns, resolution of aroma concentrates from wine is still difficult today. Acids have been subjected to separation and determination as methylesters [14,33,123], and fractional preseparations for better gas chromatographic separation have been carried out by several authors. Drawert et al. [144] and Schreier et al. [26,27,29,125,145] separated aroma concentrates on silica gel into five fractions, according to the polarity of the individual constituents using a discontinuous pentane-diethyl-ether gradient (10% diethyl ether in pentane to 100% diethyl ether). The increasing polarity of the substances led to repeated overlap between individual fractions, complicating in particular the quantitative aspects of the determination. Hardy et al. [74] reduced the alcohols in the Freon concentrations by a propylene glycol extraction, and achieved a better separation on Ucon 550 capillary column. The propylene-glycol fractionation yielded 40 components; otherwise, only 20 components are separated.

Because these prefractionations result in artifacts as well as losses in some of the aroma substances (especially when separating by column chromatography on silica gel), it is desirable to separate the original extract by direct gas chromatography, if possible. We have used several glass capillary columns [70-m LAC-3R-738, 50-m FFAP, 90-m Reoplex 400 (ID 0.25 mm), 140-m Reoplex 400 (ID 0.4 mm), and 90-m Carbowax 20M], which allow sufficient separation of the aroma concentrates, even without preliminary separation of the carboxylic acids. Figure 16 shows a section of 33 min of a total aromagram (on 140-m Reoplex 400) of a Morio-Muskat wine (headspace enrichment).

The following capillary columns—mainly glass capillary columns—have been used by several authors for the separation of aroma constituents from wine: 150-ft Carbowax 20M [29], 152-m SF96 [97], 50-m SCOT Carbowax 20M [103,146-148], Ucon 550 [74], 500-ft OV-101 [142], 50-m × 0.75-mm ID Ucon [98].

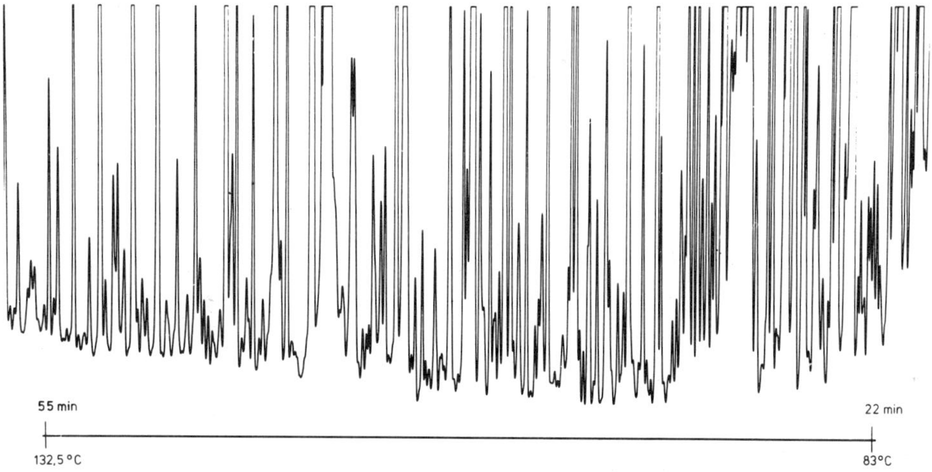

Figure 16. Aromagram of Morio-Muskat wine: headspace extraction; 100-m Reoplex glass capillary column. Sections from the 22nd to 55th min (83 to 132.5°C, 1.5°C min^{-1}).

A comparison of the aromagrams obtained from wines of different varieties (Fig. 17) reveals that these have significant differences in the aroma composition ("fingerprint patterns"). Investigations of wines obtained from grape berries of similar maturity and produced under the same technological conditions (separating, pure culture yeast, time of fermentation, etc.) indicate that these differences are due to variety-specific characteristics. The key substances typical of Riesling are marked with ⟶ , those of the cv. Morio-Muskat with ⟶⟶ . The calculation of these key substances results in relative values which differ significantly for the individual varieties. As can be seen from Table 4, the first group shows some ratios typical of Morio-Muskat and the last group represents those in which B-7-2 differs from the other varieties. Further, the aromagrams of different wines of varying quality levels (from table wine to berry selection) and vintage years reveal that some ratios increase with increasing maturity, others decrease, and some remain constant. Further, however, it turns out that some of the relative values established as typical of Morio-Muskat differ markedly from those of other varieties, in spite of these quantitative changes conditioned by vintage year and quality level. Thus, values between 6 and 65 have been found in the ratio of components 26.1 and 31.8 in Morio-Muskat (Table 5, middle line), whereas this ratio is 3500 for B-7-2, 1750 for Silvaner, and 273 for Reisling. From this it is apparent that, based on a great

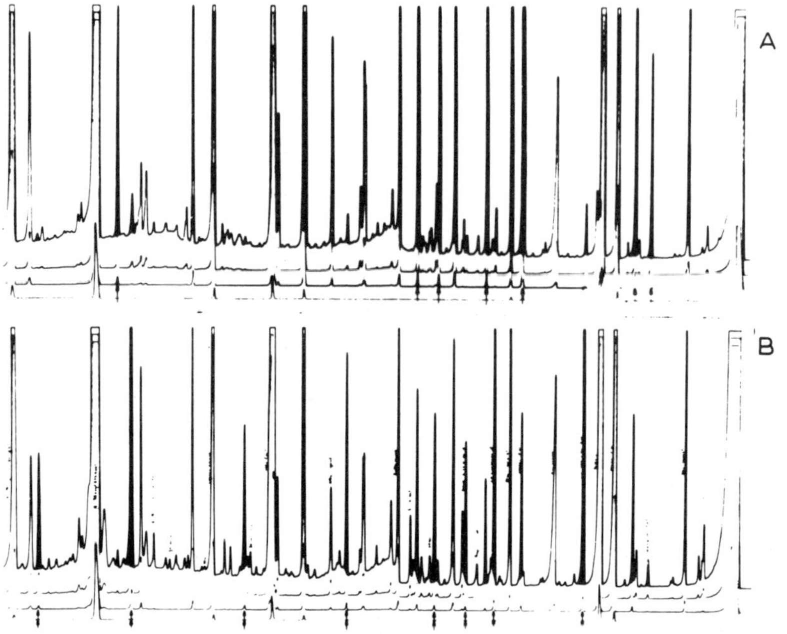

Figure 17. Aromagram sections of wines of Riesling (A) and Morio-Muskat (B): ⟶ key substances of Riesling; ⟶⟶ key substances of Morio-Muskat.

number of key substances, characterization of all varieties might become possible, quite independently of vintage year and quality level.

Subsequent to extractive enrichment (pentane/dichloro-methane) and preseparation by column chromatography on silica gel, Drawert et al. [144] and Schreier et al. [27,29] quantitatively determined the content of 11 esters and 10 terpene compounds and found distinct differences in the content of these components of the varieties investigated (Ruländer, Scheurebe, Gewürztraminer, Morio-Muskat, Riesling, and Müller-Thurgau). The order of the characteristics as shown in Table 6 (classified according to their importance to the analytical separation) resulted from the calculation of the analysis values for the differentiation by discriminant analysis. It turns out that the content of the isomeric rose oxides, nerol oxides, geraniol, trans-linalool oxide, linalool, and hotrienol are important to the differentiation of the wine varieties using discriminant analysis. These selected components allow a clear differentiation between the cvs. Gewürztraminer, Morio-Muskat, and Ruländer, whereas significant differentiation could not be achieved for the cvs. Riesling, Scheurebe, and Müller-Thurgau, which are very closely related genetically.

Table 4. Aroma Compounds of Wines of Different Grape Varieties—Relative Values (Vintage Year 1973)

Relation of components	Riesling	Silvaner	B-7-2	Morio-Muskat
26-1/29.0	93	538	233	3
26-1/31.8	273	1750	3500	6
26-1/35.8	56	97	13	2
35-9/31.8	5	18	258	3
35.9/29.0	2	6	17	2

When using the analytical data of fermentation by-products (e.g., malonic acid ethylester, lauric acid ethylester, acetic acid-i-phenylethylester) the differentiation for Gewürztraminer and Tiesling was only at the 95% level [27,29,144]. Cordner et al. [146] found a significant correlation between the yield and several gross constituents (e.g., pH, proline, sugar, acid) of grape must and wine. They also succeeded in detecting a highly significant correlation with the several headspace components determined by means of capillary chromatography. Stepwise discriminant analysis was used to show that the wines could be correctly sorted on the basis of analysis of either must, wine, or headspace. Selective must, wine, and headspace data were correlated linearly with crop levels.

Noble [97] and others determined the headspace profiles from 11 wines (9 Cabernet-Sauvignons). Initially, the aroma components of 200-ml wine were sorbed on 20-mg Tenax GC and then separated using capillary chromatography (77 components), followed by an odor description. By calculating

Table 5. Aroma Compounds of Wines of Different Grape Varieties—Relative Values

Relation of components	Riesling (1973)	Silvaner (1973)	B-7-2 1973	Morio-Muskat (1971-1974) 54 to 123° Oechsle[a]
26-1/29.0	93	538	233	3--34
26-1/31.8	273	1750	3500	6--65
26-1/35.9	56	97	13	2--28

[a]Table wine—top-quality wine.

Table 6. F Values and Sequence of the Marks for the Discriminant Analytical Separation by the Contents of the Original Substances of Grapes

Step	Mark	F Value
1	Roseoxide (cis + trans)	39.3871***
2	Neroloxide	21.6445***
3	Geraniol	8.2183**
4	Trans-linalooloxide	5.3712**
5	Linalool	3.4888*
6	Ho-trienol	2.6691*
7	α-Terpineol	1.0854
8	Cis-linalooloxide	1.0391
9	Krist. linalooloxide	2.1286
10	Nerol	.4814

Source: Ref. 27.
*SE < .01.
**SE < .001.
***SE < .0001.

the headspace components, a correlation to the attributes "vegetative," "woody," "raisin," "fruity," and "varietal character or Cabernet-like" was investigated. Although significant equations were derived within the data set, the equations cannot be used for prediction of wines outside it. Further investigation of a larger number of wines would be required for development of equations which can predict intensity values in untested wines.

Simpson [147] investigated the aroma and compositional changes in wine with oxidation storage and ageing. Using headspace techniques and capillary chromatography, the quantitative changes in some aroma components of Riesling wines were investigated, e.g., the components TDN (1,1,6-trimethyl-1,2-dihydronaphthalene) and Vitispirane, which possess intense tastes. Both are components typical of bottle-aged Riesling, and probably develop from carotinoids. The formation of these components is favored by both increased temperature and oxygen.

Benzaldehyde and furfural, however, are only developed under oxidative conditions. Changes in the concentrations of the ethylesters of C_6 to C_{10} fatty acids as well as of diethyl succinate, ethylfuroate, and 2-phenyl ethanol were followed for all treatment variations (influence of temperature, oxidative and nonoxidative storage). It was established that only the changes in the concentration of the hydrocarbon TDN might be responsible for the changes in the odor and taste. Decreases in the fruity Riesling bouquet might be explained by a decrease in monoterpenes, which strongly influence the aroma, and by an increase in the total content of the aroma components formed by oxidation and adulteration (TDN, Vitispirane) [147].

After enriching by means of the headspace technique, Williams et al. [148,149], investigated the aroma composition of Australian Flor-Sherries using capillary chromatography. Ethylesters of the C_6 to C_8 and C_{10} fatty acids and hexanol-1 were established as typical aroma components. When a film yeast had grown, produced by *Saccharomyces capensis*, significant changes in the aroma composition could be detected: the content of ethyl capronate and ethyl caprinate decreased, whereas that of 2-phenyl ethanol and diethyl succinate increased. From the sherry samples investigated, four glycerine derivates—1,3-dioxanes and 1,3-dioxolanes—were identified [148,150].

4.3 Wine Distillates

Using direct injection of brandies, cognacs, etc. on packed columns, a few alcohols and esters (8 to 15 components) have been determined [24,45,48,

151-156]. This number of detectable compounds can be increased without pretreatment or stepwise enrichment by direct injection of the samples on several different packed columns. Using this method, Postel et al. [18, 159] identified and determined quantitatively 29 volatile compounds in brandy from wine, brandy, and cognac; 37 in spirits distilled from fruit; 29 in whisky; 13 in grain spirits; and 6 in vodka. The brandies and cognacs covered 11 alcohols, 13 esters, and 5 carbonyl compounds.

Subsequent to water separation, 20 ml cognac were mixed with 50 g water-free Na_2SO_4, and the solution was drawn off after 16 hr at 0°C. Bricout [158] injected samples of wine distillates directly onto capillary columns (100-m Carbowax 20M, 0.5-mm ID) and determined quantitatively 19 components. This method permitted enrichment of the higher esters (ethylester of C_8 to C_{12} fatty acids) to such an extent that their quantitative determination was possible. The content of these compounds reveals distinct differences between the samples of cognac and armagnac.

Using a 90-m × 0.4-mm glass capillary column coated with Carbowax 400, we were able to separate 15 to 29 compounds by direct injection and to determine them quantitatively without pretreatment of the samples of wine distillates, brandies, and spirits distilled from fruit (Fig. 18). Connell and Strauss [159] also introduced wine distillates directly onto capillary columns (for alcohols 50-m Igepal CO630, for esters 100-m Apiezon-M and 30-m Ucon LB 550; 0.5 mm ID) and determined some alcohols and esters in quantity. In order to achieve better detection of the esters, the samples were mixed with carbon tetrachloride and propane-1,2-diol, which fractionated the constituents according to their polarity. The nonpolar solvent resulted in an ester concentrate which was largely free of low-molecular-weight alcohols and permitted determination of 15 esters. These results indicate that there are quantitative differences in the contents of volatile constituents that are detectable by direct injection and which allow further differentiation within the individual brandy groups. French cognacs, for example, can be differentiated from German brandies—cognac samples contain more 3-methyl butanol-1 and propanol-1 than do German brandies [24,155,157] —and cherry wine from the other spirits distilled from fruit [157]. When comparing samples of cognac and armagnac, Reinhard [24] found significant differences in the content of several components, particularly, the isopentanol/isobutanol, at 10, which was more significant in samples of armagnac (by 5.1) than in those of cognac (2.8 to 3.4). These differences might be

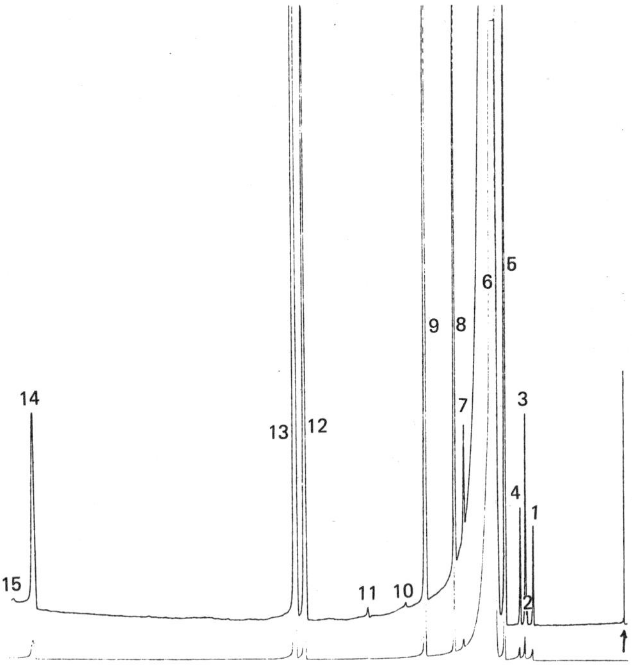

Figure 18. Volatiles from cognac. Direct injection on 90-m Carbowax 400 (0.4 mm ID) glass capillary column. Sample: cognac; 0.5 µl (1:25). Temperature: 50 to 80°C at 1°C min^{-1}. Carrier gas: hydrogen at 2 ml min^{-1}; chart speed 1 cm min^{-1}. Peaks: 1—acetaldehyde, 2—ethyl formate, 3—acetone, 4—ethyl acetate, 5—methanol, 6—ethanol, 7—butanol-2, 8—n-propanol, 9—2-methylpropanol-1, 11—n-butanol, 12—2-methylbutanol-1, 13—3-methylbutanol-1, 14—ethyl lactate, 15—hexanol.

due to different distillation methods—cognacs being produced by means of the Charentaiser technique, armagnacs by continuous distillation. In addition, there are various aroma components typical of certain brandies, e.g., benzaldehyde and benzyl alcohol for the stone fruit brandies. Compared with cognac and brandy, pomace brandy has a considerably higher content of methanol (500 mg per 100 ml; cognac 70 mg per 100 ml) and butanol-2 (35 mg per 100 ml; cognac below 5 mg per 100 ml) [157]. Barnett and Einsmann [151] found great differences between the individual brandies, mainly in the hexanol-1 content (whiskey ∿70 mg $liter^{-1}$, brandy 10 mg $liter^{-1}$, cherry wine 1 mg $liter^{-1}$). There were also indications that different distillation methods had an influence on the content of several aroma components [66, 153].

Investigations in this direction on the quantitative determination of several volatile compounds indicated that the number of components detectable by direct injection is not sufficient to obtain data indicative of purity and quality. When selecting a method for the enrichment of aroma substances from distillates (e.g., brandy, spirits), the large volumes of ethanol must be considered in addition to quantities of water. Before the extraction, several workers dilute the distillation samples with water to 20 to 27 vol % ethanol [156,160]. This method proved necessary for several solvents, as too much ethanol is otherwise coextracted. Liquid-liquid extraction [105, 109,149] and gas extraction [98,103,161], have been used to obtain aroma concentrates from distillates. When using gas extraction, the aroma substances are extracted from the gas stream mainly by adsorption on solid carriers.

From these aroma concentrates, Prillinger and Horwatitsch [67] separated 23 components on packed columns and Tsantalis [156] also identified 64 components on packed columns after decomposition analyses. Using packed columns, Miglio [162] determined 82 components from aroma concentrations of Italian marc brandies (Grappa); among these, there were several odor- and taste-intensive S compounds (e.g., dimethyl sulfide, dimethyl disulfide, diethyl disulfite). Ter Heide et al. [98] extracted 465 liters cognac using pentane/dichloropentane (2:1) and obtained 650 g of an aroma concentrate. After the separation of acids and phenols, the remaining neutral extract (550 g) was separated into a total of 22 fractions on silica gel, using a discontinuous pentane-isopropyl-chloride or a pentane-ether gradient. Each fraction was subjected to olfactory analysis by a panel of experts. All of the characteristic flavor elements of cognac were found in fractions 16 through 19. All fractions were investigated by capillary chromatography (50-m × 0.75-mm ID glass SCOT column: Ucon 50 LB-550X; 70 to 170°C at 1°C min^{-1}). In the application of the headspace technique to cognac, a N_2 gas stream (100 ml to liter m^{-1}) is passed through 2 liters cognac, and the aroma substances are adsorbed on Tenax GC or Chromosorb 102. From liquid-liquid and headspace concentrations, 339 compounds were detected and identified by means of GC/MS. Of these, 193 were identified for the first time in cognac [98]. They included 77 esters, 28 aldehydes, 24 acetals, 63 alcohols (21 terpenic alcohols), 27 acids (identified as methylesters), 19 ethers, 23 ketones (including for the first time γ-dehydroionone), 3 lactones, and 18 compounds with furan ring (2-acetylfuran, 5-methyl-2-acetylfuran, 2-acetyl-4,5-dimethylfuran, 2-propionyl-furan). Of these components, 32 were detected and identified for the first time using headspace enrichment, including 1,1-diethoxymethane, ethyl-2-methyl-butanoate, and 3-methyl-2-butanone.

Williams et al. [161] enriched the aroma substances derived from spirit produced by heap-fermented grape marc using a two-level headspace technique [103]. The first step of this method was to adsorb the volatile compounds (in a N_2 gas stream) on Chromosorb 105. From here, they were trapped in a second step on a short column with SF-96 on Chromosorb A. The compounds were then passed by reverse-flushing onto the capillary column (SCOT glass column Carbowax 20M) for gas chromatographic analysis. The investigations revealed that the undesirable odor is due to unsaturated carbonyl compounds present in trace amounts. The principal odoriferous aldehydes in the spirit included trans-2-octenal, trans,trans-2,4-heptadienal, trans-2-nonenal, trans,trans-2,4-nonadienal, and trans,trans-2,4-decadienal. The same compounds were also detected in a sample of autoxidized grape seed oil. Thus it appears that the off-odors of marc spirits are due to the oxidation of polyunsaturated fatty acids of grape seeds [161]. The acetals identified from pentane concentrates (GC/MS, SCOT Carbowax 20M) included 3,3-diethoxy-butan-2-one and 1,1,3-triethoxypropane; these are products of acetalization of diacetyl and acrolein. They are formed when distilling products with low pH values [106].

5. SUMMARY

The gas chromatographic analysis of the volatile components of grape berries, wine, and wine distillates requires initial enrichment. In spite of the large quantities of water and ethanol, successful enrichments can be achieved by either liquid-liquid extraction using various solvents (Freon-11, pentane, pentane/dichloromethane) or gas extraction. When using gas extraction--the aroma substances being enriched by refrigeration in traps or by adsorption on solid carriers from the gas stream--difficulties arise which are due mainly to the large quantities of water and ethanol. In addition, depending on the carrier and the subsequent desorption, the composition can change significantly. We have developed a new technique which allows concentration of the trace components from a steam phase of aqueous and aqueous-alcoholic solutions, the aroma components being enriched (minus water and ethanol) by introducing the gas stream into Freon-11.

Since the enrichment of the aroma substances from grape berries results in decomposition of the cell structure, it is possible that changes in the original aroma pattern occur by enzymatic processes. In order to compare the variety-specific primary aroma substances of grape berries, the fruit enzymes must be inhibited during the concentration steop, (e.g., with heavy metal salts or alcohols).

The enrichment of aroma substances leads to aroma concentrates which are composed of several hundred individual components including compounds of different classes of substances. The separation of this complex mixture requires columns of high efficiency. Only the application of capillary chromatography, especially glass capillary columns, allows sufficient separation of the volatile aroma components. In the course of an analysis time of 2 to 3 hr, aroma concentrates can be separated into 300 to 400 individual compounds. This number of separated components should be sufficient for characterizing the varieties and evaluating the qualtiy of wines from different cultivars.

Comparison of the aromagrams produced from wines or grape juices of different varieties indicates that there are distinct differences in the aroma composition ("fingerprint patterns"). In these "fingerprint patterns" of various cultivars, components occur which differ markedly in quantity (key substances). Several of these key substances are suitable for characterizing a variety independent of maturity and location. As can be seen from the multiple discriminant analysis, the use of several of these key substances allows a highly significant distinction between the investigated cultivars (Riesling, Morio-Muskat, Sylvaner, B-6-18), independent of the degree of maturity and the location (Germany). Some of these variety-typical key substances are terpene compounds, including, e.g., rose oxide, linalool oxide, nerol, linalool, and geraniol.

The analytical method used results in significant differences in the fingerprint patterns of new varieties of similar origin which are sufficient for a characterization. Subsequent to prefractionation using column chromatography, Ter Heide et al. separated and identified 339 compounds from liquid-liquid or headspace enrichments of cognac samples, of which 193 components (77 esters, 28 aldehydes, 24 acetals, 63 alcohols, 19 ethers, 18 compounds with furan rings) were determined for the first time in cognac.

The changes in the aroma concentration of sherry wines caused by film yeast (Williams) and during storage (bottle ageing) depending on oxygen and temperature (Simpson) were also investigated by means of capillary chromatography.

REFERENCES

1. Bayonove, C., and Cordonnier, R., *Ann. Technol. Agric.*, 19:95 (1970).
2. Bayonove, C., and Cordonnier, R., *C.R. Sécances Acad. Agric.*, 57:1374 (1971).
3. Bertrand, A., *Connais. Vigne et Vin*, 2:175 (1968).
4. Braun, G., and Hieke, E., *Deutsch. Lebensm. Rdsch.*, 70:165 (1974).
5. Charalambous, G., and Inglett, G. E., *Flavor of Foods and Beverages*, Academic Press, New York, 1978.
6. Charalambous, G., *Analysis of Foods and Beverages. Headspace Techniques*, Academic Press, New York, 1978.
7. Cordonnier, R., *Progr. Agric. Vitic.*, 91:320 (1974).
8. Dickes, G. J., and Nicholas, P. V., *Gas Chromatography in Food Analysis*, Butterworths, London, Boston, 1976.
9. Drawert, F., and Rapp, A., *Chromatographia*, 1:446 (1968).
10. Drawert, F., Schreier, P., and Scherer, W., *Z. Lebensm. Unters. und Forschung*, 155:342 (1974).
11. Drawert, F., *Geruchs- und Geschmacksstoffe*, Carl, Nürnberg, 1975.
12. Hieke, E., *Dt. Weinfachzeitung*, 108:24 (1972).
13. Hieke, E., *Wein Wiss.*, 28:285 (1973).
14. Kepner, R. E., Webb, A. D., and Maggiora, L., *Amer. J. Enol. Vitic.*, 19:116 (1968).
15. Koch, J., Hess, D., and Gruss, R., *Z. Lebensm. Unters. und Forschung*, 147:207 (1971).
16. Meyer, H. G., *Angew. Chemie*, 82:965 (1970).
17. Muller, C. J., Kepner, R. E., and Webb, A. D., *Amer. J. Enol. Vitic.*, 24:5 (1973).
18. Postel, W., Drawert, F., and Adam, L., *Chem. Mikrobiol. Technol. Lebensm.*, 1:224 (1972).
19. Prillinger, F., Horwatitsch, H., and Madner, A., *Mitt. Klosterneuburg*, 17:271 (1967).
20. Prillinger, F., and Madner, A., *Mitt. Klosterneuburg*, 20:202 (1970).
21. Rapp, A., Uber Inhaltsstoffe von Traubenmosten und Weinen, Doctoral Thesis, Univ. Mainz, 1965.
22. Rapp, A., *Bull. OIV.*, 492:151 (1972).
23. Reinhard, C., *Wein Wiss.*, 23:475 (1968).
24. Reinhard, C., *Dtsch. Lebensm. Rdsch.*, 65:223 (1969); 67:349 (1971).
25. Solms, J., and Neukom, H., *Aroma- und Geschmacksstoffe in Lebensmitteln*, Proceedings, symposium Technical University of Zürich, 1967. Published by Forster Verlag, Zürich, 1967.
26. Schreier, P., and Drawert, F., *Chem. Mikrobiol. Technol. Lebensm.*, 3:154 (1974).

27. Schreier, P., Drawert, F., Junker, A., and Reiner, L., *Mitt. Klosterneuburg*, 26:225 (1976).

28. Schreier, P., Drawert, F., and Junker, A., *Chem. Mikrobiol. Technol. Lebensm.*, 4:154 (1976).

29. Schreier, P., Drawert, F., and Junker, A., *Chem. Mikrobiol. Technol. Lebensm.*, 5:45 (1977).

30. Terrier, A., Boidron, J. N., and Ribéreau-Gayon, P., *C.R. Acad. Sci. (Paris)*, 275:941 (1972).

31. Usseglio-Tomasset, L., *Riv. Vitic. Enol.* (Conegliano), 22:223 (1969).

32. Webb, A. D., Kepner, R. E., and Maggiora, L., *Amer. J. Enol. Vitic.*, 17:247 (1968).

33. van Wyk, C. J., Webb, A. D., and Kepner, R. E., *J. Food Sci.*, 32:660 (1967).

34. Boeckh, J., Die Chemischen Sinne Geruch und Geschmack, in *Handbook Physiologie des Menschen*, (O. H. Gauer, K. Kramer, R. Jung, eds.), Urban und Schwarzenberg, München, 1972.

35. Guadagni, D. G., Buttery, R. G., and Okano, S., *J. Sci. Food Agric.*, 14:761 (1963).

36. Guadagni, D. G., Buttery, R. G., and Harries, J., *J. Sci. Food Agric.*, 17:142 (1966).

37. Meilgaard, M. C., Aroma volatiles in beer, in: *Geruchs- und Geschmacksstoffe*, (F. Drawert, ed.), Carl, Nürnberg, pp. 211-254, 1972.

38. Salo, P., Nykänen, N., and Suomalainen, H., *J. Food Sci.*, 37:394 (1972).

39. Hennig, K., and Villforth, F., *Vorratspflanze und Lebensm. Forsch.*, 5:181 (1942).

40. Bayer, E., Kupfer, G., and Reuther, K. H., *Zt. Anal. Chemie*, 164:1 (1958).

41. Bayer, E., and Bässler, E., *Zt. Anal. Chemie*, 181:418 (1961).

42. Bandion, F., *Mitteilung Rebe und Wein*, 22:253 (1972).

43. Baraud, J., and Genevois, C., *C.R. Acad. Sci.*, 247:2479 (1958).

44. Bertrand, A., and Ribéreau-Gayon, P., *Ann. Fals. Expert. Chim.*, 63:148 (1970).

45. Christensen, E. N., and Caputti, A., *Amer. J. Enol. Vitic.*, 19:155 (1968).

46. Hupf, H., *Wein Wiss.*, 29:1 (1974).

47. Kamibayaschi, A., Miki, M., and Duo, H., *J. Agric. Chem. Soc. Jap.*, 35:968 (1961).

48. Litschev, D. V., and Goranov, N., *Bull. OIV.*, 45:317 (1972).

49. Rapp, A., Hövermann, W., Jecht, U., Franck, H., and Ullemeyer, H., *Chemiker-Zeitung*, 97:29 (1973).

50. Reinhard, C., *Chem. Mikrobiol. Technol. Lebensm.*, 1:236 (1972).

51. Sihto, E., Nykänen, L., and Suomalainen, H., *Qual. Plant. Veg.*, 11:211 (1964).

52. Woidich, H., and Pfannhauser, W., *Mitt. Klosterneuburg*, 23:117 (1973).

53. Reinhard, C., *Branntweinwirtschaft*, 110:333 (1970).

54. Äyräpää, T., *Brauwiss.*, 18:161 (1965).

55. Flanzy, M., *Ann. Technol. Agric.*, 17:207 (1968).

56. Hardy, P. J., *Phytochem.*, 9:709 (1970).

57. Rapp, A., *Chemie und Landwirtsch. Produkten*, 201 (1970).

58. Rapp, A., and Franck, H., *Vitis*, 9:299 (1971).

59. Yoshizawa, K., *Agr. Biol. Chem.*, (Japan), 30:634 (1966).

60. Moutounet, M., *Ann. Technol. Agric.*, 18:249 (1969).

61. Rapp, A., Steffan, H., Hastrich, H., and Ullemeyer, H., *Mitt. Fachgr. Lebensm. und Gerichtl. Chemie*, 29:33 (1975).

62. Yoshizawa, K., *Agr. Biol. Chem.*, 28:279 (1964).

63. Bayonove, C., *Ind. Aliment. Agric.*, 89:275 (1972).

64. Chaudhary, S. S., Kepner, R. E., and Webb, A. D., *Amer. J. Enol. Vitic.*, 15:190 (1964).

65. Galetto, W. G., Webb, A. D., and Kepner, R. E., *Amer. J. Enol. Vitic.*, 17:11 (1966).

66. Mecke, R., Schindler, R., and de Vries, M., *Wein Wiss.*, 15:183 (1960).

67. Prillinger, F., and Horwatitsch, H., *Mitt. Klosterneuburg*, 16:115 (1966).

68. Prillinger, F., and Madner, A., *Mitt. Klosterneuburg*, 19:361 (1969).

69. Sakato, K. H., Hoekman, M., Kepner, R. E., Webb, A. D., and Muller, C. J., *Amer. J. Enol. Vitic.*, 26:70 (1975).

70. Terrier, A., and Boidron, J. N., *Conn. Vigne et Vin*, 6:69 (1972).

71. Terrier, A., Boidron, J. N., and Ribéreau-Gayon, P., *C.R. Acad. Sci.*, (Paris), 275:495 (1972).

72. Webb, A. D., *Conn. Vigne et Vin*, 2:13 (1968).

73. Chaudhary, S. S., Webb, A. D., and Kepner, R. E., *Amer. J. Enol. Vitic.*, 19:6 (1968).

74. Hardy, P. L., and Ramshaw, E. H., *J. Sci. Food Agric.*, 21:39 (1970).

75. Lemperle, E., and Mecke, R., *Wein Wiss.*, 19:210 (1964).

76. Prillinger, F., and Madner, A., *Mitt. Klosterneuburg*, 18:1 (1968).

77. Stern, D. J., Lee, A., McFadden, W. H., and Stevens, K. L., *J. Agric. Food Chem.*, 15:1100 (1967).

78. Stevens, K. L., Bomben, J. L., Lee, A., and McFadden, H. W., *J. Agric. Food Chem.*, 14:249 (1966).

79. Stevens, K. L., Bomben, J. L., and McFadden, H. W., *J. Agric. Food Chem.*, 15:378 (1967).

80. Stevens, K. L., Flath, R. A., Lee, A., and Stern, D. J., *J. Agric. Food Chem.*, 17:1102 (1969).

81. Drawert, F., Heimann, W., Emberger, R., and Tressl, R., *Zeitsch. Naturforschung*, 20b:497 (1965).

82. Emberger, R., Beiträge zur Biogenese Pflanzlicher Aromastoffe. Doctoral Thesis, Univ. Karlsruhe, 1967.

83. Rapp, A., Hastrich, H., and Engel, L., *Vitis*, 15:29 (1976).

84. Tressl, R., Über die Bildung und Enzymatische Veränderung einiger Fruchtaromastoffe. Doctoral Thesis, Univ. Karlsruhe, 1967.

85. Rapp, A., Hastrich, H., and Engel, L., *Mitt. Klosterneuburg*, 27:74 (1977).

86. Schreier, P., Drawert, F., and Junker, A., *Zeitschr. Lebensm. Unters. und Forschung*, 160:271 (1976).

87. Wagner, R., Dirninger, W., Fuchs, V., and Bronner, A., Etude des differences intervarietales de la teneur en constituants volatiles de l'arome du raisin. Proceedings of the *Int. Symp. on the Quality of the Vintage*, Cape Town, South Africa, pp. 137-142, 1977.

88. Webb, A. D., and Kepner, R. E., *Food Res.*, 22:384 (1967).

89. Wenzel, K. W. O., and Vries, M. J., *S. Afr. J. Agric. Sci.*, 11:273 (1968).

90. Bayonove, C., and Cordonnier, R., *Ann. Technol. Agric.*, 19:79 (1970).

91. Bertrend, A., Boidron, J. N., and Ribéreau-Gayon, P., *Bull. Soc. Chim. France*, 5:3149 (1967).

92. Drawert, F., Rapp, A., and Bachmann, O., in: *Gaschromatographische Untersuchung der Aromastoffe und Alkohole von Früchten*, in "Fruchtaromen", Proceedings, IVth Symposium of International Fruit Juice Producers, Berne, 1962. Published by Juris-Verlag, Zürich, pp. 235-243, 1962.

93. Kepner, R. E., and Webb, A. D., *Amer. J. Enol. Vitic.*, 7:8 (1956).

94. Prillinger, F., and Madner, A., *Weinberg und Keller*, 15:435 (1968).

95. Rapp, A., and Knipser, W., in preparation.

96. Jennings, W. G., and Filsoof, M., *Agric. and Food Chem.*, 25:440 (1977).

97. Noble, A. C., Sensory and instrumental evaluation of wine aroma. In: *Analysis of Foods and Beverages; Headspace Techniques* (G. Charalambous, ed.), Academic Press, New York, pp. 203-228, 1978.

98. ter Heide, R., de Vries, P. J., Visser, J., Jaeger, R. P., and Timmer, R., Concentration and identification of trace constituents in alcoholic beverages. In: *Analysis of Foods and Beverages; Headspace Techniques* (G. Charalambous, ed.), Academic Press, New York, pp. 249-281, 1978.

99. Jennings, W. G., Wohleb, R., and Lewis, M. J., *J. Food Sci.*, 37:69 (1972).

100. Klimes, J., and Lamparsky, D., Headspace Techniques. In: *Analysis of Foods and Beverages; Headspace Techniques* (G. Charalambous, ed.), Academic Press, New York, pp. 95-114, 1978.

101. Wyllie, S. G., Alves, S., Filsoof, M., and Jennings, W. G., Headspace Sampling: Use and abuse. In: *Analysis of Foods and Beverages; Headspace Techniques* (G. Charalambous, ed.), Academic Press, New York, pp. 1-16, 1978.

102. Heinz, D. E., Sevenants, M. R., and Jennings, W. G., *J. Food Sci.*, 31:63 (1966).

103. Williams, P. J., and Strauss, C. R., *J. Inst. Brewing*, 83:213 (1977).

104. Williams, A. A., *Flavor. Ind.*, 3:604 (1972).

105. Williams, A. A., and Tucknott, O. G., *J. Sci. Food Agric.*, 24:863 (1973).

106. Williams, P. J., and Strauss, C. R., *J. Sci. Food Agric.*, 26:1127 (1975).

107. Kepner, R. E., Webb, A. D., and Maggiora, L., *Amer. J. Enol. Vitic.*, 20:25 (1969).

108. Nelson, R. H., Acree, T. E., Lee, C. Y., and Butts, R. M., *J. Assoc. Offic. Anal. Chem.*, 59:1387 (1976).

109. Hardy, P. J., *J. Agric. Food Chem.*, 17:656 (1969).

110. Ramshaw, E. H., and Hardy, P. J., *J. Sci. Food Agric.*, 20:619 (1969).

111. Rapp, A., Hastrich, H., Engel, L., and Knipser, W., Sorten- und Qualitätsbeurteilung am Beispiel der Traubeninhaltsstoffe. Proceedings symposium of International Federation of Fruit Juice Producers, Bern, 1978. Published by Juris Verlag, Bern, pp. 39-63, 1978.

112. Rapp, A., Hastrich, H., Engel, L., and Knipser, W., Possibilities of characterizing wine quality and wine varieties by means of capillary chromatography. In: *Flavor of Foods and Beverages* (G. Charalambous, ed.), Academic Press, New York, pp. 391-417, 1978.

113. Cobb, C. S., and Bursey, M. M., *J. Agric. Food Chem.*, 26:197 (1978).

114. Bertrand, A., *Chim. Anal.*, 53:577 (1971).

115. Woidich, H., and Pfannhauser, W., *Dtsch. Lebensm. Rundschau*, 74:397 (1978).

116. Beernaert, H., and Gossele, J., *Trav. Chim. Aliment. Hyg.*, 69:516 (1978).

117. Grob, K., *Helv. Chim. Acta*, 48:1362 (1965).

118. Jennings, W. G., Wyllie, S. G., and Alves, S., *Chromatographia*, 10:426 (1977).

119. Jennings, W. G., *Gas Chromatography with Glass Capillary Columns*, 2nd edit., Academic Press, New York, 1980.

120. Kaiser, R., *Chromatographie in der Gasphase. Kapillarchromatographie.* Bibliograph. Inst., Mannheim (1975).

121. Rapp, A., Hastrich, H., and Engel, L., *Vitis*, 15:183 (1976).

122. Schomburg, G., and Husmann, H., *Chromatographia*, 8:517 (1975).

123. Drawert, F., Schreier, P., Leopold, G., Kerenyi, Z., Lessing, V., and Junker, A., *Z. Lebensm.-Unters. und Forschung*, 162:11 (1976).

124. Schreier, P., and Drawert, F., *Z. Lebensm.-Unters. und Forschung*, 154:27 (1974).

125. Schreier, P., Drawert, F., and Junker, A., *Z. Lebensm.-Unters. und Forschung*, 157:34 (1979).

126. Bertsch, W., Anderson, E., and Holzer, G., *Chromatographia*, 10:449 (1977).

127. Bertsch, W., *HRC&CC*, 1:289 (1978).

128. Deans, D. R., *Chromatographia*, 1:18 (1968).

129. Drawert, F., and Rapp, A., *Chemiker-Ztg.*, 88:267 (1964).

130. Miller, J. R., Stearns, S. D., and Freeman, R. R., *HRC&CC*, 2:55 (1979).

131. Clark, R. G., and Cronin, D. A., *J. Sci. Food Agric.*, 26:1009 (1975).

132. Schomburg, G., Husmann, H., and Weeke, F., *J. Chromatogr.*, 112:205 (1975).

133. Teitelbaum, C. L., *J. Agric. Food Chem.*, 25:466 (1977).

134. Rapp, A., and Hastrich, H., *Vitis*, 17:288 (1978).

135. Ribéreau-Gayon, P., Boidron, J. N., and Terrier, A., *J. Agric. Food Chem.*, 23:1042 (1975).

136. Rapp, A., Hastrich, H., and Engel, L., Einfluss von Reife und Standort auf die Aromastoffe. In: *Moderne Aspekte der Qualitätsweinerzeugung*, Symposium, Freiburg (Breisgau), 1977. Published by Staatliches Weinbauinstitut, Freiburg, pp. 53-66, 1977.

137. MacLeod, W. D., MacFadden W. H., and Buigues, N. M., *J. Food Sci.*, 31:591 (1966).

138. Slater, C. A., *J. Sci. Food Agric.*, 12:257 (1961).

139. Bayonove, C., Cordonnier, R., and Ratier, R., *Ann. Techn. Agric.*, 20:347 (1970).

140. Terrier, A., and Boidron, J. N., *Conn. Vigne et Vin*, 6:147 (1972).

141. Wagner, R., Dirninger, N., Fuchs, V., and Bronner, B., Premiers résultats concernant l'étude génétique de constituants volatiles importants de l'arome des raisins dans deux descendances de V. vinifera L. In: *Génétique et Amélioration de la Vigne*, Institut National de la Recherche Agronomique, pp. 429-434. Paris, 1978.

142. Stern, D. J., Guadagni, D., and Stevens, K. L., *Amer. J. Enol. Vitic.*, 26:208 (1975).

143. Parliment, T. M., and Scarpellino, R., *J. Agric. Food Chem.*, 25:97 (1977).

144. Drawert, F., and Schreier, P., *Ann. Technol.*, 27:367 (1978).

145. Schreier, P., Drawert, F., Kerenyi, Z., and Junker, A., *Zeitschr. Lebensm.-Unters. und Forschung*, 161:249 (1976).

146. Cordner, C. W., Ough, C. S., Kasimatis, A. N., and Kissler, J. J., *Amer. J. Enol. Vitic.*, 29:247 (1978).

147. Simpson, R. F., *Vitis*, 17:274 (1978).

148. Williams, P. J., and Strauss, C. R., *J. Inst. Brewing*, 84:144 (1978).

149. Williams, P. J., and Strauss, C. R., *J. Inst. Brewing*, 84:148 (1978).

150. Muller, C. J., Kepner, R. E., and Webb, A. D., *Amer. J. Enol. Vitic.*, 29:207 (1978).

151. Barnett, J. H., and Einsmann, J. R., *J. Assoc. Offic. Analyt. Chem.*, 60:297 (1977).

152. Drawert, F., Heimann, W., and Tsantalis, G., *Z. Anal. Chemie*, 228:170 (1967).

153. Gadzhiev, D. M., Nefedov, M. P., and Korbyakov, G. A., *Sadovod. Vinograd. Vinodel. Mold.*, 27:35 (1972).

154. Guymon, J. F., *Amer. J. Enol. Vitic.*, 21:61 (1970).

155. Hogben, R., and Mulav, M., *J. Sci. Food Agric.*, 27:1108 (1976).

156. Tsantalis, G., Gaschromatographische Untersuchungen der Aromastoffe von Branntweinen. Doctoral Thesis, Univ. Karlsruhe, 1968.

157. Postel, W., Drawert, F., and Adam, L., Aromastoffe in Branntweinen. In: Geruch- und Geschmacksstoffe (F. Drawert, ed.), Carl, Nürnberg, pp. 99-111, 1975.

158. Bricout, J., *Ann. Technol. Agric.*, 19:197 (1972).

159. Connell, D. W., and Strauss, C. R., *J. Sci. Food Agric.*, 25:31 (1974).

160. Guymon, J. F., and Crowell, E. A., *Amer. J. Enol. Vitic.*, 20:76 (1969).

161. Williams, P. J., and Strauss, C. R., *J. Sci. Agric.*, 29:527 (1973).

162. Miglio, G., *Branntweinwirtsch.*, 112:29 (1972).

INDEX

Addison's disease, 215, 217
Adrenal carcinoma, steroid relationships in, 321
Adsorbent capacity, determination of, 81
Adsorptive activity of column, 87
Air:
  cigarette smoke in, 97-98
  compounds identified in, 92-96
  fuel constituents in, 97
Air analysis,
  analytical techniques for, 73-84
  artifact formation in, 74-75
  column selection for, 87
  sampling methods for, 75-84
Air pollution and mortality rates, 73
Air sampling:
  highly volatile substances, 76-77
  intermediate range volatiles, 77-84
  water exclusion during, 77-78
Alcoholism and fatty acid distribution, 400
Alditols, *see* Aldoses
Aldoses:
  derivatization of, 253-258
  metabolic profiles of, 250, 259-262
  retention characteristics for derivatives of, 261
  Wohl derivatives of, 254-255
Aldosterone, 216-217
Amino acid derivatization, 332-339
  amino group blockage, 337
  goals, 336
  preferred methods, 339
Amino acids:
  beer, 350, 570, 572
  column considerations for, 340-342
  detector considerations for, 342
  diastereoisomer separations, 333, 336, 341
  general and historic considerations, 331-334
  human serum, 358-359
  injector considerations for, 340
  ion exchange contrasted with GC, 332, 334
  methyl esters of, 332
  sample preparation, 334-335
  trifluoroacetyl methylesters of, 332
Amino acid esterification, choice of alcohol, 337, 338
Amino acid quantification:
  internal standards for, 356-357
  special problems in, 352
Amino acid separation, 33
  from biological fluids, 334-335
  ion exchange vs. GC, 346
  standards, 345
Anderson's disease, 253
Androgenization, 322
Anorexia nervosa, 323
Arctic bramble, volatile composition of, 522

Argon ionization detector, *see* Detection
Aroclor, 105
Aroma enrichment, hazards in, 581
Aroma extraction of grapes and wines, 582, 584, 586-587
Arthritis:
  metabolic profiles in, 206-207
  organic acid profile of sinovial fluid, 213
  plasma acid profiles in, 213
Artichoke, volatile composition of, 517, 520
Artifacts:
  grape processing, 583, 589
  wine processing, 604
Atapet, 28, 80
Atherogenesis, fatty acid metabolism in, 399-400
Auxiliary gas, *see* Detection

Bacteriocides, 192
Banana, volatile composition of, 516, 518
Beer:
  bitter acids in, 570-574
  extraction with CS_2, 557, 559-560
  fatty acids of, 566, 567, 569-570
  fermentation effects on flavor, 555-556
  hop resins in, 569
  sunstruck flavor of, 557
  terpene hydrocarbons in, 566
Beer aroma, 562, 566, 568
Beer volatiles, 552, 554-575
Bile acid derivatives:
  acyl esters, 379-380
  alkyl esters, 378-379, 385, 386
  ethers, 380, 382, 384
  trifluoroacetate esters, 386, 387-389, 390
Bile acid extraction:
  bile, 373
  feces, 376
  gallstones, 377-378
  preferred methods of, 373, 374-377
  serum, 371-373, 374-375
  tissue cells, 376-377
  urine, 376
Bile acids:
  clinical considerations, 369-370
  dehydroxylated, 368
  general and historical aspects, 370
[Bile acids]
  hydrolysis of, 377
  metabolism of, 366-367
  occurrence and distribution of, 367-368, 370
  sample preparation, 370-380
  specific properties of, 368-369
  structure and function of, 366
  toxic effects of, 368
Biomedical applications of GC, general considerations, 204-206
Blood plasma:
  arthritic acid profiles, 213
  human acid profile, 212
Borontrifluoride in fatty acid methylation, 406-407
Brandies, *see* Wine distillates
Breakthrough, in adsorbent traps, 81-82
Breast lesions, steroidal hormonal patterns in, 244, 248-250

Cab-O-Sil, 61
Cancer maps, 99
Cantaloupe, volatile composition of, 488, 491-492, 515, 516
Capacity of column, 62
Capacity ratio, 31
Capillary columns, bores contrasted, 536-538
Carbohydrate metabolism, 252
Carbopack, trapping efficiency of, 83
Carrier gas:
  choice of, 52-53
  precautions, 52
Cataracts, human lens, 253, 267-269
Charcoal:
  recovery of adsorbed sample, 79
  use in air sampling, 78-80
Chemical ionization MS, 89
Chiral phase, 333, 334, 338, 341, 353-357, *see also* cis-trans separations
Chloracne, 192
Chlordane, 189
Chlorinated pesticides, 180
Chlorination, effect on water, 163-166
Chloromethyl ether, bis, 102-103
Chlorophenols, detection of, 193-195
Cholesterol synthesis, bile acid inhibition of, 368

cis-trans separations, *see also* Chiral phases, Dipeptide liquid phases, Enantiomers
  fatty acids, 413, 417, 419, 420, 422, 424-425, 434, 436, 437, 440
Coating efficiency, 36-37, 514-515
Coffee, volatile composition of, 521, 528
Cognacs, *see* Wine distillates
Column coating:
  dynamic method, 21-22
  historical aspects of, 20-22
  mercury plug method, 22
  static method, 20-21
Column chromatography:
  essential oil samples, 462-463, 468
Column selection, 514-515
  fatty acid analysis, 411-413
  steroid analysis, 302-306
  water analysis, 137-138
Columns:
  attachment of, 53
  coatability of, 27, 61
  deactivation of, 61
  end orientation of, 52, 53
  installation of, 53
  PLOT, *see* Columns, SCOT
  quartz, 29, 32
  SCOT, 22, 23, 25, 26, 414-418
  silica, 29
  stainless steel, 27-28
  wall modification of, 29
  whiskered, 26
Comparison of packed and capillary columns, 126, 414, 417
Cori's disease, 253
Cortisol, 321-322
  effect of level, 214-215
  general aspects of, 214-215
Crude oil, separation of, 30
Cushing's syndrome, 217, 311

DDE, 187
Deactivation of columns, 88
Desorption of trapped sample directly to column, 86
Detection:
  argon ionization, 12-13
  flame ionization, 15-16
  makeup gas, 51-52, 62
  thermal conductivity, 12
Detector considerations:
  amino acids, 342
[Detector considerations]
  fatty acids, 413, 414
  pesticides, 181, 182, 193, 195
  polynuclear aromatic hydrocarbons, 105, 106
Diabetes, 253
  cataract formation and, 269, 271
  urinary steroid profiles in, 270
Diarrhea, bile acid relationships in, 368
Dichloroaniline, 192
Dieldrin:
  in drinking water, 189
Dioxin, 176, 177
  sediment contamination by, 161
  tetrachlorodibenzo, 104
Dipeptide liquid phases, 345, 348-352, *see also* cis-trans separations
Distillation, essential oils, 461-462
Drinking water, *see* Water, drinking
Drug metabolites, 273-276
Dust, carrier gas contamination by, 52
Duter, detection of, 193

Edman degradation, 334
Effluents:
  domestic, 152-157
  industrial, 146-153
Enantiomeric labeling, 338-339
Enantiomers, *see also* Chiral phases
  amino acids, 333, 338, 346, 348-360
Endrin, 184, 185
Enfleurage, 457
Essential oils, *see also* specific product heading
  angelica root, 485, 486
  castanopis flower, 493
  commercial preparation of various, 456-458
  general considerations, 455-456
  guava, 478-479, 482
  historical aspects of, 455-456
  hop, 539, 541-552
  hop head space, 543
  isomerized hop, 547, 550, 553, 554
  jasmin, 495
  juniper, 485, 487, 488, 489-490
  lavender, 398-399
  orange peel, 588
  pepper, 479, 483-484, 485
  peppermint, 504-505

[Essential oils]
rose, 496-497
Estradiol, 215
Ethylenethiourea, fungicide contamination by, 193
Extraction, liquid carbon dioxide, 516-517

Falling needle injector, *see* Injection
Fatty acid analysis:
applications, 398-401
nutritional relationships of, 400-401
Fatty acid derivatization, 404-408
comparison of methods, 408-409
Fatty acid extraction, 402-404
Fatty acid metabolism, 397-398
Fatty acid nomenclature, 396-397
Fatty acids:
essential, 401
function, 397-398
half life in man, 397
human plasma, 277-278
methylation of, 405-407
occurrence of, 402
structures of, 396-397
Fatty acids, beer, 566, 567, 569-570
Fig volatiles, 516-517
Flame ionization detector, *see* Detection
Flow programming, 184, 187
Flophemsyl-derivatives of steroids, 300
Fluorinated derivatives:
amino acids, 337, 338, 343, 344
steroids, 300
Food analysis, general, 511, 512
Fruit juices, chromatograms of, 592
Fungicides, 193

Gallstone therapy, bile acid, 368, 369
Gas chromatography as an analytical technique, advantages of, 49-50
GC/MS, 540
interface considerations in, 136-137
Ghost peaks, 58
Golay equation, *see also* van Deemter equation, 8, 26-27

Grape aroma, 579
Grape varieties, chromatographic differentiation, 590-603
Grape volatiles:
maturity effects on, 592-596
Gueuze, 559

Halogenated organics in air, 100-101
Headspace:
food, 512
essential oils, 459-460
Heptoma cells, fatty acid metabolism of, 399
Herbicides, 191
Histidine, analytical problems with, 337-338, 344
Hop oil, *see* Essential oils
Hops, aroma of, 540-541
Hormonal steroids, *see* Steroids
Hydrocarbons:
in urban air, 90
separation of low boilers, 34
Hydrogenation, on line, 65

Industrial effluents, 146-152
Injection, *see* Sample introduction
Insect growth regulators, 195-196
Insecticides, *see also* Pesticides, and individual compound
carbamate, 176
organochlorine, 175, 182-189
organophosphorus, 176, 189-191
Insulin, 252-253, 343-344
Internal standards, water analysis, 129

Jalapeña pepper, volatile composition of, 517, 519

k, *see* Capacity ratio
Key substances:
grape varital characterizations via, 593, 595, 596-602
wine, 605
Kidney, organic acid profiles in human, 210
Kovats index, *see* Retention index

Linolenic acid, metabolic tracing with, 399
Liver disease, bile acid relationships in, 369

Liver metabolism, steroid relationships in, 318, 319, 323, 324
Liver, organic acid profiles in human, 208
Liquid crystal columns, 87
Liquid phase film, thickness of, 31

Makeup gas, *see also* Detection
  choice of, 52
McCardles's disease, 253
Meat, volatile composition of, 521, 524, 525
Menhaden oil, separation of, 15
Metabolic profile concept, 206, 207
Metal analysis by GC, 529
Milkfat:
  bovine, 428
  human, 427, 432
Mirex, 186-188
Morrel, volatile composition of, 528, 530
Muscle, organic acid profiles of human, 211

Nitrogen, as carrier or make-up gas, 52
Nylon, as column material, 27, 28

Odor of GC fractions, *see* Sniff testing
Oechsle scale, conversion factors for, 593
Oil spills, 162
Oleic acid, incorporation by liver, 399
On-column injection, *see* Sample introduction
Open tubular column, *see also* Capillary columns
  invention of, 4-6
OPGV, 537
Oven considerations, 51
Oxazolinones, resulting from amino acid derivatization, 337
Oxygen, contamination of carrier gas by, 52
Ozone, use in water treatment, 167

Pancreas, organic acid profile in human, 209
Passion fruit volatiles, 515
Pear volatiles, 519
Parathion, 190, *see also* Paraoxon
  detoxification of, 176
Paraoxon, from parathion oxidation, 176
Parenchymal necrosis, bile acid role, 368
Pear, volatile composition of, 519
Pentachloroaniline, 193, 195
Pentachlorothioanisole, 195
Peppermint oil volatiles, 13, 474, 476, 477-479
Performance index, 6, 22
Perfume:
  Arpege, 502-503
  Chanel No. 5, 500-501
Pesticides, *see also* Insecticides
  chemical modification of, 176
  detectors for, 181-182
  isolation from water or air, 179
  precautions in analysis of, 63-64
Phase ratio, 31
  in static coating, 21
Pheremones, 195-197
Photodieldrin, 184, 185
Phthalates, occurrence of, 112, 114
Phytanic acid, 400
Polynuclear aromatic hydrocarbons, air particulate occurrence, 72, 84
  analysis of, 104-114
  EC detection of, 105-106
  formation of, 104
  prefractionation for air analysis, 106-107
  recovery of from traps, 84
  syringe errors in injection of, 86
  work environment occurrence, 107-108
Polyols:
  metabolic profiles, 262-267
  retention characteristics for derivatives of, 261
Pompe's disease, 253
Porapak, *see* Porous polymers
Porous polymers:
  air sampling use, 79-82
  beer flavor enrichment by, 556-557
  essential oil analysis, 460
  grape and wine aromas, 582
  wine distillate sampling, 613
Pregnanolones, 318
Progesterone, 216
Purge and trap, in water sampling, 133-134, 135-136, 147, 150, 156
Purge gas, *see* Detection

Quintozene, 195

Radic GC, 198
Radioimmunoassay of steroids contrasted with GC, 290
Recorder, linear contrasted with logarithmic response, 432
Recycle chromatography, 515
Refsum's syndrome, 400
Retention indices, 475-476
  peppermint oil constituents, 481

Saccharide separations, 524, 529, 530
Salting out:
  aroma isolation via, 586
Sample capacity, wide bore columns, 538
Sample introduction, 85
  cold trapping, 50
  falling needle injector, 50, 59, 197
  fatty acids, 409-410
  general, 50
  insecticides, 184
  modes, 18
  on-column injection, 19-20, 86
  pesticides, 180-181
  split, 16, 19, 50-51, 57-58
  splitless, 19, 58, 86
  splitter linearity, 17
  steroids, 306-307
Sample preparation, *see also* specific item
  special considerations in, 512
SCOT columns:
  in pesticide analysis, 185
Separation number, 342, 384, 435
Sediments, contaminants on, 160-161
  sampling methods for, 160
Septum purge, in splitless injection, 19
Sewage, *see* Effluent
Sheep, peritoneal fat of, 433
Signal handling, 53
Silanox, 61
  in column coating, 25
Sinovial fluid,
  human acid profiles, 212, *see also* Arthritis
Sniff testing, 538, 540, 541
  hop oil, 547, 550, 553
Splitter, *see* Sample introduction

Steroidal hormones:
  animal sexual characteristics, 237-239
  human sexual characteristics, 233-236, 240-242
Steroid analysis:
  history of, 290-291
  sample preparation, 291-301
Steroid derivatization, 294-302
Steroid profiles:
  clinical diagnosis via, 223-231
  metabolic, 206-210
  methods for, 217, 223
  urinary, 218, 225-227, 230, 231, 245, 247, 248, 301
Steroids, *see also* Urinary steroids and specific steroids
  extraction of, 292
  derivatization of, 221-222, 244-245
  hydrolysis of conjugated, 292, 293
  ion exchange clean-up of, 293-294
  plasma, 316, 317
  production of, 223, 224, 233
  rat fecal, 273, 274
  retention characteristics of, 232
  significance of, 289-290
  urinary, 299
S-triazines, *see* Herbicides
Sugar acids and lactones, 266
Sulfur compounds, 521
  analytical requirements for, 505
  cantaloupe, 488

Tailing, due to metal contact, 28
Tea volatiles, 521, 527
Tenax, *see also* Porous polymers
  air sampling, 78
  retention data on, 82
  trapping efficiency of, 83
Terpenes, interconversion of, 466-467
Testosterone:
  analogs of, 213-214
  effect of plasma level, 214
  functions of, 212-214
Tetrahydrocortisone:
  hydrolysis and derivatization scheme, 298
Thermal conductivity, *see* Detection
Thin layer chromatography, in essential oil analysis, 462-463
Thiophosphinic derivatives of steroids, 300

Thyroid hormones, 342-344
Toluene, in urban air, 90
Toxaphene, 176
  separation of, 189
Trapping, from capillary systems, 65
Tumor cell, fatty acid metabolism, 399
Turner's syndrome, 241
Two dimensional GC, 66-67, 179-180, 184
Tygon, as column material, 27
TZ, wide bore and narrow bore columns, 537

Uremea, steroids associated with, 314
Urinary steroids, *see also* Steroids
  general, 207
  hirsute females, 312, 322
  metabolite excretory patterns, 237
  normal female, 314, 318, 319
  pathological female profiles, 245-250
  pregnant women, 312, 319, 320
  profile procedures, 301-302
  profile studies, 307-325
  uremia, 314
UTE, 32, 35-36

Vaccenic acid, liver cell incorporation of, 399
van Deemter curve, 52, 53
van Deemter equation, 3, *see also* Golay equation
Vanilla bean volatiles, 527
von Gierke's disease, 253

Water:
  air analysis complications, 74
  carrier gas considerations, 52
  volatiles in, 124
Water analysis:
  analytical scheme selection, 125
  bioaccumulation of contaminants in, 129
  extraction methods for, 129-131
  macroreticular resin sampling of, 128, 130-131
  potable, 162-163
  reverse osmosis, 128, 133
  sample collection, 127
  sample preparation, 129-134
  sample preservation, 127-128
  sampling methods, 145, 158-160
  sampling site and frequency, 126
Water pollutants, occurrence in animals, 162
Water pollution, sources of, 140-141, 143
Whisker-walled columns, 184
Wine:
  aroma of, 579
  composition of, 603-604
Wine distillates:
  cognacs-brandies differentiated by volatiles, 610-611
  extraction methods for, 611-612
  volatiles of, 609-613
Wine volatiles, varietal differences in, 605-607
Wort, *see* Beer

Xenobiotic, definition of, 56